Real Analysis
with
Real Applications

Real Analysis
with
Real Applications

Kenneth R. Davidson

University of Waterloo

Allan P. Donsig

University of Nebraska

Prentice Hall
Upper Saddle River, NJ 07458

Library of Congress Cataloging–in–Publication Data

Davidson, Kenneth R.
    Real analysis with real applications / Kenneth R. Davidson, Allan P. Donsig.
        p. cm.
    Includes bibliographical references and index.
    ISBN 0-13-041647-9
    1. Mathematical analysis. I. Donsig, Allan P. II. Title.

    QA 300 .D342 2002
    515–dc21

2001052318

Acquisitions Editor: George Lobell
Editor-in-Chief: Sally Yagan
Vice President/Director Production and Manufacturing: David W. Riccardi
Executive Managing Editor: Kathleen Schiaparelli
Senior Managing Editor: Linda Mihatov Behrens
Assistant Managing Editor: Bayani Mendoza de Leon
Production Editor: Steven S. Pawlowski
Manufacturing Buyer: Alan Fischer
Manufacturing Manager: Trudy Pisciotti
Director of Marketing: John Tweeddale
Marketing Manager: Angela Battle
Marketing Assistant: Rachel Beckman
Editorial Assistant: Melanie Van Benthuysen
Art Director: Jayne Conte
Cover Designer: Bruce Kenselaar
Cover Image: A.Y. Jackson, "Labrador Coast" (1928). Oil on canvas 50" x 61".
        Hart House Permanent Collection, University of Toronto.

ⓒ2002 by Prentice Hall, Inc.
Upper Saddle River, NJ 07458

Printed in the United States of America.
10   9   8   7   6   5   4   3   2   1

ISBN 0-13-041647-9

Pearson Education LTD.
Pearson Education Australia PTY, Limited
Pearson Education Singapose, Pte. Ltd
Pearson Education North Asia, Ltd
Pearson Education Canada Ltd.
Pearson Education de Mexico, S.A. de C.V.
Pearson Education – Japan
Pearson Education Malaysia, Pte. Ltd

To Virginia and Stephanie

CONTENTS

Part B Applications

PREFACE

This book provides an introduction both to real analysis and to a range of important applications that require this material. More than half the book is a series of essentially independent chapters covering topics from Fourier series and polynomial approximation to discrete dynamical systems and convex optimization. Studying these applications can, we believe, both improve understanding of real analysis and prepare for more intensive work in each topic. There is enough material to allow a choice of applications and to support courses at a variety of levels.

The first part of the book covers the basic machinery of real analysis, focusing on that part needed to treat the applications. This material is organized to allow a streamlined approach that gets to the applications quickly, or a more wide-ranging introduction. To this end, certain sections have been marked as enrichment topics or as advanced topics to suggest that they might be omitted. It is our intent that the instructor will choose topics judiciously in order to leave sufficient time for material in the second part of the book.

A quick look at the table of contents should convince the reader that applications are more than a passing fancy in this book. Material has been chosen from both classical and modern topics of interest in applied mathematics and related fields. Our goal is to discuss the theoretical underpinnings of these applied areas concentrating on the role of fundamental principles of analysis. This is not a methods course, although some familiarity with the computational or methods-oriented aspects of these topics may help the student appreciate how the topics are developed. In each application, we have attempted to get to a number of substantial results and to show how these results depend on the fundamental ideas of real analysis. In particular, the notions of limit and approximation are two sides of the same coin, and this interplay is central to the whole book.

We emphasize the role of normed vector spaces in analysis, as they provide a natural framework for most of the applications. This begins early with a separate treatment of $\mathbb{R}^n$. Normed vector spaces are introduced to study completeness and limits of functions. There is a separate chapter on metric spaces that we use as an opportunity to put in a few more sophisticated ideas. This format allows its omission, if need be.

The basic ideas of calculus are covered carefully, as this level of rigour is not generally possible in a first calculus course. One could spend a whole semester doing this material, which forms the basis of many standard analysis courses today. When we have taught a course from these notes, however, we have often chosen to omit topics such as the basics of differentiation and integration on the grounds that these topics have been covered adequately for many students. The goal of getting further into the applications chapters may make it worth cutting here.

We have treated only tangentially some topics commonly covered in real analysis texts, such as multivariate calculus or a brief development of the Lebesgue integral. To cover this material in an accessible way would have left no time, even in a one-year course, for the real goal of the book. Nevertheless, we deal throughout with functions on domains in $\mathbb{R}^n$, and we do manage to deal with issues of higher dimensions without differentiability. For example, the chapter on convexity and optimization yields some deep results on "nonsmooth" analysis that contain the standard differentiable results such as Lagrange multipliers. This is possible because the subject is based on directional derivatives, an essentially one-variable idea. Ideas from multivariate calculus appear once or twice in the advanced sections, such as the use of Green's Theorem in the section on the isoperimetric inequality.

Not covering measure theory was another conscious decision to keep the material accessible and to keep the size of the book under control. True, we do make use of the L^2 norm and do mention the L^p spaces because these are important ideas. We feel, however, that the basics of Fourier series, approximation theory, and even wavelets can be developed while keeping measure theory to a minimum. Of course, this does not mean we think that the subject is unimportant. Rather we wished to aim the book at an undergraduate audience. To deal partially with some of the issues that arise here, we have included a section on metric space completion. This allows a treatment of L^p spaces as complete spaces of bona fide functions, by means of the Daniell integral. This is certainly an enrichment topic, which can be used to motivate the need for measure theory and to satisfy curious students.

This book began in 1984 when the first author wrote a short set of course notes (120 pages) for a real analysis class at the University of Waterloo designed for students who came primarily from applied math and computer science. The idea was to get to the basic results of analysis quickly and then illustrate their role in a variety of applications. At that time, the applications were limited to polynomial approximation, Newton's method, differential equations, and Fourier series.

A plan evolved to expand these notes into a textbook suitable for one semester or a year-long course. We expanded both the theoretical section and the choice of applications in order to make the text more flexible. As a consequence, the text is not uniformly difficult. The material is arranged by topic, and generally each chapter gets more difficult as one progresses through it. The instructor can choose to omit some more difficult topics in the chapters on abstract analysis if they will not be needed later. We provide a flow chart indicating the topics in abstract analysis required for each part of the applications chapters. For example, the chapter on limits of functions begins with the basic notion of uniform convergence and the fundamental result that the uniform limit of continuous functions is continuous. It ends with much more difficult material, such as the Arzela–Ascoli Theorem. Even if one plans to do the chapter on differential equations, it is possible to stop before the last section on Peano's Theorem, where the Arzela–Ascoli Theorem is needed. So both topics can be conveniently omitted. Although one cannot proceed linearly through the text, we hope there is some compensation in demonstrating that, even at a high level, there is a continued interplay between theory and application.

The background assumed for using this text is decent courses in both calculus and linear algebra. What we expect is outlined in the background chapter. A student should have a reasonable working knowledge of differential and integral calculus.

Multivariable calculus is an asset because of the increased level of sophistication and the incorporation of linear algebra; it is not essential. We certainly expect that the student is used to working with exponentials, logarithms, and trigonometric functions. Linear algebra is needed because we treat $\mathbb{R}^n$, $C(X)$, and $L^2(-\pi, \pi)$ as vector spaces. We develop the notion of norms on vector spaces as an important tool for measuring convergence. As such, the reader should be comfortable with the notion of a basis in finite-dimensional spaces. Familiarity with linear transformations is also sometimes useful. A course that introduces the student to proofs would also be an asset. Although we have attempted to address this in the background chapter (Chapter 1), we have no illusions that this text would be easy for a student having no prior experience with writing proofs.

While this background is in principle enough for the whole book, sections marked with a • require additional mathematical maturity or are not central to the main development, and sections marked with a ⋆ are more difficult yet. By and large, the various applications are independent of each other. However, there are references to material in other chapters. For example, in the wavelets chapter (Chapter 15), it seems essential to make comparisons with the classical approximation results for Fourier series and for polynomials.

It is also possible to use an application chapter on its own for a student seminar or other topics course. We have included several modern topics of interest in addition to the classical subjects of applied mathematics. The chapter on discrete dynamical systems (Chapter 11) introduces the notions of chaos and fractals and develops a number of examples. The chapter on wavelets (Chapter 15) illustrates the ideas with the Haar wavelet. It continues with a construction of wavelets of compact support, and gives a complete treatment of a somewhat easier continuous wavelet. In the final chapter (Chapter 16), we study convex optimization and convex programming. Both of these latter chapters require more linear algebra than the others.

We would like to thank various people who worked with early versions of this book for their helpful comments; in particular, Robert André, John Baker, Brian Forrest, John Holbrook, David Seigel, and Frank Zorzitto. We also thank various people who offered us assistance in various ways, including Jon Borwein, Stephen Krantz, Justin Peters, and Ed Vrscay. We also thank our student Masoud Kamgarpour for working through parts of the book. We would particularly like to thank the students in various classes, at the University of Waterloo and at the University of Nebraska, where early versions of the text were used.

We welcome comments on this book.

Pure Mathematics Department, University of Waterloo, Waterloo, ON N2L 3G1 Canada
krdavids@math.uwaterloo.ca

Mathematics and Statistics Department, University of Nebraska–Lincoln, Lincoln, NE 68588-0323, United States
adonsig@math.unl.edu

DEPENDENCE ON EARLIER SECTIONS

The description of dependence on earlier material will be simplified by assuming a core of material. Some parts of this may be reasonably omitted by relying on the student's background. For example, consider omitting Background, Series, and Differentiation and Integration.

Core Material
1. Background, 1.1–1.4
2. The Real Numbers, 2.1–2.7
3. Series, 3.1–3.2
4. Topology of $\mathbb{R}^n$, 4.1–4.4
5. Functions, 5.1–5.6
6. Differentiation and Integration 6.1–6.4
7. Normed Spaces, 7.1–7.4
8. Limits of Functions, 8.1–8.4

Enrichment Topics. Special topics in Part A, marked with a bullet • for enrichment or a star ⋆ for advanced, generally may be done with the core material (to that point) with a few exceptions, which are noted below.

Enrichment Topic	Section	Dependence
• Equivalence relations	1.5	
• Cardinality	2.8	1.5
• e	3.3	
• Absolute and conditional convergence	3.4	
• Monotone functions	5.7	
• Stirling's formula	6.5	
⋆ Measure zero	6.6	
• Finite-dimensional normed spaces	7.6	
⋆ The L^p norms	7.7	
• Power series	8.5	3.4
⋆ Arzela–Ascoli theorem	8.6	
• Topology in metric spaces	9.2–9.4	
⋆ Metric completion	9.5	
⋆ The L^p spaces	9.6	1.5, 7.7, 9.5

Applied topics assume the core material and the standard sections preceding in their own chapter.

Applied Topic	Sections	Background Required
Polynomial approximation	10.1–10.4	8.5
• Best approximation	10.5–10.6	7.6
⋆ Chebychev expansions	10.7	7.6, 10.6
⋆ Splines	10.8–10.9	linear algebra
⋆ Stone–Weierstrass theorem	10.10	9.1–9.3
Discrete dynamical systems		
Contraction principle	11.1	
Newton's method	11.2	10.1,11.1
• Chaos	11.3–11.5	
⋆ Topological conjugacy	11.6	5.7
⋆ Fractals	11.7	9.1, 9.2, only 11.1
ODEs	12.1–12.5	11.1
• Linear DEs	12.6	
• Perturbations of DEs	12.7	
⋆ Peano's theorem	12.8	8.6
Fourier series and physics	13.1–13.7	
• Vibrating String	13.8–13.9	
• Complex Exponential	13.10	8.5
Fourier series and approximation	14.1–14.6	7.6, 13.3, 13.5, 13.6
⋆ Best approximation	14.7–14.9	10.1–10.5, 10.7
Wavelets	15.1–15.4	7.6, 10.4
• Daubechies wavelet	15.5–15.7	linear algebra
• Franklin wavelet	15.8–15.9	7.6, 10.8, 13.10, linear algebra
Convexity and optimization	16.1–16.5	5.7, linear algebra
• Constrained optimization	16.6–16.9	multivariable calculus

POSSIBLE COURSE OUTLINES

Core Material (Abridged)

The Real Numbers, 2.1–2.7
Series, 3.1–3.2
Topology of $\mathbb{R}^n$, 4.1–4.4
Functions, 5.1–5.6
Normed Spaces, 7.1–7.4
Limits of Functions, 8.1–8.4

A One-Semester Applied Course

Core Material (abridged)
Polynomial Approximation, 10.1–10.5
Contraction Maps and Newton's Method, 11.1, 11.2
Differential Equations, 12.1–12.4
Fourier Series, 13.1–13.7, 14.1, 14.2

A One-Semester Course with Approximation

Core Material (abridged) Plus 3.4, 7.6, 8.5
Polynomial Approximation, 10.1–10.6, 10.8
Fourier Series and Approximation, 13.3, 13.5–6, Chapter 14
Wavelets, 15.1–15.4, if time permits

A One-Semester Course with Dynamics

Core Material (abridged) Plus 3.4, 9.1
Dynamical Systems, Chapter 11
Differential Equations, Chapter 12
Fourier Series and Physics, Chapter 13

A One-Semester Course with Optimization

Core Material (abridged) Plus 9.1, 8.5, 8.6
Approximation by Polynomials, 10.1–10.6
Convexity and Optimization, Chapter 16
Differential Equations, Chapter 12

A One-Semester Standard Analysis Course

Reals, Series, Topology, Functions, Chapters 2–5
Differentiation and Integration, 6.1–6.4
Normed Spaces, 7.1, 7.2
Limits of Functions, Chapter 8
Metric Spaces, 9.1–9.4
Polynomial Approximation, 10.1–10.3
One application topic, if time permits

A Sample One-Year Course

Reals, Series, Topology, Functions, Chapters 2–5
Differentiation and Integration, 6.1–6.5
Normed Spaces, 7.1–7.6
Limits of Functions, Chapter 8
Metric Spaces, 9.1–9.4
Polynomial Approximation, 10.1–10.7
Discrete Dynamical Systems, 11.1–11.5
Fourier Series and Physics, 13.1–13.7
Fourier Series and Approximation, 14.1–14.8
Convexity and Optimization, Chapter 16, if time permits

CHAPTER 1

Background

This chapter covers the language of mathematics and the basic objects, such as functions, that underlie everything that we do later. It also summarizes what the reader is expected to know from calculus and linear algebra, the material that real analysis builds upon. The most important prerequisite is that amorphous attribute "mathematical maturity," which means being comfortable with abstraction, being able to connect abstract statements with concrete examples, and having experience in reading and writing proofs. Working through this chapter will force you to practice these skills and will help prepare you for the rest of the book.

1.1. The Language of Mathematics

The language of mathematics has to be precise, because mathematical statements must be interpreted with as little ambiguity as possible. Indeed, the rigour in mathematics is much greater than in law. There should be no doubts, reasonable or otherwise, when a theorem is proved. It is either completely correct, or it is wrong. Consequently, mathematicians have adopted a very precise language so that statements may not be misconstrued.

In complicated situations, it is easy to fool yourself. By being very precise and formal now, we can build up a set of tools that will help prevent mistakes later. The history of mathematics is full of stories in which mathematicians have fooled themselves with incorrect proofs. Clarity in mathematical language, like clarity in all other kinds of writing, is essential to communicating your ideas.

We begin with a brief discussion of the logical usage of certain innocuous words *if, then, only if, and, or* and *not*. Let A, B, C represent statements that may or may not be true in a specific instance. For example, consider the statements

A. It is raining.

B. The sidewalk is wet.

The statement "If A, then B" means that whenever A is true, it follows that B must also be true. We also formulate this as "A **implies** B." This statement does not claim either that the sidewalk is wet or that it is not. It tells you that if you

1

look outside and see that it is raining, then without looking at the sidewalk, you will know that the sidewalk is wet as a result. As in the English language, "if A, then B" is a conditional statement meaning that only when the hypothesis A is verified can you deduce that B is valid. One also writes "Suppose A. Then B" with essentially the same meaning.

On the other hand, A implies B is quite different from B implies A. For example, the sidewalk may be wet because

C. The lawn sprinkler is on.

The statement "if B, then A" is known as the **converse** of "if A, then B." This amounts to reversing the direction of the implication. As you can see from this example, one may be true but not the other.

We can also say "A if B" to mean "if B, then A." The statement "A only if B" means that in order that B be true, it is necessary that A be true. A bit of thought reveals that this is yet another reformulation of "if A, then B." For reasons of clarity, these two expressions are rarely used alone and are generally restricted to the combined statement "A if and only if B." Parsing this sentence, we arrive at two statements "A if B" and "A only if B." The former means "B implies A" and the latter means "A implies B." Together they mean that either both statements are true or both are false. In this case, we say that statements A and B are **equivalent**.

The words *and*, *or*, and *not* are used with a precise mathematical meaning that does not always coincide with English usage. It is easy to be tripped up by these changes in meaning; be careful. "Not A" is the **negation** of the statement A. So "not A" is true if and only if A is false. To say that "A and B" is true, we mean that both A is true and B is true. On the other hand, "A or B" is true when at least one is true, but both being true is also possible. For example, the statement "if A or C, then B" means that if either A is true or C is true, then B is true.

Consider these statements about an integer n:

D. n is even.

E. n is a multiple of 4.

F. There is an integer k so that $n = 4k + 2$.

The statement "not F" is "there is no integer k so that $n = 4k + 2$." The statement "D and not F" says that "n is even, and there is no integer k so that $n = 4k + 2$." One can easily check that this is equivalent to statement E. Here are some valid statements:

(1) D if and only if (E or F).

(2) If (D and not E), then F.

(3) If F, then D.

In the usual logical system of mathematics, a statement is either true or false, even if one cannot determine which is valid. A statement that is always true is a **tautology**. For example, "A or not A" is a tautology. A more complicated tautology known as **modus ponens** is "If A is true, and A implies B, then B is true." It is more common that a statement may be true or false depending on the situation (e.g., statement D may be true or false depending on the value assigned to n).

The words *not* and *and* can be used together, but you must be careful to interpret statements accurately. The statement "not (A and B)" is true if (A and B) is false. If A is false, then (A and B) is false. Likewise if B is false, then (A and B) is false. While if both A and B are true, then (A and B) is true. So "not (A and B)" is true if either A is false or B is false. Equivalently, one of "not A" or "not B" is true. Thus "not (A and B)" means the same thing as "(not A) or (not B)."

This kind of thinking may sound pedantic, but it is an important way of looking at a problem from another angle. The statement "A implies B" means that B is true whenever A is true. Thus if B is false, A cannot be true, and thus A is false. That is, "not B implies not A." For example, if the sidewalk is not wet, then it is not raining. Conversely, if "not B implies not A", then "A implies B." Go through the same reasoning to see this through. You may have to use that "not (not A)" is equivalent to A. The statement "not B implies not A" is called the **contrapositive** of "A implies B." This discussion shows that the two statements are equivalent.

In addition to the converse and contrapositive of the statement "A implies B," there is the negation, "not (A implies B)." For "A implies B" to be false, there must be *some instance* in which A is true and B is false. Such an instance is called a **counterexample** to the claim that "A implies B." So the truth of A has no direct implication on the truth of B. For example, "not (C implies B)" means that it is possible for the lawn sprinkler to be on, yet the sidewalk remains dry. Perhaps the sprinkler is in the backyard, well out of reach of the sidewalk. It does not allow one to deduce any sensible conclusion about the relationship between B and C *except* that there are counterexamples to the statement "C implies B."

G. If 2 divides 3, then 10 is prime.

H. If 2 divides n, then $n^2 + 1$ is prime.

One common point of confusion is the fact that false statements can imply anything. For example, statement G is a tautology because the condition "2 divides 3" is never satisfied, so one never arrives at the false conclusion. One the other hand, H is sometimes false (e.g., when $n = 8$).

Another important use of precise language in mathematics is the phrases **for every** (or **for all**) and **there exists**, which are known as **quantifiers**. For example,

I. For every integer n, the integer $n^2 - n$ is even.

This statement means that every substitution of an integer for n in $n^2 - n$ yields an even integer. This is correct because $n^2 - n = n(n - 1)$ is the product of the two integers n and $n - 1$, and one of them is even.

On the other hand, look at

J. For every integer $n \geq 0$, the integer $n^2 + n + 41$ is prime.

The first few terms 41, 43, 47, 53, 61, 71, 83, 97, 113, 131 are all prime. But to disprove this statement, it only takes a single instance where the statement fails. Indeed, $40^2 + 40 + 41 = 41^2$ is not prime. So this statement is false. We established this by demonstrating instead that

K. There is an integer n so that $n^2 + n + 41$ is not prime.

This is the negation of statement J, and exactly one of them is true.

Things can get tricky when several quantifiers are used together. Consider

L. For every integer m, there is an integer n so that 13 divides $m^2 + n^2$.

To verify this, one needs to take each m and prove that n exists. This can be done by noting that $n = 5m$ does the job since $m^2 + (5m)^2 = 13(2m^2)$. On the other hand, consider

M. For every integer m, there is an integer n so that 7 divides $m^2 + n^2$.

To disprove this, one needs to find just one m for which this statement is false. Take $m = 1$. To show that this statement is false for $m = 1$, it is necessary to check *every* n to make sure that $n^2 + 1$ is not a multiple of 7. This could take a rather long time by brute force. However observe that every integer may be written as $n = 7k \pm j$ where j is 0, 1, 2 or 3. Therefore

$$n^2 + 1 = (7k \pm j)^2 + 1 = 7(7k^2 \pm 2j) + j^2 + 1.$$

Note that $j^2 + 1$ takes the values 1, 2, 5 and 10. None of these is a multiple of 7, and thus all of these possibilities are eliminated.

The order in which quantifiers is critical. Suppose the words in the statement L are reordered as

N. There is an integer n so that for every integer m, 13 divides $m^2 + n^2$.

This has exactly the same words as statement L, but it claims the existence of an integer n that works with *every* choice of m. We can dispose of this by showing that for every possible n, there is at least one value of m for which the statement is false. Let us consider $m = 0$ and $m = 1$. If N is true, then for the number n satisfying the statement, we would have that both $n^2 + 1$ and $n^2 + 0$ are multiples of 13. But then 13 would divide the difference, which is 1. This contradiction shows that n does not validate statement N. As n was arbitrary, we conclude that N is false.

Exercises for Section 1.1

A. Which of the following are statements? That is, can they be true or false?
   (a) Are all cats black?
   (b) All integers are prime.
   (c) $x + y$.
   (d) $|x|$ is continuous.
   (e) Don't divide by zero.

B. Which of the following statements implies which others?
   (1) X is a quadrilateral.
   (2) X is a square.
   (3) X is a parallelogram.
   (4) X is a trapezoid.
   (5) X is a rhombus.

C. Give the converse and contrapositive statements of the following:
   (a) An equilateral triangle is isosceles.
   (b) If the wind blows, the cradle will rock.
   (c) If Jack Sprat could eat no fat and his wife could eat no lean, then together they can lick the platter clean.
   (d) (A and B) implies (C or D).

D. Three young hoodlums accused of stealing CDs make the following statements:
  (1) Ed: "Fred did it, and Ted is innocent."
  (2) Fred: "If Ed is guilty, then so is Ted."
  (3) Ted: "I'm innocent, but at least one of the others is guilty."
  (a) If they are all innocent, who is lying?
  (b) If all these statements are true, who is guilty?
  (c) If the innocent told the truth and the guilty lied, who is guilty?
    HINT: Remember that false statements imply anything.

E. Which of the following statements is true? For those that are false, write down the negation of the statement.
  (a) For every $n \in \mathbb{N}$, there is an $m \in \mathbb{N}$ so that $m > n$.
  (b) For every $m \in \mathbb{N}$, there is an $n \in \mathbb{N}$ so that $m > n$.
  (c) There is an $m \in \mathbb{N}$ so that for every $n \in \mathbb{N}$, $m \geq n$.
  (d) There is an $n \in \mathbb{N}$ so that for every $m \in \mathbb{N}$, $m \geq n$.

F. Let A, B, C, D, E be statements. Make the following inferences.
  (a) Suppose that $(A \text{ or } B)$ and $(A \text{ implies } B)$. Prove B.
  (b) Suppose that $((\text{not } A) \text{ implies } B)$ and $(B \text{ implies } (\text{not } C))$ and C. Prove A.
  (c) Suppose that $(A \text{ or } (\text{not } D))$, $((A \text{ and } B) \text{ implies } C)$, $((\text{not } E) \text{ implies } B)$, and D. Prove $(C \text{ or } E)$.

1.2. Sets and Functions

Set theory is a large subject in its own right. We assume without discussion the existence of a sensible theory of sets and leave a full and rigourous development to books devoted to the subject. Our goal here to summarize the "intuitive" parts of set theory that we need for real analysis.

Sets. A **set** is a collection of elements; for example, $A = \{0, 1, 2, 3\}$ is a set. This set has four elements, 0, 1, 2, and 3. The order in which they are listed is not relevant. A set can have other sets as elements. For example, $B = \{0, \{1, 2\}, 3\}$ has three elements, one of which is the set $\{1, 2\}$. Note that 1 is *not* an element of B, and that A and B are different.

We use $a \in A$ to denote that a is an element of the set A and $a \notin A$ to denote "not $(a \in A)$." The empty set $\varnothing$ is the set with no elements. We use the words *collection* and *family* as synonyms for sets. It is often clearer to talk about "a collection of sets" or "a family of sets" instead of "a set of sets." We say that two sets are equal if they have the same elements.

Given two sets A and B, we say A is a **subset** of B if every element of A is also an element of B. Formally, A is a subset of B if "$a \in A$ implies $a \in B$," or equivalently using quantifiers, $a \in B$ for all $a \in A$. If A is a subset of B, then we write $A \subset B$. This allows the possibility that $A = B$. It also allows the possibility that A has no elements, that is, $A = \varnothing$. We say A is a **proper subset** of B is $A \subset B$ and $A \neq B$. Notice that "$A \subset B$ and $B \subset A$" if and only if "$A = B$." Thus, if we want to prove that two sets, A and B, are equal, it is equivalent to prove the two statements $A \subset B$ and $B \subset A$.

You should recognize that there is a distinction between membership in a set and a subset of a set. For the sets A and B defined at the beginning of this section, observe that $\{1, 2\} \subset A$ and $\{1, 2\} \in B$. The set $\{1, 2\}$ is not a subset of B nor an element of A.

There are a number of ways to combine sets to obtain new sets. The two most important are **union** and **intersection**. The union of two sets A and B is the set of all elements that are in A or in B, and it is denoted $A \cup B$. Formally, $x \in A \cup B$ if and only if $x \in A$ or $x \in B$. The intersection of two sets A and B is the set of all elements that are both in A and in B, and it is denoted $A \cap B$. Formally, $x \in A \cap B$ if and only if $x \in A$ and $x \in B$. Using our example, we have

$$A \cup B = \{0, 1, 2, \{1, 2\}, 3\} \quad \text{and} \quad A \cap B = \{0, 3\}.$$

Similarly, we may have an infinite family of sets A_γ indexed by another set Γ. What this means is that for every element γ of the set Γ, we have a set A_γ indexed by that element. For example, for n a positive integer, let A_n be the set of positive numbers that divide n, so that $A_{12} = \{1, 2, 3, 4, 6, 12\}$ and $A_{13} = \{1, 13\}$. Then this collection A_n is an infinite family of sets indexed by the positive integers, $\mathbb{N}$.

For infinite families of sets, intersection and union are defined formally in the same way. The union is

$$\bigcup_{\gamma \in \Gamma} A_\gamma = \{x : \text{there is a } \gamma \in \Gamma \text{ such that } x \in A_\gamma\}$$

and the intersection is

$$\bigcap_{\gamma \in \Gamma} A_\gamma = \{x : x \in A_\gamma \text{ for every } \gamma \in \Gamma\}.$$

In a particular situation, we are often working with a given set and subsets of it, such as the set of integers $\mathbb{Z} = \{\ldots, -2, -1, 0, 1, 2, \ldots\}$ and its subsets. We call this set our **universal set**. Once we have a universal set, say U, and a subset, say $A \subset U$, we can define the **complement** of A to be the collection of all elements of U that are not in A. The complement is denoted A'. Notice that the universal set can change from problem to problem, and that this will change the complement.

Given a universal set U, we can specify a subset of U as all elements of U with a certain property. For example, we may define the set of all the integers that are divisible by two. We write this formally as $\{x \in \mathbb{Z} \text{ so that } 2 \text{ divides } x\}$. It is traditional to use a vertical bar $|$ or a colon $:$ for "so that," so that we can write the set of even integers as $2\mathbb{Z} = \{x \in \mathbb{Z} \mid 2 \text{ divides } x\}$. Similarly, we can write the complement of A in a universal set U as

$$A' = \{x \in U : x \notin A\}.$$

Given two sets A and B, we define the **relative complement** of B in A, denoted $A \setminus B$, to be

$$A \setminus B = \{x \in A : x \notin B\}.$$

Notice that B need not be a subset of A. Thus, we can talk about the relative complement of $2\mathbb{Z}$ in $\{0, 1, 2, 3\}$, namely

$$\{0, 1, 2, 3\} \setminus 2\mathbb{Z} = \{1, 3\}.$$

In our example, $A \setminus B = \{1, 2\}$. Curiously, $B \setminus A = \{\{1, 2\}\}$, the set consisting of the single element $\{1, 2\}$.

Finally, we need the idea of the **Cartesian product** of two sets, denoted $A \times B$. This is the set of ordered pairs $\{(a, b) : a \in A \text{ and } b \in B\}$. For example,

$$\{0, 1, 2\} \times \{2, 4\} = \{(0, 2), (1, 2), (2, 2), (0, 4), (1, 4), (2, 4)\}.$$

More generally, if $A_1, \ldots, A_n$ is a finite collection of sets, the Cartesian product is written $A_1 \times \cdots \times A_n$ or $\prod_{i=1}^{n} A_i$, and consists of all n-**tuples** $a = (a_1, a_2, \ldots, a_n)$ such that $a_i \in A_i$ for $1 \leq i \leq n$. If $A_i = A$ is the same set for each i, then we write A^n for the product of n copies of A. For example, $\mathbb{R}^3$ consists of all triples (x, y, z) with arbitrary real coefficients x, y, z. There is also a notion of the product of an infinite family of sets. We will not have any need of it, but we warn the reader that such infinite products raise subtle questions about the nature of sets.

Functions. In practice, a function f from A to B is a rule that assigns an element $f(a) \in B$ to each element $a \in A$. Such a rule may be very complicated with many different cases. In set theory, a very general definition of function is given that does not require the use of undefined terms such as *rule*. This definition specifies a function in terms of its graph, which is a subset of $A \times B$ with a special property. We provide the definition here. However, we will usually define functions by rules in the standard fashion.

1.2.1. DEFINITION. Given two nonempty sets A and B, a **function** f from A to B is a subset of $A \times B$, denoted $G(f)$, so that

   (1) for each $a \in A$, there is some $b \in B$ so that $(a, b) \in G(f)$,

   (2) for each $a \in A$, there is only one $b \in B$ so that $(a, b) \in G(f)$.

That is, for each $a \in A$, there is *exactly one* element $b \in B$ with $(a, b) \in G(f)$. We then write $f(a) = b$. A concise way to specify the function f and the sets A and B all at once is to write $f : A \to B$. We call $G(f)$ the **graph of the function** f.

The property of a subset of $A \times B$ that makes it the graph of a function is that $\{b \in B : (a, b) \in G(f)\}$ has precisely one element for each $a \in A$. This is the "vertical line test" for functions.

We can think of f as the rule that sends $a \in A$ to the unique point $b \in B$ such that $(a, b) \in G(f)$; and we write $f(a) = b$. Notice that $f(a)$ is an element of B while f is the name of the function as a whole. Sometimes we will use such convenient expressions as "the function x^2." This really means "the function that sends x to x^2 for all x such that x^2 makes sense."

We call A the **domain** of the function $f : A \to B$ and B is the **codomain**. Far more important than the codomain is the **range** of f, which is

$$\text{Ran}(f) := \{b \in B : b = f(a) \text{ for some } a \in A\}.$$

If f is a function from A into B and $C \subset A$, the **image** of C under f is

$$f(C) := \{b \in B : \text{ there is some } c \in C \text{ so that } f(c) = b\}.$$

The range of f is $f(A)$.

Notice that that the notation $f(r)$ has two possible meanings, depending on whether r is an element of A or a subset of A. The standard practice of using lowercase letters for elements and uppercase letters for sets makes this notation clear in practice.

The same caveat is applied to the notation f^{-1}. If f maps A into B, the **inverse image** of $C \subset B$ under f is

$$f^{-1}(C) = \{a \in A : f(a) \in C\}.$$

Observe that f^{-1} is not used here as a function from B to A. Indeed, the domain of f^{-1} is the set of all subsets of B, and the codomain consists of all subsets of A. Even if $C = \{b\}$ is a single point, $f^{-1}(\{b\})$ may be the empty set or it may be very large. For example, if $f(x) = \sin x$, then $f^{-1}(\{0\}) = \{n\pi : n \in \mathbb{Z}\}$ and $f^{-1}(\{y : |y| > 1\}) = \varnothing$.

1.2.2. DEFINITION. A function f of A into B maps A **onto** B or f is **surjective** if $\mathrm{Ran}(f) = B$. In other words, for each $b \in B$, there is *at least one* $a \in A$ such that $f(a) = b$. Similarly, if $D \subset B$, say that f maps A **onto** D if $D \subset \mathrm{Ran}(f)$.

A function f of A into B is **one-to-one** or **injective** if $f(a_1) = f(a_2)$ implies that $a_1 = a_2$ for $a_1, a_2 \in A$. In other words, for each b in the range of f, there is *at most one* $a \in A$ such that $f(a) = b$.

A function from A to B that is both one-to-one and onto is called a **bijection**.

Suppose that $f : A \to B$, $\mathrm{Ran}(f) \subset B_0 \subset B$ and $g : B_0 \to C$; then the **composition** of g and f is the function $g \circ f(a) = g(f(a))$ from A into C.

A function is one-to-one if it passes a "horizontal line test." This provides a context in which we can interpret f^{-1} as a function from B to A. This notion has a number of important consequences. The most important is that when the ordered pairs in $G(f)$ are interchanged, the new set is the graph of a function known as the **inverse function** of f.

1.2.3. LEMMA. *If $f : A \to B$ is a one-to-one function, then there is a unique one-to-one function $g : f(A) \to A$ so that $g \circ f = \mathrm{id}_A$; that is,*

$$g(f(a)) = a \text{ for all } a \in A \quad \text{and} \quad f(g(b)) = b \text{ for all } b \in f(A).$$

We call g the inverse function of f and denote it by f^{-1}.

PROOF. Let $H \subset f(A) \times A$ be defined by

$$H = \{(b, a) \in f(A) \times A : (a, b) \in G(f)\}.$$

By the definition of $f(A)$, for each $b \in f(A)$, there is an $a \in A$ with $(a, b) \in G(f)$. Thus, $(b, a) \in H$, showing H satisfies property (1) of a function.

Suppose (b, a_1) and (b, a_2) are in H. Then (a_1, b) and (a_2, b) are in $G(f)$; that is, $f(a_1) = b$ and $f(a_2) = b$. Since f is one-to-one, $a_1 = a_2$. This confirms that H has property (2); and so H is the graph of a function $g : f(A) \to A$.

Suppose that $g(b_1) = g(b_2)$. Then there is some $a \in A$ so that (b_1, a) and (b_2, a) are in $G(g) = H$. Thus, (a, b_1) and (a, b_2) are in $G(f)$. But f is a function, so by property (2) for $G(f)$, $b_1 = b_2$. Thus, g is one-to-one.

Finally, observe that if $a \in A$, and $b = f(a)$, then $(b, a) \in G(g)$, so $g(b) = a$ and thus $g(f(a)) = g(b) = a$. Similarly, $f(g(b)) = b$ for all $b \in f(A)$. ∎

We say that two functions are equal if they have the same domains and the same codomains and if they agree at every point. So $f : A \to B$ and $g : A \to B$ are equal if $f(a) = g(a)$ for all $a \in A$.

We can express the relation between a one-to-one function and its inverse in terms of the identity maps. The **identity map** on a set A is $\mathrm{id}_A(a) = a$ for $a \in A$. When only one set A is involved, we use id instead of id_A.

1.2.4. COROLLARY. *If $f : A \to B$ is a bijection, then f^{-1} is a bijection and it is the unique function $g : B \to A$ so that $g \circ f = \mathrm{id}_A$ and $f \circ g = \mathrm{id}_B$. That is, $f^{-1}(f(a)) = a$ for all $a \in A$ and $f(f^{-1}(b)) = b$ for all $b \in B$.*

Exercises for Section 1.2

A. Which of the following statements is true? Prove or give a counterexample.
  (a) $(A \cap B) \subset (B \cup C)$
  (b) $(A \cup B') \cap B = A \cap B$
  (c) $(A \cap B') \cup B = A \cup B$
  (d) $A \setminus B = B \setminus A$
  (e) $(A \cup B) \setminus (A \cap B) = (A \setminus B) \cup (B \setminus A)$
  (f) $A \cap (B \cup C) = (A \cap B) \cup (A \cap C)$
  (g) $A \cup (B \cap C) = (A \cup B) \cap (A \cup C)$
  (h) If $(A \cap C) \subset (B \cap C)$, then $(A \cup C) \subset (B \cup C)$.

B. How many different sets are there that may be described using two sets A and B and as many intersections, unions, complements and parentheses as desired?
  HINT: First show that there are four minimal nonempty sets of this type.

C. What is the Cartesian product of the empty set with another set?

D. The **power set** $P(X)$ of a set X is the set consisting of all subsets of X, including $\varnothing$.
  (a) Find a bijection between $P(X)$ and the set of all functions $f : X \to \{0, 1\}$.
  (b) How many different subsets of $\{1, 2, 3, \ldots, n\}$ are there?
    HINT: Count the functions in part (a).

E. Let f be a function from A into X, and let $Y, Z \subset X$. Prove the following:
  (a) $f^{-1}(Y \cap Z) = f^{-1}(Y) \cap f^{-1}(Z)$
  (b) $f^{-1}(Y \cup Z) = f^{-1}(Y) \cup f^{-1}(Z)$
  (c) $f^{-1}(X) = A$
  (d) $f^{-1}(Y') = f^{-1}(Y)'$

F. Let f be a function from A into X, and let $B, C \subset A$. Prove the following statements. One of these statements may be sharpened to an equality. Prove it, and show by example that the others may be proper inclusions.
  (a) $f(B \cap C) \subset f(B) \cap f(C)$

(b) $f(B \cup C) \subset f(B) \cup f(C)$
(c) $f(B) \subset X$
(d) If f is one-to-one, then $f(B') \subset f(B)'$.

G. (a) What should a *two-to-one* function be?
(b) Give an example of a two-to-one function from $\mathbb{Z}$ onto $\mathbb{Z}$.

H. Suppose that f, g, h are functions from $\mathbb{R}$ into $\mathbb{R}$. Prove or give a counterexample to each of the following statements.   HINT: Only one is true.
(a) $f \circ g = g \circ f$
(b) $f \circ (g + h) = f \circ g + f \circ h$
(c) $(f + g) \circ h = f \circ h + g \circ h$

I. Suppose that $f : A \to B$ and $g : B \to A$ satisfy $g \circ f = \mathrm{id}_A$. Show that f is one-to-one and g is onto.

1.3. Calculus

To read and understand this book, you are expected to have taken (and understood most of) a full course on calculus, although it need not be a proof-oriented course. In general, you should have an understanding of functions, the mechanics of differentiation, and the mechanics of integration. We will make use of these tools to analyze examples before we get to Chapter 6, where the theory of differentiation and integration are developed carefully, with complete proofs.

Here is a checklist of the essential skills and concepts. Although they are not taught in this book, they are used frequently throughout the text to illustrate new ideas, to build interesting examples, and to do the exercises.

- Be familiar with the standard functions, especially the log and exponential functions, trig and inverse trig functions.

- Have experience in graphing functions and in recognizing the interesting features of graphs, such as local extrema, asymptotes, and inflection points.

- Have a working knowledge of how to compute basic limits.

- Have a reasonable knowledge of how to differentiate most functions.

- Be able to connect the algorithmic technique of differentiation with the geometric idea of a tangent line.

- Know how to solve typical "max–min" problems and, more important, to *recognize* such problems when they arise.

- Have a working knowledge of integration, including the ability to compute antiderivatives using the various tricks of the trade such as substitution and integration by parts.

- Know the geometric idea that integration is a computation of area by a limiting process that approximates a region under a curve by the sum of areas of rectangles that cover it (Riemann sums).

Of course, we not only use these ideas, we also extend them. The notion of limit is developed carefully in Chapter 2, as it underlies everything done in this book. A theoretical development of both differentiation and integration is carried out in Chapter 6. If you have seen a proof-oriented development of calculus, then most of Chapter 6 may safely be omitted.

There are two ideas from calculus that you need to be aware of now, to understand some exercises and material in the first few chapters. The first is a useful theorem connecting derivatives and functions; the second is a common misconception about integration.

One central fact from differential calculus that we make use of frequently is the Mean Value Theorem, Theorem 6.2.4. Intuitively, this says that if f is a differentiable function on $[a, b]$, then the line through the endpoints is parallel to a tangent line to the curve at some interior point. More precisely, it asserts the existence of a point $c \in (a, b)$ so that

$$\frac{f(b) - f(a)}{b - a} = f'(c).$$

This yields an important quantitative estimate that will be used often: *If f is a differentiable function on $[a, b]$, then*

$$|f(b) - f(a)| \leq (b - a) \max\{|f'(c)| : c \in (a, b)\}.$$

In a calculus course, most integrals are actually computed by finding antiderivatives. But if you think that integration *is* antidifferentiation, then Chapter 6 will show you what integration *really* is. It is the computation of area, and the connection to antidifferentiation is a *theorem*. This is the Fundamental Theorem of Calculus, Theorem 6.4.1, that connects the notions of tangent line and area in a surprising way. You should remember that integration is the computation of area , and that we can compute integrals even when no simple antiderivative can be found.

1.4. Linear Algebra

Readers of this book are assumed to know some (finite-dimensional) linear algebra. Various vector spaces occur throughout this book, and norms on vector spaces is a central theme. Linear algebra is used but not developed in this book. So we take a couple of pages to outline the notions that will be used. The reader is referred to a good linear algebra book such as [12, 13, 14] for details.

1.4.1. DEFINITION. A (real) **vector space** consists of a set V with elements called **vectors** and two operations with the following properties:

vector addition: for each pair $\mathbf{u}, \mathbf{v} \in V$, there is a vector $\mathbf{u} + \mathbf{v} \in V$. This satisfies

(1) **commutativity:** $\mathbf{u} + \mathbf{v} = \mathbf{v} + \mathbf{u}$ for all $\mathbf{u}, \mathbf{v} \in V$

(2) **associativity:** $\mathbf{u} + (\mathbf{v} + \mathbf{w}) = (\mathbf{u} + \mathbf{v}) + \mathbf{w}$ for all $\mathbf{u}, \mathbf{v}, \mathbf{w} \in V$

(3) **zero:** there is a vector $\mathbf{0}$ such that $\mathbf{0} + \mathbf{u} = \mathbf{u} = \mathbf{u} + \mathbf{0}$ for all $\mathbf{u} \in V$

(4) **inverses:** for each $\mathbf{u} \in V$, there is a vector $-\mathbf{u}$ such that $\mathbf{u} + (-\mathbf{u}) = \mathbf{0}$

scalar multiplication: for each vector $\mathbf{v} \in V$ and real number $r \in \mathbb{R}$, there is a vector $r\mathbf{v} \in V$. This satisfies

(1) $(r + s)\mathbf{v} = r\mathbf{v} + s\mathbf{v}$ for all $r, s \in \mathbb{R}$ and $\mathbf{v} \in V$

(2) $r(s\mathbf{v}) = (rs)\mathbf{v}$ for all $r, s \in \mathbb{R}$ and $\mathbf{v} \in V$

(3) $r(\mathbf{u} + \mathbf{v}) = r\mathbf{u} + r\mathbf{v}$ for all $r \in \mathbb{R}$ and $\mathbf{u}, \mathbf{v} \in V$

(4) $1\mathbf{v} = \mathbf{v}$ for all $\mathbf{v} \in V$

(5) $0\mathbf{v} = \mathbf{0}$ for all $\mathbf{v} \in V$

(6) $(-1)\mathbf{v} = -\mathbf{v}$ for all $\mathbf{v} \in V$

This long list of properties for a vector space is somewhat redundant, but there is no reason for us to worry about a minimal list.

1.4.2. EXAMPLES.

(1) The space $\mathbb{R}^n$ consists of all n-tuples $\mathbf{v} = (v_1, \ldots, v_n)$, where $v_i \in \mathbb{R}$ for $1 \leq i \leq n$. Addition and scalar multiplication are defined by

$$(u_1, \ldots, u_n) + (v_1, \ldots, v_n) = (u_1 + v_1, \ldots, u_n + v_n)$$

and

$$r(v_1, \ldots, v_n) = (rv_1, \ldots, rv_n).$$

(2) The space $C[0, 1]$ of all continuous functions $f : [0, 1] \to \mathbb{R}$ is a vector space with operations

$$(f + g)(x) = f(x) + g(x) \quad \text{and} \quad (rf)(x) = rf(x) \quad \text{for} \quad 0 \leq x \leq 1.$$

(3) The set $\mathbb{P}$ consists of all polynomials $p(x) = a_0 + a_1 x + \cdots + a_n x^n$, where n is an arbitrary positive integer and $a_i \in \mathbb{R}$ for $0 \leq i \leq n$. One can always consider a polynomial as having higher coefficients by setting $a_i = 0$ for $i > n$. The operations are

$$p + q = (a_0 + a_1 x + \cdots + a_n x^n) + (b_0 + b_1 x + \cdots + b_n x^n)$$
$$= (a_0 + b_0) + (a_1 + b_1)x + \cdots + (a_n + b_n)x^n$$

and

$$rp = r(a_0 + a_1 x + \cdots + a_n x^n) = ra_0 + ra_1 x + \cdots + ra_n x^n.$$

1.4.3. DEFINITION.

A **subspace** of a vector space V is a nonempty subset W of V that is a vector space using the operations of V.

A subset $W \subset V$ is **closed under addition and scalar multiplication** if whenever $\mathbf{w}_1, \mathbf{w}_2 \in W$ and $r \in \mathbb{R}$, then $\mathbf{w}_1 + \mathbf{w}_2$ and $r\mathbf{w}_1$ belong to W.

Most of the properties in our list are automatic in a subset of a vector space, because they are known to hold in the vector space. Thus, it is much easier to check whether a subset is a subspace. For example, addition is commutative and associative. Moreover, if a subspace W contains a vector $\mathbf{w}$, it also contains the vectors $0\mathbf{w} = \mathbf{0}$ and $(-1)\mathbf{w} = -\mathbf{w}$.

1.4.4. PROPOSITION. *A nonempty subset W of a vector space V is a subspace if and only it is closed under addition and scalar multiplication.*

1.4.5. EXAMPLES.

(1) A line is a the set of multiples of a single nonzero vector $\mathbf{v}$, namely $\mathbb{R}\mathbf{v}$. It is easy to check that a line is a subspace. An affine line is the translate of a line (i.e. $A = \mathbf{u} + \mathbb{R}\mathbf{v}$). This is not a subspace unless $\mathbf{u}$ is a multiple of $\mathbf{v}$, in which case $A = \mathbb{R}\mathbf{v}$, for otherwise A does not contain the origin $\mathbf{0}$.

(2) The set $\mathbb{P}_n$ of all polynomials of degree no greater than n is a subspace of the vector space $\mathbb{P}$ of all polynomials.

(3) The set S of all functions $f(x)$ on $[0, 1]$ satisfying the differential equation $f''(x) + e^x f'(x) - x f(x) = 0$ is a subspace of $C[0, 1]$. If f and g are solutions, then

$$(f + g)''(x) + e^x (f + g)'(x) - x(f + g)(x)$$
$$= (f''(x) + e^x f'(x) - x f(x)) + (g''(x) + e^x g'(x) - x g(x)) = 0$$

and

$$(rf)''(x) + e^x (rf)'(x) - x(rf)(x) = r(f''(x) + e^x f'(x) - x f(x)) = 0.$$

So $f + g$ and rf are solutions, showing that S is closed under addition and scalar multiplication.

(4) The subset $\mathbb{R}_+^2$ consisting of all vectors (x, y) in $\mathbb{R}^2$ such that $x \geq 0$ and $y \geq 0$ is not a subspace. It *is* closed under addition, but not scalar multiplication. Indeed, $(1, 0) \in \mathbb{R}_+^2$ but $(-1)(1, 0) = (-1, 0)$ does not belong.

(5) $\{\mathbf{0}\}$ is the smallest subspace of V.

1.4.6. DEFINITION. If S is a subset of a vector space V, the **span** of S is the smallest subspace containing S. It is denoted by span S. A vector $\mathbf{w}$ is a **linear combination** of S if there are vectors $\mathbf{v}_1, \ldots \mathbf{v}_k \in S$ and scalars $r_1, \ldots, r_k$ such that $\mathbf{w} = r_1 \mathbf{v}_1 + \cdots + r_k \mathbf{v}_k$.

Phrases like *the smallest* are dangerous, because they assume that there is a unique smallest subspace. We must prove that this object exists; fortunately, it is easy to do so. First, verify that that the collection of all linear combinations of S is a subspace. Second, verify that any subspace containing S will contain all linear combinations of S, and hence will contain this unique minimal subspace. This process of showing that the definition of an object makes sense is known as showing that the object is **well defined**. It comes up often.

Besides showing that our definition of span makes sense, this also proves the following result.

1.4.7. PROPOSITION. *The subspace spanned by a nonempty set $S \subset V$ consists of all linear combinations of S.*

We now consider one of the central notions of linear algebra.

1.4.8. DEFINITION. A subset S of a vector space V is **linearly dependent** if there are vectors $\mathbf{v}_1, \ldots \mathbf{v}_k \in S$ and scalars $r_1, \ldots, r_k$, which are not all 0, such that $r_1\mathbf{v}_1 + \cdots + r_k\mathbf{v}_k = \mathbf{0}$.

A subset S of a vector space V is **linearly independent** if whenever vectors $\mathbf{v}_1, \ldots \mathbf{v}_k \in S$ and scalars $r_1, \ldots, r_k \in \mathbb{R}$ satisfy $r_1\mathbf{v}_1 + \cdots + r_k\mathbf{v}_k = \mathbf{0}$, this implies that $r_1 = \cdots = r_k = 0$.

A **basis** for a vector space V is a linearly independent set that spans V. We say that V is **finite dimensional** if it has a finite basis.

A set S is linearly dependent if there is a nontrivial linear relationship between certain vectors in S. If $r_1\mathbf{v}_1 + \cdots + r_k\mathbf{v}_k = \mathbf{0}$ and $r_j \neq 0$, you can solve for $\mathbf{v}_j$ as

$$\mathbf{v}_j = \tfrac{r_1}{r_j}\mathbf{v}_1 + \cdots + \tfrac{r_{j-1}}{r_j}\mathbf{v}_{j-1} + \tfrac{r_{j+1}}{r_j}\mathbf{v}_{j+1} + \ldots \tfrac{r_k}{r_j}\mathbf{v}_k.$$

Thus $\mathbf{v}_j$ belongs to $\mathrm{span}\{S \setminus \{\mathbf{v}_j\}\}$. Therefore, it can be deleted from S without changing the span.

On the other hand, there is no such relation among vectors in a linearly independent set. Consequently, different linear combinations yield different vectors. That is, if $S = \{\mathbf{v}_1, \ldots, \mathbf{v}_k\}$ is linearly independent and

$$r_1\mathbf{v}_1 + \cdots + r_k\mathbf{v}_k = s_1\mathbf{v}_1 + \cdots + s_k\mathbf{v}_k,$$

then

$$(r_1 - s_1)\mathbf{v}_1 + \cdots + (r_k - s_k)\mathbf{v}_k = \mathbf{0}.$$

Linear independence forces $r_i = s_i$ for $1 \leq i \leq k$. So each vector $\mathbf{v}$ in $\mathrm{span}\, S$ can be written in exactly one way (*uniquely*) as a linear combination of vectors in S.

So if S is linearly independent and $\mathrm{span}\, S = V$ and thus S is a basis for V, then every vector in V has a unique expression in terms of this basis. Thus the choice of a basis provides a coordinate system for V. The expression of each vector as a linear combination of the basis can be interpreted as coordinates with respect to this basis.

1.4.9. EXAMPLES.

(1) Consider the space $\mathbb{R}^n$. The vectors $\mathbf{e}_1 = (1, 0, 0, \ldots, 0)$, $\mathbf{e}_2 = (0, 1, 0, \ldots, 0)$, $\ldots$, $\mathbf{e}_n = (0, 0, \ldots, 0, 1)$ are called the **standard basis** for $\mathbb{R}^n$. The vector $\mathbf{v} = (v_1, \ldots, v_n) = \sum\limits_{i=1}^{n} v_i\mathbf{e}_i$ is in $\mathrm{span}\{\mathbf{e}_1, \ldots, \mathbf{e}_n\}$. Conversely, suppose that $\mathbf{0} = \sum\limits_{i=1}^{n} r_i\mathbf{e}_i = (r_1, \ldots, r_n)$. Then $r_1 = \cdots = r_n = 0$. So this set is linearly independent. Thus it is a basis.

(2) The vectors $(1, 2)$, $(2, 3)$ and $(3, 4)$ are linearly dependent because

$$(1, 2) - 2(2, 3) + (3, 4) = (0, 0)$$

is a nontrivial relation.

(3) The set $\{1, x, x^2, \ldots, x^n\}$ is a basis for $\mathbb{P}_n$. The infinite set $\{1, x^j : j \geq 1\}$ is a basis for $\mathbb{P}$. Note that in algebra (as opposed to analysis), linear combination means a *finite* linear combination. Algebraists do not have to worry about convergence.

1.4.10. THEOREM. *Let V be a finite-dimensional vector space. Then every basis for V has the same finite number of elements. This number is the **dimension** of V, written $\dim V$.*

It follows that $\mathbb{R}^n$ is n-dimensional and $\mathbb{P}_n$ is $(n+1)$-dimensional. Neither the polynomials $\mathbb{P}$ nor the space of continuous functions $C[0, 1]$ is finite dimensional. Indeed, many infinite-dimensional vector spaces arise in analysis.

Linear Transformations. In this book, we will be concerned with functions on vector spaces. Linear algebra is focused on the special case of linear functions.

1.4.11. DEFINITION. A **linear transformation** A from a vector space V to a vector space W is a function with domain V and codomain W satisfying

$$A(r_1\mathbf{v}_1 + r_2\mathbf{v}_2) = r_1 A\mathbf{v}_1 + r_2 A\mathbf{v}_2 \quad \text{for all } \mathbf{v}_1, \mathbf{v}_2 \in V \text{ and } r_1, r_2 \in \mathbb{R}.$$

We use $\mathcal{L}(V, W)$ to denote the set of all linear transformations from V to W. When $W = V$, we write $\mathcal{L}(V)$ rather than $\mathcal{L}(V, V)$.

A linear transformation is determined by what it does to a basis. Suppose that $\mathbf{e}_1, \ldots, \mathbf{e}_m$ is a basis for V and $\mathbf{f}_1, \ldots, \mathbf{f}_n$ is a basis for W. If $A\mathbf{e}_j = \mathbf{w}_j$ for $1 \leq j \leq m$, then we can compute

$$A \sum_{j=1}^{m} r_j\mathbf{e}_j = \sum_{j=1}^{m} r_j A\mathbf{e}_j = \sum_{j=1}^{m} r_j\mathbf{w}_j.$$

Moreover if each $\mathbf{w}_j$ is expressed in the basis $\mathbf{f}_1, \ldots, \mathbf{f}_n$, say $A\mathbf{e}_j = \mathbf{w}_j = \sum_{i=1}^{n} a_{ij}\mathbf{f}_i$, then we may compute

$$A \sum_{j=1}^{m} r_j\mathbf{e}_j = \sum_{j=1}^{m} r_j \sum_{i=1}^{n} a_{ij}\mathbf{f}_i = \sum_{i=1}^{n} \left(\sum_{j=1}^{m} a_{ij}r_j \right) \mathbf{f}_i.$$

The $n \times m$ matrix $\left[a_{ij} \right]$ is called the **matrix representation** of A, with respect to the bases $\mathbf{e}_1, \ldots, \mathbf{e}_m$ and $\mathbf{f}_1, \ldots, \mathbf{f}_n$. Notice the meaning of a coefficient a_{ij} in this representation. It tells us how much of $A\mathbf{e}_j$ is in the direction of $\mathbf{f}_i$. By writing vectors in terms of the two bases and representing the linear transformation by its matrix, we can turn the application of the transformation into multiplication by its matrix, using the preceding formula.

1.4.12. EXAMPLE. Consider the function taking $\mathbb{P}_4$ into $\mathbb{P}_3$ by differentiation, $Dp = p'$. Since $(r_1 p_1 + r_2 p_2)' = r_1 p_1' + r_2 p_2'$, it is evident that this is a linear map. Let us use the standard basis $1, x, x^2, x^3, x^4$ for $\mathbb{P}_4$ and use $1, x, x^2, x^3$ for $\mathbb{P}_3$.

Observe that $Dx^k = kx^{k-1}$ for $0 \le k \le 4$. If $p = a_0 + a_1x + a_2x^2 + a_3x^3 + a_4x^4$, then $Dp = a_1 + 2a_2x + 3a_3x^2 + 4a_4x^3$ is given by the matrix computation

$$\begin{bmatrix} 0 & 1 & 0 & 0 & 0 \\ 0 & 0 & 2 & 0 & 0 \\ 0 & 0 & 0 & 3 & 0 \\ 0 & 0 & 0 & 0 & 4 \end{bmatrix} \begin{bmatrix} a_0 \\ a_1 \\ a_2 \\ a_3 \\ a_4 \end{bmatrix} = \begin{bmatrix} a_1 \\ 2a_2 \\ 3a_3 \\ 4a_4 \end{bmatrix}.$$

The space $\mathcal{L}(V, W)$ is a vector space with the two operations $A + B$ and rA for A and B in $\mathcal{L}(V, W)$ and scalars r, defined by

$$(A + B)\mathbf{v} = A\mathbf{v} + B\mathbf{v} \quad \text{and} \quad (rA)\mathbf{v} = r(A\mathbf{v}) \quad \text{for } \mathbf{v} \in V.$$

However, there is another important operation that is defined because linear maps are functions, namely composition. If A is a linear map from V to W and B is a linear map of W into a third vector space X, then BA is defined as the map $(BA)\mathbf{v} = B(A\mathbf{v})$. Convince yourself that this is indeed just composition. To see that BA is linear, we compute

$$(BA)(r_1\mathbf{v}_1 + r_2\mathbf{v}_2) = B(A(r_1\mathbf{v}_1 + r_2\mathbf{v}_2)) = B(r_1A\mathbf{v}_1 + r_2A\mathbf{v}_2)$$
$$= r_1B(A\mathbf{v}_1) + r_2B(A\mathbf{v}_2) = r_1(BA)\mathbf{v}_1 + r_2(BA)\mathbf{v}_2.$$

The more interesting computation to do is the matrix representation of BA. Suppose that bases have been chosen for V, W and X, say $\mathbf{e}_1, \ldots, \mathbf{e}_m$ for V, $\mathbf{f}_1, \ldots, \mathbf{f}_n$ for W and $\mathbf{g}_1, \ldots, \mathbf{g}_p$ for X. Then there will be matrices representing A and B. As before, we see that $A = \begin{bmatrix} a_{ij} \end{bmatrix}$ is an $n \times m$ matrix. Similarly, $B = \begin{bmatrix} b_{kl} \end{bmatrix}$ is an $p \times n$ matrix. Compute

$$BA\mathbf{e}_j = B\sum_{i=1}^{n} a_{ij}\mathbf{f}_i = \sum_{i=1}^{n} a_{ij}B\mathbf{f}_i$$

$$= \sum_{i=1}^{n} a_{ij} \sum_{k=1}^{p} b_{ki}\mathbf{g}_k = \sum_{k=1}^{p} \left(\sum_{i=1}^{n} b_{ki}a_{ij} \right) \mathbf{g}_k.$$

So BA has a matrix representation for the bases $\mathbf{e}_1, \ldots, \mathbf{e}_m$ and $\mathbf{g}_1, \ldots, \mathbf{g}_p$ as $\begin{bmatrix} c_{kj} \end{bmatrix}$ where $c_{kj} = \sum_{i=1}^{n} b_{ki}a_{ij}$. This formula is known as **matrix multiplication**.

1.4.13. DEFINITION. The **kernel** of a linear transformation $A \in \mathcal{L}(V, W)$ is the set $\ker A = \{\mathbf{v} \in V : A\mathbf{v} = \mathbf{0}\}$. The **nullity** of A is $\dim \ker A$, the dimension of the kernel. The **rank** of A is $\operatorname{rank} A = \dim \operatorname{Ran} A$, the dimension of the range.

Implicit in this definition is the observation that the kernel and range of A are subspaces, so that dimension is defined. This is easy to verify. The kernel of A measures how far A is from being one-to-one. Indeed, $A\mathbf{u} = A\mathbf{v}$ if and only if $A(\mathbf{u} - \mathbf{v}) = \mathbf{0}$ (i.e., $\mathbf{u} - \mathbf{v} \in \ker A$). So A is one-to-one if and only if $\ker A = \{\mathbf{0}\}$.

The range of A is the subspace $A(V)$. The main result relating the two dimensions is the following.

1.4.14. THEOREM. *Let V be a finite-dimensional vector space, and let A be a linear transformation of V into W. Then*

$$\dim \ker A + \operatorname{rank} A = \dim V.$$

An important corollary of this applies to maps from V into itself.

1.4.15. COROLLARY. *Let $A \in \mathcal{L}(V)$, where V is a finite-dimensional vector space. Then the following are equivalent:*

(1) *A is invertible.*

(2) *A is one-to-one (i.e., $\ker A = \{0\}$).*

(3) *A is onto (i.e., $\operatorname{Ran} A = V$).*

PROOF. The map A is invertible if and only it is one-to-one and onto. So (1) implies (2) and (3). However, (2) means that $\dim \ker A = 0$ and (3) means that $\dim \operatorname{Ran} A = \dim V$. Theorem 1.4.14 says that $\dim \ker A = \dim V - \dim \operatorname{Ran} A$, so (2) and (3) are equivalent. Together they say that A is one-to-one and onto, and thus is invertible. ∎

While it is not stated in the theorem, it is important to observe that the inverse of an invertible linear transformation is also linear. Let A^{-1} be the inverse map. Consider $\mathbf{v}_1, \mathbf{v}_2 \in V$ and $r_1, r_2 \in \mathbb{R}$. Suppose that $A^{-1}\mathbf{v}_1 = \mathbf{u}_1$ and $A^{-1}\mathbf{v}_2 = \mathbf{u}_2$. Then

$$A(r_1\mathbf{u}_1 + r_2\mathbf{u}_2) = r_1 A\mathbf{u}_1 + r_2 A\mathbf{u}_2 = r_1\mathbf{v}_1 + r_2\mathbf{v}_2.$$

Therefore, linearity follows from

$$A^{-1}(r_1\mathbf{v}_1 + r_2\mathbf{v}_2) = r_1\mathbf{u}_1 + r_2\mathbf{u}_2 = r_1 A^{-1}\mathbf{v}_1 + r_2 A^{-1}\mathbf{v}_2.$$

Use I for the identity map on the vector space (i.e., I sends each vector to itself). Thus in any basis, I is represented as the diagonal matrix with 1s down the diagonal. The composition of A and A^{-1} is the identity map. This is expressed algebraically as $A^{-1}A = I = AA^{-1}$.

Systems of Linear Equations. The most basic application of the ideas of linear algebra is in the solution of a system of linear equations such as

$$
\begin{aligned}
a_{11}x_1 + a_{12}x_2 + \cdots + a_{1m}x_m &= b_1, \\
a_{21}x_1 + a_{22}x_2 + \cdots + a_{2m}x_m &= b_2, \\
&\vdots \\
a_{n1}x_1 + a_{n2}x_2 + \cdots + a_{nm}x_m &= b_n.
\end{aligned}
$$

We define an $n \times m$ matrix $A = [a_{ij}]$, a vector $\mathbf{b} = (b_1, \ldots, b_n)$ in $\mathbb{R}^n$, and an unknown vector $\mathbf{x} = (x_1, \ldots, x_m)$ in $\mathbb{R}^m$. Then this system can be succinctly written as $A\mathbf{x} = \mathbf{b}$.

You should be familiar with the Gaussian elimination algorithm for solving a system of this kind. A solution exists precisely when $\mathbf{b}$ belongs to Ran A. If $\mathbf{x}$ is a solution, then every vector in $\mathbf{x} + \ker A$ is also a solution, and all solutions are in this set. This gives necessary and sufficient conditions for solving the problem. In particular, if $m = n$, Corollary 1.4.15 can be used to show that a solution exists for every vector $\mathbf{b}$ if and only if A is invertible if and only if the solution is unique.

Exercises for Section 1.4

A. Show that the space S of all infinite sequences of real numbers, $\mathbf{x} = (x_1, x_2, x_3, \dots)$, satisfies the axioms of a vector space.

B. (a) Verify that $\mathcal{L}(V, W)$ satisfies the axioms of a vector space.
   (b) If V and W are finite-dimensional vector spaces, show that $\dim \mathcal{L}(V, W)$ is equal to $(\dim V)(\dim W)$.

C. (a) Let V and W be vector spaces. Define $V \oplus W$ to be the set of vectors $(\mathbf{v}, \mathbf{w})$ with $\mathbf{v} \in V$ and $\mathbf{w} \in W$. Define $(\mathbf{v}_1, \mathbf{w}_1) + (\mathbf{v}_2, \mathbf{w}_2) = (\mathbf{v}_1 + \mathbf{v}_2, \mathbf{w}_1 + \mathbf{w}_2)$ and $r(\mathbf{v}_1, \mathbf{w}_1) = (r\mathbf{v}_1, r\mathbf{w}_1)$ for $\mathbf{v}_i \in V$, $\mathbf{w}_i \in W$ and $r \in \mathbb{R}$. Show that $V \oplus W$ is a vector space.
   (b) Show that $\dim(V \oplus W) = \dim V + \dim W$.

D. Show that the set of all vectors $\mathbf{x} = (w, x, y, z)$ such that $2w - x + 3y + 5z = 0$ is a subspace of $\mathbb{R}^4$. Express this subspace as the kernel of a linear map.

E. Which of the following subsets of $C[0, 1]$ are subspaces?
   (a) $W = \{f : f(0) = f(1)\}$
   (b) $X = \{f : f(0) + f(1) = 2\}$
   (c) $Y = \{f : f \text{ is linear on } [0, .5]\}$
   (d) $Z = \{f : |f(0)| \le |f(1)|\}$

F. Let W_1 and W_2 be subspaces of a vector space V.
   (a) Show that $W_1 + W_2 = \{\mathbf{w}_1 + \mathbf{w}_2 : \mathbf{w}_1 \in W_1, \mathbf{w}_2 \in W_2\}$ is a subspace of V.
   (b) Show that $W_1 \cap W_2$ is a subspace.
   (c) Define $A \in \mathcal{L}(W_1 \oplus W_2, V)$ by $A(\mathbf{w}_1, \mathbf{w}_2) = \mathbf{w}_1 + \mathbf{w}_2$. Show Ran $A = W_1 + W_2$.
   (d) Compute $\ker A$.
   (e) Prove that $\dim(W_1 + W_2) + \dim(W_1 \cap W_2) = \dim W_1 + \dim W_2$.

G. Given a subset S of a vector space V, consider the collection of all subspaces of V containing S. Show that there is a smallest subspace of V containing S, using the intersection of this collection.

H. Show that if $\text{span}\{\mathbf{v}_1, \dots, \mathbf{v}_m\} = V$, then a basis for V may be obtained by deleting every $\mathbf{v}_j$ that belongs to $\text{span}\{\mathbf{v}_i : i < j\}$.

I. Let W be a subspace of a finite-dimensional vector space V.
   (a) Show that a basis for W may be extended to a basis for V.
      HINT: Start with a basis for W, add a basis for V, and use the previous exercise.
   (b) Hence deduce that $\dim W \le \dim V$, with equality only when $W = V$.

J. Let $A \in \mathcal{L}(V, W)$, where V has a basis $\mathbf{v}_1, \dots, \mathbf{v}_n$. Show that A is one-to-one if and only if $A\mathbf{v}_1, \dots, A\mathbf{v}_n$ are linearly independent.

K. Let $A = \begin{bmatrix} a_{ij} \end{bmatrix}$ be an upper triangular $n \times n$ matrix, meaning that $a_{ij} = 0$ if $i > j$. Prove that A is invertible if and only if $a_{ii} \neq 0$ for $1 \leq i \leq n$.

L. Consider an $n \times n$ system of equations $A\mathbf{x} = \mathbf{b}$. Show that a solution exists for every vector $\mathbf{b}$ if and only if $A\mathbf{x} = \mathbf{0}$ has a unique solution.

1.5. The Role of Proofs

Mathematics is all about proofs. Mathematicians are not as much interested in *what* is true as in *why* it is true. For example, you were taught in high school that the roots of the quadratic equation $ax^2 + bx + c = 0$ are $\dfrac{-b \pm \sqrt{b^2 - 4ac}}{2a}$ provided that $a \neq 0$. A serious class would not have been given this as a fact to be memorized. It would have been justified by the technique of *completing the square*. This raises the formula from the realm of magic to the realm of understanding.

There are several important reasons for teaching this argument. The first goes beyond intellectual honesty and addresses the real point, which is that you shouldn't accept mathematics (or science) on faith. The essence of scientific thought is understanding why things work out the way they do.

Second, the formula itself does not help you do anything beyond what it is designed to accomplish. It is no better than a quadratic solver button that could be built into your calculator. The numbers a, b, c go into a black box and two numbers come out or they don't—you might get an error message if $b^2 - 4ac < 0$. At this stage, you have no way of knowing if the calculator gave you a reasonable answer, or why it might give an error. If you know where the formula comes from, you can analyze all of these issues clearly.

Third, knowledge of the proof makes further progress a possibility. The creation of a new proof about something that you don't yet know is much more difficult than understanding the arguments someone else has written down. Moreover, understanding these arguments makes it easier to push further. It is for this reason that we can make progress. As Isaac Newton once said, "If I have seen further than others, it is by standing on the shoulders of giants." The first step toward proving things for yourself is to understand how others have done it before.

Fourth, if you understand that the *idea* behind the quadratic formula is completing the square, then you can always recover the quadratic formula whenever you forget it. This nugget of the proof is a useful method of data compression that saves you the trouble of memorizing a bunch of arcane formulae.

It is our hope that most students reading this book already have had some introduction to proofs in their earlier courses. If this is not the case, the examples in this section will help. This may be sufficient to tackle the basic material in this book. But be warned that some parts of this book require significant sophistication on the part of the reader.

Direct Proofs. We illustrate several proof techniques that occur frequently. The first is **direct proof**. In this technique, one takes a statement, usually one asserting the existence of some mathematical object, and proceeds to verify it. Such an argument may amount to a computation of the answer. On the other hand, it might just show the existence of the object without actually computing it. The crucial distinction for existence proofs is between those that are **constructive proofs**—that is, those that give you a method or algorithm for finding the object—and those that are **nonconstructive proofs**—that is, they don't tell you how to find it. Needless to say, constructive proofs do something more than nonconstructive ones, but they sometimes take more work.

Every real number x has a decimal expansion $x = a_0.a_1a_2a_3\ldots$, where a_i are integers and $0 \le a_i \le 9$ for all $i \ge 1$. This will be discussed thoroughly in Chapter 2. This expansion is **eventually periodic** if there are integers N and d so that $a_{n+d} = a_n$ for $n \ge N$.

Occasionally a direct proof is just a straightforward calculation or verification.

1.5.1. THEOREM. *If the decimal expansion of a real number x is eventually periodic, then x is rational.*

PROOF. Suppose that N and d are given so that $a_{n+d} = a_n$ for $n \ge N$. Compute $10^N x$ and $10^{N+d}x$ and observe that

$$10^{N+d}x = b.a_{N+1+d}a_{N+2+d}a_{N+3+d}a_{N+4+d}\cdots$$
$$= b.a_{N+1}a_{N+2}a_{N+3}a_{N+4}\cdots$$
$$10^N x = c.a_{N+1}a_{N+2}a_{N+3}a_{N+4}\cdots,$$

where b and c are integers that you can easily compute. Subtracting the second equation from the first yields

$$(10^{N+d} - 10^N)x = b - c.$$

Therefore, $x = \dfrac{b - c}{10^{N+d} - 10^N}$ is a rational number. ∎

The converse of this statement is also true. We will prove it by an existential argument that does not actually exhibit the exact answer, although the argument does provide a method for finding the exact answer. The next proof is definitely more sophisticated than a computational proof. It still, like the last proof, has the advantage of being constructive.

We need a simple but very useful fact.

1.5.2. PIGEONHOLE PRINCIPLE.
If $n + 1$ items are divided into n categories, then at least two of the items are in the same category.

This is evident after a little thought, and we do not attempt to provide a formal proof. Note that it has variants that may also be useful. If $nd + 1$ objects are divided into n categories, then at least one category contains $d + 1$ items. Also, if infinitely many items are divided into finitely many categories, then at least one category has infinitely many items.

1.5.3. THEOREM. *If x is rational, then the decimal expansion of x is eventually periodic.*

PROOF. Since x is rational, we may write it as $x = \frac{p}{q}$, where p, q are integers and $q > 0$. When an integer is divided by q, we obtain another integer with a remainder in the set $\{0, 1, \ldots, q - 1\}$. Consider the remainders r_k when 10^k is divided by q for $0 \le k \le q$. There are $q + 1$ numbers r_k, but only q possible remainders. By the Pigeonhole Principle, there are two integers $0 \le k < k + d \le q$ so that $r_k = r_{k+d}$. Therefore, q divides $10^{k+d} - 10^k$ exactly, say $qm = 10^{k+d} - 10^k$.

Now compute

$$\frac{p}{q} = \frac{pm}{qm} = \frac{pm}{10^{k+d} - 10^k} = 10^{-k}\frac{pm}{10^d - 1}.$$

Divide $10^d - 1$ into pm to obtain quotient a with remainder b, $0 \le b < 10^d - 1$. So

$$x = \frac{p}{q} = 10^{-k}\Big(a + \frac{b}{10^d - 1}\Big),$$

where $0 \le b < 10^d - 1$. Write $b = b_1 b_2 \ldots b_d$ as a decimal number with exactly d digits even if the first few are zero. For example, if $d = 4$ and $b = 13$, we will write $b = 0013$. Then consider the periodic (or repeating) decimal

$$r = 0.b_1 b_2 \ldots b_d b_1 b_2 \ldots b_d b_1 b_2 \ldots b_d \ldots.$$

Using the proof of Theorem 1.5.1, we find that $(10^d - 1)r = b$ and thus $r = \frac{b}{10^d - 1}$. Observe that $10^k x = a + r = a.b_1 b_2 \ldots b_d \ldots$ has a repeating decimal expansion. The decimal expansion of $x = 10^{-k}(a + r)$ begins repeating every d terms after the first k. Therefore, this expansion is eventually periodic. ∎

Proof by Contradiction. The second common proof technique is generally called **proof by contradiction**. Suppose that we wish to verify statement A. Now either A is true or it is false. We assume that A is false and make a number of logical deductions until we establish as true something that is clearly false. No false statement can be deduced from a logical sequence of deductions based on a valid hypothesis. So our hypothesis that A is false must be incorrect, whence A is true.

Here is a well-known example of this type.

1.5.4. THEOREM. $\sqrt{3}$ *is an irrational number.*

PROOF. Suppose to the contrary that $\sqrt{3} = a/b$, where a, b are positive integers with no common factor. (This proviso of no common factor is crucial to setting the stage correctly. Watch for where it gets used.) Manipulating the equation, we obtain

$$a^2 = 3b^2.$$

When the number a is divided by 3, it leaves a remainder $r \in \{0, 1, 2\}$. Let us write $a = 3k + r$. Then

$$a^2 = (3k + r)^2 = 3(3k^2 + 2kr) + r^2 = \begin{cases} 9k^2 & \text{if } r = 0 \\ 3(3k^2 + 2k) + 1 & \text{if } r = 1 \\ 3(3k^2 + 4k + 1) + 1 & \text{if } r = 2 \end{cases}$$

Observe that a^2 is a multiple of 3 only when a is a multiple of 3. Therefore, we can write $a = 3c$ for some integer c. So $9c^2 = 3b^2$. Dividing by 3 yields $b^2 = 3c^2$.

Repeating exactly the same reasoning, we deduce that $b = 3d$ for some integer d. It follows that a and b do have a common factor 3, contrary to our assumption. The reason for this contradiction was the incorrect assumption that $\sqrt{3}$ was rational. Therefore, $\sqrt{3}$ is irrational. ∎

The astute reader might question why a fraction may be expressed in lowest terms. This is an easy fact that does not depend on deeper facts such as unique factorization into primes. It is merely the observation that if a and b have a common factor, then after it is factored out, one obtains a new fraction a_1/b_1 with a smaller denominator. This procedure must terminate by the time the denominator is reduced to 1, if not sooner. A very crude estimate of how many times the denominator can be factored is b itself.

The same reasoning is commonly applied to verify "A implies B." It is enough to show that "A and not B" is always false. For then if A is true, it follows that not B is false, whence B is true. This is usually phrased as follows: A is given as true. Assume that B is false. If we can make a sequence of logical deductions leading to a statement that is evidently false, then given that A is true, our assumption that B was false is itself incorrect. Thus B is true.

Proof by Induction. The Principle of Induction is the mathematical version of the domino effect.

1.5.5 PRINCIPLE OF INDUCTION. Let $P(n)$, $n \geq 1$, be a sequence of statements. Suppose that we can verify the following two statements:

   (1) $P(1)$ is true.

   (2) If $n > 1$ and $P(k)$ is true for $1 \leq k < n$, then $P(n)$ is true.

Then $P(n)$ is true for each $n \geq 1$.

We note that there is nothing special about starting at $n = 1$. For example, we can also start at $n = 0$ if the statements are numbered beginning at 0. You may have seen step (2) replaced by

   (2′) If $n > 1$ and $P(n - 1)$ is true, then $P(n)$ is true.

This requires a stronger dependence on the previous statements and thus is a somewhat weaker principle. However, it is frequently sufficient.

Most students reading this book will have seen how to verify statements like $\sum_{k=1}^{n} k^3 = \left(\sum_{k=1}^{n} k \right)^2$ by induction. As a quick warmup, we outline the proof that the sum of the first n odd numbers is n^2, that is, $\sum_{k=1}^{n} (2k - 1) = n^2$. If $n = 1$, then both sides are 1 and hence equal. Suppose the statement if true for $n - 1$, so that $\sum_{k=1}^{n-1} (2k - 1) = (n - 1)^2$. Then

$$\sum_{k=1}^{n} (2k - 1) = (2n - 1) + \sum_{k=1}^{n-1} (2k - 1) = 2n - 1 + (n - 1)^2 = n^2.$$

By induction, the statement is true for all integers $n \geq 1$.

Next, we provide an example that requires a bit more work and relies on the stronger version of induction. In fact, this example require two steps to get going, not just one.

1.5.6. THEOREM. *The **Fibonacci sequence** is given recursively by*

$$F(0) = F(1) = 1 \quad and \quad F(n) = F(n - 1) + F(n - 2) \quad for \ all \quad n \geq 2.$$

Let $\tau = \dfrac{1 + \sqrt{5}}{2}$. Then $F(n) = \dfrac{\tau^{n+1} - (-\tau)^{-n-1}}{\sqrt{5}}$ for all $n \geq 0$.

PROOF. The statements are $P(n)$: $F(n) = \dfrac{\tau^{n+1} - (-\tau)^{-n-1}}{\sqrt{5}}$. Before we begin, observe that

$$\tau^2 = \left(\frac{1 + \sqrt{5}}{2} \right)^2 = \frac{6 + 2\sqrt{5}}{4} = \frac{3 + \sqrt{5}}{2} = \tau + 1.$$

Therefore, τ is a root of $x^2 - x - 1 = 0$. Now dividing by τ and rearranging yields

$$-\frac{1}{\tau} = 1 - \tau = \frac{1 - \sqrt{5}}{2}.$$

Consider $n = 0$. It is generally better to begin with the complicated side of the equation and simplify it.

$$\frac{\tau^1 - (-\tau)^{-1}}{\sqrt{5}} = \frac{1}{\sqrt{5}} \left(\frac{1 + \sqrt{5}}{2} - \frac{1 - \sqrt{5}}{2} \right) = \frac{2\sqrt{5}}{2\sqrt{5}} = 1$$

This verifies the first step $P(0)$.

Right away we have a snag compared with a standard induction. Each $F(n)$ for $n \geq 2$ is determined by the two previous terms. But $F(1)$ does not fit into this pattern. It must also be verified separately.

$$\frac{\tau^2 - (-\tau)^{-2}}{\sqrt{5}} = \frac{1}{\sqrt{5}} \frac{(6 + 2\sqrt{5}) - (6 - 2\sqrt{5})}{4} = \frac{4\sqrt{5}}{4\sqrt{5}} = 1$$

This verifies statement $P(1)$.

Now consider the case $P(n)$ for $n \geq 2$, assuming that the statements $P(k)$ are known to be true for $0 \leq k < n$. In particular, they are valid for $k = n - 1$ and $k = n - 2$. Therefore,

$$
\begin{aligned}
F(n) &= F(n-1) + F(n-2) \\
&= \frac{\tau^n - (-\tau)^{-n}}{\sqrt{5}} + \frac{\tau^{n-1} - (-\tau)^{1-n}}{\sqrt{5}} \\
&= \frac{\tau^{n-1}(\tau + 1) - (-\tau)^{-n}(1 - \tau)}{\sqrt{5}} \\
&= \frac{\tau^{n-1}(\tau^2) - (-\tau)^{-n}(-\tau^{-1})}{\sqrt{5}} = \frac{\tau^{n+1} - (-\tau)^{-n-1}}{\sqrt{5}}.
\end{aligned}
$$

Thus $P(n)$ follows from knowing $P(n-1)$ and $P(n-2)$. The Principle of Induction now establishes that $P(n)$ is valid for each $n \geq 0$. ∎

We will several times need a slightly stronger form of induction known as **recursion**. Simply put, the Principle of Recursion states that after an induction argument has been established, one has *all* of the statements $P(n)$. This undoubtedly seems to be what induction says. The difference is a subtle point of logic. Induction guarantees that each statement $P(n)$ is true, one at a time. To take all infinitely many of them at once requires a bit more. In order to deal with this rigourously, one needs to discuss the axioms of set theory, which takes us outside of the scope of this book. However it is intuitively believable, and we will take this as valid.

Exercises for Section 1.5

A. Let $a \neq 0$. Prove that the quadratic equation $ax^2 + bx + c = 0$ has real solutions if and only if the discriminant $b^2 - 4ac$ is nonnegative.   HINT: Complete the square.

B. Prove that the following numbers are irrational.
   (a) $\sqrt[3]{2}$ (b) $\log_{10} 3$ (c) $\sqrt{3} + \sqrt[3]{7}$ (d) $\sqrt{6} - \sqrt{2} - \sqrt{3}$

C. Prove by induction that $\displaystyle\sum_{k=1}^{n} k^3 = \left(\sum_{k=1}^{n} k\right)^2 = \left(\frac{n(n+1)}{2}\right)^2$.

D. Recall that the **binomial coefficient** $\binom{n}{k}$ is $n!/(k!(n-k)!)$. Prove by induction that
$$\sum_{k=0}^{n} \binom{n}{k} = 2^n.$$ HINT: First prove that $\displaystyle\binom{n+1}{k} = \binom{n}{k} + \binom{n}{k-1}$.

E. Let A and B be $n \times n$ matrices. Prove that AB is invertible if and only if both A and B are invertible.   HINT: Use direct algebraic calculations.

F. Prove by induction that every integer $n \geq 2$ factors as the product of prime numbers. HINT: You need the statements $P(k)$ for all $2 \leq k < n$ here.

G. (a) Prove directly that if $a, b \geq 0$, then $\dfrac{a+b}{2} \geq \sqrt{ab}$.

   (b) If $a_1, \ldots, a_{2^n} \geq 0$, show by induction that $\dfrac{a_1 + \cdots + a_{2^n}}{2^n} \geq \sqrt[2^n]{a_1 a_2 \ldots a_{2^n}}$.

(c) If $a_1, \ldots, a_m$ are positive numbers, choose $2^n \geq m$ and set $a_i = \dfrac{a_1 + \cdots + a_m}{m}$ for $m < i \leq 2^n$. Apply part (b) to deduce the **arithmetic mean–geometric mean inequality**, $\dfrac{a_1 + \cdots + a_m}{m} \geq \sqrt[m]{a_1 a_2 \ldots a_m}$.

H. Fix an integer $N \geq 2$. Consider the remainders $q(n)$ obtained by dividing the Fibonacci number $F(n)$ by N, so that $0 \leq q(n) < N$. Prove that this sequence is periodic with period $d \leq N^2$ as follows:
   (a) Show that there are integers $0 \leq i < j \leq N^2$ such that $q(i) = q(j)$ and $q(i+1) = q(j+1)$.   HINT: Pigeonhole.
   (b) Show that if $q(i+d) = q(i)$ and $q(i+1+d) = q(i+1)$, then $q(n+d) = q(n)$ for all $n \geq i$.
   HINT: Use the recurrence relation for $F(n)$ and induction.
   (c) Show that if $q(i+d) = q(i)$ and $q(i+1+d) = q(i+1)$, then $q(n+d) = q(n)$ for all $n \geq 0$.
   HINT: Work backward using the recurrence relation.

I. **The Binomial Theorem.** By induction on n, prove that for all real numbers x and y,

$$(x+y)^n = \sum_{k=0}^{n} \binom{n}{k} x^k y^{n-k}.$$

HINT: Exercise D is the special case $x = y = 1$. Imitate its proof.

J. Consider the following "proof" by induction. We will argue that all students receive the same mark in calculus. Let $P(n)$ be the statement that every set of n students receives the same mark. This is evidently valid for $n = 1$. Now look at larger n. Suppose that $P(n-1)$ is true. Given a group of n people, apply the induction hypothesis to all but the last person in the group. The students in this smaller group all have the same mark. Now repeat this argument with all but the first person. Combining these two facts, we find that all n students have the same mark. By induction, all students have the same mark.
This is patently absurd, and you are undoubtedly ready to refute this by saying that Paul has a much lower mark than Mary. But you must find the mistake in the induction argument, not just in the conclusion.
HINT: The mistake is not $P(1)$, and $P(73)$ does imply $P(74)$.

1.6. Appendix: Equivalence Relations

We make a short diversion to introduce a basic mathematical construction known as an equivalence relation. Equivalence relations occur frequently in mathematics and will appear occasionally later in this book. The reader may have seen other types of relations such as orderings.

1.6.1. DEFINITION. Let X be a set, and let R be a subset of $X \times X$. Then R is a **relation** on X. Let us write $x \sim y$ if $(x, y) \in R$. We say that R or $\sim$ is an **equivalence relation** if it is
   (1) (**reflexive**) $x \sim x$ for all $x \in X$.
   (2) (**symmetric**) if $x \sim y$ for any $x, y \in X$, then $y \sim x$.
   (3) (**transitive**) if $x \sim y$ and $y \sim z$ for any $x, y, z \in X$, then $x \sim z$.

If $\sim$ is an equivalence relation on X and $x \in X$, then the **equivalence class** $[x]$ is the set $\{y \in X : y \sim x\}$. By $X/\sim$ we mean the collection of all equivalence classes.

1.6.2. EXAMPLES.

(1) Equality is an equivalence relation on any set. Verify this.

(2) Consider the integers $\mathbb{Z}$. Say that $m \equiv n \pmod{12}$ if 12 divides $m - n$. Note that 12 divides $n - n = 0$ for any n, and thus $n \equiv n \pmod{12}$. So it is reflexive. Also if 12 divides $m - n$, then it divides $n - m = -(m - n)$. So $m \equiv n \pmod{12}$ implies that $n \equiv m \pmod{12}$ (i.e., symmetry). Finally, if $l \equiv m \pmod{12}$ and $m \equiv n \pmod{12}$, then we may write $l - m = 12a$ and $m - n = 12b$ for certain integers a, b. Thus $l - n = (l - m) + (m - n) = 12(a + b)$ is also a multiple of 12. Therefore, $l \equiv n \pmod{12}$, which is transitivity.

There are twelve equivalence classes $[r]$ for $0 \le r < 12$ determined by the remainder r obtained when n is divided by 12. So $[r] = \{12a + r : a \in \mathbb{Z}\}$.

(3) Consider the set $\mathbb{R}$ with the relation $x \le y$. This relation is reflexive ($x \le x$) and transitive $x \le y$ and $y \le z$ implies $x \le z$. However, it is **antisymmetric**: $x \le y$ and $y \le x$ both occur if and only if $x = y$. This is not an equivalence relation.

When dealing with functions defined on equivalence classes, we often define the function on an equivalence class in terms of a representative. In order for the function to be well defined, that is, for the definition of the function to make sense, we must check that we get same value regardless of which representative is used.

1.6.3. EXAMPLES.

(1) Consider the set of real numbers $\mathbb{R}$. Say that $x \equiv y \pmod{2\pi}$ if $x - y$ is an integer multiple of 2π. Verify that this is an equivalence relation. Define a function $f([x]) = (\cos x, \sin x)$. We are really defining a function $F(x) = (\cos x, \sin x)$ on $\mathbb{R}$ and asserting that $F(x) = F(y)$ when $x \equiv y \pmod{2\pi}$. Indeed, we then have $y = x + 2\pi n$ for some $n \in \mathbb{Z}$. As sin and cos are 2π-periodic, we have

$$
\begin{aligned}
F(y) &= (\cos y, \sin y) \\
&= (\cos(x + 2\pi n), \sin(x + 2\pi n)) \\
&= (\cos x, \sin x) = F(x).
\end{aligned}
$$

It follows that the function $f([x]) = F(x)$ yields the same answer for every $y \in [x]$. So f is well defined. One can imagine the function f as wrapping the real line around the circle infinitely often, matching up equivalent points.

(2) Consider $\mathbb{R}$ modulo 2π again, and look at $f([x]) = e^x$. Then $0 \equiv 2\pi \pmod{2\pi}$ but $e^0 = 1 \ne e^{2\pi}$. So f is not well defined on equivalence classes.

(3) Now consider Example 1.6.2(2). We wish to define multiplication modulo 12 by $[n][m] = [nm]$. To check that this is well defined, consider two representatives

$n_1, n_2 \in [n]$ and two representatives $m_1, m_2 \in [m]$. Then there are integers a and b so that $n_2 = n_1 + 12a$ and $m_2 = m_1 + 12b$. Then

$$n_2 m_2 = (n_1 + 12a)(m_1 + 12b)$$
$$= n_1 m_1 + 12(am_1 + n_1 b + 12ab).$$

Therefore, $n_2 m_2 \equiv n_1 m_1 \pmod{12}$. Consequently, multiplication modulo 12 is well defined.

Exercises for Section 1.6

A. Put a relation on $C[0, 1]$ by $f \sim g$ if $f(k/10) = g(k/10)$ for $0 \le k \le 10$.
   (a) Verify that this is an equivalence relation.
   (b) Describe the equivalence classes.
   (c) Show that $[f] + [g] = [f + g]$ is a well-defined operation.
   (d) Show that $t[f] = [tf]$ is well defined for all $t \in \mathbb{R}$ and $f \in C[0, 1]$.
   (e) Show that these operations make $C[0, 1]/\sim$ into a vector space of dimension 11.

B. Consider the set of all infinite decimal expansions $x = a_0.a_1 a_2 a_3 \ldots$, where a_0 is any integer and a_i are digits between 0 and 9 for $i \ge 1$. Say that $x \sim y$ if x and y represent the same real number. That is, if $y = b_0.b_1 b_2 b_3 \ldots$, then $x \sim y$ if (1) $x = y$, or (2) there is an integer $m \ge 1$ so that $a_i = b_i$ for $i < m - 1$, $a_{m-1} = b_{m-1} + 1$, $b_i = 9$ for $i \ge m$ and $a_i = 0$ for $i \ge m$, or (3) there is an integer $m \ge 1$ so that $a_i = b_i$ for $i < m - 1$, $a_{m-1} + 1 = b_{m-1}$, $a_i = 9$ for $i \ge m$ and $b_i = 0$ for $i \ge m$. Prove that this is an equivalence relation.

C. Define a relation on the set $PC[0, 1]$ of all piecewise continuous functions on $[0, 1]$ (see Definition 5.2.4) by $f \approx g$ if $\{x \in [0, 1] : f(x) \ne g(x)\}$ is finite.
   (a) Prove that this is an equivalence relation.
   (b) Decide which of the following functions are well defined.

   (i) $\varphi([f]) = f(0)$ (ii) $\psi([f]) = \int_0^1 f(t)\, dt$ (iii) $\gamma([f]) = \lim_{x \to 1^-} f(x)$

D. Let $d \ge 2$ be an integer. Define a relation on $\mathbb{Z}$ by $m \equiv n \pmod{d}$ if d divides $m - n$.
   (a) Verify that this is an equivalence relation, and describe the equivalence classes.
   (b) Show that $[m] + [n] = [m + n]$ is a well-defined addition.
   (c) Show that $[m][n] = [mn]$ is a well-defined multiplication.
   (d) Let $\mathbb{Z}_d$ denote the equivalence classes modulo d. Prove the distributive law:
      $$[k]([m] + [n]) = [k][m] + [k][n].$$

E. Say that two real vector spaces V and W are **isomorphic** if there is an invertible linear map T of V onto W.
   (a) Prove that this is an equivalence relation on the collection of all vector spaces.
   (b) When are two finite-dimensional vector spaces isomorphic?

Part A

Abstract Analysis

CHAPTER 2

The Real Numbers

2.1. An Overview of the Real Numbers

This section describes the history and motivation behind the development of the real number system. Readers will be familiar, in some sense, with the real numbers from studying calculus. A rigorous development of the real numbers requires both checking many details and working through some subtle properties. Instead, we will describe the real numbers in a way that suffices to establish the crucial properties without belabouring the more foundational issues.

Intuitively, we think of the real numbers as the points on a line stretching off to infinity in both directions. However, to make any sense of this, we must label all the points on this line and determine the relationship between them from different points of view. First, the real numbers form an algebraic object known as a field, meaning that one may add, subtract, and multiply real numbers and divide by nonzero real numbers. Moreover, there are well-known relationships between addition and multiplication. There is also an order on the real numbers compatible with these algebraic properties, and there is the notion of distance between two points.

All of these nice properties are also shared by the set of rational numbers:

$$\mathbb{Q} = \left\{ \frac{a}{b} : a, b \in \mathbb{Z}, b \neq 0 \right\}.$$

The ancient Greeks understood how to construct all fractions geometrically and knew that they satisfied all the properties that we alluded to in the previous paragraph. However, they were also aware that there were other points on the line that could be constructed but were not rational, such as $\sqrt{3}$. See Theorem 1.5.4 for the easy argument that shows that $\sqrt{3}$ cannot be expressed as a fraction.

This immediately raises the question of why the rationals are inadequate and what larger set of numbers fills all the apparent gaps. While the Greeks were focussed on those numbers that could be obtained by geometric construction, we have since found other reasonable numbers that do not fit this restrictive definition. The simplest example of such a number is perhaps π, the circumference of a circle of diameter 1 (and the area of a circle of radius 1). Again the Greeks knew of this quantity but were not able to find a construction of it. This was the famous problem of *squaring the circle*. Lambert showed that e and π were irrational in 1761. In the nineteenth century, Hermite showed that e was transcendental (not the root of any polynomial with integer coefficients). A number of years later, Lindemann generalized Hermite's argument to deduce that π is transcendental, and thus in particular is not constructible. This solved the Greeks' famous puzzle. In fact, a major achievement of nineteenth-century algebra explained exactly that numbers can be constructed. It was Abel who first showed that there were roots of polynomials of degree five that could not be described by any process of taking kth roots. Galois developed a beautiful connection between roots of polynomials and group theory that provided a method to analyze any polynomial. Galois's work also explains another famous Greek problem, the trisection of an angle. Indeed, a $20°$ angle cannot be constructed with a straightedge and compass.

Like the Greeks, we accept the fact that $\sqrt{3}$ and π are bona fide numbers that must be included on our real line. The approach most suited to our analytic viewpoint also goes back to the Greeks—successively better approximation. Archimedes, who lived in the third century B.C., was the first to obtain a method for computing π by inscribing 2^n-gons inside a circle and computing their perimeters, which converge (slowly) to the desired answer. Today, more sophisticated formulae for π and supercomputers have allowed mathematicians to compute the decimal expansion of π to over 2 billion digits. Perhaps even more remarkable is a new formula for π due to Bailey, P. Borwein, and Plouffe that allows them to compute any digit in the hexadecimal expansion of π *without* computing the earlier digits.

The answer to the question of what the real numbers are came as a result of the development of calculus. It turns out to be closely tied up with the notion of a limit. Later in this chapter, we will see that the crucial properties that distinguish the real numbers from the rational numbers are formulated using limits. So the definitions of the real numbers and of limits had to be developed together.

The theory of Newton and Leibniz in the seventeenth century relied on a very vague notion of the limiting process. It was not until the early nineteenth century that Cauchy made the notion of a limit precise. A large part of the difficulty was a failure to recognize what the problem was. The notion of limit is evidently an extremely subtle one. Fortunately, it is a concept that is much easier to understand than it is to formulate in the first place. Nevertheless, it is not an obvious one, and the ability to make use of it requires some hard work.

The notion of limit will be explored carefully in this book. It is the most important concept in analysis. It goes hand in hand with the more computational viewpoint of approximation. While one point of view may be abstract and existential and the other implies a more algorithmic view aimed at calculation, they are really two sides of the same coin. Implicit in the notion of limit is the estimation of the error of successive approximations. Only by showing that the error can be made arbitrarily small can we establish that a limit exists.

Even once the notion of limit was made precise, the development of the real numbers took quite a long time. It wasn't until about the 1850s that mathematicians such as Cauchy and Weierstrass realized that a formal treatment of the real numbers was necessary. Once one recognizes that it is important to consider limits of real numbers, it becomes crucial that there are enough real numbers to include the limits of all convergent sequences of real numbers. Implicit in this statement is some method for determining whether a sequence is supposed to converge without having to name the limit point.

Since the notion of infinite decimal expansions is taught from a very early stage, we will take this as our definition of the real numbers. A subtle point of our definition is that an infinite decimal expansion is just an object and does not imply the need to sum an infinite series. We do not want to use the notion of limit to define the real numbers. In the next section, we outline how to define infinite decimals precisely and how to order, add, and multiply them. After that is done, we can safely define the notion of limit, using the order and arithmetical properties of real numbers.

Our construction of the real numbers appears to be strongly dependent on the choice of 10 as the base. For this reason, purists prefer a base independent method of defining the real numbers, albeit a more abstract one. We are left with the nagging question of whether the number line we construct depends on the number of digits on our hands. Fortunately, this is not the case but proving it requires considerably more work than we wish to do at present. Our main goal is to get on with the study of analysis. There is a proof much later in the book, in Example 9.5.6, that our construction does not depend on the choice of 10 as the base.

As was implied previously, infinite decimals are not the only way to define the real numbers. To satisfy the curious, we now sketch one of the base independent definitions of the real numbers. At end of this chapter, we show some of the ingredients needed for yet another definition of the real numbers, in Exercise 2.7.J.

In 1858, Dedekind described a formal construction of the real numbers that did not require the use of any base nor any notion of limit at all. He noticed that for each real number x, there was an associated set $S_x = \{r \in \mathbb{Q} : r < x\}$ of rational numbers. This determines a different set of rational numbers for each real x. Of course, we defined these sets using the real numbers. But we can turn it around. Dedekind considered all sets S of rational numbers that have the properties

(1) S is a nonempty subset of $\mathbb{Q}$ that is bounded above,

(2) S does not contain its upper bound, and

(3) if $s \in S$ and $r < s$ for $r \in \mathbb{Q}$, then $r \in S$.

These sets are known as **Dedekind cuts**. He then associated a point x to each of these sets. In particular, each rational number r is associated to the set S_r described previously. We can then go on to define order by inclusion of sets, arithmetic operations, and limits. This somewhat artificial construction finally freed the definition of $\mathbb{R}$ from reliance on intuitive notions and put analysis on a firm footing at last.

Exercises for Section 2.1

A. Using Dedekind's notion of the real numbers, show that addition of two Dedekind cuts can be defined easily by $S + T = \{s + t : s \in S, t \in T\}$. Verify that $S + T$ is a Dedekind cut.

B. Define $-S$, ST and $1/S$, in terms of Dedekind cuts S and T.

2.2. Infinite Decimals

As discussed in the previous section, we will define a real number by using an infinite decimal expansion such as

$$\frac{1}{3} = 0.3333333333333333333333333333333333333333333333333333\ldots$$

$$\frac{1}{4} = 0.25000000000000000000000000000000000000000000000000000\ldots$$

$$\sqrt{3} = 1.7320508075688772935274463415058723669428052538 1038\ldots$$

$$\pi = 3.1415926535897932384662643383279502884197169399375 10\ldots$$

$$e = 2.7182818284590452353602874713526624977572470936999 59\ldots$$

and, in general,

$$x = a_0.a_1a_2a_3a_4a_5a_6a_7a_8a_9a_{10}a_{11}a_{12}a_{10}a_{11}a_{12}a_{13}a_{14}a_{15}a_{16}a_{17}a_{18}\cdots.$$

To be formal, an **infinite decimal expansion** is a function, say x, from the set $\{0, 1, 2, \ldots\}$ to the integers $\mathbb{Z}$. We require that for all $n \geq 1$, $x(n)$ is in the set $\{0, 1, 2, 3, 4, 5, 6, 7, 8, 9\}$. The point is that this function in no way depends on the concept of a limit. Of course, it doesn't seem related to our intuitive picture of the real line, either.

To relate infinite decimal expansions to the real line, start with a line and mark two points on the line; and call the left-hand one 0 and the right-hand one 1. Then we can easily construct points for every integer $\mathbb{Z}$, equally spaced along the line. Now divide each interval from an integer n to $n + 1$ into 10 equal pieces, marking the cuts as $n.1$, $n.2$, ..., $n.9$. Proceed in this way, cutting each interval of length 10^{-k} into 10 equal intervals of length 10^{-k-1} and mark the endpoints by the corresponding number with $k + 1$ decimals. In this way, all finite decimals are placed on the line.

It seems clear that for every infinite decimal expansion $x = a_0.a_1a_2a_3\ldots$, there will be a point on this line called a **real number** x with the property that for each positive integer k, x lies in the interval between the two decimal numbers $y = a_0.a_1\ldots a_k$ and $y + 10^{-k}$. For example,

(2.2.1) $3.141592653589 \leq \pi \leq 3.141592653590.$

In other words, the decimal expansion of x up to the kth decimal approximates x to an accuracy of at least 10^{-k}. We also have, implicitly, the notion of convergence of these finite decimal approximants, which will allow us to make sense of limits of real numbers.

The astute reader will be aware that there is a problem with this as a definition of the real numbers. What is the relationship between the numbers 1 and $z = 0.9999999999999\ldots$? Clearly these are *different* infinite decimal expansions. However, for each positive integer k, we have

$$1 - 10^{-k} = 0.\underbrace{99999999999999}_{k} \leq z \leq 1.$$

Thus the difference between this number z and 1 is arbitrarily small. It would create quite a nonintuitive line if these were allowed to represent different points. To fit in with our intuition, we must agree that $z = 1$. That means that some real numbers (precisely all those numbers with a finite decimal expansion) have two different expansions, one ending in an infinite string of zeros, and the other ending with an infinite string of nines. For example, $0.125 = 0.12499999\ldots$. If you have read Section 1.6, you will recognize this as an equivalence relation (see Exercise 1.6.B).

The set of all real numbers is denoted by $\mathbb{R}$. Technically, this is the set of all infinite decimal numbers with the identifications described in the previous paragraph. This achieves a definition of a large set of points that, on the surface, appears to make sense and to be adequate for our purposes.

The rational numbers are distinguished among all real numbers by the fact that their decimal expansions are eventually periodic. See Theorems 1.5.1 and 1.5.3 in Chapter 1 if this is unfamiliar. What we need to do next is to extend the operations on $\mathbb{Q}$ to all of $\mathbb{R}$. However, there are many details to check. We will not carry out all of these necessary verifications, but we will at least outline what needs to be done.

First, we have a built-in order on the line given by the placement of the points. This extends the natural order on the finite decimals. Notice that between any two distinct finite decimal numbers, there are (infinitely many) other finite decimal numbers. Now if x and y are distinct real numbers given by infinite decimal expansions, these expansions will differ at some finite point. This enables us to find finite decimals in between them. Because we know how to compare an infinite decimal expansion to its finite decimal approximants [using equations such as (2.2.1)], we can determine which of x or y is larger.

Second, we must extend the arithmetic properties of the rational numbers to all real numbers—namely addition, multiplication, and their inverse operations—and verify all the field axioms. This is done by making all the operations consistent with the order. For example, if x and y are real numbers and k is a positive integer, then we have the finite decimal approximants

$$a = a_0.a_1 \ldots a_k \leq x \leq a + 10^{-k} \quad \text{and} \quad b = b_0.b_1 \ldots b_k \leq y \leq b + 10^{-k},$$

and so

$$a + b \leq x + y \leq a + b + 2 \cdot 10^{-k}.$$

Clearly this determines the sum $x + y$ to an accuracy of $2 \cdot 10^{-k}$ for each k. In this way, the sum is "determined" for all real numbers.

The computation of the sum of two infinite decimals is subtle and cannot be done exactly by a computer program. The reason is that the first digit of $x + y$ may not be determined exactly after any fixed finite number of steps, even though the sum can be known to any desired accuracy. To see why this is the case, consider

$$x = 0.\overbrace{999999\ldots999999}^{10^{15}\text{ nines}}\overbrace{0123456789\ldots0123456789}^{10^4\text{ repetitions}}31415\ldots$$

$$y = 0.\underbrace{999999\ldots999999}_{10^{15}\text{ nines}}\underbrace{9876543210\ldots9876543210}_{10^4\text{ repetitions}}a9066\ldots.$$

When we add $x + y$ using the first k decimal digits for any $k \leq 10^{15}$, we obtain

$$1.\overbrace{999999\ldots999999}^{k-1 \text{ nines}}8 \leq x + y \leq 2.\overbrace{000000\ldots000000}^{k \text{ zeros}}.$$

If we take $k = 10^{15}$, we have computed $x + y$ to an accuracy of $2 \cdot 10^{-10^{15}}$ and still we cannot say for sure whether the first digit of the sum is 1 or 2. When we proceed with the computation using one more digit, we obtain

$$1.\overbrace{999999\ldots999999}^{10^{15}-1 \text{ nines}}89 \leq x + y \leq 1.\overbrace{999999\ldots999999}^{10^{15}-1 \text{ nines}}91.$$

All of a sudden, not only is the first digit certainly a 1, but the next $10^{15} - 1$ digits are all nines.

A new period of uncertainty now occurs, again because of the problem that a long string of nines can *roll over* to a string of zeros like the odometer in a car. After using another 10^4 digits, we obtain a different result depending on whether $a \leq 4$, $a = 5$ or 6, or $a \geq 7$. When $a = 4$, we get

$$1.\overbrace{9999\ldots9999}^{10^{15}-1 \text{ nines}}8\overbrace{9999\ldots9999}^{10^4 \text{ nines}}7 \leq x + y \leq 1.\overbrace{9999\ldots9999}^{10^{15}-1 \text{ nines}}8\overbrace{9999\ldots9999}^{10^4 \text{ nines}}9.$$

So the digits are now determined for another $10^4 + 1$ places. When $a = 7$, we obtain

$$1.\overbrace{9999\ldots9999}^{10^{15}-1 \text{ nines}}9\overbrace{000\ldots0000}^{10^4 \text{ zeros}}0 \leq x + y \leq 1.\overbrace{9999\ldots9999}^{10^{15}-1 \text{ nines}}9\overbrace{000\ldots0000}^{10^4 \text{ zeros}}2.$$

Again, the next $10^4 + 1$ digits are now determined. However, when $a = 5$, these digits of the sum are still ambiguous:

$$1.\overbrace{9999\ldots9999}^{10^{15}-1 \text{ nines}}8\overbrace{9999\ldots9999}^{10^4 \text{ nines}}8 \leq x + y \leq 1.\overbrace{9999\ldots9999}^{10^{15}-1 \text{ nines}}9\overbrace{000\ldots0000}^{10^4 \text{ zeros}}0.$$

The 10^{15}-th decimal digit is still not known.

The important thing to recognize is that these difficulties are not a serious impediment to defining the real numbers as infinite decimals. In theory, we know that the digits in the sum are *eventually* resolved. It may be true that no matter how large k is, looking at the first k digits of x and y does not tell us if the first digit of $x + y$ is a 1 or a 2. But then we must have, for all k, $2 - 10^{-k} \leq x + y \leq 2$. So we end up with two possible answers, one ending in an infinite string of nines and the other with an infinite string of zeros. Since we identify these two expansions as the same real number, this settles the first digit of $x + y$ (and, in fact, all of them).

In real life, knowing the sum to, say, within $2 \cdot 10^{-15}$ is much the same as knowing it to $10^{15} - 1$ decimal places (in fact marginally better). So we are content, on both theoretical and practical grounds, that we have an acceptable working model of addition.

The issues are similar for the other arithmetic operations: multiplication, additive inverses, and multiplicative inverses. It is crucial that these operations are consistent with order, as this means that they are also continuous (respect limits). Carrying out all the details of this program is tedious but not especially difficult.

The key points of this section are that we can define real numbers as infinite decimal expansions (with some identifications) and can rigourously define all the field operations; the result fits our intuitive picture of the real line. Moreover, we have the order, distance, and arithmetic properties that we expect. Once we have developed the notion of limit in the next two sections, we will use infinite decimal expansions in Section 2.5 to prove the Least Upper Bound Principle (2.5.3). This principle says, in effect, that the real number system has no gaps, resolving the problems discussed in the last section and giving us a solid foundation for the rest of our work.

Exercises for Section 2.2

A. If $x \neq y$, explain an algorithm to decide if $x < y$ or $y < x$. Does your method break down if $x = 0.9999\ldots$ and $y = 1.0000\ldots$?

B. If $a < b$ and $x < y$, is $ax < by$? What additional order hypotheses make the conclusion correct?

C. Define $|x| = \max\{x, -x\}$.
   (a) Prove that $|xy| = |x|\,|y|$ and $|x^{-1}| = |x|^{-1}$.
   (b) Prove the triangle inequality $|x + y| \leq |x| + |y|$.
       HINT: Consider x and y of the same sign and different signs as separate cases.

D. Prove by induction that $|x_1 + x_2 + \cdots + x_n| \leq |x_1| + |x_2| + \cdots + |x_n|$.

E. Prove that $\big||x| - |y|\big| \leq |x - y|$.

F. (a) Prove that if $x < y$, then there is a rational number r with a finite decimal expansion such that $x < r < y$.
   (b) Prove that if $x < y$, then there is an irrational number z such that $x < z < y$.
       HINT: Use (a) and add a small multiple of $\sqrt{2}$ to r.

G. Suppose that $r \neq 0$ is a rational number and that x is irrational. Show that $r + x$ and rx are irrational.

H. If m and n are integers, show that $\left|\sqrt{3} - \dfrac{m}{n}\right| \geq \dfrac{1}{5n^2}$.
   HINT: Rationalize the numerator and use the irrationality of $\sqrt{3}$.

2.3. Limits

The notion of a limit is *the* basic notion of analysis. Limits are the culmination of an infinite process; and it is the concern with limits in particular that separates analysis from algebra. In this section, we will deal with limits of a sequence of real numbers. Later we will concern ourselves with limits of functions, possibly with values in other spaces.

In the 1680s, Newton and Leibniz independently developed calculus. But it is not calculus as we know it today. Their writings about limits were vague and depended on physical reasoning that was somewhat circular and certainly was imprecise. In the late eighteenth century, some mathematicians, such as d'Alembert, saw the need to develop a precise notion of limit, while other great mathematicians, such as Lagrange, tried to develop calculus without dependence on this notion. Gauss in 1812 was the first mathematician to concern himself with tests for convergence of infinite series as necessary before attempting to evaluate the limit. It was not until 1829 that Cauchy gave a definition of limit that is close to the modern one we use today.

Intuitively, to say that a sequence a_n converges to a limit L means that eventually *all* the terms of the (tail of the) sequence approximate the limit value L to *any* desired accuracy. To make this precise, we introduce a subtle definition.

2.3.1. DEFINITION. A real number L is the **limit** of a sequence of real numbers $(a_n)_{n=1}^{\infty}$ if for *every* $\varepsilon > 0$, there is an integer $N = N(\varepsilon) > 0$ so that

$$|a_n - L| < \varepsilon \quad \text{for all} \quad n \geq N.$$

We say that the sequence $(a_n)_{n=1}^{\infty}$ **converges** to L, and we write $\lim\limits_{n \to \infty} a_n = L$.

The important issue in this definition is that, from some point on, *every* element of the sequence approximates the limit L to any desired accuracy. A little thought shows that we could consider only values for ε of the form $\frac{1}{2}10^{-k}$. The statement $|a_n - L| < \frac{1}{2}10^{-k}$ means that a_n and L agree to k decimal places. Thus a sequence converges to L precisely when eventually all the terms of the sequence agree with L to k decimals of accuracy for every k, no matter how large.

2.3.2. EXAMPLE. Consider the sequence $(a_n) = (n/(n+1))_{n=1}^{\infty}$, which we claim converges to 1. If the definition agrees with our intuitive idea of convergence, we should be able to pick N for any ε. Suppose $\varepsilon = .05$. We need to find some N so that

$$\left| \frac{n}{n+1} - 1 \right| < .05 \quad \text{for all} \quad n \geq N.$$

First we simplify the left-hand side of this equation: $\left| \frac{n}{n+1} - 1 \right| = \frac{1}{n+1}$. If $n \geq 20$, then

$$\left| \frac{n}{n+1} - 1 \right| = \frac{1}{n+1} \leq \frac{1}{21} < .05.$$

So it is enough to choose $N = 20$.

We could also choose $N = 73$. It is not necessary to find the best choice for N. However, as we shall see in connection with the analysis of numerical methods, better estimates can lead to better algorithms for computation.

If $\varepsilon = \frac{1}{2}10^{-k}$, what should we choose for N? Arguing as before, we see that if $n \geq 2 \cdot 10^k$, then

$$\left| \frac{n}{n+1} - 1 \right| = \frac{1}{n+1} \leq \frac{1}{2 \cdot 10^k + 1} < \frac{1}{2}10^{-k}.$$

So we can choose $N = 2 \cdot 10^k$.

2.3.3. EXAMPLE. Consider $a_{2n-1} = \pi + \frac{1}{n}$ and $a_{2n} = \pi$ for $n \geq 1$. This sequence converges to π. Indeed, given $\varepsilon > 0$, choose a large positive integer N so that $\frac{1}{N} < \varepsilon$. Then if $n > 2N$, we may write $n = 2k - 1$ or $n = 2k$ for some $k > N$. In the first case,

$$|a_n - \pi| = |a_{2k-1} - \pi| = \frac{1}{k} < \frac{1}{N} < \varepsilon,$$

while in the second case,

$$|a_n - \pi| = |a_{2k} - \pi| = 0 < \varepsilon.$$

Note that some terms of a convergent sequence may actually equal the limit exactly.

2.3.4. EXAMPLE. Consider the sequence $(a_n) = ((-1)^n)_{n=1}^{\infty}$. Since this flips back and forth between two values that are far apart, it evidently does not converge in any intuitive sense. To show this using our definition, we need to show that the definition of limit fails for *any* choice of L. However, for this L, we need find *only one* value of ε that violates the definition.

Consider the fact that

$$|a_n - a_{n+1}| = |(-1)^n - (-1)^{n+1}| = 2$$

for all n, no matter how large. So let L be any real number. We notice that L cannot be close to both 1 and -1. To turn this into a quantitative statement that avoids cases, we use a trick. For any real number L,

$$|a_n - L| + |a_{n+1} - L| \geq |(a_n - L) - (a_{n+1} - L)| = |a_n - a_{n+1}| = 2.$$

Thus

$$\max\{|a_n - L|, |a_{n+1} - L|\} \geq 1.$$

So now take $\varepsilon = 1$. If this sequence *did* converge, there would be an integer N so that $|a_n - L| < 1$ for all $n \geq N$. But this is not true for both N and $N + 1$. Consequently, this sequence does not converge.

2.3.5. EXAMPLE. Consider the sequence $\left(\dfrac{\sin n}{n}\right)_{n=1}^{\infty}$. The numerator oscillates wildly, but it remains bounded between ± 1 while the denominator goes off to infinity. We obtain the estimates

$$-\frac{1}{n} \leq \frac{\sin n}{n} \leq \frac{1}{n}.$$

We know that $\lim\limits_{n \to \infty} \dfrac{1}{n} = 0 = \lim\limits_{n \to \infty} -\dfrac{1}{n}$ as this is exactly like Example 2.3.2. Therefore, the limit can be computed using a familiar principle from calculus:

2.3.6. THE SQUEEZE THEOREM.
Suppose that three sequences (a_n), (b_n) and (c_n) satisfy

$$a_n \leq b_n \leq c_n \quad \text{for all} \quad n \geq 1 \qquad \text{and} \qquad \lim_{n \to \infty} a_n = \lim_{n \to \infty} c_n = L.$$

Then $\lim\limits_{n \to \infty} b_n = L$.

PROOF. Let $\varepsilon > 0$. There is some N_1 so that

$$|a_n - L| < \varepsilon \quad \text{for all} \quad n \geq N_1$$

or equivalently, $L - \varepsilon < a_n < L + \varepsilon$ for all $n \geq N_1$. There is also some N_2 so that

$$|c_n - L| < \varepsilon \quad \text{for all} \quad n \geq N_2$$

or $L - \varepsilon < c_n < L + \varepsilon$ for all $n \geq N_2$. Then, if $n \geq \max\{N_1, N_2\}$, we have

$$L - \varepsilon < a_n \leq b_n \leq c_n < L + \varepsilon.$$

Thus $|b_n - L| < \varepsilon$ for $n \geq \max\{N_1, N_2\}$, as required.   ∎

Returning to our example $\left(\dfrac{\sin n}{n} \right)_{n=1}^{\infty}$, we have

$$\lim_{n \to \infty} \frac{1}{n} = \lim_{n \to \infty} \frac{-1}{n} = 0.$$

By the Squeeze Theorem,

$$\lim_{n \to \infty} \frac{\sin n}{n} = 0.$$

2.3.7. EXAMPLE. A more sophisticated example comes from calculus. Consider the sequence $\left(n \sin\left(\frac{1}{n}\right) \right)_{n=1}^{\infty}$. To apply the Squeeze Theorem, we need to obtain an estimate for $\sin \theta$ when the angle θ is small. Consider a sector of the circle of radius 1 with angle θ and the two triangles, as shown in Figure 2.1.

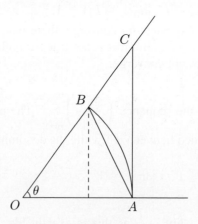

FIGURE 2.1. Sector OAB between $\triangle OAB$ and $\triangle OAC$.

Since

$$\triangle OAB \subset \text{sector } OAB \subset \triangle OAC,$$

we have the same relationship for their areas:

$$\frac{\sin \theta}{2} < \frac{\theta}{2} < \frac{\tan \theta}{2} = \frac{\sin \theta}{2 \cos \theta}.$$

A manipulation of these inequalities yields

$$\cos\theta < \frac{\sin\theta}{\theta} < 1.$$

In particular, $\cos\frac{1}{n} < n\sin\frac{1}{n} < 1$. Moreover,

$$\cos\left(\tfrac{1}{n}\right) = \sqrt{1 - \sin^2\left(\tfrac{1}{n}\right)} > \sqrt{1 - \left(\tfrac{1}{n}\right)^2} > 1 - \frac{1}{n^2}.$$

However,

$$\lim_{n\to\infty} 1 - \tfrac{1}{n^2} = 1 = \lim_{n\to\infty} 1.$$

Therefore, by the Squeeze Theorem, $\lim\limits_{n\to\infty} n\sin\frac{1}{n} = 1$.

Exercises for Section 2.3

A. In each of the following, compute the limit. Then, using $\varepsilon = 10^{-6}$, find an integer N that satisfies the limit definition.

(a) $\lim\limits_{n\to\infty} \dfrac{\sin n^2}{\sqrt{n}}$ (b) $\lim\limits_{n\to\infty} \cos\frac{1}{n}$ (c) $\lim\limits_{n\to\infty} \dfrac{1}{\log\log n}$

(d) $\lim\limits_{n\to\infty} \dfrac{3^n}{n!}$ (e) $\lim\limits_{n\to\infty} \dfrac{n^2 + 2n + 1}{2n^2 - n + 2}$ (f) $\lim\limits_{n\to\infty} \sqrt{n^2 + n} - n$

B. Prove from the definition that the sequence $a_n = L$ for $n \geq 1$ has a limit.

C. Show that $\lim\limits_{n\to\infty} \sin\frac{n\pi}{2}$ does not exist using the definition of limit.

D. Prove that if $a_n \leq b_n$ for $n \geq 1$, $L = \lim\limits_{n\to\infty} a_n$ and $M = \lim\limits_{n\to\infty} b_n$, then $L \leq M$.

E. Prove that if $L = \lim\limits_{n\to\infty} a_n$, then $L = \lim\limits_{n\to\infty} a_{2n}$ and $L = \lim\limits_{n\to\infty} a_{n^2}$.

F. Sometimes, a limit is defined informally as follows: "As n goes to infinity, a_n gets closer and closer to L." Find as many faults with this definition as you can.
(a) Can a sequence satisfy this definition and still fail to converge?
(b) Can a sequence converge yet fail to satisfy this definition?

G. Define a sequence $(a_n)_{n=1}^\infty$ so that $\lim\limits_{n\to\infty} a_{n^2}$ exists but $\lim\limits_{n\to\infty} a_n$ does not exist.

H. Suppose that $\lim\limits_{n\to\infty} a_n = L$ and $L \neq 0$. Prove that $a_n \neq 0$ with only finitely many exceptions.

I. Let a_0 and a_1 be positive real numbers, and set $a_{n+2} = \sqrt{a_{n+1}} + \sqrt{a_n}$ for $n \geq 0$.
(a) Show that $a_n \geq 1$ for n sufficiently large. (That is, there is some N so that this holds for all $n \geq N$.)
(b) Let $\varepsilon_n = a_n - 4$. Show that $\varepsilon_{n+2} \leq (\varepsilon_{n+1} + \varepsilon_n)/3$ for $n \geq N$.
(c) Prove that this sequence converges.

J. For each real number x, determine if the sequence $\left(\frac{1}{1+x^n}\right)_{n=1}^\infty$ has a limit, and compute it when it exists.

K. Show that the sequence $(\log n)_{n=1}^\infty$ does not converge.

L. Provide an example of sequences with $a_n \leq b_n \leq c_n$ such that both $L = \lim\limits_{n\to\infty} a_n$ and $M = \lim\limits_{n\to\infty} c_n$ exist, but $\lim\limits_{n\to\infty} b_n$ does not exist.

2.4. Basic Properties of Limits

We have already developed a number of basic properties of limits in the examples and exercises of the previous section. For example, the Squeeze Theorem and Exercise 2.3.D show that limits respect order. Another simple but important observation is that convergent sequences are bounded.

2.4.1. DEFINITION. A set $A \subset \mathbb{R}$ is **bounded above** if there is a real number M so that $a \leq M$ for all $a \in A$. Similarly, the set A is **bounded below** if there is a real number m such that $a \geq m$ for all $a \in A$. A set that is bounded above and below is called **bounded**. Equivalently, A is bounded if there is one real number B so that $|a| \leq B$ for all $a \in A$.

2.4.2. PROPOSITION. *If $(a_n)_{n=1}^{\infty}$ is a convergent sequence of real numbers, then the set $\{a_n : n \in \mathbb{N}\}$ is bounded.*

PROOF. Let $L = \lim\limits_{n\to\infty} a_n$. If we set $\varepsilon = 1$, then by the definition of limit, there is some $N > 0$ so that $|a_n - L| < 1$ for all $n \geq N$. In other words,

$$L - 1 < a_n < L + 1 \quad \text{for all} \quad n \geq N.$$

Let

$$M = \max\{a_1, a_2, \ldots, a_{N-1}, L+1\}$$

and

$$m = \min\{a_1, a_2, \ldots, a_{N-1}, L-1\}.$$

Clearly, for all n, we have $m \leq a_n \leq M$.   ∎

Note that there is no special reason to use 1 in this proof except convenience. We could have picked $\varepsilon = 1/2$ or $\varepsilon = 42$ and the argument would still work.

It is also crucial that the arithmetic operations respect limits. Proving this is straightforward. For completeness, here are the details.

2.4.3. THEOREM. *If $\lim\limits_{n\to\infty} a_n = L$, $\lim\limits_{n\to\infty} b_n = M$ and $\alpha \in \mathbb{R}$, then*

(1) $\lim\limits_{n\to\infty} a_n + b_n = L + M,$

(2) $\lim\limits_{n\to\infty} \alpha a_n = \alpha L,$

(3) $\lim\limits_{n\to\infty} a_n b_n = LM,$ *and*

(4) $\lim\limits_{n\to\infty} \dfrac{a_n}{b_n} = \dfrac{L}{M}$ *if $M \neq 0$.*

In the sequence $(a_n/b_n)_{n=1}^{\infty}$, we ignore terms where $b_n = 0$. There is no problem doing this because $M \neq 0$ implies that $b_n \neq 0$ for all sufficiently large n. (We use "for all sufficiently large n" as shorthand for saying there is some N so that this holds for all $n \geq N$.) We will return to this point in the proof of (4).

PROOF. (1) Notice that

$$|(a_n + b_n) - (L + M)| = |a_n - L + b_n - M| \leq |a_n - L| + |b_n - M|.$$

Since $\lim_{n \to \infty} a_n = L$, we can find $N_1 > 0$ so that

$$|a_n - L| < \frac{\varepsilon}{2} \quad \text{for all} \quad n \geq N_1.$$

Similarly, we can find $N_2 > 0$ so that

$$|b_n - M| < \frac{\varepsilon}{2} \quad \text{for all} \quad n \geq N_2.$$

Thus, if $n \geq \max\{N_1, N_2\}$, then

$$|(a_n + b_n) - (L + M)| \leq |a_n - L| + |b_n - M|$$
$$< \frac{\varepsilon}{2} + \frac{\varepsilon}{2} = \varepsilon$$

(2) Here, we have

$$|\alpha a_n - \alpha L| = |\alpha||a_n - L|.$$

We employ a simple trick to avoid dividing by 0. Using the definition of limit, we can find N so that

$$|a_n - L| < \frac{\varepsilon}{|\alpha| + 1} \quad \text{for all} \quad n \geq N.$$

Thus, if $n \geq N$,

$$|\alpha a_n - \alpha L| \leq |\alpha||a_n - L| < |\alpha|\frac{\varepsilon}{|\alpha| + 1} < \varepsilon.$$

So $(\alpha a_n)_{n=1}^{\infty}$ converges to αL, as required.

(3) First we simplify the difference we are trying to control. Observe that

$$|a_n b_n - LM| = |a_n b_n - Lb_n + Lb_n - LM|$$
$$\leq |a_n b_n - Lb_n| + |Lb_n - LM|$$
$$= |a_n - L||b_n| + |L||b_n - M|.$$

We would like to repeat the method of part (2) for each term on the right-hand side. For the second term, there is no problem. Choose N_2 so that

$$|b_n - M| < \frac{\varepsilon}{2|L| + 1} \quad \text{for all} \quad n \geq N_2.$$

For the first term, there is a problem: $|b_n|$ is not a constant. However, since $(b_n)_{n=1}^{\infty}$ converges to M, Proposition 2.4.2 implies there is a number B so that $|b_n| \leq B$ for all n. Thus,

$$|a_n b_n - LM| \leq |a_n - L||b_n| + |L||b_n - M| \leq |a_n - L|B + |L||b_n - M|.$$

Now we can choose N_1 so that

$$|a_n - L| < \frac{\varepsilon}{2B + 1} \quad \text{for all} \quad n \geq N_1.$$

Putting everything together, if $n \geq \max\{N_1, N_2\}$, then

$$|a_n b_n - LM| < |a_n - L|B + |L||b_n - M|$$

$$< \frac{\varepsilon}{2B + 1}B + |L|\frac{\varepsilon}{2|L| + 1}$$

$$< \frac{\varepsilon}{2} + \frac{\varepsilon}{2} = \varepsilon.$$

(4) Although the algebra is more complicated, the same strategy works here. Assuming $b_n \neq 0$, we can write

(2.4.4)

$$\left|\frac{a_n}{b_n} - \frac{L}{M}\right| = \left|\frac{a_n M - Lb_n}{b_n M}\right|$$

$$= \left|\frac{a_n M - LM + LM - Lb_n}{b_n M}\right|$$

$$\leq \left|\frac{a_n M - LM}{b_n M}\right| + \left|\frac{LM - Lb_n}{b_n M}\right|$$

$$= |a_n - L|\left|\frac{1}{b_n}\right| + |M - b_n|\left|\frac{L}{b_n M}\right|.$$

We would like to repeat the arguments of the previous case. As in case (3), we must replace $1/|b_n|$ and $|L/(Mb_n)|$ with constant upper bounds.

Since $M \neq 0$, $|M|/2 > 0$. Using $|M|/2$ as our ε, we can find some N_1 so that

$$|b_n - M| < \frac{|M|}{2} \quad \text{for all} \quad n \geq N_1.$$

The key point is that $|b_n| \geq |M|/2$. Taking the reciprocals of both sides, we get

$$\frac{1}{|b_n|} \leq \frac{2}{|M|} \quad \text{for all} \quad n \geq N_1.$$

So continuing from (2.4.4), we have

$$\left|\frac{a_n}{b_n} - \frac{L}{M}\right| \leq |a_n - L|\left|\frac{1}{b_n}\right| + |M - b_n|\left|\frac{L}{b_n M}\right|$$

$$\leq |a_n - L|\left|\frac{2}{M}\right| + |M - b_n|\left|\frac{2L}{M^2}\right|.$$

Now we are set to use the ideas of the previous part. Choose N_2 so that

$$|a_n - L| < \frac{\varepsilon|M|}{4} \quad \text{for all} \quad n \geq N_2;$$

and choose N_3 so that

$$|b_n - M| < \frac{\varepsilon|M|^2}{4|L| + 1} \quad \text{for all} \quad n \geq N_3.$$

Now if $n \geq \max\{N_1, N_2, N_3\}$, then since $b_n \geq |M|/2 > 0$, we know b_n is not zero and we can use (2.4.4) and previous paragraph to conclude

$$\left| \frac{a_n}{b_n} - \frac{L}{M} \right| \leq |a_n - L| \frac{2}{|M|} + |M - b_n| \frac{2|L|}{|M|^2}$$

$$< \frac{\varepsilon |M|}{4} \frac{2}{|M|} + \frac{\varepsilon |M|^2}{4|L| + 1} \frac{2|L|}{|M|^2}$$

$$< \frac{\varepsilon}{2} + \frac{\varepsilon}{2} = \varepsilon.$$

■

Exercises for Section 2.4

A. Give a careful proof using the definition of limit of the fact that when $\lim\limits_{n \to \infty} a_n = L$ and $\lim\limits_{n \to \infty} b_n = M$, then $\lim\limits_{n \to \infty} 2a_n + 3b_n = 2L + 3M$.

B. Compute the following limits.

(a) $\lim\limits_{n \to \infty} \dfrac{\tan \frac{\pi}{n}}{n \sin^2 \frac{2}{n}}$ (b) $\lim\limits_{n \to \infty} \dfrac{2^{100+5n}}{e^{4n-10}}$ (c) $\lim\limits_{n \to \infty} \dfrac{\csc \frac{1}{n}}{n} + \dfrac{2 \tan^{-1} x}{\log n}$

C. If $\lim\limits_{n \to \infty} a_n = L > 0$, prove that $\lim\limits_{n \to \infty} \sqrt{a_n} = \sqrt{L}$. Be sure to discuss the issue of when $\sqrt{a_n}$ makes sense.
HINT: Express $|\sqrt{a_n} - \sqrt{L}|$ in terms of $|a_n - L|$.

D. Let $(a_n)_{n=1}^{\infty}$ and $(b_n)_{n=1}^{\infty}$ be two sequences of real numbers such that $|a_n - b_n| < \frac{1}{n}$. Suppose that $L = \lim\limits_{n \to \infty} a_n$ exists. Show that $(b_n)_{n=1}^{\infty}$ converges to L also.

E. Find $\lim\limits_{n \to \infty} \dfrac{\log(2 + 3^n)}{2n}$.   HINT: $\log(2 + 3^n) = \log 3^n + \log \frac{2+3^n}{3^n}$

F. (a) Let $x_n = \sqrt[n]{n} - 1$. Use the fact that $(1 + x_n)^n = n$ to show that $x_n^2 \leq 2/n$.
      HINT: Use the Binomial Theorem and throw away most terms.
    (b) Hence compute $\lim\limits_{n \to \infty} n^{1/n}$.

G. Show that the set of rational numbers is **dense** in $\mathbb{R}$, meaning that every real number is a limit of rational numbers.

H. (a) Show that $\frac{b-1}{b} \leq \log b \leq b - 1$.   HINT: Integrate $1/x$ from 1 to b.
    (b) Apply this to $b = \sqrt[n]{a}$ to show that $\log a \leq n(\sqrt[n]{a} - 1) \leq \sqrt[n]{a} \log a$.
    (c) Hence evaluate $\lim\limits_{n \to \infty} n(\sqrt[n]{a} - 1)$.

I. Suppose that $\lim\limits_{n \to \infty} a_n = L$. Show that $\lim\limits_{n \to \infty} \dfrac{a_1 + a_2 + \cdots + a_n}{n} = L$.

J. Show that the set $S = \{n + m\sqrt{2} : m, n \in \mathbb{Z}\}$ is dense in $\mathbb{R}$.
HINT: Find infinitely many of these numbers in $[0, 1]$ and use the Pigeonhole Principle to find two which are close within 10^{-k}.

2.5. Upper and Lower Bounds

2.5.1. DEFINITION. If S is a nonempty subset of $\mathbb{R}$ that is bounded above, the **supremum** or **least upper bound** is the number L such that

$$s \leq L \quad \text{for all} \quad s \in S$$

and whenever L' is another upper bound for S, then $L' \geq L$. This value is denoted $\sup S$.

Similarly, if S is a nonempty subset of $\mathbb{R}$ that is bounded below, the **infimum** or **greatest lower bound** is the number L such that

$$s \geq L \quad \text{for all} \quad s \in S$$

and whenever L' is another lower bound for S, then $L' \leq L$. This value is denoted $\inf S$.

It is not obvious that supremums and infimums will always exist. The next result will describe precisely when they do exist. However, a finite set will always have a supremum and an infimum—just pick the biggest element and the smallest element of the set respectively. An *infinite* set may not have biggest and smallest elements.

2.5.2. EXAMPLES.

(1) If $A = \{4, -2, 5, 7\}$, then $\inf A = -2$ and $\sup A = 7$. Both supremum and infimum belong to A.

(2) If $B = \{2, 4, 6, \ldots\}$, then $\inf B = 2$ and $\sup B$ does not exist. Sometimes we write $\sup B = +\infty$ to mean there are arbitrarily large elements of B. The infimum belongs to B.

(3) If $C = \{\pi/n : n \in \mathbb{N}\}$, then $\sup C = \pi$ and $\inf C = 0$. The supremum is the maximum and belongs to C. The infimum does not.

(4) If $D = \left\{\dfrac{(-1)^n n}{n+1} : n \in \mathbb{N}\right\}$, then $\inf D = -1$ and $\sup D = 1$. Neither 1 nor -1 belong to D.

In the next result, our definition of the real numbers as *all* infinite decimals plays a crucial role. This result is not true for subsets of rational numbers for this very reason. For example, the set $\{s \in \mathbb{Q} : s^2 < 2\}$ has no least upper bound in $\mathbb{Q}$, but $\sqrt{2}$ is the supremum in $\mathbb{R}$.

2.5.3. LEAST UPPER BOUND PRINCIPLE.
Every nonempty subset S of $\mathbb{R}$ that is bounded above has a supremum. Similarly, every nonempty subset S of $\mathbb{R}$ that is bounded below has an infimum.

PROOF. We prove the second statement first, as it is more convenient. Let M be some lower bound for S with decimal expansion $M = m_0.m_1m_2\dots$. Let s be some element of S with decimal expansion $s = s_0.s_1s_2\dots$. Note that m_0 is a lower bound for S but $s_0 + 1$ is not. There are only finitely many integers between m_0 and s_0. Pick the largest of these that is still a lower bound for S, and call it a_0. Since $a_0 + 1$ is not a lower bound, we also choose an element x_0 in S such that $x_0 < a_0 + 1$.

Next pick the greatest integer a_1 such that $y_1 = a_0 + 10^{-1}a_1$ is a lower bound for S. Since $a_1 = 0$ works and $a_1 = 10$ does not, a_1 belongs to $\{0, 1, \dots, 9\}$. To verify our choice, pick an element x_1 in S such that

$$x_1 < a_0 + 10^{-1}(a_1 + 1).$$

Continue in this way recursively. At the kth stage, we find the largest integer a_k in $\{0, 1, \dots, 9\}$ such that $y_k = a_0.a_1a_2\dots a_k = \sum_{i=0}^{k} 10^{-i}a_i$ is a lower bound for S. Since $y_k + 10^{-k}$ is not a lower bound, we also pick an element x_k in S such that $x_k < y_k + 10^{-k}$ to verify our choice. The integer a_{k+1} is now chosen by induction. Figure 2.2 shows how a_2 and x_2 would be chosen.

FIGURE 2.2. The second stage ($k = 2$) in the proof.

Let $L = a_0.a_1a_2 \cdots = \lim_{k\to\infty} y_k$. We claim that L is the infimum of S. First note that if $s \in S$, then $y_k \le s$ for all $k \ge 0$. Therefore

$$L = \lim_{k\to\infty} y_k \le \lim_{k\to\infty} s = s.$$

So L is a lower bound.

Now if $L' = b_0.b_1b_2 \cdots > L$, there is some first integer k such that $b_k > a_k$ and $b_i = a_i$ for $0 \le i < k$. But then

$$L' = a_0.a_1\dots a_{k-1}b_k \ge y_k + 10^{-k} > x_k.$$

Thus L' is not a lower bound for S. Hence L is the greatest lower bound.

To deal with upper bounds, notice that L is an upper bound for S precisely when $-L$ is a lower bound for $-S = \{-s : s \in S\}$. Thus $\sup S = -\inf(-S)$ exists. ∎

A sequence (a_n) is **(strictly) monotone increasing** if $a_n \le a_{n+1}$ ($a_n < a_{n+1}$) for all $n \ge 1$. Similarly, we define (strictly) monotone decreasing sequences.

2.5.4. MONOTONE CONVERGENCE THEOREM.
A monotone increasing sequence that is bounded above converges. A monotone decreasing sequence that is bounded below converges.

PROOF. Suppose $(a_n)_{n=1}^{\infty}$ is an increasing sequence that is bounded above. Then by the Least Upper Bound Principle, there is a number

$$L = \sup\{a_n : n \in \mathbb{N}\}.$$

We will show $\lim_{n\to\infty} a_n = L$.

Let $\varepsilon > 0$ be given. Since $L - \varepsilon$ is not an upper bound for A, there is some integer N so that $a_N > L - \varepsilon$. Then because the sequence is monotone increasing,

$$L - \varepsilon < a_N \le a_n \le L \quad \text{for all} \quad n \ge N.$$

So $|a_n - L| < \varepsilon$ for all $n \ge N$ as required. Therefore, $\lim_{n\to\infty} a_n = L$.

The case of a decreasing sequence is similar. It is less work to deduce it as a consequence of the result for increasing sequences. That is, if a_n is decreasing and bounded below by B, then the sequence $-a_n$ is increasing and bounded above by $-B$. Thus the sequence $(-a_n)_{n=1}^{\infty}$ has a limit $L = \lim_{n\to\infty} - a_n$. Therefore $-L = \lim_{n\to\infty} a_n$. ∎

2.5.5. EXAMPLE. Consider the sequence given recursively by

$$a_1 = 1 \quad \text{and} \quad a_{n+1} = \sqrt{2 + \sqrt{a_n}} \quad \text{for all} \quad n \ge 1.$$

Evaluating the first few terms, we obtain

$$1, \quad 1.7320508076, \quad 1.8210090645, \quad 1.8301496356, \quad 1.8310735189,$$
$$1.831166746, \quad 1.8311761518, \quad 1.8311771007, \quad 1.8311771965, \quad \dots.$$

It appears that this sequence increases to some limit.

First we show by induction that

$$1 \le a_n < a_{n+1} < 2 \quad \text{for all} \quad n \ge 1.$$

Since $1 = a_1 < \sqrt{3} = a_2 < 2$, this is valid for $n = 1$. Suppose that it holds for some n. Then

$$a_{n+2} = \sqrt{2 + \sqrt{a_{n+1}}} > \sqrt{2 + \sqrt{a_n}} = a_{n+1} \ge 1,$$

and

$$a_{n+2} = \sqrt{2 + \sqrt{a_{n+1}}} < \sqrt{2 + \sqrt{2}} < 2.$$

This verifies our claim at the next step, and hence by induction, it is valid for each $n \ge 1$.

Therefore, (a_n) is a monotone increasing sequence. So by the Monotone Convergence Theorem (Theorem 2.5.4), it follows that there is a limit $L = \lim_{n\to\infty} a_n$. It is not clear that there is a nice expression for L. However, once we know the sequence converges, it is not hard to find a formula for L. Notice that

$$L = \lim_{n\to\infty} a_{n+1} = \lim_{n\to\infty} \sqrt{2 + \sqrt{a_n}}$$

$$= \sqrt{2 + \sqrt{\lim_{n\to\infty} a_n}} = \sqrt{2 + \sqrt{L}}.$$

Here we use the fact that the limit of square roots is the square root of the limit (see Exercise 2.4.C). Squaring both sides gives $L^2 - 2 = \sqrt{L}$, and further squaring yields

$$0 = L^4 - 4L^2 - L + 4 = (L-1)(L^3 + L^2 - 3L - 4).$$

Since $L > 1$, it must be a root of the cubic $p(x) = x^3 + x^2 - 3x - 4$ in the interval $(1, 2)$. There is only one such root, as graphing the curve shows (see Figure 2.3). Indeed,

$$p'(x) = 3x^2 + 2x - 3 = 3(x^2 - 1) + 2x$$

is positive on $[1, 2]$, and so p is strictly increasing. As $p(1) = -5$ and $p(2) = 2$, it follows that p has exactly one root in between. (See the Intermediate Value Theorem, Theorem 5.6.1.)

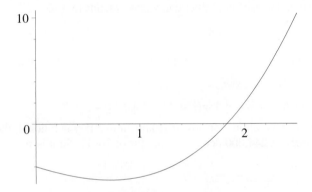

FIGURE 2.3. Graph of $x^3 + x^2 - 3x - 4$.

For the amusement of the reader, we give an explicit algebraic formula:

$$L = \frac{1}{3}\left(\sqrt[3]{\tfrac{79+\sqrt{2241}}{2}} + \sqrt[3]{\tfrac{79-\sqrt{2241}}{2}} - 1 \right).$$

Notice that we proved that the sequence converged first and then evaluated the limit afterwards. This is important, for consider the sequence given by $a_1 = 2$ and $a_{n+1} = (a_n^2 + 1)/2$. This is a monotone increasing sequence. Suppose we let L denote the limit and compute

$$L = \lim_{n\to\infty} a_{n+1} = \lim_{n\to\infty} (a_n^2 + 1)/2 = (L^2 + 1)/2.$$

Thus $(L-1)^2 = 0$, which means that $L = 1$. This is an absurd conclusion because this sequence is monotone increasing and greater than 2. The fault lay in assuming that the limit L actually exists, because instead it diverges to $+\infty$.

Exercises for Section 2.5

A. We say that $\lim_{n\to\infty} a_n = +\infty$ if for every real number R, there is an integer N so that $a_n > R$ for all $n \geq N$. Show that a divergent monotone increasing sequence converges to infinity in this sense.

B. Let $a_1 = 0$ and $a_{n+1} = \sqrt{5 + 2a_n}$ for $n \geq 1$. Show that this sequence converges and find the limit.

C. Is the set $S = \{x \in \mathbb{R} : 0 < \sin(\frac{1}{x}) < \frac{1}{2}\}$ bounded above (below)? Find $\sup S$ and $\inf S$.

D. (a) Evaluate $\lim\limits_{n \to \infty} \sqrt[n]{3^n + 5^n}$.
(b) Show that this sequence is monotone decreasing.

E. Let a, b be positive real numbers. Set $x_0 = a$ and $x_{n+1} = \dfrac{1}{x_n^{-1} + b}$ for $n \geq 0$.

(a) Prove that x_n is monotone decreasing.
(b) Prove that the limit exists and find it.

F. Suppose (a_n) is a sequence of positive real numbers such that $a_{n+1} - 2a_n + a_{n-1} > 0$ for all $n \geq 1$. Prove that the sequence either converges or tends to $+\infty$.

G. **Euler's constant**, denoted γ, is defined as $\gamma = \lim\limits_{n \to \infty} 1 + \frac{1}{2} + \frac{1}{3} + \cdots + \frac{1}{n} - \log n$.

Let $a_n = \left(\sum\limits_{k=1}^{n} \frac{1}{k}\right) - \log n$ for $n \geq 1$. Show that $(a_n)_{n=1}^{\infty}$ is decreasing and bounded below by zero, and so this limit exists.

HINT: Prove the inequality $\dfrac{1}{n+1} \leq \log(n+1) - \log n \leq \dfrac{1}{n}$.
(It is unknown whether γ is rational or not. It is known that if γ is rational, then the denominator has more than 244,000 decimal digits [as of 2001]. So it is *suspected* to be irrational.)

H. Let $x_n = \sqrt{1 + \sqrt{2 + \sqrt{3 + \cdots + \sqrt{n}}}}$.

(a) Show that $x_n < x_{n+1}$.
(b) Show that $x_{n+1}^2 \leq 1 + \sqrt{2}x_n$.
    HINT: Square x_{n+1} and factor a 2 out of the square root.
(c) Hence show that x_n is bounded above by 2. Deduce that $\lim\limits_{n \to \infty} x_n$ exists.

I. (a) Let $(a_n)_{n=1}^{\infty}$ be a bounded sequence. Define a sequence $b_n = \sup\{a_k : k \geq n\}$ for $n \geq 1$. Prove that (b_n) converges. This limit is called the **limit superior** of (a_n), almost always abreviated to $\limsup a_n$.
(b) Without redoing the proof, do the same for the **limit inferior** of (a_n), which is defined as $\liminf a_n := \lim\limits_{n \to \infty} \left(\inf\limits_{k \geq n} a_k\right)$.

J. Show that a sequence $(a_n)_{n=1}^{\infty}$ converges if and only if $\limsup a_n = \liminf a_n$.

K. Suppose that $(a_n)_{n=1}^{\infty}$ is a sequence of positive real numbers. Show that

$$\limsup \frac{1}{a_n} = \frac{1}{\liminf a_n}.$$

L. Suppose that $(a_n)_{n=1}^{\infty}$ and $(b_n)_{n=1}^{\infty}$ are sequences of positive real numbers and that $\limsup \dfrac{a_n}{b_n} < \infty$. Prove that there is a constant M so that $a_n \leq Mb_n$ for all $n \geq 1$.

M. Suppose that the real numbers were defined using Dedekind cuts at the end of Section 2.2. Show that if S is a family of Dedekind cuts that is bounded above, then the union of all of these cuts is also a cut and represents the least upper bound.

2.6. Subsequences

Even if a sequence does not converge, by picking out certain terms of the sequence, it may be possible to find a new convergent sequence.

2.6.1. DEFINITION. A **subsequence** of a sequence $(a_n)_{n=1}^{\infty}$ is a new sequence $(a_{n_k})_{k=1}^{\infty} = (a_{n_1}, a_{n_2}, a_{n_3}, \ldots)$, where $n_1 < n_2 < n_3 < \ldots$.

For example, $(a_{2n})_{n=1}^{\infty}$ and $(a_{n^3})_{n=1}^{\infty}$ are subsequences. It is easy to verify that if $(a_n)_{n=1}^{\infty}$ converges to a limit L, then $(a_{n_k})_{k=1}^{\infty}$ also converges to the same limit.

The sequence $a_n = n$ does not have a limit, nor does any subsequence because any subsequence tends to $+\infty$. However, we will show that as long as a sequence remains bounded, it has subsequences that converge.

2.6.2. NESTED INTERVALS LEMMA.
Suppose that
$$I_n = [a_n, b_n] = \{x \in \mathbb{R} : a_n \leq x \leq b_n\}$$
are nonempty closed intervals such that I_{n+1} is contained in I_n for each $n \geq 1$. Then the intersection $\bigcap_{n \geq 1} I_n$ is nonempty.

2.6.3. REMARK. This proof of this result is so easy that a couple of warning remarks in advance are appropriate. This result depends in a crucial way on the completeness of the reals implicit in the existence of a real number for every infinite decimal expansion. This was used in establishing the Least Upper Bound Principle, which we use here in the guise of the convergence of monotone sequences.

The result is false for intervals of rational numbers. In Example 2.6.8, we construct a sequence of closed intervals with rational endpoints that intersect exactly in the point $\sqrt{8}$, which is not rational. So the corresponding intervals of rational numbers intersects in the empty set.

Likewise, this result is false for open intervals. Indeed, $\bigcap_{n \geq 1}(0, \frac{1}{n}) = \varnothing$.

PROOF. Notice that since I_{n+1} is contained in I_n, it follows that
$$a_n \leq a_{n+1} \leq b_{n+1} \leq b_n.$$
Thus (a_n) is a monotone increasing sequence bounded above by any b_k; and likewise (b_n) is a monotone decreasing sequence bounded below by any a_k. Hence by Corollary 2.5.4, $a = \lim_{n \to \infty} a_n$ exists; as does $b = \lim_{n \to \infty} b_n$. By Exercise 2.3.D, $a \leq b$. Thus
$$a_k \leq a \leq b \leq b_k.$$
Consequently, the point a belongs to each I_k for each $k \geq 1$. ∎

This is just the tool needed to establish the key result of this section.

2.6.4. BOLZANO–WEIERSTRASS THEOREM.
Every bounded sequence of real numbers has a convergent subsequence.

PROOF. Let (a_n) be a sequence bounded by B. Thus the interval $[-B, B]$ contains the whole (infinite) sequence. Now if I is an interval containing infinitely many points of the sequence (a_n), and $I = J_1 \cup J_2$ is the union of two smaller intervals, then at least one of them contains infinitely many points of the sequence too.

So let $I_1 = [-B, B]$. Split it into two closed intervals of length B, namely $[-B, 0]$ and $[0, B]$. One of these halves contains infinitely many points of (a_n); call it I_2. Similarly, divide I_2 into two closed intervals of length $B/2$. Again pick one, called I_3, that contains infinitely many points of our sequence. Recursively, we construct a decreasing sequence I_k of closed intervals of length $2^{2-k}B$ so that each contains infinitely many points of our sequence. Figure 2.4 shows the choice of I_3 and I_4, where the terms of the sequence are indicated by vertical lines.

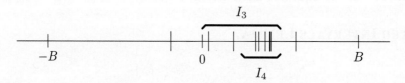

FIGURE 2.4. Choice of intervals I_3 and I_4.

By the Nested Interval Lemma, we know that $\bigcap_{k \geq 1} I_k$ contains a point L. Now choose an increasing sequence n_k such that a_{n_k} belongs to I_k. This is possible since it contains infinitely many points in the sequence, and only finitely many are less than n_{k-1}. It will be shown that

$$\lim_{k \to \infty} a_{n_k} = L.$$

Indeed, both a_{n_k} and L belong to I_k, and hence

$$|a_{n_k} - L| \leq |I_k| = 2^{-k}(4B).$$

The right-hand side tends to 0, and thus $\lim_{k \to \infty} a_{n_k} = L$. ∎

2.6.5. EXAMPLE. Consider the sequence $(a_n) = (\text{sign}(\sin n))_{n=1}^{\infty}$, where the sign function takes values ± 1 depending on the sign of x except for $\text{sign} \, 0 = 0$. Without knowing anything about the properties of the sin function, we can observe that the sequence (a_n) takes at most three different values. At least one of these values is taken infinitely often. Thus it is possible to deduce the existence of a subsequence that is constant and therefore converges.

Using our knowledge of sin allows us to get somewhat more specific. Now $\sin x = 0$ exactly when x is an integer multiple of π. Since π is irrational, $k\pi$ is never an integer for $k > 0$. Therefore, a_n takes only the values ± 1. Note that

$\sin x > 0$ if there is an integer k so that $2k\pi < x < (2k + 1)\pi$; and $\sin x < 0$ if there is an integer k so that $(2k - 1)\pi < x < 2k\pi$. Observe that n increases by steps of length 1 while the intervals on which $\sin x$ takes positive or negative values has length $\pi \approx 3.14$. Consequently, a_n takes the value $+1$ for three or four terms in a row, followed by three or four terms taking the value -1.

Consequently, both 1 and -1 are limits of certain subsequences of (a_n). To compute a particular sequence n_k for which $a_{n_k} = 1$ for all k requires a much more delicate analysis depending on π. One of the nice things about analysis is that one can often make significant use of such a sequence *without* knowing the details of which subsequence is used.

2.6.6. EXAMPLE. Consider the sequence $(a_n) = (\sin n)_{n=1}^{\infty}$. As the angles n radians for $n \geq 1$ are marked on a circle, they appear gradually to fill in a dense subset. If this can be demonstrated, we should be able to show that $\sin\theta$ is a limit of a subsequence of our sequence for any θ in $[0, 2\pi]$. The key is to approximate the angle 0 modulo 2π by integers.

This can be done using the Pigeonhole Principle. Let m be a positive integer and $\varepsilon > 0$ be any positive real number. Choose an integer N so large that $N\varepsilon > 2\pi$. Divide the circle into N arcs of length $2\pi/N$ radians each. Then consider the $N+1$ points $0, m, 2m, \ldots, Nm$ modulo 2π on the circle. Since there are $N + 1$ points distributed into only N arcs, at least one of them contains two points, say im and jm, where $i < j$. Then $n = jm - im$ represents an angle of at most $2\pi/N < \varepsilon$ radians up to a multiple of 2π. That is, $n = \psi + 2\pi s$ for some integer s and real number $|\psi| < \varepsilon$. In particular, $|\sin n| < \varepsilon$ and $n \geq m$. Moreover, since π is not rational, n is not an exact multiple of 2π.

So given $\theta \in [0, 2\pi]$, construct a subsequence as follows. Let $n_1 = 1$. Recursively we construct an increasing sequence n_k such that

$$|\sin n_k - \sin\theta| < \frac{1}{k}.$$

Once n_k is defined, take $\varepsilon = \frac{1}{k+1}$ and $m = n_k + 1$ in the previous paragraph. This provides an integer $n > n_k$ such that $n = \psi + 2\pi s$ and $|\psi| < \frac{1}{k+1}$. Thus there is a positive integer t such that $|\theta - t\psi| < \frac{1}{k+1}$. Therefore

$$(2.6.7) \qquad |\sin(tn) - \sin(\theta)| = |\sin(t\psi) - \sin(0)| < |t\psi - \theta| < \frac{1}{k+1}.$$

Set $n_{k+1} = tn$. This completes the induction. The result is a subsequence such that

$$\lim_{k\to\infty} \sin(n_k) = \sin\theta.$$

To verify equation (2.6.7), we could use the Mean Value Theorem or we can use trig identities. See the Exercises for the latter, more computational argument. If $f(x) = \sin x$, then $f'(x) = \cos x$. So by the Mean Value Theorem (Theorem 6.2.4), there is a point ξ between $t\psi$ and θ such that

$$\left|\frac{\sin(t\psi) - \sin(\theta)}{t\psi - \theta}\right| = |\cos\xi| \leq 1.$$

Rearranging yields $|\sin(t\psi) - \sin(\theta)| < |t\psi - \theta|$.

Therefore, we have shown that every value in the interval $[-1, 1]$ is the limit of some subsequence of the sequence $(\sin n)_{n=1}^{\infty}$.

2.6.8. EXAMPLE. Consider the sequence $b_1 = 3$ and $b_{n+1} = (b_n + 8/b_n)/2$. Notice that

$$
\begin{aligned}
b_{n+1}^2 - 8 &= \frac{b_n^2 + 16 + (64/b_n^2) - 32}{4} \\
&= \frac{b_n^2 - 16 + (64/b_n^2)}{4} \\
&= \frac{(b_n - 8/b_n)^2}{4} = \frac{(b_n^2 - 8)^2}{4b_n^2}.
\end{aligned}
$$

It follows that $b_n^2 > 8$ for all $n \geq 2$, and $b_1^2 - 8 = 1 > 0$ also. Thus

$$
0 < b_{n+1}^2 - 8 < \frac{(b_n^2 - 8)^2}{32}.
$$

Iterating this, we obtain $b_2^2 - 8 < 32^{-1}$, $b_3 < 32^{-3}$ and $b_4 < 32^{-7}$. In general, we establish by induction that

$$
0 < b_n^2 - 8 < 32^{1-2^{n-1}}.
$$

Since b_n is positive and $b^2 - 8 = (b - \sqrt{8})(b + \sqrt{8})$, it follows that

$$
\begin{aligned}
0 < b_n - \sqrt{8} &= \frac{b_n^2 - 8}{b_n + \sqrt{8}} \\
&< \frac{32^{1-2^{n-1}}}{2\sqrt{8}} < 6(32^{-2^{n-1}}).
\end{aligned}
$$

Lastly, using the fact that $32^2 = 1024 > 10^3$, we obtain

$$
0 < b_n - \sqrt{8} < 10 \cdot 10^{-3 \cdot 2^{n-2}}.
$$

In particular, $\lim_{n \to \infty} b_n = \sqrt{8}$. In fact, the convergence is so rapid that b_{10} approximates $\sqrt{8}$ to more than 750 digits of accuracy. See Example 11.2.2 for a more general analysis using Newton's method.

Now let $a_n = 8/b_n$. Then a_n is monotone increasing to $\sqrt{8}$. Both a_n and b_n are rational numbers. Thus the sets $J_n = \{x \in \mathbb{Q} : a_n \leq x \leq b_n\}$ form a decreasing sequence of nonempty intervals of rational numbers with empty intersection.

Exercises for Section 2.6

A. Show that $(a_n) = \left(\frac{n \cos^n(n)}{\sqrt{n^2 + 2n}} \right)_{n=1}^{\infty}$ has a convergent subsequence.

B. Does the sequence $(b_n) = \left(n + \cos(n\pi)\sqrt{n^2 + 1} \right)_{n=1}^{\infty}$ have a convergent subsequence?

C. Does the sequence $(a_n) = \left(\cos \log n\right)_{n=1}^{\infty}$ converge?

D. Show that every sequence has a monotone subsequence.

E. Use trig identities to show that $|\sin x - \sin y| \leq |x - y|$.
HINT: Let $a = (x + y)/2$ and $b = (x - y)/2$. Use the addition formula for $\sin(a \pm b)$.

F. Define $x_1 = 2$ and $x_{n+1} = \frac{1}{2}(x_n + 5/x_n)$ for $n \geq 1$.
(a) Find a formula for $x_{n+1}^2 - 5$ in terms of $x_n^2 - 5$.
(b) Hence evaluate $\lim_{n \to \infty} x_n$.
(c) Compute the first ten terms on a calculator.
(d) Show that the tenth term approximates the limit to over 200 decimal places.

G. Let $(x_n)_{n=1}^{\infty}$ be a sequence of real numbers. Suppose that there is a real number L so that $L = \lim_{n \to \infty} x_{3n-1} = \lim_{n \to \infty} x_{3n+1} = \lim_{n \to \infty} x_{3n}$. Show that $\lim_{n \to \infty} x_n$ exists and equals L.

H. Let $(x_n)_{n=1}^{\infty}$ be a sequence of real numbers. Suppose that there is a real number L with the property that every subsequence $(x_{n_k})_{k=1}^{\infty}$ has a subsubsequence $(x_{n_{k(l)}})_{l=1}^{\infty}$ with $\lim_{l \to \infty} x_{n_{k(l)}} = L$. Show that the whole sequence converges to L.
HINT: If it were false, you could find a subsequence bounded away from L.

I. Suppose that $(x_n)_{n=1}^{\infty}$ is a sequence of real numbers. Also suppose that L_k are real numbers with $\lim_{k \to \infty} L_k = L$. If for each $k \geq 1$, there is a subsequence of $(x_n)_{n=1}^{\infty}$ converging to L_k, show that some subsequence converges to L.
HINT: Find an increasing sequence n_k so that $|x_{n_k} - L| < 1/k$.

J. Suppose that $(x_n)_{n=1}^{\infty}$ is a sequence of real numbers.
(a) If $L = \liminf x_n$, show that there is a subsequence $(x_{n_k})_{k=1}^{\infty}$ so that $\lim_{k \to \infty} x_{n_k} = L$.
(b) Similarly, prove that there is a subsequence $(x_{n_l})_{l=1}^{\infty}$ so that $\lim_{l \to \infty} x_{n_l} = \limsup x_n$.

K. Let $(x_n)_{n=1}^{\infty}$ be an arbitrary sequence of real numbers. Prove that there is a subsequence $(x_{n_k})_{k=1}^{\infty}$ so that either $\lim_{k \to \infty} x_{n_k} = \infty$ or $\lim_{k \to \infty} x_{n_k} = -\infty$, or there is a real number L so that $\lim_{k \to \infty} x_{n_k} = L$.

L. Construct a sequence $(x_n)_{n=1}^{\infty}$ so that for every real number L, there is a subsequence $(x_{n_k})_{k=1}^{\infty}$ with $\lim_{k \to \infty} x_{n_k} = L$.

2.7. Cauchy Sequences

Can we decide if a sequence converges *without* finding the potential limit value? We are looking for an intrinsic property of a sequence that is equivalent to convergence and that doesn't use any information about the limit. This leads us to the notion of a space being *complete* if all sequences that are "supposed" to converge actually do. As we shall see, this completeness property has been built into the real numbers by our construction of all infinite decimals.

To obtain an appropriate condition, notice that if a sequence (a_n) converges to L, then as well as getting close to the limit, the terms of the sequence are getting close to each other. To be precise,

2.7.1. PROPOSITION. *Let $(a_n)_{n=1}^{\infty}$ be a sequence converging to a limit L. Then for every $\varepsilon > 0$, there is an integer N such that*

$$|a_n - a_m| < \varepsilon \quad \text{for all} \quad m, n \geq N.$$

PROOF. Fix $\varepsilon > 0$ and use the value $\varepsilon/2$ in the definition of limit. It follows that there is an integer N so that

$$|a_n - L| < \frac{\varepsilon}{2} \quad \text{for all} \quad n \geq N.$$

Thus if $m, n \geq N$, we obtain

$$|a_n - a_m| \leq |a_n - L| + |L - a_m| < \frac{\varepsilon}{2} + \frac{\varepsilon}{2} = \varepsilon. \qquad \blacksquare$$

We make the conclusion of this proposition into a definition.

2.7.2. DEFINITION. A sequence $(a_n)_{n=1}^{\infty}$ of real numbers is called a **Cauchy sequence** provided that for *every $\varepsilon > 0$*, there is an integer N so that

$$|a_m - a_n| < \varepsilon \quad \text{for all} \quad m, n \geq N.$$

This definition retains the flavour of the definition of a limit, in that it has the same logical structure: *For all $\varepsilon > 0$, there is an integer N ....* However, it does not require the use of a potential limit L. This permits the following definition.

2.7.3. DEFINITION. A subset S of $\mathbb{R}$ is said to be **complete** if every Cauchy sequence (a_n) in S (that is, $a_n \in S$) converges to a point in S.

This brings us to an important conclusion about the real numbers themselves.

2.7.4. COMPLETENESS THEOREM.
A sequence of real numbers converges if and only if it is a Cauchy sequence. In particular, $\mathbb{R}$ is complete.

PROOF. Proposition 2.7.1 shows that convergent sequences are Cauchy.

Conversely, suppose that $(a_n)_{n=1}^{\infty}$ is a Cauchy sequence. First we show that the set $\{a_n : n \geq 1\}$ is bounded. The proof is basically the same as Proposition 2.4.2. Indeed, take $\varepsilon = 1$ and find N sufficiently large that

$$|a_n - a_N| < 1 \quad \text{for all} \quad n \geq N.$$

It follows that the sequence is bounded by

$$\max\{|a_1|, |a_2|, \ldots, |a_{N-1}|, |a_N| + 1\}.$$

By the Bolzano–Weierstrass Theorem 2.6.4, this sequence has a convergent subsequence, say

$$\lim_{k \to \infty} a_{n_k} = L.$$

Now let $\varepsilon > 0$. From the definition of Cauchy sequence for $\varepsilon/2$, there is an integer N so that

$$|a_m - a_n| < \frac{\varepsilon}{2} \quad \text{for all} \quad m, n \geq N.$$

And from the definition of limit using $\varepsilon/2$, there is an integer K so that

$$|a_{n_k} - L| < \frac{\varepsilon}{2} \quad \text{for all} \quad k \geq K.$$

Pick any $k \geq K$ such that $n_k \geq N$. Then for every $n \geq N$,

$$|a_n - L| = |a_n - a_{n_k}| + |a_{n_k} - L| < \frac{\varepsilon}{2} + \frac{\varepsilon}{2} = \varepsilon.$$

So $\lim_{n \to \infty} a_n = L$. ∎

2.7.5. REMARK. This theorem is not true for the rational numbers. Here is an example of a Cauchy sequence of rational numbers that does not converge to a rational number. Define the sequence $(a_n)_{n=1}^{\infty}$ by

$$a_1 = 1.4, \quad a_2 = 1.41, \quad a_3 = 1.414, \quad a_4 = 1.4142, \quad a_5 = 1.41421, \ldots$$

and in general, a_n is the first $n + 1$ digits in the decimal expansion of $\sqrt{2}$. If n and m are greater than N, then a_n and a_m agree for at least first $N + 1$ digits. Thus

$$|a_n - a_m| < 10^{-N} \quad \text{for all} \quad m, n \geq N.$$

This shows that $(a_n)_{n=1}^{\infty}$ is a Cauchy sequence. (Why?)

However, this sequence has no limit *in the rationals*. In our terminology, $\mathbb{Q}$ is not complete. Of course, this sequence does converge to a real number, namely $\sqrt{2}$. This is the essential difference between $\mathbb{R}$ and $\mathbb{Q}$: The set of real numbers is complete and $\mathbb{Q}$ is not.

2.7.6. EXAMPLE. Let α be an arbitrary real number. Define $a_n = [n\alpha]/n$, where $[x]$ is the nearest integer to x. Then $\big|[n\alpha] - n\alpha\big| \leq 1/2$. So

$$|a_n - \alpha| = \frac{\big|[n\alpha] - n\alpha\big|}{n} \leq \frac{1}{2n}.$$

Therefore, $\lim_{n \to \infty} a_n = \alpha$. Indeed, given $\varepsilon > 0$, choose N so large that $\frac{1}{N} < \varepsilon$. Then for $n \geq N, |a_n - \alpha| < \varepsilon/2$. Moreover if $m, n \geq N$,

$$|a_n - a_m| \leq |a_n - \alpha| + |\alpha - a_m| < \frac{\varepsilon}{2} + \frac{\varepsilon}{2} = \varepsilon.$$

Thus this sequence is Cauchy.

2.7.7. EXAMPLE. Consider the infinite **continued fraction**

$$\cfrac{1}{2 + \cfrac{1}{2 + \cfrac{1}{2 + \cfrac{1}{2 + \cfrac{1}{2 + \cdots}}}}}.$$

Does this make any sense? It has to be interpreted as the limit of the finite fractions

$$a_1 = \frac{1}{2} \qquad a_2 = \cfrac{1}{2 + \cfrac{1}{2}} \qquad a_3 = \cfrac{1}{2 + \cfrac{1}{2 + \cfrac{1}{2}}} \qquad \cdots .$$

This formulation of the sequence is not very useful, so we should look for a better way of defining the general term. In this case, there is a recursion formula for obtaining one term from the preceding one:

$$a_1 = \frac{1}{2}, \qquad a_{n+1} = \frac{1}{2 + a_n} \quad \text{for} \quad n \geq 1.$$

In order to establish convergence, we will show that (a_n) is Cauchy. Consider the difference

$$a_{n+1} - a_{n+2} = \frac{1}{2 + a_n} - \frac{1}{2 + a_{n+1}} = \frac{a_{n+1} - a_n}{(2 + a_n)(2 + a_{n+1})}.$$

Now $a_1 > 0$ and it is readily shown that $a_n > 0$ for all $n \geq 2$ by induction. Hence the denominator $(2 + a_n)(2 + a_{n+1}) > 4$. So we obtain

$$|a_{n+1} - a_{n+2}| < \frac{|a_n - a_{n+1}|}{4} \quad \text{for all} \quad n \geq 1.$$

Since $|a_1 - a_2| = 1/10$, we may iterate this inequality to estimate

$$|a_2 - a_3| < \frac{1}{10 \cdot 4}$$

$$|a_3 - a_4| < \frac{1}{10 \cdot 4^2}$$

$$|a_4 - a_5| < \frac{1}{10 \cdot 4^3}$$

$$|a_n - a_{n+1}| < \frac{1}{10 \cdot 4^{n-1}} = \tfrac{2}{5}(4^{-n}).$$

The general formula estimating the difference may be verified by induction. Now it is straightforward to estimate the difference between arbitrary terms a_m and

a_n for $m < n$:

$$\begin{aligned}
|a_m - a_n| &= \left|(a_m - a_{m+1}) + (a_{m+1} - a_{m+2}) + \cdots + (a_{n-1} - a_n)\right| \\
&\leq |a_m - a_{m+1}| + |a_{m+1} - a_{m+2}| + \cdots + |a_{n-1} - a_n| \\
&< \tfrac{2}{5}(4^{-m} + 4^{-m-1} + \cdots + 4^{1-n}) \\
&< \frac{2 \cdot 4^{-m}}{5(1 - \frac{1}{4})} = \frac{8}{15}4^{-m} < 4^{-m}.
\end{aligned}$$

This tells us that our sequence is Cauchy. Indeed, if $\varepsilon > 0$, choose N so large that $4^{-N} < \varepsilon$. Then for all $m, n \geq N$,

$$|a_m - a_n| < 4^{-m} \leq 4^{-N} < \varepsilon.$$

Thus by our Completeness Theorem (Theorem 2.7.4), it follows that $(a_n)_{n=1}^{\infty}$ converges; say, $\lim\limits_{n \to \infty} a_n = L$. To calculate L, use the recursion relation

$$L = \lim_{n \to \infty} a_n = \lim_{n \to \infty} a_{n+1} = \lim_{n \to \infty} \frac{1}{2 + a_n} = \frac{1}{2 + L}.$$

It follows that $L^2 + 2L - 1 = 0$. Solving yields $L = \pm\sqrt{2} - 1$. Since $L > 0$, we see that $L = \sqrt{2} - 1$.

Exercises for Section 2.7

A. Give an example of a sequence (a_n) such that $\lim\limits_{n \to \infty} |a_n - a_{n+1}| = 0$, but the sequence does not converge.

B. Let (a_n) be a sequence with the property that $\lim\limits_{N \to \infty} \sum\limits_{n=1}^{N} |a_n - a_{n+1}| < \infty$. Show that (a_n) is Cauchy.

C. Show that if $(x_n)_{n=1}^{\infty}$ is a Cauchy sequence, then it has a subsequence (x_{n_k}) such that $\lim\limits_{K \to \infty} \sum\limits_{k=1}^{K} |x_{n_k} - x_{n_{k+1}}| < \infty.$

D. Suppose that (a_n) is a sequence such that $a_{2n} \leq a_{2n+2} \leq a_{2n+3} \leq a_{2n+1}$ for all $n \geq 0$. Show that this sequence is Cauchy if and only if $\lim\limits_{n \to \infty} |a_n - a_{n+1}| = 0$.

E. Give an example of a sequence (a_n) such that $a_{2n} \leq a_{2n+2} \leq a_{2n+3} \leq a_{2n+1}$ for all $n \geq 0$ that does not converge.

F. Let $a_0 = 0$ and set $a_{n+1} = \cos(a_n)$ for $n \geq 0$.
(a) Try this on your calculator. (Remember to use radians mode.)
(b) Show that $a_{2n} \leq a_{2n+2} \leq a_{2n+3} \leq a_{2n+1}$ for all $n \geq 0$.
(c) Use the Mean Value Theorem to find an explicit number $r < 1$ such that $|a_{n+2} - a_{n+1}| \leq r|a_n - a_{n+1}|$ for all $n \geq 0$. Hence show that this sequence is Cauchy.
(d) Describe the limit geometrically as the intersection point of two curves.

G. Evaluate the continued fraction (Example 2.7.7)

$$1 + \cfrac{1}{1 + \cfrac{1}{1 + \cfrac{1}{1 + \cfrac{1}{1 + \cfrac{1}{1 + \cdots}}}}}.$$

H. Let $x_0 = 0$ and $x_{n+1} = \sqrt{5 - 2x_n}$ for $n \geq 0$. Show that this sequence converges and compute the limit.
HINT: Show that the even terms increase and the odd terms decrease to the same limit.

I. Consider an infinite binary expansion $(0.e_1 e_2 e_3 \dots)_{\text{base } 2}$, where each $e_i \in \{0, 1\}$ and this expansion is defined to be the real number that is the limit of the sequence $a_n = \sum_{i=1}^{n} 2^{-i} e_i$ when the limit exists. Prove that for every choice of zeros and ones, this sequence is Cauchy and therefore defines a unique real number.

J. We can develop a construction of the real numbers from the rational numbers using Cauchy sequences. This exercise presents the definitions that go into such a proof.
   (a) Associate a "point" to each Cauchy sequence of rational numbers. Find a way to decide when two Cauchy sequences should determine the same point without using limits.   HINT: Combine the two sequences into one.
   (b) Your definition in (a) should be an equivalence relation. Is it? (See Appendix 1.6.)
   (c) How can addition and multiplication be defined?
   (d) How is order defined?

2.8. Appendix: Cardinality

Cardinality is the notion that measures the size of a set in the crudest of ways—by counting the numbers of elements. Obviously, the number of elements in a set could be 0, 1, 2, 3, 4, or some other finite number. Or a set can have infinitely many elements. Perhaps surprisingly, not all infinite sets have the same cardinality. For our purposes, infinite sets have two possible sizes: countable and uncountable (the uncountable ones are larger). In this section, the most important ideas to understand are what *countable* means and what distinguishes countable sets from those with larger cardinality. We include a few more sophisticated arguments for enrichment purposes.

2.8.1. DEFINITION. Two sets A and B have the same **cardinality** if there is a *bijection* f from A onto B. We write $|A| = |B|$ in this case. Similarly, we say that the cardinality of A is less than that of B ($|A| \leq |B|$) if there is an *injection* f from A into B.

The definition says simply that if all of the elements of A can be paired, one-to-one, with all of the elements of B, then A and B have the same size. If A fits inside B in a one-to-one manner, then A is smaller than B. One of the subtleties that we address later is whether $|A| \leq |B|$ and $|B| \leq |A|$ mean that $|A| = |B|$. The answer is yes, but this is not obvious for infinite sets.

2.8.2. EXAMPLES.

(1) The cardinality of any finite set is the number of elements, and this number belongs to $\mathbb{N}_0 = \{0, 1, 2, 3, 4, \dots\}$. Set theorists go to some trouble to define the natural numbers too. But we will take for granted that the reader is familiar with the notion of a finite set.

(2) Most sets encountered in analysis are infinite, meaning that they are not finite. The sets of natural numbers $\mathbb{N}$, integers $\mathbb{Z}$, rational numbers $\mathbb{Q}$, and real numbers $\mathbb{R}$ are all infinite. Moreover, we have the natural containments $\mathbb{N} \subset \mathbb{Z} \subset \mathbb{Q} \subset \mathbb{R}$. So $|\mathbb{N}| \leq |\mathbb{Z}| \leq |\mathbb{Q}| \leq |\mathbb{R}|$. Notice that the integers can be written as a list $0, 1, -1, 2, -2, 3, -3, \dots$. This amounts to defining a bijection $f : \mathbb{N} \to \mathbb{Z}$ by

$$f(n) = \begin{cases} (1-n)/2 & \text{if} \quad n \text{ is odd} \\ n/2 & \text{if} \quad n \text{ is even.} \end{cases}$$

Therefore, $|\mathbb{N}| = |\mathbb{Z}|$.

2.8.3. DEFINITION. A set A is a **countable set** is it is finite or if $|A| = |\mathbb{N}|$. The cardinal $|\mathbb{N}|$ is also denoted by $\aleph_0$. This is the first letter of the Hebrew alphabet, aleph, with subscript zero. It is pronounced **aleph nought**.

An infinite set that is not countable is called an **uncountable set**.

Notice that two uncountable sets could have different cardinalities. We refer the reader interested in the possible cardinalities of uncountable sets to [1].

Equivalently, A is countable if the elements of A may be listed as $a_1, a_2, a_3, \dots$. Indeed, the list itself determines a bijection from $\mathbb{N}$ to A by $f(k) = a_k$. It is a basic fact that countable sets are the smallest infinite sets.

2.8.4. LEMMA. *Every infinite subset of $\mathbb{N}$ is countable. Moreover, if A is an infinite set such that $|A| \leq |\mathbb{N}|$, then $|A| = |\mathbb{N}|$.*

PROOF. Any nonempty subset X of $\mathbb{N}$ has a smallest element. Indeed, as X is nonempty, it contains an integer n. Consider the elements of the finite set $\{1, 2, \dots, n\}$ in order and pick the first one that belongs to X—that is, the smallest.

Let B be an infinite subset of $\mathbb{N}$. List the elements of B in increasing order as $b_1 < b_2 < b_3 < \dots$. This is done by choosing the smallest element b_1, then the smallest of the remaining set $B \setminus \{b_1\}$, then the smallest of $B \setminus \{b_1, b_2\}$ and so on. The result is an infinite list of elements of B in increasing order. It must include every element $b \in B$ because $\{n \in B : n \leq b\}$ is finite, containing say k elements. Then $b_k = b$. As noted before the proof, this implies that $|B| = |\mathbb{N}|$.

Now consider a set A with $|A| \leq |\mathbb{N}|$. By definition, there is a injection f of A into $\mathbb{N}$. Let $B = f(A)$. Note that f is a bijection of A onto B. Then B is an infinite subset of $\mathbb{N}$. So $|A| = |B| = |\mathbb{N}|$. ∎

2.8.5. PROPOSITION. *The countable union of countable sets is countable.*

PROOF. By the previous lemma, we may assume that there is a countably infinite collection of sets $A_1, A_2, A_3, \ldots$ that are each countably infinite. Write the elements of A_i as a list $a_{i,1}, a_{i,2}, a_{i,3}, \ldots$. Then we may write $A = \bigcup_{i \geq 1} A_i$ as a list as follows:

$$a_{1,1}, a_{1,2}, a_{2,1}, a_{1,3}, a_{2,2}, a_{3,1}, a_{1,4}, a_{2,3}, a_{3,2}, a_{4,1}, \ldots,$$

where the elements $a_{i,j}$ are written so that $i + j$ is monotone increasing, and within the set of pairs (i, j) with $i + j = n$, the terms are written with the i's in increasing order. See Figure 2.5. Thus A is countable. ∎

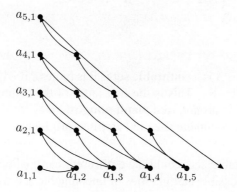

FIGURE 2.5. The set $\mathbb{N} \times \mathbb{N}$ is countable.

2.8.6. COROLLARY. *The set $\mathbb{Q}$ of rational numbers is countable.*

PROOF. The set $\mathbb{Z} \times \mathbb{N} = \{(i, j) : i \in \mathbb{Z}, j \in \mathbb{N}\}$ is the disjoint union of the sets $A_i = \{(i, j) : j \in \mathbb{N}\}$ for $i \in \mathbb{Z}$. Each A_i is evidently countable. By Example 2.8.2(2), $\mathbb{Z}$ is countable. Hence $\mathbb{Z} \times \mathbb{N}$ is the countable union of countable sets, and hence is countable by Proposition 2.8.5.

Define a map from $\mathbb{Q}$ into $\mathbb{Z} \times \mathbb{N}$ by $f(r) = (a, b)$ if $r = a/b$, where a and b are integers with no common factor and $b > 0$. These conditions uniquely determine the pair (a, b) for each rational r, and so f is a function. Clearly, f is injective since r is recovered from (a, b) by division. Therefore, f is an injection of $\mathbb{Q}$ into a countable set. Hence $\mathbb{Q}$ is an infinite set with $|\mathbb{Q}| \leq |\mathbb{N}|$. So $\mathbb{Q}$ is countable by Lemma 2.8.4. ∎

There are infinite sets that are not countable.

2.8.7. THEOREM. *The set $\mathbb{R}$ of real numbers is uncountable.*

PROOF. The proof uses a **diagonalization** argument due to Cantor. Suppose to the contrary that $\mathbb{R}$ is countable. Then all real numbers may be written as a list $x_1, x_2, x_3, \ldots$. Express each x_i as an infinite decimal, which we write as $x_i = x_{i0}.x_{i1}x_{i2}x_{i3}\ldots$, where x_{i0} is any integer and x_{ik} is an integer from 0 to 9 for each $k \geq 1$. Our goal is to write down another real number that does not appear in this (supposedly exhaustive) list. Let $a_0 = 0$ and define $a_k = 7$ if $x_{ik} \in \{0, 1, 2, 3, 4\}$ and $a_k = 2$ if $x_{ik} \in \{5, 6, 7, 8, 9\}$. Define a real number $a = a_0.a_1a_2a_3\ldots$.

Since a is a real number, it must appear somewhere in this list, say $a = x_k$. However, the kth decimal place a_k of a and $x_{k,k}$ of x_k differ by at least 3. This cannot be accounted for by the fact that certain real numbers have two decimal expansions, one ending in zeros and the other ending in nines because this changes any digit by no more than 1 (counting 9 and 0 as being within 1). So $a \neq x_k$, and hence a does not occur in this list. It follows that there is no list containing all real numbers, and thus $\mathbb{R}$ is uncountable. ∎

We conclude with the result promised in the start of this section.

2.8.8. SCHROEDER–BERNSTEIN THEOREM.
If A and B are sets with $|A| \leq |B|$ and $|B| \leq |A|$, then $|A| = |B|$.

PROOF. The proof is surprisingly simple. Since $|A| \leq |B|$, there is an injection f mapping A into B. Likewise, as $|B| \leq |A|$, there is an injection g mapping B into A. Let $B_1 = B \setminus f(A)$. Recursively define $A_i = g(B_i)$ and $B_{i+1} = f(A_i)$ for $i \geq 1$. Define $A_0 = A \setminus \bigcup_{i \geq 1} A_i$ and $B_0 = B \setminus \bigcup_{i \geq 1} B_i$. We will show that the actions of f and g fit the scheme of Figure 2.6.

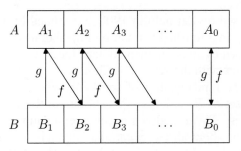

FIGURE 2.6. Schematic of action of f and g on A and B.

First we show that the B_i's are disjoint. Clearly each B_i for $i \geq 2$ is in the range of f and hence does not intersect B_1. Suppose that $1 < i < j$. Then $(fg)^{i-1}$ is an injection of B into itself that carries B_k onto B_{k+i-1} for every $k \geq 1$. In particular, B_1 is mapped onto B_i and B_{j-i+1} is mapped onto B_j. Since $B_1 \cap B_{j-i+1} = \varnothing$ and $(fg)^{i-1}$ is one-to-one, it follows that $B_i \cap B_j = \varnothing$.

By construction, g^{-1} is a bijection of each A_i onto B_i for $i \geq 1$. We claim that f maps A_0 onto B_0. Observe that f maps A_i onto B_{i+1} for each $i \geq 1$. Thus the remainder of A, namely A_0, is mapped onto the remainder of the image. Thus

$$f(A_0) = f(A) \setminus \bigcup_{i \geq 1} f(A_i) = (B \setminus B_1) \setminus \bigcup_{i \geq 1} B_{i+1} = B \setminus \bigcup_{i \geq 1} B_i = B_0.$$

This means that the function

$$h(a) = \begin{cases} g^{-1}(a) & \text{if} \quad a \in \bigcup_{i \geq 1} A_i \\ f(a) & \text{if} \quad a \in A_0 \end{cases}$$

is a bijection between A and B. So $|A| = |B|$. ∎

The whole subject of cardinality is wrapped up in the subtleties of set theory. The commonly used axioms of set theory include the **Axiom of Choice**, which has the simple sounding statement that given any collection of nonempty sets, one can select an element from each of them. This has many significant ramifications that are beyond the scope of this book. One of these is that if A and B are any sets, then either $|A| \leq |B|$ or $|B| \leq |A|$.

Another aspect of set theory that arises in this context is that sets must be built up from smaller sets only in certain allowable ways. This prevents the universe of all sets from being a set itself. The reason behind this is a famous contradiction, known as **Russell's Paradox**, to a more casually defined theory of sets.

Russell's argument is like Cantor's diagonalization argument. Let X be the set consisting of all sets A that do not contain themselves as an element. Intuition suggests that no set contains itself. However, the set of all sets (were it a set) would have to contain itself, and thus is not an element of X. The question is, Does X contain itself? If X does not contain itself, then by definition, it does belongs to X. Conversely, if X is an element of X, then by definition it would not be a member. So neither possibility is logical.

The solution was proposed by Zermelo in 1908 and refined by various other mathematicians, culminating in a finished version by Fraenkel in 1922. The standard axioms of set theory used by most mathematicians today are called ZFC, for **Zermelo–Fraenkel set theory** with the Axiom of Choice.

A curious question in the fundamentals of set theory was raised by Cantor. If A is an uncountable subset of $\mathbb{R}$, is $|A| = |\mathbb{R}|$? The **continuum hypothesis** asserts that the answer is yes. There is also a generalized continuum hypothesis that makes a parallel assertion about all larger cardinal numbers, not just $|\mathbb{R}|$. The Austrian mathematician Gödel established several deep results about the foundations of mathematics. One of these was that, if there is an inconsistency of ZFC together with the generalized continuum hypothesis, then there is an inconsistency in the Zermelo–Fraenkel axioms themselves. This means that there is no additional danger of an inconsistency in assuming the Axiom of Choice or the generalized continuum hypothesis. In 1965, Cohen showed that the generalized continuum hypothesis does not follow from ZFC, so we are also free to not assume the generalized continuum hypothesis, if we so choose.

Any further discussion of these issues leads far away from analysis and the purpose of this book. We refer the interested reader to [1] as a starting point for the study of set theory.

Exercises for Section 2.8

A. Prove that the set $\mathbb{Z}^n$, consisting of all n-tuples $\mathbf{a} = (a_1, a_2, \ldots, a_n)$, where $a_i \in \mathbb{Z}$, is countable.

B. Do $(0, 1)$ and $[0, 1]$ have the same cardinality as $\mathbb{R}$? Do not use the Schroeder–Bernstein Theorem.

C. Show that if $|A| \leq |B|$ and $|B| \leq |C|$, then $|A| \leq |C|$.

D. Show that the relation of equal cardinality $|A| = |B|$ is an equivalence relation.

E. Show that $|\mathbb{R}^2| = |\mathbb{R}|$.

F. Prove that the set of all infinite sequences of integers is uncountable.
HINT: Modify the diagonalization argument.

G. A real number α is called an **algebraic number** if there is a polynomial with integer coefficients with α as a root. Prove that the set of all algebraic numbers is countable.
HINT: First count the set of all polynomials with integer coefficients.

H. A real number that is not algebraic is called a **transcendental number**. Prove that the set of transcendental numbers has the same cardinality as $\mathbb{R}$.

I. Show that the set of all *finite* subsets of $\mathbb{N}$ is countable.

J. Prove **Cantor's Theorem**: that for any set X, the power set $P(X)$ of all subsets of X satisfies $|X| \neq |P(X)|$.
HINT: If f is an injection from X into $P(X)$, consider $A = \{x \in X : x \notin f(x)\}$.

K. If A is an infinite set, show that A has a countable subset.
HINT: Use recursion to choose a sequence a_n of distinct points in A.

L. Show that A is infinite if and only if there is a proper subset B of A such that $|B| = |A|$.
HINT: Use the previous exercise and let $B = A \setminus \{a_1\}$.

CHAPTER 3

Series

3.1. Convergent Series

We turn now to the problem of adding up an infinite series of numbers. As we shall quickly see, this is really no different from dealing with the sequence of partial sums of the series. However, there are tests for convergence that are more conveniently expressed for series than for sequences.

3.1.1. DEFINITION. If $(a_n)_{n=1}^{\infty}$ is a sequence of numbers, the **infinite series** with terms a_n is the formal expression $\sum_{n=1}^{\infty} a_n$. This series **converges** or, equivalently, the sequence is **summable** if the sequence of partial sums $(s_n)_{n=1}^{\infty}$ defined by $s_n = \sum_{k=1}^{n} a_k$ converges. If $L = \lim_{n \to \infty} s_n$, then L is the sum of the series, and we write $L = \sum_{n=1}^{\infty} a_n$. If the series does not converge, then it is said to **diverge**.

It is worth pointing out that the convergence of the series $\sum_{n=1}^{\infty} a_n$ and of the sequence $(a_n)_{n=1}^{\infty}$ are two very different questions. We will work out the exact relation between these concepts in this chapter.

It can be fairly difficult or even virtually impossible to find the sum of a series. However, it is not nearly as hard to determine whether or not a series converges. We devote this chapter to examples of series and to tests for convergence of series. While these tests may be familiar to you from calculus, the proofs may not be. And of course, you should learn both how to use these tests and how to prove that the tests are correct.

3.1.2. EXAMPLE. First consider $\sum_{k=1}^{\infty} \frac{1}{k}$, which is known as the **harmonic series**. We will show that this series diverges. The idea is to group the terms cleverly.

Suppose that n satisfies $2^k \leq n < 2^{k+1}$. Then

$$s_n = 1 + \frac{1}{2} + \cdots + \frac{1}{n}$$

$$> 1 + \frac{1}{2} + (\frac{1}{3} + \frac{1}{4}) + \cdots + (\frac{1}{2^{k-1}+1} + \cdots + \frac{1}{2^k})$$

$$\geq 1 + \frac{1}{2} + 2\frac{1}{4} + \cdots + 2^{k-1}\frac{1}{2^k} = 1 + \frac{k}{2}.$$

Thus $\lim\limits_{n \to \infty} s_n = +\infty$.

There is another way to estimate the terms s_n that gives a more precise idea of the rate of divergence of the harmonic series. Consider the graph of $y = 1/x$, as given in Figure 3.1.

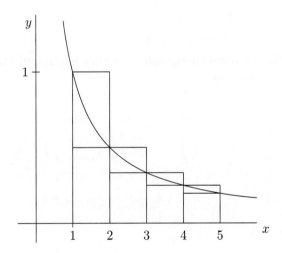

FIGURE 3.1. The graph of $1/x$ with bounding rectangles.

It is clear that

$$\frac{1}{k+1} = \int_k^{k+1} \frac{1}{k+1}\, dx < \int_k^{k+1} \frac{1}{x}\, dx < \int_k^{k+1} \frac{1}{k}\, dx = \frac{1}{k}.$$

Notice that s_n is the upper Riemann sum estimate for the integral of $1/x$ from 1 to $n + 1$ using the integer partition.

$$s_n = \sum_{k=1}^{n} \frac{1}{k} > \sum_{k=1}^{n} \int_k^{k+1} \frac{1}{x}\, dx$$

$$= \int_1^{n+1} \frac{1}{x}\, dx = \log x \Big|_1^{n+1} = \log(n+1).$$

Similarly, $s_n - 1$ is the lower Riemann sum estimate for the integral of $1/x$ from 1 to n using the integer partition.

$$s_n - 1 = \sum_{k=2}^{n} \frac{1}{k} < \sum_{k=2}^{n} \int_{k-1}^{k} \frac{1}{x}\, dx$$

$$= \int_{1}^{n} \frac{1}{x}\, dx = \log x \Big|_{1}^{n} = \log(n)$$

Therefore,

$$\log(n+1) < s_n < 1 + \log(n) \qquad \text{for all} \quad n \geq 1.$$

Hence s_n diverges to infinity roughly at the same rate as the log function.

3.1.3. EXAMPLE. On the other hand, consider $\displaystyle\sum_{n=1}^{\infty} \frac{1}{n(n+3)}$. First observe that

$$\frac{3}{n(n+3)} = \frac{1}{n} - \frac{1}{n+3}$$

and so we have an example of a **telescoping sum** (so named because of the convenient cancellation in the following sum):

$$3s_n = \frac{3}{4} + \frac{3}{10} + \cdots + \frac{3}{n(n+3)}$$

$$= \left(1 - \frac{1}{4}\right) + \left(\frac{1}{2} - \frac{1}{5}\right) + \cdots + \left(\frac{1}{n} - \frac{1}{n+3}\right)$$

$$= \left(1 + \frac{1}{2} + \cdots + \frac{1}{n}\right) - \left(\frac{1}{4} + \frac{1}{5} + \cdots + \frac{1}{n+3}\right)$$

$$= 1 + \frac{1}{2} + \frac{1}{3} - \frac{1}{n+1} - \frac{1}{n+2} - \frac{1}{n+3}.$$

Thus,

$$\sum_{n=1}^{\infty} \frac{1}{n(n+3)} = \lim_{n\to\infty} s_n = \frac{1 + 1/2 + 1/3}{3} = \frac{11}{18}.$$

The harmonic series shows that a series $\displaystyle\sum_{n=1}^{\infty} a_n$ can diverge even if the a_n go to zero. However, if a series $\displaystyle\sum_{n=1}^{\infty} a_n$ does converge, then $\displaystyle\lim_{n\to\infty} a_n$ must be zero. That is, $\displaystyle\lim_{n\to\infty} a_n = 0$ is necessary for $\displaystyle\sum_{n=1}^{\infty} a_n$ to converge, but it is not sufficient to make $\displaystyle\sum_{n=1}^{\infty} a_n$ converge.

3.1.4. THEOREM. *If the series* $\displaystyle\sum_{n=1}^{\infty} a_n$ *is convergent, then* $\displaystyle\lim_{n\to\infty} a_n = 0.$

PROOF. If $(s_n)_{n=1}^{\infty}$ is the sequence of partial sums, then $a_n = s_n - s_{n-1}$ for $n \geq 2$. Using the properties of limits, we have $\lim\limits_{n \to \infty} s_n = \lim\limits_{n \to \infty} s_{n-1}$, and thus

$$\lim_{n \to \infty} a_n = \lim_{n \to \infty} s_n - s_{n-1} = \lim_{n \to \infty} s_n - \lim_{n \to \infty} s_{n-1} = 0. \qquad \blacksquare$$

The rigourous ε–N definition of convergence and the Cauchy criterion have a nice form for series.

3.1.5. CAUCHY CRITERION FOR SERIES.

The following are equivalent for a series $\sum\limits_{n=1}^{\infty} a_n$.

(1) *The series converges.*

(2) *For every $\varepsilon > 0$, there is an $N \in \mathbb{N}$ so that $\left| \sum\limits_{k=n+1}^{\infty} a_k \right| < \varepsilon$ for all $n \geq N$.*

(3) *For every $\varepsilon > 0$, there is an $N \in \mathbb{N}$ so that $\left| \sum\limits_{k=n+1}^{m} a_k \right| < \varepsilon$ if $n, m \geq N$.*

PROOF. Let s_n be the sequence of partial sums of the series. If the series converges to a limit L, then for every $\varepsilon > 0$ there is an integer N such that

$$|L - s_n| < \varepsilon \quad \text{for all} \quad n \geq N.$$

Moreover,

$$L - s_n = \lim_{m \to \infty} s_m - s_n = \lim_{m \to \infty} \sum_{k=n+1}^{m} a_k = \sum_{k=n+1}^{\infty} a_k.$$

This shows that (1) implies (2).

If (2) holds, then the previous paragraph shows that there is an integer N so that (for $\varepsilon/2$)

$$|L - s_n| < \frac{\varepsilon}{2} \quad \text{for all} \quad n \geq N.$$

Thus

$$\left| \sum_{k=n+1}^{m} a_k \right| = |s_m - s_n| \leq |s_m - L| + |L - s_n| < \frac{\varepsilon}{2} + \frac{\varepsilon}{2} = \varepsilon.$$

So (3) holds.

Finally, if (3) holds, this says that (s_n) is a Cauchy sequence, and therefore it converges by the completeness of the real numbers. $\qquad \blacksquare$

Exercises for Section 3.1

A. Sum the series $\sum\limits_{n=1}^{\infty} \dfrac{1}{n(n+2)}$.

B. Sum the series $\sum\limits_{n=1}^{\infty} \dfrac{1}{n(n+1)(n+3)(n+4)}$.

HINT: Show that $\dfrac{12}{n(n+1)(n+3)(n+4)} = \dfrac{1}{n} - \dfrac{2}{n+1} + \dfrac{2}{n+3} - \dfrac{1}{n+4}$.

C. Prove that if $\sum\limits_{k=1}^{\infty} t_k$ is a convergent series of nonnegative numbers and $p > 1$, then $\sum\limits_{k=1}^{\infty} t_k^p$ converges.

D. Let $(a_n)_{n=1}^{\infty}$ be a sequence such that $\lim\limits_{n\to\infty} |a_n| = 0$. Prove that there is a subsequence a_{n_k} so that $\sum\limits_{k=1}^{\infty} a_{n_k}$ converges.

E. Compute $\sum\limits_{n=1}^{\infty} \dfrac{1}{(n+1)\sqrt{n} + n\sqrt{n+1}}$.

HINT: Multiply the nth term by by $1 = \dfrac{\sqrt{n+1} - \sqrt{n}}{\sqrt{n+1} - \sqrt{n}}$ and simplify.

F. Let $|a| < 1$ and set $S_n = \sum\limits_{k=0}^{n} a^k$ and $T_n = \sum\limits_{k=0}^{n} (k+1)a^k$.

(a) Show that $S_n^2 = \sum\limits_{k=0}^{n} (k+1)a^k + \sum\limits_{k=1}^{n} (n+1-k)a^{n+k}$.

(b) Hence show that $|T_n - S_n^2| \le \frac{n(n+1)}{2}|a|^{n+1}$.

(c) Show that $\lim\limits_{n\to\infty} T_n = \left(\lim\limits_{n\to\infty} S_n\right)^2$. Hence obtain a formula for this sum.

(d) Evaluate $\sum\limits_{k=0}^{\infty} \dfrac{n+1}{3^n}$.

G. Let $x_0 = 1$ and $x_{n+1} = x_n + 1/x_n$.
(a) Find $\lim\limits_{n\to\infty} x_n$.
(b) Let $y_n = x_n^2 - 2n$. Find a recursion formula for y_{n+1} in terms of y_n only.
(c) Show that y_n is monotone increasing and $y_n < \frac{1}{2}\log n$.
(d) Hence show that $\lim\limits_{n\to\infty} x_n - \sqrt{2n} = 0$.

3.2. Convergence Tests for Series

We start by considering infinite series with positive terms. If each $a_n \ge 0$, then $s_{n+1} = s_n + a_{n+1} \ge s_n$, so the sequence of partial sums is increasing. This fact and the Monotone Convergence Theorem (Theorem 2.5.4) show that (s_n) converges if and only if it is bounded above. We have established the following proposition.

3.2.1. Proposition. *If $a_k \geq 0$ for $k \geq 1$ and $s_n = \sum\limits_{k=1}^{n} a_k$, then either*

(1) $(s_n)_{n=1}^{\infty}$ *is bounded above, in which case $\sum\limits_{n=1}^{\infty} a_n$ converges,*

or

(2) $(s_n)_{n=1}^{\infty}$ *is unbounded, in which case $\sum\limits_{n=1}^{\infty} a_n$ diverges.*

A sequence $(a_n)_{n=0}^{\infty}$ is a **geometric sequence** with ratio r if $a_{n+1} = r a_n$ for all $n \geq 0$ or, equivalently, $a_n = a_0 r^n$ for all $n \geq 0$. Finding the sum of a geometric sequence is a standard result from the calculus, so we content ourselves with stating the result and leave the proof as an exercise.

3.2.2. Geometric Series.
A geometric series converges if $|r| < 1$. Moreover, $\sum\limits_{n=0}^{\infty} ar^n = \dfrac{a}{1-r}$.

Of course, if $a \neq 0$ and $|r| \geq 1$, then the terms ar^n do not converge to 0. In this case, the geometric sequence $(a_n)_{n=0}^{\infty}$ is not summable.

Another possibly familiar test is the Comparison Test.

3.2.3. The Comparison Test.
Consider two sequences of real numbers (a_n) and (b_n) with $|a_n| \leq b_n$ for all $n \geq 1$. If (b_n) is summable, then (a_n) is summable and

$$\left| \sum_{n=1}^{\infty} a_n \right| \leq \sum_{n=1}^{\infty} b_n.$$

If (a_n) is not summable, then (b_n) is not summable.

Proof. Let $\varepsilon > 0$ be given. Since (b_n) is summable, Lemma 3.1.5 yields an integer N so that

$$\sum_{k=n+1}^{m} b_k < \varepsilon \quad \text{for all} \quad N \leq n \leq m.$$

Therefore,

$$\left| \sum_{k=n+1}^{m} a_k \right| \leq \sum_{k=n+1}^{m} |a_k| \leq \sum_{k=n+1}^{m} b_k < \varepsilon.$$

Therefore, applying Lemma 3.1.5 again shows that $\sum_{n=1}^{\infty} a_n$ converges.

Conversely, if (a_n) is not summable, then neither is (b_n), for the summability of (b_n) would imply that (a_n) was also summable. This is the contrapositive of the previous paragraph. ∎

The root test can decide the summability of sequences that are dominated by a geometric sequence "at infinity."

3.2.4. THE ROOT TEST.

Suppose that $a_n \geq 0$ for all n and let $\ell = \limsup \sqrt[n]{a_n}$. If $\ell < 1$, then $\sum_{n=1}^{\infty} a_n$

converges, and if $\ell > 1$, then $\sum_{n=1}^{\infty} a_n$ diverges.

NOTE: If $\limsup \sqrt[n]{a_n} = 1$, the series may or may not be converge (see Exercise 3.2.M).

PROOF. Suppose that $\limsup_{n \to \infty} \sqrt[n]{a_n} = \ell < 1$. To show the series converges, we need to show that the sequence of partial sums is bounded above. Pick a number r with $\ell < r < 1$ and let $\varepsilon = r - \ell$. Since $\varepsilon > 0$, we can find an integer $N > 0$ so that

$$a_n^{1/n} < \ell + \varepsilon = r \quad \text{for all} \quad n \geq N.$$

Therefore, $a_n < r^n$ for all $n \geq N$.

Consider the sequence $(b_n)_{n=1}^{\infty}$ given by

$$b_n = a_n, \quad 1 \leq n < N \qquad\qquad b_n = r^n, \quad n \geq N.$$

This sequence is summable by Theorem 3.2.2.

$$\sum_{n=1}^{\infty} b_n = \sum_{n=1}^{N-1} b_n + \sum_{n=N}^{\infty} b_n = \sum_{n=1}^{N-1} b_n + \frac{r^N}{1-r}$$

Since $|a_n| \leq b_n$ for $n \geq 1$, the Comparison Test (3.2.3) shows that the sequence $(a_n)_{n=1}^{\infty}$ is also summable.

Conversely, if $\liminf_{n \to \infty} \sqrt[n]{a_n} = \ell > 1$, then we can choose r with $1 < r < \ell$ and let $\varepsilon = r - \ell$. From the definition of liminf, there is an integer $N > 0$ so that

$$a_n^{1/n} > \ell - \varepsilon = r \quad \text{for all} \quad n \geq N.$$

Therefore, $a_n \geq r^n$ for all $n \geq N$. Hence

$$\lim_{n \to \infty} a_n = \lim_{n \to \infty} r^n = \infty.$$

Consequently, $\sum_{n=1}^{\infty} a_n$ diverges.   ■

3.2.5. DEFINITION.
A sequence is **alternating** if it has the form $((-1)^n a_n)$ or $((-1)^{n+1} a_n)$, where $a_n \geq 0$ for all $n \geq 1$.

3.2.6. LEIBNIZ ALTERNATING SERIES TEST.
Suppose that $(a_n)_{n=1}^{\infty}$ is a monotone decreasing sequence $a_1 \geq a_2 \geq a_3 \geq \cdots \geq 0$ and that $\lim_{n \to \infty} a_n = 0$. Then the alternating series $\sum_{n=1}^{\infty} (-1)^n a_n$ converges.

PROOF. Let $s_n = \sum_{k=1}^{n} (-1)^k a_k$. Intuitively, (s_n) behaves as in Figure 3.2.

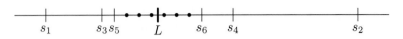

FIGURE 3.2. Behaviour of partial sums.

Making this formal, we claim that

(1) $s_2 \geq s_4 \geq s_6 \cdots$,

(2) $s_1 \leq s_3 \leq s_5 \cdots$, and

(3) $s_{2m-1} \leq s_{2n}$ for all $m, n \geq 1$.

To prove (1), notice that $s_{2n} - s_{2n-2} = a_{2n} - a_{2n-1} \leq 0$ since $a_{2n} \leq a_{2n-1}$. For (2), $s_{2n+1} - s_{2n-1} = a_{2n} - a_{2n+1} \geq 0$.

The inequalities in (3) follow simply from (1) and (2). For any positive integer N, $s_{2N} = s_{2N-1} + a_{2N} \geq s_{2N-1}$. If m and n are integers, then for $N = \max\{m, n\}$, we have

$$s_{2m-1} \leq s_{2N-1} \leq s_{2N} \leq s_{2n}.$$

The first inequality follows from (1), we have just proved the second, and the third follows from (2).

Since the sequence $(s_2, s_4, \ldots)$ is bounded below by s_1, it converges to some number L by Theorem 2.5.4. Similarly, since $(s_1, s_3, \ldots)$ is bounded above by s_2, it converges to some number M.

All that remains is to show $L = M$. To do this, we use the properties of limits,

$$L - M = \lim_{n \to \infty} s_{2n} - \lim_{n \to \infty} s_{2n-1}$$
$$= \lim_{n \to \infty} s_{2n} - s_{2n-1} = \lim_{n \to \infty} a_{2n} = 0. \qquad \blacksquare$$

Note that the proof gives upper and lower bounds on $\sum_{n=1}^{\infty} (-1)^n a_n$, namely the even and odd partial sums, respectively. Precisely, we have the following statement.

3.2.7. COROLLARY. *Suppose that $(a_n)_{n=1}^{\infty}$ is a monotone decreasing sequence $a_1 \geq a_2 \geq a_3 \geq \cdots \geq 0$ and that $\lim_{n \to \infty} a_n = 0$. Then the difference between the sum of the alternating series $\sum_{n=1}^{\infty} (-1)^n a_n$ and the Nth partial sum is at most $|a_N|$.*

3.2.8. EXAMPLE. Consider the alternating harmonic series

$$\sum_{n=1}^{\infty} \frac{(-1)^{n-1}}{n} = 1 - \frac{1}{2} + \frac{1}{3} - \frac{1}{4} + \ldots.$$

Since this series is alternating and $\frac{1}{n}$ is monotone decreasing to 0, the series must converge. Compare this with the harmonic series, which has the same terms without the sign changes.

It is possible to sum this series in several ways. All rely on calculus in some way. Notice that

$$s_{2n} = \sum_{k=1}^{2n} \frac{(-1)^{k-1}}{k} = 1 - \frac{1}{2} + \cdots + \frac{1}{2n-1} - \frac{1}{2n}$$

$$= \left(1 + \frac{1}{2} + \cdots + \frac{1}{2n-1} + \frac{1}{2n}\right) - 2\left(\frac{1}{2} + \frac{1}{4} + \cdots + \frac{1}{2n}\right)$$

$$= \sum_{k=1}^{2n} \frac{1}{k} - 2\sum_{k=1}^{n} \frac{1}{2k} = \sum_{k=1}^{2n} \frac{1}{k} - \sum_{k=1}^{n} \frac{1}{k} = \sum_{k=1}^{n} \frac{1}{n+k}.$$

We have to recognize this as a Riemann sum approximating an integral. Indeed, consider the integral $\int_1^2 \frac{1}{x}\,dx$. Partition the interval $[1, 2]$ into n equal pieces. Then

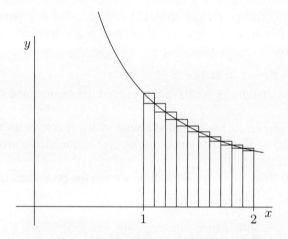

FIGURE 3.3. Riemann sum for $\int_1^2 \frac{1}{x}\,dx$.

from Figure 3.3, we see that the Riemann (lower) sum for $f(x) = \frac{1}{x}$ is

$$\frac{1}{n}\sum_{k=1}^{n} f\left(1 + \tfrac{k}{n}\right) = \sum_{k=1}^{n} \frac{1}{n+k} = s_{2n}.$$

From the calculus, we obtain

$$\sum_{n=1}^{\infty} \frac{(-1)^{n-1}}{n} = \lim_{n\to\infty} s_{2n} = \int_1^2 \frac{1}{x}\,dx = \log x \Big|_1^2 = \log 2.$$

Exercises for Section 3.2

A. Prove Theorem 3.2.2.

B. Show that if $(|a_n|)_{n=1}^{\infty}$ is summable, then so is $(a_n)_{n=1}^{\infty}$.

C. Euler proposed that $1 - 2 + 4 - 8 + \cdots = \sum_{n=0}^{\infty} (-2)^n = \dfrac{1}{1 - (-2)} = \dfrac{1}{3}$.
What is wrong with this argument?

D. Let $(a_n)_{n=1}^{\infty}$ be a monotone decreasing sequence of positive real numbers. Show that the series $\sum_{n=1}^{\infty} a_n$ converges if and only if the series $\sum_{k=0}^{\infty} 2^k a_{2^k}$ converges.

E. Apply the previous exercise to the series $\sum_{n=1}^{\infty} \dfrac{1}{n^p}$ for $p > 0$. For which values of p does this converge?

F. Let $\sum_{n=1}^{\infty} a_n$ be a convergent series of positive terms.

(a) Show that $\sum_{n=1}^{\infty} \dfrac{\sqrt{a_n}}{n^p}$ converges for $p > 1/2$.   HINT: Schwarz inequality

(b) Find an example for which the series $\sum_{n=1}^{\infty} \sqrt{\dfrac{a_n}{n}}$ diverges.

G. If $\sum_{k=1}^{\infty} a_k^2$ and $\sum_{k=1}^{\infty} b_k^2$ both converge, then prove that $\sum_{k=1}^{\infty} a_k b_k$ converges.

H. (THE LIMIT COMPARISON TEST) Show that if $\sum_{n=1}^{\infty} a_n$ and $\sum_{n=1}^{\infty} b_n$ are series with $b_n \geq 0$ such that $\limsup_{n \to \infty} \dfrac{a_n}{b_n} < \infty$ and $\sum_{n=1}^{\infty} b_n < \infty$, then the series $\sum_{n=1}^{\infty} a_n$ converges.

I. (THE RATIO TEST) Suppose that $(a_n)_{n=1}^{\infty}$ is a sequence of positive terms. Show that if $\limsup_{n \to \infty} \dfrac{a_{n+1}}{a_n} < 1$, then $\sum_{n=1}^{\infty} a_n$ converges. Conversely, show that if $\liminf_{n \to \infty} \dfrac{a_{n+1}}{a_n} > 1$, then $\sum_{n=1}^{\infty} a_n$ diverges.
HINT: Imitate the proof of the Root Test (i.e., find a suitable r and integer $N > 0$ and compare a_n with $a_N r^{n-N}$ for all $n \geq N$).

J. (a) Find a convergent series $\sum_{n=1}^{\infty} a_n$, with positive entries, so that $\lim_{n \to \infty} \dfrac{a_{n+1}}{a_n} = 1$.
(b) Find a divergent series with the same property.

K. Construct a convergent series of positive terms with $\limsup_n \dfrac{a_{n+1}}{a_n} = \infty$.

L. Prove that for a sequence of positive terms (a_n), if there is N with $a_{n+1}/a_n \geq 1$ for all $n \geq N$, then the series $\sum_{n=1}^{\infty} a_n$ diverges. Show that this result implies one direction of the Ratio Test (Exercise I).

M. (a) Find a convergent series $\sum_{n=1}^{\infty} a_n$, with positive entries, so that $\lim_{n \to \infty} \sqrt[n]{a_n} = 1$.
(b) Find a divergent series with the same property.

N. If $a_n \geq 0$ for all n, prove that $\sum_{n=1}^{\infty} a_n$ converges if and only if $\sum_{n=1}^{\infty} \dfrac{a_n}{1 + a_n}$ converges.

O. (THE INTEGRAL TEST) Let $f(x)$ be a positive monotone decreasing function on $[1, \infty)$. Show that the sequence $(f(n))$ is summable if and only if $\int_1^\infty f(x)\, dx < \infty$.

HINT: Show that $\sum_{n=2}^{k+1} f(n) < \int_1^k f(x)\, dx < \sum_{n=1}^{k} f(n)$.

P. Apply the previous exercise to the series $\sum_{n=2}^{\infty} \dfrac{1}{n(\log n)^p}$ for $p > 0$.

Q. Find two convergent series $\sum_{k=1}^{\infty} a_k$ and $\sum_{k=1}^{\infty} b_k$ so that $\sum_{k=1}^{\infty} a_k b_k$ diverges.

R. Let $\sum_{n\geq 1} a_n$ be a *divergent* series of positive terms. Set $s_n = \sum_{i=1}^{n} a_i$.

    (a) Prove that $\sum_{n\geq 1} \dfrac{a_n}{s_n^p}$ converges for $p > 1$.

        HINT: Rewrite the sum as $\sum_{n\geq 1} s_n\left(s_n^{-p} - s_{n+1}^{-p}\right) = \sum_{n\geq 1} t_n^{-1/p}(t_n - t_{n+1})$.
        Bound the latter sum by an integral.

    (b) Prove that $\sum_{n\geq 1} \dfrac{a_n}{s_n}$ diverges.

        HINT: First consider $a_n \geq s_{n-1}$. Then group terms to achieve this.

S. Determine if the following series converge or diverge.

(a) $\sum_{n=2}^{\infty} \dfrac{3n}{n^3 + 1}$

(b) $\sum_{n=1}^{\infty} \dfrac{n}{2^n}$

(c) $\sum_{n=2}^{\infty} \dfrac{(-1)^n \log n}{n}$

(d) $\sum_{n=1}^{\infty} \sqrt{n+1} - \sqrt{n}$

(e) $\sum_{n=1}^{\infty} e^{-n^2}$

(f) $\sum_{n=1}^{\infty} \sin(n\pi/4)$

(g) $\sum_{n=1}^{\infty} (-1)^n \sin(1/n)$

(h) $\sum_{n=1}^{\infty} \dfrac{1}{\sqrt{n^3 + 4}}$

(i) $\sum_{n=1}^{\infty} \left(\sqrt[n]{n} - 1 \right)^n$

(j) $\sum_{n=2}^{\infty} \dfrac{\sqrt{n+1} - \sqrt{n}}{n}$

(k) $\sum_{n=2}^{\infty} \dfrac{(-1)^k}{\sqrt{n}\log n}$

(l) $\sum_{n=2}^{\infty} \dfrac{(-1)^n}{\sqrt[n]{n}}$

(m) $\sum_{n=2}^{\infty} \dfrac{1}{(\log n)^k}$

(n) $\sum_{n=1}^{\infty} \dfrac{n!}{n^n}$

(o) $\sum_{n=1}^{\infty} \dfrac{(-1)^n \arctan(n)}{n}$

(p) $\sum_{n=2}^{\infty} \dfrac{(-1)^n}{\sqrt{n} + (-1)^n}$

(q) $\sum_{n=1}^{\infty} (-1)^n (e^{1/n} - 1)$

(r) $\sum_{n=1}^{\infty} (-1)^n \dfrac{n^{42}}{(n+1)!}$

(s) $\sum_{n=1}^{\infty} \dfrac{1}{1 + n^2}$

(t) $\sum_{n=1}^{\infty} \dfrac{1}{\log(e^n + e^{-n})}$

(u) $\sum_{n=1}^{\infty} \dfrac{\sin(\pi n/3)}{n}$

(v) $\sum_{n=1}^{\infty} \dfrac{n^{10}}{10^n}$

(w) $\sum_{n=2}^{\infty} \dfrac{1}{(\log n)^n}$

(x) $\sum_{n=2}^{\infty} \dfrac{1}{n\log n}$

(y) $\sum_{n=1}^{\infty} \sin\left(\tfrac{1}{n}\right)$

(z) $\sum_{n=1}^{\infty} \dfrac{1}{n^{1+1/n}}$

3.3. The Number *e*

Recall from calculus the formula

$$e = 1 + 1 + \frac{1}{2} + \frac{1}{6} + \frac{1}{24} + \cdots = \sum_{k=0}^{\infty} \frac{1}{k!} \approx 2.7182818285\ldots$$

If you don't recall this, you can review the section on Taylor series in your calculus book or study Example 10.1.4 later in this book. With $a_k = 1/k!$, we obtain

$$\lim_{k\to\infty} \frac{a_{k+1}}{a_k} = \lim_{k\to\infty} \frac{1}{k+1} = 0.$$

Thus this series converges by the Ratio Test. The limit is called *e*.

There are other ways to compute *e*. We give one such well-known formula and verify that it has the same limit.

3.3.1. PROPOSITION. *Consider the sequences*

$$b_n = \left(1 + \frac{1}{n}\right)^n \quad and \quad c_n = \left(1 + \frac{1}{n}\right)^{n+1}.$$

These sequences are monotone increasing and decreasing, respectively, and

$$\lim_{n\to\infty} \left(1 + \frac{1}{n}\right)^n = \lim_{n\to\infty} \left(1 + \frac{1}{n}\right)^{n+1} = e.$$

PROOF. We need the inequality:

$$(1+x)^n > (1+nx) \quad \text{for} \quad x > -1 \quad \text{and} \quad n \geq 1.$$

To see this, let $f(t) = (1+t)^n - (1+nt)$. Notice that $f(0) = 0$ and

$$f'(t) = n\big((1+t)^{n-1} - 1\big) \quad \text{for} \quad t > -1.$$

Thus f is decreasing on $(-1, 0]$ and increasing on $[0, \infty)$, and so takes its minimum value at 0.

As $c_n = \left(1 + \frac{1}{n}\right)b_n$, we have $b_n < c_n$ for $n \geq 1$. Now compute

$$\frac{b_{n+1}}{c_n} = \left(\frac{1 + \frac{1}{n+1}}{1 + \frac{1}{n}}\right)^{n+1} = \left(\frac{n^2 + 2n}{n^2 + 2n + 1}\right)^{n+1}$$

$$= \left(1 - \frac{1}{(n+1)^2}\right)^{n+1} > 1 - \frac{n+1}{(n+1)^2} = \frac{n}{n+1}.$$

Therefore,

$$b_{n+1} > \frac{n}{n+1}c_n > \frac{n-1}{n}c_n = b_n.$$

Similarly, inverting the preceding identity,

$$\frac{c_n}{b_{n+1}} = \left(1 + \frac{1}{n^2 + 2n}\right)^{n+1} > 1 + \frac{n+1}{n^2 + 2n} > \frac{n+2}{n+1}$$

whence $c_{n+1} = \frac{n+2}{n+1}b_{n+1} < c_n$.

It follows that $(b_n)_{n=1}^{\infty}$ is a monotone increasing sequence bounded above by c_1. Thus it converges to some limit L by Theorem 2.5.4. Similarly, $(c_n)_{n=1}^{\infty}$ is a monotone decreasing sequence bounded below by b_1, and

$$\lim_{n\to\infty} c_n = \lim_{n\to\infty} \left(1+\tfrac{1}{n}\right) b_n = L.$$

Next we use the Binomial Theorem to estimate the terms b_n:

$$b_n = \left(1+\frac{1}{n}\right)^n = \sum_{k=0}^{n} \binom{n}{k} \frac{1}{n^k}$$

$$= \sum_{k=0}^{n} \frac{1}{k!} \frac{n}{n} \frac{(n-1)}{n} \cdots \frac{(n-k+1)}{n}$$

$$= \sum_{k=0}^{n} \frac{1}{k!} \left(1-\frac{1}{n}\right) \left(1-\frac{2}{n}\right) \cdots \left(1-\frac{k-1}{n}\right).$$

It follows that $b_n < s_n = \sum_{k=0}^{n} \frac{1}{k!}$, and therefore

$$L = \lim_{n\to\infty} b_n \le \lim_{n\to\infty} s_n = e.$$

On the other hand, this also provides the lower bound. If p is a fixed integer, then

$$b_n > \sum_{k=0}^{p} \frac{1}{k!} \left(1-\frac{1}{n}\right) \left(1-\frac{2}{n}\right) \cdots \left(1-\frac{k-1}{n}\right)$$

$$> \left(1-\frac{1}{n}\right) \left(1-\frac{2}{n}\right) \cdots \left(1-\frac{p-1}{n}\right) s_p.$$

Thus

$$L = \lim_{n\to\infty} b_n \ge \lim_{n\to\infty} \left(1-\frac{1}{n}\right) \left(1-\frac{2}{n}\right) \cdots \left(1-\frac{p-1}{n}\right) s_p = s_p$$

for all $p \ge 1$. So

$$L \ge \lim_{p\to\infty} s_p = e.$$

We obtain

$$\lim_{n\to\infty} \left(1+\frac{1}{n}\right)^n = \lim_{n\to\infty} \left(1+\frac{1}{n}\right)^{n+1} = e. \qquad \blacksquare$$

Infinite series have unexpected uses. We use the series for e to prove that it is not a rational number.

3.3.2. THEOREM. *e is irrational.*

PROOF. Suppose that e is rational and can be written as $e = p/q$ in lowest terms, where p and q are positive integers. We look for a contradiction. Do the following calculation:

$$(q-1)!\,p = q!\frac{p}{q} = q!\,e = q! \sum_{k=0}^{\infty} \frac{1}{k!} = \sum_{k=0}^{q} \frac{q!}{k!} + \sum_{k=q+1}^{\infty} \frac{q!}{k!}.$$

Rearranging, we get

$$\sum_{k=q+1}^{\infty} \frac{q!}{k!} = (q-1)!p - \sum_{k=1}^{q} \frac{q!}{k!}.$$

Since both terms on the right-hand are integers, the left-hand side must also be an integer. However, using the properties of limits, we have

$$0 < \sum_{k=q+1}^{\infty} \frac{q!}{k!} = \frac{1}{q+1} + \frac{1}{(q+1)(q+2)} + \frac{1}{(q+1)(q+2)(q+3)} + \cdots$$

$$< \frac{1}{q+1} + \frac{1}{(q+1)^2} + \frac{1}{(q+1)^3} + \cdots$$

$$= \sum_{n \geq 1} \left(\frac{1}{q+1} \right)^n = \frac{1/(q+1)}{1 - 1/(q+1)} = \frac{1}{q} < 1.$$

Therefore, $0 < \sum_{k=q+1}^{\infty} \frac{q!}{k!} < 1$, contradicting the conclusion that this summation is an integer. ∎

Exercises for Section 3.3

A. (a) Show that $\sum_{k=0}^{\infty} \frac{x^k}{k!}$ converges for each $x \in \mathbb{R}$. The limit is called e^x.

  (b) Show that $\lim_{n \to \infty} \left(1 + \frac{x}{n} \right)^n = e^x$.   HINT: Compare with the series in part (a).

B. Compute the sum of the series $\sum_{n=1}^{\infty} \frac{1^2 + 2^2 + \cdots + n^2}{n!}$.

  HINT: Use induction to show that $1^2 + 2^2 + \cdots + n^2 = n(n+1)(2n+1)/6$. Then express this in the form $An(n-1)(n-2) + Bn(n-1) + Cn + D$.

C. Compute the sum of the series $\sum_{n=0}^{\infty} \frac{1}{(2n+1)!}$.

  HINT: Express this as the difference of two known series.

D. Evaluate $\lim_{n \to \infty} \left(1 + \frac{1}{n} \right)^{n^2} \left(1 + \frac{1}{n+1} \right)^{-(n+1)^2}$.

E. Decide whether $\sum_{n=1}^{\infty} \frac{n!}{n^n}$ converges.

F. Show that the Root Test implies the Ratio Test by proving that if $\lim_{n \to \infty} \frac{a_{n+1}}{a_n} = r$, then $\lim_{n \to \infty} (a_n)^{1/n} = r$.

G. Combine the two previous exercises to compute $\lim_{n \to \infty} \frac{\sqrt[n]{n!}}{n}$.

H. Decide if $\sum_{n=1}^{\infty} \frac{1}{(\log n)^{\log n}}$ converges.   HINT: Show that $(\log n)^{\log n} = n^{\log \log n}$.

I.   If a is a real number, consider the limit $\lim\limits_{n\to\infty} \sin(n!\pi a)$.

    (a) If a is rational, find the limit.

    (b) Show that $n! \sum\limits_{k=n+1}^{\infty} \dfrac{1}{k!} < \dfrac{1}{n}$.

    (c) If $a = e/2$, compute $\lim\limits_{n\to\infty} \sin((2n)!\pi a)$.

3.4. Absolute and Conditional Convergence

In this section, we further investigate convergence properties of series. The Alternating Series Test shows that badly behaved series such as the harmonic series become more tractable when appropriate signs for the terms keep the partial close together. However, the following variant on Example 3.2.8 shows that considerable care must be taken when adding this type of series.

3.4.1. EXAMPLE. Consider the series

$$1 - \frac{1}{2} - \frac{1}{4} + \frac{1}{3} - \frac{1}{6} - \frac{1}{8} + \cdots,$$

where $a_{3n-2} = \frac{1}{2n-1}$, $a_{3n-1} = -\frac{1}{4n-2}$ and $a_{3n} = -\frac{1}{4n}$. This has exactly the same terms as the alternating harmonic series except that the negative terms are coming twice as fast as the positive ones.

First let's convince ourselves that this series converges. Notice that

$$
\begin{aligned}
a_{3n-2} + a_{3n-1} + a_{3n} &= \frac{1}{2n-1} - \frac{1}{4n-2} - \frac{1}{4n} \\
&= \frac{4n - 2n - (2n-1)}{4n(2n-1)} = \frac{1}{4n(2n-1)}.
\end{aligned}
$$

Therefore, $s_{3n} = \sum\limits_{k=1}^{n} \dfrac{1}{4k(2k-1)}$. The terms of this series are dominated by the

series $\sum\limits_{k=1}^{\infty} \dfrac{1}{4k^2}$. This latter series converges by the Integral Test (see Exercise 3.2.O) because

$$\int_{1}^{\infty} \frac{1}{4x^2}\, dx = -\frac{1}{4x}\Big|_{1}^{\infty} = \frac{1}{4} < \infty.$$

Therefore, $\lim\limits_{n\to\infty} s_{3n} = \sum\limits_{k=1}^{\infty} \dfrac{1}{4k(2k-1)}$ converges by the Comparison Test (3.2.3).

However, $|s_{3n} - s_{3n\pm1}| < \frac{1}{2n}$. With a little bit of work (see Exercise 2.6.G), we can show that the limit exists, using

$$\lim_{n\to\infty} s_{3n-1} = \lim_{n\to\infty} s_{3n+1} = \lim_{n\to\infty} s_{3n} = \lim_{n\to\infty} s_n.$$

The proof of the Integral Test yields the estimate

$$\sum_{n=1}^{\infty} \frac{1}{4n^2} < \frac{1}{4} + \int_{1}^{\infty} \frac{1}{4x^2}\, dx = \frac{1}{2}.$$

Hence our limit is also less than $\frac{1}{2}$. However, the alternating harmonic series has the limit $\log 2 \approx 0.6931471806$. So these two series have *different* sums even though they have the same terms. We can actually sum this series exactly because

$$\frac{1}{4k(2k-1)} = \frac{1}{2}\left(\frac{1}{2k-1} - \frac{1}{2k}\right).$$

Therefore,

$$s_{3n} = \frac{1}{2}\sum_{k=1}^{2n} \frac{(-1)^{n+1}}{k}.$$

By Example 3.2.8, we conclude that the series converges to $\frac{1}{2}\log 2$.

3.4.2. DEFINITION. A series $\sum_{n=1}^{\infty} a_n$ is called **absolutely convergent** if the series $\sum_{n=1}^{\infty} |a_n|$ converges. A series that converges but is not absolutely convergent is called **conditionally convergent**.

Example 3.2.8 shows that a convergent series need not be absolutely convergent. The next simple fact is that absolute convergence is a stronger notion than convergence. The proof is left to Exercise 3.2.B, which is an easy application of the Comparison Test to the series $\sum_{n=1}^{\infty} a_n$ and $\sum_{n=1}^{\infty} |a_n|$.

3.4.3. PROPOSITION. *An absolutely convergent series is convergent.*

Consider the question of what happens when the terms of a series are permuted.

3.4.4. DEFINITION. A **rearrangement** of a series $\sum_{n=1}^{\infty} a_n$ is another series with the same terms in a different order. This can be described by a permutation π of the natural numbers $\mathbb{N}$ determining the series $\sum_{n=1}^{\infty} a_{\pi(n)}$.

The situation for absolutely convergent series is good.

3.4.5. THEOREM. *For an absolutely convergent series, every rearrangement converges to the same limit.*

PROOF. Let $\sum_{n=1}^{\infty} a_n$ be an absolutely convergent series that converges to L. Suppose that π is a permutation of $\mathbb{N}$ and that $\varepsilon > 0$ is given. Then there is an integer N such that $\sum_{k=N+1}^{\infty} |a_k| < \varepsilon/2$.

Since the rearrangement contains exactly the same terms in a different order, the first N terms $a_1, \ldots, a_N$ eventually occur in the rearranged series. Thus there is an integer M so that all of these terms occur in the first M terms of the rearrangement. Hence for $m \geq M$,

$$\left| \sum_{k=1}^{m} a_{\pi(k)} - L \right| \leq \left| \sum_{k=1}^{m} a_{\pi(k)} - \sum_{k=1}^{N} a_k \right| + \left| \sum_{k=1}^{N} a_k - L \right|$$

$$\leq 2 \sum_{k=N+1}^{\infty} |a_k| < \varepsilon.$$

Therefore, $\sum_{k=1}^{\infty} a_{\pi(k)} = L$. ∎

3.4.6. EXAMPLE. Consider the series $\sum_{n=1}^{\infty} \dfrac{(-1)^{n+1}}{n^4}$. This converges by the Alternating Series Test. In fact, it is absolutely convergent since the series $\sum_{n=1}^{\infty} \dfrac{1}{n^4}$ converges by the Integral Test (verify). Hence we may manipulate the terms freely. Therefore,

$$\sum_{n=1}^{\infty} \frac{(-1)^{n+1}}{n^4} = \sum_{n=1}^{\infty} \frac{1}{n^4} - 2 \sum_{n=1}^{\infty} \frac{1}{(2n)^4} = \frac{7}{8} \sum_{n=1}^{\infty} \frac{1}{n^4}.$$

Using techniques from Fourier series (see Chapter 14), we will be able to show that $\sum_{n=1}^{\infty} 1/n^4 = \pi^4/90$. From this, it follows that the preceding summation equals $7\pi^4/720$.

On the other hand, the worst possible scenario holds for the rearrangements of conditionally convergent series. First, we need the following dichotomy.

3.4.7. LEMMA. *Let $\sum_{n=1}^{\infty} a_n$ be a convergent series. Denote the positive terms as $b_1, b_2, b_3, \ldots$ and the other terms as $c_1, c_2, c_3, \ldots$.*

(1) *If $\sum_{n=1}^{\infty} a_n$ is absolutely convergent, then so are both $\sum_{n=1}^{\infty} b_n$ and $\sum_{n=1}^{\infty} |c_n|$, and $\sum_{n=1}^{\infty} a_n = \sum_{n=1}^{\infty} b_n - \sum_{n=1}^{\infty} |c_n|$.*

(2) *If $\sum_{n=1}^{\infty} a_n$ is conditionally convergent, then $\sum_{n=1}^{\infty} b_n$ and $\sum_{n=1}^{\infty} |c_n|$ both diverge.*

PROOF. If the series (a_n) is absolutely convergent, consider the new sequence given by $x_n = \max\{a_n, 0\}$. This is just the sequence (b_k) in order with zeros in place of the c_ks. Thus $\sum_{n=1}^{\infty} b_n$ converges if $\sum_{n=1}^{\infty} x_n$ does. However, $0 \le x_n \le |a_n|$. So by the Comparison Test (3.2.3), this series converges. The series for (c_n) is handled in the same way.

Now suppose that the sequence converges conditionally. Hence

$$\sum_{n=1}^{\infty} |a_n| = \sum_{n=1}^{\infty} b_n + \sum_{n=1}^{\infty} |c_n| = +\infty.$$

Therefore, at least one of the series $\sum_{n=1}^{\infty} b_n$ and $\sum_{n=1}^{\infty} |c_n|$ diverges.

However, if one of these two series is convergent, we can obtain a contradiction. For convenience, suppose that $\sum_{n=1}^{\infty} |c_n|$ converges to L. Since $\sum_{n=1}^{\infty} b_n$ diverges, for each $R > 0$, there is an integer N so that

$$\sum_{n=1}^{N} b_n > R + L.$$

Choose M sufficiently large so that the first M terms of $(a_n)_{n=1}^{\infty}$ contains the first N terms of (b_i). Then for all $m \ge M$

$$\sum_{k=1}^{m} a_k \ge \sum_{n=1}^{N} b_n - \sum_{k=1}^{\infty} |c_k| \ge (R + L) - L = R.$$

Therefore, the series $\sum_{n=1}^{\infty} a_n$ diverges, contrary to fact. This contradiction shows that $\sum_{n=1}^{\infty} |c_n|$ must diverge. The case of $\sum_{n=1}^{\infty} b_n$ converging is similar. ∎

3.4.8. REARRANGEMENT THEOREM.

If $\sum_{n=1}^{\infty} a_n$ is a conditionally convergent series, then for every real number L, there is a rearrangement that converges to L.

PROOF. Let us write the positive terms as $b_1, b_2, b_3, \ldots$ and the negative terms as $c_1, c_2, c_3, \ldots$. By Lemma 3.4.7,

$$\sum_{k=1}^{\infty} b_k = +\infty \qquad \text{and} \qquad \sum_{k=1}^{\infty} c_k = -\infty.$$

Also by Theorem 3.1.4,

$$\lim_{n \to \infty} a_n = 0 = \lim_{n \to \infty} b_n = \lim_{n \to \infty} c_n.$$

Now fix a real number L. Choose the least positive integer m_1 such that

$$u_1 = \sum_{i=1}^{m_1} b_i > L.$$

Then choose the least positive integer n_1 such that

$$v_1 = \sum_{i=1}^{m_1} b_i + \sum_{j=1}^{n_1} c_j < L.$$

We continue in this way, adding just enough positive terms to make the total greater than L and then switching to negative terms until the total is less than L. In this way, we define increasing sequences m_k and n_k to be the least positive integers greater than m_{k-1} and n_{k-1}, respectively, such that

$$u_k = \sum_{i=1}^{m_k} b_i + \sum_{j=1}^{n_{k-1}} c_j > L > \sum_{i=1}^{m_k} b_i + \sum_{j=1}^{n_k} c_j = v_k.$$

This new series is

$$b_1 + \cdots + b_{m_1} + c_1 + \cdots + c_{n_1} + b_{m_1+1} + \cdots + b_{m_2} + c_{n_1+1} + \cdots + c_{n_2} + \cdots.$$

We shall show that this series has limit L. From the construction of the new series, we have the inequalities

$$u_i - b_{m_i} \leq L < u_i \qquad \text{and} \qquad v_j < L \leq v_j - c_{n_j}.$$

Therefore,

$$L + c_{n_j} \leq v_j < L < u_i \leq L + b_{m_i}.$$

Since $\lim_{n \to \infty} L + b_n = \lim_{n \to \infty} L + c_n = L$, the Squeeze Theorem (Theorem 2.3.6) shows that

$$\lim_{i \to \infty} u_i = L = \lim_{j \to \infty} v_j.$$

Finally, let s_k be the partial sums of the new series. It is evident that

$$v_{i-1} \leq s_k \leq u_i \quad \text{for} \quad \sum_{j=1}^{i-1}(m_j + n_j) \leq k \leq m_i + \sum_{j=1}^{i-1}(m_j + n_j)$$

and

$$v_i \leq s_k \leq u_i \quad \text{for} \quad m_i + \sum_{j=1}^{i-1}(m_j + n_j) \leq k \leq \sum_{j=1}^{i}(m_j + n_j).$$

A second application of the Squeeze Theorem shows that our series converges to L as desired. ∎

We complete this section with yet another convergence test. The proof utilizes a rearrangement technique called **summation by parts**, which is analogous to integration by parts.

3.4.9. SUMMATION BY PARTS LEMMA.

Suppose (x_n) and (y_n) are sequences of real numbers and define $X_n = \sum_{k=1}^{n} x_k$ and $Y_n = \sum_{k=1}^{n} y_k$. Then

$$\sum_{n=1}^{m} x_n Y_n + \sum_{n=1}^{m} X_n y_{n+1} = X_m Y_{m+1}.$$

PROOF. The argument is essentially an exercise in reindexing summations. Let $X_0 = 0$ and notice that the left-hand side (LHS) equals

$$\text{LHS} = \sum_{n=1}^{m} (X_n - X_{n-1})Y_n + \sum_{n=1}^{m} X_n (Y_{n+1} - Y_n)$$
$$= \sum_{n=1}^{m} X_n Y_n - \sum_{n=1}^{m} X_{n-1} Y_n + \sum_{n=1}^{m} X_n Y_{n+1} - \sum_{n=1}^{m} X_n Y_n.$$
$$= -X_0 Y_1 + X_m Y_{m+1} = X_m Y_{m+1}.$$

■

Thus, provided that $\lim_{m \to \infty} X_m Y_{m+1}$ exists, the two series $\sum x_n Y_n$ and $\sum X_n y_n$ either both converge or both diverge.

3.4.10. DIRICHLET'S TEST.

Suppose that $(a_n)_{n=1}^{\infty}$ is a sequence of real numbers with bounded partial sums:

$$\left| \sum_{k=1}^{n} a_k \right| \leq M < \infty \quad \text{for all} \quad n \geq 1.$$

If $(b_n)_{n=1}^{\infty}$ is a sequence of positive numbers decreasing monotonically to 0, then the series $\sum_{n=1}^{\infty} a_n b_n$ converges.

PROOF. We use the Summation by Parts Lemma to rewrite $a_n b_n$. Let $y_1 = b_1$, $y_n = b_n - b_{n-1}$ for $n > 1$, and let $x_n = a_n$ for all n. Define X_n and Y_n as in the lemma. Observe that $Y_n = b_n$ and so $a_n b_n = x_n Y_n$.

Notice that $|X_n| = \left| \sum_{k=1}^{n} a_k \right| \leq M$ for all n. Since $|X_n Y_{n+1}| \leq M b_{n+1}$, the Squeeze Theorem shows that $X_n Y_{n+1}$ converges to zero. Furthermore,

$$\sum_{k=1}^{n} |X_k y_{k+1}| \leq \sum_{k=1}^{n} M y_{k+1} \leq M b_{n+1}.$$

Thus $\sum_{k=1}^{\infty} X_k y_{k+1}$ converges absolutely. Using the Summation by Parts Lemma, convergence follows from

$$\sum_{n=1}^{\infty} a_n b_n = \lim_{m\to\infty} \sum_{n=1}^{m} x_n Y_n = \lim_{m\to\infty} X_m Y_m - \sum_{n=1}^{m} X_n y_{n+1} = \sum_{k=1}^{\infty} X_k y_{k+1}. \quad\blacksquare$$

3.4.11. EXAMPLE. Consider the series $\sum_{n=1}^{\infty} \dfrac{\sin n\theta}{n}$ for $0 \le \theta \le 2\pi$. At the endpoints $\theta = 0$ and 2π, the series is just 0. For θ in $(0, 2\pi)$, we will show that this series converges conditionally.

Let $a_n = \sin n\theta$ and $b_n = \frac{1}{n}$. To evaluate the partial sums of the a_ns, use the following trigonometric identities:

$$\cos(k \pm \tfrac{1}{2})\theta = \cos k\theta \cos \tfrac{\theta}{2} \mp \sin k\theta \sin \tfrac{\theta}{2}$$

whence

$$2 \sin k\theta \sin \tfrac{\theta}{2} = \cos(k - \tfrac{1}{2})\theta - \cos(k + \tfrac{1}{2})\theta.$$

Therefore, we obtain a telescoping sum

$$\sum_{k=1}^{n} a_k = \sum_{k=1}^{n} \sin k\theta = \frac{1}{2 \sin \frac{\theta}{2}} \sum_{k=1}^{n} \cos(k - \tfrac{1}{2})\theta - \cos(k + \tfrac{1}{2})\theta$$

$$= \frac{1}{2} \csc \tfrac{\theta}{2} \left(\cos \tfrac{\theta}{2} - \cos(n + \tfrac{1}{2})\theta \right).$$

Consequently,

$$\left| \sum_{k=1}^{n} a_k \right| \le \csc \tfrac{1}{2}\theta < \infty.$$

So Dirichlet's Test applies and the series converges.

To see that convergence is conditional, it suffices to show that most of the terms behave like the harmonic series. Out of every two consecutive terms $\sin k\theta$, at most one term has absolute value less than $\sin \frac{\theta}{2}$. Therefore, we obtain

$$\frac{|\sin(2k-1)\theta|}{2k-1} + \frac{|\sin 2k\theta|}{2k} \ge \frac{\sin \frac{\theta}{2}}{2k}.$$

By comparison with a multiple of the harmonic series, $\sum_{k=1}^{\infty} \dfrac{|\sin k\theta|}{k}$ diverges.

Exercises for Section 3.4

A. Find the series in Exercise 3.2.S that converge conditionally but not absolutely.

B. Decide which of the following series converge absolutely, conditionally, or not at all.

(a) $\displaystyle\sum_{n=1}^{\infty} \frac{(-1)^n}{n \log(n+1)}$ (b) $\displaystyle\sum_{n=1}^{\infty} \frac{(-1)^n}{(2 + (-1)^n)n}$ (c) $\displaystyle\sum_{n=1}^{\infty} \frac{(-1)^n \sin(\frac{1}{n})}{n}$

C. Compute the sum of the series $\sum_{n=1}^{\infty} \dfrac{1}{n^2(2n-1)}$ given that $\sum_{n=1}^{\infty} \dfrac{1}{n^2} = \dfrac{\pi^2}{6}$.

HINT: $\frac{1}{n^2(2n-1)} = \frac{4}{2n(2n-1)} - \frac{1}{n^2}$.

D. Show that the series $\sum_{n=1}^{\infty} \dfrac{\sin(\frac{2n\pi}{3})}{n^2}$ converges absolutely. Find the sum, given that

$\sum_{n=1}^{\infty} \dfrac{1}{n^2} = \pi^2/6$. (See Example 13.6.5.)

E. Show that a conditionally convergent series has a rearrangement converging to $+\infty$.

F. Show that the Alternating Series Test is a special case of Dirichlet's Test.

G. Use summation by parts to prove **Abel's Test**: Suppose that $\sum_{n=1}^{\infty} a_n$ converges and (b_n) is a monotonic convergent sequence. Show that $\sum_{n=1}^{\infty} a_n b_n$ converges.

H. Determine the values of θ for which the series $\sum_{n=1}^{\infty} \dfrac{e^{in\theta}}{\log(n+1)}$ converges.

I. Let $a_n = \dfrac{(-1)^k}{n}$ for $(k-1)^2 < n \le k^2$ and $k \ge 1$. Decide if the series $\sum_{n=1}^{\infty} a_n$ converges.

CHAPTER 4

Topology of $\mathbb{R}^n$

The space $\mathbb{R}^n$ is the right setting for many problems in real analysis. For example, in many situations, functions of interest depend on several variables. This puts us into the realm of multivariable calculus, which is naturally set in $\mathbb{R}^n$. We will study normed vector spaces further in Chapter 7, building on the properties and concepts we study here. The space $\mathbb{R}^n$ is the most important normed vector space, after the real numbers themselves.

4.1. *n*-Dimensional Space

The space $\mathbb{R}^n$ is the set of n-vectors $\mathbf{x} = (x_1, x_2, \ldots, x_n)$ with arbitrary real coefficients x_i for $1 \leq i \leq n$. Generally, vectors in $\mathbb{R}^n$ will be referred to as **points**. This space has a lot of structure, most of which should be familiar from advanced calculus or linear algebra courses. In particular, we should mention that the **zero vector** is $(0, 0, \ldots, 0)$, which we denote as $\mathbf{0}$.

First, it is a *vector space*. Recall (see Section 1.4) that vectors may be added and subtracted and multiplied by (real) scalars. Indeed, for any $\mathbf{x}$ and $\mathbf{y}$ in $\mathbb{R}^n$ and scalars $t \in \mathbb{R}$,

$$\mathbf{x} + \mathbf{y} = (x_1, x_2, \ldots, x_n) + (y_1, y_2, \ldots, y_n)$$
$$= (x_1 + y_1, x_2 + y_2, \ldots, x_n + y_n)$$

and

$$t\mathbf{x} = t(x_1, x_2, \ldots, x_n) = (tx_1, tx_2, \ldots, tx_n).$$

Vector spaces are reviewed in Section 1.4, but we are assuming that you know the basics of linear algebra. Instead, we concentrate on the properties of $\mathbb{R}^n$ that build on the ideas of distance and convergence.

There is the notion of length of a vector, given by

$$\|\mathbf{x}\| = \|(x_1, x_2, \ldots, x_n)\| = \left(\sum_{i=1}^{n} |x_i|^2 \right)^{1/2}.$$

This is called the **Euclidean norm** on $\mathbb{R}^n$, and $\|\mathbf{x}\|$ is the **norm** of $\mathbf{x}$. This conforms to our usual notion of distance in the plane and in space. Moreover, it is the natural consequence of the Euclidean distance in the plane using the Pythagorean formula and induction on the number of variables. (See Exercise 4.1.A.) The distance between two points $\mathbf{x}$ and $\mathbf{y}$ is then determined by

$$\|\mathbf{x} - \mathbf{y}\| = \left(\sum_{i=1}^{n} |x_i - y_i|^2\right)^{1/2}.$$

An important property of distance is the **triangle inequality**: The distance from point A to point B and then on to a point C is at least as great as the direct distance from A to C. This is interpreted geometrically as saying that the sum of the lengths of two sides of a triangle is greater than the length of the third side (Figure 4.1). (Equality can occur if the triangle has no area.)

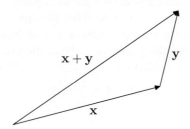

FIGURE 4.1. The triangle inequality.

To verify this algebraically, we need an inequality involving the dot product, which is useful in its own right. Recall that the **dot product** or **inner product** of two vectors $\mathbf{x}$ and $\mathbf{y}$ is given by

$$\langle \mathbf{x}, \mathbf{y} \rangle = \langle (x_1, \ldots, x_n), (y_1, \ldots, y_n) \rangle = \sum_{i=1}^{n} x_i y_i.$$

There is a close connection between the inner product and the Euclidean norm because of the evident identity

$$\langle \mathbf{x}, \mathbf{x} \rangle = \|\mathbf{x}\|^2.$$

The inner product is linear in both variables:

$$\langle r\mathbf{x} + s\mathbf{y}, \mathbf{z} \rangle = r\langle \mathbf{x}, \mathbf{z} \rangle + s\langle \mathbf{y}, \mathbf{z} \rangle \quad \text{for all} \quad \mathbf{x}, \mathbf{y}, \mathbf{z} \in \mathbb{R}^n \text{ and } r, s \in \mathbb{R}$$

and

$$\langle \mathbf{x}, s\mathbf{y} + t\mathbf{z} \rangle = s\langle \mathbf{x}, \mathbf{y} \rangle + t\langle \mathbf{x}, \mathbf{z} \rangle \quad \text{for all} \quad \mathbf{x}, \mathbf{y}, \mathbf{z} \in \mathbb{R}^n \text{ and } s, t \in \mathbb{R}$$

4.1.1. SCHWARZ INEQUALITY.
For all $\mathbf{x}$ and $\mathbf{y}$ in $\mathbb{R}^n$,

$$|\langle \mathbf{x}, \mathbf{y} \rangle| \leq \|\mathbf{x}\| \, \|\mathbf{y}\|.$$

Equality holds if and only if $\mathbf{x}$ and $\mathbf{y}$ are collinear.

PROOF. Let $\mathbf{x} = (x_1, \ldots, x_n)$ and $\mathbf{y} = (y_1, \ldots, y_n)$. Then

$$2\|\mathbf{x}\|^2\|\mathbf{y}\|^2 - 2|\langle \mathbf{x}, \mathbf{y}\rangle|^2 = 2\sum_{i=1}^{n}\sum_{j=1}^{n} x_i^2 y_j^2 - 2\Big(\sum_{i=1}^{n} x_i y_i\Big)^2$$

$$= \sum_{i=1}^{n}\sum_{j=1}^{n} x_i^2 y_j^2 + x_j^2 y_i^2 - \sum_{i=1}^{n}\sum_{j=1}^{n} 2x_i y_i x_j y_j$$

$$= \sum_{i=1}^{n}\sum_{j=1}^{n} x_i^2 y_j^2 - 2x_i y_j x_j y_i + x_j^2 y_i^2$$

$$= \sum_{i=1}^{n}\sum_{j=1}^{n} (x_i y_j - x_j y_i)^2 \geq 0.$$

This establishes the inequality because a sum of squares is positive.

Equality holds precisely when $x_i y_j - x_j y_i = 0$ for all i and j. If both $\mathbf{x}$ and $\mathbf{y}$ equal $\mathbf{0}$, there is nothing to prove. So we may suppose that at least one coefficient is nonzero. There is no harm in assuming that $x_1 \neq 0$, as the proof is the same in all other cases. Then

$$y_j = \frac{y_1}{x_1} x_j \quad \text{for all} \quad 1 \leq j \leq n.$$

Hence $\mathbf{y} = \frac{y_1}{x_1}\mathbf{x}$.   ∎

4.1.2. TRIANGLE INEQUALITY.
The triangle inequality holds for the Euclidean norm on $\mathbb{R}^n$:

$$\|\mathbf{x} + \mathbf{y}\| \leq \|\mathbf{x}\| + \|\mathbf{y}\| \quad \text{for all} \quad \mathbf{x}, \mathbf{y} \in \mathbb{R}^n.$$

Moreover, equality holds if and only if either $\mathbf{x} = \mathbf{0}$ *or* $\mathbf{y} = c\mathbf{x}$ *with* $c \geq 0$.

PROOF. Use the relationship between the inner product and norm to compute

$$\begin{aligned}
\|\mathbf{x} + \mathbf{y}\|^2 &= \langle \mathbf{x} + \mathbf{y}, \mathbf{x} + \mathbf{y}\rangle \\
&= \langle \mathbf{x}, \mathbf{x}\rangle + \langle \mathbf{x}, \mathbf{y}\rangle + \langle \mathbf{y}, \mathbf{x}\rangle + \langle \mathbf{y}, \mathbf{y}\rangle \\
&\leq \langle \mathbf{x}, \mathbf{x}\rangle + |\langle \mathbf{x}, \mathbf{y}\rangle| + |\langle \mathbf{y}, \mathbf{x}\rangle| + \langle \mathbf{y}, \mathbf{y}\rangle \\
&\leq \|\mathbf{x}\|^2 + \|\mathbf{x}\|\,\|\mathbf{y}\| + \|\mathbf{x}\|\,\|\mathbf{y}\| + \|\mathbf{y}\|^2 \\
&= (\|\mathbf{x}\| + \|\mathbf{y}\|)^2.
\end{aligned}$$

If equality holds, then we must have $\langle \mathbf{x}, \mathbf{y}\rangle = \|\mathbf{x}\|\|\mathbf{y}\|$. In particular, the Schwarz inequality holds. So either $\mathbf{x} = \mathbf{0}$ or $\mathbf{y} = c\mathbf{x}$. Substituting $\mathbf{y} = c\mathbf{x}$ into $\langle \mathbf{x}, \mathbf{y}\rangle = \|\mathbf{x}\|\,\|\mathbf{y}\|$ gives $c = \|\mathbf{y}\| \geq 0$.   ∎

Colinearity does not imply equality for the triangle inequality in all cases because $\langle x, y\rangle$ could be negative. For example, $\mathbf{x}$ and $-\mathbf{x}$ are collinear for any nonzero vector $\mathbf{x}$, but

$$0 = \|\mathbf{x} + (-\mathbf{x})\| < \|\mathbf{x}\| + \|-\mathbf{x}\| = 2\|\mathbf{x}\|.$$

When we write elements of $\mathbb{R}^n$ in vector notation, we are implicitly using the standard basis $\{\mathbf{e}_i : 1 \le i \le n\}$, as defined in Section 1.4. (Recall that $\mathbf{e}_i$ is the vector with a single 1 in the ith position and zeros in the other coordinates.) A set $\{\mathbf{v}_1, \dots, \mathbf{v}_m\}$ in $\mathbb{R}^n$ is **orthonormal** if $\langle \mathbf{v}_i, \mathbf{v}_j \rangle = \delta_{ij}$ for $1 \le i, j \le m$, where $\delta_{ij} = 0$ when $i \ne j$ and $\delta_{ii} = 1$. If, in addition, $\{\mathbf{v}_1, \dots, \mathbf{v}_m\}$ spans $\mathbb{R}^n$, it is called an **orthonormal basis**. In particular, $\{\mathbf{e}_1, \dots, \mathbf{e}_n\}$ is an orthonormal basis for $\mathbb{R}^n$.

4.1.3. LEMMA. *Let $\{\mathbf{v}_1, \dots, \mathbf{v}_m\}$ be an orthonormal set in $\mathbb{R}^n$. Then*

$$\left\| \sum_{i=1}^m a_i \mathbf{v}_i \right\| = \left(\sum_{i=1}^m |a_i|^2 \right)^{1/2}.$$

An orthonormal set in $\mathbb{R}^n$ is linearly independent. So an orthonormal basis for $\mathbb{R}^n$ is a basis and has exactly n elements.

PROOF. Use the inner product to compute

$$\left\| \sum_{i=1}^m a_i \mathbf{v}_i \right\|^2 = \sum_{i=1}^m \sum_{j=1}^m \langle a_i \mathbf{v}_i, a_j \mathbf{v}_j \rangle$$

$$= \sum_{i=1}^m \sum_{j=1}^m a_i a_j \delta_{ij} = \sum_{i=1}^m |a_i|^2.$$

In particular, if $\sum_{i=1}^m a_i \mathbf{v}_i = \mathbf{0}$, we find that $\sum_{i=1}^m |a_i|^2 = 0$ and thus $a_i = 0$ for $1 \le i \le m$. This shows that $\{\mathbf{v}_1, \dots, \mathbf{v}_m\}$ is linearly independent. Finally, a basis for $\mathbb{R}^n$ is a linearly independent set of vectors that spans $\mathbb{R}^n$. An orthonormal basis spans by definition and is independent, as previously. A basic result of linear algebra shows that every basis for $\mathbb{R}^n$ has exactly n elements. ∎

Exercises for Section 4.1

A. Establish the **Pythagorean formula**: If $\mathbf{x}$ and $\mathbf{y}$ are orthogonal vectors, prove that $\|\mathbf{x} + \mathbf{y}\| = \left(\|\mathbf{x}\|^2 + \|\mathbf{y}\|^2 \right)^{1/2}$.

B. (a) Suppose that $\mathbf{x} = \sum_{i=1}^j x_i \mathbf{e}_i$ is a vector in $\mathbb{R}^n$ with nonzero coefficients only in the first j positions. Apply the Pythagorean formula to the orthogonal vectors $\mathbf{x}$ and $\mathbf{y} = x_{j+1} \mathbf{e}_{j+1}$.

(b) Show by induction that the Pythagorean formula yields the norm of a vector in all dimensions.

C. Show that $\|\mathbf{x} + \mathbf{y}\|^2 + \|\mathbf{x} - \mathbf{y}\|^2 = 2\|\mathbf{x}\|^2 + 2\|\mathbf{y}\|^2$ for all vectors $\mathbf{x}$ and $\mathbf{y}$ in $\mathbb{R}^n$. This is called the **parallelogram law**. What does it mean geometrically?

D. Prove that if $\mathbf{x}$ and $\mathbf{y}$ are vectors in $\mathbb{R}^n$, then $\big| \|\mathbf{x}\| - \|\mathbf{y}\| \big| \le \|\mathbf{x} - \mathbf{y}\|$.

E. Prove by induction that $\|\mathbf{x}_1 + \cdots + \mathbf{x}_k\| \le \|\mathbf{x}_1\| + \cdots + \|\mathbf{x}_k\|$ for vectors $\mathbf{x}_i$ in $\mathbb{R}^n$.

F. Suppose that $\mathbf{x}$ and $\mathbf{y}$ are unit vectors in $\mathbb{R}^n$. Show that if $\left\| \frac{\mathbf{x}+\mathbf{y}}{2} \right\| = 1$, then $\mathbf{x} = \mathbf{y}$.

G. Let $\mathbf{x}$ and $\mathbf{y}$ be two nonzero vectors in $\mathbb{R}^2$ such that the angle between them is θ. Prove that $\langle \mathbf{x}, \mathbf{y} \rangle = \|\mathbf{x}\| \, \|\mathbf{y}\| \cos \theta$.
HINT: If $\mathbf{x}$ makes the angle α to the positive x-axis, then $\mathbf{x} = (\|\mathbf{x}\| \cos \alpha, \|\mathbf{x}\| \sin \alpha)$.

H. For nonzero vectors $\mathbf{x}$ and $\mathbf{y}$ in $\mathbb{R}^n$, define θ by $\|\mathbf{x}\| \, \|\mathbf{y}\| \cos \theta = \langle \mathbf{x}, \mathbf{y} \rangle$, and call this the angle between them.
(a) Prove the **cosine law**: If $\mathbf{x}$ and $\mathbf{y}$ are vectors and θ is the angle between them, then $\|\mathbf{x} + \mathbf{y}\|^2 = \|\mathbf{x}\|^2 + 2\|\mathbf{x}\| \, \|\mathbf{y}\| \cos \theta + \|\mathbf{y}\|^2$.
(b) Prove that $\langle \mathbf{x}, \mathbf{y} \rangle$ can be defined only in terms of norms of related vectors.

I. Suppose that U is a linear transformation from $\mathbb{R}^n$ into $\mathbb{R}^m$ that is **isometric**, meaning that $\|U\mathbf{x}\| = \|\mathbf{x}\|$ for all $\mathbf{x} \in \mathbb{R}^n$.
(a) Prove that $\langle U\mathbf{x}, U\mathbf{y} \rangle = \langle \mathbf{x}, \mathbf{y} \rangle$ for all $\mathbf{x}, \mathbf{y} \in \mathbb{R}^n$.
(b) If $\{\mathbf{v}_1, \ldots, \mathbf{v}_k\}$ is an orthonormal set in $\mathbb{R}^n$, show that $\{U\mathbf{v}_1, \ldots, U\mathbf{v}_m\}$ is also orthonormal.

J. (a) Let U be an isometric linear transformation of $\mathbb{R}^n$ onto itself. Show that the n columns of the matrix of U form an orthonormal basis for $\mathbb{R}^n$.
(b) Conversely, if $\{\mathbf{v}_1, \ldots, \mathbf{v}_n\}$ is an orthonormal basis for $\mathbb{R}^n$, show that the linear transformation $U\mathbf{x} = \sum_{i=1}^{n} x_i \mathbf{v}_i$ is isometric.

K. Let M be a subspace of $\mathbb{R}^n$ with an orthonormal basis $\{\mathbf{v}_1, \ldots, \mathbf{v}_k\}$. Define a linear transformation on $\mathbb{R}_n$ by $P\mathbf{x} = \sum_{j=1}^{k} \langle \mathbf{x}, \mathbf{v}_i \rangle \mathbf{v}_i$.
(a) Show that $P\mathbf{x}$ belongs to M, and $P\mathbf{y} = \mathbf{y}$ for all $\mathbf{y} \in M$. Hence show $P^2 = P$.
(b) Show that $\langle P\mathbf{x}, \mathbf{x} - P\mathbf{x} \rangle = 0$.
(c) Hence show that $\|\mathbf{x}\|^2 = \|P\mathbf{x}\|^2 + \|\mathbf{x} - P\mathbf{x}\|^2$.
(d) If $\mathbf{y}$ belongs to M, show that $\|\mathbf{x} - \mathbf{y}\|^2 = \|\mathbf{y} - P\mathbf{x}\|^2 + \|\mathbf{x} - P\mathbf{x}\|^2$.
(e) Hence show that $P\mathbf{x}$ is the closest point in M to $\mathbf{x}$.

4.2. Convergence and Completeness in $\mathbb{R}^n$

The notion of norm for points in $\mathbb{R}^n$ immediately allows us to discuss convergence of sequences in this context. The definition of limit of a sequence of points $\mathbf{x}_k$ in $\mathbb{R}^n$ is virtually identical to the definition of convergence in $\mathbb{R}$. The only change is to replace absolute value, which is the measure of distance in the reals, with the Euclidean norm in n-space.

4.2.1. DEFINITION. A sequence of points $(\mathbf{x}_k)$ in $\mathbb{R}^n$ **converges** to a point $\mathbf{a}$ if for *every* $\varepsilon > 0$, there is an integer $N = N(\varepsilon)$ so that

$$\|\mathbf{x}_k - \mathbf{a}\| < \varepsilon \quad \text{for all} \quad k \geq N.$$

In this case, we write $\lim_{k \to \infty} \mathbf{x}_k = \mathbf{a}$.

The parallel between the two definitions of convergence allows us to reformulate the definition of limit of a sequence of points in n-space to the consideration of a sequence of real numbers, namely the Euclidean norms of the points.

4.2.2. LEMMA. *Let* $(\mathbf{x}_k)$ *be a sequence in* $\mathbb{R}^n$. *Then* $\lim_{k\to\infty} \mathbf{x}_k = \mathbf{a}$ *if and only if* $\lim_{k\to\infty} \|\mathbf{x}_k - \mathbf{a}\| = 0.$

The second limit is a sequence of real numbers, and thus it may be understood using only ideas from Chapter 2.

Just as important is the relation between convergence in $\mathbb{R}^n$ and convergence of the coefficients. The following lemma is conceptually quite easy (draw a picture in $\mathbb{R}^2$), but the proof requires careful bookkeeping.

4.2.3. LEMMA. *A sequence* $\mathbf{x}_k = (x_{k,1}, \ldots, x_{k,n})$ *in* $\mathbb{R}^n$ *converges to a point* $\mathbf{a} = (a_1, \ldots, a_n)$ *if and only if each coefficient converges:*

$$\lim_{k\to\infty} x_{k,i} = a_i \quad \textit{for} \quad 1 \le i \le n.$$

PROOF. First suppose that $\lim_{k\to\infty} \mathbf{x}_k = \mathbf{a}$. Then given $\varepsilon > 0$, we obtain an integer N so that $\|\mathbf{x}_k - \mathbf{a}\| < \varepsilon$ for all $k \ge N$. Then for each $1 \le i \le n$ and all $k \ge N$,

$$|x_{k,i} - a_i| \le \left(\sum_{j=1}^{n} |x_{k,j} - a_j|^2 \right)^{1/2} = \|\mathbf{x}_k - \mathbf{a}\| < \varepsilon.$$

Therefore, $\lim_{k\to\infty} x_{k,i} = a_i$ for all $1 \le i \le n$.

Conversely, suppose that each coordinate sequence $x_{k,i}$ converges to a real number a_i for $1 \le i \le n$. Then given $\varepsilon > 0$, use ε/n in the definition of limit and choose N_i so large that

$$|x_{k,i} - a_i| < \frac{\varepsilon}{n} \quad \text{for all} \quad k \ge N_i.$$

Then using $N = \max\{N_i : 1 \le i \le n\}$, all n of these inequalities are valid for $k \ge N$. Hence

$$\|\mathbf{x}_k - \mathbf{a}\| = \left(\sum_{i=1}^{n} |x_{k,i} - a_i|^2 \right)^{1/2} < \left(\sum_{i=1}^{n} \left(\frac{\varepsilon}{n}\right)^2 \right)^{1/2} < \varepsilon.$$

Therefore, $\lim_{k\to\infty} \mathbf{x}_k = \mathbf{a}$. ■

Following the same route as for the line, we will define Cauchy sequences and completeness in the higher-dimensional context. For the real line, it was necessary to build the completeness of $\mathbb{R}$ into its construction. However, the completeness for $\mathbb{R}^n$ will be a consequence of the completeness of $\mathbb{R}$.

4.2.4. DEFINITION. A sequence $\mathbf{x}_k$ in $\mathbb{R}^n$ is **Cauchy** if for every $\varepsilon > 0$, there is an integer N so that

$$\|\mathbf{x}_k - \mathbf{x}_l\| < \varepsilon \quad \text{for all} \quad k, l \ge N.$$

A set $S \subset \mathbb{R}^n$ is **complete** if every Cauchy sequence of points in S converges to a point in S.

As in Proposition 2.7.1, it is easy to show that a convergent sequence is Cauchy. Indeed, suppose that $\lim_{k\to\infty} \mathbf{x}_k = \mathbf{a}$ and $\varepsilon > 0$. Then, using $\varepsilon/2$ in the definition of limit, choose an integer N so that $\|\mathbf{x}_k - \mathbf{a}\| < \varepsilon/2$ for all $k \geq N$. Then for all $k, l \geq N$, use the triangle inequality to obtain

$$\|\mathbf{x}_k - \mathbf{x}_l\| \leq \|\mathbf{x}_k - \mathbf{a}\| + \|\mathbf{a} - \mathbf{x}_l\| < \frac{\varepsilon}{2} + \frac{\varepsilon}{2} = \varepsilon.$$

The converse lies deeper and is more important.

4.2.5. COMPLETENESS THEOREM FOR $\mathbb{R}^n$.
Every Cauchy sequence in $\mathbb{R}^n$ *converges. Thus,* $\mathbb{R}^n$ *is complete.*

PROOF. Let $\mathbf{x}_k$ be a Cauchy sequence in $\mathbb{R}^n$. The proof is accomplished by reducing the problem to each coordinate. Let us write the elements of the sequence as $\mathbf{x}_k = (x_{k,1}, \ldots, x_{k,n})$. We will show that the sequences $(x_{k,i})_{k=1}^{\infty}$ are Cauchy for each $1 \leq i \leq n$. Indeed, if $\varepsilon > 0$, choose N so large that

$$\|\mathbf{x}_k - \mathbf{x}_l\| < \varepsilon \quad \text{for all} \quad k, l \geq N.$$

Then

$$|x_{k,i} - x_{l,i}| \leq \|\mathbf{x}_k - \mathbf{x}_l\| < \varepsilon \quad \text{for all} \quad k, l \geq N.$$

Thus $(x_{k,i})_{k=1}^{\infty}$ are Cauchy for $1 \leq i \leq n$.

By the completeness of $\mathbb{R}$, Theorem 2.7.4, each of these sequences has a limit, say

$$\lim_{k\to\infty} x_{k,i} = a_i \quad \text{for} \quad 1 \leq i \leq n.$$

Define a vector $\mathbf{a} \in \mathbb{R}^n$ by $\mathbf{a} = (a_1, \ldots, a_n)$. By Lemma 4.2.3, $\lim_{k\to\infty} \mathbf{x}_k = \mathbf{a}$ and hence $\mathbb{R}^n$ is complete. ∎

4.2.6. EXAMPLE. Let $\mathbf{v}_0 = (0,0)$, and define a sequence $\mathbf{v}_n = (x_n, y_n)$ in $\mathbb{R}^2$ recursively by

$$x_{n+1} = \frac{x_n + y_n + 1}{2} \qquad y_{n+1} = \frac{x_n - y_n + 1}{2}.$$

The first few terms are

$$(0,0), \ (\tfrac{1}{2}, \tfrac{1}{2}), \ (1, \tfrac{1}{2}), \ (\tfrac{5}{4}, \tfrac{3}{4}), \ (\tfrac{3}{2}, \tfrac{3}{4}), \ (\tfrac{13}{8}, \tfrac{7}{8}), \ (\tfrac{7}{4}, \tfrac{7}{8}), \ \ldots.$$

To get an idea of what the limit might be (if it exists), look for fixed points of the map

$$T(x, y) = \left(\frac{x + y + 1}{2}, \frac{x - y + 1}{2} \right).$$

In other words, solve the equation $T\mathbf{u} = \mathbf{u}$. This is a linear system

$$x = \tfrac{1}{2}x + \tfrac{1}{2}y + \tfrac{1}{2}$$
$$y = \tfrac{1}{2}x - \tfrac{1}{2}y + \tfrac{1}{2}.$$

Solving, we find the solution $\mathbf{u} = (2, 1)$.

This leads us to consider the distance of $\mathbf{v}_n$ to $\mathbf{u}$.

$$\|\mathbf{v}_{n+1} - \mathbf{u}\|^2 = \|(x_{n+1} - 2, y_{n+1} - 1)\|^2$$

$$= \left\| \left(\frac{x_n + y_n - 3}{2}, \frac{x_n - y_n - 1}{2} \right) \right\|^2$$

$$= \frac{(x_n + y_n - 3)^2 + (x_n - y_n - 1)^2}{4}$$

$$= \frac{2x_n^2 + 2y_n^2 - 8x_n - 4y_n + 10}{4}$$

$$= \frac{(x_n - 2)^2 + (y_n - 1)^2}{2} = \tfrac{1}{2}\|\mathbf{v}_n - \mathbf{u}\|^2$$

By induction, it follows that

$$\|\mathbf{v}_n - \mathbf{u}\| = 2^{-n/2}\|\mathbf{v}_0 - \mathbf{u}\| = 2^{-n/2}\sqrt{5}.$$

Hence $\lim_{n\to\infty} \|\mathbf{v}_n - \mathbf{u}\| = 0$, which means that $\lim_{n\to\infty} \mathbf{v}_n = (2,1)$.

Exercises for Section 4.2

A. (a) If $(\mathbf{x}_n)_{n=1}^{\infty}$ is a sequence in $\mathbb{R}^n$ with $\lim_{n\to\infty} \mathbf{x}_n = \mathbf{a}$, show that $\lim_{n\to\infty} \|\mathbf{x}_n\| = \|\mathbf{a}\|$.
   (b) Show by example that the converse is false.

B. Show that if $(\mathbf{x}_n)_{n=1}^{\infty}$ is a sequence in $\mathbb{R}^n$ such that $\sum_{n\geq 1} \|\mathbf{x}_n - \mathbf{x}_{n+1}\| < \infty$, then $(\mathbf{x}_n)$ is a Cauchy sequence.

C. (a) Give an example of a Cauchy sequence for which the condition of the previous exercise fails.
   (b) However, show that every Cauchy sequence $(\mathbf{x}_n)_{n=1}^{\infty}$ has a subsequence $(\mathbf{x}_{n_i})_{i=1}^{\infty}$ such that $\sum_{i\geq 1} \|\mathbf{x}_{n_i} - \mathbf{x}_{n_{i+1}}\| < \infty$.

D. Let $\mathbf{x}_0 \in \mathbb{R}^n$ and $R > 0$. Prove that $B_R(\mathbf{x}_0) = \{\mathbf{x} \in \mathbb{R}^n : \|\mathbf{x} - \mathbf{x}_0\| \leq R\}$ is complete.

E. Let M be a subspace of $\mathbb{R}^n$.
   (a) Let $\{\mathbf{v}_1, \dots, \mathbf{v}_m\}$ be an orthonormal basis for M. Formulate an analogue of Lemma 4.2.3 for M and prove it.
   (b) Prove that M is complete.

F. Let $\mathbf{v}_0 = (x_0, y_0)$ with $0 < x_0 < y_0$. Define $\mathbf{v}_{n+1} = (x_{n+1}, y_{n+1}) = \left(\sqrt{x_n y_n}, \frac{x_n + y_n}{2} \right)$ for all $n \geq 0$.
   (a) Show by induction that $0 < x_n < x_{n+1} < y_{n+1} < y_n$.
   (b) Then estimate $y_{n+1} - x_{n+1}$ in terms of $y_n - x_n$.
   (c) Thereby show that there is a number c such that $\lim_{n\to\infty} \mathbf{v}_n = (c, c)$. This value c is known as the **arithmetic–geometric mean** of x_0 and y_0.

G. Let $\mathbf{v}_0 = (x_0, y_0) = (0, 0)$, and for $n \geq 0$ define
$$\mathbf{v}_{n+1} = (x_{n+1}, y_{n+1}) = \left(\sqrt{\frac{x_n^2 + 2y_n^2}{4}}, \frac{x_n + y_n + 1}{3} \right).$$
   (a) Show that x_n and y_n are increasing sequences that are bounded above.

(b) Prove that $\lim_{n \to \infty} \mathbf{v}_n$ exists, and find the limit.

H. Let $T = \begin{bmatrix} 5/4 & -1/4 \\ 3/4 & 1/4 \end{bmatrix}$. Set $\mathbf{x}_n = T^n(1,0)$ for $n \geq 1$.

   (a) Prove that $(\mathbf{x}_n)$ converges and find the limit $\mathbf{y}$.

   (b) Find an explicit N so that $\|\mathbf{x}_n - \mathbf{y}\| < \frac{1}{2}10^{-100}$ for all $n \geq N$.

      HINT: Show by induction that $\mathbf{x}_n = \left(\frac{3-2^{-n}}{2}, \frac{3(1-2^{-n})}{2} \right)$.

4.3. Closed and Open Subsets of $\mathbb{R}^n$

Two classes of subsets play a crucial role in analysis: the closed sets, which contain all of their limit points, and the open sets, which contain small balls around each point. These notions will be made precise in this section. From the point of view thus far, closed sets seem more natural because they are directly connected to limiting procedures. Later, however, we shall see that open sets play at least as important a role when considering continuous functions. They are intimately related in any case. We develop closed sets first.

4.3.1. DEFINITION. A point $\mathbf{x}$ is a **limit point** of a subset A of $\mathbb{R}^n$ if there is a sequence $(\mathbf{a}_n)_{n=1}^{\infty}$ with $\mathbf{a}_n \in A$ such that $\mathbf{x} = \lim_{n \to \infty} \mathbf{a}_n$. A set $A \subset \mathbb{R}^n$ is **closed** if it contains all of its limit points.

NOTE: Be warned that some other books define limit point to be a slightly more complicated concept, which we call a cluster point (see Exercise 4.3.N).

4.3.2. EXAMPLES.

(1) $[a, b] = \{x \in \mathbb{R} : a \leq x \leq b\}$ is closed.

(2) $\varnothing$ and $\mathbb{R}^n$ are both closed.

(3) $[0, +\infty)$ is closed in $\mathbb{R}$.

(4) $(0, 1]$ and $(0, 1)$ are not closed.

(5) $\{\mathbf{x} \in \mathbb{R}^n : \|\mathbf{x}\| \leq 1\}$ is closed.

(6) $\{\mathbf{x} \in \mathbb{R}^n : \|\mathbf{x}\| < 1\}$ is not closed.

(7) $\{(x, y) \in \mathbb{R}^2 : xy \geq 1\}$ is closed.

(8) Finite sets are closed.

In the following proposition, I denotes an arbitrary index set. This may be an infinite set of very large cardinality (such as the real line) or a countably infinite set (like $\mathbb{N}$) or even a finite set.

4.3.3. PROPOSITION. *If* $A, B \subset \mathbb{R}^n$ *are closed, then* $A \cup B$ *is closed.*
If $\{A_i : i \in I\}$ *is a family of closed subsets of* $\mathbb{R}^n$, *then* $\bigcap_{i \in I} A_i$ *is closed.*

PROOF. Suppose that $(\mathbf{x}_n)_{n=1}^{\infty}$ is a sequence in $A \cup B$ with limit $\mathbf{x}$. Clearly, either infinitely many of the $\mathbf{x}_n$'s belong to A or infinitely many belong to B. Without loss of generality, we may suppose that A has this property. Hence there is a subsequence $(\mathbf{x}_{n_i})_{i=1}^{\infty}$ of $(\mathbf{x}_n)_{n=1}^{\infty}$ such that each $\mathbf{x}_{n_i}$ belongs to A. But this subsequence has limit x. Since A is closed, we deduce that $\mathbf{x}$ belongs to A, and thus belongs to $A \cup B$. So $A \cup B$ is closed.

Now suppose that $(\mathbf{x}_n)_{n=1}^{\infty}$ is a sequence in $\bigcap_{i \in I} A_i$ with limit $\mathbf{x}$. For each $i \in I$, the sequence $(\mathbf{x}_n)_{n=1}^{\infty}$ belongs to A_i, which is closed. Therefore, the limit $\mathbf{x}$ also belongs to A_i. Since this holds for every $i \in I$, it follows that $\mathbf{x}$ also belongs to the intersection. Hence the intersection is closed. ∎

Since a closed set has the very useful property of containing all of its limit points, it is natural to want to construct closed sets from other, less well-behaved sets. This is the motivation for the following definition.

4.3.4. DEFINITION. If A is a subset of $\mathbb{R}^n$, the **closure** of A is the set $\overline{A}$ consisting of all limit points of A.

To justify the name, we establish some basic properties of the closure operation.

4.3.5. PROPOSITION. *Let* A *be a subset of* $\mathbb{R}^n$. *Then* $\overline{A}$ *is the smallest closed set containing* A. *In particular,* $\overline{\overline{A}} = \overline{A}$.

PROOF. First notice that for each $\mathbf{a}$ in A, we may consider the sequence $\mathbf{x}_n = \mathbf{a}$ for all $n \geq 1$. This has limit $\mathbf{a}$, and thus A is contained in $\overline{A}$.

To show that $\overline{A}$ is closed, consider a sequence $(\mathbf{x}_n)_{n=1}^{\infty}$, where each $\mathbf{x}_n$ belongs to $\overline{A}$ with limit $\mathbf{x} = \lim_{n \to \infty} \mathbf{x}_n$. For each n, there is a sequence of points in A converging to $\mathbf{x}_n$. Hence we may choose an element $\mathbf{a}_n \in A$ from this sequence such that $\|\mathbf{x}_n - \mathbf{a}_n\| < \frac{1}{n}$. Then

$$\lim_{n \to \infty} \mathbf{a}_n = \lim_{n \to \infty} \mathbf{x}_n + (\mathbf{a}_n - \mathbf{x}_n)$$
$$= \lim_{n \to \infty} \mathbf{x}_n + \lim_{n \to \infty} \mathbf{a}_n - \mathbf{x}_n = \mathbf{x} + 0 = \mathbf{x}.$$

Thus $\mathbf{x}$ is also a limit of points in A, whence it belongs to $\overline{A}$. So $\overline{A}$ is closed.

If C is a closed set containing A, then it also contains all limits of sequences in A and therefore contains $\overline{A}$. So $\overline{A}$ is the smallest closed set containing A.

Now $\overline{\overline{A}}$ is the smallest closed set containing $\overline{A}$. Since $\overline{A}$ is already closed, $\overline{\overline{A}} = \overline{A}$. ∎

4.3.6. DEFINITION. The ball about **a** in $\mathbb{R}^n$ of radius r is the set

$$B_r(\mathbf{a}) = \{\mathbf{x} \in \mathbb{R}^n : \|\mathbf{x} - \mathbf{a}\| < r\}.$$

A subset U of $\mathbb{R}^n$ is **open** if for every $\mathbf{a} \in U$, there is some $r = r(\mathbf{a}) > 0$ so that the ball $B_r(\mathbf{a})$ is contained in U.

Figure 4.2 illustrates the definitions of open and closed sets.

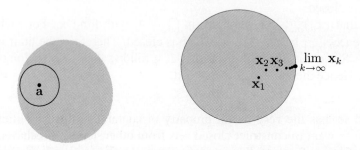

FIGURE 4.2. Open and closed sets

4.3.7. EXAMPLES.

(1) $(a, b) = \{x \in \mathbb{R} : a < x < b\}$ is open.

(2) $\varnothing$ and $\mathbb{R}^n$ are both open.

(3) $(0, +\infty)$ is open in $\mathbb{R}$.

(4) $(0, 1]$ and $[0, 1]$ are not open.

(5) $B_r(a)$ is open.

(6) $\overline{B_r(a)} = \{x \in \mathbb{R}^n : \|x - a\| \le r\}$ is not open.

(7) $\{(x, y) \in \mathbb{R}^2 : xy < 1\}$ is open.

(8) $\{(x, 0) \in \mathbb{R}^2 : 0 < x < 1\}$ is not open.

It is important to remember that while a door must be either open or closed, a set can be neither. The following connection between open and closed sets makes the relation clear.

4.3.8. THEOREM. *A set $A \subset \mathbb{R}^n$ is open if and only if the complement of A, $A' = \{\mathbf{x} \in \mathbb{R}^n : \mathbf{x} \notin A\}$, is closed.*

PROOF. Let A be open. Let $(\mathbf{x}_n)_{n=1}^{\infty}$ be a sequence in A' with limit $\mathbf{x}$. If $\mathbf{a}$ is any point in A, there is a positive number $r > 0$ such that $B_r(\mathbf{a})$ is contained in A.

Hence $\|\mathbf{a} - \mathbf{x}_n\| \geq r$ for all $n \geq 1$. Therefore,

$$\|\mathbf{a} - \mathbf{x}\| = \lim_{n \to \infty} \|\mathbf{a} - \mathbf{x}_n\| \geq r.$$

In particular, $\mathbf{x} \neq \mathbf{a}$. This is true for every point in A, and hence $\mathbf{x}$ belongs to A'. That is, A' is closed.

Conversely, suppose that A is not open. Then there is some $\mathbf{a} \in A$ so that for every $r > 0$, the ball $B_r(\mathbf{a})$ is not contained in A. In particular, if we let $r = \frac{1}{n}$, we can find $\mathbf{x}_n \in A'$ so that $\|\mathbf{a} - \mathbf{x}_n\| < \frac{1}{n}$. Then $\mathbf{a} = \lim_{n \to \infty} \mathbf{x}_n$ is a limit point of A' belonging to A. Hence A' is not closed. ∎

We have the following proposition, which is dual to Proposition 4.3.3. The proof is left as an exercise.

4.3.9. PROPOSITION. *If U and V are open subsets of $\mathbb{R}^n$, then $U \cap V$ is an open subset of $\mathbb{R}^n$. If $\{U_i : i \in I\}$ is a family of open subsets of $\mathbb{R}^n$, then $\bigcup_{i \in I} U_i$ is open.*

There is also a notion for open sets that is dual to the closure. The **interior** int X of a set X is the largest open set contained inside X (see the Exercises). If the interior of a set X is the empty set, then we say X has **empty interior**.

4.3.10. EXAMPLE. Let $A = \{(x, y) : x \in \mathbb{Q}, y > x^3\}$. This set is neither open nor closed. Indeed, the point $(0, 0) = \lim_{n \to \infty} (0, \frac{1}{n})$ is a limit point not contained in A; so A is not closed. And the point $(0, 1) = \lim_{n \to \infty} (\sqrt{2}/n, 1)$ belongs to A, yet it is the limit of points in A'. So A is not open either.

The closure of A is the set

$$\overline{A} = \{(x, y) : y \geq x^3\}.$$

To see this, let (x, y) be given such that $y \geq x^3$. Let x_n be an increasing sequence of rationals converging to x (such as the finite decimal approximations of x). Set $y_n = y + \frac{1}{n}$. Then it is clear that $y_n > x^3 \geq x_n^3$ and thus $\mathbf{a}_n = (x_n, y_n)$ belongs to A. Now

$$\lim_{n \to \infty} x_n = x \qquad \text{and} \qquad \lim_{n \to \infty} y_n = \lim_{n \to \infty} y + \frac{1}{n} = y.$$

Hence $\lim_{n \to \infty} \mathbf{a}_n = (x, y)$. Therefore, $\overline{A}$ contains $\{(x, y) : y \geq x^3\}$. Conversely, if $(x, y) = \lim_{n \to \infty} \mathbf{a}_n$ for any sequence $\mathbf{a}_n = (x_n, y_n)$ in A, it follows that

$$y = \lim_{n \to \infty} y_n \geq \lim_{n \to \infty} x_n^3 = x^3.$$

Thus $\{(x, y) : y \geq x^3\}$ contains $\overline{A}$.

In this case, A has empty interior. The reason is that every open ball contains points with irrational coordinates, and A does not.

The interior of $\overline{A}$ is the set $U = \{(x, y) : y > x^3\}$. First we show that U is open. If $\mathbf{a} = (x, y)$ belongs to U, then $s = y - x^3 > 0$. We need to determine a value for r so that $B_r(\mathbf{a})$ is contained in U. Some calculation is needed to determine the proper choice. Suppose that a point (u, v) satisfies $\|(u, v) - (x, y)\| < r$. Then, in particular, $|u - x| < r$ and $|v - y| < r$. Hence

$$v - u^3 > (y - r) - (x + r)^3 = y - r - (x^3 + 3rx^2 + 3r^2x + r^3)$$
$$= y - x^3 - r - 3rx^2 - 3r^2x - r^3$$
$$= s - (r + 3rx^2 + 3r^2x + r^3).$$

To make the right-hand side positive, which we require, a choice must be made for r so that $r + 3rx^2 + 3r^2x + r^3 \leq s$. Let us decide that $r \leq 1$ (our choice—and it can't hurt). Then

$$r + 3rx^2 + 3r^2x + r^3 \leq r(2 + 3x^2 + 3|x|).$$

Define $r = \min\{1, s/(2 + 3x^2 + 3|x|)\}$. Then it follows that

$$v - u^3 > s - (r + 3rx^2 + 3r^2x + r^3) > 0.$$

This shows that $B_r(\mathbf{a})$ is contained in U. Thus U is open.

Now suppose that $\mathbf{a} = (x, y)$ belongs to $\overline{A}$ but is not in U. Then $y \geq x^3$ but $y \not> x^3$, whence $y = x^3$. To see that $\mathbf{a}$ is not in the interior of $\overline{A}$, it must be shown that whenever $r > 0$, the ball $B_r(\mathbf{a})$ intersects $\overline{A}'$. This is easy, since the point $(x, x^3 - r/2)$ belongs to this ball and does not belong to $\overline{A}$. So $\operatorname{int}\overline{A} = U$.

Exercises for Section 4.3

A. Find the closure of the following sets:
  (a) $\mathbb{Q}$
  (b) $\{(x, y) \in \mathbb{R}^2 : xy < 1\}$
  (c) $\{(x, \sin(\frac{1}{x})) : x > 0\}$
  (d) $\{(x, y) : x, y \in \mathbb{Q},\ x^2 + y^2 < 1\}$

B. Let $(\mathbf{a}_n)_{n=1}^{\infty}$ be a sequence in $\mathbb{R}^n$ with $\lim_{n \to \infty} \mathbf{a}_n = \mathbf{a}$. Show that $A = \{\mathbf{a}_n : n \geq 1\} \cup \{\mathbf{a}\}$ is a closed set.

C. Show that $U = \{(x, y) \in \mathbb{R}^2 : x^2 + 4y^2 < 4\}$ is open by explicitly finding a ball around each point which is contained in U.

D. If A is a bounded subset of $\mathbb{R}$, show that $\sup A$ and $\inf A$ belong to $\overline{A}$.

E. Show that the interior satisfies $\operatorname{int} A = (\overline{A'})'$.

F. Find the interior of $A \cup B$, where $A = \{(x, y) : x \in \mathbb{Q},\ y^2 \geq x\}$ and $B = \{(x, y) : x \notin \mathbb{Q},\ y \geq x^2\}$.

G. If a subset A of $\mathbb{R}^n$ has no interior, must it be closed?

H. Show that a subset of $\mathbb{R}^n$ is complete if and only if it is closed.

I. Prove Proposition 4.3.9 using Theorem 4.3.8 and Proposition 4.3.3.

J. Show that if U is open and A is closed, then $U \setminus A = \{\mathbf{x} \in U : \mathbf{x} \notin A\}$ is open. What can be said about $A \setminus U$?

K. Suppose that A and B are closed subsets of $\mathbb{R}$.
  (a) Show that the product set $A \times B = \{(x, y) \in \mathbb{R}^2 : x \in A \text{ and } y \in B\}$ is closed.
  (b) Likewise show that if both A and B are open, then $A \times B$ is open.

L. A set A is **dense** in B if B is contained in $\overline{A}$.
  (a) Show that the set of irrational numbers is dense in $\mathbb{R}$.
  (b) Hence show that $\mathbb{Q}$ has empty interior.

M. Suppose that A is a dense subset of $\mathbb{R}^n$.
  (a) Show that if U is open in $\mathbb{R}^n$, then $A \cap U$ is dense in U.
  (b) Show by example that this may fail for sets that are not open.

N. A point $\mathbf{x}$ is a **cluster point** of a subset A of $\mathbb{R}^n$ if there is a sequence $(\mathbf{a}_n)_{n=1}^\infty$ with $\mathbf{a}_n \in A \backslash \{\mathbf{x}\}$ such that $\mathbf{x} = \lim_{n \to \infty} \mathbf{a}_n$. Thus, every cluster point is a limit point but not conversely.
  (a) Show that if $\mathbf{x}$ is a limit point of A, then either $\mathbf{x}$ is a cluster point of A or $\mathbf{x} \in A$.
  (b) Hence show that a set is closed if it contains all of its cluster points.
  (c) Find all cluster points of (i) $\mathbb{Q}$, (ii) $\mathbb{Z}$, (iii) $(0, 1)$.

O. Starting with a subset A of $\mathbb{R}^n$, form all the possible sets that may be obtained by repeated use of the operations of closure and complement. Up to 14 different sets can be obtained in this way. Find such a subset of $\mathbb{R}$.

4.4. Compact Sets and the Heine–Borel Theorem

Now we turn to the notion of compactness. At this stage, compactness seems like a convenience and may not appear to be much more useful than completeness. However, when we study continuous functions, compactness will be very useful and then its full power will become apparent.

4.4.1. DEFINITION. A subset A of $\mathbb{R}^n$ is **compact** if every sequence $(\mathbf{a}_k)_{k=1}^\infty$ of points in A has a convergent subsequence $(\mathbf{a}_{k_i})_{i=1}^\infty$ with limit $\mathbf{a} = \lim_{i \to \infty} \mathbf{a}_{k_i}$ in A.

Recall that the Bolzano–Weierstrass Theorem (Theorem 2.6.4) states that every bounded sequence has a convergent subsequence. Using this new language, we may deduce that every subset of $\mathbb{R}$ that is both closed and bounded is compact. This rephrasing naturally suggests the question, Which subsets of $\mathbb{R}^n$ are compact? Before answering this question, we consider a few examples.

4.4.2. EXAMPLES. Consider the set $(0, 1]$. The sequence $1, 1/2, 1/3, \ldots$ is in this set but converges to 0, which is not in the set. Since any subsequence will also converge to zero, there is no subsequence of $1, 1/2, 1/3, \ldots$ that converges to a number in $(0, 1]$. So this set is not compact.

Next, consider the set $\mathbb{N}$. The sequence $1, 2, 3, \ldots$ is in $\mathbb{N}$. However, no subsequence converges (because each subsequence is unbounded and being bounded is a necessary condition for convergence by Proposition 2.4.2). So $\mathbb{N}$ is not compact.

A subset S of $\mathbb{R}^n$ is called **bounded** provided that there is a real number R such that S is contained in the ball $B_R(0)$. Equivalently, S is bounded if $\sup_{x \in S} \|x\| < \infty$.(Notice that when $n = 1$, this definition of bounded agrees with our old definition of bounded subsets of $\mathbb{R}$.)

The previous examples suggest that sets that are not closed or not bounded cannot be compact. This is true, and the proofs are an abstraction of the arguments for these examples.

4.4.3. LEMMA. *A compact subset of* $\mathbb{R}^n$ *is closed and bounded.*

PROOF. Let C be a compact subset of $\mathbb{R}^n$. Suppose that $\mathbf{x}$ is a limit point of C, say $\mathbf{x} = \lim_{n \to \infty} \mathbf{c}_n$ for a sequence $(\mathbf{c}_n)$ in C. Then this sequence has a subsequence $(\mathbf{c}_{n_i})$ converging to a point $\mathbf{c}$ in C. Therefore,

$$\mathbf{x} = \lim_{n \to \infty} \mathbf{c}_n = \lim_{i \to \infty} \mathbf{c}_{n_i} = \mathbf{c} \in C.$$

Thus C is closed.

To show that C is bounded, suppose that it were unbounded. That means that there is a sequence $\mathbf{c}_n \in C$ such that $\|\mathbf{c}_n\| > n$ for each $n \geq 1$. Consider the sequence $(\mathbf{c}_n)$. If there were a convergent subsequence $(\mathbf{c}_{n_i})$ with limit $\mathbf{c}$, it would follow that

$$\|\mathbf{c}\| = \lim_{i \to \infty} \|\mathbf{c}_{n_i}\| \geq \lim_{i \to \infty} n_i = +\infty.$$

This is an absurd conclusion, and thus C must be bounded. $\blacksquare$

To establish the converse, we build up a couple of partial results.

4.4.4. LEMMA. *If C is a closed subset of a compact subset of* $\mathbb{R}^n$*, then C is compact.*

PROOF. Let K be the compact set containing C. Suppose $(\mathbf{x}_n)_{n=1}^{\infty}$ is a sequence in C. To show that C is compact, we must find a subsequence that converges to an element of C.

However, $(\mathbf{x}_n)_{n=1}^{\infty}$ is contained in the compact set K. So it has a subsequence that converges to a number $\mathbf{x}$ in K, say $\mathbf{x} = \lim_{k \to \infty} \mathbf{x}_{n_k}$. Since $(\mathbf{x}_{n_k})_{k=1}^{\infty}$ is contained in C and C is closed, it follows that $\mathbf{x}$ belongs to C as required. $\blacksquare$

4.4.5. LEMMA. *The cube $[a, b]^n$ is a compact subset of* $\mathbb{R}^n$*.*

PROOF. Let $\mathbf{x}_k = (x_{k,1}, \ldots, x_{k,n})$ for $k \geq 1$ be a sequence in $\mathbb{R}^n$ such that the coefficients satisfy $a \leq x_{k,i} \leq b$ for all $k \geq 1$ and $1 \leq i \leq n$. Consider the sequence $(x_{k,1})_{k=1}^{\infty}$ of first coordinates. By the Bolzano–Weierstrass Theorem (Theorem 2.6.4), there is a subsequence $(x_{k_j,1})_{j=1}^{\infty}$ converging to a point z_1 in $[a, b]$,

$$\lim_{j \to \infty} x_{k_j,1} = z_1.$$

Next consider the sequence $y_j = x_{k_j,2}$ for $j \geq 1$. This sequence is contained in the closed interval $[a, b]$. Thus a second application of the Bolzano–Weierstrass Theorem yields a subsequence $(y_{j_l}) = (x_{k_{j_l},2})$ such that $\lim_{l \to \infty} x_{k_{j_l},2} = z_2$. We still have $\lim_{l \to \infty} x_{k_{j_l},1} = z_1$ since every subsequence of a convergent sequence has the same limit.

Proceeding in this way, finding n consecutive sub-subsequences, we obtain a subsequence $p_1 < p_2 < \dots$ and z_i in $[a, b]$ such that

$$\lim_{j \to \infty} x_{p_j,i} = z_i \quad \text{for} \quad 1 \leq i \leq n.$$

Thus $\lim_{j \to \infty} \mathbf{x}_{p_j} = \mathbf{z} := (z_1, \dots, z_n)$ by Lemma 4.2.3. ∎

4.4.6. THE HEINE–BOREL THEOREM.
A subset of $\mathbb{R}^n$ is compact if and only if it is closed and bounded.

PROOF. The easy direction is given by Lemma 4.4.3.

For the other direction, suppose that C is a closed and bounded subset of $\mathbb{R}^n$. Since it is bounded, there is some $M > 0$ so that $\|x\| \leq M$ for all $x \in C$. In particular, C is contained in the cube $[-M, M]^n$. Now $[-M, M]^n$ is compact by Lemma 4.4.5. So C is a closed subset of a compact set and therefore is compact by Lemma 4.4.4. ∎

This leads to an important generalization of the Nested Interval Theorem.

4.4.7. CANTOR'S INTERSECTION THEOREM.
If $A_1 \supset A_2 \supset A_3 \cdots$ is a decreasing sequence of nonempty *compact subsets of $\mathbb{R}^n$, then $\bigcap_{k \geq 1} A_k$ is not empty.*

PROOF. Since A_n is not empty, we may choose a point $\mathbf{a}_n$ in A_n for each $n \geq 1$. Then the sequence $(\mathbf{a}_n)_{n=1}^{\infty}$ belongs to the compact set A_1. By compactness, there is a subsequence $(\mathbf{a}_{n_k})_{k=1}^{\infty}$ that converges to a limit point $\mathbf{x}$. For each i, the terms a_{n_k} belong to A_i for all $k \geq i$. Thus $\mathbf{x}$ is the limit of points in A_i, whence $\mathbf{x}$ belongs to A_i for all $i \geq 1$. Therefore, $\mathbf{x}$ belongs to their intersection. ∎

4.4.8. EXAMPLE: THE CANTOR SET.
We now give a more subtle example of a compact set in $\mathbb{R}$. The **Cantor set** is a fractal subset of the real line. Let $S_0 = [0, 1]$, and construct S_{i+1} from S_i recursively by removing the *middle third* from each interval in S_i. For example, the first three terms are

$$S_1 = [0, 1/3] \cup [2/3, 1]$$
$$S_2 = [0, 1/9] \cup [2/9, 1/3] \cup [2/3, 7/9] \cup [8/9, 1]$$
$$S_3 = [0, 1/27] \cup [2/27, 1/9] \cup [2/9, 7/27] \cup [8/27, 1/3]$$
$$\cup [2/3, 19/27] \cup [20/27, 7/9] \cup [8/9, 25/27] \cup [26/27, 1]$$

By Proposition 4.3.3, the intersection $C = \cap_{i \geq 1} S_i$ is a closed set. It is bounded and hence compact. By Cantor's Intersection Theorem, this intersection is not empty. Figure 4.3 shows an approximation to the Cantor set, namely S_5.

FIGURE 4.3. The Cantor set (actually S_5).

Every endpoint of an interval in one of the sets S_n belongs to C. But in fact, C contains many other points. Each point in C is determined by a binary decision tree. At the first stage, pick one of the two intervals of S_1 of length $1/3$, which we label 0 and 2. This interval is split into two in S_2 by removing the middle third. Choose either the left (label 0) or right (label 2) to obtain an interval labeled 00, 02, 20, or 22. Continuing in this way, we choose a decreasing sequence of intervals determined by an infinite sequence of 0s and 2s. By Cantor's Intersection Theorem (Theorem 4.4.7), every choice determines a point of intersection. There is only one point in each of these intersections because the length of the intervals tends to 0. We leave it to you to describe the (proper) set of decision trees that correspond to left or right endpoints.

The Cantor set has empty interior. For if C contained an open interval (a, b) with $a < b$, it would also be contained in each S_n. This forces $b - a \leq 3^{-n}$ for every n, whence $a = b$. So the interior of C is empty. A set whose closure has no interior is **nowhere dense**.

Yet C has no isolated points. A point x of a set A is **isolated** if there is an $\varepsilon > 0$ such that the ball $B_\varepsilon(x)$ intersects A only in the singleton $\{x\}$. In fact, C is a **perfect set**, meaning that every point of C is the limit of a sequence of other points in C. In other words, every point is a cluster point. To see this, suppose first that x is not the right endpoint of one of the intervals of some S_n. For each n, let x_n be the right endpoint of the interval of S_n containing x. Then $x_n \neq x$ and $|x_n - x| \leq 3^{-n}$. So $x = \lim_{n \to \infty} x_n$. If x is the right endpoint of one of these intervals, use the left endpoints instead to define the sequence x_n.

The set C is very large, the same size as $[0, 1]$, in the sense of cardinality from Appendix 2.8. Consider the numbers in $[0, 1]$ expanded as infinite "decimals" in base 3 (the **ternary expansion**). That is, each number may be expressed as

$$x = (x_0.x_1x_2x_3\ldots)_{\text{base 3}} = \sum_{k \geq 0} 3^{-k} x_k,$$

where x_i belong to $\{0, 1, 2\}$ for $i \geq 1$. Note that S_1 consists of all numbers in $[0, 1]$ that have an expansion with the first digit equal to 0 or 2. In particular,

$$\tfrac{1}{3} = (.1)_{\text{base 3}} = (.02222222\ldots)_{\text{base 3}} \quad \text{and} \quad 1 = (.22222222\ldots)_{\text{base 3}}.$$

Likewise, S_i consists of all numbers in $[0, 1]$ such that the first i terms of some ternary expansion are all 0s and 2s. Since C is the intersection of all the S_i, it consists of precisely all the numbers in $[0, 1]$ that have a ternary expansion using only 0s and 2s.

As C is a subset of $[0, 1]$, it is clear that C can have no more points than $[0, 1]$. To see that $[0, 1]$ can have no more points than C, we construct a one-to-one map from $[0, 1]$ into C. Think of the points in $[0, 1]$ in terms of their binary expansion (base 2). These are all the "decimal" expansions

$$y = (y_0.y_1y_2y_3\cdots)_{\text{base 2}} = \sum_{k\geq 0} 2^{-k}y_k,$$

where $y_k \in \{0, 1\}$. For each point, pick one binary expansion. (Some numbers like $\frac{1}{2}$ have two possible expansions, one ending in an infinite string of 0s and the other ending in an infinite string of 1s. In this case, pick the expansion ending in 0s.) Send it to the corresponding point in base 3 using 0s and 2s by changing each 1 to 2. Since this corresponding point is in C, we have a map of $[0, 1]$ into C. This map is one-to-one because the only duplication of ternary expansions comes from a sequence ending in all 0s corresponding to another ending with all 2s. But we do not send any number to a ternary expansion ending in all 2s.

Warning: This map is not onto because of numbers with two expansions in base 2 such as $\frac{1}{2}$, which in base two equals both $(.1)_{\text{base 2}}$ and $(.01111\ldots)_{\text{base 2}}$. We only used the first one, which we send to $(.2)_{\text{base 3}}$, namely $\frac{2}{3}$. But the other expansion would go to $(.02222\ldots)_{\text{base 3}}$, which equals $(.1)_{\text{base 3}} = \frac{1}{3}$.

At this point, it seems obvious that there are as many points in C as in $[0, 1]$—it can't have any more, and it can't have any less. However, with infinite sets, proving this is a subtle business. The Schroeder–Bernstein Theorem (Theorem 2.8.8) *could* be invoked to obtain a bijection between C and $[0, 1]$. However, the special nature of our setup allows a bijection between C and $[0, 1]$ to be constructed fairly easily; see Exercise 4.4.K. Therefore, C and $[0, 1]$ have the same cardinality, which is uncountable by Theorem 2.8.7.

On the other hand, using a different notion of "size," the Cantor set is very small. We can measure how much of the interval has been removed at each step. The set S_n contains 2^n intervals of length 3^{-n}. The middle third of length 3^{-n-1} is removed from each of these 2^n intervals to obtain S_{n+1}. The total length of the pieces removed is computed by adding an infinite geometric series

$$\sum_{n=0}^{\infty} \frac{2^n}{3^{n+1}} = \frac{1/3}{1-(2/3)} = 1.$$

Thus the Cantor set has measure zero, a notion that is discussed in Section 6.6. In some sense, C squeezes its very large number of points into a very small space.

Exercises for Section 4.4

A. Which of the following sets are compact?
   (a) $\{(x, y) \in \mathbb{R}^2 : 2x^2 - y^2 \leq 1\}$
   (b) $\{\mathbf{x} \in \mathbb{R}^n : 2 \leq \|\mathbf{x}\| \leq 4\}$
   (c) $\{(e^{-x}\cos x, e^{-x}\sin x) : x \geq 0\} \cup \{(x, 0) : 0 \leq x \leq 1\}$
   (d) $\{(e^{-x}\cos\theta, e^{-x}\sin\theta) : x \geq 0,\ 0 \leq \theta \leq 2\pi\}$

B. Give an example to show that Cantor's Intersection Theorem would not be true if compact sets were replaced by closed sets.

C. Show that the union of finitely many compact sets is compact.

D. Show that the intersection of any family of compact sets is compact.

E. (a) Show that the sum of a closed subset and a compact subset of $\mathbb{R}^n$ is closed. Recall that $A + B = \{\mathbf{a} + \mathbf{b} : \mathbf{a} \in A \text{ and } \mathbf{b} \in B\}$.
(b) Is this true for the sum of two compact sets and a closed set?
(c) Is this true for the sum of two closed sets?

F. Let $(\mathbf{x}_n)_{n=1}^{\infty}$ be a sequence in a compact set K that is *not* convergent. Show that there are two subsequences of this sequence that are convergent to *different* limit points.

G. Prove that a set $S \subset \mathbb{R}$ has no cluster points if and only if $S \cap [-n, n]$ is a finite set for each $n \geq 1$.

H. Describe all subsets of $\mathbb{R}^n$ that have no cluster points at all.

I. Let A and B be *disjoint* closed subsets of $\mathbb{R}^n$. Define

$$d(A, B) = \inf\{\|\mathbf{a} - \mathbf{b}\| : \mathbf{a} \in A, \ \mathbf{b} \in B\}.$$

(a) If $A = \{\mathbf{a}\}$ is a singleton, show that $d(A, B) > 0$.
(b) If A is compact, show that $d(A, B) > 0$.
(c) Find an example of two disjoint closed sets in $\mathbb{R}^2$ with $d(A, B) = 0$.

J. The **Sierpinski snowflake** is constructed in the plane as follows. Start with a solid equilateral triangle. Remove the open middle triangle with vertices at the midpoint of each side of the larger triangle, leaving three solid triangles with half the side length of the original. From each of these three, remove the open middle triangle, leaving 9 triangles of one fourth the original side lengths. Proceed in this process ad infinitum. Let S denote the intersection of all the finite stages.

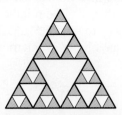

FIGURE 4.4. The third stage in constructing the Sierpinski snowflake.

(a) Show that S is a nonempty compact set.
(b) Show that S has no interior.
(c) Show that the boundaries of the triangles at the nth stage belong to S. Hence show that there is a path in S from the top vertex of the original triangle that gets as close as desired (within ε) to any point in S.
(d) Compute the area of the material removed from the triangle to leave S behind.
(e) Construct a decision tree for S. Does each decision tree correspond to exactly one point in the set? Show that S is uncountable.

K. Show that there is a bijection from $[0, 1]$ onto the Cantor set.
   HINT: Adjust the map constructed in the text by redefining it on a countable sequence
   to include the missing points in the range.

L. Prove that a countable compact set $X = \{x_n : n \geq 1\}$ cannot be perfect.
   HINT: Use a decreasing family X_n of closed nonempty subsets of X with $x_n \notin X_n$.

CHAPTER 5

Functions

The main purpose of this chapter is to introduce the notion of a continuous function. Continuity is a basic notion of analysis. It is only with continuous functions that there can be any reasonable approximation or estimation of values at specific points. Most physical phenomena are continuous over most of their domain. The ideas we study in this chapter are sufficiently powerful that even discrete phenomena are sometimes best understood by using continuous approximations, where these ideas can be used.

5.1. Limits and Continuity

General functions, between any two sets, were defined in Section 1.2. In this chapter, and indeed in most of this book, functions will be defined on some subset S of $\mathbb{R}^n$ with range contained in $\mathbb{R}^m$. Everything is based on the notion of limit, which is a natural variant of the definition for limit of a sequence.

5.1.1. DEFINITION OF LIMIT. Let $S \subset \mathbb{R}^n$ and let f be a function from S into $\mathbb{R}^m$. If $\mathbf{a}$ is a limit point of S, then a point $\mathbf{v} \in \mathbb{R}^m$ is the **limit** of f at $\mathbf{a}$ if for every $\varepsilon > 0$, there is an $r > 0$ so that

$$\|f(\mathbf{x}) - \mathbf{v}\| < \varepsilon \quad \text{whenever} \quad 0 < \|\mathbf{x} - \mathbf{a}\| < r \text{ and } \mathbf{x} \in S.$$

We write $\lim_{\mathbf{x} \to \mathbf{a}} f(\mathbf{x}) = \mathbf{v}$.

Geometrically, we have a picture such as Figure 5.1. Notice that $f(\mathbf{a})$ itself need not be defined. Certainly, saying that $\lim_{\mathbf{x} \to \mathbf{a}} f(\mathbf{x}) = \mathbf{v}$ does not tell us anything about $f(\mathbf{a})$.

Let us specialize to the case of a real-valued function f defined on an interval (a, b) and let c be a point in this interval. Then $\lim_{x \to c} f(x) = L$ means that for every $\varepsilon > 0$, there is an $r > 0$ so that

$$|f(x) - L| < \varepsilon \quad \text{for all} \quad 0 < |x - c| < r.$$

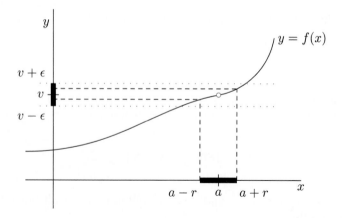

FIGURE 5.1. Limit for a function $f : \mathbb{R} \to \mathbb{R}$.

5.1.2. DEFINITION. Let $S \subset \mathbb{R}^n$ and let f be a function from S into $\mathbb{R}^m$. We say that f is **continuous at a** $\in S$ if $\lim_{\mathbf{x} \to \mathbf{a}} f(\mathbf{x}) = f(\mathbf{a})$. Moreover, f is **continuous on** S if it is continuous at each point $\mathbf{a} \in S$.

If f is not continuous at $\mathbf{a}$, we say that f is **discontinuous** at $\mathbf{a}$.

In contrast to limits, for continuity it is crucial that $f(\mathbf{a})$ be defined and equal to the limit of $f(\mathbf{x})$ as $\mathbf{x}$ approaches $\mathbf{a}$. In terms of the epsilon form, we have that f is continuous at $\mathbf{a} \in S$ provided that for every $\varepsilon > 0$, there is an $r > 0$ such that

$$\|f(\mathbf{x}) - f(\mathbf{a})\| < \varepsilon \quad \text{whenever} \quad \|\mathbf{x} - \mathbf{a}\| < r \text{ and } \mathbf{x} \in S.$$

5.1.3. EXAMPLE. Consider the function $f : \mathbb{R}^n \setminus \{\mathbf{0}\} \to \mathbb{R}$ given by

$$f(\mathbf{x}) = 1/\|\mathbf{x}\|.$$

Let us show that this is continuous on its domain. Fix a point $\mathbf{a} \in \mathbb{R}^n \setminus \{\mathbf{0}\}$. Then

$$|f(\mathbf{x}) - f(\mathbf{a})| = \left| \frac{1}{\|\mathbf{x}\|} - \frac{1}{\|\mathbf{a}\|} \right| = \frac{\big|\|\mathbf{a}\| - \|\mathbf{x}\|\big|}{\|\mathbf{x}\| \, \|\mathbf{a}\|}.$$

Our goal is to make this difference small only by controlling the distance from $\mathbf{x}$ to $\mathbf{a}$, namely $\|\mathbf{x} - \mathbf{a}\|$.

The first step is to show that the numerator is controllable. Indeed, what we need follows from the triangle inequality.

$$\|\mathbf{x}\| \le \|\mathbf{a}\| + \|\mathbf{x} - \mathbf{a}\| \quad \text{and} \quad \|\mathbf{a}\| \le \|\mathbf{x}\| + \|\mathbf{x} - \mathbf{a}\|.$$

Manipulating these inequalities yields

$$-\|\mathbf{x} - \mathbf{a}\| \le \|\mathbf{x}\| - \|\mathbf{a}\| \le \|\mathbf{x} - \mathbf{a}\|.$$

Hence $\big|\|\mathbf{x}\| - \|\mathbf{a}\|\big| \le \|\mathbf{x} - \mathbf{a}\|$. This takes care of the numerator.

Now let's worry about the denominator $\|\mathbf{x}\| \, \|\mathbf{a}\|$. Since $\|\mathbf{a}\|$ is a positive constant, it creates no problems. However, $\|\mathbf{x}\|$ must be kept away from 0 to keep the quotient in control. This is accomplished by making **x** sufficiently close to **a**. The previous paragraph shows that

$$\|\mathbf{x}\| \geq \|\mathbf{a}\| - \|\mathbf{a} - \mathbf{x}\|,$$

so provided that $\|\mathbf{a} - \mathbf{x}\| < \|\mathbf{a}\|/2$, we obtain $\|\mathbf{x}\| > \|\mathbf{a}\|/2$.

Putting all of this together, choose $r \leq \|\mathbf{a}\|/2$ and consider all **x** so that $\|\mathbf{x} - \mathbf{a}\| < r$. Then, remembering that making the denominator smaller makes the quotient larger, we have

$$|f(\mathbf{x}) - f(\mathbf{a})| = \frac{|\|\mathbf{a}\| - \|\mathbf{x}\||}{\|\mathbf{x}\|\,\|\mathbf{a}\|} \leq \frac{\|\mathbf{x} - \mathbf{a}\|}{\|\mathbf{a}\|^2/2} < \frac{2r}{\|\mathbf{a}\|^2}.$$

To make this less than ε, choose $r \leq \varepsilon\|\mathbf{a}\|^2/2$. Combining this with our other condition on r, it follows that

$$r = \min\left\{ \frac{\|\mathbf{a}\|}{2}, \frac{\varepsilon\|\mathbf{a}\|^2}{2} \right\}$$

is sufficient to ensure that $\|f(\mathbf{x}) - f(\mathbf{a})\| < \varepsilon$.

This shows that f is a continuous function. Notice that the function goes to infinity at 0. So the limit at 0 does not exist.

5.1.4. EXAMPLE. A function does not need to have a simple analytic expression to be continuous. However, extra care needs to be taken at the interface. Consider the function graphed in Figure 5.2, given by

$$f(x) = \begin{cases} 0 & \text{if} \quad x \leq 0 \\ e^{-1/x} & \text{if} \quad x > 0. \end{cases}$$

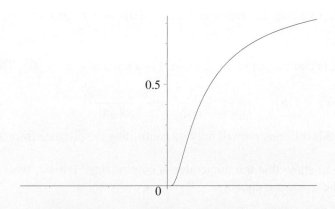

FIGURE 5.2. Graph of $e^{-1/x}$ for $x > 0$.

When $a < 0$ and $\varepsilon > 0$ is any positive number, we may take $r = |a|$. For if $|x - a| < |a|$, then $x < 0$ and thus

$$|f(x) - f(a)| = |0 - 0| = 0 < \varepsilon.$$

Therefore, f is continuous at a.

Now if $a > 0$, finding an appropriate r for each ε is similar to the previous example. This is an example of a function that is the composition of the more elementary functions $g(x) = e^x$ and $h(x) = -1/x$, so that $f(x) = g(h(x))$. We shall see shortly that the composition of continuous functions is continuous. This simplifies the exercise to showing that both g and h are continuous. We leave these details to the reader.

Finally, we must consider $a = 0$ separately because the function f has different definitions on each side of 0. Fix $\varepsilon > 0$. We use the fact that e^x is an increasing function such that $\lim_{x \to -\infty} e^x = 0$, which follows from the basic properties of the exponential function established in calculus. It follows that there is a large negative number $-N$ such that $e^{-N} < \varepsilon$. Therefore, if $0 < x < 1/N$, it follows that $-1/x < -N$ and thus

$$0 < f(x) = e^{-1/x} < e^{-N} < \varepsilon.$$

So take $r = 1/N$. We obtain

$$|f(x) - f(0)| = \begin{cases} 0 & < \varepsilon \quad \text{if} \quad -r < x \le 0 \\ e^{-1/x} & < \varepsilon \quad \text{if} \quad 0 < x < r. \end{cases}$$

Hence f is continuous at $a = 0$.

As a further example, we treat a general class of useful functions that are automatically continuous.

5.1.5. DEFINITION. A function f from $S \subset \mathbb{R}^n$ into $\mathbb{R}^m$ is called a **Lipschitz function** if there is a constant C such that

$$\|f(\mathbf{x}) - f(\mathbf{y})\| \le C\|\mathbf{x} - \mathbf{y}\| \quad \text{for all} \quad \mathbf{x}, \mathbf{y} \in S.$$

The **Lipschitz constant** of f is the smallest choice of C for which the previous condition holds.

As a matter of course, you should check that the Lipschitz constant is well defined. The following easy result will have several important consequences.

5.1.6. PROPOSITION. *Every Lipschitz function is continuous.*

PROOF. Let f be a Lipschitz function with constant C. Given $\varepsilon > 0$, let $r = \varepsilon/C$. Then if $\|\mathbf{x} - \mathbf{y}\| < r$,

$$\|f(\mathbf{x}) - f(\mathbf{y})\| \le C\|\mathbf{x} - \mathbf{y}\| < Cr = \varepsilon.$$

Therefore, f is continuous. ∎

5.1.7. COROLLARY. *Every linear transformation A from $\mathbb{R}^n$ to $\mathbb{R}^m$ is Lipschitz, and therefore is continuous.*

PROOF. Recall that every linear transformation is given by an $m \times n$ matrix, which we also call $A = [a_{ij}]$. The function is then given by the matrix multiplication formula

$$A(x_1, x_2, \ldots, x_n) = \left(\sum_{j=1}^{n} a_{1j} x_j, \ldots, \sum_{j=1}^{n} a_{mj} x_j \right).$$

Let $\mathbf{x} = (x_1, x_2, \ldots, x_n)$ and $\mathbf{y} = (y_1, y_2, \ldots, y_m)$. Compute

$$\|A\mathbf{x} - A\mathbf{y}\| = \|A(\mathbf{x} - \mathbf{y})\|$$
$$= \|A(x_1 - y_1, x_2 - y_2, \ldots, x_n - y_n)\|$$
$$= \left(\sum_{i=1}^{m} \left(\sum_{j=1}^{n} a_{ij}(x_j - y_j) \right)^2 \right)^{1/2}.$$

Apply the Schwarz inequality to obtain that

$$\left| \sum_{j=1}^{n} a_{ij}(x_j - y_j) \right|^2 \le \sum_{j=1}^{n} |a_{ij}|^2 \sum_{j=1}^{n} |x_j - y_j|^2$$
$$= \sum_{j=1}^{n} |a_{ij}|^2 \, \|\mathbf{x} - \mathbf{y}\|^2.$$

Putting this into our computation, we get

$$\|A\mathbf{x} - A\mathbf{y}\| \le \left(\sum_{i=1}^{m} \sum_{j=1}^{n} |a_{ij}|^2 \right)^{1/2} \|\mathbf{x} - \mathbf{y}\| = C \|\mathbf{x} - \mathbf{y}\|,$$

where

$$C = \left(\sum_{i=1}^{m} \sum_{j=1}^{n} |a_{ij}|^2 \right)^{1/2}.$$

Therefore, linear maps are Lipschitz, and hence are continuous. ∎

There are two basic linear functions, called **coordinate functions**, which we will use regularly. There is the map $\pi_j(x_1, \ldots, x_n) = x_j$, which maps $\mathbb{R}^n$ into $\mathbb{R}$ by reading off the jth coordinate. And there is the map $\epsilon_i(t) = t\mathbf{e}_i$, which maps $\mathbb{R}$ into $\mathbb{R}^m$ by sending $\mathbb{R}$ onto the ith coordinate axis. In Exercise 5.1.K, you are asked to show that every linear map can be built up as linear combinations of $\epsilon_i \pi_j$.

Exercises for Section 5.1

A. Use the ϵ–r definition of the limit of a function to show $\lim_{x \to 2} x^2 = 4$.

B. Show that the function

$$f(x) = \begin{cases} \dfrac{x}{\sin x} & \text{if } 0 < |x| < \pi/2 \\ 1 & \text{if } x = 0 \end{cases}$$

is continuous at 0. Find an $r > 0$ such that $|f(x) - 1| < 10^{-6}$ for all $|x| < r$.
HINT: Use inequalities from Example 2.3.7.

C. Show that the sawtooth function

$$f(x) = \begin{cases} x - 2n & \text{if} \quad 2n \le x \le 2n+1, \ n \in \mathbb{Z} \\ 2n - x & \text{if} \quad 2n-1 \le x \le 2n, \ n \in \mathbb{Z} \end{cases}$$

is continuous.

D. Consider a function defined on $\mathbb{R}^2$ by

$$f(x,y) = \begin{cases} (1+xy)^{1/x} & \text{if} \quad x \ne 0 \\ e^y & \text{if} \quad x = 0. \end{cases}$$

Prove carefully that f is continuous at $(0, y_0)$.

E. Consider a function defined on $\mathbb{R}^2$ by

$$f(x,y) = \begin{cases} 0 & \text{if} \quad y \le 0 \ \text{ or if } \ y \ge x^2 \\ \sin\left(\frac{\pi y}{x^2}\right) & \text{if} \quad 0 < y < x^2. \end{cases}$$

(a) Show that f is not continuous at the origin.
(b) Show that the restriction of f to any straight line through the origin is continuous.

F. (a) Show that the definition of limit can be reformulated using open balls instead of norms as follows: A function f mapping a subset $S \subset \mathbb{R}^n$ into $\mathbb{R}^m$ has limit $\mathbf{v}$ as $\mathbf{x}$ approaches $\mathbf{a}$ provided that for every $\varepsilon > 0$, there is an $r > 0$ such that $f\big(B_r(\mathbf{a}) \cap S \setminus \{\mathbf{a}\}\big) \subset B_\varepsilon(\mathbf{v})$.
(b) Provide a similar reformulation of the statement that f is continuous at $\mathbf{a}$.

G. Suppose that functions f, g, h mapping $S \subset \mathbb{R}^n$ into $\mathbb{R}$ satisfy $f(\mathbf{x}) \le g(\mathbf{x}) \le h(\mathbf{x})$ for $\mathbf{x} \in S$. Suppose that $\mathbf{c}$ is a limit point of S and

$$\lim_{\mathbf{x}\to\mathbf{c}} f(\mathbf{x}) = \lim_{\mathbf{x}\to\mathbf{c}} h(\mathbf{x}) = L.$$

Show that $\lim_{\mathbf{x}\to\mathbf{c}} g(\mathbf{x}) = L$.

H. Define a function on the set $S = \{0\} \cup \{\frac{1}{n} : n \ge 1\}$ by $f(\frac{1}{n}) = a_n$ and $f(0) = L$. Prove that f is continuous on S if and only if $\lim_{n\to\infty} a_n = L$.

I. Show that if $f : [a,b] \to \mathbb{R}$ is a differentiable function such that $|f'(x)| \le M$ on $[a,b]$, then f is Lipschitz. HINT: Mean Value Theorem.

J. Find a bounded continuous function on $\mathbb{R}$ that is not Lipschitz.
HINT: The derivative should blow up somewhere.

K. (a) Show that a linear transformation $A : \mathbb{R}^n \to \mathbb{R}^m$ with $m \times n$ matrix $\big[a_{ij}\big]$ can be written as $A = \sum_{i=1}^{m}\sum_{j=1}^{n} a_{ij}\epsilon_i\pi_j$.
(b) Show that $\epsilon_i\pi_j$ is Lipschitz with constant 1.
(c) Hence deduce that A is Lipschitz with constant $\sum_{i=1}^{n}\sum_{j=1}^{m} |a_{ij}|$.

NOTE: This estimate is weaker than the one obtained in Corollary 5.1.7 but does not require the Schwarz inequality.

L. Consider the linear transformation on $\mathbb{R}^4$ given by the matrix

$$A = \begin{bmatrix} \frac{1}{2} & \frac{1}{2} & \frac{1}{2} & \frac{1}{2} \\ \frac{1}{2} & -\frac{1}{2} & \frac{1}{2} & -\frac{1}{2} \\ \frac{1}{2} & \frac{1}{2} & -\frac{1}{2} & -\frac{1}{2} \\ \frac{1}{2} & -\frac{1}{2} & -\frac{1}{2} & \frac{1}{2} \end{bmatrix}.$$

(a) Compute the Lipschitz constant obtained in Corollary 5.1.7.
(b) Show that $\|A\mathbf{x}\| = \|\mathbf{x}\|$ for all $\mathbf{x} \in \mathbb{R}^4$. Hence deduce that the optimal Lipschitz constant is 1. HINT: The columns of A form an orthonormal basis for $\mathbb{R}^4$.

M. At some point, you may have been told that a continuous function is one that can be drawn without lifting your pencil off the paper. Is this actually true?

5.2. Discontinuous Functions

The purpose of this section is to show, through a variety of examples, some of the pathologies that can occur in discontinuous functions. We make no serious attempt to classify discontinuities, although we give names to some of the simpler kinds, the ones that have useful descriptions.

5.2.1. EXAMPLE. Consider

$$f(x) = \begin{cases} 0 & \text{if} \quad x \neq 0 \\ 1 & \text{if} \quad x = 0. \end{cases}$$

This function is discontinuous at 0 because

$$\lim_{x \to 0} f(x) = 0 \neq 1 = f(0).$$

This is the simplest kind of discontinuity, known as a **removable singularity** because the function may be altered at the point of discontinuity in order to fix the problem. Changing $f(0)$ to 0 makes the function continuous.

5.2.2. EXAMPLE. For our next example of a discontinuous function, consider the **Heaviside function**, which is much used in engineering. Its graph is given in Figure 5.3. Define H on $\mathbb{R}$ by

$$H(x) = \begin{cases} 0 & \text{if} \quad x < 0 \\ 1 & \text{if} \quad x \geq 0. \end{cases}$$

We claim that $\lim_{x \to 0} H(x)$ does not exist. Suppose $\lim_{x \to 0} H(x) = L$, for some number L. Then for any $\varepsilon > 0$, there would be some $r > 0$ so that

$$|H(x) - L| < \varepsilon \quad \text{whenever} \quad 0 < |x - 0| < r.$$

Let $\varepsilon = 1/2$ and let r be any positive real number. The values $\pm r/2$ both satisfy $|\pm r/2 - 0| = r/2 < r$. But for any choice of L, the triangle inequality yields

$$\left| H(\tfrac{r}{2}) - L \right| + \left| H(\tfrac{-r}{2}) - L \right| \geq \left| H(\tfrac{r}{2}) - H(\tfrac{-r}{2}) \right| = 1.$$

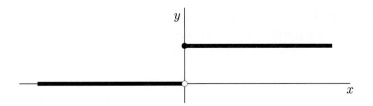

FIGURE 5.3. The Heaviside function.

Therefore, we conclude that

$$\max\left\{\left|H(\tfrac{r}{2})-L\right|,\left|H(\tfrac{-r}{2})-L\right|\right\}\geq\tfrac{1}{2}=\varepsilon.$$

So no possible limit can exist. This trick has the effect of removing consideration of what the limit L might be and eliminates all possible limit values at the same time.

However, H is continuous at every $a\neq 0$. Indeed, if $|x-a|<|a|$, then a and x have the same sign. Hence

$$|H(x)-H(a)|=0<\varepsilon$$

for any $\varepsilon>0$.

There is no way to "repair" this discontinuity by redefining H at the origin. However, it is not difficult to understand this function's behaviour. We make the following definition of one-sided limits.

5.2.3. DEFINITION. Say that the limit of f as x approaches a from the right is L if for every $\varepsilon>0$, there is an $r>0$ so that

$$|f(x)-L|<\varepsilon\quad\text{for all}\quad a<x<a+\varepsilon.$$

We write $\lim\limits_{x\to a^+}f(x)=L$.

Similarly, we define a limit from the left [written $\lim\limits_{x\to a^-}f(x)=L$].

5.2.4. DEFINITION. When a function f on $\mathbb{R}$ has limits from the left and right that are different, we say that f has a **jump discontinuity**. A function on an interval is called **piecewise continuous** if on every finite subinterval, it has only a finite number of jump discontinuities.

Indeed, the restriction of H to $[0,\infty)$ is constant and therefore continuous. What happens at $a=0$ is that

$$\lim\limits_{x\to 0^+}H(x)=1=H(0).$$

Also on $(-\infty,0)$ the function is continuous and moreover has a limit at 0. That is,

$$\lim\limits_{x\to 0^-}H(x)=0.$$

Thus H is a piecewise continuous function with a jump discontinuity at 0.

Piecewise continuity allows a function to have infinitely many jump discontinuities, provided the set of jump discontinuities does not "bunch up." For example,
the **ceiling function** on $\mathbb{R}$, defined by letting $\lceil x \rceil$ be the least integer greater than
or equal to x, is piecewise continuous on $\mathbb{R}$.

We also consider what it means to have an infinite limit.

5.2.5. DEFINITION. Say that the limit of a function $f(x)$ as x approaches a is
$+\infty$ if for every positive integer N, there is an $r > 0$ so that

$$f(x) > N \quad \text{for all} \quad 0 < |x - a| < r.$$

We write $\lim_{x \to a} f(x) = +\infty$. We define the limit $\lim_{x \to a} f(x) = -\infty$ similarly.

5.2.6. EXAMPLE. Another simple type of discontinuity was already observed
in Example 5.1.3. Define f on $\mathbb{R}^n$ by

$$f(\mathbf{x}) = \begin{cases} 1/\|\mathbf{x}\| & \text{if} \quad \mathbf{x} \neq \mathbf{0} \\ 0 & \text{if} \quad \mathbf{x} = \mathbf{0}. \end{cases}$$

This was shown to be continuous on $\mathbb{R}^n \setminus \{\mathbf{0}\}$. However,

$$\lim_{\mathbf{x} \to \mathbf{0}} f(\mathbf{x}) = +\infty.$$

Indeed, for each positive integer N, take $r = 1/N$. Then for $0 < \|\mathbf{x}\| < 1/N$, we
have $f(\mathbf{x}) > N$ as desired.

No redefinition of the value of f at the origin can make this function continuous. Nevertheless, this can be seen as a straightforward kind of behaviour.

Let us look at two examples that behave more wildly.

5.2.7. EXAMPLE. Consider the function defined on $\mathbb{R}^2$ by

$$f(x, y) = \begin{cases} \dfrac{x^2}{x^2 + y^2} & \text{if} \quad (x, y) \neq (0, 0) \\ 0 & \text{if} \quad (x, y) = (0, 0). \end{cases}$$

It is easy to verify that f is continuous on $\mathbb{R}^2 \setminus \{(0, 0)\}$. However, at the origin,
this function behaves in a nasty fashion. To understand this function, we convert to
polar coordinates.

Recall that a vector $(x, y) \neq (0, 0)$ is determined by its length $r = \sqrt{x^2 + y^2}$
and the angle θ that the vector makes to the positive real axis determined up to a
multiple of 2π by

$$x = r \cos \theta \quad \text{and} \quad y = r \sin \theta.$$

With this notation, we may compute

$$f(x, y) = \frac{x^2}{x^2 + y^2} = \frac{r^2 \cos^2 \theta}{r^2} = \cos^2 \theta.$$

Now it is clear that this function is constant on rays from the origin (those points with a fixed angle in polar coordinates). However, even though the function remains bounded, its value, $f(x, y)$, oscillates between 0 and 1 as (x, y) progresses around the circle. There is no limit as one approaches the origin from an arbitrary direction as every value in $[0, 1]$ is a limit value of some sequence.

5.2.8. EXAMPLE. A similar phenomenon can be seen in functions on the real line. Consider the function graphed in Figure 5.4, which is given by

$$f(x) = \begin{cases} \sin \frac{1}{x} & \text{if } x \neq 0 \\ 0 & \text{if } x = 0. \end{cases}$$

First think about the problem of graphing this function. Evidently, $f(-x) = -f(x)$ so it suffices to consider $x > 0$ and reflect to the left half line using the fact that this is an odd function. Now as x tends to $+\infty$, the reciprocal $1/x$ tends monotonically to zero. As $\sin \theta \approx \theta$ for small values of θ, our function is **asymptotic to the curve** $y = 1/x$ as x tends to $+\infty$, meaning $\lim_{x \to \infty} |f(x) - 1/x| = 0$.

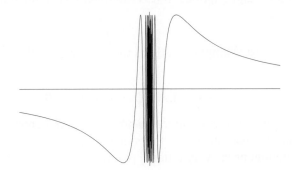

FIGURE 5.4. A partial graph of $\sin(1/x)$.

On the other hand, as x tends to 0^+, $1/x$ goes off to $+\infty$. Indeed, it passes through values from $2n\pi$ to $2(n+1)\pi$ as x passes from $\frac{1}{2n\pi}$ to $\frac{1}{2(n+1)\pi}$. Hence $\sin$ takes values running from 0 up to 1, down to -1, and back up to 0. This happens infinitely often as x approaches the origin. So the curve oscillates rapidly up and down between 1 and -1. No limit is possible.

General arguments show that this function is continuous on $\mathbb{R} \setminus \{0\}$. However, it has a nasty discontinuity at 0. Let us show that every value in $[-1, 1]$ is a limit value along some subsequence. Consider a number $t = \sin \theta$. Notice that

$$\begin{aligned} f(x) = t \quad &\text{if and only if} \quad \sin \tfrac{1}{x} = \sin \theta \\ &\text{if and only if} \quad \tfrac{1}{x} = \theta + 2n\pi \text{ or } (\pi - \theta) + 2n\pi, \quad n \in \mathbb{Z} \\ &\text{if and only if} \quad x = \tfrac{1}{\theta + 2n\pi} \text{ or } \tfrac{1}{(\pi - \theta) + 2n\pi}, \quad n \in \mathbb{Z}. \end{aligned}$$

In particular,

$$\lim_{n \to \infty} f\left(\tfrac{1}{\theta + 2n\pi}\right) = \sin \theta = t.$$

This shows that every point $(0, t)$ for $|t| \leq 1$ lies in the closure of the graph of f. It is not difficult to see that the closure of the graph is precisely this line segment together with the graph itself.

Finally, we look at a couple of bizarre examples.

5.2.9. EXAMPLE. For any subset A of $\mathbb{R}^n$, the **characteristic function** of A is

$$\chi_A(\mathbf{x}) = \begin{cases} 1 & \text{if} \quad \mathbf{x} \in A \\ 0 & \text{if} \quad \mathbf{x} \notin A. \end{cases}$$

The behaviour of this function depends on the character of the set A. See Exercises 5.2.A.

Let us take A to be the set $\mathbb{Q}$ of rationals in $\mathbb{R}$. The function $\chi_\mathbb{Q}$ takes the values 0 and 1 on every open interval, no matter how small, because the sets of the rational and the irrational numbers are both dense in the line. Thus for every $a \in \mathbb{R}$ and any $r > 0$ there is a point x with $|x - a| < r$ such that $|f(x) - f(a)| = 1$. (Just take x irrational if a is rational and x rational if a is irrational.) This function is not continuous at any point!

5.2.10. EXAMPLE. This last example is perhaps the strangest of all. Let

$$f(x) = \begin{cases} 0 & \text{if} \quad x \notin \mathbb{Q} \\ \frac{1}{q} & \text{if} \quad x = \frac{p}{q} \text{ in lowest terms and } q > 0 \end{cases}$$

meaning that p, q are integers with no common factor. Figure 5.5 shows part of the graph of this function. We will show that this function is continuous at every irrational point, and discontinuous at every rational point.

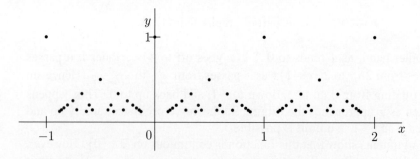

FIGURE 5.5. Partial graph of function $f(p/q) = 1/q$.

First suppose that a is irrational and $\varepsilon > 0$. Fix an integer $M > |a|$. There is an integer N sufficiently large that $1/N < \varepsilon$. The set

$$S = \left\{ \frac{p}{q} : 1 \leq q \leq N, \ -Mq \leq p \leq Mq \right\}$$

is finite and thus is closed. Since S does not contain the point a and its complement S' is open, there is a real number $r > 0$ such that the ball $B_r(a)$ is contained in

$S' \cap (-M, M)$, and so is disjoint from S. Now if $x \in (-M, M)$ is not in S, then either it is irrational, whence $f(x) = 0$, or it is a rational $\frac{p}{q}$ with $q > N$, whence $f(x) < \frac{1}{N} < \varepsilon$. Hence

$$|f(x) - f(a)| = |f(x)| < \varepsilon \quad \text{for all} \quad |x - a| < r.$$

This shows that f is continuous at each irrational point.

Now consider a rational $a = \frac{p}{q}$ in lowest terms. Then $f(a) = \frac{1}{q} > 0$. Take $\varepsilon = \frac{1}{q}$. For every $r > 0$, there are irrational numbers x with $|x - a| < r$. Thus

$$|f(x) - f(a)| = \frac{1}{q} \not< \varepsilon.$$

Consequently, f is not continuous at a.

In fact, f does have a limit at a and so the discontiniuity at a is removable. Defining the set S as in the irrational case, there is a real number $r > 0$ such that $B_r(a) \setminus \{a\}$ is disjoint from the set S. Hence we conclude that

$$|f(x)| < \varepsilon \quad \text{for all} \quad 0 < |x - a| < r.$$

Therefore, $\lim_{x \to a} f(x) = 0$ for *every* point a in $\mathbb{R}$. Since $f(a) \neq 0$ when a is rational, this shows by another method that f is discontinuous there. The really amazing fact is that f has a limit at every point in $\mathbb{R}$ yet is discontinuous on a dense set.

Exercises for Section 5.2

A. Let A be a subset of $\mathbb{R}^n$. Show that the characteristic function χ_A is continuous on the interior of A and on the interior of its complement A' but is discontinuous on the boundary $\partial A = \overline{A} \cap \overline{A'}$.

B. Show that $f(x) = x \log x^2$ for $x \in \mathbb{R} \setminus \{0\}$ has a removable singularity at $x = 0$.

C. Give an example of a bounded function $f : [-1, 1] \to \mathbb{R}$ that has only jump discontinuities but is not piecewise continuous.

D. What is the nature of the singularity at $x = 1$ of the function $f(x) = x^{\frac{1}{1-x}}$ defined for $x \geq 0$, $x \neq 1$?   HINT: $\lim_{x \to 1} \frac{\log x}{x - 1}$ is a derivative.

E. Let $f(x) = \sin^{-1}(\sin x)$, where $\sin^{-1}(y)$ is the unique value $\theta \in [-\frac{\pi}{2}, \frac{\pi}{2}]$ such that $\sin \theta = y$.
(a) Show that f has left and right limits at every point.
(b) Where is f discontinuous?

F. Prove that $L = \lim_{x \to a} f(x)$ if and only if both $\lim_{x \to a-} f(x) = L$ and $\lim_{x \to a+} f(x) = L$.

G. (A monotone convergence test for functions.) Suppose that f is an increasing function on (a, b) that is bounded above. Prove that the one-sided limit $\lim_{x \to b-} f(x)$ exists.

H. Define f on $\mathbb{R}$ by $f(x) = x\chi_\mathbb{Q}(x)$. Show that f is continuous at 0 and that this is the *only* point where f is continuous.

5.3. Properties of Continuous Functions

We start with several properties equivalent to continuity. Then we will record various consequences of continuity, most of which are easy to verify. As the domain of a function is often a proper subset of $\mathbb{R}^n$, we introduce another topological notion. A subset $V \subset S \subset \mathbb{R}^n$ is **open in** S or **relatively open** (with respect to S) if there is an open set U in $\mathbb{R}^n$ such that $U \cap S = V$. In other words, V is open in S if for every $\mathbf{v} \in V$, there is an $\varepsilon > 0$ so that $B_\varepsilon(\mathbf{v}) \cap S \subset V$.

5.3.1. THEOREM. *For a function f mapping $S \subset \mathbb{R}^n$ into $\mathbb{R}^m$, the following are equivalent:*

(1) *f is continuous on S.*

(2) *For every convergent sequence $(\mathbf{x}_n)_{n=1}^{\infty}$ with $\lim_{n \to \infty} \mathbf{x}_n = \mathbf{a}$ in S,*

$$\lim_{n \to \infty} f(\mathbf{x}_n) = f(\mathbf{a}).$$

(3) *For every open set U in $\mathbb{R}^m$, the set*

$$f^{-1}(U) = \{\mathbf{x} \in S : f(\mathbf{x}) \in U\}$$

is open in S.

PROOF. If we assume (1), that is, f is continuous on S, then clearly (2) holds. Conversely, assume that (1) is false, and f is not continuous at some point $\mathbf{a} \in S$. Then reversing the definition of continuity, we can find some positive number $\varepsilon > 0$ for which the definition fails, meaning that there is no value of $r > 0$ that works. That is, fixing this ε, for *every* $r > 0$ there is some point $\mathbf{x} \in S$ (depending on r) such that

$$\|\mathbf{x} - \mathbf{a}\| < r \quad \text{and} \quad \|f(\mathbf{x}) - f(\mathbf{a})\| \geq \varepsilon.$$

So take $r = 1/n$ and find an $\mathbf{x}_n \in S$ with

$$\|\mathbf{x}_n - \mathbf{a}\| < \tfrac{1}{n} \quad \text{and} \quad \|f(\mathbf{x}_n) - f(\mathbf{a})\| \geq \varepsilon.$$

It follows that $\lim_{n \to \infty} \mathbf{x}_n = \mathbf{a}$ and $f(\mathbf{x}_n)$ does not converge to $f(\mathbf{a})$. This shows that if (1) fails, then (2) is false also. Therefore, (1) and (2) are equivalent.

Suppose that f is continuous and U is an open subset of $\mathbb{R}^m$. Pick any point $\mathbf{a}$ in $f^{-1}(U)$. Since U is open and contains $\mathbf{u} = f(\mathbf{a})$, there is an $\varepsilon > 0$ such that $B_\varepsilon(\mathbf{u})$ is contained in U. From the continuity of f, there is a real number $r > 0$ such that

$$\|f(\mathbf{x}) - \mathbf{u}\| < \varepsilon \quad \text{for all} \quad \mathbf{x} \in S, \ \|\mathbf{x} - \mathbf{a}\| < r.$$

This means that $f(B_r(\mathbf{a}) \cap S)$ is contained in $B_\varepsilon(\mathbf{u})$ and thus in U. Hence $f^{-1}(U)$ contains $B_r(\mathbf{a}) \cap S$. Consequently, $f^{-1}(U)$ is open in S.

Conversely, suppose that (3) holds. Fix $\mathbf{a}$ in S and $\varepsilon > 0$. Using the open set $U = B_\varepsilon(f(\mathbf{a}))$, we obtain an open set $f^{-1}(U)$ in S containing $\mathbf{a}$. Therefore, there is a real number $r > 0$ such that

$$B_r(\mathbf{a}) \cap S \subset f^{-1}(U).$$

In other words, $\|f(\mathbf{x}) - f(\mathbf{a})\| < \varepsilon$ for all $\mathbf{x} \in S$ such that $\|\mathbf{x} - \mathbf{a}\| < r$. So f is continuous. ∎

Property (2) could be called the **sequential characterization of continuity**. It will often be more convenient to work with a sequence and this property rather than finding some r for each ε, as in the original definition. Property (3) could be called the **topological charactizeration of continuity**. This is a formulation that readily generalizes to settings in which there is no appropriate distance function. In certain ways, this version is more powerful than the others, but it is only valid for continuity on a set, not continuity at a point.

Recall that if f, g are functions from a common domain $S \subset \mathbb{R}^m$ into $\mathbb{R}^m$ and $\alpha, \beta \in \mathbb{R}$, then $\alpha f + \beta g$ denotes the function that sends $\mathbf{x}$ to $\alpha f(\mathbf{x}) + \beta g(\mathbf{x})$. When the range is $\mathbb{R}$, fg denotes the function sending $\mathbf{x}$ to $f(\mathbf{x})g(\mathbf{x})$. From the properties of limits and the definition of continuity, we obtain the following basic properties.

5.3.2. THEOREM. *If f, g are functions from a common domain S into $\mathbb{R}^m$ and $\mathbf{a} \in S$ such that $\lim_{\mathbf{x}\to\mathbf{a}} f(\mathbf{x}) = \mathbf{u}$ and $\lim_{\mathbf{x}\to\mathbf{a}} g(\mathbf{x}) = \mathbf{v}$, then*

(1) $\lim_{\mathbf{x}\to\mathbf{a}} f(\mathbf{x}) + g(\mathbf{x}) = \mathbf{u} + \mathbf{v}$,

(2) $\lim_{\mathbf{x}\to\mathbf{a}} \alpha f(\mathbf{x}) = \alpha\mathbf{u}$ *for any* $\alpha \in \mathbb{R}$.

When the range is contained in $\mathbb{R}$, say $\lim_{\mathbf{x}\to\mathbf{a}} f(\mathbf{x}) = u$ and $\lim_{\mathbf{x}\to\mathbf{a}} g(\mathbf{x}) = v$, then

(3) $\lim_{\mathbf{x}\to\mathbf{a}} f(\mathbf{x})g(\mathbf{x}) = uv$, *and*

(4) $\lim_{\mathbf{x}\to\mathbf{a}} \dfrac{f(\mathbf{x})}{g(\mathbf{x})} = \dfrac{u}{v}$ *provided that $v \neq 0$.*

PROOF. The proofs are analogous to the corresponding fact for limits of sequences. We will prove (1) from the definition of limit. The others will be left as exercises.

Let $\varepsilon > 0$ be given. From $\lim_{\mathbf{x}\to\mathbf{a}} f(\mathbf{x}) = \mathbf{u}$ and using $\varepsilon/2$ in the definition of limit, there is an $r_1 > 0$ such that

$$\|f(\mathbf{x}) - \mathbf{u}\| < \frac{\varepsilon}{2} \quad \text{for all} \quad \mathbf{x} \in S \text{ with } \|\mathbf{x} - \mathbf{a}\| < r_1.$$

Similarly, for $\lim_{\mathbf{x}\to\mathbf{a}} g(\mathbf{x}) = \mathbf{v}$, there is an $r_2 > 0$ such that

$$\|g(\mathbf{x}) - \mathbf{v}\| < \frac{\varepsilon}{2} \quad \text{for all} \quad \mathbf{x} \in S \text{ with } \|\mathbf{x} - \mathbf{a}\| < r_2.$$

Take $r = \min\{r_1, r_2\}$, which is positive. Then for $\|\mathbf{x} - \mathbf{a}\| < r$,

$$\begin{aligned}
\|(f + g)(\mathbf{x}) - (\mathbf{u} + \mathbf{v})\| &= \|(f(\mathbf{x}) - \mathbf{u}) + (g(\mathbf{x}) - \mathbf{v})\| \\
&\leq \|(f(\mathbf{x}) - \mathbf{u})\| + \|(g(\mathbf{x}) - \mathbf{v})\| \\
&< \frac{\varepsilon}{2} + \frac{\varepsilon}{2} = \varepsilon.
\end{aligned}$$

This establishes that $\lim_{\mathbf{x}\to\mathbf{a}} f(\mathbf{x}) + g(\mathbf{x}) = \mathbf{u} + \mathbf{v}$. ∎

5.3.3. THEOREM. *If f, g are functions from a common domain S into $\mathbb{R}^m$ that are continuous at $\mathbf{a} \in S$, and $\alpha \in \mathbb{R}$, then*

    (1) $f + g$ *is continuous at* $\mathbf{a}$,

    (2) αf *is continuous at* $\mathbf{a}$,

and when the range is contained in $\mathbb{R}$,

    (3) fg *is continuous at* $\mathbf{a}$, *and*

    (4) f/g *is continuous at* $\mathbf{a}$ *provided that* $g(\mathbf{a}) \neq 0$.

PROOF. We prove (3) and leave the others as exercises. To show that fg is continuous at $\mathbf{a}$, we must show that $\lim_{\mathbf{x} \to \mathbf{a}} fg(\mathbf{x}) = f(\mathbf{a})g(\mathbf{a})$. By Theorem 5.3.2 (3),

$$\lim_{\mathbf{x} \to \mathbf{a}} fg(\mathbf{x}) = \lim_{\mathbf{x} \to \mathbf{a}} f(\mathbf{x})g(\mathbf{x}) = \lim_{\mathbf{x} \to \mathbf{a}} f(\mathbf{x}) \lim_{\mathbf{x} \to \mathbf{a}} g(\mathbf{x}) = f(\mathbf{a})g(\mathbf{a}),$$

where the last equality follows from f and g each being continuous. ∎

5.3.4. EXAMPLE. Observe that the function $f(x) = x$ is continuous at every $a \in \mathbb{R}$, since $\lim_{x \to a} f(x) = \lim_{x \to a} x = a$. By Theorem 5.3.3 (2), products of this function are continuous, so $g(x) = x^2$, $h(x) = x^3$, and in general $k(x) = x^n$ for every positive integer n are all continuous functions. By Theorem 5.3.3 (1) and (3), linear combinations of these functions are continuous, and so we conclude that every polynomial is continuous on $\mathbb{R}$.

    If f is a **rational function**—that is, $f(x) = p(x)/q(x)$, where p and q are polynomials—then f is continuous at all $a \in \mathbb{R}$, where $q(a) \neq 0$. This follows from the previous paragraph and Theorem 5.3.3 (4).

    Recall that if f maps a domain $S \subset \mathbb{R}^n$ into a set $T \subset \mathbb{R}^m$, and g maps T into $\mathbb{R}^l$, then the **composition** of g and f, denoted $g \circ f$, is the function that sends x to $g(f(x))$. For example, if $f(x, y) = x^2 + y^2$ is defined on $\mathbb{R}^2$ and $g(x) = \sqrt{x}$ for $x \in [0, \infty)$, then $g \circ f(x, y) = \sqrt{x^2 + y^2}$.

5.3.5. THEOREM. *Suppose that f maps a domain S contained in $\mathbb{R}^n$ into a subset T of $\mathbb{R}^m$, and g maps T into $\mathbb{R}^l$. If f is continuous at $\mathbf{a} \in S$ and g is continuous at $f(\mathbf{a}) \in T$, then the function $g \circ f$ is continuous at $\mathbf{a}$. Thus if f and g are continuous, then so is $g \circ f$.*

PROOF. We will use the sequential characterization of continuity. Let $(\mathbf{x}_n)_{n=1}^{\infty}$ be any sequence of points in S with $\lim_{n \to \infty} \mathbf{x}_n = \mathbf{a}$. Since f is continuous at $\mathbf{a}$, we know that $\lim_{n \to \infty} f(\mathbf{x}_n) = f(\mathbf{a})$. Thus $(f(\mathbf{x}_n))_{n=1}^{\infty}$ is a sequence in T with limit $f(\mathbf{a})$, and since g is continuous at $f(\mathbf{a})$, we conclude that

$$\lim_{n \to \infty} g(f(\mathbf{x}_n)) = g(f(\mathbf{a})).$$

Therefore, by Theorem 5.3.1, $g \circ f$ is continuous at $\mathbf{a}$. ∎

5.3.6. EXAMPLE. If f maps $S \subset \mathbb{R}^n$ into $\mathbb{R}^m$, then f_i, the ith coordinate of $f(\mathbf{x})$, is a real-valued function on S. Using this notation, we may write

$$f(\mathbf{x}) = (f_1(\mathbf{x}), \ldots, f_m(\mathbf{x})).$$

We claim that f is continuous if and only if each function f_i is continuous for $1 \le i \le m$.

One way to see this is to argue exactly as in Lemma 4.2.3. Instead, we will use Corollary 5.1.7 and the (continuous) coordinate functions $\pi_i : \mathbb{R}^m \to \mathbb{R}$ and $\epsilon_i : \mathbb{R} \to \mathbb{R}^m$ given by

$$\pi_i(x_1, \ldots, x_m) = x_i \quad \text{and} \quad \epsilon_i(t) = t e_i.$$

Notice that $f_i(\mathbf{x}) = \pi_i \circ f(\mathbf{x})$. Thus if f is continuous, each f_i is continuous. Conversely,

$$f(\mathbf{x}) = \sum_{i=1}^{m} \epsilon_i \circ f_i(\mathbf{x}).$$

Hence if each f_i is continuous, then each $\epsilon_i \circ f_i$ is continuous by Theorem 5.3.5; and their sum is continuous by Theorem 5.3.3 (1).

Exercises for Section 5.3

A. Show that the function defined on $\mathbb{R}^2 \setminus \{(0,0)\}$ by $f(x, y) = \dfrac{\sin(\log(x^2 + y^2))}{\cos^2 y + y^2 e^x}$ is continuous.

B. Use Lemma 4.2.3 and the sequential characterization of continuity to give a second proof for Example 5.3.6.

C. Consider f mapping $S \subset \mathbb{R}^n$ into $\mathbb{R}^m$ and two points: $\mathbf{a}$ a limit point of S and $\mathbf{v} \in \mathbb{R}^m$. Show that $\lim\limits_{\mathbf{x} \to \mathbf{a}} f(\mathbf{x}) = \mathbf{v}$ if and only if for each sequence $(\mathbf{x}_n) \in S \setminus \{\mathbf{a}\}$ with $\lim\limits_{n \to \infty} \mathbf{x}_n = \mathbf{a}$, we have $\lim\limits_{n \to \infty} f(\mathbf{x}_n) = \mathbf{v}$.

D. Suppose that f mapping $S \subset \mathbb{R}^n$ into $\mathbb{R}^m$ is given by $f(\mathbf{x}) = (f_1(\mathbf{x}), \ldots, f_n(\mathbf{x}))$ with $f_i : S \to \mathbb{R}$ for each i. Show that $\lim\limits_{\mathbf{x} \to \mathbf{a}} f(\mathbf{x}) = (u_1, \ldots, u_m)$ if and only if $\lim\limits_{\mathbf{x} \to \mathbf{a}} f_i(\mathbf{x}) = u_i$ for each i. HINT: Use the previous exercise and Lemma 4.2.3.

E. Let f and g be continuous mapping of $S \subset \mathbb{R}^n$ into $\mathbb{R}^m$. Show that the inner product $h(\mathbf{x}) = \langle f(\mathbf{x}), g(\mathbf{x}) \rangle$ is continuous.

F. Finish the proof of Theorem 5.3.2.

G. Suppose that f is a continuous function on $[a, b]$ and g is a continuous function on $[b, c]$ such that $f(b) = g(b)$. Show that

$$h(x) = \begin{cases} f(x) & \text{if} \quad a \le x \le b \\ g(x) & \text{if} \quad b \le x \le c \end{cases}$$

is continuous on $[a, c]$.

H. Let f be a continuous real-valued function defined on an open subset U of $\mathbb{R}^n$. Show that $\{(\mathbf{x}, y) : \mathbf{x} \in U, \ y > f(\mathbf{x})\}$ is an open subset of $\mathbb{R}^{n+1}$.

I.   (a) Show that $m(x, y) = \max\{x, y\}$ is continuous on $\mathbb{R}^2$.
(b) Hence show that if f and g are continuous real-valued functions on a set $S \subset \mathbb{R}^n$, then $h(x) = \max\{f(x), g(x)\}$ is continuous on S.
(c) Use induction to show that if f_i are continuous real-valued functions on S for $1 \le i \le k$, then $h(x) = \max\limits_{1 \le i \le k} f_i(x)$ is continuous.

J.   Show that a function f from $\mathbb{R}^n$ into $\mathbb{R}^m$ is continuous if and only if $f^{-1}(C)$ is closed for every closed set $C \subset \mathbb{R}^m$.

K.   Suppose that A and B are subsets of $\mathbb{R}^n$. Find necessary and sufficient conditions for there to be a continuous function f on $\mathbb{R}^n$ with $f|_A = 1$ and $f|_B = 0$.
HINT: Consider $g(x) = \text{dist}(x, A)$ and $h(x) = \text{dist}(x, B)$.

L.   Give example of a continuous function f and an open set U such that $f(U)$ is not open.

M.   Suppose that $f : \mathbb{R} \to \mathbb{R}$ satisfies the functional equation

$$f(u + v) = f(u) + f(v) \quad \text{for all} \quad u, v \in \mathbb{R}.$$

(a) Prove that $f(mx) = mf(x)$ for all $x \in \mathbb{R}$ and $m \in \mathbb{Z}$.
HINT: Use induction for $m \ge 1$ and show that $f(-x) = -f(x)$.
(b) Prove that $f(x) = mx$ for all $x \in \mathbb{Q}$, where $m = f(1)$.
HINT: Use (a) to solve for $f(p/q)$ when $p, q \in \mathbb{Z}$.
(c) Use (b) to prove that if f is continuous on $\mathbb{R}$, then $f(x) = mx$ for all $x \in \mathbb{R}$.

5.4. Compactness and Extreme Values

In every calculus course, a lot of effort is spent finding the maximum or minimum of various functions. Sometimes there were physical reasons why such a point should exist. However, generally it was taken on blind faith and the student dutifully differentiates the function to find critical points. Even when the function is not differentiable, the function may attain its maximum value. On the other hand, many very nice functions do not attain maxima. In this section, we will see how our new topological tools can explain this phenomenon.

First consider a couple of easy examples in which there are no maxima.

5.4.1. EXAMPLE. Consider $f(\mathbf{x}) = \dfrac{-1}{1 + \|\mathbf{x}\|^2}$ for $\mathbf{x} \in \mathbb{R}^n$. This function is bounded above, yet the supremum $0 = \lim\limits_{\|\mathbf{x}\| \to \infty} f(\mathbf{x})$ is never attained. The function $g(\mathbf{x}) = \|\mathbf{x}\|$ is unbounded and thus also does not attain its supremum. These problems can occur whenever the domain of the function is unbounded.

5.4.2. EXAMPLE. Consider $f(x) = -x$ for $x \in (0, 1]$. This function is bounded above yet does not attain its supremum $0 = \lim\limits_{x \to 0^+} f(x)$ because the limit point 0 is missing from the domain. Similarly, the function $f(x) = \frac{1}{x}$ for $x \in (0, 1]$ is unbounded and thus does not attain its supremum. A modification of this example would show that the same problem results whenever the domain is not closed.

Both of these difficulties may be avoided if the domain is compact. It turns out that this is all that we need. As in most proofs using compactness, the aim is to find an appropriate sequence in the compact set C so that a convergent subsequence can be obtained with good properties.

5.4.3. THEOREM. *Let C be a compact subset of $\mathbb{R}^n$, and let f be a continuous function from C into $\mathbb{R}^m$. Then the image set $f(C)$ is compact.*

PROOF. Let $(\mathbf{y}_n)_{n=1}^{\infty}$ be a sequence in $f(C)$. We must find a subsequence converging to a point in the image. First choose points $\mathbf{x}_n$ in C such that $\mathbf{y}_n = f(\mathbf{x}_n)$. Now $(\mathbf{x}_n)_{n=1}^{\infty}$ is a sequence in the compact set C. Therefore, there is a subsequence $(\mathbf{x}_{n_i})$ such that

$$\lim_{i \to \infty} \mathbf{x}_{n_i} = \mathbf{c} \in C.$$

Therefore, by the continuity of f,

$$\lim_{i \to \infty} \mathbf{y}_{n_i} = \lim_{i \to \infty} f(\mathbf{x}_{n_i}) = f(\mathbf{c}).$$

Thus the subsequence $(\mathbf{y}_{n_i})$ has the limit $f(\mathbf{c})$ in $f(C)$. Consequently, $f(C)$ is compact. ■

This immediately yields a result often used (without proof) in calculus.

5.4.4. EXTREME VALUE THEOREM.
Let C be a compact subset of $\mathbb{R}^n$, and let f be a continuous function from C into $\mathbb{R}$. Then there are points $\mathbf{a}$ and $\mathbf{b}$ in C attaining the minimum and maximum values of f on C. That is,

$$f(\mathbf{a}) \le f(\mathbf{x}) \le f(\mathbf{b}) \quad \text{for all} \quad \mathbf{x} \in C.$$

PROOF. Since C is compact, Theorem 5.4.3 shows that $f(C)$ is compact. Hence it is closed and bounded in $\mathbb{R}$. Boundedness shows that

$$m = \inf_{\mathbf{x} \in C} f(\mathbf{x}) \quad \text{and} \quad M = \sup_{\mathbf{x} \in C} f(\mathbf{x})$$

are both finite. From the definition of supremum, M is a limit of values in $f(C)$. Thus since $f(C)$ is closed, $M \in f(C)$. This means that there is a point $\mathbf{b} \in C$ such that $f(\mathbf{b}) = M$. Similarly, the infimum is attained at some point $\mathbf{a} \in C$. ■

Exercises for Section 5.4

A. If A is a noncompact subset of $\mathbb{R}^n$, show that there is a bounded continuous real-valued function on A that does not attain its maximum.

B. Find a *discontinuous* function on $[0, 1]$ that is bounded but does not achieve its supremum.

C. Suppose that f is a continuous function on $[a, b]$ with no local maximum or local minimum. Prove that f is monotone.

D. Find a linear transformation T on $\mathbb{R}^2$ and a closed subset C of $\mathbb{R}^2$ such that $T(C)$ is not closed.

E. Show that a function f mapping a *compact* set $S \subset \mathbb{R}^n$ into $\mathbb{R}^m$ is continuous if and only if its graph $G(f) = \{(\mathbf{x}, f(\mathbf{x})) : \mathbf{x} \in S\}$ is compact.
HINT: For $\Rightarrow$, use Theorem 5.4.3. For $\Leftarrow$, use Theorem 5.3.1(2).

F. Give a function defined on $[0, 1]$ that has a closed graph but is not continuous.

G. Suppose that f is a *positive* continuous function on $\mathbb{R}^n$ such that $\lim\limits_{\|\mathbf{x}\| \to \infty} f(\mathbf{x}) = 0$
[i.e., for all $\varepsilon > 0$, there is an N so that $|f(\mathbf{x})| < \varepsilon$ for all $\mathbf{x}$ with $\|\mathbf{x}\| > N$]. Show that f attains its maximum.

H. Let f be a **periodic function** on $\mathbb{R}$, meaning that there is $d \in \mathbb{R}$ with $f(x+d) = f(x)$ for all $x \in \mathbb{R}$. (We call d the **period** of f if it is the least positive number with this property and say that f is **d-periodic**.) Show that if f is continuous, then f attains its maximum and minimum on $\mathbb{R}$.

I. (a) Give an example of a continuous function on $\mathbb{R}^2$ satisfying $f(x + 1, y) = f(x, y)$ for all $x, y \in \mathbb{R}$ that does not attain its maximum.
(b) Find and prove a variant of the previous exercise that is valid for functions on $\mathbb{R}^2$.

J. Let A be a compact subset of $\mathbb{R}^n$. Show that for any point $\mathbf{x} \in \mathbb{R}^n$, there is a closest point $\mathbf{a}$ in A to $\mathbf{x}$. (This means that the point $\mathbf{a} \in A$ satisfies $\|\mathbf{x} - \mathbf{a}\| \leq \|\mathbf{x} - \mathbf{b}\|$ for all $\mathbf{b} \in A$. It does not imply that $\mathbf{a}$ is unique—it may not be.)
HINT: Fix $\mathbf{x}$ and define a useful continuous function on A.

K. For a function f on $[0, \infty)$, we say that $\lim\limits_{x \to \infty} f(x) = L$ if for every $\varepsilon > 0$, there is some $N > 0$ so that $\|f(x) - L\| < \epsilon$ for all $x > N$. Suppose that $\lim\limits_{x \to \infty} f(x) = f(0)$. Prove that f attains its maximum and minimum values.

L. Suppose that C is a compact subset of $\mathbb{R}^n$ and that f is a continuous, one-to-one function of C onto $D \subset \mathbb{R}^m$. Prove that the inverse function f^{-1} is continuous.
HINT: Fix $d_0 \in D$, $c_0 = f^{-1}(d_0)$ and $\varepsilon > 0$. Show that there is a $r > 0$ so that $B_r(d_0)$ is disjoint from $f(C \setminus B_\varepsilon(c_0))$.

M. *A space-filling curve.* Let T be a right triangle with side lengths 3, 4, and 5. Drop a perpendicular line from the right angle to the opposite side, splitting the triangle into two congruent pieces. Label the smaller triangle $T(0)$ and the larger one $T(1)$. Then divide each $T(\varepsilon)$ into two pieces in the same way, labeling the smaller $T(\varepsilon 0)$ and the larger $T(\varepsilon 1)$. Recursively divide each triangle $T(\varepsilon_1 \ldots \varepsilon_n)$ into two smaller congruent triangles labeled $T(\varepsilon_1 \ldots \varepsilon_n 0)$ and $T(\varepsilon_1 \ldots \varepsilon_n 1)$. Now consider each point $x \in [0, 1]$ in its base 2 (binary) expansion $x = 0.\varepsilon_1 \varepsilon_2 \varepsilon_3 \ldots$, where ε_i is 0 or 1. Define a function $f : [0, 1] \to T$ by defining $f(x)$ to be the point in $\bigcap_{n \geq 1} T(\varepsilon_1 \ldots \varepsilon_n)$.
(a) Prove that $T(\varepsilon_1 \ldots \varepsilon_n)$ has diameter at most $5(.8)^n$.
(b) If $x = 0.\varepsilon_1 \ldots \varepsilon_{n-1} 100000 \ldots$ has a finite binary expansion, then it has a second binary representation $x = 0.\varepsilon_1 \ldots \varepsilon_{n-1} 011111 \ldots$ ending in ones. Prove that both expansions yield the same value for $f(x)$.
(c) Hence prove that $f(x)$ is well defined for each $x \in [0, 1]$.
(d) Prove that f is continuous. HINT: If x and y agree to the nth decimal, what do $f(x)$ and $f(y)$ have in common?
(e) Prove that f maps $[0, 1]$ onto T.

N. *A space-filling curve II.* Adapt the triangle filling function of the previous exercise to construct a continuous function on $\mathbb{R}$ that maps onto the entire plane.
HINT: Consider covering the plane by triangles somehow.

5.5. Uniform Continuity

Mathematical terminology is not always consistent, but the adjective *uniform* is (almost) always used the same way. A property is uniform on a set if that property holds at every point in the set with common estimates. Uniform estimates often lead to more powerful conclusions.

5.5.1. DEFINITION. A function f from $S \subset \mathbb{R}^n$ into $\mathbb{R}^m$ is **uniformly continuous** if for every $\varepsilon > 0$, there is a positive real number $r > 0$ so that

$$\|f(\mathbf{x}) - f(\mathbf{a})\| < \varepsilon \quad \text{whenever} \quad \|\mathbf{x} - \mathbf{a}\| < r, \ \mathbf{x}, \mathbf{a} \in S.$$

Read the definition carefully to note what makes it different from continuity at each point $\mathbf{a} \in S$. For f to be continuous, we fix both $\varepsilon > 0$ and $\mathbf{a} \in S$ before obtaining the value of r. So the choice of r might depend on $\mathbf{a}$ as well as on ε. Uniform continuity means that for each $\varepsilon > 0$, the value $r > 0$ that we obtain can be chosen independently of the point $\mathbf{a}$. This is a subtle difference, so we look at some examples.

5.5.2. EXAMPLE. Consider the function $f(x) = x^2$ defined on the bounded interval $[c, d]$. Let us try to obtain a common estimate for r for each $\varepsilon > 0$. Remember that we are trying to control the difference $|f(x) - f(a)|$ *only* by controlling $|x - a|$. Hence we always look for a method of getting a factor close to $|x - a|$ into our estimate while gaining some (perhaps crude) control over the rest. Compute

$$|f(x) - f(a)| = |x^2 - a^2| = |x + a|\,|x - a|.$$

In this case, the factor of $|x - a|$ comes out naturally. A bound must be found for $|x + a|$. Note that for any $x \in [c, d]$, we have

$$|x| \le \max\{|c|, |d|\} =: M.$$

Hence we can show that

$$|x + a| \le |x| + |a| \le 2M.$$

We are allowed to choose a positive number r and insist that $|x - a| < r$. With this choice, the combined estimates yield

$$|f(x) - f(a)| = |x + a|\,|x - a| < 2Mr.$$

To make this at most ε, it suffices to choose $r = \varepsilon/(2M)$, whence

$$|f(x) - f(a)| < 2Mr = \varepsilon \quad \text{for all} \quad |x - a| < r, \ x, a \in [c, d].$$

Hence f is uniformly continuous.

On the other hand, consider $f(x) = x^2$ defined on the whole real line. The preceding argument doesn't work because M would be infinite.

Intuitively, we can put it this way: Suppose, for a fixed $\varepsilon > 0$, we had some $r > 0$ that satisfies the definition of uniform continuity for $f(x) = x^2$ on $\mathbb{R}$. As Figure 5.6 suggests, as a goes to infinity, the interval between $f(a - r)$ and $f(a + r)$ will no longer be contained in the interval from $f(a) - \varepsilon$ to $f(a) + \varepsilon$, giving a contradiction. Since this happens for all $r > 0$, there is no choice of r that "works."

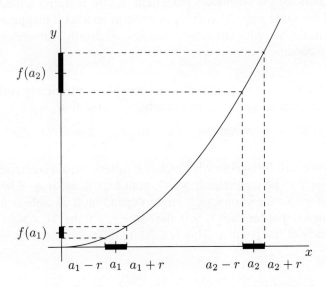

FIGURE 5.6. The function $f(x) = x^2$ on $\mathbb{R}$.

In fact, the estimates used before show us how to make this precise. Suppose that $|x - a| = r$. Then

$$|f(x) - f(a)| = |x^2 - a^2| = |x + a|\,|x - a| = |x + a|\, r.$$

When $|x + a|$ is very large compared with $1/r$, this difference will be large. It suggests that for any choice of r, values of x and a can be found to violate our hoped-for estimates. Search for a sequence that shows that $r = 1/n$ won't work for each $n \geq 1$. For example, let

$$a_n = n \quad \text{and} \quad x_n = n + \tfrac{1}{n}.$$

Then

$$|f(x_n) - f(a_n)| = \left(n + \frac{1}{n}\right)^2 - n^2 = 2 + \frac{1}{n^2}.$$

To violate the definition of uniform continuity, only one value of ε is needed provided that it is shown that *no* value of $r > 0$ will work for that ε. In this case, we may take $\varepsilon = 1$. For any $r > 0$, there is some n so large that $\frac{1}{n} < r$. Then $|x_n - a_n| = \frac{1}{n} < r$, yet $|f(x_n) - f(a_n)| > 2 > \varepsilon$. Therefore, f is not uniformly continuous on $\mathbb{R}$.

5.5.3. EXAMPLE. Consider the function $f(x) = 1/x$ on $(0, 1]$. Notice that the graph blows up at the origin and becomes very steep. This is the same property that we just exploited for x^2 as x goes off to infinity. Very close values in the domain are mapped to points which are far apart. Let $x_n = \frac{1}{n}$. Then

$$|f(x_{n+1}) - f(x_n)| = (n+1) - n = 1.$$

However, $|x_{n+1} - x_n| = \frac{1}{n+n^2}$ tends to 0. So let $\varepsilon = 1$ and consider any $r > 0$. For n large enough, $|x_{n+1} - x_n| < r$, but $|f(x_{n+1}) - f(x_n)| = 1 = \varepsilon$. So f is not uniformly continuous.

A number of properties of functions imply uniform continuity.

5.5.4. PROPOSITION. *Every Lipschitz function is uniformly continuous.*

PROOF. Recall that f is Lipschitz on S means that there is a constant C such that

$$\|f(\mathbf{x}) - f(\mathbf{y})\| \le C\|\mathbf{x} - \mathbf{y}\|.$$

Given $\varepsilon > 0$, choose $r = \varepsilon/C$. Then if $\mathbf{x}, \mathbf{a} \in S$ and $\|\mathbf{x} - \mathbf{a}\| < r$,

$$\|f(\mathbf{x}) - f(\mathbf{a})\| \le C\|\mathbf{x} - \mathbf{a}\| < C\frac{\varepsilon}{C} = \varepsilon.$$

Thus uniform continuity is established (almost by definition). ∎

Corollary 5.1.7 shows that every linear transformation is a Lipschitz function and Exercise 5.1.I shows that every function $f : [a, b] \to \mathbb{R}$ with a bounded derivative is a Lipschitz function. Hence we obtain the following:

5.5.5. COROLLARY. *Every linear transformation from $\mathbb{R}^n$ to $\mathbb{R}^m$ is uniformly continuous.*

5.5.6. COROLLARY. *Let f be a differentiable real-valued function on $[a, b]$ with a bounded derivative; that is, there is $M > 0$ so that $|f'(x)| \le M$ for all $a \le x \le b$. Then f is uniformly continuous on $[a, b]$.*

Before getting to our main result, we look at two more examples. The first is to show that a function does not need to be unbounded in order to fail uniform continuity. However, the previous theorem shows us that the function should be very steep (large derivative) frequently. This suggests returning to one of our favourite functions.

5.5.7. EXAMPLE. Let $f(x) = \sin\frac{1}{x}$ on $(0, 1]$. A computation of the derivative is not necessary since the qualitative features of this function have been considered before. See Example 5.2.8. The function oscillates wildly between $+1$ and -1 as

x approaches 0. In particular,

$$\lim_{n\to\infty} f\left(\frac{1}{(2n+\frac{1}{2})\pi}\right) = 1$$

and

$$\lim_{n\to\infty} f\left(\frac{1}{(2n-\frac{1}{2})\pi}\right) = -1.$$

Let

$$x_n = \frac{1}{(2n+\frac{1}{2})\pi} \quad \text{and} \quad a_n = \frac{1}{(2n-\frac{1}{2})\pi}.$$

Then $f(x_n) - f(a_n) = 2$ for all $n \geq 1$ while

$$\lim_{n\to\infty} |x_n - a_n| = \lim_{n\to\infty} \frac{1}{(4n^2 - \frac{1}{4})\pi} = 0.$$

As before, this means that f is not uniformly continuous.

Finally, let us look at an example in which the derivative is unbounded yet the function is still uniformly continuous.

5.5.8. EXAMPLE. Let $f(x) = x \sin \frac{1}{x}$ on $(0, \infty)$, which we graph in Figure 5.7. What makes this different from the previous examples is the behaviour at the endpoints. At 0, the Squeeze Theorem shows that

$$\lim_{x\to 0} x \sin \frac{1}{x} = 0.$$

Thus we may define $f(0) = 0$ and obtain a continuous function on $[0, \infty)$. At infinity, the substitution $y = \frac{1}{x}$ yields

$$\lim_{x\to\infty} x \sin \frac{1}{x} = \lim_{y\to 0} \frac{\sin y}{y} = 1.$$

This latter limit is established in Example 2.3.7. In fact, we obtained the explicit estimates

$$1 - \frac{1}{x^2} < x \sin \frac{1}{x} < 1 \quad \text{for} \quad x \geq 1.$$

We will show that f is uniformly continuous. The two limits will be used to deal with points near 0 and near infinity (sufficiently large). Let $0 < \varepsilon < 1$ be given. First consider values near the origin. If $|x| < \varepsilon/2$ and $|y| < \varepsilon/2$, then

$$|f(x) - f(y)| \leq |x||\sin \frac{1}{x}| + |y||\sin \frac{1}{y}| \leq |x| + |y| < \varepsilon.$$

Thus if $x \in [0, \varepsilon/4]$ and $|x - y| < \varepsilon/4$, then this estimate holds.

Now do the same thing near infinity. Pick an integer $N > \varepsilon^{-1/2}$. If x and y are greater than N, then

$$1 - \varepsilon < 1 - \frac{1}{N^2} < f(x) < 1.$$

The same is true for y, and thus if $x \geq N + 1$ and $|x - y| < 1$, then

$$|f(y) - f(x)| < \varepsilon.$$

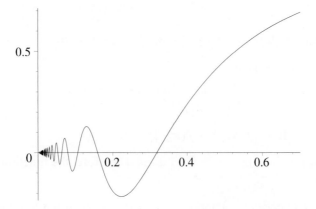

FIGURE 5.7. Partial graph of $x \sin \frac{1}{x}$.

These two estimates show that $|f(y) - f(x)| < \varepsilon$ if $|x - y| < \varepsilon/4$ and either x or y lies in either $[0, \varepsilon/4]$ or $[N + 1, \infty)$.

Now consider the case in which both x and y in the interval $[\varepsilon/4, N + 1]$. On this interval, the function f has a continuous derivative

$$f'(x) = \sin \frac{1}{x} - \frac{1}{x} \cos \frac{1}{x}.$$

An easy estimate shows that

$$|f'(x)| \leq 1 + \frac{1}{x} \leq 1 + \frac{4}{\varepsilon}.$$

Let $M = 1 + 4/\varepsilon$. Hence by Corollary 5.5.6, if we take $|x - y| < \varepsilon/M$ and $\varepsilon/4 \leq x, y \leq N + 1$, then $|f(x) - f(y)| < \varepsilon$. (We could also establish this case using Theorem 5.5.9, but then we would not get an explicit estimate for r in terms of ε.)

Finally, as $\varepsilon/M = \varepsilon^2/(4+\varepsilon)$, we can choose $r = \min\{\varepsilon/4, \varepsilon^2/(4+\varepsilon)\}$. Then if $|x-y| < r$, one of the preceding estimates applies to show that $|f(y)-f(x)| < \varepsilon$. Therefore, f is uniformly continuous.

We conclude this section with an important result relating compactness to uniform continuity.

5.5.9. THEOREM. *Suppose that $C \subset \mathbb{R}^n$ is compact and $f : C \to \mathbb{R}^n$ is countinuous. Then f is uniformly continuous on C.*

PROOF. Suppose that f were not uniformly continuous. Then there would be some $\varepsilon > 0$ for which no $r > 0$ satisfies the definition. That is, for each $r = 1/n$, there are points $\mathbf{a}_n$ and $\mathbf{x}_n$ in C such that $\|\mathbf{x}_n - \mathbf{a}_n\| < 1/n$ but $\|f(\mathbf{x}_n) - f(\mathbf{a}_n)\| \geq \varepsilon$.

Since C is compact and $(\mathbf{a}_n)_{n=1}^{\infty}$ is a sequence in C, there is convergent subsequence $(\mathbf{a}_{n_k})$ with

$$\lim_{k \to \infty} \mathbf{a}_{n_k} = \mathbf{a} \in C.$$

Thus we also have

$$\lim_{k \to \infty} \mathbf{x}_{n_k} = \lim_{k \to \infty} \mathbf{a}_{n_k} + (\mathbf{x}_{n_k} - \mathbf{a}_{n_k}) = \mathbf{a} + \mathbf{0} = \mathbf{a}.$$

By the continuity of f, we have

$$\lim_{k \to \infty} f(\mathbf{a}_{n_k}) = f(\mathbf{a}) \quad \text{and} \quad \lim_{k \to \infty} f(\mathbf{x}_{n_k}) = f(\mathbf{a}).$$

Consequently,

$$\lim_{k \to \infty} \left\| f(\mathbf{a}_{n_k}) - f(\mathbf{x}_{n_k}) \right\| = 0.$$

This contradicts $\| f(\mathbf{a}_n) - f(\mathbf{x}_n) \| \geq \varepsilon > 0$ for all n. Therefore, the function must be uniformly continuous. ∎

Compare how the sequences $(\mathbf{a}_n)$ and $(\mathbf{x}_n)$ are used in this proof to how similar sequences are used to show specific functions are not uniformly continuous in Examples 5.5.3, 5.5.2, and 5.5.7.

Exercises for Section 5.5

A. Show that $g(x) = \sqrt{x}$ is uniformly continuous on $[0, +\infty)$.
HINT: Show that $\sqrt{a - b} \geq \sqrt{a} - \sqrt{b}$ and $\sqrt{a + b} \leq \sqrt{a} + \sqrt{b}$.

B. Given a polynomial $p(x, y) = \sum_{m,n=0}^{N} a_{mn} x^m y^n$ in variables x and y and an $\varepsilon > 0$, find an explicit $\delta > 0$ establishing uniform continuity on the square $[-R, R]^2$.
HINT: Try $\delta = \varepsilon / C$, where $C = \sum_{m,n=0}^{N} |a_{mn}| (m + n) R^{m+n-1}$.

C. (a) Show that the function $f(x) = \frac{1}{x} \sin x$ for $x \neq 0$ can be extended to a continuous function on all of $\mathbb{R}$.
(b) Prove that it is uniformly continuous on $\mathbb{R}$.

D. Show that $f(x) = x^p$ is not uniformly continuous on $\mathbb{R}$ if $p > 1$.

E. Show that if f is continuous on $(0, 1)$ and $\lim_{x \to 0^+} f(x) = +\infty$, then f is not uniformly continuous.

F. Show that a periodic continuous function on $\mathbb{R}$ is bounded and uniformly continuous.

G. Suppose that $f = (f_1, \dots, f_m)$ is a continuous function from an open subset S of $\mathbb{R}^n$ into $\mathbb{R}^m$ that has partial derivatives $\left| \frac{\partial f_i}{\partial x_j}(\mathbf{x}) \right| \leq M < \infty$ for all $\mathbf{x} \in S$. Prove that f is uniformly continuous.

H. Suppose that f is continuous on (a, b) and that $a < b < c$. Show that if f is uniformly continuous on both $(a, b]$ and $[b, c)$, then f is uniformly continuous on (a, b).
HINT: Use $\varepsilon / 2$ to get a δ for each interval separately.

I. Let f be a continuous function on $(0, 1]$. Show that f is uniformly continuous if and only if $\lim_{x \to 0^+} f(x)$ exists.

J. For which real values of α is the function $g_\alpha(x) = x^\alpha \log(x)$ uniformly continuous on $(0, \infty)$? HINT: Use Exercises D, E, and H as a guide. For $0 < \alpha < 1$, consider $[0, 1]$ and $[1, \infty)$ separately.

K. We say a function $f : [a, b] \to \mathbb{R}$ satisfies a **Lipschitz condition of order** $\alpha > 0$ if there is some positive constant M so that

$$|f(x_1) - f(x_2)| \le M|x_1 - x_2|^\alpha.$$

Let Lip α denote the set of all functions satisfying a Lipschitz condition of order α.
(a) Prove that if $f \in \text{Lip } \alpha$, then f is uniformly continuous.
(b) Prove that if $f \in \text{Lip } \alpha$ and $\alpha > 1$, then f is constant.
(c) For $\alpha \in (0, 1)$, show that $f(x) = x^\alpha$ belongs to Lip α.

5.6. The Intermediate Value Theorem

The Intermediate Value Theorem (IVT) is another fundamental result often used in the calculus without proof.

5.6.1. INTERMEDIATE VALUE THEOREM.
If f is a continuous real-valued function on $[a, b]$ with $f(a) < 0 < f(b)$, then there exists a point $c \in (a, b)$ such that $f(c) = 0$.

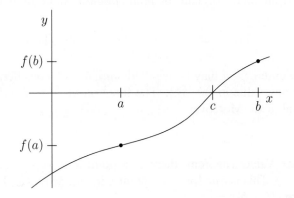

FIGURE 5.8. Applying IVT to a function $f : \mathbb{R} \to \mathbb{R}$.

Figure 5.8 shows the meaning of this theorem graphically. Like the Extreme Value Theorem, this result seems intuitively clear. However, it also depends on the completeness of the real numbers. For example, if our functions were only defined on the set of rational numbers, then the function $f(x) = x^3 - 2$ for $x \in \mathbb{Q}$ would satisfy $f(1) = -1 < 0$ and $f(2) = 6 > 0$, but $f(x)$ would never take the value 0 for any *rational* value of x.

PROOF. Define a subset A of $[a, b]$ by

$$A = \{x \in [a, b] : f(x) < 0\}.$$

Since $a \in A$, A is not empty. And since b is an upper bound for A, the Least Upper Bound Principle allows us to define $c = \sup A$; and it belongs to $[a, b]$.

We claim that $f(c) = 0$. First, since c is the least upper bound for A, there is a sequence (a_n) in A such that $c - \frac{1}{n} < a_n \leq c$. So $c = \lim_{n \to \infty} a_n$. Therefore, since $f(a_n) < 0$ for all $n \geq 1$ and f is continuous at c,

$$f(c) = \lim_{n \to \infty} f(a_n) \leq 0.$$

This means that $c \neq b$, and thus $c < b$. Choose any sequence $c < b_n \leq b$ such that $c = \lim_{n \to \infty} b_n$. Since c is the upper bound for A, it follows that $b_n \notin A$ and so $f(b_n) \geq 0$. Consequently,

$$f(c) = \lim_{n \to \infty} f(b_n) \geq 0.$$

Putting the two inequalities together yields $f(c) = 0$.   ∎

5.6.2. COROLLARY. *If f is a continuous real-valued function on $[a, b]$, then $f([a, b])$ is a closed interval.*

PROOF. The Extreme Value Theorem (Theorem 5.4.4) shows that the range of f is bounded, and the extrema are attained. Thus there are points c and d in $[a, b]$ such that

$$f(c) = m := \inf_{x \in [a,b]} f(x) \quad \text{and} \quad f(d) = M := \sup_{x \in [a,b]} f(x).$$

Suppose that $c \leq d$ (the case $c > d$ may be handled similarly by considering the function $-f$). Let y be any value in (m, M). The function $g(x) = f(x) - y$ is continuous on the interval $[c, d]$. Moreover,

$$g(c) = m - y < 0 < M - y = g(d).$$

Thus by the Intermediate Value Theorem, there is a point x in (c, d) such that $g(x) = 0$, and so $f(x) = y$. This is true for every point y in (m, M), as well as the two endpoints. Therefore, $f([a, b]) = [m, M]$.   ∎

A **path** in $S \subset \mathbb{R}^n$ from $\mathbf{a}$ to $\mathbf{b}$, both points in S, is the image of a continuous function γ from $[0, 1]$ into S such that $\gamma(0) = \mathbf{a}$ and $\gamma(1) = \mathbf{b}$.

5.6.3. COROLLARY. *Suppose that $S \subset \mathbb{R}^n$ and f is a continuous real-valued function on S. If there is a path from $\mathbf{a}$ to $\mathbf{b}$ in S and $f(\mathbf{a}) < 0 < f(\mathbf{b})$, then there is a point $\mathbf{c}$ on the path so that $f(\mathbf{c}) = 0$.*

PROOF. Let $\gamma : [0, 1] \to S$ define the path from $\mathbf{a}$ to $\mathbf{b}$. Consider the continuous function $g = f \circ \gamma$. Then $g(0) < 0 < g(1)$. By the Intermediate Value Theorem, there is a point x in $(0, 1)$ so that $g(x) = 0$. Then $\mathbf{c} = \gamma(x)$ is the desired point.   ∎

Exercises for Section 5.6

A. (a) Show that there is some $x \in (0, \pi/2)$ so that $\cos x = x$.

(b) Prove that this is the only real solution.

B. How many solutions are there to $\tan x = x$ in $[0, 11]$?

C. Show that $2 \sin x + 3 \cos x = x$ has three solutions.

D. Show that a polynomial of odd degree has at least one real root.

E. The temperature $T(\mathbf{x})$ at each point $\mathbf{x}$ on the surface of Mars (a sphere) is a continuous function. Show that there is a point $\mathbf{x}$ on the surface such that $T(\mathbf{x}) = T(-\mathbf{x})$.
HINT: Represent the surface of Mars as $\{\mathbf{x} \in \mathbb{R}^3 : \|\mathbf{x}\| = 1\}$. Consider the function $f(\mathbf{x}) = T(\mathbf{x}) - T(-\mathbf{x})$, and compare $f(\mathbf{x})$ and $f(-\mathbf{x})$.

F. Let f be a continuous function from $B = \{\mathbf{x} \in \mathbb{R}^2 : \|\mathbf{x}\| \leq 1\}$, into $\mathbb{R}$, the closed ball in $\mathbb{R}^2$, into $\mathbb{R}$. Show that f cannot be one-to-one.

G. If f is a continuous real-valued function on $(0, 1)$, what are the possibilities for the range of f? Give examples for each possibility, and prove that your list is complete.

H. (a) Show that a continuous function on $(-\infty, +\infty)$ cannot take *every* real value *exactly twice*.

(b) Find a continuous function on $(-\infty, +\infty)$ that takes *every* real value *exactly three times*.

5.7. Monotone Functions

Since most functions we encounter are increasing or decreasing on intervals contained in their domain, it makes sense to spend a short time tying down some of the special properties of functions on such intervals. You have probably seen the following definitions in calculus.

5.7.1. DEFINITION. A function f is called **increasing** on an interval (a, b) if $f(x) \leq f(y)$ whenever $a < x \leq y < b$. It is **strictly increasing** on (a, b) if $f(x) < f(y)$ whenever $a < x \leq y < b$. Similarly, we define **decreasing** and **strictly decreasing** functions. All of these functions are called **monotone**.

Sometimes, **monotone increasing** and **monotone decreasing** are used as synonyms for increasing and decreasing.

5.7.2. PROPOSITION. *If f is an increasing function on the interval (a, b), then the one-sided limits of f exist at each point $c \in (a, b)$, and*

$$\lim_{x \to c^-} f(x) = L \leq f(c) \leq \lim_{x \to c^+} f(x) = M.$$

For decreasing functions, the inequalities are reversed.

PROOF. Consider the case of f increasing. The set $F = \{f(x) : a < x < c\}$ is a nonempty set of real numbers bounded above by $f(c)$. Therefore, $L = \sup\limits_{a<x<c} f(x)$ is defined by the Least Upper Bound Principle (2.5.3), and $L \leq f(c)$. We will show that L is the left limit.

Indeed, let $\varepsilon > 0$. Since $L - \varepsilon$ is not an upper bound for F, there is a point x_0 with $a < x_0 < c$ such that $f(x_0) > L - \varepsilon$. By definition of the supremum, we have $f(x) \leq L$ for all x in (a, c). Hence for all $x_0 < x < c$, we have

$$L - \varepsilon < f(x_0) \leq f(x) \leq L < L + \varepsilon.$$

From the definition of limit, it now follows that $\lim\limits_{x \to c^-} f(x) = L$.

The limit from the right and the decreasing case are handled similarly. ∎

Since monotone functions have limits from each side, this restricts the possible discontinuities is several ways. The first thing to observe is that if the function does not jump at c, it is necessarily continuous there.

5.7.3. COROLLARY. *The only type of discontinuity that a monotone function on an interval can have is a jump discontinuity.*

PROOF. Let $\lim\limits_{x \to c^-} f(x) = L$ and $\lim\limits_{x \to c^+} f(x) = M$. Since $L \leq f(c) \leq M$, the equality $L = M$ implies that

$$\lim\limits_{x \to c} f(x) = L = M = f(c)$$

and thus f is continuous at c. ∎

5.7.4. COROLLARY. *If f is a monotone function on $[a, b]$ and the range of f intersects every nonempty open interval in $[f(a), f(b)]$, then f is continuous.*

PROOF. Suppose that f is increasing and has a jump discontinuity at c. Then (with notation as previously) the range of f is contained in $(-\infty, L] \cup [M, \infty)$ with the exception of at most one point, $f(c)$, in between. Thus either $(L, f(c))$ or $(f(c), M)$ is a nonempty interval in $[f(a), f(b)]$, which is disjoint from the range of f. Consequently, if the range of f meets every open interval in $[f(a), f(b)]$, then f must be continuous. ∎

Here is a stronger conclusion. Recall that a set is countable if it is finite or can be written as a list indexed by $\mathbb{N}$ (see Appendix 2.8).

5.7.5. THEOREM. *A monotone function on $[a, b]$ has at most countably many discontinuities.*

PROOF. Assume that f is increasing, and let $A = f(b) - f(a)$. If f is discontinuous at c, define the *jump* at c to be $J(c) = f(c^+) - f(c^-)$. This is the length of the gap $(f(c^-), f(c^+))$ in the range of f. Since these intervals are disjoint, the sum of all the jumps is at most A.

Let's estimate how many discontinuities with jump at least 2^{-k} are possible. If there are N_k of them, we see that $2^{-k} N_k \leq A$ or $N_k \leq 2^k A$. In particular, N_k is finite. Therefore the points with $J(c) \geq 1$ can be listed as $c_1, \ldots, c_{k_0}$, where $k_0 \leq A$. Then the points with $2^{-1} \leq J(c) < 1$ may be listed as $c_{k_0+1}, \ldots, c_{k_1}$; and $k_1 \leq 2A$. Continuing in this manner, the discontinuities with $J(c) \in [2^{-k}, 2^{1-k})$ may be listed as $c_{k_{n-1}+1}, \ldots, c_{k_n}$ with $k_n \leq 2^n A$. In this way, we construct a sequence containing all of the discontinuities of f. Thus the set of discontinuities is countable. ∎

The relation to inverse functions is important. Recall, from Corollary 1.2.4, that if f maps a set S one-to-one and onto a set T, then there is a unique function, denoted f^{-1}, mapping T back to S such that $f^{-1}(f(s)) = s$ for all $s \in S$ and $f(f^{-1}(t)) = t$ for all $t \in T$.

5.7.6. THEOREM. *Let f be a continuous strictly increasing function on $[a, b]$. Then f maps $[a, b]$ one-to-one and onto $[f(a), f(b)]$. Moreover the inverse function f^{-1} is also continuous and strictly increasing.*

PROOF. Since f is strictly increasing, it is clearly one-to-one. By the Intermediate Value Theorem (Theorem 5.6.1), the range of f contains $[f(a), f(b)]$. Again, since f is monotone, there are no other values in the range. So f maps $[a, b]$ one-to-one and onto $[f(a), f(b)]$.

Let g be the inverse function. Suppose that $f(a) \leq s < t \leq f(b)$, and let $x = g(s)$ and $y = g(t)$. Then we must have $x < y$, for if $x \geq y$ then $f(x) \geq f(y)$, contrary to fact. Hence g is strictly increasing. The range of g is $[a, b]$. Thus by Corollary 5.7.4, g is continuous. ∎

5.7.7. EXAMPLE. Consider the function $\tan x$. This function has period π and so clearly is not monotone. However it is monotone on subintervals of its range. A natural choice is the interval $(-\frac{\pi}{2}, \frac{\pi}{2})$. Note that $\tan x$ is strictly increasing on this interval and that $\lim_{x \to -\pi/2^+} \tan x = -\infty$ and $\lim_{x \to \pi/2^-} \tan x = +\infty$. So $\tan x$ maps $(-\frac{\pi}{2}, \frac{\pi}{2})$ one-to-one and onto $\mathbb{R}$. The inverse function $\tan^{-1}$ is the uniquely defined function that assigns to each real y the value $x \in (-\frac{\pi}{2}, \frac{\pi}{2})$, satisfying $\tan x = y$. Hence $\lim_{y \to +\infty} \tan^{-1}(y) = \frac{\pi}{2}$ and $\lim_{y \to -\infty} \tan^{-1}(y) = -\frac{\pi}{2}$. Figures 5.9 and 5.10 give the graphs of these two functions.

5.7.8. EXAMPLE. We construct a function associated to the Cantor set known as the **Cantor function**. Recall from Example 4.4.8 that the Cantor set is constructed by successively removing the middle thirds from the unit interval and obtaining the

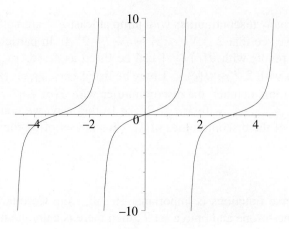

FIGURE 5.9. The graph of $\tan(x)$.

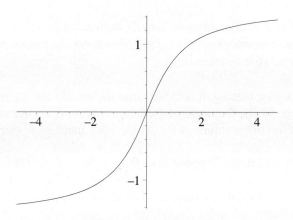

FIGURE 5.10. The graph of $\tan^{-1}(x)$.

set as the intersection of the sets S_k obtained from this procedure:

$$S_1 = [0, 1/3] \cup [2/3, 1],$$
$$S_2 = [0, 1/9] \cup [2/9, 1/3] \cup [2/3, 7/9] \cup [8/9, 1],$$
$$S_3 = [0, 1/27] \cup [2/27, 1/9] \cup [2/9, 7/27] \cup [8/27, 1/3]$$
$$\cup [2/3, 19/27] \cup [20/27, 7/9] \cup [8/9, 25/27] \cup [26/27, 1].$$

We define a function on $[0, 1]$ as follows. Set $f(0) = 0$ and $f(1) = 1$; next set $f(x) = 1/2$ on $[1/3, 2/3]$; then $f(x) = 1/4$ on $[1/9, 2/9]$ and $3/4$ on $[7/9, 8/9]$; and so on. The road to understanding this function is the ternary (base 3) expansion. We may write a number in $[0, 1]$ in base 3 as

$$x = (0.x_1 x_2 x_3 \ldots)_{\text{base 3}} = \sum_{k \geq 1} 3^{-k} x_k,$$

where each x_i belongs to $\{0, 1, 2\}$. The first time a 1 appears as some x_k, the point x_k belongs to one of the $2^k - 1$ closed intervals on which f is assigned a value of the form $m/2^k$.

More precisely, for $x = (0.x_1 \ldots x_{k-1} 1 x_{k+1} \ldots)_{\text{base 3}}$, where $x_i \in \{0, 2\}$ for $1 \leq i < k$, we set $f(x) = (0.\frac{x_1}{2} \ldots \frac{x_{k-1}}{2} 1000 \ldots)_{\text{base 2}}$. For the remaining points in C that have a ternary expansion using only 0s and 2s, we have $f(x) = (0.\frac{x_1}{2} \ldots \frac{x_k}{2} \ldots)_{\text{base 2}}$. The restriction of f to C was considered in Example 4.4.8 in order to show that C was very large. Some numbers have two ternary expansions. Rather than verify that both expansions lead to the same function value (which they do, as the interested reader can verify), we merely choose the expansion which contains a 1. This leads to a well-defined function.

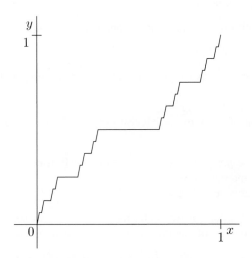

FIGURE 5.11. An approximation to the Cantor function.

It is evident from our construction that f is increasing. Moreover, it is constant on each of the intervals removed from C. So the function has a rather flat appearance. The interesting thing is that f is continuous. To see this, it suffices by Corollary 5.7.4 to show that the range of f has no gaps coming from a jump discontinuity. This follows since every number of the form $m/2^k$ for $0 \leq m \leq 2^k$ belongs to the range of f. Any interval contains many of these points. In particular, there is no gap in the range of f, and thus f is continuous and maps $[0, 1]$ onto itself.

By construction, f is constant on each interval component of $[0, 1] \setminus S_i$. Thus f is differentiable with derivative 0 on each of these intervals. In Example 4.4.8, we showed that the lengths of the intervals removed from $[0, 1]$ to form C summed to 1. So f is continuous on $[0, 1]$ and has derivative 0 at "almost all" points. That is, f has derivative 0 except on a set of measure zero. Nevertheless, it cannot be differentiable everywhere because the Mean Value Theorem would then show it to be constant, which it clearly is not. In fact, it can be shown that f fails to be differentiable at every point of C.

For each component interval of $[0, 1] \setminus C$, the two endpoints lie in C and f takes the same value at both. Thus it follows that $f(C) = [0, 1]$. Recall the discussion of cardinality in Appendix 2.8 and at the end of Example 4.4.8. This function f provides another way to show that the cardinality of C is greater than (and thus the same as) the real line.

Exercises for Section 5.7

A. Let f and g be decreasing functions defined on $\mathbb{R}$.
   (a) Is the composition $f(g(x))$ monotone?
   (b) Is the sum $f(x) + g(x)$ monotone?
   (c) Is the product $f(x)g(x)$ monotone?

B. Show that if f is continuous on $[0, 1]$ and one-to-one, then it is monotone.

C. What is the inverse function of $f(x) = x^2$ on $(-\infty, 0]$?

D. (a) Show that the restriction f of $\cos x$ to $[0, \pi]$ has an inverse function, and graph them both.
   (b) Why do we choose the interval $[0, \pi]$?
   (c) Let g be the restriction of $\cos x$ to the interval $[3\pi, 4\pi]$. What is the relationship between f^{-1} and g^{-1}?

E. Show that the cubic $f(x) = ax^3 + bx^2 + cx + d$, with $a \neq 0$, is one-to-one and thus has an inverse function if and only if $3ac \geq b^2$.
   HINT: Compute the derivative and its discriminant.

F. Verify that the formula for Cantor function in terms of the ternary expansion yields the same answer for both expansions of a point x when two expansions exist.

G. Define f on $S = [0, 1] \cup (2, 3]$ by $f(x) = \begin{cases} x & \text{for } 0 \leq x \leq 1 \\ x - 1 & \text{for } 2 < x \leq 3. \end{cases}$
   (a) Show that f is continuous, strictly increasing on S.
   (b) Show that f maps S one-to-one and onto $[0, 2]$.
   (c) Show that f^{-1} is not continuous.
   (d) Why is this not a contradiction to Theorem 5.7.6?

H. For $x \in [0, 1]$, express it as a decimal $x = x_0.x_1x_2x_3 \ldots$. Use a finite decimal expansion without repeating 9s when there is a choice. Then define a function f by $f(x) = x_0.0x_10x_20x_3 \ldots$.
   (a) Show that f is strictly increasing.
   (b) Compute $\lim_{x \to 1^-} f(x)$.
   (c) Show that $\lim_{x \to a^+} f(x) = f(a)$ for $0 \leq a < 1$.
   (d) Find all discontinuities of f.

CHAPTER 6

Differentiation and Integration

In this chapter, we examine the mathematical foundations of differentiation and integration. The theorems of this chapter are useful not only to make calculus work but also to study functions in many other contexts. We do not spend any time on the important applications that typically appear in courses devoted to calculus, such as optimization problems. Rather we will highlight those aspects that either depend on or apply to results in real analysis.

After developing the basic properties of differentiation, we focus on the most important approximation theorem of differential calculus, the Mean Value Theorem. Integration, although based on the natural idea of area, requires some care to define precisely. We concentrate on the proof of existence of the Riemann integral and on the appropriately named Fundamental Theorem of Calculus. This theorem, which ties differential and integral calculus together, makes the calculation of integrals an algorithmic (if sometimes tricky) process.

As we assume that the reader has already seen calculus, we dive right in with key definitions, assuming that the motivating examples are familiar.

6.1. Differentiable Functions

6.1.1. DEFINITION. A real-valued function $f : (a, b) \to \mathbb{R}$ is **differentiable at a point** $x_0 \in (a, b)$ if

$$\lim_{h \to 0} \frac{f(x_0 + h) - f(x_0)}{h} = \lim_{x_0 \to x} \frac{f(x) - f(x_0)}{x - x_0}$$

exists. In this case, we write $f'(x_0)$ for this limit.

If a function is defined on a closed interval $[a, b]$, then we say it is differentiable at a or b if the appropriate one-sided limit exists. The function f is **differentiable on an interval** $[a, b]$ if it is differentiable at each point x_0 in the interval.

When f is differentiable at x_0, we define the **tangent line** to f at x_0 to be the linear function $T(x) = f(x_0) + f'(x_0)(x - x_0)$. Figure 6.1 shows a typical tangent line.

The phrase *linear function* has two different meanings. In linear algebra, a linear function is one satisfying $f(\alpha\mathbf{x}+\beta\mathbf{y}) = \alpha f(\mathbf{x})+\beta f(\mathbf{y})$; in calculus, a linear function is one whose graph is a line [i.e., a linear function (in the linear algebra sense) plus a constant]. To keep the language unambiguous, sometimes this second family of functions are called *affine functions*, and we do this in studying convexity (Chapter 16). In this chapter, we use linear function in its usual calculus sense.

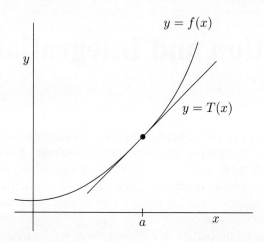

FIGURE 6.1. The tangent line T to f at a.

An immediate consequence of differentiability is continuity.

6.1.2. PROPOSITION. *If f is differentiable at x_0, then it is continuous at x_0. So every differentiable function is continuous.*

PROOF. We compute

$$\lim_{x \to x_0} f(x) = \lim_{x \to x_0} f(x_0) + (x - x_0)\frac{f(x) - f(x_0)}{x - x_0}$$
$$= f(x_0) + 0f'(x_0) = f(x_0). \qquad \blacksquare$$

Notice that when f is differentiable at x_0, the tangent line T passes through the point $(x_0, f(x_0))$ with slope $f'(x_0)$. The point of the derivative is that, locally, the line T has the same slope as the function f. Thus $T(x)$ is the best linear approximation to $f(x)$ for x very close to x_0. Before making this precise, notice that T is the best approximation to f near x using a *first order* polynomial. In Chapter 10, we will look at approximations using higher-order polynomials.

6.1.3. LEMMA. *Let f be a function on $[a, b]$ that is differentiable at x_0. Let $T(x)$ be the tangent line to f at x_0. Then T is the unique linear function with property that*

$$\lim_{x \to x_0} \frac{f(x) - T(x)}{x - x_0} = 0.$$

PROOF. First we compute the limit by substituting $h = x - x_0$,

$$\lim_{x \to x_0} \frac{f(x) - T(x)}{x - x_0} = \lim_{h \to 0} \frac{f(x_0 + h) - (f(x_0) + f'(x_0)h)}{h}$$

$$= \lim_{h \to 0} \frac{f(x_0 + h) - f(x_0)}{h} - f'(x_0) = 0.$$

Now if another linear function $L(x)$ satisfies this limit, then by the continuity of L and f, we have

$$f(x_0) - L(x_0) = \lim_{x \to x_0} f(x) - L(x)$$

$$= \lim_{x \to x_0} (x - x_0) \frac{f(x) - L(x)}{x - x_0} = 0.$$

So $L(x) = f(x_0) + m(x - x_0)$, where m is its slope, and

$$m = L'(x_0) = \lim_{h \to 0} \frac{L(x_0 + h) - L(x_0)}{h}$$

$$= \lim_{h \to 0} \frac{L(x_0 + h) - f(x_0 + h)}{h} + \frac{f(x_0 + h) - f(x_0)}{h}$$

$$= 0 + f'(x_0) = f'(x_0).$$

Thus the line L goes through the point $(x_0, f(x_0))$ and has the same slope as T. Consequently, $L = T$. ∎

An immediate consequence is that the tangent line is a good approximant to f near x_0 in the sense of (2) in Corollary 6.1.4.

6.1.4. COROLLARY. *If $f(x)$ is a function on (a, b) and $x_0 \in (a, b)$, then the following are equivalent:*

(1) *f is differentiable at x_0.*

(2) *There is a linear function $T(x)$ and a function $\varepsilon(x)$ on (a, b) such that $\lim_{x \to x_0} \varepsilon(x) = 0$ and $f(x) = T(x) + \varepsilon(x)(x - x_0)$.*

(3) *There is a function $\varphi(x)$ on (a, b) such that $f(x) = f(x_0) + \varphi(x)(x - x_0)$ and $\lim_{x \to x_0} \varphi(x)$ exists.*

If these hold, then the linear function $T(x)$ in (2) is the tangent line, and the limit $\lim_{x \to x_0} \varphi(x)$ in (3) equals $f'(x_0)$.

PROOF. Clearly, we must define $\varepsilon(x) = \dfrac{f(x) - T(x)}{x - x_0}$ and $\varphi(x) = \dfrac{f(x)}{x - x_0}$ for $x \neq x_0$. Lemma 6.1.3 shows that (1) and (2) are equivalent and that the linear function T is necessarily the tangent line $T(x) = f(x_0) + f'(x_0)(x - x_0)$. Thus if (2) holds, then $\varphi(x) = f'(x_0) + \varepsilon(x)$ satisfies $f(x) = f(x_0) + \varphi(x)(x - x_0)$ and $\lim_{x \to x_0} \varphi(x) = f'(x_0)$. So (3) holds and limit equals $f'(x_0)$. Conversely, if (3) holds, then let $L = \lim\limits_{x \to x_0} \varphi(x)$. It is easy to check that $T(x) = f(x_0) + L(x - x_0)$ and $\varepsilon(x) = \varphi(x) - L$ satisfy (2). ∎

6.1.5. REMARK. It is easy to describe continuous functions that are not differentiable at every point. The prototypical example is $f(x) = |x|$. This function is differentiable at every point except $x = 0$. Here it has left and right derivatives -1 and 1, respectively. As this function comes to a point at the origin, it is intuitively clear that no straight line is a good approximant near $x = 0$.

A more subtle example is $g(x) = \sqrt[3]{x}$, graphed in Figure 6.2 This has derivative $g'(x) = \frac{1}{3}x^{-2/3}$ for all $x \neq 0$. But at $x = 0$, we have

$$\lim_{h \to 0} \frac{g(h) - g(0)}{h} = \lim_{h \to 0} h^{-2/3} = +\infty.$$

So this function is not differentiable because it has a vertical tangent at the origin.

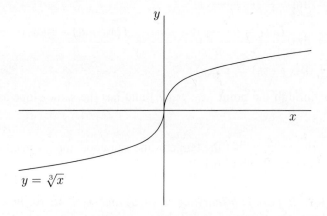

$y = \sqrt[3]{x}$

FIGURE 6.2. The graph of $\sqrt[3]{x}$.

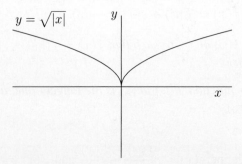

$y = \sqrt{|x|}$

FIGURE 6.3. The graph of $\sqrt{|x|}$.

A related example is the function $h(x) = \sqrt{|x|}$, graphed in Figure 6.3. This function has a cusp, with right derivative $+\infty$ and left derivative $-\infty$.

Looking only at these examples, it is intuitively plausible that a continuous function has to be differentiable at "most" points, but this is not true. These functions do not give the full extent of nasty behaviour. There are continuous functions that are not differentiable at any point! We will construct such a function in Example 8.4.9. Unsurprisingly, this function is not given by a simple formula, but rather is expressed as a convergent infinite series of functions. In spite of the difficulties in writing down such functions, there are actually many of them, in a certain sense; see Proposition 9.3.3.

The linearity of differentiation [i.e., that $(\alpha f + \beta g)' = \alpha f' + \beta g'$ for functions f and g and scalars α and β] follows from the linearity of limits. The product and quotient rules require more care; they are left as exercises with hints. However, the chain rule is yet more subtle.

6.1.6. THE CHAIN RULE.
Suppose that f is defined on $[a, b]$ and has range contained in $[c, d]$. Let g be defined on $[c, d]$. Suppose that f is differentiable at $x_0 \in [a, b]$ and g is differentiable at $f(x_0)$. Then the composition $h(x) = g(f(x))$ is defined, and

$$h'(x_0) = g'(f(x_0))f'(x_0).$$

PROOF. Using part (3) of Lemma 6.1.4, write $f(x) = f(x_0) + \varphi(x)(x - x_0)$, where $\lim_{x \to x_0} \varphi(x) = f'(x_0)$. Similarly, $g(y) = g(f(x_0)) + \psi(y)(y - f(x_0))$, where $\lim_{y \to f(x_0)} \psi(y) = g'(f(x_0))$. Then

$$h(x) = g(f(x)) = g(f(x_0)) + \psi(f(x))(f(x) - f(x_0))$$
$$= g(f(x_0)) + \psi(f(x))\varphi(x)(x - x_0)$$

and by Lemma 6.1.4,

$$\lim_{x \to x_0} \psi(f(x))\varphi(x) = g'(f(x_0))f'(x_0).$$

By Lemma 6.1.4 again, h is differentiable at x_0 and $h'(x_0) = g'(f(x_0))f'(x_0)$. ∎

6.1.7. EXAMPLE.
There are functions that are differentiable but do not have a continuous derivative. Let $a > 0$ and consider $f(x) = x^a \sin(1/x)$ for $x > 0$ and set $f(0) = 0$. This is evidently continuous at every point except possibly 0. However, since $|\sin(1/x)| \le 1$ and $\lim_{x \to 0^+} x^a = 0$, the Squeeze Theorem shows that f is also continuous at $x = 0$.

Consider the derivative of f. For $x > 0$, we invoke the product and chain rules to compute

$$f'(x) = ax^{a-1} \sin(1/x) - x^{a-2} \cos(1/x).$$

The point $x = 0$ must be handled separately using the definition of derivative.

$$f'(0) = \lim_{h \to 0^+} \frac{h^a \sin(1/h)}{h} = \lim_{h \to 0^+} h^{a-1} \sin(1/h)$$

If $a > 1$, the Squeeze Theorem yields $f'(0) = 0$; while for $a \leq 1$, the values of $h^{a-1}\sin(1/h)$ oscillate wildly between large positive and negative values, and the limit does not exist.

So for $0 < a \leq 1$, f is continuous but not differentiable at $x = 0$. A key point of interest is the values $1 < a \leq 2$. For these values, $f(x)$ is differentiable at every point but the derivative $f'(x)$ is not continuous at $x = 0$. Thus a function may be differentiable but not C^1.

Exercises for Section 6.1

A. Prove that $(\alpha f + \beta g)'(x_0) = \alpha f'(x_0) + \beta g'(x_0)$ for all functions f and g on $[a, b]$ that are differentiable at x_0.

B. If f is differentiable at x_0 and $a \in \mathbb{R}$, show that $\lim_{h\to 0} \dfrac{f(x_0 + ah) - f(x_0)}{h} = af'(x_0)$.

C. Explain what goes wrong with the proof of Proposition 6.1.2 if f is not differentiable.

D. Let f and g be differentiable functions on (a, b). Suppose there is a point x_0 in (a, b) with $f(x_0) = g(x_0)$ and $f(x) \leq g(x)$ for $a < x < b$. Prove that $f'(x_0) = g'(x_0)$.

E. Show that the derivative of an even function is odd, and the derivative of an odd function is even. Recall that a function f is **even** if $f(-x) = f(x)$ and is **odd** if $f(-x) = -f(x)$.

F. Prove the product rule: $(fg)'(x_0) = f'(x_0)g(x_0) + f(x_0)g'(x_0)$ for all functions f and g on $[a, b]$ that are differentiable at x_0.
HINT: Express $f(x_0 + h)g(x_0 + h) - f(x_0)g(x_0)$ as
$$\big(f(x_0 + h) - f(x_0)\big)g(x_0 + h) + f(x_0)\big(g(x_0 + h) - g(x_0)\big).$$

G. For each positive integer n, give an example of a function which is C^n but not C^{n+1}.
HINT: Look at Example 6.1.7.

H. Prove the quotient rule: $(f/g)'(x_0) = \dfrac{f'(x_0)g(x_0) - f(x_0)g'(x_0)}{g(x_0)^2}$ for all functions f and g on $[a, b]$ that are differentiable at x_0 and $g(x_0) \neq 0$.
HINT: Let $h = f/g$ and use the product rule on $f = gh$. Solve for $h'(x_0)$.

I. Suppose that f is a continuous, one-to-one function from the interval $[a, b]$ onto $[c, d]$. By Theorem 5.7.6, the inverse function $f^{-1}(x)$ is also continuous. Prove that if $f(x_0) = y_0$ and $f'(x_0) \neq 0$, then $(f^{-1})'(y_0) = \dfrac{1}{f'(x_0)} = \dfrac{1}{f'(f^{-1}(y_0))}$.
HINT: Use Theorem 6.1.4(3) to establish that $f^{-1}(y) - f^{-1}(y_0) = \dfrac{1}{\varphi(f^{-1}(y))}(y - y_0)$.

J. We say f is **left-differentiable** at x_0 if $\lim_{h\to 0^+} [f(x_0 + h) - f(x_0)]/h$ exists. In this case, the value is denoted $f'(x_0+)$.
(a) Define **right-differentiable** at x_0 and assign a meaning to $f'(x_0-)$.
(b) Show that f is differentiable at x_0 if and only if it is both left-differentiable and right-differentiable at x_0 and $f'(x_0+) = f'(x_0-)$.

K. Find left and right derivatives of $f(x) = \sqrt{1 - \sin x}$ at every point. Where does f fail to be differentiable?

L. Suppose $f : [a, b] \to \mathbb{R}$ is differentiable on (a, b) and continuous on $[a, b]$. Does it follow that f is left-differentiable at a and right-differentiable at b?

M. Some calculus text books write "$(f^{-1})' = \dfrac{1}{f'}$". Find several things wrong with this statement. Which error is the most egregious?

N. The function $\sin x$ is 2π-periodic and consequently is definitely not one-to-one.
   (a) How do we define the inverse function $\arcsin y$?
   (b) How does the choice you make in part (a) affect the graph of $\arcsin y$? What is the effect on the derivative?

O. If f is periodic with period T, show that f' is also T-periodic.

P. Recall that we said a function $f(x)$ is asymptotic to a curve $c(x) = ax + b$ as x tends to $+\infty$ if $\lim\limits_{x \to +\infty} |f(x) - c(x)| = 0$.
   (a) Show that if $f(x)$ is asymptotic to a line $L(x) = ax + b$ as x tends to $+\infty$, then
   $$a = \lim_{x \to +\infty} \frac{f(x)}{x} \text{ and } b = \lim_{x \to +\infty} f(x) - ax. \text{ (As usual, this includes showing that}$$
   the limits exist.)
   (b) Find all of the asymptotes (including horizontal and vertical ones) for
   $$f(x) = \frac{(x - 2)^3}{(x + 1)^2}. \text{ Sketch the graph.}$$

Q. Sketch the curve $f(x) = xe^{-\frac{5}{x} - \frac{2}{x^2}}$ for $x \neq 0$. Pay attention to the following:
   (a) asymptotic behaviour at $\pm\infty$
   (b) find all critical points
   (c) limits of f and f' at 0^+
   (d) points of inflection

R. Sketch the curve $f(x) = \dfrac{(\log x)^2 + 4\log x}{(1 + \log x)^2}$ for $x > 0$. Pay attention to the following:
   (a) points where f is zero or undefined
   (b) limits of f and f' at 0^+
   (c) all asymptotes and local extrema
   (d) Indicate all points of inflection on the graph, but do not compute the second derivative to compute them exactly

S. Suppose that f is continuous on (a, b).
   (a) Show that f is differentiable at $x_0 \in (a, b)$ if and only if $\lim\limits_{\substack{x \to x_0^+ \\ y \to x_0^-}} \dfrac{f(x) - f(y)}{x - y}$ exists.
   (b) Why is part (a) false without the continuity hypothesis?

T. Establish the Leibniz formula that the nth derivative of a product $f(x)g(x)$ is given by
   $$\sum_{k=0}^{n} \binom{n}{k} f^{(k)}(x) g^{(n-k)}(x). \quad \text{HINT: Use induction.}$$

U. (a) Suppose that g is continuous at $x = 0$. Prove that $f(x) = xg(x)$ is differentiable at $x = 0$.
   (b) Conversely, suppose that $f(0) = 0$ and f is differentiable at $x = 0$. Prove that there is a function g that is continuous at $x = 0$ and satisfies $f(x) = xg(x)$.

6.2. The Mean Value Theorem

The Mean Value Theorem is the fundamental approximation result of differential calculus. It is endlessly applied and, indeed, is the most useful analytic tool that calculus provides. However, the proof depends on the Extreme Value Theorem, putting a valid proof beyond many calculus courses. We devote this section to proving the Mean Value Theorem.

The starting point is the following basic result on which all of calculus relies—the location of possible extrema. Simply stated, it says that extrema of continuous functions occur at the endpoints or at critical points, where the derivative is either undefined or equal to 0. Sometimes this result is credited to Fermat.

6.2.1. FERMAT'S THEOREM.
Let f be a continuous function on an interval $[a, b]$ that takes its maximum or minimum value at a point x_0. Then

(1) *x_0 is an endpoint a or b, or*

(2) *f is not differentiable at x_0, or*

(3) *f is differentiable at x_0 and $f'(x_0) = 0$.*

PROOF. Suppose that the first two options do not apply. Then x_0 is an interior point at which f is differentiable. For convenience, assume that x_0 is a maximum. Since $f(x_0 + h) - f(x_0) \leq 0$, the limit from the right yields

$$f'(x_0) = \lim_{h \to 0^+} \frac{f(x_0 + h) - f(x_0)}{h} \leq 0$$

and the limit from the left yields

$$f'(x_0) = \lim_{h \to 0^-} \frac{f(x_0 + h) - f(x_0)}{h} \geq 0.$$

Hence $f'(x_0) = 0$. ∎

Next, we prove a special case of Mean Value Theorem, traditionally called Rolle's Theorem, which we can then use to deduce the general case.

6.2.2. ROLLE'S THEOREM.
Suppose that f is a function that is continuous on $[a, b]$ and differentiable on (a, b) such that $f(a) = f(b) = 0$. Then there is a point $c \in (a, b)$ such that $f'(c) = 0$.

PROOF. If the maximum value and minimum value of f are both 0, then f is constant and $f'(c) = 0$ for every $c \in (a, b)$. Otherwise, either the maximum is greater than 0, or the minimum is less. For convenience, assume the former.

By the Extreme Value Theorem (Theorem 5.4.4), there is a point c at which f attains its maximum value. Evidently, c is an interior point. So by Fermat's Theorem, we have $f'(c) = 0$. ∎

6.2.3. EXAMPLE. Differentiability is necessary at every interior point for this theorem to be valid. For example, the continuous function on $[-1, 1]$ given by $f(x) = 1 - |x|$ does not have any point at which $f'(x) = 0$. Here f is differentiable at every point except the point $x = 0$ at which the maximum occurs.

Now a rescaling trick yields the general result. Actually, this idea can used to solve other problems as well, so perhaps we should call it a method.

6.2.4. MEAN VALUE THEOREM.
Suppose that f is a function that is continuous on $[a, b]$ and differentiable on (a, b). Then there is a point $c \in (a, b)$ such that

$$f'(c) = \frac{f(b) - f(a)}{b - a}.$$

Before proving the theorem, it is worth pointing out its significance. The key point is that we have a connection between the derivative of the function and its average slope. It is this connection that allows us to deduce properties of f from properties of its derivative. Figure 6.4 shows the theorem graphically.

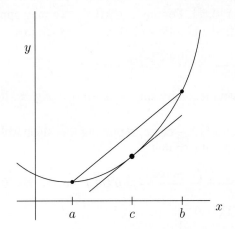

FIGURE 6.4. Graph for the Mean Value Theorem.

PROOF. Let $L(x)$ be the linear function through $(a, f(a))$ and $(b, f(b))$, namely

$$L(x) = f(a) + \frac{f(b) - f(a)}{b - a}(x - a).$$

Consider the function $g(x) = f(x) - L(x)$. Then g is continuous on $[a, b]$ and differentiable on $[a, b]$ since f and L are. Moreover, $g(a) = g(b) = 0$. So by Rolle's Theorem, there is a point $c \in (a, b)$ at which

$$0 = g'(c) = f'(c) - \frac{f(b) - f(a)}{b - a}.$$

This is the desired point. ∎

The first important consequence is used repeatedly in calculus arguments.

6.2.5. COROLLARY. *Let f be a differentiable function on* $[a, b]$.
(1) *If* $f'(x)$ *is (strictly) positive, then* f *is (strictly) increasing.*
(2) *If* $f'(x)$ *is (strictly) negative, then* f *is (strictly) decreasing.*
(3) *If* $f'(x) = 0$ *at every* $x \in (a, b)$, *then* f *is constant.*

PROOF. For any $a \leq x < y \leq b$, apply the Mean Value Theorem on $[x, y]$ to obtain a point c in between so that

$$f(y) - f(x) = f'(c)(y - x) \geq 0.$$

So f is increasing. If $f' > 0$, then the same argument yields a strict inequality. Likewise, the decreasing case follows. A function that is both increasing and decreasing at the same time is constant. ∎

6.2.6. EXAMPLE. We show how the Mean Value Theorem may be used to obtain useful approximations.

Consider $f(x) = \sin x$ on $[0, \frac{\pi}{2}]$. For any $x \in (0, \frac{\pi}{2}]$, we may apply the Mean Value Theorem on $[0, x]$ and find a point c with $0 < c < x$ such that

$$\frac{f(x) - f(0)}{x - 0} = \frac{\sin x}{x} = f'(c) = \cos c < 1.$$

Thus we obtain the well-known inequality $\sin x < x$ for $0 < x \leq \frac{\pi}{2}$. It is evidently valid for $x > \frac{\pi}{2}$ as well.

Now consider $g(x) = 1 - x^2/2 - \cos x$. Applying the Mean Value Theorem again, we obtain a (different) point c so that

$$\frac{g(x) - g(0)}{x - 0} = \frac{1 - x^2/2 - \cos x}{x} = g'(c) = \sin c - c < 0.$$

So

$$\cos x > 1 - \frac{x^2}{2} \quad \text{for} \quad 0 < x \leq \frac{\pi}{2}.$$

Then consider the function $h(x) = \sin x - x + x^3/6$. Once again, apply the Mean Value Theorem on $[0, x]$ with $0 < x \leq \pi/2$:

$$\frac{\sin x - x + x^3/6}{x} = h'(c) = \cos c - 1 + \frac{c^2}{2} > 0.$$

Hence

$$x - \frac{x^3}{6} < \sin x < x \qquad \text{on} \ (0, \pi/2].$$

A fourth application using $k(x) = 1 - x^2/2 + x^4/24 - \cos x$ yields

$$\frac{1 - x^2/2 + x^4/24 - \cos x}{x} = k'(c) = \sin c - (c - c^3/6) > 0.$$

Thus

$$1 - \frac{x^2}{2} < \cos x < 1 - \frac{x^2}{2} + \frac{x^4}{24} \qquad \text{on } (0, \pi/2].$$

These approximations are reasonably good when x is near zero. They are not much use for x away from zero. If you were interested in approximating $\cos x$ near $x = \pi/4$, you would begin with $a = \pi/4$ rather than 0.

Many functions that occur in practice can be differentiated several times. We define the higher-order derivatives recursively by $f^{(n+1)}(x) = \left(f^{(n)}\right)'(x)$, assuming that $f^{(n)}$ turns out to be differentiable.

It stands to reason that information about higher order derivatives should add to our information about the behaviour of the function f. The simplest example of the usefulness of higher derivatives is the sign of the second derivative. If $f''(x)$ is positive on an interval $[a, b]$, then the derivative $f'(x)$ is increasing over this range. Thus the graph of f is seen to curve upward [even if $f'(x) < 0$] and f is said to be **convex** or **concave up**. See Exercise 6.2.J. Likewise, if $f''(x)$ is negative over an interval, then the derivative of f is decreasing and the graph of f curves downward. Functions with such graphs are called **concave** or **concave down**. For this reason, the points at which $f''(x)$ changes sign are called **inflection points** to indicate that the curvature of the graph changes direction. These points are easily identified by eye; see Figure 6.5. Similar changes in higher derivatives are not so easily recognized, as they have a more subtle effect.

Notice that the ideas of convex and concave also make sense for functions that are not differentiable. We study such nondifferentiable convex functions in Chapter 16.

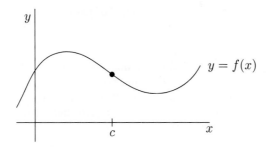

FIGURE 6.5. An inflection point: f'' changes sign at c.

Exercises for Section 6.2

A. If f and g are differentiable on $[a, b]$ and $f'(x) = g'(x)$ for all $a < x < b$, show that $g(x) = f(x) + C$ for some constant C.

B. If f is strictly increasing on $[a, b]$, is $f'(x) > 0$ for all $x \in (a, b)$?

C. Suppose that f is C^3 on (a, b), and f has four zeros in (a, b). Show that $f^{(3)}$ has a zero.

D. (a) Show that $\tan x > x + \dfrac{x^3}{3} + \dfrac{2x^5}{15}$ for $0 < x < \dfrac{\pi}{2}$.

(b) Show that $\tan x < x + \dfrac{x^3}{3} + \dfrac{2x^5}{5}$ for $0 < x < 1$.

E. Suppose that f is continuous on an interval $[a, b]$ and is differentiable at all points of (a, b) except possibly at a single point $x_0 \in (a, b)$. If $\lim\limits_{x \to x_0} f'(x)$ exists, show that $f'(x_0)$ exists and $f'(x_0) = \lim\limits_{x \to x_0} f'(x)$.

HINT: Consider the intervals $[x_0, x_0 + h]$ and $[x_0 - h, x_0]$.

F. (a) Show that the error between a differentiable function $f(x)$ on $[a, b]$ and the tangent line $T(x)$ to f at a may be estimated by $|f(x) - T(x)| \le C|x - a|$, where $C = \sup_{a \le y \le x} |f'(y) - f'(a)|$.

(b) Hence if f is C^2, refine this error estimate to $|f(x) - T(x)| \le D|x - a|^2$, where $D = \sup_{a \le y \le x} |f''(y)|$.

G. (a) Let $f(x) = x^2 \sin(1/x)$ for $x \ne 0$ and $f(0) = 0$ as in Example 6.1.7. Show that 0 is a critical point of f that is not a local maximum nor a local minimum nor an inflection point.

(b) Let $g(x) = 2x^2 + f(x)$. Show that g does have a global minimum at 0, but $g'(x)$ changes sign infinitely often on both $(0, \varepsilon)$ and $(-\varepsilon, 0)$ for any $\varepsilon > 0$.

(c) Let $h(x) = x + 2f(x)$. Show that $h'(0) > 0$ but h is not monotone increasing on any interval including 0.

H. (a) Suppose that g is C^1 on $[a, b]$. Prove that for every $\epsilon > 0$, there is $\delta > 0$ so that
$$\left| g'(c) - \frac{g(d) - g(c)}{d - c} \right| < \varepsilon \text{ for all points } c, d \in [a, b] \text{ with } 0 < |d - c| < \delta.$$
HINT: Use the Mean Value Theorem (MVT) and the fact that g' is uniformly continuous.

(b) Use this to give a second proof that the function f of Example 6.1.7 is not C^1 on $[0, 1]$.

I. Suppose that f is differentiable on $[a, b]$ and $f'(a) < 0 < f'(b)$.

(a) Show that there are points $a < c < d < b$ such that $f(c) < f(a)$ and $f(d) < f(b)$.

(b) Show that the minimum on $[a, b]$ occurs at an interior point.

(c) Hence show that there is a point x_0 in (a, b) such that $f'(x_0) = 0$.

(d) Prove **Darboux's Theorem**: If f is differentiable on $[a, b]$ and $f'(a) < L < f'(b)$, then there is a point x_0 in (a, b) at which $f'(x_0) = L$.

J. A function f is **convex** on an interval $[a, b]$ if $f(tx + (1 - t)y) \le tf(x) + (1 - t)f(y)$ for all $x, y \in [a, b]$ and all $t \in [0, 1]$. In other words, the graph of f is below the line segment joining $(x, f(x))$ and and $(y, f(y))$ for all $x, y \in [a, b]$.

(a) If f is differentiable on $[a, b]$ and f' is increasing, then f is convex on $[a, b]$.
HINT: If $x < y$ and $z = tx + (1 - t)y$, apply the MVT to $[x, z]$ and $[z, y]$.

(b) If f is C^2 on an interval $[a, b]$ and for some $x_0 \in [a, b]$, $f''(x_0) > 0$, then f is convex in some interval about x_0.

(c) If f is C^2 on an interval $[a, b]$ and $f''(x) \ge 0$ for all $x \in (a, b)$, show that f is convex on $[a, b]$.

K. Suppose that f is differentiable on $[0, \infty)$ and f' is strictly increasing.

(a) Show that $f'(x)$ is continuous.

(b) Suppose that $f(0) = 0$, and let $g(0) = f'(0)$ and $g(x) = f(x)/x$ for $x > 0$. Show that g is continuous and strictly increasing.

L. Suppose that f is a continuous function on $\mathbb{R}$ such that $\displaystyle\lim_{h \to 0} \frac{f(x+h) - f(x-h)}{h} = 0$
for every $x \in \mathbb{R}$. Prove that f is constant.
HINT: Fix $\varepsilon > 0$. For each x, find a $\delta > 0$ so that $|f(x+h) - f(x-h)| \le \varepsilon h$ for $0 \le h \le \delta$. Let Δ be the supremum of all such δ. Show that $\Delta = \infty$.

M. Find a discontinuous function f on $\mathbb{R}$ such that $\displaystyle\lim_{h \to 0} \frac{f(x+h) - f(x-h)}{h} = 0$ for every $x \in \mathbb{R}$.

N. Suppose that f is C^1 on $[0, \infty)$, and let $C = \{x : f'(x) = 0\}$.
(a) If C is bounded, show that $\displaystyle\lim_{x \to \infty} f(x)$ exists or is $\pm\infty$.
(b) If C is unbounded and $\displaystyle\lim_{x \in C,\, x \to \infty} f(x) = L$, prove that $\displaystyle\lim_{x \to \infty} f(x) = L$.
HINT: Compare $f(x)$ to the value of f at the nearest critical points on either side.

O. Suppose that f is C^1 on $[0, \infty)$, and $\displaystyle\lim_{x \to \infty} f(x) + f'(x) = 0$. Prove that $\displaystyle\lim_{x \to \infty} f(x) = 0$.
HINT: Use the previous exercise.

6.3. Riemann Integration

We turn now to integration. A crucial point of this section is that the word *derivative* never appears—well, only twice. An integral is *defined* as a limit related to area, not as an antiderivative. In the next section, we establish the Fundamental Theorem of Calculus, which shows that integration and differentiation are, in some sense, inverse operations. This theorem *sometimes* allows us to replace the complicated limit calculation with a simpler calculation using differential calculus. However, we cannot prove this theorem until we first define clearly what an integral is. This necessarily involves a limiting process.

Besides carefully defining integrals, we will establish that two classes of functions are always integrable: continuous functions and monotone functions. This is sufficient for many applications. There are more powerful kinds of integrals that, at the price of much more technical machinery, can integrate more functions and satisfy very strong limit theorems. We describe one approach to such "better, stronger, faster" integrals in Section 9.6. But for the applications we study in this book, the Riemann integral will do all we need.

The goal is to define the Riemann integral of a function f defined on an interval $[a, b]$. The idea is simple and geometric. Chop the interval up into a partition consisting of a number of smaller subintervals. Approximate f as well as possible above and below by functions that are constant on each subinterval. This approximates the region bounded by f above and below by the union of a number of rectangles. We know the areas of these upper and lower approximations, and they are called the upper and lower sums for this partition. For a 'reasonable' function, we will show that as the partition is made finer and finer, these upper and lower estimates will converge to a common value. This limit, when it exists, will be called the integral of f from a to b.

Now we make precise what we just described.

6.3.1. DEFINITION. Let f be a *bounded* function defined on an interval $[a, b]$. A **partition** of $[a, b]$ is a finite set $P = \{a = x_0 < x_1 < \ldots < x_{n-1} < x_n = b\}$. Set $\Delta_j = x_j - x_{j-1}$ and define the **mesh** of a partition P as $\text{mesh}(P) = \max_{1 \le j \le n} \Delta_j$. For each interval $[x_{j-1}, x_j]$ of this partition, we define the maximum and minimum of f on this interval by

$$M_j(f, P) = \sup_{x_{j-1} \le x \le x_j} f(x) \quad \text{and} \quad m_j(f, P) = \inf_{x_{j-1} \le x \le x_j} f(x).$$

Then define the **upper and lower sums** of f with respect to the partition P by

$$U(f, P) = \sum_{j=1}^{n} M_j(f, P)\Delta_j \quad \text{and} \quad L(f, P) = \sum_{j=1}^{n} m_j(f, P)\Delta_j.$$

If, in addition, we are given a set of points $X = \{x_j' : 1 \le j \le n\}$, where $x_j' \in [x_{j-1}, x_j]$ for $1 \le j \le n$, we define the **Riemann sum**

$$I(f, P, X) = \sum_{j=1}^{n} f(x_j')\Delta_j.$$

A partition R is a **refinement** of a partition P provided that $P \subset R$. If P and Q are two partitions, then R is a **common refinement** of P and Q provided that $P \cup Q \subset R$.

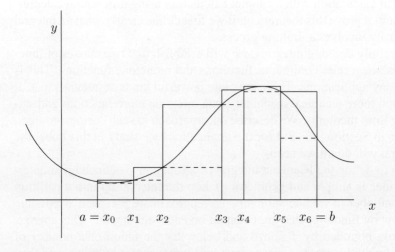

FIGURE 6.6. Example of upper and lower sums

Figure 6.6 illustrates upper and lower sums. We always have

$$L(f, P) \leq I(f, P, X) \leq U(f, P).$$

If R refines P, then each interval I of R is contained in an interval J of P. Now it is evident that if I is a subinterval of J, then the maximum over I is smaller than the corresponding max over J, and the minimum over I is larger than the corresponding min over J. This is only depth to the following estimate.

6.3.2. LEMMA. *If R is a refinement of P, then*

$$L(f, P) \leq L(f, R) \leq U(f, R) \leq U(f, P).$$

PROOF. Consider an interval $[x_{j-1}, x_j]$ of P, and note that this may be subdivided by R into $x_{j-1} = t_k < \cdots < t_l = x_j$. As noted previously, we have the estimate for $k + 1 \leq i \leq l$,

$$m_j(f, P) = \inf_{x_{j-1} \leq x \leq x_j} f(x) \leq \inf_{t_{i-1} \leq t \leq t_i} f(t) = m_i(f, R)$$

and

$$M_i(f, R) = \sup_{t_{i-1} \leq t \leq t_i} f(t) \leq \sup_{x_{j-1} \leq x \leq x_j} f(x) = M_j(f, P).$$

Thus

$$m_j(f, P)(x_j - x_{j-1}) \leq \sum_{i=k+1}^{l} m_i(f, R)(t_i - t_{i-1})$$

and

$$\sum_{i=k+1}^{l} m_i(f, R)(t_i - t_{i-1}) \leq M_j(f, P)(x_j - x_{j-1}).$$

Summing over all the intervals of P now yields

$$L(f, P) \leq L(f, R) \leq U(f, R) \leq U(f, P). \qquad \blacksquare$$

We easily obtain another useful inequality from this:

6.3.3. COROLLARY. *If P and Q are any two partitions of $[a, b]$,*

$$L(f, P) \leq U(f, Q).$$

PROOF. Let R be the common refinement of P and Q. Then by the previous lemma, $L(f, P) \leq L(f, R) \leq U(f, R) \leq U(f, Q)$. $\qquad \blacksquare$

In particular, we see that the set of numbers $\{L(f, P)\}$ is bounded above by any $U(f, Q)$. Hence by the completeness of $\mathbb{R}$, $\sup_P L(f, P)$ is defined. Moreover, $\sup_P L(f, P) \leq U(f, Q)$ for every partition Q. Therefore, $\inf_P U(f, P)$ is defined and $\sup_P L(f, P) \leq \inf_P U(f, P)$.

6.3.4. DEFINITION. Define $L(f) = \sup_P L(f, P)$ and $U(f) = \inf_P U(f, P)$. As pointed out previously, $L(f) \leq U(f)$. A bounded function f on a finite interval $[a, b]$ is called **Riemann integrable** if $L(f) = U(f)$. In this case, we write $\int_a^b f(x)\,dx$ for the common value.

We establish Riemann's Condition for integrability, which follows easily from our definition.

6.3.5. RIEMANN'S CONDITION.
Let $f(x)$ be a bounded function on $[a, b]$. The following are equivalent:

  (1) *f is Riemann integrable.*

  (2) *For each $\varepsilon > 0$, there is a partition P so that $U(f, P) - L(f, P) < \varepsilon$.*

PROOF. Assume that (2) holds. Then for any $\varepsilon > 0$, using the given P we have
$$L(f, P) \leq L(f) \leq U(f) \leq U(f, P).$$
Hence $0 \leq U(f) - L(f) < U(f, P) - L(f, P) < \varepsilon$. So $U(f) = L(f)$ and (1) holds.

If f is Riemann integrable, let $L = L(f) = U(f)$. Let $\varepsilon > 0$. We can find two partitions P_1 and P_2 so that $U(f, P_1) < L + \varepsilon/2$ and $L(f, P_2) > L - \varepsilon/2$. Let P be their common refinement, $P_1 \cup P_2$. By Lemma 6.3.2,
$$L - \frac{\varepsilon}{2} < L(f, P) \leq U(f, P) < L + \frac{\varepsilon}{2}$$
and so $U(f, P) - L(f, P) \leq \varepsilon$, proving (2).   ∎

6.3.6. COROLLARY. *Let f be a bounded real-valued function $[a, b]$. If there is a sequence of partitions of $[a, b]$, P_n, so that*
$$\lim_{n \to 0} U(f, P_n) - L(f, P_n) = 0,$$
then f is Riemann integrable. Moreover, if X_n is any choice of points $x'_{n,j}$ selected from each interval of P_n, then
$$\lim_{n \to \infty} I(f, P_n, X_n) = \int_a^b f(x)\,dx.$$

PROOF. Riemann's condition is verified by hypothesis. Hence by Theorem 6.3.5, f is Riemann integrable. Therefore, both $I(f, P_n, X_n)$ and $\int_a^b f(x)\,dx$ lie in the intervals $[L(f, P_n), U(f, P_n)]$. Consequently,
$$\left| I(f, P_n, X_n) - \int_a^b f(x)\,dx \right| \leq U(f, P_n) - L(f, P_n).$$

The right-hand side is less than any given $\varepsilon > 0$ for n sufficiently large, and there-fore $\lim_{n\to\infty} I(f, P_n, X_n) = \int_a^b f(x)\, dx.$ ∎

A typical choice is the evenly spaced partition

$$P = \{x_j = a + j(b-a)/n : 0 \le j \le n\}.$$

However, it is sometimes convenient to choose a partition better suited to the function f such as the next example, where constant ratios are used rather than constant widths.

6.3.7. EXAMPLE. Consider the function $f(x) = x^p$ on $[a, b]$, where $p \ne -1$ and $0 < a < b$. Take the partition $P_n = \{a = x_0 < x_1 < \ldots < x_n = b\}$, where $x_j = a\left(\frac{b}{a}\right)^{j/n}$ for $0 \le j \le n$.

To keep the notation under control, let $R = (b/a)^{1/n}$. For example, $x_j = aR^j$ and $\Delta_j = x_j - x_{j-1} = aR^{j-1}(R-1)$. Since f is monotone increasing, we easily compute

$$m_j(f, P_n) = \inf_{x_{j-1} \le x \le x_j} f(x) = x_{j-1}^p = a^p R^{p(j-1)}$$

and

$$M_j(f, P_n) = \sup_{x_{j-1} \le x \le x_j} f(x) = x_j^p = a^p R^{pj} = R^p m_j(f, P_n).$$

So $U(f, P_n) = R^p L(f, P_n)$, and

$$L(f, P_n) = \sum_{j=1}^n m_j(f, P_n)\Delta_j$$

$$= \sum_{j=1}^n a^p R^{p(j-1)} a R^{j-1}(R-1)$$

$$= a^{p+1}(R-1)\sum_{j=0}^{n-1} R^{(p+1)j}.$$

Summing the geometric series and rearranging, we have

$$L(f, P_n) = a^{p+1}(R-1)\frac{R^{n(p+1)} - 1}{R^{p+1} - 1}$$

$$= a^{p+1}(R-1)\frac{\left(\frac{b}{a}\right)^{p+1} - 1}{R^{p+1} - 1}$$

$$= (b^{p+1} - a^{p+1})\frac{R-1}{R^{p+1} - 1}.$$

We will take a limit as $n \to +\infty$. To show the role of n clearly, we set $r = \frac{b}{a}$ and $h = 1/n$, so that $R = r^{1/n} = r^h$. The key is to recognize the limit as the computation of two derivatives.

$$\lim_{n\to\infty} L(f, P_n) = (b^{p+1} - a^{p+1}) \lim_{n\to\infty} \frac{r^{1/n} - 1}{r^{(p+1)/n} - 1}$$

$$= (b^{p+1} - a^{p+1}) \lim_{h\to0} \frac{r^h - 1}{h} \frac{h}{r^{(p+1)h} - 1}$$

$$= (b^{p+1} - a^{p+1}) \cdot \frac{\frac{d}{dx}(r^x)|_{x=0}}{\frac{d}{dx}(r^{(p+1)x})|_{x=0}}$$

$$= (b^{p+1} - a^{p+1}) \frac{\log r}{(p+1)\log r} = \frac{b^{p+1} - a^{p+1}}{p+1}.$$

Since $U(f, P_n) = \left(\frac{a}{b}\right)^{p/n} L(f, P_n)$ has the same limit, we conclude that

$$\frac{b^{p+1} - a^{p+1}}{p+1} \le L(f) \le U(f) \le \frac{b^{p+1} - a^{p+1}}{p+1}.$$

So this function is Riemann integrable with $\int_a^b x^p \, dx = \dfrac{b^{p+1} - a^{p+1}}{p+1}$.

Given the many concepts we have already presented using an ϵ–δ formulation, it is natural to ask if Riemann integrable functions can be described in this way. There is such a formulation, given as Condition (3) in the next theorem, but it is harder to verify than the definition or Riemann's condition. Condition (4) shows that we can skip the step of finding the maximum and minimum values on each interval and instead use the Riemann sum for arbitrarily chosen points. Frequently, the choice of the left or right endpoint of the interval is both natural and convenient.

6.3.8. THEOREM. *Let $f(x)$ be a bounded function on $[a, b]$. The following are equivalent:*

(1) *f is Riemann integrable.*

(2) *For each $\varepsilon > 0$, there is a partition P so that $U(f, P) - L(f, P) < \varepsilon$.*

(3) *For every $\varepsilon > 0$, there is a $\delta > 0$ so that every partition Q such that $\text{mesh}(Q) < \delta$ satisfies $U(f, Q) - L(f, Q) < \varepsilon$.*

(4) *For every $\varepsilon > 0$, there is a $\delta > 0$ so that every partition Q such that $\text{mesh}(Q) < \delta$ and every choice of set $X = \{x'_j : 1 \le j \le n\}$, where $x'_j \in [x_{j-1}, x_j]$ satisfies $\left| I(f, Q, X) - \int_a^b f(x)\, dx \right| < \varepsilon$.*

PROOF. We have already verified that (1) and (2) are equivalent. Clearly, (3) implies (2). Let us prove that (1) implies (3).

If f is Riemann integrable, let $L = L(f) = U(f)$. Let $\varepsilon > 0$. We can find two partitions P_1 and P_2 so that $U(f, P_1) < L + \varepsilon/4$ and $L(f, P_2) > L - \varepsilon/4$. Let P

be their common refinement, $P_1 \cup P_2$. By Lemma 6.3.2,

$$L - \frac{\varepsilon}{4} < L(f, P) \le U(f, P) < L + \frac{\varepsilon}{4}.$$

Let n be the number of points in P, and set $\delta = \dfrac{\varepsilon}{8n\|f\|_\infty}$.

Now suppose that Q is any partition with mesh$(Q) < \delta$. Define $R = P \cup Q$ to be the common refinement of P and Q. By Lemma 6.3.2 again, we obtain

$$L - \frac{\varepsilon}{4} < L(f, R) \le U(f, R) < L + \frac{\varepsilon}{4}.$$

The intervals of R coincide with the intervals of Q except for at most $n-1$ intervals of Q which are split in two by the points in P. Thus in the sums determining $L(f, R)$ and $L(f, Q)$, all the terms are the same except for terms from these $n - 1$ intervals. Fix one of these intervals of Q, say I. The infimum of f over I is no smaller than $-\|f\|_\infty$, and the infimum of f over any subinterval of R contained in I is no more than $+\|f\|_\infty$. Adding up the differences over the $n - 1$ such intervals of Q, the total can be no more than

$$L(f, R) - L(f, Q) \le (n - 1)(2\|f\|_\infty) \, \mathrm{mesh}(Q)$$

$$< 2n\|f\|_\infty \frac{\varepsilon}{8n\|f\|_\infty} = \frac{\varepsilon}{4}.$$

Hence $L(f, Q) > L - \frac{\varepsilon}{2}$. Likewise, $U(f, Q) < L + \frac{\varepsilon}{2}$. So $U(f, Q) - L(f, Q) < \varepsilon$ and (3) is valid.

To see that (3) implies (4), fix a partition Q and a set $X = \{x'_j : 1 \le j \le n\}$, where $x'_j \in [x_{j-1}, x_j]$. Then

$$L(f, Q) \le I(f, Q, X) \le U(f, Q) < L(f, Q) + \varepsilon.$$

Since $L(f, Q) \le \displaystyle\int_a^b f(x)\, dx \le U(f, Q) < L(f, Q) + \varepsilon$ also, it follows that

$$\left| I(f, Q, \{x'_j\}) - \int_a^b f(x)\, dx \right| < \varepsilon.$$

Conversely, if (4) holds, then *every* choice of X satisfies this inequality for $\varepsilon/3$. If x'_j satisfy $f(x'_j) = \displaystyle\inf_{x_{j-1} \le x \le x_j} f(x)$, then $I(f, Q, X) = L(f, Q)$. Hence $\left| L(f, Q) - \displaystyle\int_a^b f(x)\, dx \right| < \varepsilon/3$. If the infimum is not attained, then we can choose X so that $f(x'_j)$ are sufficiently close to this infimum to obtain the inequality $\left| L(f, Q) - \displaystyle\int_a^b f(x)\, dx \right| < \varepsilon/2$. The details of this argument are left as an exercise. Similarly, $\left| U(f, Q) - \displaystyle\int_a^b f(x)\, dx \right| < \varepsilon/2$. Hence $U(f, Q) - L(f, Q) < \varepsilon$. So (3) holds. ∎

Using Riemann's condition, we can now show that many functions are integrable.

6.3.9. THEOREM. *Every monotone function on $[a, b]$ is Riemann integrable.*

PROOF. We will assume that f is monotone increasing, but the reader can easily convert this to a proof for decreasing functions. Consider the uniform partition P given by $x_j = a + \frac{j(b-a)}{n}$ for $0 \le j \le n$. Notice that $m_j(f, P) = f(x_{j-1})$ and $M_j(f, P) = f(x_j)$. Thus we obtain a telescoping sum

$$U(f, P) - L(f, P) = \sum_{j=1}^{n} f(x_j) \frac{b-a}{n} - \sum_{j=1}^{n} f(x_{j-1}) \frac{b-a}{n}$$

$$= \frac{\big(f(b) - f(a)\big)(b-a)}{n}.$$

It is evident that given any $\varepsilon > 0$, we may choose an integer n sufficiently large (i.e. $n > \big(f(b) - f(a)\big)(b-a)\varepsilon^{-1}$) so that $U(f, P) - L(f, P) < \varepsilon$. This verifies Riemann's condition, and therefore f is integrable. ∎

6.3.10. THEOREM. *Every continuous function on $[a, b]$ is integrable.*

PROOF. This result is deeper than the result for monotone functions because we must use Theorem 5.5.9 to deduce that a continuous function f on $[a, b]$ is uniformly continuous.

Let $\varepsilon > 0$. By uniform continuity, there is a $\delta > 0$ so that for $x, y \in [a, b]$ with $|x - y| < \delta$, we have $|f(x) - f(y)| < \varepsilon/(b - a)$. Let P be any partition with $\text{mesh}(P) < \delta$. Then for any points x, y in a common interval $[x_{j-1}, x_j]$, we have $|f(x) - f(y)| < \varepsilon/(b - a)$. Hence $M_j(f, P) - m_j(f, P) \le \varepsilon/(b - a)$. Consequently,

$$U(f, P) - L(f, P) = \sum_{j=1}^{n} \big(M_j(f, P) - m_j(f, P)\big)\Delta_j \le \frac{\varepsilon}{b - a} \sum_{j=1}^{n} \Delta_j = \varepsilon.$$

Thus Riemann's condition is satisfied, and f is Riemann integrable. ∎

6.3.11. EXAMPLE. There do exist functions that are not Riemann integrable. For example, consider $f : [0, 1] \to \mathbb{R}$ defined by

$$f(x) = \begin{cases} 1 & \text{if} \quad x \in \mathbb{Q} \\ 0 & \text{if} \quad x \notin \mathbb{Q}. \end{cases}$$

Let P be any partition. Notice that $M_j(f, P) = 1$ and $m_j(f, P) = 0$ for all j. Thus we see that

$$U(f, P) = \sum_{j=1}^{n} x_j - x_{j-1} = 1 \qquad \text{and} \qquad L(f, P) = 0.$$

This holds for all P. Thus it follows that $L(f) = 0$ and $U(f) = 1$, and so f is not Riemann integrable. The reason for this failure is that f is discontinuous at every point in $[0, 1]$.

6.3.12. EXAMPLE. On the other hand, there are discontinuous functions that are Riemann integrable. For example, the characteristic function $\chi_{(.5, 1]}$ is Riemann integrable on $[0, 1]$ because it is monotone. However, the discontinuity is rather banal.

Consider $f(x) = \sin(1/x)$ on $(0, 1]$, and set $f(0) = 0$. This function has a nasty discontinuity at the origin, as shown in Figure 6.7. But it turns out that provided that the function remains bounded, even bad discontinuities at a few points do not prevent integrability.

Let $\varepsilon > 0$ be given. We will choose a partition P with $x_1 = \varepsilon/4$. Notice that since f is continuous on $[\varepsilon/4, 1]$, it is integrable there. Thus there is a partition $Q = \{x_1 = \varepsilon/4 < \cdots < x_n = 1\}$ of $[\varepsilon/4, 1]$ with

$$U(f|_{[x_1, 1]}, Q) - L(f|_{[x_1, 1]}, Q) < \frac{\varepsilon}{2}.$$

Now take $P = \{x_0\} \cup Q$ as a partition of $[0, 1]$. Then since $\sin(1/x)$ oscillates wildly between ± 1 near $x = 0$, it follows that $M_1(f, P) = 1$ and $m_1(f, P) = -1$. So

$$U(f, P) = \Delta_1 + U(f|_{[x_1, 1]}, Q) = \frac{\varepsilon}{4} + U(f|_{[x_1, 1]}, Q)$$

and

$$L(f, P) = -\Delta_1 + L(f|_{[x_1, 1]}, Q) = -\frac{\varepsilon}{4} + L(f|_{[x_1, 1]}, Q).$$

Therefore,

$$U(f, P) - L(f, P) = \frac{\varepsilon}{2} + U(f|_{[x_1, 1]}, Q) - L(f|_{[x_1, 1]}, Q) < \varepsilon.$$

So f is integrable.

A similar analysis applies to the function $g(x) = \sin(\csc(1/x))$, which is graphed in Figure 6.8. This function behaves badly at those x where $\csc(1/x)$ is undefined, namely where $\sin(1/x) = 0$. Since $\sin(t) = 0$ for $t = n\pi$, these discontinuities occur at $x = 1/(n\pi)$ for $n \geq 1$ and at the endpoint $x = 0$ where g is undefined. Moreover, on an interval around one of these discontinuities, say $[1/(x\pi), 1/((x+1)\pi)]$, where $x = n+1/2$, g has, qualitatively, the same behaviour as $\sin(1/x)$ has around the origin.

Nonetheless, this function is integrable on $[0, 1]$. A small choice of x_1 takes care of all but finitely many of these bad points, much as before. The remaining points can be handled by ensuring that the partition contains each remaining discontinuity in a very small interval. The details are left to the reader as an exercise.

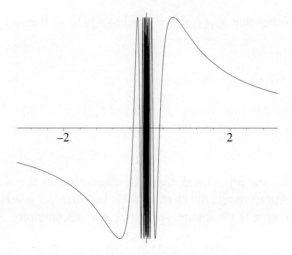

FIGURE 6.7. The graph of $\sin(1/x)$.

.

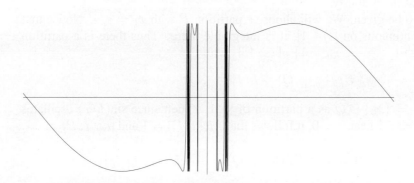

FIGURE 6.8. Partial graph of $\sin(\csc(1/x))$ from $-\pi$ to π.

.

Exercises for Section 6.3

A. (a) Compute the upper Riemann sum for $f(x) = x^{-1}$ on $[a, b]$ using the partition $P_n = \{x_j = a(b/a)^{j/n} : 0 \le j \le n\}$.

    (b) Evaluate the integral $\int_a^b \frac{1}{x}\, dx$.   HINT: Recognize $\lim\limits_{n \to \infty} U(f, P_n)$ as a derivative.

B. (a) Compute the upper Riemann sum for $f(x) = x^2$ on $[a, b]$ using the uniform partition $P_n = \{x_j = a + j(b - a)/n : 0 \le j \le n\}$.   HINT: $\sum\limits_{j=1}^{n} j^2 = \frac{n(n+1)(2n+1)}{6}$.

    (b) Hence evaluate the integral $\int_a^b x^2\, dx$.

C. Show that if a function $f : [a, b] \to \mathbb{R}$ is Lipschitz with constant C, then for any partition P of $[a, b]$, we have $U(f, P) - L(f, P) \le C\,\mathrm{mesh}(P)$.

D. Show that if f and g are Riemann integrable on $[a, b]$, then so is $\alpha f + \beta g$; and

$$\int_a^b \alpha f(x) + \beta g(x)\, dx = \alpha \int_a^b f(x)\, dx + \beta \int_a^b g(x)\, dx.$$

E. We call $f : [a, b] \to \mathbb{R}$ a **step function** if there is a partition of $[a, b]$ so that f is constant on each interval of the partition. Show that every step function is Riemann integrable using two arguments: first, using Riemann's condition (Theorem 6.3.5) and second, using Condition (3) from Theorem 6.3.8.

F. Show that every piecewise continuous function is Riemann integrable. (This subsumes the previous exercise.)

G. Show that if f is Riemann integrable on $[a, b]$, then so is $|f|$.
HINT: Show that $M_i(|f|, P) - m_i(|f|, P) \le M_i(f, P) - m_i(f, P)$.

H. Show that f is Riemann integrable if and only if for each $\varepsilon > 0$, there are step functions f_1 and f_2 on $[a, b]$ with $f_1(x) \le f(x) \le f_2(x)$ such that $\int_a^b f_2(x) - f_1(x)\, dx < \varepsilon$.

I. (a) Show that if $f \ge 0$ is Riemann integrable on $[a, b]$, then $\int_a^b f(x)\, dx \ge 0$.

 (b) Hence show that if f and g are Riemann integrable on $[a, b]$ and $f(x) \le g(x)$ for $a \le x \le b$, then $\int_a^b f(x)\, dx \le \int_a^b g(x)\, dx$.

 (c) Show that $\left| \int_a^b f(x)\, dx \right| \le \int_a^b |f(x)|\, dx$.

J. Show that if f is Riemann integrable on $[a, b]$ and $c \in \mathbb{R}$, then the function g defined on $[a + c, b + c]$ by $g(x) = f(x - c)$ is also Riemann integrable and

$$\int_{a+c}^{b+c} g(x)\, dx = \int_a^b f(x)\, dx.$$

This property is called **translation invariance**.

K. Suppose that f is Lipschitz with constant L on $[a, b]$. Prove that

$$\left| \int_0^1 f(x)\, dx - \frac{1}{n} \sum_{j=1}^n f\left(\tfrac{j}{n}\right) \right| \le \frac{L}{n}.$$

L. If f is Riemann integrable on $[a, b]$, show that $F(x) = c + \int_a^x f(t)\, dt$ is Lipschitz.

M. Show that if f is integrable on $[a, b]$, then it is integrable on each interval $[c, d] \subset [a, b]$ as well. Moreover, for $a < c < b$,

$$\int_a^b f(x)\, dx = \int_a^c f(x)\, dx + \int_c^b f(x)\, dx.$$

N. If $b < a$, define $\int_a^b f(x)\, dx = -\int_b^a f(x)\, dx$. Show that the formula of the previous exercise also holds for c outside $[a, b]$. On what interval should f be Riemann integrable for the formula to make sense in this context?

O. Show that $\sin(\csc(1/x))$ is integrable.   HINT: See the discussion of Example 6.3.12.

P. Verify the implication (4) implies (3) of Theorem 6.3.8 when the infimum and supremum over the intervals $[x_{j-1}, x_j]$ are not attained.

Q. If f and g are both Riemann integrable on $[a, b]$, show that fg is also integrable.
HINT: Use the identity $f(x)g(x) - f(t)g(t) = f(x)\big(g(x) - g(t)\big) + \big(f(x) - f(t)\big)g(t)$
to show that $M_i(fg, P) - m_i(fg, P)$ is bounded by

$$\|f\|_\infty \big(M_i(g, P) - m_i(g, P)\big) + \|g\|_\infty \big(M_i(f, P) - m_i(f, P)\big).$$

R. Show that the function of Example 5.2.10 is Riemann integrable, even though it is discontinuous at every rational number.
HINT: For $\varepsilon > 0$, there are only finitely many points taking values greater than ε. Choose a partition that includes those points in very small intervals.

S. **Improper Integrals.** Say that f is Riemann integrable on $[a, \infty)$ if $\displaystyle\lim_{b \to \infty} \int_a^b f(x)\,dx$
exists. Use $\displaystyle\int_a^\infty f(x)\,dx$ to denote this limit.

(a) For which real values of p does $\displaystyle\int_e^\infty \frac{(\log x)^p}{x}\,dx$ exist?

(b) Show that $\displaystyle\int_0^\infty \frac{\sin x}{x}\,dx$ exists.

HINT: Consider the alternating series $\displaystyle\sum_{n \geq 0} \int_{n\pi}^{(n+1)\pi} \frac{\sin x}{x}\,dx$.

T. If f is unbounded as it approaches a, we define an improper integral by
$\displaystyle\lim_{\varepsilon \to 0} \int_{a+\varepsilon}^b f(x)\,dx$, when the limit exists. Of course, $\displaystyle\int_a^b f(x)\,dx$ is used to denote this
limit, so you have to keep an eye out for unbounded functions.

(a) For which real values of p does $\displaystyle\int_0^1 x^p\,dx$ exist?

(b) Show by example that $\displaystyle\lim_{\varepsilon \to 0} \int_{-b}^{a-\varepsilon} f(x)\,dx + \int_{a+\varepsilon}^b f(x)\,dx$ can exist even though

$\displaystyle\lim_{\varepsilon \to 0} \int_{a+\varepsilon}^b f(x)\,dx$ does not.

6.4. The Fundamental Theorem of Calculus

The calculation of integrals via Riemann sums is of theoretical importance, but it is not a simple method for evaluating integrals. Fortunately, there is a crucial connection between integrals and derivatives that makes evaluating integrals by hand practical and efficient. We stress that this is *not* the definition of integral. The word *fundamental* is used to emphasize that this is the central result connecting differential and integral calculus. Nevertheless, it is not difficult to prove.

6.4.1. FUNDAMENTAL THEOREM OF CALCULUS.
Let f be a bounded Riemann integrable function on $[a, b]$, and let

$$F(x) = \int_a^x f(t)\,dt \quad for \quad a \le x \le b.$$

Then F is a continuous function. If f is continuous at a point x_0, then F is differentiable at x_0 and $F'(x_0) = f(x_0)$.

Before proving this, we record the most useful consequence. A function f on $[a, b]$ has an **antiderivative** if there is a continuous function $F(x)$ on $[a, b]$ such that $F'(x) = f(x)$ for *every* point $x \in (a, b)$. The second part of following theorem is in fact valid for the larger class of functions that are Riemann integrable and have antiderivatives, even though they are not continuous. This more technical result is left to Exercise 6.4.J.

6.4.2. COROLLARY. *Let f be a continuous function on $[a, b]$. Then f has an antiderivative. Moreover, if G is any antiderivative of f, then*

$$\int_a^b f(x)\,dx = G(b) - G(a).$$

PROOF. The Fundamental Theorem of Calculus shows that $F(x) = \int_a^x f(t)\,dt$ is a differentiable function with $F' = f$. If G is any antiderivative of f, then $(G-F)'(x) = 0$ for all $x \in (a, b)$. By Corollary 6.2.5 of the Mean Value Theorem, this means that $G - F$ is some constant c, whence $G = F + c$. Finally,

$$G(b) - G(a) = F(b) - F(a) = \int_a^b f(x)\,dx. \qquad \blacksquare$$

We require a simple estimate.

6.4.3. LEMMA. *Suppose that f is an integrable function on $[a, b]$ bounded by M. Then*

$$\left| \int_a^b f(t)\,dt \right| \le M(b - a).$$

PROOF. It is evident that for any partition P and any $1 \le j \le n$,

$$-M \le m_j(f, P) \le M_j(f, P) \le M.$$

Hence $U(f, P) \le \sum_{j=1}^n M(x_j - x_{j-1}) = M(b - a)$. Therefore,

$$\int_a^b f(t)\,dt = \inf_P U(f, P) \le M(b - a).$$

Similarly, we obtain $\int_a^b f(t)\,dt = \sup_P L(f, P) \ge -M(b - a).$ $\qquad \blacksquare$

PROOF OF THE FUNDAMENTAL THEOREM. Let f be bounded by M. For x, y in $[a, b]$, use the lemma to compute

$$|F(x) - F(y)| = \left| \int_a^x f(t)\, dt - \int_a^y f(t)\, dt \right|$$

$$= \left| \int_y^x f(t)\, dt \right| \leq M|x - y|.$$

Hence F is Lipshitz with constant M and thus is continuous.

Now suppose that f is continuous at x_0. Given $\varepsilon > 0$, choose $\delta > 0$ so that $|y - x_0| < \delta$ implies that $|f(y) - f(x_0)| < \varepsilon$. Then for $|h| < \delta$, compute

$$\left| \frac{F(x_0 + h) - F(x_0)}{h} - f(x_0) \right| = \left| \frac{1}{h} \int_{x_0}^{x_0 + h} f(t)\, dt - \frac{1}{h} \int_{x_0}^{x_0 + h} f(x_0)\, dt \right|$$

$$= \left| \frac{1}{h} \int_{x_0}^{x_0 + h} f(t) - f(x_0)\, dt \right| \leq \varepsilon.$$

The lemma was used again for the final inequality. Thus

$$F'(x_0) = \lim_{h \to 0} \frac{F(x_0 + h) - F(x_0)}{h} = f(x_0).$$

So F has the desired derivative. ∎

6.4.4. REMARK. A jump discontinuity in the integrand f can result in a point where the integral is not differentiable. For example, take $f(x) = 1$ for $0 \leq x \leq 1$ and 2 for $1 < x \leq 2$. Then

$$F(x) = \int_0^x f(t)\, dt = \begin{cases} x & \text{for } 0 \leq x \leq 1 \\ 2x - 1 & \text{for } 1 \leq x \leq 2. \end{cases}$$

This function is not differentiable at $x = 1$, but it does have a left derivative of 1 and a right derivative 2.

Nor is it the case that every differentiable function is an integral. An easy way for this to fail is when the derivative is unbounded. Recall from Example 6.1.7, the function $F(x) = x^a \sin(1/x)$ on $[0, 1]$ for some constant a in $(1, 2)$. Then $F'(x) = ax^{a-1} \sin(1/x) - x^{a-2} \cos(1/x)$ for $x > 0$ and $F'(0) = 0$. Thus F is differentiable, but the derivative is an unbounded function and thus is not Riemann integrable.

We can take various formulas for differentiation and, using the Fundamental Theorem, integrate them to obtain useful integration techniques. We are not concerned here with the all tricks of the trade, but just a glance at the major methods. Integration rules are more subtle than differentiation because it can be difficult to recognize which method to apply.

The product rule translates into integration by parts. We will assume that F and G are C^1 to avoid pathology. Since $(FG)'(x) = F'(x)G(x) + F(x)G'(x)$, we obtain

$$\int_a^b F'(x)G(x) + F(x)G'(x)\, dx = FG \Big|_a^b = F(b)G(b) - F(a)G(a).$$

Rearranging, we obtain the usual formulation

$$\int_a^b F'(x)G(x)\,dx = FG\Big|_a^b - \int_a^b F(x)G'(x)\,dx.$$

The chain rule corresponds to the substitution rule, also known as the change of variable formula. Let u be a C^1 function on $[a, b]$, and let F be C^1 on an interval $[c, d]$ containing the range of u. Then if $G(x) = F(u(x))$, the chain rule states that $G'(x) = F'(u(x))u'(x)$. Thus if we set $f = F'$,

$$(6.4.5) \qquad \int_a^b f(u(x))u'(x)\,dx = G(b) - G(a)$$

$$= F(u(b)) - F(u(a)) = \int_{u(a)}^{u(b)} f(t)\,dt.$$

We interpret this as making the substitution $t = u(x)$ and think of dt as $u'(x)\,dx$.

This change of variables is sometimes formulated somewhat differently. Suppose that the function u satisfies $u'(x) \neq 0$ for all $x \in [a, b]$. If we set $c = u(a)$ and $d = u(b)$, we obtain

$$(6.4.6) \qquad \int_c^d f(x)\,dx = \int_{u^{-1}(c)}^{u^{-1}(d)} f(u(t))u'(t)\,dt.$$

This corresponds to the substitution $x = u(t)$.

Without any attempt to be complete, we give a couple of examples of integration technique to refresh the reader's memory. Introductory calculus textbooks are full of further examples.

6.4.7. EXAMPLE. Consider $\int_0^1 \tan^{-1}(x)\,dx$. You will likely recall that the derivative of $\tan^{-1}(x)$ is $\dfrac{1}{1+x^2}$, but the integral is probably not stored in memory. This suggests that an integration by parts approach might help. Of course, you need something to integrate, and so we put in a factor of 1, which integrates to x. Thus

$$\int \tan^{-1}(x)\,dx = x \tan^{-1}(x) - \int \frac{x}{1+x^2}\,dx.$$

Now substitute $u = 1 + x^2$, which has derivative $du = 2x\,dx$, to obtain

$$= x \tan^{-1}(x) - \int \frac{du}{2u} = x \tan^{-1}(x) - \frac{1}{2}\log u$$

$$= x \tan^{-1}(x) - \frac{1}{2}\log(1+x^2).$$

Thus

$$\int_0^1 \tan^{-1}(x)\,dx = x \tan^{-1}(x) - \frac{1}{2}\log(1+x^2)\Big|_0^1 = \frac{\pi}{4} - \frac{1}{2}\log 2.$$

6.4.8. EXAMPLE. Now consider the integral $\int_0^8 e^{\sqrt[3]{x}}\, dx$. This integrand has a complicated exponent, $\sqrt[3]{x}$, which can be simplified by substituting $x = u^3$. Then $dx = 3u^2\, du$ and $u = \sqrt[3]{x}$ runs from 0 to 2. So we obtain

$$\int_0^8 e^{\sqrt[3]{x}}\, dx = \int_0^2 e^u 3u^2\, du.$$

This now can be integrated by parts twice by integrating e^u and differentiating $3u^2$.

$$= 3u^2 e^u \Big|_0^2 - \int_0^2 6u e^u\, du$$

$$= (3u^2 - 6u)e^u \Big|_0^2 + \int_0^2 6e^u\, du$$

$$= (3u^2 - 6u + 6)e^u \Big|_0^2 = 6(e^2 - 1)$$

Exercises for Section 6.4

A. Evaluate the following:

(a) $\displaystyle \lim_{n\to\infty} \sum_{j=1}^n \frac{1}{n + jc}$ for $c > 1$, and

(b) $\displaystyle \lim_{n\to\infty} \frac{1}{n^{a+1}} + \frac{2^a}{n^{a+1}} + \cdots + \frac{n-1^a}{n^{a+1}}$ for $a > -1$.

HINT: Recognize these as Riemann sums.

B. Define $\log(x)$, for $x > 0$, to be $\int_1^x \frac{1}{x}\, dx$. By manipulating integrals, show that $\log(ab) = \log(a) + \log(b)$.

C. (a) Prove the **Mean Value Theorem for Integrals**: If f is a continuous function on $[a, b]$, then there is a point $c \in (a, b)$ such that $\frac{1}{b-a} \int_a^b f(x)\, dx = f(c)$.

(b) Show by example that this may fail for a Riemann integrable function that is not continuous.

D. Let $f(x) = \text{sign}(x)$, and $F(x) = |x|$. Show that f is Riemann integrable on $[a, b]$ and that $\int_a^b f(x)\, dx = F(b) - F(a)$ for any $a < b$. Why is F not an antiderivative of f?

E. Let f be a continuous function on $\mathbb{R}$, and suppose that $b(x)$ is a C^1 function. Define $G(x) = \int_a^{b(x)} f(t)\, dt$. Compute $G'(x)$.

HINT: Let $F(x) = \int_a^x f(t)\, dt$ and note that $G(x) = F(b(x))$.

F. Compute the following integrals:

(a) $\displaystyle\int_1^e \left(\log x\right)^2 dx$ (b) $\displaystyle\int_0^{\pi/2} \frac{\sin^3 x}{\sqrt{\cos x}}\, dx$ (c) $\displaystyle\int_1^{125} \frac{dt}{\sqrt{t} + \sqrt[3]{t}}$

G. Let f be a continuous function on $\mathbb{R}$, and fix $\varepsilon > 0$. Define a function G by

$$G(x) = \frac{1}{\varepsilon} \int_x^{x+\varepsilon} f(t)\, dt.$$

Show that G is C^1 and compute G'.

H. Let u be a strictly increasing C^1 function on $[a, b]$.
(a) By considering the area under the graph plus the area between the graph and the
y-axis, establish the formula $\displaystyle\int_a^b u(x)\, dx + \int_{u(a)}^{u(b)} u^{-1}(t)\, dt = bu(b) - au(a).$
(b) Use the substitution formula (6.4.6) using $g(x) = x$ and integrate the second expression by parts to derive the same formula as in part (a).

I. (a) Let $x(t)$ and $y(t)$ be C^1 functions on $[0, 1]$ such that $x'(t) \geq 0$. Prove that the area
under the curve $C = \{(x(t), y(t)) : 0 \leq t \leq 1\}$ is $\displaystyle\int_0^1 y(t)x'(t)\, dt.$
(b) Now suppose that C is a closed curve [i.e., $(x(0), y(0)) = (x(1), y(1))$] that
doesn't intersect itself and that $x'(t)$ changes sign only a finite number of times.
Prove that the area enclosed by C is $\left| \displaystyle\int_0^1 y(t)x'(t)\, dt \right|.$

J. Suppose that a function f is Riemann integrable and has an antiderivative F. Prove
that $\displaystyle\int_a^b f(x)\, dx = F(b) - F(a).$
HINT: Apply the Mean Value Theorem to obtain $F(x_i) - F(x_{i-1}) = f(c_i)(x_i - x_{i-1})$
for some $c_i \in (x_{i-1}, x_i)$. Then apply Theorem 6.3.8 (4).

K. Suppose that f is twice differentiable on $\mathbb{R}$ and $\|f\|_\infty = A$ and $\|f''\|_\infty = C$. Prove
that $\|f'\|_\infty \leq \sqrt{2AC}.$
HINT: If $f'(x_0) = b > 0$, show that $f'(x_0 + t) \geq b - C|t|$. Integrate from $x_0 - b/C$
to $x_0 + b/C$.

6.5. Wallis's Product and Stirling's Formula

In this section, we will obtain Stirling's formula, an elegant asymptotic formula
for $n!$, using only basic calculus. However, to get a sharp result, the estimates must
be done quite carefully. By a **sharp inequality**, we mean an inequality that cannot
be improved. For example, $|\sin(x)/x| \leq 1$ for $x > 0$ is sharp but $\tan^{-1}(x) < 2$ is
not. These estimates lead us to a general method of numerical integration. The first
step is an exercise in integration that leads to a useful formula for π.

6.5.1. EXAMPLE. To begin, we wish to compute $I_n = \int_0^\pi \sin^n x \, dx$. We use integration by parts and induction. If $n \geq 2$,

$$I_n = \int_0^\pi \sin^n x \, dx = \int_0^\pi \sin^{n-1} x \sin x \, dx$$

$$= -\sin^{n-1} x \cos x \Big|_0^\pi + \int_0^\pi (n-1) \sin^{n-2} x \cos^2 x \, dx$$

$$= (n-1) \int_0^\pi \sin^{n-2} x (1 - \sin^2 x) \, dx = (n-1)(I_{n-2} - I_n).$$

Solving for I_n, we obtain a **recursion formula**

$$I_n = \frac{n-1}{n} I_{n-2}.$$

Rather than repeatedly integrating by parts, we use this formula. For example,

$$I_6 = \frac{5}{6} I_4 = \frac{5}{6} \frac{3}{4} I_2 = \frac{5}{6} \frac{3}{4} \frac{1}{2} I_0 = \frac{5\pi}{16}.$$

Since the formula drops the index by 2 each time, we end up at a multiple of $I_0 = \pi$ if n is even and a multiple of $I_1 = 2$ when n is odd. Indeed,

$$I_{2n} = \frac{1}{2} \cdot \frac{3}{4} \cdot \frac{5}{6} \cdots \frac{2n-3}{2n-2} \cdot \frac{2n-1}{2n} \pi$$

$$I_{2n+1} = \frac{2}{3} \cdot \frac{4}{5} \cdot \frac{6}{7} \cdots \frac{2n-2}{2n-1} \cdot \frac{2n}{2n+1} 2.$$

Since $0 \leq \sin x \leq 1$ for $x \in [0, \pi]$, we have $\sin^{2n+2} x \leq \sin^{2n+1} x \leq \sin^{2n} x$. Therefore, $I_{2n+2} \leq I_{2n+1} \leq I_{2n}$. Divide through by I_{2n} and use the preceding formula to obtain

$$\frac{2n+1}{2n+2} \leq \frac{2^2 \cdot 4^2 \cdot 6^2 \cdots (2n-2)^2 \cdot (2n)^2}{1 \cdot 3^2 \cdot 5^2 \cdots (2n-1)^2 \cdot (2n+1)} \frac{2}{\pi} \leq 1.$$

Rearranging this slightly and taking the limit, we obtain **Wallis's product**

$$\frac{\pi}{2} = \lim_{n \to \infty} \frac{2^2 \cdot 4^2 \cdot 6^2 \cdots (2n-2)^2 \cdot (2n)^2}{1 \cdot 3^2 \cdot 5^2 \cdots (2n-1)^2 \cdot (2n+1)}.$$

We estimate $n!$ by approximating the integral of $\log x$ using the **trapezoidal rule**, a modification of Riemann sums. First verify that $\int \log x \, dx = x \log x - x$ by differentiating. Notice that the second derivative of $\log x$ is $-x^{-2}$, which is negative for all $x > 0$. So the graph of $\log x$ is curving downward (i.e., this function is convex). Rather than using the rectangular approximants used in Riemann sums, we can do significantly better by using a trapezoid. In other words, we approximate the curve $\log x$ for x between $k-1$ and k by the straight line segment connecting $(k-1, \log(k-1))$ to $(k, \log k)$. By the convexity, this line lies strictly below $\log x$ and thus has a smaller area.

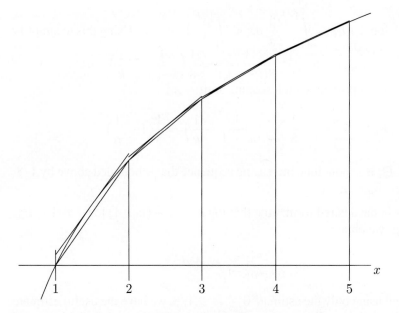

FIGURE 6.9. The graph of $\log x$ with the tangent approximation above and the trapezoidal approximation below.

The area of a trapezoid is the base times the average height. Therefore

$$\frac{\log(k-1) + \log k}{2} < \int_{k-1}^{k} \log x \, dx.$$

Sum this from 2 to n to obtain

$$\sum_{k=2}^{n-1} \log k + \tfrac{1}{2} \log n < \int_{1}^{n} \log x \, dx = n \log n - (n-1).$$

This may be rearranged to compute the error as

$$E_n := (n + \tfrac{1}{2}) \log n - (n-1) - \log n!.$$

To bound this error, let us estimate $\int_{k-1}^{k} \log x \, dx$ from above by another trapezoid, as shown in Figure 6.9. The tangent line to $\log x$ at $x = k - \frac{1}{2}$ lies strictly above the curve because $\log x$ is concave. This yields a trapezoid with average height $\log(k - \frac{1}{2})$. Therefore,

$$\varepsilon_k := \int_{k-1}^{k} \log x \, dx - \frac{\log(k-1) + \log k}{2}$$

$$< \log(k - \tfrac{1}{2}) - \frac{1}{2} \log(k-1)k$$

$$= \frac{1}{2} \log \frac{(k - \tfrac{1}{2})^2}{k^2 - k} = \frac{1}{2} \log \left(1 + \tfrac{1}{4(k^2 - k)}\right).$$

Now for $h > 0$, $\log(1+h) = \int_1^{1+h} \frac{1}{x} \, dx < \int_1^{1+h} 1 \, dx = h$. Using this to simplify our formula for ε_k, we obtain $\varepsilon_k < \dfrac{1}{8(k^2 - k)} = \dfrac{1}{8}\left(\dfrac{1}{k-1} - \dfrac{1}{k}\right)$.

Observe that this produces a telescoping sum

$$E_n = \sum_{k=2}^{n} \varepsilon_k < \frac{1}{8} \sum_{k=2}^{n} \left(\frac{1}{k-1} - \frac{1}{k}\right) = \frac{1}{8}\left(1 - \frac{1}{n}\right).$$

Consequently, E_n is a monotone increasing sequence that is bounded above by $1/8$. Therefore, $\lim\limits_{n \to \infty} E_n = E$ exists.

To put this in the desired form, note that $\log n! - n - (n+\frac{1}{2})\log n = 1 - E_n$. Exponentiating, we obtain

$$\lim_{n \to \infty} \frac{n!}{n^n \sqrt{n} e^{-n}} = e^{1-E}.$$

Rearranging and using only the estimate $0 \le E \le 1/8$, we have the useful estimate known as **Stirling's inequality**:

$$e^{\frac{7}{8}}\left(\frac{n}{e}\right)^n \sqrt{n} < n! < e\left(\frac{n}{e}\right)^n \sqrt{n}.$$

We will evaluate e^{1-E} exactly. To this end, set $a_n = \dfrac{n!}{n^n \sqrt{n} e^{-n}}$. Then

$$\frac{a_n^2}{a_{2n}} = \frac{(n!)^2}{n^{2n+1}e^{-2n}} \frac{(2n)^{2n}\sqrt{2n}e^{-2n}}{(2n)!} = \sqrt{\frac{2}{n}}\frac{(2^n n!)^2}{(2n)!}$$

$$= \sqrt{\frac{2}{n}}\frac{2^2 \cdot 4^2 \cdot 6^2 \cdots (2n-4)^2 \cdot (2n-2)^2 \cdot (2n)^2}{1 \cdot 2 \cdot 3 \cdot 4 \cdots (2n-2) \cdot (2n-1) \cdot (2n)}$$

$$= \sqrt{\frac{2}{n}}\frac{2 \cdot 4 \cdot 6 \cdots (2n-4) \cdot (2n-2) \cdot (2n)}{1 \cdot 3 \cdot 5 \cdots (2n-3) \cdot (2n-1)}$$

$$= \sqrt{\frac{2(2n+1)}{n}}\sqrt{\frac{2^2 \cdot 4^2 \cdot 6^2 \cdots (2n-4)^2 \cdot (2n-2)^2 \cdot (2n)^2}{3^2 \cdot 5^2 \cdot 7^2 \cdots (2n-3)^2 \cdot (2n-1)^2 \cdot (2n+1)}}.$$

Now combining this with our knowledge of Wallis's product,

$$e^{1-E} = \lim_{n \to \infty} \frac{a_n^2}{a_{2n}} = 2\sqrt{\frac{\pi}{2}} = \sqrt{2\pi}.$$

So we obtain the following:

6.5.2. STIRLING'S FORMULA.

$$\lim_{n \to \infty} \frac{n!}{n^n e^{-n} \sqrt{2\pi n}} = 1$$

Finally, let us look more closely at the trapezoidal approximation as a method for computing integrals. Consider a continuous function $f(x)$ on $[a, b]$ and a uniform partition $x_k = a + \frac{k(b-a)}{n}$ for $0 \leq k \leq n$. As previously, we estimate the area $\int_{x_{k-1}}^{x_k} f(x)\, dx$ by the trapezoid with vertices $(x_{k-1}, f(x_{k-1}))$ and $(x_k, f(x_k))$. This has base $\dfrac{b-a}{n}$ and average height $\dfrac{f(x_{k-1}) + f(x_k)}{2}$. The approximation is therefore

$$A_n = \sum_{k=1}^{n} \left(\frac{b-a}{n}\right)\left(\frac{f(x_{k-1}) + f(x_k)}{2}\right)$$

$$= \frac{b-a}{n}\left(\tfrac{1}{2}f(x_0) + f(x_1) + f(x_2) + \cdots + f(x_{n-1}) + \tfrac{1}{2}f(x_n)\right).$$

The crucial questions about this, or any approximation, are, first, How close is it to the true answer, and second, How much easier is it to work with? Ideally, an approximation is both very close to the true answer and much easier to work with. In practice, there is usually a trade-off between these two properties. We return to this issue in detail when we consider approximating functions by polynomials (Chapter 10) and by Fourier series (Chapter 14).

For the trapezoidal rule, so long as the function is C^2, we can use the Mean Value Theorem twice (once disguised as Rolle's Theorem) to obtain an estimate for the error.

6.5.3. TRAPEZOIDAL RULE.
Suppose that f is a C^2 function on $[a, b]$. Then the trapezoidal approximants A_n satisfy

$$\left| \int_a^b f(x)\, dx - A_n \right| \leq \frac{(b-a)^3 \|f''\|_\infty}{12n^2}.$$

PROOF. Let $F(x) = \int_a^x f(t)\, dt$. For the interval $[x_{k-1}, x_k]$, set

$$\varepsilon_k = F(x_k) - F(x_{k-1}) - \frac{b-a}{2n}\big(f(x_k) - f(x_{k-1})\big).$$

Define $c = \tfrac{1}{2}(x_{k-1} + x_k)$ and consider the function

$$G(t) = F(c + t) - F(c - t) - t\big(f(c + t) + f(c - t)\big) - Bt^3.$$

Notice that $G(0) = 0$. We are of course interested in $t_0 = (b-a)/2n$. So we choose the constant B so that $G(t_0) = 0$, namely $B = \varepsilon_k / t_0^3$.

By Rolle's Theorem, there is a point $t_1 \in (0, t_0)$ so that $G'(t_1) = 0$. By the Fundamental Theorem of Calculus and the chain rule,

$$\big(F(c + t) - F(c - t)\big)' = f(c + t) + f(c - t).$$

Therefore,

$$G'(t) = t\big(f'(c + t) - f'(c - t)\big) - 3Bt^2.$$

Substituting t_1 and solving for B yields

$$B = \frac{f'(c + t_1) - f'(c - t_1)}{3t_1}.$$

An application of the Mean Value Theorem produces a point $t_2 \in (0, t_1)$ so that

$$\frac{f'(c + t_1) - f'(c - t_1)}{2t_1} = f''(t_2)$$

and hence $B = 2f''(t_2)/3$. Consequently,

$$|\varepsilon_k| = |Bt_0^3| = \left| \frac{(b - a)^3 f''(t_2)}{12n^3} \right| \le \frac{(b - a)^3 \|f''\|_\infty}{12n^3}.$$

Summing from 1 to n yields

$$\left| \int_a^b f(x)\, dx - A_n \right| \le \sum_{k=1}^n |\varepsilon_k| \le \frac{(b - a)^3 \|f''\|_\infty}{12n^2}. \qquad \blacksquare$$

Exercises for Section 6.5

A. Evaluate $\lim\limits_{n \to \infty} \dfrac{\sqrt[n]{n!}}{n}$.

B. Estimate the choice of n to guarantee an approximation to $\int_0^1 e^{x^2}\, dx$ to 4 decimals accuracy using the trapezoidal rule.

C. For every Riemann integrable function, f, use the trapezoidal rule to obtain a sequence converging $\int_a^b f(x)\, dx$.

D. **Simpson's Rule**. This method for estimating integrals is based on approximating the function by a parabola passing through three points on the graph of f. Given a partition $P = \{a = x_0 < \ldots < x_{2n} = b\}$, let $y_k = f(x_k)$ for all k.
(a) Find the parabola passing through (x_{2k-2}, y_{2k-2}), (x_{2k-1}, y_{2k-1}) and (x_{2k}, y_{2k}).
(b) Find the integral of this parabola from x_{2k-2} to x_{2k}.
(c) Sum these areas to obtain Simpson's rule for approximating $\int_a^b f(x)\, dx$:

$$A_n = \frac{b - a}{3n} \left(y_0 + 4y_1 + 2y_2 + 4y_3 + \cdots + 2y_{2n-2} + 4y_{2n-1} + y_{2n} \right).$$

E. Decide whether $\displaystyle\sum_{n \ge 0} \frac{1}{2^{2n+k}\sqrt{n}} \binom{2n + k}{n}$ converges.

F. The **gamma function** is defined by $\Gamma(x) = \displaystyle\int_0^\infty t^{x-1} e^{-t}\, dt$ for all $x > 0$.
(a) Prove that this improper integral has a finite value for all $x > 0$.
(b) Prove that $\Gamma(x + 1) = x\Gamma(x)$. HINT: Integrate by parts.
(c) Prove by induction that $\Gamma(n + 1) = n!$ for $n \ge 1$.
(d) Calculate $\Gamma(\frac{1}{2})$. HINT: Substitute $t = u^2$, write the square as a double integral, and convert to polar coordinates.

6.6. Measure Zero and Lebesgue's Theorem

As we have already said, it is possible to build a more powerful theory of integration, based on the ideas of measure theory. The first step toward this theory, and the crucial notion for this section, is the idea of a set of measure zero. This is a reasonable condition for a set to be "small," although the condition can have surprising and unintuitive properties. Using this idea, we obtain a characterization of precisely which functions are Riemann integrable.

If $U = (c, d)$ is an open interval, let us write $|U| = d - c$ for its length.

6.6.1. DEFINITION. A subset A of $\mathbb{R}$ has **measure zero** if for every $\varepsilon > 0$, there is a countable family of intervals $\{U_n = (c_n, d_n) : n \geq 1\}$ such that $A \subset \bigcup_{n \geq 1} U_n$ and $\sum_{n \geq 1} |U_n| = \sum_{n \geq 1} d_n - c_n < \varepsilon$.

A subset A of (a, b) has **content zero** if for every $\varepsilon > 0$, there is a *finite* family of intervals $\{U_n = (c_n, d_n) : 1 \leq n \leq N\}$ such that $A \subset \bigcup_{n=1}^{N} U_n$ and

$$\sum_{n=1}^{N} |U_n| < \varepsilon.$$

6.6.2. EXAMPLES.

(1) The set $\mathbb{Q}$ of all rational numbers has measure zero. To see this, write $\mathbb{Q}$ as a list $r_1, r_2, r_3, \ldots$. Given $\varepsilon > 0$, let $U_n = (r_n - 2^{-n-1}\varepsilon, r_n + 2^{-n-1}\varepsilon)$. Evidently, $\bigcup_{n \geq 1} U_n$ contains $\mathbb{Q}$ and $\sum_{n \geq 1} |U_n| = \sum_{n \geq 1} 2^{-n}\varepsilon = \varepsilon$. So $\mathbb{Q}$ has measure zero.

However, $\mathbb{Q}$ does not have content zero because any *finite* collection of intervals covering $\mathbb{Q}$ can miss only finitely many points in $\mathbb{R}$. Consequently, at least one would be infinite! In fact, the set of rational points in $[0, 1]$ will not have content zero either for the same reason combined with the next example.

(2) Suppose that an interval $[a, b]$ is covered by open intervals $\{U_n\}$. Now $[a, b]$ is compact by the Heine–Borel Theorem (Theorem 4.4.6). Moreover by the Borel–Lebesgue Theorem (Theorem 9.2.3), the open cover $\{U_n\}$ has a finite subcover, say $U_1, \ldots, U_N$. Now this *finite* set of intervals pieces together to cover an interval of length $b - a$. From this we can easily show that $\sum_{n=1}^{N} d_n - c_n \geq b - a$. Consequently, $[a, b]$ does not have measure zero. (It has measure $b - a$.)

(3) In Example 4.4.8, we constructed the Cantor set. We showed there that it has measure zero and indeed has content zero. It also shows that an uncountable set can have measure zero.

6.6.3. PROPOSITION.

(1) *If A has measure zero, and $B \subset A$, then B has measure zero.*

(2) *If A_n are sets of measure zero for $n \geq 1$, then $\bigcup_{n \geq 1} A_n$ has measure zero.*

(3) *Every countable set has measure zero.*

(4) *If A is compact and has measure zero, then it has content zero.*

PROOF. (1) is trivial. For (2), suppose that $\varepsilon > 0$. We modify the argument in Example 6.6.2 (1). Choose a collection of open intervals $\{U_{nm} : m \geq 1\}$ covering A_n such that $\sum_{m \geq 1} |U_{nm}| < 2^{-n}\varepsilon$. Then $\{U_{nm} : m, n \geq 1\}$ covers $\bigcup_{n \geq 1} A_n$ and the combined lengths of these intervals is $\sum_{n \geq 1} \sum_{m \geq 1} |U_{nm}| < \sum_{n \geq 1} 2^{-n}\varepsilon = \varepsilon$. (3) now follows from (2) and the observation that a single point has measure zero.

(4) Suppose that A is compact and $\{U_n\}$ is an open cover of intervals with $\sum_{n \geq 1} |U_n| < \varepsilon$. By the Borel–Lebesgue Theorem, this cover has a finite subcover. This finite cover has total length less than ε also. Therefore, A has content zero. ∎

6.6.4. DEFINITION. A property is valid **almost everywhere** (a.e.) if the set of points where it fails has measure zero.

For example, $f = g$ a.e. means that $\{x : f(x) \neq g(x)\}$ has measure zero. And $f(x) = \lim_{n \to \infty} f_n(x)$ a.e. means that this limit exists and equals $f(x)$ except on a set of measure zero.

It is important to distinguish between two similar but distinct statements. If f is continuous almost everywhere, then the set of points of discontinuity has measure zero. If f equals a continuous function almost everywhere, then there would be a continuous function g so that $\{x : f(x) \neq g(x)\}$ is measure zero. This is a much stronger property. For example, the characteristic function of $[0, 1]$ has only two points of discontinuity in $\mathbb{R}$ and thus is continuous almost everywhere. But any continuous function will differ from this on a whole interval near 0 and another near 1. Since these intervals are not measure zero (see Example 6.6.2), the continuous function is not equal to the characteristic function almost everywhere.

The next result is very appealing because it provides a simple description of exactly which functions are integrable. But in practice, it is often enough to know that piecewise continuous functions are integrable. It will be more useful to build a more powerful integral, as we do in Section 9.6.

We need a notion of the "size" of a discontinuity. There is a global version of this concept, known as the modulus of continuity, which we will develop later (see Definition 10.4.2).

6.6.5. DEFINITION. The **oscillation** of a function f over an interval I is defined as $\mathrm{osc}(f, I) = \sup\{|f(x) - f(y)| : x, y \in I\}$. Then set

$$\mathrm{osc}(f, x) = \inf_{r > 0} \mathrm{osc}\left(f, (x - r, x + r)\right).$$

It is easy (Exercise 6.6.A) to prove that f is continuous at x if and only if $\mathrm{osc}(f, x) = 0$.

6.6.6. LEBESGUE'S THEOREM.
A bounded function on $[a, b]$ is Riemann integrable if and only if it is continuous almost everywhere.

PROOF. First suppose that f is Riemann integrable. Then for each $k \geq 1$, there is a finite partition P_k of $[a, b]$ so that $U(f, P_k), -L(f, P_k) < 4^{-k}$. Let u_k and l_k be the step functions that are constant on the intervals of P_k and bound f from above and below (as in Exercise 6.3.H) so that $\int_a^b u_k(x) - l_k(x) \, dx < 4^{-k}$.

The set $B_k = \{x : u_k(x) - l_k(x) \geq 2^{-k}\}$ is the union of certain intervals of the partition P_k, say $J_{k,i}$ for $i \in S_k$. Compute

$$4^{-k} > \int_a^b u_k(x) - l_k(x) \, dx \geq \sum_{i \in S_k} 2^{-k} |J_{k,i}|.$$

Then $\sum_{i \in S_k} |J_{k,i}| < 2^{-k}$. Let $A_1 = \bigcap_{k \geq 1} \text{int } B_k$ and let $A_2 = \bigcup_{k \geq 1} P_k$. Observe that $\{\text{int } J_{k,i} : i \in S_k\}$ covers A_1 for each k. As these intervals have length summing to less than 2^{-k}, it follows that A_1 has measure zero. Since A_2 is countable, it also has measure zero. Thus the set $A = A_1 \cup A_2$ has measure zero.

For any $x \notin A$ and any $\varepsilon > 0$, choose k so that $2^{-k} < \varepsilon$ and $x \notin B_k$. Then $u_k(x) - l_k(x) < 2^{-k} < \varepsilon$. As x is not a point in P_k, it is an interior point of some interval J of this partition. Choose $r > 0$ so that $(x - r, x + r) \subset J$. For any y with $|x - y| < r$, we have $l_k(x) \leq f(y), f(x) \leq u_k(x)$. Thus $|f(x) - f(y)| < \varepsilon$ and so f is continuous at x.

Conversely, suppose that f is continuous almost everywhere on $[a, b]$ and is bounded by M. Let

$$A_k = \{x \in [a, b] : \text{osc}(f, x) \geq 2^{-k}\}.$$

Then each A_k has measure zero, and the set of points of discontinuity is $A = \bigcup_{k \geq 1} A_k$. Observe that A_k is closed. Indeed suppose that x is the limit of a sequence (x_n) with all x_n in A_k. By definition, there are points y_n and z_n with $|x_n - y_n| < 1/n$ and $|x_n - z_n| < 1/n$ such that $|f(y_n) - f(z_n)| \geq 2^{-k} - 2^{-n}$. Therefore,

$$\lim_{n \to \infty} y_n = \lim_{n \to \infty} z_n = x.$$

Consequently, $\text{osc}(f, x) \geq 2^{-k}$.

By Proposition 6.6.3 (4), each A_k has content zero. Cover A_k with a finite number of open intervals $J_{k,i}$ such that $\sum_i |J_{k,i}| < 2^{-k}$. The complement X consists of a finite number of closed intervals on which $\text{osc}(f, x) < 2^{-k}$ for all $x \in X$. Thus there is an open interval J_x containing x so that $\text{osc}(f, J_x) < 2^{-k}$. Notice that X is a closed and bounded subset of $[a, b]$ and so is compact. The collection $\{J_x : x \in X\}$ is an open cover of X. By the Borel–Lebesgue Theorem, there is a finite subcover $J_{x_1}, \ldots, J_{x_p}$. Let P_k be the finite partition consisting of all the endpoints of all of these intervals together with the endpoints of each $J_{k,i}$.

Let us estimate the upper and lower sums for this partition. As usual, let $M_j(f, P_k)$ and $m_j(f, P_k)$ be the supremum and infimum of f over the jth interval, namely $I_{k,j} = [x_{k,j-1}, x_{k,j}]$. These intervals split into two groups, those contained in X and those contained in $U_k = \bigcup_i J_{k,i}$. For the first group, the oscillation is less than 2^{-k} and thus $M_j - m_j \leq 2^{-k}$. For the second group, the total length of the intervals is at most 2^{-k}. Combining these estimates, we obtain

$$U(f, P_k) - L(f, P_k) = \sum_{I_{k,i} \subset X} (M_i - m_i)\Delta_i + \sum_{I_{k,i} \subset U_k} (M_i - m_i)\Delta_i$$

$$\leq \sum_{I_{k,i} \subset X} 2^{-k}\Delta_i + \sum_{I_{k,i} \subset U_k} 2M\Delta_i$$

$$\leq 2^{-k}(b - a) + 2M2^{-k} = 2^{-k}(b - a + 2M).$$

Therefore, $\lim_{k \to \infty} U(f, P_k) - L(f, P_k) = 0$, and so f is integrable by Riemann's condition. ∎

You may want to review the examples of discontinuous functions in Section 5.2 to see which are equal to a continuous function almost everywhere.

Exercises for Section 6.6

A. Prove that f is continuous at x if and only if $\text{osc}(f, x) = 0$.

B. If measure zero sets were defined using closed intervals instead of open intervals, show that one obtains the same sets.

C. If $A \subset \mathbb{R}$ has measure zero, what is $\text{int}(A)$? Is $\overline{A}$ also measure zero?

D. If A has measure zero and B is countable, show that $A + B = \{a + b : a \in A, b \in B\}$ has measure zero.

E. Consider the set C' obtained using the construction of the Cantor set in Example 4.4.8 but removing 2^{n-1} intervals of length 4^{-n} (instead of length 3^{-n}) at the nth stage.
(a) Show that C' is closed and has no interior.
(b) Show that C' is not measure zero.
    HINT: Any cover of C' together with the intervals removed covers $[0, 1]$.

F. Let D be the set of numbers $x \in [0, 1]$ with a decimal expansion containing no odd digits. Prove that D has measure zero.
HINT: Cover D with some intervals of length 10^{-n}.

G. Suppose that f and g are both Riemann integrable on $[a, b]$. Use Lebesgue's Theorem to prove that fg is Riemann integrable. (Compare with Exercise 6.3.Q.)

H. Show that $A \subset \mathbb{R}$ has content zero if and only if $\overline{A}$ is compact and has measure zero.
HINT: Show that an unbounded set cannot have content zero. Note that a finite open cover of A contains most of $\overline{A}$.

I. Define a relation on functions on $[a, b]$ by $f \sim g$ if the set $\{x : f(x) \neq g(x)\}$ has measure zero. Prove that this is an equivalence relation.

CHAPTER 7

Normed Vector Spaces

In this chapter, we generalize to vector spaces the absolute value function on $\mathbb{R}$ and the norm of a vector in $\mathbb{R}^n$. From this perspective, convergence of functions is similar to convergence of points in $\mathbb{R}$. The additional complication comes from the fact that these vector spaces are generally infinite dimensional. The notions of topology go through with almost no change in the definitions. However, the theorems can be quite different. For example, being closed and bounded is not sufficient to imply compactness in infinite-dimensional spaces such as $C[a, b]$. In particular, we develop the properties of inner product spaces in detail because these spaces are especially useful and tractable.

7.1. Definition and Examples

Recall that a vector space over $\mathbb{R}$ is a set V together with two operations, vector addition and scalar multiplication. Vector spaces are discussed in more detail in Section 1.4; an important example of a vector space, $\mathbb{R}^n$, is the focus of Chapter 4. We write $x + y$ for vector addition and αx for scalar multiplication, where $x, y \in V$ and $\alpha \in \mathbb{R}$.

7.1.1. DEFINITION. Let V be a vector space over $\mathbb{R}$. A **norm** on V is a function $\| \cdot \|$ on V taking values in $[0, +\infty)$ with the following properties:

(1) (positive definite) $\|x\| = 0$ if and only if $x = 0$,

(2) (homogeneous) $\|\alpha x\| = |\alpha| \|x\|$ for all $x \in V$ and $\alpha \in \mathbb{R}$, and

(3) (triangle inequality) $\|x + y\| \le \|x\| + \|y\|$ for all $x, y \in V$.

We call the pair $(V, \| \cdot \|)$ a **normed vector space**.

The first two properties are usually easy to verify. The positive definite property just says that nonzero vectors have nonzero length. And the homogeneous property says that the norm is scalable. The important property, which often requires some cleverness to verify, is the triangle inequality. It says that the path from point A to B and on to C is at least as long as the direct route from A to C. As we indicated in Figure 4.1 in Chapter 4, this algebraic inequality is equivalent to the geometric statement that the length of one side of a triangle is at most sum of the lengths of the other two sides.

7.1.2. EXAMPLE. Consider the vector space $\mathbb{R}^n$. In Chapter 4, we showed that the Euclidean norm

$$\|\mathbf{x}\| = \|(x_1, \ldots, x_n)\|_2 = \left(\sum_{i=1}^{n} |x_i|^2 \right)^{1/2}$$

is a norm. Indeed, properties (1) and (2) are evident, and the triangle inequality was a consequence of Schwarz's inequality.

Consider two other functions:

$$\|\mathbf{x}\|_1 = \|(x_1, \ldots, x_n)\|_1 = \sum_{i=1}^{n} |x_i|$$

$$\|\mathbf{x}\|_\infty = \|(x_1, \ldots, x_n)\|_\infty = \max_{1 \le i \le n} |x_i|.$$

Again it is easy to see that they are positive definite and homogeneous. The key is the triangle inequality. But for these functions, even that is straightforward. Compute

$$\|\mathbf{x} + \mathbf{y}\|_1 = \sum_{i=1}^{n} |x_i + y_i|$$

$$\le \sum_{i=1}^{n} |x_i| + |y_i| = \|\mathbf{x}\|_1 + \|\mathbf{y}\|_1$$

and

$$\|\mathbf{x} + \mathbf{y}\|_\infty = \max_{1 \le i \le n} |x_i + y_i|$$

$$\le \max_{1 \le i \le n} |x_i| + \max_{1 \le i \le n} |y_i| = \|\mathbf{x}\|_\infty + \|\mathbf{y}\|_\infty.$$

To illustrate the differences between these norms, consider the vectors of norm at most 1 in $\mathbb{R}^2$ for these three norms, as given in Figure 7.1. This set is sufficiently useful that we give it a name. For any normed vector space $(V, \| \cdot \|)$, the **unit ball** of V is the set $\{x \in V : \|x\| \le 1\}$.

The next example is very important for our applications as it allows us to apply vector space methods to collections of functions; that is, to think of functions as vectors.

7.1.3. EXAMPLE. Let K be a compact subset of $\mathbb{R}^n$, and let $C(K)$ denote the vector space of all continuous real-valued functions on K. If $f, g \in C(K)$ and $\alpha \in \mathbb{R}$, then $f + g$ is the function given by $(f + g)(x) := f(x) + g(x)$ and $(\alpha f)(x) := \alpha f(x)$. There are several different possible norms on $C(K)$. The most

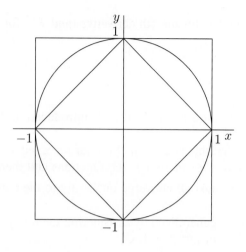

FIGURE 7.1. The unit balls $\{(x, y) : \|(x, y)\|_p \le 1\}$ for $p = 1, 2, \infty$.

natural and most important is the uniform norm, given by

$$\|f\|_\infty = \sup_{x \in K} |f(x)|.$$

By the Extreme Value Theorem (Theorem 5.4.4), $|f|$ achieves its maximum at some point $x_0 \in K$. So using this point x_0, we have $\|f\|_\infty = |f(x_0)| < \infty$.

Clearly, this function is nonnegative. To see that it is really a norm, observe first that if $\|f\|_\infty = 0$, then $|f(x)| = 0$ for all $x \in K$ and so $f = 0$. For homogeneity, we have

$$\|\alpha f\|_\infty = \sup_{x \in K} |\alpha f(x)| = |\alpha| \sup_{x \in K} |f(x)| = |\alpha| \|f\|_\infty.$$

Finally, the triangle inequality is proved as follows:

$$\begin{aligned}
\|f + g\|_\infty &= \sup_{x \in K} |f(x) + g(x)| \\
&\le \sup_{x \in K} |f(x)| + |g(x)| \\
&\le \sup_{x \in K} |f(x)| + \sup_{x \in K} |g(x)| = \|f\|_\infty + \|g\|_\infty.
\end{aligned}$$

We shall see in Chapter 8 that a sequence of functions f_n converges to a function f in $\big(C(K), \|\cdot\|_\infty\big)$ if and only if the sequence converges uniformly. Since uniform convergence is often the "right" notion of convergence for many applications, we will use this normed vector space often.

Controlling the derivatives of a function, as well as the function itself, is often important. Fortunately, this is easy to do, and we will regularly use normed vector spaces like the following example.

7.1.4. EXAMPLE. For simplicity, we restrict our attention to an interval of $\mathbb{R}$, but the same idea readily generalizes. Let $C^3[a, b]$ denote the vector space of all functions $f : [a, b] \to \mathbb{R}$ such that f and its first 3 derivatives f', f'', f''' are all defined and continuous. Using $f^{(j)}$ for the jth derivative (and $f^{(0)}$ for f), we can define a new norm $\|\cdot\|_{C^3}$ by

$$\|f\|_{C^3} = \max_{0 \leq j \leq 3} \|f^{(j)}\|_\infty,$$

where $\|\cdot\|_\infty$ is the uniform norm on the interval $[a, b]$ introduced in the previous example.

It is an exercise to verify that this is a norm. In order for a sequence of functions f_n in $C^3[a, b]$ to converge to a function f, the functions and their first three derivatives $f_n^{(j)}$ for $0 \leq j \leq 3$ must all converge uniformly to the corresponding derivative $f^{(j)}$.

Clearly, this example can be generalized by considering the first p derivatives, for any positive integer p instead of just $p = 3$.

7.1.5. EXAMPLE. For certain applications, we'll need the L^p norms on $C[a, b]$. Fix a real number p in $[1, \infty)$. The $L^p[a, b]$ norm is defined on $C[a, b]$ by

$$\|f\|_p = \left(\int_a^b |f(x)|^p \, dx \right)^{1/p}.$$

First notice that $\|f\|_p \geq 0$. Moreover, if $f \neq 0$, then there is a point $x_0 \in [a, b]$ such that $f(x_0) \neq 0$. Take $\varepsilon = |f(x_0)|/2$ and use the continuity of f to find an $r > 0$ so that

$$|f(x) - f(x_0)| < \frac{|f(x_0)|}{2} \quad \text{for} \quad x_0 - r < x < x_0 + r.$$

Hence

$$|f(x)| \geq |f(x_0)| - |f(x) - f(x_0)| > \frac{|f(x_0)|}{2}$$

for $x_0 - r < x < x_0 + r$. We may suppose that r is small enough so that $a \leq x_0 - r$ and $x_0 + r \leq b$. (If $x_0 = a$ or b, the simple modification is left to the reader.) Consequently,

$$\|f\|_p \geq \left(\int_{x_0-r}^{x_0+r} \left(\frac{|f(x_0)|}{2} \right)^p \, dx \right)^{1/p} \geq \frac{(2r)^{1/p}|f(x_0)|}{2} > 0.$$

So the p-norms are positive definite.

Homogeneity is easy to verify from the definition. However, the triangle inequality is tricky, unless p is 1 or 2. We leave the general case to Section 7.7, where we study L^p spaces in more detail. The case $p = 2$ is easier because it arises from an inner product (much as the Euclidean norm on $\mathbb{R}^n$ is obtained from the dot product). We prove it in Section 7.3.

Here we will establish the triangle inequality for $p = 1$. If f and g are in $C[a, b]$, compute

$$\|f + g\|_1 = \int_a^b |f(x) + g(x)| \, dx$$

$$\leq \int_a^b |f(x)| + |g(x)| \, dx$$

$$= \int_a^b |f(x)| \, dx + \int_a^b |g(x)| \, dx = \|f\|_1 + \|g\|_1.$$

Exercises for Section 7.1

A. Show that $\|(x, y, z)\| = |x| + 2\sqrt{y^2 + z^2}$ is a norm on $\mathbb{R}^3$. Sketch the unit ball.

B. Is $\|(x, y)\| = \left(|x|^{1/2} + |y|^{1/2}\right)^2$ a norm on $\mathbb{R}^2$?

C. For f in $C^1[a, b]$, define $\rho(f) = \|f'\|_\infty$. Show that ρ is nonnegative, homogeneous, and satisfies the triangle inequality. Why is it not a norm?

D. If $(V, \|\cdot\|)$ is a normed vector space, show that $\big| \|x\| - \|y\| \big| \leq \|x - y\|$ for all $x, y \in V$.

E. Show that the unit ball of a normed vector space, $(V, \|\cdot\|)$, is convex, meaning that if $\|x\| \leq 1$ and $\|y\| \leq 1$, then every point on the line segment between x and y has norm at most 1.
HINT: Describe the line segment algebraically in terms of x and y and a parameter t.

F. Let K be a compact subset of $\mathbb{R}^n$, and let $C(K, \mathbb{R}^m)$ denote the vector space of all continuous functions from K into $\mathbb{R}^m$. Show that for f in $C(K, \mathbb{R}^m)$, the quantity $\|f\|_\infty = \sup_{x \in K} \|f(x)\|_2$ is finite and $\|\cdot\|_\infty$ is a norm on $C(K, \mathbb{R}^m)$.

G. Define $C^p[a, b]$ for $p \in \mathbb{N}$ and verify that the $C^p[a, b]$ norm is indeed a norm. Is there a reasonable definition for $C^0[a, b]$?

H. (a) Show that if $\|\cdot\|$ and $\|\!|\cdot|\!\|$ are both norms on V, then $\|v\|_m := \max\{\|v\|, \|\!|v|\!\|\}$ is also a norm on V.
(b) Take $V = \mathbb{R}^2$ and $\|(x, y)\| = \sqrt{x^2 + y^2}$ and $\|\!|(x, y)|\!\| = \frac{3}{2}|x| + |y|$. Then define $\|(x, y)\|_m$ as previously. Draw a sketch of the unit balls for these three norms.

I. Let S be any subset of $\mathbb{R}^n$. Let $C_b(S)$ denote the vector space of all *bounded* continuous functions on S. For $f \in C(S)$, define $\|f\|_\infty = \sup_{x \in S} |f(x)|$.
(a) Show that this is a norm on $C_b(S)$.
(b) When is this a norm on the vector space of *all* continuous functions on S?

J. (a) Let $x_1, \ldots, x_{n+1}$ be distinct points in $[a, b]$. Show that there are polynomials p_i of degree n so that $p_i(x_j) = \delta_{ij}$ for $1 \leq i, j \leq n + 1$.
(b) Hence show that the subspace of $C[a, b]$ of all polynomials of degree at most n is $(n + 1)$-dimensional.
(c) Deduce that $C[a, b]$ is infinite dimensional.

K. (a) Let $a = x_0 < \cdots < x_n = b$ be distinct points in a compact subset K of $\mathbb{R}$. Let

$$h_0(x) = \begin{cases} \frac{x-a}{x_1-a} & a \leq x \leq x_1 \\ 0 & x_1 \leq x \leq b \end{cases} \qquad h_n = \begin{cases} 0 & a \leq x \leq x_{n-1} \\ \frac{x-x_{n-1}}{b-x_{n-1}} & x_{n-1} \leq x \leq b \end{cases}$$

and for $1 \leq j \leq n-1$, let

$$h_j = \begin{cases} 0 & a \leq x \leq x_{k-1} \text{ and } x_{k+1} \leq x \leq b \\ \frac{x-x_{k-1}}{x_k-x_{k-1}} & x_{k-1} \leq x \leq x_k \\ \frac{x_{k+1}-x}{x_{k+1}-x_k} & x_k \leq x \leq x_{k+1}. \end{cases}$$

Describe the linear span of $\{h_0, h_1, \ldots, h_n\}$ as a subspace of $C(K)$.

(b) Hence show that $C(K)$ is infinite dimensional if K is an infinite set.

7.2. Topology in Normed Spaces

The point of this section is to show how the notions of convergence and topology, which we developed in $\mathbb{R}$ and in $\mathbb{R}^n$, can be generalized to any normed vector space. We give the definitions and some simple properties where everything is almost identical to the treatment of $\mathbb{R}^n$. Other properties, such as the Heine–Borel Theorem, do not hold in general. For such properties, we will have to study specific kinds of normed vector spaces individually using their special properties.

The notion of convergence is the most fundamental and is exactly the same as our definition for $\mathbb{R}^n$.

7.2.1. DEFINITION. In a normed vector space $(V, \|\cdot\|)$, we say that a sequence $(v_n)_{n=1}^{\infty}$ **converges** to $v \in V$ if $\lim\limits_{n \to \infty} \|v_n - v\| = 0$. Equivalently, for every $\varepsilon > 0$, there is an integer $N > 0$ so that $\|v_n - v\| < \varepsilon$ for all $n \geq N$. This is written $\lim\limits_{n \to \infty} v_n = v$.

As in $\mathbb{R}^n$, we can decide if a sequence is attempting to converge without mentioning a limit point.

7.2.2. DEFINITION. Call $(v_n)_{n=1}^{\infty}$ a **Cauchy sequence** if for every $\varepsilon > 0$, there is an integer $N > 0$ so that $\|v_n - v_m\| < \varepsilon$ for all $n, m \geq N$.

This leads to the notion of completeness in this context. Completeness is the fundamental property that distinguishes the real numbers from the rational numbers. We have already seen several important theorems that depend on this notion, such as the Least Upper Bound Principle (2.5.3) and the completeness both of $\mathbb{R}$ (Theorem 2.7.4) and of $\mathbb{R}^n$ (Theorem 4.2.5). So it should not surprise the reader to find out that this is also a fundamental property of bigger normed spaces such as $(C(K), \|\cdot\|_\infty)$. That $C(K)$ is complete will be established in the next chapter; see Theorem 8.2.2.

7.2.3. DEFINITION. Say that $(V, \| \cdot \|)$ is **complete** if every Cauchy sequence in V converges to some vector $v \in V$. A complete normed space is called a **Banach space**.

Now we can reformulate convergence in terms of open and closed sets, again exactly as for $\mathbb{R}^n$.

7.2.4. DEFINITION. For a normed vector space $(V, \| \cdot \|)$, we define the **open ball** with centre $a \in V$ and radius $r > 0$ to be $B_r(a) = \{v \in V : \|v - a\| < r\}$.

A subset U of V is **open** if for every $a \in U$, there is some $r > 0$ so that $B_r(a) \subset U$.

A subset C of V is **closed** if it contains all of its limit points. That is, whenever (x_n) is a convergent sequence of points in C with limit $x = \lim_{n \to \infty} x_n$, then x belongs to C.

Proposition 4.3.8 works just as well for any normed space. So the open sets are precisely the complements of closed sets. Here is a sample result in showing the relationship between convergence and topology.

7.2.5. PROPOSITION. *A sequence x_n in a normed vector space V converges to a vector x if and only if for each open set U containing x, there is an integer N so that $x_n \in U$ for all $n \geq N$.*

PROOF. Suppose that $x = \lim_{n \to \infty} x_n$ and U is an open set containing x. Then there is an $r > 0$ so that $B_r(x)$ is contained in U. From the definition of limit, there is an integer N so that

$$\|x - x_n\| < r \quad \text{for all} \quad n \geq N.$$

This just says that $x_n \in B_r(x) \subset U$ for all $n \geq N$.

Conversely, suppose that the latter condition holds. In order to establish that $x = \lim_{n \to \infty} x_n$, let $r > 0$ be given. Take the open set $U = B_r(x)$. By hypothesis, there is an integer N so that $x_n \in U$ for all $n \geq N$. As before, this just means that

$$\|x - x_n\| < r \quad \text{for all} \quad n \geq N.$$

Hence $\lim_{n \to \infty} x_n = x$. ∎

There is one more fundamental property that we can define in this general context, compactness. However, the reader should be warned that the main theorem about compactness in $\mathbb{R}^n$, the Heine–Borel Theorem, is *not valid* in infinite-dimensional spaces. The correct characterization of compact sets is given by the Borel–Lebesgue Theorem (Theorem 9.2.3), which we will prove in the context of metric spaces in Chapter 9.

7.2.6. DEFINITION. A subset K of a normed vector space V is **compact** if every sequence (x_n) of points in K has a subsequence (x_{n_i}) which converges to a point in K.

Exercises for Section 7.2

A. If $x = \lim\limits_{n \to \infty} x_n$ and $y = \lim\limits_{n \to \infty} y_n$ in a normed space V and $\alpha = \lim\limits_{n \to \infty} \alpha_n$, show that
$x + y = \lim\limits_{n \to \infty} x_n + y_n$ and $\alpha x = \lim\limits_{n \to \infty} \alpha_n x_n$.

B. Show that every convergent sequence in any normed space is a Cauchy sequence.

C. If A is a subset of $(V, \|\cdot\|)$, let $\bar{A}$ denote its closure. Show that if $x \in V$ and $\alpha \in \mathbb{R}$, then $\overline{x + A} = x + \bar{A}$ and $\alpha\bar{A} = \overline{\alpha A}$.

D. Show that if A is an arbitrary subset of a normed space V and U is an open subset, then $A + U = \{a + u : a \in A, \ u \in U\}$ is open.

E. Prove that if two norms on V have the same unit ball, then the norms are equal.

F. Which of the following sets are open in $C^2[0, 1]$? Explain.
  (a) $A = \{f \in C^2[0, 1] : f(x) > 0, \ \|f'\|_\infty < 1, \ |f''(0)| > 2\}$
  (b) $B = \{f \in C^2[0, 1] : f(1) < 0, \ f'(1) = 0, \ f''(1) > 0\}$
  (c) $C = \{f \in C^2[0, 1] : f(x)f'(x) > 0 \text{ for } 0 \le x \le 1\}$.
       HINT: Extreme Value Theorem and Intermediate Value Theorem
  (d) $D = \{f \in C^2[0, 1] : f(x)f'(x) > 0 \text{ for } 0 < x < 1\}$.
       HINT: Why is this different from the previous example?

G. Prove that a compact subset of a normed vector space is closed and bounded.

H. (a) Prove that a compact subset of a normed vector space is complete.
  (b) Prove that a closed subset of a complete normed vector space is complete.

I. Consider the functions in $C[-1, 1]$ given by $f_n(x) = \begin{cases} 0 & -1 \le x \le 0 \\ nx & 0 \le x \le \frac{1}{n} \\ 1 & \frac{1}{n} \le x \le 1 \end{cases}$.
  (a) Show that $\|f_n - f_m\|_\infty \ge \frac{1}{2}$ if $m \ge 2n$.
  (b) Hence show that no subsequence of $(f_n)_{n=1}^\infty$ converges.
  (c) Conclude that the unit ball of $C[-1, 1]$ is not compact.
  (d) Show that the unit ball of $C[-1, 1]$ is closed and bounded and complete.

J. Prove that the following are equivalent for a normed vector space $(V, \|\cdot\|)$.
  (1) $(V, \|\cdot\|)$ is complete.
  (2) Every decreasing sequence of closed balls has a nonempty intersection.
       Note that the balls need not be concentric.
  (3) Every decreasing sequence of closed balls with radii going to zero has nonempty intersection.
  HINT: (1) $\implies$ (2) show that the centres of the balls form a Cauchy sequence.

K. (a) Show that if A is a closed subset of a normed vector space V and C is a compact subset, then $A + C = \{a + c : a \in A, \ c \in C\}$ is closed.
  (b) Is it enough for C to also be closed, or is compactness necessary?
  (c) If A and C are both compact, show that $A + C$ is compact.

L. Let $X_n = \{f \in C[0,1] : f(0) = 0, \|f\|_\infty \leq 1, \text{ and } f(x) \geq \frac{1}{2} \text{ for } x \geq \frac{1}{n}\}$.
  (a) Show that X_n is a closed bounded subset of $C[0,1]$.
  (b) Show that X_{n+1} is a proper subset of X_n for $n \geq 1$, and compute $\bigcap_{n \geq 1} X_n$.
  (c) Compare this with the Cantor Intersection Theorem (Theorem 4.4.7). Why does the theorem fail in this context?

M. Let c_0 be the vector space of all sequences $\mathbf{x} = (x_n)_{n=1}^\infty$ such that $\lim_{n \to \infty} x_n = 0$. Define a norm on c_0 by $\|\mathbf{x}\|_\infty = \sup_{n \geq 1} |x_n|$. Prove that c_0 is complete.
  HINT: Let $\mathbf{x}_k = (x_{k,n})_{n=1}^\infty$ be a Cauchy sequence in c_0.
  (a) Show that $(x_{k,n})_{k=1}^\infty$ is Cauchy for each $n \geq 1$. Hence define $\mathbf{x} = (x_n)$ by $x_n = \lim_{k \to \infty} x_{k,n}$.
  (b) Show that $\|\mathbf{x}_k\|_\infty$ is Cauchy and $\|\mathbf{x}\| \leq \lim_{k \to \infty} \|\mathbf{x}_k\|_\infty$.
  (c) Given $\varepsilon > 0$, apply the Cauchy criterion. Show that there is an integer K so that $|x_n - x_{k,n}| \leq \varepsilon$ for all $n \geq 1$ and all $k \geq K$.
  (d) Conclude that $\mathbf{x}$ belongs to c_0 and that $\lim_{k \to \infty} \mathbf{x}_k = \mathbf{x}$.

N. Consider the sequence f_n in $C[-1,1]$ from Exercise 7.2.I, but use the $L^1[-1,1]$ norm.
  (a) Show that f_n is Cauchy in the L^1 norm.
  (b) Show that f_n converges to $\chi_{(0,1]}$, the characteristic function of $(0,1]$, in the L^1 norm.
  (c) Show that $\|\chi_{(0,1]} - h\|_1 > 0$ for every h in $C[-1,1]$.
  (d) Conclude that $C[-1,1]$ is not complete in the L^1 norm.

7.3. Inner Product Spaces

In studying $\mathbb{R}^n$, we constructed the Euclidean norm using the dot product. An inner product on a vector space is a generalization of the dot product. It is one of the most important sources of norms, and the norms obtained from inner products are particularly tractable. For example, the L^2 norm on $C[a,b]$ arises this way.

7.3.1. DEFINITION. An **inner product** on a vector space V is a function $\langle x, y \rangle$ on pairs (x,y) of vectors in $V \times V$ taking values in $\mathbb{R}$ satisfying the following properties:

(1) (positive definiteness) $\langle x, x \rangle \geq 0$ for all $x \in V$ and $\langle x, x \rangle = 0$ only if $x = 0$.

(2) (symmetry) $\langle x, y \rangle = \langle y, x \rangle$ for all $x, y \in V$.

(3) (bilinearity)   For all $x, y, z \in V$ and scalars $\alpha, \beta \in \mathbb{R}$,
$$\langle \alpha x + \beta y, z \rangle = \alpha \langle x, z \rangle + \beta \langle y, z \rangle.$$

Given an inner product space, it is easy to check that the following definition gives us a norm:
$$\|x\| = \langle x, x \rangle^{1/2}.$$

The dot product on $\mathbb{R}^n$ is an inner product and the norm obtained from it is the usual Euclidean norm.

The bilinearity condition just given is linearity in the first variable. But as the term suggests, it really means a twofold linearity because combining it with symmetry yields linearity in the second variable as well. For x, y in V and scalars α, β in $\mathbb{R}$,

$$\langle z, \alpha x + \beta y \rangle = \langle \alpha x + \beta y, z \rangle = \alpha \langle x, z \rangle + \beta \langle y, z \rangle = \alpha \langle z, x \rangle + \beta \langle z, y \rangle.$$

7.3.2. EXAMPLE. The space $C[a, b]$ can be given an inner product

$$\langle f, g \rangle = \int_a^b f(x) g(x) \, dx.$$

This gives rise to the L^2 norm, which we defined in Example 7.1.5. Positive definiteness of this norm was established in Example 7.1.5 for arbitrary p, including $p = 2$. The other two properties follow from the linearity of the integral and are left as an exercise for the reader.

7.3.3. EXAMPLE. The space $\mathbb{R}^n$ can be given other inner product structures by weighting the vectors by a matrix $A = \begin{bmatrix} a_{ij} \end{bmatrix}$ by

$$\langle \mathbf{x}, \mathbf{y} \rangle_A = \langle A\mathbf{x}, \mathbf{y} \rangle = \sum_{i=1}^n \sum_{j=1}^n a_{ij} x_i y_j.$$

This is easily seen to be bilinear because A is linear and the standard inner product is bilinear. To be symmetric, we must require that $a_{ij} = a_{ji}$, so A is a symmetric matrix. Moreover, we need an additional condition to ensure that the inner product is positive definite. It turns out that the necessary condition is that the eigenvalues of A are all strictly positive. The proof of this is an important result from linear algebra known as the Spectral Theorem for Symmetric Matrices or as the Principal Axis Theorem.

For the purposes of this example, consider the 2×2 matrix $A = \begin{bmatrix} 3 & 1 \\ 1 & 2 \end{bmatrix}$. We noted that since A is symmetric, the inner product $\langle \cdot, \cdot \rangle_A$ is symmetric and bilinear. Let us establish directly that it is positive definite. Take a vector $\mathbf{x} = (x, y)$.

$$\langle \mathbf{x}, \mathbf{x} \rangle_A = 3x^2 + xy + yx + 2y^2 = 2x^2 + (x + y)^2 + y^2$$

From this identity, it is clear that $\langle \mathbf{x}, \mathbf{x} \rangle_A \geq 0$. Moreover, equality requires that x, y and $x + y$ all be 0, whence $\mathbf{x} = 0$. So it is positive definite.

It is a fundamental fact that every inner product space satisfies the Schwarz inequality. Our proof for $\mathbb{R}^n$ was special, using the specific formula for the dot product. We now show that it follows just from the basic properties of an inner product discussed previously.

7.3.4. CAUCHY–SCHWARZ INEQUALITY.

For all vectors x, y in an inner product space V,

$$|\langle x, y \rangle| \le \|x\| \, \|y\|.$$

Equality holds if and only if x and y are colinear.

PROOF. If either x or y is 0, both sides of the inequality are 0. Equality holds here, and these vectors are colinear. So we may assume that x and y are nonzero.

Apply the positive definite property to the vector $x - ty$ for $t \in \mathbb{R}$.

$$
\begin{aligned}
0 \le \langle x - ty, x - ty \rangle &= \langle x, x - ty \rangle - t \langle y, x - ty \rangle \\
&= \langle x, x \rangle - t \langle x, y \rangle - t \langle x, y \rangle + t^2 \langle y, y \rangle \\
&= \|x\|^2 - 2t \langle x, y \rangle + t^2 \|y\|^2
\end{aligned}
$$

We may rearrange this to obtain

$$2t \langle x, y \rangle \le \|x\|^2 + t^2 \|y\|^2.$$

Choose $t = \pm \|x\| / \|y\|$. Divide by $2|t|$ to obtain

$$\pm \langle x, y \rangle \le \frac{\|x\|^2 + (\|x\|/\|y\|)^2 \|y\|^2}{2(\|x\|/\|y\|)} = \frac{2\|x\|^2}{2(\|x\|/\|y\|)} = \|x\| \, \|y\|.$$

This establishes the inequality.

For equality to hold, the vector $x - ty$ must have norm 0. By the positive definite property, this means that $x = ty$ and so they are colinear. Conversely, if $x = ty$, then $\|x\|^2 = \langle ty, ty \rangle = t^2 \|y\|^2$ and

$$|\langle x, y \rangle| = |t| \langle y, y \rangle = \sqrt{t^2 \langle y, y \rangle} \sqrt{\langle y, y \rangle} = \|x\| \, \|y\|. \qquad \blacksquare$$

7.3.5. COROLLARY. *For $f, g \in C[a, b]$, we have*

$$\left| \int_a^b f(x) g(x) \, dx \right| \le \left(\int_a^b f(x)^2 \, dx \right)^{1/2} \left(\int_a^b g(x)^2 \, dx \right)^{1/2}.$$

As for $\mathbb{R}^n$, the triangle inequality in an immediate consequence. In particular, the L^2 norms on $C[a, b]$ are indeed norms.

7.3.6. COROLLARY. *An inner product space V satisfies the triangle inequality*

$$\|x + y\| \le \|x\| + \|y\| \quad \text{for all} \quad x, y \in V.$$

Moreover, if equality occurs, then x and y are colinear.

PROOF. This proof is identical to the $\mathbb{R}^n$ case. The key is the use of Cauchy–Schwarz inequality.

$$\begin{aligned} \|x + y\|^2 &= \langle x + y, x + y \rangle \\ &= \langle x, x \rangle + 2\langle x, y \rangle + \langle y, y \rangle \\ &\leq \|x\|^2 + 2\|x\|\,\|y\| + \|y\|^2 = \left(\|x\| + \|y\| \right)^2 \end{aligned}$$

This establishes the inequality.

Moreover, equality only occurs if $\langle x, y \rangle = \|x\|\,\|y\|$, which by the Cauchy–Schwarz inequality can only happen when x and y are colinear. ∎

Another fundamental consequence of the Cauchy–Schwarz inequality is that the inner product is continuous with respect to the induced norm. We leave the proof as an exercise.

7.3.7. COROLLARY. *Let V be an inner product space with induced norm $\|\cdot\|$. Then the inner product is continuous (i.e., if x_n converges to x and y_n converges to y, then $\langle x_n, y_n \rangle$ converges to $\langle x, y \rangle$).*

Exercises for Section 7.3

A. Let $A = \begin{bmatrix} 3 & 1 & 2 \\ 1 & 2 & 1 \\ 2 & 1 & 4 \end{bmatrix}$. Show that the form $\langle \cdot, \cdot \rangle_A$ is positive definite on $\mathbb{R}^3$.

B. Minimize the quantity $\|x\|^2 - 2t\langle x, y \rangle + t^2\|y\|^2$ over $t \in \mathbb{R}$. You will see why we chose t as we did in the proof of the Cauchy–Schwarz inequality.

C. Show that it is possible for x and y to be colinear yet the triangle inequality is still a strict inequality.

D. Prove Corollary 7.3.7.

E. Let $w(x)$ be a strictly positive continuous function on $[a, b]$. Define a form on $C[a, b]$ by the formula $\langle f, g \rangle_w = \displaystyle\int_a^b f(x)g(x)w(x)\,dx$ for $f, g \in C[a, b]$. Show that this is an inner product.

F. A normed vector space V is **strictly convex** if $\|u\| = \|v\| = \|(u + v)/2\| = 1$ for vectors $u, v \in V$ implies that $u = v$.
(a) Show that an inner product space is always strictly convex.
(b) Show that $\mathbb{R}^2$ with the norm $\|(x, y)\|_\infty = \max\{|x|, |y|\}$ is not strictly convex.

G. Let T be an $n \times n$ matrix. Define a form on $\mathbb{R}^n$ by $\langle \mathbf{x}, \mathbf{y} \rangle_T = \langle T\mathbf{x}, T\mathbf{y} \rangle$ for $\mathbf{x}, \mathbf{y} \in \mathbb{R}^n$. Show that this is an inner product if and only if T is invertible.

H. Let T be an invertible $n \times n$ matrix. Prove that a sequence of vectors $\mathbf{x}_k$ in $\mathbb{R}^n$ converges to a vector $\mathbf{x}$ in the usual Euclidean norm if and only if it converges to $\mathbf{x}$ in the norm of Exercise G, namely $\|\mathbf{x}\|_T := \|T\mathbf{x}\|$.
HINT: Use the previous exercise to show that $\dfrac{\|\mathbf{x}\|}{\|T^{-1}\|_2} \leq \|T\mathbf{x}\| \leq \|T\|_2\,\|\mathbf{x}\|$.

I. Show that there is an inner product on the space $\mathcal{M}_n$ of all $n \times n$ matrices given by
$$\langle A, B \rangle = \operatorname{Tr}(AB^t) = \sum_{i=1}^{n} \sum_{j=1}^{n} a_{ij} b_{ij} \text{ where } \operatorname{Tr}(A) = \sum_{i=1}^{n} a_{ii}$$ denotes the trace. The
norm $\|A\|_2 = \langle A, A \rangle^{1/2}$ is called the **Hilbert–Schmidt norm**.

J. Let $A = \begin{bmatrix} a_{ij} \end{bmatrix}$ be an $n \times n$ matrix, and let $\mathbf{x} = (x_1, \ldots, x_n)$ be a vector in $\mathbb{R}^n$. Show
that $\|A\mathbf{x}\| \le \|A\|_2 \|\mathbf{x}\|$.
    HINT: Compute $\|A\mathbf{x}\|^2$ using coordinates, and apply the Cauchy–Schwarz inequality.

K. Given an inner product space V, define a function on $V \setminus \{0\}$ by $Rx = x/\|x\|^2$. This
R is called *inversion* with respect to the unit sphere $\{x \in V : \|x\| = 1\}$.
    (a) Prove that $\|Rx - Ry\| = \dfrac{\|x - y\|}{\|x\|\,\|y\|}$.
    (b) Hence show that the inversion R is continuous.
    (c) Show that for all $w, x, y, z \in V$,

$$\|w - y\|\,\|x - z\| \le \|w - x\|\,\|y - z\| + \|w - z\|\,\|x - y\|.$$

    HINT: Reduce to the case $w = 0$, and reinterpret the inequality using inversion.

7.4. Orthonormal Sets

7.4.1. DEFINITION. Two vectors x and y are called **orthogonal** if $\langle x, y \rangle = 0$.
A collection of vectors $\{e_n : n \in S\}$ in V is called **orthonormal** if $\|e_n\| = 1$ for
all $n \in S$ and $\langle e_n, e_m \rangle = 0$ for $n \ne m \in S$. This set is called an **orthonormal
basis** if in addition this set is maximal with respect to being an orthonormal set.

The space $\mathbb{R}^n$ has many orthonormal bases, including the canonical one given
by $\mathbf{e}_1 = (1, 0, \ldots, 0), \ldots, \mathbf{e}_n = (0, \ldots, 0, 1)$.

7.4.2. PROPOSITION. *An orthonormal set is linearly independent.*
    An orthonormal basis for a finite-dimensional inner product space is a basis.

PROOF. Suppose that $\{e_n : n \in S\}$ is orthonormal and that there is a relation
$\sum\limits_{n \in S} \alpha_n e_n = \mathbf{0}$ with only finitely many $\alpha_n \ne 0$. Then

$$0 = \langle \mathbf{0}, e_k \rangle = \sum_{n \in S} \langle \alpha_n e_n, e_k \rangle = \alpha_k.$$

So $\alpha_k = 0$ for all $k \in S$ and this set is linearly independent.
    If V is a finite-dimensional inner product space and $\{e_n : n \in S\}$ is an or-
thonormal basis, then by the first paragraph it follows that $\{e_n : n \in S\}$ is linearly
independent. Let $V_0 = \operatorname{span}\{e_n : n \in S\}$. If $V_0 = V$, then $\{e_n : n \in S\}$ is a
spanning linearly independent set, and hence a basis.

On the other hand, if V_0 is a proper subspace, then there is a vector $v \in V$ that is not in V_0. Define

$$w = v - \sum_{n \in S} \langle v, e_n \rangle e_n.$$

Observe that $w \neq 0$ because $w = 0$ would mean that v belonged to the span of $\{e_n : n \in S\}$. Moreover,

$$\langle w, e_k \rangle = \langle v, e_k \rangle - \sum_{n \in S} \langle v, e_n \rangle \langle e_n, e_k \rangle = \langle v, e_k \rangle - \langle v, e_k \rangle = 0.$$

Thus w is orthogonal to $\{e_n : n \in S\}$. Therefore, $\{e_n : n \in S\} \cup \{w/\|w\|\}$ is a larger orthonormal set. This contradiction establishes that $V_0 = V$ as desired. ∎

7.4.3 THE GRAM–SCHMIDT PROCESS. The idea used in the proof of the proposition can be turned into an algorithm for producing orthonormal sets with nice properties. It provides an orthonormal set with the same span as some initial set of vectors $\{x_1, \ldots, x_n\}$. Set $y_1 = x_1$ and inductively define

$$y_{k+1} = x_{k+1} - \sum_{i=1}^{k} \langle x_{k+1}, y_i \rangle \frac{y_i}{\|y_i\|^2} \quad \text{for} \quad 1 \le k < n,$$

where we interpret $y_i/\|y_i\|^2$ to be 0 if $y_i = 0$. Now delete those terms y_i that are zero, and for the rest define $f_i = y_i/\|y_i\|$. This yields a set with three properties:

(1) $\mathrm{span}\{f_1, \ldots, f_k\} = \mathrm{span}\{x_1, \ldots, x_k\}$ for $1 \le k \le n$,

(2) $y_k = 0$ (and f_k is not defined) if and only if $x_k \in \mathrm{span}\{x_i : i < k\}$, and

(3) $\{f_1, \ldots, f_n\}$ is an orthonormal basis for $\mathrm{span}\{x_1, \ldots, x_n\}$.

The proof of these facts is left to the exercises. By applying this algorithm to any basis for a finite-dimensional inner product space, it follows that an inner product space of dimension n always has an orthonormal basis consisting of n vectors.

Let us look at a fundamental example that underlies Fourier series, a subject to which we devote two chapters later on.

7.4.4. EXAMPLE. We use the natural inner product on $C[-\pi, \pi]$ given by

$$\langle f, g \rangle = \frac{1}{2\pi} \int_{-\pi}^{\pi} f(\theta) g(\theta) \, d\theta.$$

The factor $\frac{1}{2\pi}$ is put in to make the constant function 1 have unit length. The norm is given by

$$\|f\|_2 := \langle f, f \rangle^{1/2} = \left(\frac{1}{2\pi} \int_{-\pi}^{\pi} |f(\theta)|^2 \, d\theta \right)^{1/2}.$$

This is called the L^2 norm.

NOTE: This inner product and L^2 norm have the factor $\frac{1}{2\pi}$ unlike the original definition of $\langle f, g \rangle$ and $\|f\|_2$ for $f \in C[a, b]$ given in the last two sections. We will *always* use these normalizations for $C[-\pi, \pi]$ and *never* for general spaces $C[a, b]$.

7.4.5. LEMMA. *The functions $\{1, \sqrt{2} \cos n\theta, \sqrt{2} \sin n\theta : n \geq 1\}$ form an orthonormal set in $C[-\pi, \pi]$ with this inner product.*

PROOF. Starting with the cosines for $n \geq m \geq 1$, we have

$$\langle \sqrt{2} \cos n\theta, \sqrt{2} \cos m\theta \rangle = \frac{1}{\pi} \int_{-\pi}^{\pi} \cos n\theta \cos m\theta \, dt$$

$$= \frac{1}{2\pi} \int_{-\pi}^{\pi} \cos(m+n)\theta + \cos(m-n)\theta \, dt,$$

where we've used the identity $2 \cos A \cos B = \cos(A + B) + \cos(A - B)$. If $n > m$, then both $m + n$ and $m - n$ are not zero and the integral is

$$\langle \sqrt{2} \cos n\theta, \sqrt{2} \cos m\theta \rangle = \frac{1}{2\pi} \left(\frac{\sin(m+n)\theta}{m+n} + \frac{\sin(m-n)\theta}{m-n} \right) \Big|_{-\pi}^{\pi} = 0.$$

If $n = m$, then

$$\langle \sqrt{2} \cos n\theta, \sqrt{2} \cos n\theta \rangle = \frac{1}{2\pi} \left(\frac{\sin 2n\theta}{2n} + \theta \right) \Big|_{-\pi}^{\pi} = 1.$$

The other cases are very similar and are left to the reader. ∎

A **trigonometric polynomial** is a finite sum

$$f(\theta) = A_0 + \sum_{k=1}^{N} A_k \cos k\theta + B_k \sin k\theta.$$

In this case, the coefficients may be recovered using these orthogonality relations. Indeed,

$$\langle f, 1 \rangle = A_0 \langle 1, 1 \rangle + \sum_{k=1}^{N} A_k \langle \cos k\theta, 1 \rangle + B_k \langle \sin k\theta, 1 \rangle = A_0,$$

$$\langle f, \cos n\theta \rangle = A_0 \langle 1, \cos n\theta \rangle + \sum_{k=1}^{N} A_k \langle \cos k\theta, \cos n\theta \rangle + B_k \langle \sin k\theta, \cos n\theta \rangle$$

$$= A_n \langle \cos n\theta, \cos n\theta \rangle = \frac{A_n}{2},$$

and

$$\langle f, \sin n\theta \rangle = A_0 \langle 1, \sin n\theta \rangle + \sum_{k=1}^{N} A_k \langle \cos k\theta, \sin n\theta \rangle + B_k \langle \sin k\theta, \sin n\theta \rangle$$

$$= B_n \langle \sin n\theta, \sin n\theta \rangle = \frac{B_n}{2}.$$

We make the following definition.

7.4.6. DEFINITION. Denote the **Fourier series** of $f \in C[-\pi, \pi]$ by

$$f \sim A_0 + \sum_{n=1}^{\infty} A_n \cos n\theta + B_n \sin n\theta,$$

where $A_0 = \dfrac{1}{2\pi} \displaystyle\int_{-\pi}^{\pi} f(t)\, dt$, and, for $n \geq 1$,

$$A_n = \frac{1}{\pi} \int_{-\pi}^{\pi} f(t) \cos nt\, dt \qquad \text{and} \qquad B_n = \frac{1}{\pi} \int_{-\pi}^{\pi} f(t) \sin nt\, dt.$$

The sequences $(A_n)_{n \geq 0}$ and $(B_n)_{n \geq 1}$ are the **Fourier coefficients** of f.

If f is an arbitrary continuous function, it is not immediately evident that the Fourier series of f will equal f. This is a serious problem, which required a lot of work and helped force mathematicians to adopt the careful definitions and concern with proofs that drive this book. The answer must involve convergence of a series of functions. The general ideas that we need are developed in Chapter 8.

7.4.7. LEMMA. *Let $\{e_1, \ldots, e_n\}$ be an orthonormal set in an inner product space V. If M is the subspace spanned by $\{e_1, \ldots, e_n\}$, then every vector $x \in M$ can be written* uniquely *as $\sum_{i=1}^{n} \alpha_i e_i$, where $\alpha_i = \langle x, e_i \rangle$. In other words, the set $\{e_1, \ldots, e_n\}$ is linearly independent.*

Moreover, for each y in V with $\langle y, e_i \rangle = \beta_i$ and each $x = \sum_{j=1}^{n} \alpha_j e_j$ in M,

$$\langle x, y \rangle = \sum_{i=1}^{n} \alpha_i \beta_i.$$

In particular, $\|x\|^2 = \sum_{j=1}^{n} \alpha_j^2.$

PROOF. If we can write a vector $x \in M$ as $\sum_{i=1}^{n} \alpha_i e_i$ and as $\sum_{i=1}^{n} \beta_i e_i$, then

$$\langle x, e_i \rangle = \Big\langle \sum_{j=1}^{n} \alpha_j e_j, e_i \Big\rangle = \sum_{j=1}^{n} \alpha_j \langle e_j, e_i \rangle = \alpha_i$$

and

$$\langle x, e_i \rangle = \Big\langle \sum_{j=1}^{n} \beta_j e_j, e_i \Big\rangle = \sum_{j=1}^{n} \beta_j \langle e_j, e_i \rangle = \beta_i.$$

Thus, $\alpha_i = \beta_i$ for all i. In other words, the coefficients of x are uniquely determined. In particular, if $x = 0$, then $\alpha_i = 0$ for $1 \le i \le n$; whence the set $\{e_1, \ldots, e_n\}$ is linearly independent.

Let y be a vector in V with $\langle y, e_i \rangle = \beta_i$ and let $x = \sum_{j=1}^{n} \alpha_j e_j$ be a vector in M. Compute

$$\langle x, y \rangle = \Big\langle \sum_{j=1}^{n} \alpha_j e_j, y \Big\rangle = \sum_{j=1}^{n} \alpha_j \langle e_j, y \rangle = \sum_{j=1}^{n} \alpha_j \beta_j$$

Taking $y = x$ gives $\|x\|^2 = \sum_{j=1}^{n} \alpha_j^2$. ∎

This lemma suffices to understand finite-dimensional inner product spaces. In particular, we see that every inner product of finite dimension n space behaves exactly as $\mathbb{R}^n$ with the dot product, once we coordinatize it using an orthonormal basis.

7.4.8. COROLLARY. *If V is an inner product space of finite dimension n, then it has an orthonormal basis $\{e_i : 1 \le i \le n\}$ and the inner product is given by*

$$\Big\langle \sum_{i=1}^{n} \alpha_i e_i, \sum_{j=1}^{n} \beta_j e_j \Big\rangle = \sum_{i=1}^{n} \alpha_i \beta_i$$

and the norm is

$$\Big\| \sum_{i=1}^{n} \alpha_i e_i \Big\| = \Big(\sum_{i=1}^{n} \alpha_i^2 \Big)^{1/2}.$$

PROOF. By definition of dimension, V has a basis consisting of n linearly independent vectors. Apply the Gram–Schmidt process (7.4.3) to this basis to obtain an orthonormal basis spanning V (Exercise 7.4.B). Now Lemma 7.4.7 provides the formulas for inner product and norm. ∎

Exercises for Section 7.4

A. Show that every inner product space satisfies the parallelogram law:
$$\|x + y\|^2 + \|x - y\|^2 = 2\|x\|^2 + 2\|y\|^2 \quad \text{for all} \quad x, y \in V.$$

B. The Gram–Schmidt process. We use the notation of (7.4.3).
   (a) Show by induction that $\operatorname{span}\{y_1, \ldots, y_k\} = \operatorname{span}\{x_1, \ldots, x_k\}$ for $1 \le k \le n$.
   (b) Show that y_{k+1} is orthogonal each y_i for $1 \le i \le k$.
   (c) Show that $y_{k+1} = \mathbf{0}$ if and only if it belongs to $\operatorname{span}\{y_1, \ldots, y_k\}$.
   (d) Hence conclude that the Gram–Schmidt process produces an orthonormal basis for $\operatorname{span}\{x_1, \ldots, x_n\}$.

C. Show that every finite-dimensional inner product space has an orthonormal basis.

D. Show that if $\{x_k : k \geq 1\}$ is a countable set of vectors in an inner product space, then the Gram–Schmidt process still produces an orthonormal set $\{f_i : i \in S\}$ for $S \subset \mathbb{N}$ with properties (1) and (2) of 7.4.3.

E. Find an orthonormal basis for the $n \times n$ matrices using the inner product of Exercise 7.3.I.

F. Complete the proof of Lemma 7.4.5.

G. Show that $f_n = \sin(n\pi x)$ for $n \geq 1$ forms an orthonormal set in $C[0, 1]$ with respect to the $L^2[0, 1]$ norm.

H. (a) Find the Fourier series for $\cos^3(\theta)$.
  (b) Use trig identities to verify that $\cos^3(\theta)$ can be expressed as the trig polynomial you found in (a).

I. If $f \in C[-\pi, \pi]$ has the Fourier series $f \sim A_0 + \sum_{n=1}^{\infty} A_n \cos n\theta + B_n \sin n\theta$, show that $A_0^2 + \dfrac{1}{2} \sum_{n=1}^{\infty} |A_n|^2 + |B_n|^2 \leq \dfrac{1}{2\pi} \int_{-\pi}^{\pi} |f(x)|^2 \, dx$.
  HINT: Consider the finite sums.
  NOTE: We will show in Example 13.6.5 that this is an equality.

J. Let $f(x) = x$ for $-\pi \leq x \leq \pi$. Compute the inner product $\langle f, \sin nx \rangle$ for $n \geq 1$. Hence show that $\sum_{n=1}^{\infty} \dfrac{1}{n^2} \leq \dfrac{\pi^2}{6}$.
  HINT: Integrate by parts. (See Example 13.6.5.)

7.5. Orthogonal Expansions in Inner Product Spaces

Given our success in the last section in understanding *finite-dimensional* inner product spaces by using orthonormal sets, it is natural to look at orthonormal sets in other inner product spaces. The first major theorem of this section, the Projection Theorem, deals with finite orthonormal sets in general inner product spaces. The rest of the section is devoted to extending the Projection Theorem to infinite orthonormal sets.

7.5.1. DEFINITION. A **projection** is a linear map P such that $P^2 = P$. In addition, say that P is an **orthogonal projection** if $\ker P = \{v \in V : Pv = 0\}$ is orthogonal to $\operatorname{Ran} P = PV$.

Observe that $\ker P = \operatorname{Ran}(I - P)$ because $Px = 0$ if and only if $(I - P)x = x$. This shows that $\ker P \subset \operatorname{Ran}(I - P)$. Conversely, $x = (I - P)y$ implies that $Px = (P - P^2)y = 0$. So when P is an orthogonal projection, the vectors Px and $(I - P)x$ are orthogonal. Therefore, we obtain the Pythagorean identity

$$\|x\|^2 = \|Px\|^2 + \|(I - P)x\|^2.$$

7.5.2. PROJECTION THEOREM.

Let $\{e_1, \ldots, e_n\}$ be an orthonormal set in an inner product space V and let M be the subspace spanned by $\{e_1, \ldots, e_n\}$. Define $P : V \to M$ by $Py = \sum_{j=1}^{n} \langle y, e_j \rangle e_j$, for each $y \in V$. Then P is the orthogonal projection onto M and

(7.5.3)
$$\|y\|^2 \geq \sum_{j=1}^{n} \langle y, e_j \rangle^2.$$

Moreover, for all $v \in M$,

(7.5.4)
$$\|y - v\|^2 = \|y - Py\|^2 + \|Py - v\|^2.$$

In particular, Py is the closest vector in M to y.

Figure 7.2 illustrates the theorem when M is a plane in $\mathbb{R}^3$.

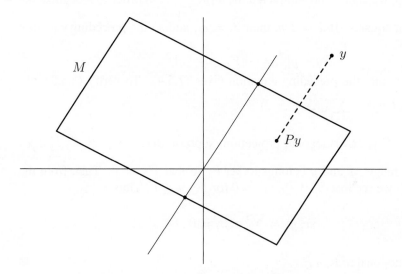

FIGURE 7.2. The projection of a point.

PROOF. We leave it to the reader to verify that P is linear by using the bilinearity of the inner product. It is evident by its definition that P maps V into M. Suppose that a typical vector in M is expressed as $x = \sum_{j=1}^{n} \alpha_j e_j$. By Lemma 7.4.7, $\langle x, e_j \rangle = \alpha_j$. So

$$Px = \sum_{j=1}^{n} \langle x, e_j \rangle e_j = \sum_{j=1}^{n} \alpha_j e_j = x.$$

For any $y \in V$, we have $Py \in M$ and so $P^2 y = P(Py) = Py$. Therefore, P is a projection of V onto M. As Py belongs to M, Lemma 7.4.7 shows that $\|Py\|^2 = \sum_{j=1}^{n} \beta_j^2$, where $\beta_j = \langle y, e_j \rangle$.

Now with $x \in M$ and $y \in V$ as in the previous lemma, we compute

$$\|x - y\|^2 = \langle x - y, x - y \rangle = \langle x, x \rangle - 2\langle x, y \rangle + \langle y, y \rangle$$

$$= \sum_{j=1}^{n} \alpha_j^2 - 2 \sum_{j=1}^{n} \alpha_j \beta_j + \|y\|^2$$

$$= \sum_{j=1}^{n} \alpha_j^2 - 2 \sum_{j=1}^{n} \alpha_j \beta_j + \sum_{j=1}^{n} \beta_j^2 - \sum_{j=1}^{n} \beta_j^2 + \|y\|^2$$

$$= \sum_{j=1}^{n} (\alpha_j - \beta_j)^2 - \|Py\|^2 + \|y\|^2$$

$$= \|x - Py\|^2 - \|Py\|^2 + \|y\|^2.$$

In the third step, we added and subtracted the term $\sum_{j=1}^{n} \beta_j^2$ in order to recognize the rest as a sum of squares. If $x = Py$, then $\beta_j = \alpha_j$ and so the preceding equation becomes

$$\|Py - y\|^2 = -\|Py\|^2 + \|y\|^2.$$

Substituting this into the preceding equation gives (7.5.4). To verify (7.5.3), observe $\|y\|^2 \geq \|Py\|^2 = \sum_{j=1}^{n} \langle y, e_i \rangle^2$.

To see that P is an orthogonal projection, suppose that $x = Px = \sum_{j=1}^{n} \alpha_j e_j$ belongs to the range M and y belongs to $\ker P$, so that $Py = 0$. Then from the definition of P, we see that $\beta_j = \langle y, e_j \rangle = 0$ for $1 \leq i \leq n$. Thus

$$\langle x, y \rangle = \sum_{j=1}^{n} \alpha_j \beta_j = 0.$$

So $\ker P$ is orthogonal to $\mathrm{Ran}\, P$. ∎

How much of the Projection Theorem extends to infinite-dimensional spaces such as $C[a, b]$? We can generalize the inequality in (7.5.3) to any orthonormal set in an inner product space.

7.5.5. BESSEL'S INEQUALITY.
Let $\{e_n : n \in S\}$ be an orthonormal set in an inner product space V. For each vector $x \in V$,

$$\sum_{n \in S} |\langle x, e_n \rangle|^2 \leq \|x\|^2.$$

PROOF. Let us write $\alpha_n = \langle x, e_n \rangle$. If S is a finite set, then the Projection Theorem applies. In particular, if Px denotes the projection onto the span of the e_n, then

$$\sum_{n \in S} |\langle x, e_n \rangle|^2 = \|Px\|^2 \leq \|x\|^2.$$

So suppose that $S = \mathbb{N}$ is infinite. Using limits and the preceding argument for the finite set $\{e_n : 1 \leq n \leq N\}$ gives

$$\sum_{n=1}^{\infty} |\langle x, e_n \rangle|^2 = \lim_{N \to \infty} \sum_{n=1}^{N} |\langle x, e_n \rangle|^2 \leq \lim_{N \to \infty} \|x\|^2 = \|x\|^2. \qquad \blacksquare$$

To extend other parts of the Projection Theorem, we need to deal with infinite series of vectors and their convergence. This can be a delicate issue, however, the problem has an accessible solution if the inner product space is complete.

7.5.6. DEFINITION. A complete inner product space is called a **Hilbert space**.

We give a Hilbert space in the next example, but it is important to know that not all "natural" inner product spaces are complete. For example, the space $C[-\pi, \pi]$ examined in the previous section is not complete (the argument is outlined in Exercise 7.5.H). An abstract way to complete $C[-\pi, \pi]$ in the L^2 norm to obtain the Hilbert space $L^2(-\pi, \pi)$ is discussed in Section 9.6. Alternatively, $L^2(-\pi, \pi)$ may be constructed by developing a more powerful theory of integration, known as the Lebesgue integral, which is a central topic in a course on measure theory.

7.5.7. EXAMPLE. The space ℓ^2 consists of all sequences $\mathbf{x} = (x_n)_{n=1}^{\infty}$ such that $\|\mathbf{x}\|_2 := \left(\sum_{n=1}^{\infty} x_n^2 \right)^{1/2}$ is finite. The inner product on ℓ^2 is given by

$$\langle \mathbf{x}, \mathbf{y} \rangle = \sum_{n=1}^{\infty} x_n y_n.$$

In order for this inner product to be well defined, we need to know that this series always converges. In fact, it always converges absolutely (see Exercise 7.5.B).

7.5.8. THEOREM. *The space ℓ^2 is complete.*

PROOF. We must show that if a sequence $\mathbf{x}_k = (x_{k,n})_{n=1}^{\infty}$ is Cauchy, then it converges to a vector $\mathbf{x}$ in ℓ^2. We know that for every $\varepsilon > 0$, there is a number K so large that $\|\mathbf{x}_k - \mathbf{x}_l\| < \varepsilon$ for all $k, l \geq K$. In particular,

$$|x_{k,n} - x_{l,n}| \leq \|\mathbf{x}_k - \mathbf{x}_l\| < \varepsilon \quad \text{for all} \quad k, l \geq K.$$

So for each coordinate n, the sequence $(x_{k,n})_{k=1}^{\infty}$ is a Cauchy sequence of real numbers. By the completeness of the real numbers (Theorem 2.7.4), there is a real number y_n, so that

$$y_n = \lim_{k \to \infty} x_{k,n} \quad \text{exists for each } n \geq 1.$$

Let $\mathbf{y} = (y_n)_{n=1}^{\infty}$. We need to show two things: first, that $\mathbf{y}$ is in ℓ^2 and, second, that $\mathbf{x}_k$ converges in ℓ^2 to $\mathbf{y}$.

It also follows from the triangle inequality that

$$\left| \|\mathbf{x}_k\| - \|\mathbf{x}_l\| \right| \le \|\mathbf{x}_k - \mathbf{x}_l\| < \varepsilon \quad \text{for all} \quad k, l \ge K.$$

Hence the sequence $(\|\mathbf{x}_k\|)_{k=1}^{\infty}$ is Cauchy. Let $L = \lim_{k \to \infty} \|\mathbf{x}_k\|$.

Fix an integer N. Then compute

$$\sum_{n=1}^{N} |y_n|^2 = \lim_{k \to \infty} \sum_{n=1}^{N} |x_{k,n}|^2 \le \lim_{k \to \infty} \|\mathbf{x}_k\|^2 = L^2.$$

Now take a limit as N tends to infinity to obtain

$$\|\mathbf{y}\|^2 = \lim_{N \to \infty} \sum_{n=1}^{N} |y_n|^2 \le L^2.$$

This shows that $\mathbf{y}$ belongs to ℓ^2.

A similar argument shows that $\mathbf{x}_k$ converges to $\mathbf{y}$. Indeed, fix $\varepsilon > 0$ and choose K as previously using the Cauchy criterion. Then fix N and compute

$$\sum_{n=1}^{N} |y_n - x_{k,n}|^2 = \lim_{l \to \infty} \sum_{n=1}^{N} |x_{l,n} - x_{k,n}|^2$$

$$\le \lim_{l \to \infty} \|\mathbf{x}_l - \mathbf{x}_k\|^2 \le \varepsilon^2.$$

Now that the right hand side is independent of N, we may let N tend to infinity to obtain

$$\|\mathbf{y} - \mathbf{x}_k\|^2 = \lim_{N \to \infty} \sum_{n=1}^{N} |y_n - x_{k,n}|^2 \le \varepsilon^2$$

for all $k \ge K$. Since $\varepsilon > 0$ is arbitrary, this establishes convergence. ∎

In a Hilbert space, the **closed span** of a set of vectors S, denoted $\overline{\text{span}\{S\}}$, is the closure of the linear subspace spanned by S. This is still a subspace. Since it is a closed subset of a complete space, it is also complete. Thus closed subspaces of Hilbert spaces are themselves Hilbert spaces in the given inner product.

We can now give a sharpening of Bessel's inequality. This theorem tells us precisely when Bessel's inequality is an equality. Solely to avoid the technicalities of uncountable bases (see Appendix 2.8), we assume that our Hilbert spaces are **separable**, meaning that every orthonormal set is countable. In other words, we assume that an orthonormal set can be indexed either by a finite set or by $\mathbb{N}$.

7.5.9. PARSEVAL'S THEOREM.

Let $S \subset \mathbb{N}$ and $E = \{e_n : n \in S\}$ be an orthonormal set in a Hilbert space H. Then the subspace $M = \overline{\text{span}\{E\}}$ consists of all vectors $x = \sum_{n \in S} \alpha_n e_n$, where the coefficient sequence $(\alpha_n)_{n=1}^{\infty}$ belongs to ℓ^2. Further, if x is a vector in H, then x belongs to M if and only if

$$\sum_{n \in S} |\langle x, e_n \rangle|^2 = \|x\|^2.$$

PROOF. When S is a finite set, this theorem follows from the Projection Theorem (Theorem 7.5.2). So suppose that $S = \mathbb{N}$.

Suppose that $(\alpha_n)_{n=1}^{\infty} \in \ell^2$. Define $x_k = \sum_{n=1}^{k} \alpha_n e_n$. We will show that this is a Cauchy sequence. Indeed, if $\varepsilon > 0$, then the convergence of $\sum_{n \geq 1} |\alpha_n|^2$ shows that there is an integer K so that $\sum_{n=K+1}^{\infty} |\alpha_n|^2 < \varepsilon^2$. Thus if $l \geq k \geq K$,

$$\|x_l - x_k\|^2 = \left\| \sum_{n=k+1}^{l} \alpha_n e_n \right\| = \sum_{n=k+1}^{l} |\alpha_n|^2 < \varepsilon^2.$$

As H is complete, this sequence converges to a vector x.

Since M is closed and each x_k lies in M, it follows that x belongs to M. Moreover, using Corollary 7.3.7,

$$\langle x, e_n \rangle = \lim_{k \to \infty} \langle x_k, e_n \rangle = \alpha_n \quad \text{for all} \quad n \geq 1.$$

So we can write without confusion $x = \sum_{n=1}^{\infty} \alpha_n e_n$. Thus M contains all of the ℓ^2 linear combinations of the basis vectors.

Now let x be an arbitrary vector in H and set $\alpha_n = \langle x, e_n \rangle$. By Bessel's inequality (7.5.5), the sequence $(\alpha_n)_{n=1}^{\infty}$ belongs to ℓ^2 and $\sum_{n \geq 1} |\alpha_n|^2 \leq \|\mathbf{x}\|^2$. Let $y = \sum_{n=1}^{\infty} \alpha_n e_n$. Compute

$$\|x - y\|^2 = \|x\|^2 - 2\langle x, y \rangle + \|y\|^2$$

$$= \|x\|^2 - 2 \sum_{n=1}^{\infty} \langle x, \alpha_n e_n \rangle + \sum_{n=1}^{\infty} |\alpha_n|^2$$

$$= \|x\|^2 - 2 \sum_{n=1}^{\infty} |\alpha_n|^2 + \sum_{n=1}^{\infty} |\alpha_n|^2$$

$$= \|x\|^2 - \sum_{n=1}^{\infty} |\alpha_n|^2.$$

So we see that if Bessel's inequality is an equality, then $x = y$ and thus it belongs to M.

Conversely, if x belongs to M, we must show that the series $\sum_{n=1}^{\infty} \alpha_n e_n$ actually converges to x itself. Since x belongs to M, it is the limit of vectors in the algebraic span of the basis vectors. So given any $\varepsilon > 0$, there is an integer N and a vector z in $\text{span}\{e_n : 1 \leq n \leq N\}$ such that $\|x - z\| < \varepsilon$. By the Projection Theorem, the vector $x_N = \sum_{n=1}^{N} \alpha_n e_n$ is closer to x:

$$\|x - x_N\| \leq \|x - z\| < \varepsilon.$$

Since this holds for all $\varepsilon > 0$, we deduce that a subsequence of the sequence $(x_k)_{n=1}^{\infty}$ converges to x. But this whole sequence converges (as shown in the second paragraph), so that $x = \sum_{n=1}^{\infty} \alpha_n e_n$. ■

7.5.10. COROLLARY. *Let $E = \{e_n : n \in S\}$ be an orthonormal set in a Hilbert space H. Then there is a continuous linear orthogonal projection P_E of H onto $M = \overline{\operatorname{span}\{E\}}$ given by $P_E x = \sum_{n \in S} \langle x, e_n \rangle e_n$.*

PROOF. The preceding proof established that $y = P_E x = \sum_{n \in S} \langle x, e_n \rangle e_n$ is defined and that
$$\|x\|^2 = \|y\|^2 + \|x - y\|^2 = \|P_E x\|^2 + \|x - P_E x\|^2.$$
Since the coefficients are determined in a linear fashion,
$$\langle \alpha x + \beta y, e_n \rangle = \alpha \langle x, e_n \rangle + \beta \langle y, e_n \rangle,$$
it follows that $P_E(\alpha x + \beta y) = \alpha P_E x + \beta P_E y$ for all $x, y \in H$ and scalars $\alpha, \beta \in \mathbb{R}$. So P_E is linear.

Also since $\|P_E x\| \leq \|x\|$, we obtain that
$$\|P_E x - P_E y\| = \|P_E(x - y)\| \leq \|x - y\|.$$
So P_E is Lipschitz and thus (uniformly) continuous. Parseval's Theorem also established that $P_E x = x$ if and only if $x \in M$. In particular, the range of P_E is precisely M.

Finally, $P_E y = 0$ if and only if $\langle y, e_n \rangle = 0$ for all $n \in S$. Thus if $x = P_E x = \sum_{n \in S} \alpha_n e_n$ and $P_E y = 0$, it follows that
$$\langle x, y \rangle = \sum_{n \in S} \alpha_n \langle e_n, y \rangle = 0.$$
Hence P_E is an orthogonal projection. ■

Recall that an orthonormal basis for a Hilbert space was defined as a maximal orthonormal set. Every Hilbert space has an orthonormal basis, but a proof of this fact requires assumptions from set theory, including the Axiom of Choice.

7.5.11. COROLLARY. *If $E = \{e_i : i \geq 1\}$ is an orthonormal basis for a Hilbert H, every vector $x \in H$ may be uniquely expressed as $x = \sum_{i=1}^{\infty} \alpha_i e_i$, where $\alpha_i = \langle x, e_i \rangle$.*

PROOF. We need to show that the closed span of a basis is H. If $M = \overline{\operatorname{span}\{E\}}$ is a proper subspace, there is a vector $x \in H$ that is not in M. Let $y = x - P_M x$ and $e = e/\|e\|$. By the previous corollary, $y \neq 0$. So e is a unit vector such that
$$P_M e = \frac{P_M(x - P_M x)}{\|y\|} = 0.$$

Thus e is orthogonal to M. In particular, $\{e, e_i : i \geq 1\}$ is orthonormal, which contradicts the maximality of E. Since E is maximal, it follows that $M = H$.

Parseval's Theorem shows that the closed span of an orthonormal set consists of all ℓ^2 combinations of this orthonormal set. So every vector $x \in H$ may be expressed as $\sum_{i=1}^{\infty} \alpha_i e_i$. The coefficients are unique because this expression for x implies that

$$\langle x, e_j \rangle = \lim_{n \to \infty} \left\langle \sum_{i=1}^{n} \alpha_i e_i, e_j \right\rangle = \sum_{i=1}^{n} \alpha_i \langle e_i, e_j \rangle = \alpha_j \qquad \blacksquare$$

Exercises for Section 7.5

A. When does equality hold in Equation (7.5.3)?

B. Let $\mathbf{x} = (x_n)_{n=1}^{\infty}$ and $\mathbf{y} = (y_n)_{n=1}^{\infty}$ be elements of ℓ^2.
   (a) Show that $\sum_{n=1}^{N} |x_n y_n| \leq \|\mathbf{x}\|\,\|\mathbf{y}\|$.   HINT: Schwarz inequality.
   (b) Hence prove that $\sum_{n=1}^{\infty} x_n y_n$ converges absolutely.

C. If M is a closed subspace of a Hilbert space H, define the **orthogonal complement** of M to be $M^{\perp} = \{x : \langle x, m \rangle = 0 \text{ for all } m \in M\}$.
   (a) Show that every vector in H can be written uniquely as $x = m + y$, where $m \in M$ and $y \in M^{\perp}$. Moreover, $\|x\|^2 = \|m\|^2 + \|y\|^2$.
   (b) Show that $M = (M^{\perp})^{\perp}$.

D. Let P be a projection on an inner product space V. Prove that the following are equivalent:
   (a) P is an orthogonal projection.
   (b) $\|v\|^2 = \|Pv\|^2 + \|v - Pv\|^2$ for all $v \in V$.
   (c) $\|Pv\| \leq \|v\|$ for all $v \in V$.
   (d) $\langle Pv, w \rangle = \langle v, Pw \rangle$ for all $v, w \in V$.
   HINT: For (c) $\Longrightarrow$ (d), show that *not* (d) implies there are vectors $v = Pv$ and $w = (I - P)w$ such that $\langle v, w \rangle > 0$. Compute $\|v - tw\|^2 - \|v\|^2$ for small $t > 0$. For (d) $\Longrightarrow$ (a), show that $\langle Pv, (I - P)w \rangle = 0$.

E. Show that the orthogonal projection P onto a subspace M is unique.
   HINT: If P and Q are orthogonal projections onto M, apply Q to $x = Px + (I - P)x$.

F. Formulate and prove a precise version of the following statement "A separable infinite-dimensional Hilbert space with an orthonormal basis $\{e_n : n \geq 1\}$ behaves like ℓ^2."
   HINT: Look at the finite-dimensional statement, Corollary 7.4.8.

G. Let M and N be closed subspaces of a Hilbert space H.
   (a) Show that $(M \cap N)^{\perp} = \overline{M^{\perp} + N^{\perp}}$.
   (b) Let $\{e_n : n \geq 1\}$ be an orthonormal basis for ℓ^2. Let $M = \overline{\text{span}}\{e_{2n} : n \geq 1\}$ and $N = \overline{\text{span}}\{e_{2n-1} + n e_{2n} : n \geq 1\}$. Show that $M + N$ is not closed.
   (c) Use (b) to show that closure is needed in part (a).

H. Consider the functions f_n in $C[-\pi, \pi]$ given by

$$f_n(x) = \begin{cases} 0 & -\pi \leq x \leq 0 \\ nx & 0 \leq x \leq \frac{1}{n} \\ 1 & \frac{1}{n} \leq x \leq \pi. \end{cases}$$

(a) Show that f_n converges in the L^2 norm to the characteristic function χ of $(0, \pi]$. In particular, $(f_n)_{n=1}^{\infty}$ is an L^2 Cauchy sequence.
(b) Show that $\|\chi - h\|_2 > 0$ for every function h in $C[-\pi, \pi]$.
(c) Hence conclude that $C[-\pi, \pi]$ is not complete in the L^2 norm.

I. Let $\{r_n : n \geq 1\}$ be a list of the rational points in $[0, 1]$. Define an inner product on $C[0, 1]$ by $\langle f, g \rangle_* = \sum_{n=1}^{\infty} 2^{-n} f(r_n) g(r_n)$.
(a) Show that this is indeed a positive definite inner product.
(b) Why is this not an inner product on the vector space of all bounded functions on $[0, 1]$?
(c) Show that $C[0, 1]$ is not complete in this inner product norm.
   HINT: Find a Cauchy sequence converging to the characteristic function of $\{0\}$.

7.6. Finite-Dimensional Normed Spaces

In this section, we will study the finite-dimensional subspaces of an arbitrary normed vector space. Many of the nice properties of $\mathbb{R}^n$ carry over to this setting. A particularly important result is Theorem 7.6.5, which will be very useful in the chapter on approximation.

7.6.1. LEMMA. *If $\{v_1, v_2, \ldots, v_n\}$ is a linearly independent set in a normed vector space $(V, \| \cdot \|)$, then there exist positive constants $0 < c < C$ so that for all $\mathbf{a} = (a_1, \ldots, a_n) \in \mathbb{R}^n$, we have*

$$c\|\mathbf{a}\|_2 \leq \left\| \sum_{i=1}^{n} a_i v_i \right\| \leq C\|\mathbf{a}\|_2$$

PROOF. By the triangle inequality and the Schwarz inequality (4.1.1),

$$\left\| \sum_{i=1}^{n} a_i v_i \right\| \leq \sum_{i=1}^{n} |a_i| \, \|v_i\|$$

$$\leq \left(\sum_{i=1}^{n} a_i^2 \right)^{1/2} \left(\sum_{i=1}^{n} \|v_i\|^2 \right)^{1/2} = C\|\mathbf{a}\|_2,$$

where $C = \left(\sum_{i=1}^{n} \|v_i\|^2 \right)^{1/2}$.

Define a function N on $\mathbb{R}^n$ by $N(\mathbf{a}) = \left\| \sum\limits_{i=1}^{n} a_i v_i \right\|$. Observe that N is a continuous function since

$$|N(\mathbf{x}) - N(\mathbf{y})| = \left| \left\| \sum_{i=1}^{n} x_i v_i \right\| - \left\| \sum_{i=1}^{n} y_i v_i \right\| \right|$$

$$\leq \left\| \sum_{i=1}^{n} (x_i - y_i) v_i \right\| \leq C \|\mathbf{x} - \mathbf{y}\|_2.$$

So N is Lipschitz and thus continuous.

Let $S = \{\mathbf{a} \in \mathbb{R}^n : \|\mathbf{a}\|_2 = 1\}$ be the unit sphere of $\mathbb{R}^n$. Since the set $\{v_1, \dots, v_n\}$ is linearly independent, it follows that $N(\mathbf{x}) > 0$ when $\mathbf{x} \neq 0$. In particular, N never vanishes on the compact set S. By the Extreme Value Theorem (Theorem 5.4.4), N must achieve its minimum value c at some point on S; whence $c > 0$. So if $\mathbf{a}$ is an arbitrary vector in $\mathbb{R}^n$, we obtain

$$\left\| \sum_{i=1}^{n} a_i v_i \right\| = \|\mathbf{a}\|_2 N(\mathbf{a}/\|\mathbf{a}\|_2) \geq c\|\mathbf{a}\|_2. \qquad \blacksquare$$

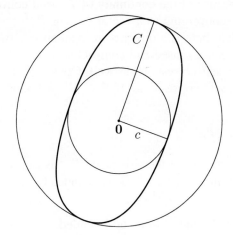

FIGURE 7.3. Euclidean balls inside and outside the unit ball of V.

The effect of this theorem is that every finite-dimensional normed space has the same topology as $(\mathbb{R}^n, \|\cdot\|_2)$, in the sense that they have the "same" convergent sequences, the "same" open sets, and so on. Let us make this more precise. If $\{v_1, v_2, \dots, v_n\}$ is a basis for an n-dimensional normed space V, define a linear transformation from $\mathbb{R}^n$ into V by $T\mathbf{a} = \sum\limits_{i=1}^{n} a_i v_i$. The map T carries $\mathbb{R}^n$ one-to-one and onto V by the definition of a basis. Since every element of V is a linear combination of $\{v_1, \dots, v_n\}$, we can define the inverse map by $T^{-1} \sum\limits_{i=1}^{n} a_i v_i = \mathbf{a}$.

7.6.2. COROLLARY. *Suppose that V is an n-dimensional normed space with basis $\{v_1, v_2, \ldots, v_n\}$. Then the maps T and T^{-1} defined previously are both Lipschitz, and therefore continuous. Thus a set A in $\mathbb{R}^n$ is closed, bounded, open, or compact if and only if $T(A)$ is closed, bounded, open, or compact in V.*

PROOF. We have for vectors $\mathbf{x}$ and $\mathbf{y}$ in $\mathbb{R}^n$,

$$\|T\mathbf{x} - T\mathbf{y}\| = N(\mathbf{x} - \mathbf{y}) \le C\|\mathbf{x} - \mathbf{y}\|_2.$$

So T is Lipschitz with constant C. Similarly, if $u = T\mathbf{x}$ and $v = T\mathbf{y}$ are typical vectors in V,

$$\|T^{-1}u - T^{-1}v\|_2 = \|T^{-1}(T\mathbf{x}) - T^{-1}(T\mathbf{y})\|_2 = \|\mathbf{x} - \mathbf{y}\|_2$$
$$\le \frac{1}{c}\|T\mathbf{x} - T\mathbf{y}\| = \frac{1}{c}\|u - v\|.$$

So T^{-1} is Lipschitz with constant $1/c$. It follows that both T and T^{-1} are continuous.

This means that if $\mathbf{x}_k$ is a sequence of vectors in $\mathbb{R}^n$ converging to a point x, then $T\mathbf{x}_k$ converges to Tx because of the continuity of T. And conversely, if v_k is a sequence of vectors in V converging to a vector v, then $T^{-1}v_k$ converges to $T^{-1}v$ in $\mathbb{R}^n$. This just says that there is a direct correspondence between convergent sequences in $\mathbb{R}^n$ and V. Since closed and compact sets are defined in terms of convergent sequences, these sets correspond as well. Open sets are the complements of closed sets, so open sets correspond. If $A \subset \mathbb{R}^n$ is bounded by L, the Lipschitz condition shows that

$$\|T\mathbf{a}\| = \|T\mathbf{a} - T\mathbf{0}\| \le C\|\mathbf{a}\|_2 \le CL$$

for every $\mathbf{a}$ in A. So $T(A)$ is bounded. Likewise, since T^{-1} is Lipschitz, if B is a subset of V bounded by L, it follows that $T^{-1}(B)$ is bounded by L/c in $\mathbb{R}^n$. ∎

Notice that we may conclude that closed and bounded sets also correspond. Since the Heine–Borel Theorem shows that closed and bounded sets are compact in $\mathbb{R}^n$, we can conclude that this is also true in all finite-dimensional normed spaces.

7.6.3. COROLLARY. *A subset of a finite-dimensional normed vector space is compact if and only if it is closed and bounded.*

Another immediate consequence refers to the way a finite-dimensional subspace sits inside an arbitrary normed space. Arbitrary normed spaces are in general not complete. However, finite-dimensional subspaces are, because we know the Heine–Borel Theorem in $\mathbb{R}^n$.

7.6.4. COROLLARY. *A finite-dimensional subspace of a normed vector space is complete, and in particular it is closed.*

PROOF. Let V be a normed vector space and let W be an n-dimensional subspace. Let T be a linear invertible map from $\mathbb{R}^n$ onto W as constructed previously. Suppose that $(w_k)_{k=1}^{\infty}$ is a Cauchy sequence in W. Then since T^{-1} is Lipschitz, the sequence $x_k = T^{-1}w_k$ for $k \geq 1$ form a Cauchy in $\mathbb{R}^n$. (Check this yourself!) Since $\mathbb{R}^n$ is complete (Theorem 4.2.5), the sequence x_k converges to a vector x. Again by Corollary 7.6.2, we see that w_k must converge to $w = Tx$. Thus W is complete. In particular, all of the limit points of W lie in W, so W is closed. ■

As an application of these corollaries, we prove the following result, which is fundamental to approximation theory.

7.6.5. THEOREM. *Let $(V, \|\cdot\|)$ be a normed a vector space, and let W be a finite dimensional subspace of V. Then for any $v \in V$, there is at least one* closest *point $w^* \in W$ so that $\|v - w^*\| = \inf\{\|v - w\| : w \in W\}$.*

PROOF. Notice that the zero vector is in W and so

$$\inf\{\|v - w\| : w \in W\} \leq \|v - 0\| = \|v\|.$$

Let $M = \|v\|$. If w satisfies $\|v - w\| \leq \|v\|$, then

$$\|w\| \leq \|w - v\| + \|v\| \leq M + M = 2M.$$

Thus if we define $K := \{w \in W : \|w\| \leq 2M\}$, then

$$\inf\{\|v - w\| : w \in K\} = \inf\{\|v - w\| : w \in W\}.$$

We will show that K is compact. Clearly K is bounded by $2M$. The norm function is Lipschitz, and hence continuous. Thus any convergent sequence of vectors in K will converge to a vector of norm at most $2M$. And since W is complete, this limit also lies in W; whence the limit lies in K. This shows that K is closed and bounded. By our corollary, it follows that K is compact.

Now define a function on K by $f(w) = \|v - w\|$. This function has Lipschitz constant 1 since

$$|f(w) - f(x)| = \big|\|v - w\| - \|v - x\|\big| \leq \|w - x\|.$$

In particular, f is continuous on the compact set K By the Extreme Value Theorem (Theorem 5.4.4), f achieves its minimum at some point w^* in K. This is a closest point to v in W. ■

Exercises for Section 7.6

A. Let V be a finite-dimensional vector space with two norms $\|\cdot\|$ and $\|\|\cdot\|\|$. Show that there are constants $0 < a < A$ such that $a\|v\| \leq \|\|v\|\| \leq A\|v\|$ for all $v \in V$.

B. Let T be the invertible linear map from Corollary 7.6.2.
  (a) Use the Lipschitz property of T and T^{-1} to show that $T(B_r(x))$ contains a ball about Tx in V; and that $T^{-1}(B_r(Tx))$ contains a ball about x in $\mathbb{R}^n$.
  (b) Hence show directly that U is open if and only if $T(U)$ is open.

C. Write out a careful proof of Corollary 7.6.3.

D. Suppose that $(w_k)_{k=1}^{\infty}$ is a Cauchy sequence in a normed space W. If $T : W \to V$ is a Lipschitz map into another normed space V, show that the sequence $v_k = Tw_k$ for $k \geq 1$ forms a Cauchy in V.

E. Show that for each integer n and each function f in $C[a, b]$, there is a polynomial of degree at most n that is closest to f in the max norm on $C[a, b]$.

F. Let $\mathbb{R}^n$ have the max norm $\|\mathbf{x}\|_{\infty} = \max\{|x_i| : 1 \leq i \leq n\}$. Let K be the unit ball of V and let $v = (2, 0, \ldots, 0)$. Find all closest points to v in K.

G. For strictly convex normed vector spaces (see Exercise 7.3.F), we have better approximation results.

   (a) Show that if W is a finite-dimensional subspace of a strictly convex normed vector space V, then each point $v \in V$ has a *unique* closest point in W.
   (b) Show that $\mathbb{R}^n$ with the standard Euclidean norm is strictly convex.
   (c) Show that $\mathbb{R}^2$ with the norm $\|(x, y)\|_1 = |x| + |y|$ is not strictly convex.
   (d) Find a subspace W of $V = (\mathbb{R}^2, \|\cdot\|_1)$ so that every point in V that is not in W has more than one closest point in W.

7.7. The L^p norms

The L^p norms on $C[a, b]$ for $1 \leq p < \infty$ were defined in Example 7.1.5 by

$$\|f\|_p = \left(\int_a^b |f(x)|^p \, dx \right)^{1/p}.$$

The point of this section is to prove that these really are norms by proving the triangle inequality, which is the only part of the definition not established in Section 7.1.

There are other variants of the L^p norms that are handled in exactly the same way. Here are two important examples.

7.7.1. EXAMPLES. (1) For $1 \leq p < \infty$, ℓ^p consists of the set of all infinite sequences $\mathbf{a} = (a_n)$ such that

$$\|\mathbf{a}\|_p = \left(\sum_{n=1}^{\infty} |a_n|^p \right)^{1/p} < \infty.$$

It is again easy to see that this measure is positive definite and homogeneous.

We can deduce the triangle inequality from the main example by identifying the sequence $\mathbf{a}$ with the function f on $[0, \infty)$ given by

$$f(x) = a_n \quad \text{for} \quad n - 1 \leq x < n, \quad n \geq 1.$$

This function has the property that $\|f\|_p = \|\mathbf{a}\|_p$. So we can deduce our inequality from the main case. You probably noticed that this function is not continuous. However, if we prove the result for piecewise continuous functions, we shall be able to apply the result here. (Alternatively, we could doctor this function so that we still obtain the correct integral and the function is continuous. But this is not as natural.)

(2) Let $w(x)$ be a strictly positive piecewise continuous function on $[a, b]$. Then define a norm on $C[a, b]$ or even on the space $PC[a, b]$ of piecewise continuous functions by

$$\|f\|_{L^p(w)} = \left(\int_a^b |f(x)|^p w(x) \, dx \right)^{1/p}.$$

We call this the $L^p(w)$ norm. The standard case takes $w(x) = 1$. So if we prove the result for $L^p(w)$, then it will follow for L^p and for ℓ^p.

First we need an easy lemma from calculus.

7.7.2. LEMMA. *Let $A, B > 0$. Then*

$$A^t B^{1-t} \le tA + (1-t)B \quad \text{for all} \quad 0 < t < 1.$$

Moreover, equality holds for some (or all) t only if $A = B$.

PROOF. Since $A > 0$, write $A = e^a$, where $a = \log A$. Similarly, $B = e^b$ for $b = \log B$. Substituting these formulas for A and B, we have to prove that

$$e^{ta} e^{(1-t)b} = e^{ta+(1-t)b} \le te^a + (1-t)e^b.$$

But the preceding inequality follows from the convexity of the exponential function on $\mathbb{R}$ (see Figure 7.4). As the exponential function *is* convex (apply Exercise 6.2.J), we are done. ∎

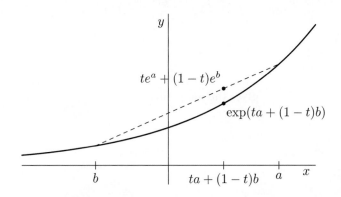

FIGURE 7.4. The convexity of e^x.

We can now prove one of the most important inequalities of analysis, the Hölder inequality.

7.7.3. HÖLDER'S INEQUALITY.

Let w be a positive function on an interval $[a, b]$. Let $f \in L^p(w)$ and $g \in L^q(w)$ where $1 < p < \infty$ and $\dfrac{1}{p} + \dfrac{1}{q} = 1$. Then

$$\left| \int_a^b f(x)g(x)w(x)\, dx \right| \leq \|f\|_{L^p(w)} \|g\|_{L^q(w)}.$$

PROOF. We may assume that f and g are nonzero because both sides are easily seen to be 0 otherwise. Let

$$A = \frac{|f(x)|^p}{\|f\|_{L^p(w)}^p} \quad \text{and} \quad B = \frac{|g(x)|^q}{\|g\|_{L^q(w)}^q}.$$

Then take $t = 1/p$, so that $1 - t = 1/q$, and apply the lemma. We obtain

$$\frac{|f(x)|}{\|f\|_{L^p(w)}} \frac{|g(x)|}{\|g\|_{L^q(w)}} = A^t B^{1-t} \leq tA + (1-t)B$$

$$= \frac{|f(x)|^p}{p\|f\|_{L^p(w)}^p} + \frac{|g(x)|^q}{q\|g\|_{L^q(w)}^q}.$$

We now multiply by $w(x)$ and integrate from a to b.

$$\left| \int_a^b \frac{f(x)g(x)}{\|f\|_{L^p(w)}\|g\|_{L^q(w)}} w(x)\, dx \right| \leq \int_a^b \frac{|f(x)|}{\|f\|_{L^p(w)}} \frac{|g(x)|}{\|g\|_{L^q(w)}} w(x)\, dx$$

$$\leq \frac{1}{p} \int_a^b \frac{|f(x)|^p}{\|f\|_{L^p(w)}^p} w(x)\, dx + \frac{1}{q} \int_a^b \frac{|g(x)|^q}{\|g\|_{L^q(w)}^q} w(x)\, dx$$

$$= \frac{1}{p} + \frac{1}{q} = 1$$

Multiplying by $\|f\|_{L^p(w)}^p \|g\|_{L^q(w)}^q$ gives the inequality. ∎

7.7.4. EXAMPLES.

(1)
$$\int_a^b f(x)g(x)\, dx \leq \left(\int_a^b f(x)^p\, dx \right)^{1/p} \left(\int_a^b g(x)^q\, dx \right)^{1/q}$$

for all continuous functions f and g on $[a, b]$. This is just the standard weight $w(x) = 1$.

(2) If we take $p = 2$, then $q = 2$ and we obtain the Cauchy–Schwarz inequality, Theorem 7.3.4, for $L^2(w)$.

(3) As in Example 7.7.1(1), take $w(x) = 1$ on $[0, \infty)$ and consider $f(x) = a_n$ and $g(x) = b_n$ on $[n-1, n]$ for $n \geq 1$, where $(a_n) \in \ell^p$ and $(b_n) \in \ell^q$. Then

$$\sum_{n=1}^{\infty} a_n b_n = \int_0^{\infty} f(x)g(x)\, dx \leq \left(\int_a^b f(x)^p\, dx \right)^{1/p} \left(\int_a^b g(x)^q\, dx \right)^{1/q}$$

$$= \left(\sum_{n=1}^{\infty} |a_n|^p \right)^{1/p} \left(\sum_{n=1}^{\infty} |b_n|^q \right)^{1/q}.$$

Again, this reduces to the Cauchy–Schwarz inequality if $p = q = 2$.

The triangle inequality now follows by another trick.

7.7.5. MINKOWSKI'S INEQUALITY.
The triangle inequality holds for $L^p(w)$, that is,

$$\left(\int_a^b |f(x) + g(x)|^p w(x)dx \right)^{1/p} \leq \left(\int_a^b |f(x)|^p w(x)dx \right)^{1/p} + \left(\int_a^b |g(x)|^p w(x)dx \right)^{1/p}.$$

PROOF. Notice that $\dfrac{1}{q} = 1 - \dfrac{1}{p} = \dfrac{p-1}{p}$. So $q = \dfrac{p}{p-1}$. Thus

$$\| |f|^{p-1} \|_{L^q(w)} = \left(\int_a^b |f(x)|^{(p-1)q} w(x)\, dx \right)^{1/q}$$

$$= \left(\int_a^b |f(x)|^p w(x)\, dx \right)^{1/q} = \|f\|_{L^p(w)}^{p/q} = \|f\|_{L^p(w)}^{p-1}.$$

We use this equality in the last line of the next calculation, which estimates the norm of $f + g$ in $L^p(w)$,

$$\|f + g\|_{L^p(w)}^p = \int_a^b |f(x) + g(x)|^{p-1} |f(x) + g(x)| w(x)\, dx$$

$$\leq \int_a^b |f(x) + g(x)|^{p-1} |f(x)| w(x)\, dx$$

$$+ \int_a^b |f(x) + g(x)|^{p-1} |g(x)| w(x)\, dx$$

$$\leq \| |f + g|^{p-1} \|_{L^q(w)} \left(\|f\|_{L^p(w)} + \|g\|_{L^p(w)} \right)$$

$$= \|f + g\|_{L^p(w)}^{p-1} \left(\|f\|_{L^p(w)} + \|g\|_{L^p(w)} \right).$$

Now divide by $\|f + g\|_{L^p(w)}^{p-1}$ to obtain

$$\|f + g\|_{L^p(w)} \leq \|f\|_{L^p(w)} + \|g\|_{L^p(w)}. \qquad \blacksquare$$

Exercises for Section 7.7

A. Rework the Hölder inequality to show: If $a_n \geq 0$ and $b_n \geq 0$ and $0 < t < 1$, then

$$\sum_{n=1}^{\infty} a_n^t b_n^{1-t} \leq \left(\sum_{n=1}^{\infty} a_n \right)^t \left(\sum_{n=1}^{\infty} b_n \right)^{1-t}.$$

B. Show that the Hölder inequality is sharp in the sense that for every $f \in L^p(w)$, there is a nonzero $g \in L^q(w)$ so that $\left| \int_a^b f(x)g(x)w(x)\,dx \right| = \|f\|_{L^p(w)}\|g\|_{L^q(w)}$.

HINT: For nonzero f, set $g(x) = |f(x)|^p/f(x)$ if $f(x) \neq 0$ and $g(x) = 0$ if $f(x) = 0$.

C. Let $f(t)$ be a continuous, strictly increasing function on $[0, \infty)$ with $f(0) = 0$. Recall that f has an inverse function g with the same properties by Theorem 5.7.6. Define
$F(x) = \int_0^x f(t)\,dt$ and $G(x) = \int_0^x g(t)\,dt$.

(a) Prove **Young's Inequality**: $xy \leq F(x) + G(y)$ for all $x, y \geq 0$, with equality if and only if $f(x) = y$.

HINT: Sketch f and find two regions with areas $F(x)$ and $G(y)$.

(b) Take $f(x) = x^{p-1}$ for $1 < p < \infty$. What inequality do you get?

D. Let $f_n = n\chi_{[\frac{1}{n}, \frac{2}{n}]}$ be defined on $[0, 1]$ for $n \geq 2$.

(a) Show that f_n converges pointwise to 0.

(b) Show that f_n does not converge in $L^p[0, 1]$ for any $1 \leq p < \infty$.

E. (a) If $f \in C[0, 1]$ and $1 \leq r \leq s < \infty$, show that $\|f\|_1 \leq \|f\|_r \leq \|f\|_s \leq \|f\|_\infty$.

HINT: Let $p = \dfrac{s}{r}$ and think of $\int_0^1 |f|^r \cdot 1\,dx$ as the product of an L^p function and an L^q function.

(b) Hence show that if $f_n \in C[0, 1]$ converge uniformly to f, then they also converge in the L^p norm for all $1 \leq p < \infty$.

F. Let $f_n = \frac{1}{n}\chi_{[0, n^n]}$ be defined on $[0, \infty)$ for $n \geq 1$.

(a) Show that f_n converges uniformly to 0.

(b) Show that f_n does not converge in $L^p[0, 1]$ for any $1 \leq p < \infty$.

G. (a) If $\mathbf{a} = (a_k) \in \ell_1$ and $1 \leq r \leq s < \infty$, show that $\|\mathbf{a}\|_\infty \leq \|\mathbf{a}\|_s \leq \|\mathbf{a}\|_r \leq \|\mathbf{a}\|_1$.

HINT: Show that for $p \geq 1$, $\sum\limits_{k=1}^{n} |b_k|^p \leq \left(\sum\limits_{k=1}^{n} |b_k| \right)^p$.

(b) Hence show that $\ell_1 \subset \ell_r \subset \ell_s \subset \ell_\infty$.

H. Find continuous functions f and g on $[0, \infty)$ so that f is in $L^1[0, \infty)$ but not in $L^2[0, \infty)$, and g is in $L^2[0, \infty)$ but not in $L^1[0, \infty)$.

CHAPTER 8

Limits of Functions

8.1. Limits of Functions

There are several reasonable definitions for the limit of a sequence of functions. Clearly the entries of the sequence should approximate the limit function f to greater and greater accuracy in some sense. But there are different ways of measuring the accuracy of an approximation, depending on the problem. Different approximation schemes generally correspond to different norms, although not all convergence criteria come from a norm. In this section, we consider two natural choices and see why the stronger notion is better for many purposes.

8.1.1. DEFINITION. Let (f_n) be a sequence of functions from $S \subset \mathbb{R}^n$ into $\mathbb{R}^m$. This sequence **converges pointwise** to a function f if

$$\lim_{n \to \infty} f_n(x) = f(x) \quad \text{for all} \quad x \in S.$$

This is the most obvious and perhaps simplest notion of convergence. It is also a rather weak concept fraught with difficulties.

8.1.2. EXAMPLE. Define piecewise linear continuous functions f_n on $[0, 1]$ by connecting the points $(0, 0)$, $(\frac{1}{n}, n)$, $(\frac{2}{n}, 0)$, and $(1, 0)$ by straight lines, namely

$$f_n(x) = \begin{cases} n^2 x & \text{for} \quad 0 \le x \le \frac{1}{n} \\ n^2(\frac{2}{n} - x) & \text{for} \quad \frac{1}{n} \le x \le \frac{2}{n} \\ 0 & \text{for} \quad \frac{2}{n} \le x \le 1. \end{cases}$$

See Figure 8.1. This sequence converges pointwise to the zero function; that is,

$$\lim_{n \to \infty} f_n(x) = 0 \quad \text{for all} \quad 0 \le x \le 1.$$

Indeed, at $x = 0$, we have $f_n(x) = 0$ for all $n \ge 1$; and if $x > 0$, then there is an integer N such that $x \ge 2/N$. Thus once $n \ge N$, we have $f_n(x) = 0$. So at every

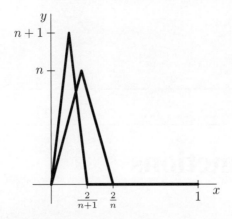

FIGURE 8.1. Graphs of f_n and f_{n+1}.

point, the functions are eventually constant. Notice, however, that the closer x is to zero, the larger the choice of N must be.

The limit is a continuous function. However, the limit of the integrals is not the integral of the limit. The area between the graph of f_n and the x-axis forms a triangle with base $2/n$ and height n and thus has area 1. Therefore,

$$\lim_{n \to \infty} \int_0^1 f_n(x)\, dx = \lim_{n \to \infty} 1 = 1 \neq 0 = \int_0^1 0\, dx.$$

In fact, an easy modification of this example would yield functions converging pointwise to 0 with integrals tending to infinity or any finite value or oscillating wildly.

The other notion of convergence that we study, uniform convergence, will demand that convergence occur at a uniform rate on the whole space S. To formulate this, we first consider the ε–N version of pointwise limit first. A sequence (f_n) converges pointwise to f if for every $x \in S$ and $\varepsilon > 0$, there is an integer N so that

$$\|f_n(x) - f(x)\| < \varepsilon \quad \text{for all} \quad n \geq N.$$

In this case, N depends on both ε and on the point x. (Think about how you would choose N for different $x \in [0, 1]$ in Example 8.1.2.) Uniform convergence demands that this choice depend only on ε, providing a common N that works for all x in S simultaneously.

8.1.3. DEFINITION. Let (f_n) be a sequence of functions from $S \subset \mathbb{R}^n$ into $\mathbb{R}^m$. This sequence **converges uniformly** to a function f if for every $\varepsilon > 0$, there is an integer N so that

$$\|f_n(x) - f(x)\| < \varepsilon \quad \text{for all} \quad x \in S \text{ and } n \geq N.$$

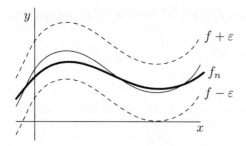

FIGURE 8.2. Graph of neighbourhood of f and a sample f_n.

To understand this definition, look at Figure 8.2. The point is that the graph of f_n must lie between the graphs of $f + \varepsilon$ and $f - \varepsilon$.

Clearly, if (f_n) converges uniformly to f, then (f_n) also converges pointwise to f. But this is not reversible.

As we have seen in Example 7.1.3, when K is a compact subset of $\mathbb{R}^n$, we may define a norm on the space $C(K)$ of all continuous functions on K by

$$\|f\|_\infty = \sup_{x \in K} |f(x)|.$$

This is defined because the Extreme Value Theorem guarantees that the supremum is finite.

When S is a subset of $\mathbb{R}^n$ that is not compact, there are unbounded continuous functions on S. Nevertheless, we may restrict ourselves to the space $C_b(S)$ consisting of all bounded continuous functions from S to $\mathbb{R}$. Then the supremum becomes a norm in the same manner. Similarly, we may consider bounded continuous functions with values in $\mathbb{R}^m$. This space is denoted $C_b(S, \mathbb{R}^m)$ and has the norm

$$\|f\|_\infty = \sup_{x \in S} \|f(x)\|_2.$$

We have the following theorem.

8.1.4. THEOREM. *For a sequence of functions (f_n) in $C_b(S, \mathbb{R}^m)$, (f_n) converges uniformly to f if and only if*

$$\lim_{n \to \infty} \|f_n - f\|_\infty = 0.$$

After the preceding discussion, the proof is immediate. Indeed, the statement $\|f_n(x) - f(x)\| \le \varepsilon$ for all $x \in S$ is equivalent to saying

$$\|f_n - f\|_\infty \le \varepsilon.$$

Returning to Example 8.1.2, the maximum of f_n occurs at $\frac{1}{n}$ with $f_n(\frac{1}{n}) = n$ and hence

$$\|f_n - 0\|_\infty = n.$$

This does not converge to 0. So f_n does not converge uniformly to the zero function (or any other bounded function, for that matter).

8.1.5. EXAMPLE. Consider $f_n(x) = x^n$ for $x \in [0, 1]$. It is easy to check that

$$\lim_{n \to \infty} f_n(x) = \lim_{n \to \infty} x^n = \begin{cases} 0 & \text{for} \quad 0 \le x < 1 \\ 1 & \text{for} \quad x = 1. \end{cases}$$

Thus the pointwise limit is the function $\chi_{\{1\}}$, the characteristic function of the point $\{1\}$. The functions f_n are polynomials, and hence not only continuous but even smooth; while the limit function has a discontinuity at the point 1.

For each $n \ge 1$, we have $f_n(1) = 1$ and so

$$\|f_n - \chi_{\{1\}}\|_\infty = \sup_{0 \le x < 1} |x^n - 0| = 1.$$

So f_n does not converge in the uniform norm. Indeed, to contradict the definition, take $\varepsilon = 1/2$. For each n, let $x_n = 2^{-1/2n}$. Then

$$\left| f_n(x_n) - \chi_{\{1\}}(x_n) \right| = \frac{1}{\sqrt{2}} > \varepsilon.$$

Hence there is no integer N satisfying the definition.

8.1.6. EXAMPLE. Consider the functions f_n on $[0, \pi]$ given by

$$f_n(x) = \frac{1}{n} \sin nx.$$

Several of the f_n are graphed in Figure 8.3. By the Squeeze Theorem,

$$\lim_{n \to \infty} \frac{1}{n} \sin nx = 0 \quad \text{for all} \quad 0 \le x \le \pi.$$

Moreover, $\|f_n\|_\infty = \sup_{0 \le x \le \pi} \frac{1}{n} |\sin nx| = \frac{1}{n}$. Thus this sequence converges uniformly to 0. If $\varepsilon > 0$, we may choose N so large that $\frac{1}{N} < \varepsilon$. Then for any $n \ge N$,

$$|f_n(x) - 0| = \left| \frac{1}{n} \sin nx \right| \le \frac{1}{N} < \varepsilon \quad \text{for all} \quad 0 \le x \le \pi.$$

This sequence does not behave well with respect to derivatives—a typical feature of uniform approximation. Compute

$$f_n'(x) = \cos nx.$$

Hence $\lim_{n \to \infty} f_n'(0) = \lim_{n \to \infty} 1 = 1 \ne 0 = f'(0)$; while $\lim_{n \to \infty} f_n'(\pi) = \lim_{n \to \infty} (-1)^n$ does not even exist. Indeed, this limit does not exist at any point except 0.

The intuition is that for any nice smooth function, there are functions that oscillate up and down very rapidly and yet remain close to the nice function, such as our previous functions. The sequence $g_n(x) = \frac{1}{n} \sin n^2 x$ converges uniformly to 0 as well, yet has derivatives, $g_n'(x) = n \cos n^2 x$, which do not converge anywhere. So uniform convergence does not give control of derivatives.

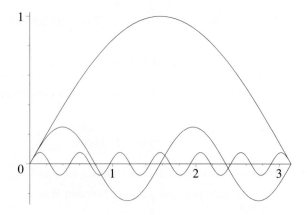

FIGURE 8.3. The graphs of f_n for $n = 1, 4$, and 13.

Exercises for Section 8.1

A. Let $f_n(x) = xne^{-nx}$ for all $x \geq 0$ and $n \geq 1$. Show that (f_n) converges to zero on $[0, \infty)$ pointwise but not uniformly.

B. Let $f_n(x) = nx(1 - x^2)^n$ on $[0, 1]$ for $n \geq 1$.
Find $f(x) = \lim\limits_{n \to \infty} f_n(x)$. Is this convergence uniform?
HINT: Recall that $\lim\limits_{n \to \infty} (1 - \frac{h}{n})^n = e^{-h}$.

C. For the sequence of functions in the previous exercise, compare the limit of the integrals (from 0 to 1) with the integral of the limit.

D. Does the sequence $f_n(x) = \dfrac{x}{1 + nx^2}$ converge uniformly on $\mathbb{R}$?

E. Show that $f_n(x) = \dfrac{\tan^{-1}(nx)}{n}$, $n \geq 1$ converges uniformly on $\mathbb{R}$.

F. Show that $f_n(x) = n\sin(x/n)$ converges uniformly on $[-R, R]$ for any finite R but does not converge uniformly on $\mathbb{R}$.

G. Find all intervals on which the sequence $f_n(x) = \dfrac{x^{2n}}{n + x^{2n}}$, $n \geq 1$, converges uniformly.

H. Suppose that $f_n : [0, 1] \to \mathbb{R}$ is a sequence of C^1 functions (i.e., functions with continuous derivatives) that converges pointwise to a function f. If there is a constant M so that $\|f_n'\|_\infty \leq M$ for all n, then prove that (f_n) converges to f uniformly.

I. Prove **Dini's Theorem**: suppose that f and f_n are continuous functions on $[a, b]$ such that $f_n \leq f_{n+1}$ for all $n \geq 1$ and (f_n) converges to f pointwise. Prove (f_n) converges to f uniformly.
HINT: Work with $g_n = f - f_n$, which decrease to 0. Show that for any point x_0 and $\varepsilon > 0$, there is an integer N and a positive $r > 0$ such that $g_N(x) \leq \varepsilon$ on $(x_0 - r, x_0 + r)$. If convergence is not uniform, say $\lim \|g_n\|_\infty = d > 0$, find x_n such that $\lim g_n(x_n) = d$. Obtain a contradiction.

J. Find an example that shows Dini's Theorem is false if $[a, b]$ is replaced with a non-compact subset of $\mathbb{R}$.

K. (a) Suppose that $f : \mathbb{R} \to \mathbb{R}$ is uniformly continuous. Let $f_n(x) = f(x + 1/n)$. Prove that f_n converges uniformly to f on $\mathbb{R}$.
(b) Does this remain true if f is just continuous? Prove it or provide a counterexample.

L. For which values of $x \geq 1$ does the expression $x^{x^{x^{\cdot^{\cdot^{\cdot}}}}}$ make sense?
HINT: Define $f_1(x) = x$ and $f_{n+1}(x) = x^{f_n(x)}$ for $n \geq 1$. Then
(a) Show that $f_{n+1}(x) \geq f_n(x)$ for all $n \geq 1$.
(b) When $L(x) = \lim\limits_{n \to \infty} f_n(x)$ exists, find optimal upper bounds for x and L.
(c) For these values of x, show by induction that $f_n(x)$ is bounded above by e for all $n \geq 1$. What can you conclude?
(d) What happens for larger x?

M. The behaviour of $x^{x^{x^{\cdot^{\cdot^{\cdot}}}}}$ when $0 < x < 1$ is more complicated and so more interesting. To get started, compute $f_n(1/16)$ for small values of n, using the functions f_n from the previous exercise, and see what occurs.

8.2. Uniform Convergence and Continuity

Our first positive result is that uniform convergence preserves continuity and so is (almost) always the right notion of convergence for continuous functions.

8.2.1. THEOREM. *Let (f_n) be a sequence of continuous functions mapping a subset S of $\mathbb{R}^n$ into $\mathbb{R}^m$ that converges uniformly to a function f. Then f is continuous.*

PROOF. Fix a point $a \in S$ and an $\varepsilon > 0$. We must control $\|f(x) - f(a)\|$ only by controlling the bound on $\|x - a\|$. To this end, we make use of the proximity of one of the continuous functions f_n and compute

$$\|f(x) - f(a)\| = \|f(x) - f_n(x) + f_n(x) - f_n(a) + f_n(a) - f(a)\|$$
$$\leq \|f(x) - f_n(x)\| + \|f_n(x) - f_n(a)\| + \|f_n(a) - f(a)\|.$$

Note that the first and last terms may be controlled by

$$\|f(x) - f_n(x)\| \leq \|f_n - f\|_\infty \quad \text{for all} \quad x \in S,$$

including the point $x = a$. The middle term may be controlled by the continuity of f_n.

To be precise, first choose n so large that

$$\|f_n - f\|_\infty < \frac{\varepsilon}{3}.$$

Then using the continuity of f_n at a, choose a positive number $r > 0$ such that

$$\|f_n(x) - f_n(a)\| < \frac{\varepsilon}{3} \quad \text{for all} \quad \|x - a\| < r.$$

Then for all $x \in S$ with $\|x - a\| < r$, we obtain

$$\|f(x) - f(a)\| \leq \|f(x) - f_n(x)\| + \|f_n(x) - f_n(a)\| + \|f_n(a) - f(a)\|$$
$$< \frac{\varepsilon}{3} + \frac{\varepsilon}{3} + \frac{\varepsilon}{3} = \varepsilon.$$

Thus f is continuous. ∎

Note that in this proof, smaller values of ε require using a closer approximant f_n in order to achieve the desired estimate.

Now we will use the compactness of K and Theorem 8.2.1 to show that $C(K)$ is complete. Just as we used completeness to understand the real line, we can use it to understand other spaces. First, of course, we have to prove that the space is complete.

8.2.2. COMPLETENESS THEOREM FOR *C(K)*.
If K is a compact set, the space $C(K)$ of all continuous functions on K with the sup norm is complete.

PROOF. A sequence (f_n) in $C(K)$ is a Cauchy sequence for the sup norm if for every $\varepsilon > 0$, there is an integer N so that

$$\|f_n - f_m\|_\infty < \varepsilon \quad \text{for all} \quad m, n \geq N.$$

We must show that every Cauchy sequence has a (uniform) limit in $C(K)$.

First consider an arbitrary point $x \in K$. We have

$$|f_n(x) - f_m(x)| \leq \|f_n - f_m\|_\infty < \varepsilon \quad \text{for all} \quad m, n \geq N.$$

Hence the sequence $(f_n(x))_{n=1}^\infty$ is a Cauchy sequence of real numbers. Since $\mathbb{R}$ is complete, this has a pointwise limit

$$f(x) := \lim_{n \to \infty} f_n(x).$$

This must be shown to converge uniformly. With ε and N as before, we obtain the estimate

$$|f(x) - f_m(x)| = \lim_{n \to \infty} |f_n(x) - f_m(x)| \leq \varepsilon \quad \text{for all} \quad m \geq N.$$

Since this holds for *all* $x \in K$, it follows that $\|f - f_n\|_\infty \leq \varepsilon$. Therefore, the limit is uniform.

By the previous theorem, the uniform limit of continuous functions is continuous. Thus f is continuous and hence belongs to $C(K)$. This establishes that $C(K)$ is complete. ∎

We have used this method of proof before and will use it again; for example, Theorem 4.2.5, showing the completeness of $\mathbb{R}^n$, and Theorem 8.4.7, later in this chapter, both follow a similar strategy.

Exercises for Section 8.2

A. Find the limits of the following functions. Find an interval on which convergence is uniform and another on which it is not. Explain.

(a) $f_n(x) = \left(\dfrac{x}{2}\right)^n + \left(\dfrac{5}{x}\right)^n$ (b) $g_n(x) = \dfrac{nx}{2 + 5nx}$

B. Show that $h_n(x) = \dfrac{n + x}{4n + x}$ converges uniformly on $[0, N]$ for any $N < \infty$ but not uniformly on $[0, \infty)$.

C. Consider a sequence of continuous functions $f_n : (0, 1) \to \mathbb{R}$. Suppose there is a function $f : (0, 1) \to \mathbb{R}$ so that whenever $0 < a < b < 1$, f_n converges uniformly on $[a, b]$ to f. Prove that f is continuous on $(0, 1)$.

D. Let f_n and g_n be continuous functions on $[a, b]$. Suppose that (f_n) converges uniformly to f and (g_n) converges uniformly to g on $[a, b]$. Prove that $(f_n g_n)$ converges uniformly to fg on $[a, b]$.

E. Suppose that (f_n) converge uniformly to f on a compact subset K of $\mathbb{R}^n$ and that (g_n) converge uniformly on K to a continuous function g such that $g(x) \neq 0$ for all $x \in K$. Prove that $f_n(x)/g_n(x)$ is everywhere defined for large n and that this quotient converges uniformly to $f(x)/g(x)$ on K.

F. Let $f_n(x) = \tan^{-1}(nx)/\sqrt{n}$.
(a) Find $f(x) = \lim\limits_{n \to \infty} f_n(x)$, and show that f_n converges uniformly to f on $\mathbb{R}$.
(b) Compute $\lim\limits_{n \to \infty} f_n'(x)$, and compare this with $f'(x)$.
(c) Where is the convergence of f_n' is uniform? Prove your answer.

G. Suppose that functions f_n defined on $\mathbb{R}^k$ converge uniformly to a function f. Suppose that each f_n is bounded, say by A_n. Prove that f is bounded.

H. Suppose that f_n in $C[0, 1]$ all have Lipschitz constant L. Show that if f_n converges pointwise to f, then the convergence is uniform and f is Lipschitz with constant L.

I. Give an example of a sequence of discontinuous functions f_n that converge uniformly to a continuous function.

8.3. Uniform Convergence and Integration

A useful feature of uniform convergence is its good behaviour with respect to limits. We now show that integration over a compact set respects uniform limits.

8.3.1. INTEGRAL CONVERGENCE THEOREM.
Let (f_n) be a sequence of continuous functions on the closed interval $[a, b]$ converging uniformly to $f(x)$ and fix $c \in [a, b]$. Then the functions

$$F_n(x) = \int_c^x f_n(t)\, dt \quad \text{for} \quad n \geq 1$$

converge uniformly on $[a, b]$ to the function $F(x) = \int_c^x f(t)\, dt.$

PROOF. The proof is straightforward:

$$|F_n(x) - F(x)| = \left| \int_c^x f_n(t) - f(t)\, dt \right|$$

$$\leq \int_c^x |f_n(t) - f(t)|\, dt \leq \int_c^x \|f_n - f\|_\infty\, dt$$

$$\leq |x - c|\,\|f_n - f\|_\infty \leq (b - a)\,\|f_n - f\|_\infty.$$

This estimate no longer depends on x. Hence

$$\|F_n - F\|_\infty \leq (b - a)\|f_n - f\|_\infty.$$

Since (f_n) converges uniformly to f,

$$\lim_{n \to \infty} \|F_n - F\|_\infty \leq (b - a) \lim_{n \to \infty} \|f_n - f\|_\infty = 0.$$

That is, (F_n) converges uniformly to F.   ∎

This can be reformulated in terms of derivatives as follows.

8.3.2. COROLLARY. *Suppose that (f_n) is a sequence of continuously differentiable functions on $[a, b]$ such that (f_n') converges uniformly to a function g and there is a point $c \in [a, b]$ so that $\lim_{n \to \infty} f_n(c) = \gamma$ exists. Then (f_n) converges uniformly to a differentiable function f with $f(c) = \gamma$ and $f' = g$.*

PROOF. By the Fundamental Theorem of Calculus, f_n is the unique antiderivative of f_n' whose value at c is $f_n(c)$. That is,

$$f_n(x) = f_n(c) + \int_c^x f_n'(t)\, dt.$$

By the previous theorem, the sequence of functions $F_n(x) = \int_c^x f_n'(t)\, dt$ for $n \geq 1$ converges uniformly to $F(x) = \int_c^x g(t)\, dt$. Since $\lim_{n \to \infty} f_n(c) = \gamma$, it follows that

$$\lim_{n \to \infty} \|f_n - (\gamma + F)\|_\infty \leq \lim_{n \to \infty} |f_n(c) - \gamma| + \|F_n - F\|_\infty = 0.$$

Therefore, (f_n) converges uniformly to

$$f(x) = \gamma + \int_c^x g(x)\, dx.$$

Finally, the Fundamental Theorem of Calculus shows that f is differentiable and $f' = g$.   ∎

Consider a function of two variables $f(x, t)$. Notice that

$$F(x) = \int_c^d f(x, t)\, dt$$

is a function of x. The previous theorem can be seen as a special case of this situation, where x is in $\mathbb{N}$ and $f(t, n)$ is written as $f_n(t)$. It turns out that $F'(x)$

equals the integral of $\partial f / \partial x$, but proving it requires some careful estimates. We begin with a continuity result that is a continuous parameter version of the Integral Convergence Theorem (Theorem 8.3.1).

8.3.3. PROPOSITION. *Let $f(x,t)$ be a continuous function on $[a,b] \times [c,d]$. Define $F(x) = \int_c^d f(x,t)\, dt$. Then F is continuous on $[a,b]$.*

PROOF. Since f is continuous on a compact set, it is uniformly continuous. Therefore, given $\varepsilon > 0$, there is a $\delta > 0$ so that $|f(x,t) - f(y,t)| < \varepsilon/(d-c)$ whenever $|x - y| < \delta$. Therefore,

$$|F(x) - F(y)| = \left| \int_c^d f(x,t) - f(y,t)\, dt \right| \leq \int_c^d |f(x,t) - f(y,t)|\, dt$$

$$\leq \int_c^d \frac{\varepsilon}{d-c}\, dt = \varepsilon.$$

Thus f is continuous.   ∎

8.3.4. LEIBNIZ'S RULE.
Suppose that $f(x,t)$ and $\frac{\partial}{\partial x} f(x,t)$ are continuous functions on $[a,b] \times [c,d]$. Then $F(x) = \int_c^d f(x,t)\, dt$ is differentiable and

$$F'(x) = \int_c^d \frac{\partial}{\partial x} f(x,t)\, dt.$$

PROOF. Fix $x_0 \in [a,b]$ and let $h \neq 0$. Observe that

$$\frac{F(x_0 + h) - F(x_0)}{h} = \int_c^d \frac{f(x_0 + h, t) - f(x_0, t)}{h}\, dt.$$

Since $f(x,t)$ is a differentiable function of x for fixed t, we may apply the Mean Value Theorem to obtain a point $x(t)$ depending on t so that

$$\frac{f(x_0 + h, t) - f(x_0, t)}{h} = \frac{\partial}{\partial x} f(x(t), t).$$

The Mean Value Theorem does not show that the function $x(t)$ is continuous. However, in this situation, the left-hand side of this identity is evidently a continuous function of t, and hence so is the right-hand side $\partial f / \partial x (x(t), t)$.

Since $f_x(x,t) = \frac{\partial}{\partial x} f(x,t)$ is continuous on a compact set, it is uniformly continuous. So for $\varepsilon > 0$, we can choose $\delta > 0$ so that

$$\left| f_x(x,t) - f_x(y,t) \right| < \frac{\varepsilon}{d-c}$$

whenever $|x - y| < \delta$. Therefore,

$$\left| \frac{F(x_0+h) - F(x_0)}{h} - \int_c^d f_x(x, t)\, dt \right| = \left| \int_c^d \frac{f(x_0+h, t) - f(x_0, t)}{h} - f_x(x, t)\, dt \right|$$

$$\le \int_c^d \left| f_x(x(t), t) - f_x(x, t) \right| dt$$

$$\le \int_c^d \frac{\varepsilon}{d - c}\, dt = \varepsilon.$$

Since $\varepsilon > 0$ was arbitrary, we obtain

$$F'(x_0) = \lim_{h \to 0} \frac{F(x_0 + h) - F(x_0)}{h} = \int_c^d \frac{\partial}{\partial x} f(x, t)\, dt. \qquad \blacksquare$$

8.3.5. EXAMPLE. We will establish the improper integral $\displaystyle\int_0^\infty e^{-x^2}\, dx = \frac{\sqrt{\pi}}{2}$.
For the definition of improper integral, see Exercise 6.3.S. It is known that the integral $g(u) = \int_0^u e^{-x^2}\, dx$ cannot be expressed in closed form in terms of the standard elementary functions. However, the definite integral can be evaluated in a number of ways. Here we exploit Leibniz's rule to accomplish this. The auxiliary function that we introduce is unmotivated, but the rest of the proof is straightforward.

Before we begin computing, observe that e^{-x^2} is positive and thus $g(u)$ is monotone increasing. Thus to prove that a limit exists as u tends to $+\infty$, it suffices to show that g is bounded. However, $e^{-x^2} \le 1$ for all x and $e^{-x^2} \le e^{-x}$ when $x \ge 1$. So

$$g(u) \le \int_0^1 1\, ds + \int_1^u e^{-s}\, ds = 1 + (e - e^{-u}) \le 1 + e.$$

Consequently, $J = \int_0^\infty e^{-x^2}\, dx$ is defined and finite.
Consider

$$f(x) = \int_0^1 \frac{e^{-x(1+t^2)}}{1 + t^2}\, dt.$$

Observe that

$$f(0) = \int_0^1 \frac{1}{1 + t^2}\, dt = \tan^{-1} \Big|_0^1 = \frac{\pi}{4}.$$

The integrand $h(x, t) = \dfrac{e^{-x(1+t^2)}}{1 + t^2}$ is continuous on $[0, 1] \times [0, \infty)$. We define $h_x(t) = h(x, t)$ and observe that $0 \le h_x(t) \le e^{-x}$. Hence h_x converges uniformly to $h_\infty(x) = 0$ on $[0, 1]$. By the Integral Convergence Theorem, we conclude that

$$\lim_{x \to \infty} f(x) = \int_0^1 \lim_{x \to \infty} h_x(t)\, dt = \int_0^1 0\, dt = 0.$$

Now apply the Leibniz Rule to compute

$$f'(x) = \int_0^1 \frac{\partial}{\partial x} \frac{e^{-x(1+t^2)}}{1+t^2}\, dt$$

$$= \int_0^1 \frac{e^{-x(1+t^2)}(-(1+t^2))}{1+t^2}\, dt = -e^{-x}\int_0^1 e^{-xt^2}\, dt$$

Make the change of variables $s = \sqrt{x}\,t$ (where x is held constant!), to obtain

$$f'(x) = -e^{-x}\int_0^{\sqrt{x}} \frac{e^{-s^2}}{\sqrt{x}}\, ds = -\frac{e^{-x}}{\sqrt{x}}g(\sqrt{x}).$$

Next evaluate $f(0)$ in two ways:

$$\frac{\pi}{4} = \lim_{n\to\infty} f(0) - f(n) = \lim_{n\to\infty} -\int_0^n f'(x)\, dx = \lim_{n\to\infty} \int_0^n \frac{e^{-x}}{\sqrt{x}}g(\sqrt{x})\, dx$$

Substitute $s = \sqrt{x}$. By the Fundamental Theorem of Calculus, $g'(s) = e^{-s^2}$. So

$$\frac{\pi}{4} = \lim_{n\to\infty} \int_0^{\sqrt{n}} 2e^{-s^2} g(s)\, ds$$

$$= \lim_{n\to\infty} \int_0^{\sqrt{n}} 2g'(s)g(s)\, ds = \lim_{n\to\infty} g^2(s)\Big|_0^n = J^2.$$

Therefore, $J = \sqrt{\pi}/2$.

Exercises for Section 8.3

A. For $x \in [-1, 1]$, let $F(x) = \int_0^1 x(1 - x^2 y^2)^{-1/2}\, dy$. Show that $F'(x) = (1 - x^2)^{-1/2}$ and deduce that $F(x) = \arcsin(x)$.

B. For $n \geq 1$, define functions f_n on $[0, \infty)$ by

$$f_n(x) = \begin{cases} e^{-x} & \text{for} \quad 0 \leq x \leq n \\ e^{-2n}(e^n + n - x) & \text{for} \quad n \leq x \leq n + e^n \\ 0 & \text{for} \quad x \geq n + e^n. \end{cases}$$

(a) Find the pointwise limit f of f_n. Show that the convergence is uniform on $[0, \infty)$.

(b) Compute $\int_0^\infty f(x)\, dx$ and $\lim_{n\to\infty} \int_0^\infty f_n(x)\, dx$.

(c) Why does this not contradict Theorem 8.3.1?

C. Suppose that $g \in C[0, 1]$ and (f_n) is a sequence in $C[0, 1]$ that converges uniformly to f. Prove that

$$\lim_{n\to\infty} \int_0^1 f_n(x)g(x)\, dx = \int_0^1 f(x)g(x)\, dx.$$

D. Compute $\lim_{n\to\infty} \int_0^\pi \frac{\sin nx}{nx}\, dx$.

HINT: Find the limit of the integral over $[\varepsilon, \pi]$ and estimate the rest.

E. Define $f(x) = \int_0^\pi \frac{\sin xt}{t} \, dt$.

    (a) Prove that this integral is defined.

    (b) Compute $f'(x)$ explicitly.

    (c) Prove that f' is continuous at 0.

F. Define the **Bessel function** J_0 by $J_0(x) = \frac{1}{\pi} \int_{-1}^1 \frac{\cos(xt)}{\sqrt{1-t^2}} \, dt$. Prove that J_0 satisfies the differential equation $y'' + y'/x + y = 0$, that is, $J_0'' + J_0'/x + J_0$ is identically zero.

G. With the setup for the Leibniz Rule, let b be a variable and set $F(x, b) = \int_a^b f(x, t) \, dt$.

Let $b(x)$ be a differentiable function, and define $G(x) = F(x, b(x)) = \int_a^{b(x)} f(x, t) \, dt$.

Show that $G'(x) = \int_a^{b(x)} \frac{\partial f}{\partial x}(x, t) \, dt + f(x, b(x))b'(x)$.

HINT: $G'(x) = \frac{\partial F}{\partial x}(x, b(x)) + \frac{\partial F}{\partial y}(x, b(x))b'(x)$

H. Suppose that $f \in C^2[0, 1]$ such that $f''(x) + bf'(x) + cf(x) = 0$, $f(0) = 0$ and $f'(0) = 1$. Let $d(x)$ be continuous on $[0, 1]$ and define $g(x) = \int_0^x f(x - t)d(t) \, dt$. Prove that $g(0) = g'(0) = 0$ and $g''(x) + bg'(x) + cg(x) = d(x)$.

I. Suppose that f_n are Riemann integrable integrable functions on $[a, b]$ that converge uniformly to a function f. Prove that f is Riemann integrable.

HINT: Use Lebesgue's Theorem (Theorem 6.6.6).

8.4. Series of Functions

By analogy with series of numbers, we define a series of functions, $\sum_{n=1}^\infty f_n(x)$, as the limit of the sequence of partial sums. Thus, we say $\sum_{n=1}^\infty f_n(x)$ converges pointwise (or uniformly) if the partial sums $\sum_{n=1}^k f_n(x)$ converge pointwise (or uniformly).

8.4.1. EXAMPLE. Consider the series of functions $\sum_{n=1}^\infty \frac{\sin(nx)}{n^2}$. To see that the partial sums converge, first observe that if $k \geq l$, then

$$\left| \sum_{n=1}^k f_n(x) - \sum_{n=1}^l f_n(x) \right| \leq \sum_{n=l+1}^k |f_n(x)| \leq \sum_{n=l+1}^k \frac{1}{n^2}.$$

As $\sum_{n=1}^{\infty} \dfrac{1}{n^2}$ is a convergent series, the Cauchy criterion shows that for any $\varepsilon > 0$, there is an integer N so that if $l, k \geq N$, then $\sum_{n=l+1}^{k} \dfrac{1}{n^2} < \varepsilon$. Thus, for $l, k \geq N$,

$$\left| \sum_{n=1}^{k} f_n(x) - \sum_{n=1}^{l} f_n(x) \right| < \varepsilon,$$

proving that the partial sums are uniformly Cauchy and so converge.

8.4.2. EXAMPLE. On the other hand, consider the sequence of functions f_n on $[0, 1]$ given by $f_n = \chi_{(0,1/n)}$. For any x in $[1/(n+1), 1/n)$, the values $f_{n+1}(x)$, $f_{n+2}(x), \ldots$ are all zero and the values $f_1(x), \ldots, f_n(x)$ are all one. Hence

$$\sum_{k=0}^{\infty} f_k(x) = n \quad \text{for} \quad \frac{1}{n+1} \leq x < \frac{1}{n}.$$

Thus, the series $\sum_{n=1}^{\infty} f_n$ converges at each point of $[0, 1]$. It does not converge uniformly, since for all $k > l$, we have

$$\left| \sum_{n=1}^{k} f_n(x) - \sum_{n=1}^{l} f_n(x) \right| \geq f_{l+1}(x) = 1 \quad \text{for all} \quad x \in (0, 1/(l+1)).$$

8.4.3. EXAMPLE. One of the most important types of series of functions is a **power series**. This is a series of the form

$$\sum_{n=1}^{\infty} a_n x^n = a_0 + a_1 x + a_2 x^2 + a_3 x^3 + \cdots.$$

We will consider these series in detail in the next section. As a starter, consider the series $\sum_{n=0}^{\infty} \dfrac{x^n}{n!}$. Fixing $x \in \mathbb{R}$, we can apply the Ratio Test, to obtain

$$\lim_{n \to \infty} \frac{x^{n+1}/(n+1)!}{x^n/n!} = \lim_{n \to \infty} \frac{x}{n} = 0.$$

So this series converges pointwise for each $x \in \mathbb{R}$. After first obtaining a few theorems, we will see that it converges uniformly on each interval $[-A, A]$. We evaluate it in Example 8.5.4.

Using the partial sums, we can translate all of the results of the previous section about sequences of functions into results about series of functions. Here are two samples. We leave the reformulation of the other theorems of Section 8.2 as exercises.

8.4.4. THEOREM. *Let* (f_n) *be a sequence of continuous functions from a subset* S *of* $\mathbb{R}^n$ *into* $\mathbb{R}^m$. *If* $\sum_{n=1}^{\infty} f_n(x)$ *converges uniformly, then it is continuous.*

8.4.5. DEFINITION. Let $S \subset \mathbb{R}^n$. We say that a series of functions f_k from S to $\mathbb{R}^m$ is **uniformly Cauchy** on S if for every $\varepsilon > 0$, there is an N so that

$$\left\| \sum_{i=k+1}^{l} f_i(x) \right\| \leq \varepsilon \quad \text{whenever} \quad x \in S \text{ and } l > k \geq N.$$

The proof that a series of real numbers converges if and only if it is Cauchy can be modified in a straightforward way to show the following.

8.4.6. THEOREM. *A series of functions converges uniformly if and only if it is uniformly Cauchy.*

PROOF. If f_n converges uniformly to f, then for each $\varepsilon > 0$, there is an integer N so that $\|f_n - f\| < \varepsilon/2$ for all $n \geq N$. Then if $m, n \geq N$,

$$\|f_m - f_n\| \leq \|f_m - f\| + \|f - f_n\| < \frac{\varepsilon}{2} + \frac{\varepsilon}{2} = \varepsilon.$$

Conversely, if (f_n) is uniformly Cauchy, then $(f_n(x))$ is Cauchy for every x, and thus $f(x) = \lim_{n \to \infty} f_n(x)$ exists as a pointwise limit. Moreover, if $\varepsilon > 0$ and $\|f_m - f_n\| < \varepsilon$ for all $m, n \geq N$, then

$$\|f - f_n\| = \lim_{m \to \infty} \|f_m - f_n\| \leq \varepsilon.$$

Thus this convergence is uniform. $\blacksquare$

There is a useful test for uniform convergence of a series of functions. The proof is easy, and the test comes up often in practice.

8.4.7. WEIERSTRASS M-TEST.
Suppose that $a_n(x)$ *is a sequence of functions on* $S \subset \mathbb{R}^k$ *into* $\mathbb{R}^m$ *and* (M_n) *is a sequence of real numbers so that*

$$\|a_n\|_{\infty} = \sup_{x \in S} \|a_n(x)\| \leq M_n \quad \text{for all} \quad x \in S.$$

If $\sum_{n=1}^{\infty} M_n$ *converges, then the series* $\sum_{n=1}^{\infty} a_n(x)$ *converges uniformly on* S.

PROOF. For each $x \in S$, the sequence $(a_n(x))$ is an absolutely convergent sequence of real numbers since

$$\sum_{n=1}^{\infty} \|a_n(x)\| < \sum_{n=1}^{\infty} \|a_n\|_{\infty} \leq \sum_{n=1}^{\infty} M_n < \infty.$$

Thus the sum exists. Define $f(x) = \sum_{n=1}^{\infty} a_n(x)$. Then for every $x \in S$,

$$\left\| f(x) - \sum_{n=1}^{k} a_n(x) \right\| = \left\| \sum_{n=k+1}^{\infty} a_n(x) \right\| \leq \sum_{n=k+1}^{\infty} \|a_n(x)\|$$

$$\leq \sum_{n=k+1}^{\infty} \|a_n\|_{\infty} \leq \sum_{n=k+1}^{\infty} M_n.$$

This estimate does not depend on x. Thus

$$\lim_{k \to \infty} \left\| f - \sum_{n=1}^{k} a_n \right\|_{\infty} \leq \lim_{k \to \infty} \sum_{n=k+1}^{\infty} M_n = 0.$$

Therefore, this series converges uniformly to f.   ■

As an application, we return to the series $\sum_{n=0}^{\infty} \dfrac{x^n}{n!}$ considered in Example 8.4.3. On any interval $[-A, A]$ with $A \geq 0$, $|x^n/n!| \leq A^n/n! =: M_n$. Applying the Ratio Test to M_n shows that $\sum_{n=0}^{\infty} M_n$ converges. Hence by the M-test, the series converges uniformly on $[-A, A]$. The series does not converge uniformly on the whole real line, but since it converges uniformly on every bounded interval, we may conclude that the limit is continuous on the whole line.

8.4.8. EXAMPLE. Consider the geometric series $\sum_{n=0}^{\infty} (-x^2)^n$. The ratio of successive terms this series at the point x is $-x^2$. Thus for $|x| < 1$, this series converges; while it diverges for $|x| > 1$. By inspection, it also diverges at $x = \pm 1$. For each x in $(-1, 1)$, we readily obtain that

$$\sum_{n=0}^{\infty} (-x^2)^n = \frac{1}{1 - (-x^2)} = \frac{1}{1 + x^2}.$$

On the interval of convergence $(-1, 1)$, the convergence is not uniform. Indeed, for any integer N, take $a = 2^{-1/2N}$ and note that the Nth term $(-a^2)^N = \frac{1}{2}$ is large. However, on the interval $[-r, r]$ for any $r < 1$, we have

$$\sup_{|x| \leq r} |(-x^2)^n| = r^{2n}.$$

Since $\sum_{n=0}^{\infty} r^{2n} = \dfrac{1}{1 - r^2} < \infty$, the Weierstrass M-test shows that the series converges uniformly to $f(x) = \dfrac{1}{1 + x^2}$ on $[-r, r]$.

Consider the functions

$$F_k(x) := \int_0^x \sum_{k=0}^n (-t^2)^n \, dt = \sum_{n=0}^k \frac{(-1)^n}{2n+1} x^{2n+1}.$$

Apply Theorem 8.3.1 to see that F_n converges uniformly on $[-r, r]$ to the function

$$F(x) = \int_0^x \frac{1}{1+t^2} \, dt = \tan^{-1}(x).$$

This yields the Taylor series for $\tan^{-1}$ about the point 0. (See Section 10.1.)

$$\tan^{-1}(x) = \sum_{n=0}^\infty \frac{(-1)^n}{2n+1} x^{2n+1}$$

The radius of convergence of this series is still 1.

This converges at $x = \pm 1$ as well because it is an alternating series in which the terms are monotone decreasing to zero.

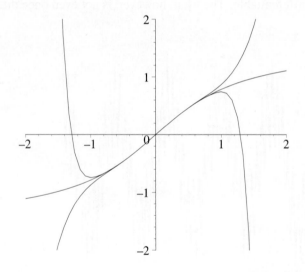

FIGURE 8.4. The function $\tan^{-1}$ with the degree 5 and 11 approximations.

It happens that this convergence is uniform on the whole interval $[-1, 1]$. To see this, we need the error estimate for alternating series that shows that the size of the next term is a bound on the error, Corollary 3.2.7. Note that the series is an alternating series for every $x \in \mathbb{R}$. The terms converge monotonely to 0 precisely when $|x| \le 1$. So the error between the nth partial sum and the limit is no greater than

$$\sup_{|x| \le 1} \left| \frac{(-1)^n}{2n+1} x^{2n+1} \right| \le \frac{1}{2n+1}.$$

Since this tends to 0, the series converges uniformly on $[-1, 1]$ to $\tan^{-1}(x)$.

8.4.9. EXAMPLE. The reason we have not proved good results about the derivative of uniformly convergent sequences is that, in general, there is nothing good to say. Indeed, there are continuous functions that are not differentiable at any point. These are called **nowhere differentiable functions**. The first example was constructed by Bolzano sometime before 1830 but was not published. Weierstrass independently discovered such functions in 1861 and he published his construction in 1872.

To construct a continuous nowhere differentiable function, let

$$f(x) = \sum_{k \geq 1} 2^{-k} \cos(10^k \pi x).$$

Set $f_k(x) = 2^{-k} \cos(10^k \pi x)$. Then $\sum_{k \geq 1} \|f_k\|_\infty = \sum_{k \geq 1} 2^{-k} = 1$ converges. Thus by the Weierstrass M-test, this series converges uniformly on the whole real line to a continuous function.

Figure 8.5 gives the graph of a partial sum $\sum_{k=1}^{n} 2^{-k} \cos(10^k \pi x)$. Bear in mind that such a function, being a finite linear combination of infinitely differentiable functions, is infinitely differentiable. The limit, however, is not even once differentiable.

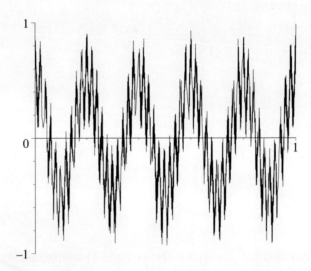

FIGURE 8.5. An approximant to Weierstrass's function.

Consider x be an arbitrary point in $\mathbb{R}$, say $x = x_0.x_1 x_2 x_3 \ldots$. We will show that f is not differentiable at x by constructing a sequence z_n converging to x so that the difference quotient $(f(z_n) - f(x))/(z_n - x)$ goes to $+\infty$.

Fix $n \geq 1$. Let $y_0 = x_0.x_1 x_2 \ldots x_n$ and $y_1 = y_0 + 10^{-n}$. So $y_0 \leq x \leq y_1$. Let us estimate $|f(y_0) - f(y_1)|$. As $10^n \pi y_0$ and $10^n \pi y_1$ are integer multiples of π, we have

$$f_n(y_0) = (-1)^{x_n} 2^{-n} \quad \text{and} \quad f_n(y_1) = (-1)^{x_n+1} 2^{-n}.$$

Hence $|f_n(y_0) - f_n(y_1)| = 2^{1-n}$. For $k > n$, $10^k y_i \pi$ is an integer multiple of 2π. So $f_k(y_0) = f_k(y_1) = 1$. And for $1 \leq k < n$, the Mean Value Theorem yields

$$|f_k(y_0) - f_k(y_1)| \leq \|f_k'\|_\infty |y_0 - y_1| = (2^{-k} 10^k \pi) 10^{-n} = 2^{-n} \pi 5^{k-n}.$$

Combining all of these estimates judiciously, we obtain

$$|f(y_0) - f(y_1)| \geq |f_n(y_0) - f_n(y_1)| - \sum_{k \neq n} |f_k(y_0) - f_k(y_1)|$$

$$\geq 2^{1-n} - \sum_{k=1}^{n-1} 2^{-n} \pi 5^{k-n}$$

$$> 2^{-n}\left(2 - \frac{\pi}{4}\right) > 2^{-n}.$$

One of these values is far from $f(a)$ since

$$|f(y_0) - f(a)| + |f(a) - f(y_1)| \geq |f(y_0) - f(y_1)| > 2^{-n}.$$

Choose $i = 0$ or 1 so that $|f(y_i) - f(a)| > 2^{-n-1}$, and set $z_n = y_i$. Clearly $|z_n - a| \leq |y_1 - y_0| = 10^{-n}$. Therefore,

$$\left| \frac{f(z_n) - f(a)}{z_n - a} \right| \geq \left| \frac{2^{-n-1}}{10^{-n}} \right| = 5^n/2.$$

As n tends to infinity, it is clear that the sequence (z_n) converges to a while the differential quotient blows up. Therefore, f is not differentiable at a.

Exercises for Section 8.4

A. Reformulate Theorem 8.3.1 and Corollary 8.3.2 in terms of series of functions.

B. Prove Theorem 8.4.6.

C. (a) Show that $\sum_{n=1}^{\infty} x^n e^{-nx}$ converges uniformly on $[0, A]$ for each $A > 0$.
   (b) Does it converge uniformly on $[0, \infty)$?

D. Does $\sum_{n=1}^{\infty} \frac{1}{x^2 + n^2}$ converge uniformly on the whole real line?

E. Show that if $\sum_{n=1}^{\infty} |a_n| < \infty$, then $\sum_{n=1}^{\infty} a_n \cos nx$ converges uniformly on $\mathbb{R}$.

F. (a) Let $f_n(x) = \frac{x^2}{(1 + x^2)^n}$ for $x \in \mathbb{R}$. Evaluate the sum $S(x) = \sum_{n=0}^{\infty} f_n(x)$.
   (b) Is this convergence uniform? For which values $a < b$ does this series converge uniformly on $[a, b]$?

G. Consider the series $\sum_{n=0}^{\infty} \left(\frac{x - 7}{x + 1}\right)^n$ for $x \geq 0$, $x \neq 7$. Find the limit, and discuss where the convergence is uniform or not.

H. Suppose that $a_k(x)$ are continuous functions on $[0, 1]$, and define $s_n(x) = \sum_{k=1}^{n} a_k(x)$.
   Show that if (s_n) converges uniformly on $[0, 1]$, then (a_n) converges uniformly to 0.

I. Prove the series version of Dini's Theorem (Exercise 8.1.I): If g_n are nonnegative continuous functions on $[a, b]$ and $\sum_{n=1}^{\infty} g_n$ converges pointwise to a *continuous* function on $[a, b]$, then it converges uniformly.

J. Let (f_n) be a sequence of functions defined on $\mathbb{N}$ such that $\lim_{k \to \infty} f_n(k) = L_n$ exists for each $n \geq 0$. Suppose that $\|f_n\|_\infty \leq M_n$ where $\sum_{n=0}^{\infty} M_n < \infty$. Define a function $F(k) = \sum_{n=0}^{\infty} f_n(k)$. Prove that $\lim_{k \to \infty} F(k) = \sum_{n=0}^{\infty} L_n$.
HINT: Think of f_n as a function g_n on $\{\frac{1}{k} : k \geq 1\} \cup \{0\}$. How will you define $g_n(0)$?

K. Apply the previous theorem to the functions $f_n(k) = \binom{k}{n} \left(\frac{x}{k}\right)^n$ for $n \geq 0$ and $k \geq 1$.
Hence show that $\lim_{k \to \infty} \left(1 + \frac{x}{k}\right)^k = e^x$.

L. In Example 8.4.9, we could use $f(x) = \sum_{k \geq 1} b^k \cos(a^k \pi x)$, where $b < 1$ and a is an even integer. Prove that if $ab > 1 + \pi/2$, then f is nowhere differentiable.

M. Let $d(x) = \operatorname{dist}(x, \mathbb{N})$ and $f_k(x) = 2^{-k}\left(d(2^k x) - 2d(2^{k-1}x)\right)$ for $k \geq 1$.
  (a) Compute $g_k(x) = d(x) + \sum_{k=1}^{n} f_k(x)$.
  (b) Where does g_k fail to be differentiable? This is an increasing sequence of sets with union dense in $\mathbb{R}$.
  (c) Find the limit of g_k. How can it turn out to be differentiable?

8.5. Power Series

As mentioned in the previous section, a power series is a series of functions of the form

$$\sum_{n=0}^{\infty} a_n x^n = a_0 + a_1 x + a_2 x^2 + a_3 x^3 + \cdots .$$

Formally, this is a power series in x and we could also consider a power series in $x - x_0$, namely

$$\sum_{n=0}^{\infty} a_n (x - x_0)^n = a_0 + a_1(x - x_0) + a_2(x - x_0)^2 + a_3(x - x_0)^3 + \cdots .$$

This increase in generality is only apparent, as we can set $y = x - x_0$ and work with a power series in y.

Clearly, a power series converges when $x = 0$. This may be the only value of x for which the series converges. For example, apply the Ratio Test to $\sum_{n=1}^{\infty} n! x^n$. The following theorem provides a full answer to the general question of when a power series converges.

8.5.1. HADAMARD'S THEOREM.

Given a power series $\sum_{n=0}^{\infty} a_n x^n$, there is R in $[0, +\infty) \cup \{+\infty\}$ so that the series converges for all x with $|x| < R$ and diverges for all x with $|x| > R$. Moreover, the series converges uniformly on each closed *interval $[a, b]$ contained in $(-R, R)$.*

Finally, if $\alpha = \limsup_{n \to \infty} |a_n|^{1/n}$, then

$$R = \begin{cases} +\infty & \text{if} \quad \alpha = 0, \\ 0 & \text{if} \quad \alpha = +\infty, \\ \frac{1}{\alpha} & \text{if} \quad \alpha \in (0, +\infty). \end{cases}$$

We call R the **radius of convergence** of the power series.

PROOF. Fixing $x \in \mathbb{R}$ and applying the Root Test to $\sum_{n=0}^{\infty} a_n x^n$ gives

$$\limsup |a_n x^n|^{1/n} = |x| \limsup |a_n|^{1/n} = |x|\alpha.$$

So if $\alpha = 0$, then $|x|\alpha < 1$ for all choices of x and so the series always converges, as claimed. If $\alpha = +\infty$, then $|x|\alpha > 1$ for all $x \neq 0$, and so the series diverges for nonzero x, again as claimed. Otherwise, $|x|\alpha < 1$ if and only if $|x| < R$ and $|x|\alpha > 1$ if and only if $|x| > R$. By the Root Test (Exercise 3.2.4), it follows that we have convergence and divergence on the required intervals.

It remains only to show uniform convergence on each interval $[a, b]$ contained in $(-R, R)$. There is some $c < R$ so that $[a, b] \subset [-c, c]$ for some $c < R$. Observe that, for $x \in [-c, c]$, $|a_n x^n| \leq |a_n| c^n$. As $c < R$, the previous paragraph shows that $\sum_{n=0}^{\infty} |a_n| c^n$ converges. By the M-test, it follows that $\sum_{n=0}^{\infty} a_n x^n$ converges uniformly on $[-c, c]$ and hence on $[a, b]$. ∎

By Exercise 3.3.F, if $\lim_{n \to \infty} \left| \frac{a_{n+1}}{a_n} \right|$ is defined, then $\lim_{n \to \infty} |a_n|^{1/n} = \lim_{n \to \infty} \left| \frac{a_{n+1}}{a_n} \right|$. Thus, we can often use ratios instead of roots to compute radii of convergence. See Exercise 8.5.C.

8.5.2. EXAMPLE.

The previous theorem contains no information about what happens if $|x| = R$? The answer, like the Ratio and Root Tests for series of numbers, is that the series may converge or diverge at these points. We consider three series,

$$\sum_{n=1}^{\infty} \frac{x^n}{2^n n^2}, \qquad \sum_{n=1}^{\infty} \frac{x^n}{2^n n}, \qquad \sum_{n=1}^{\infty} \frac{x^{2n}}{2^n n},$$

to illustrate this. For the first series, the limit ratio of successive coefficients is

$$\lim_{n \to \infty} \frac{2^n n^2}{2^{n+1}(n+1)^2} = \lim_{n \to \infty} \frac{n^2}{2(n+1)^2} = \frac{1}{2}.$$

Therefore, the radius of convergence is 2. Similarly, the second series also has radius of convergence 2.

Consider the first series at $x = \pm 2$. We have $\sum_{n=1}^{\infty} 1/n^2$ and $\sum_{n=1}^{\infty} (-1)^n/n^2$, both of which converge. For the second series at $x = \pm 2$, we have $\sum_{n=1}^{\infty} 1/n$ and $\sum_{n=1}^{\infty} (-1)^n/n$, which diverge and converge, respectively. The first series has interval of convergence $[-2, 2]$ while the second has $[-2, 2)$.

The third series differs crucially from the other two. To write this series in the form $\sum_{n=1}^{\infty} a_n x^n$, we cannot define a_n to be $1/(2^n n)$, but rather $a_{2k+1} = 0$ and $a_{2k} = 2^{-k}/k$ for $k \geq 0$. Using this formula for a_n, only the even terms matter:

$$\limsup_{n \to \infty} |a_n|^{1/n} = \lim_{k \to \infty} \left| \frac{1}{2^k k} \right|^{1/2k} = \frac{1}{\sqrt{2}} \lim_{k \to \infty} \left(\frac{1}{k} \right)^{1/2k} = \frac{1}{\sqrt{2}}.$$

So the series has radius of convergence $\sqrt{2}$. At $x = \pm\sqrt{2}$, this series is $\sum_{n=1}^{\infty} \frac{1}{n}$, which diverges. Therefore, the interval of convergence is $(-\sqrt{2}, \sqrt{2})$.

It seems natural to differentiate and integrate a power series term-by-term. That is, the derivative and indefinite integral of $f(x) = \sum_{n=0}^{\infty} a_n x^n$ should be the sum of the terms $n a_n x^{n-1}$ and $a_n x^{n+1}/(n+1)$, respectively. The badly behaved examples of previous sections show that this intuition cannot be trusted for arbitrary series of functions. However, we will prove that this is true for power series.

This pair of properties make power series particularly useful. For example, the following theorem implies that if a power series has radius of convergence $R > 0$, then it is infinitely differentiable on $(-R, R)$.

8.5.3. TERM-BY-TERM DIFFERENTIATION OF SERIES.

If $f(x) = \sum_{n=0}^{\infty} a_n x^n$ has radius of convergence $R > 0$, then $\sum_{n=1}^{\infty} n a_n x^{n-1}$ has radius of convergence R, f is differentiable on $(-R, R)$ and, for $x \in (-R, R)$,

$$f'(x) = \sum_{n=1}^{\infty} n a_n x^{n-1}.$$

Further, $\sum_{n=0}^{\infty} \frac{a_n}{n+1} x^{n+1}$ has radius of convergence R and, for $x \in (-R, R)$,

$$\int_0^x f(t)\, dt = \sum_{n=0}^{\infty} \frac{a_n}{n+1} x^{n+1}.$$

PROOF. Observe that $\sum_{n=0}^{\infty} na_n x^{n-1}$ and $\sum_{n=0}^{\infty} na_n x^n$ have the same radius of convergence. We have

$$\limsup_{n\to\infty} |na_n|^{1/n} = \lim_{n\to\infty} n^{1/n} \limsup_{n\to\infty} |a_n|^{1/n} = \frac{1}{R}.$$

Thus $\sum_{n=0}^{\infty} na_n x^n$ has radius of convergence R. As the partial sums $\sum_{n=0}^{k} na_n x^n$ converge uniformly on each interval $[-a, a] \subset (-R, R)$, we can apply Corollary 8.3.2 (with $c = 0$) to show that f is differentiable and $f'(x) = \sum_{n=0}^{\infty} na_n x^{n-1}$.

Similarly, $\sum_{n=0}^{\infty} \frac{a_n}{n+1} x^{n+1}$ and $\sum_{n=0}^{\infty} \frac{a_n}{n+1} x^n$ have the same radius of convergence and

$$\limsup_{n\to\infty} \left| \frac{a_n}{n+1} \right|^{1/n} = \lim_{n\to\infty} \frac{1}{(n+1)^{1/n}} \limsup_{n\to\infty} |a_n|^{1/n} = \frac{1}{R}.$$

So $\sum_{n=0}^{\infty} \frac{a_n}{n+1} x^{n+1}$ has radius of convergence R. If $x \in (-R, R)$, then since $\sum_{n=0}^{k} a_n t^n$ converges uniformly to $f(t)$ on the interval $[0, x]$, Theorem 8.3.1 implies that the sequence of integrals

$$F_n(x) = \int_0^x \sum_{n=0}^{k} a_n t^n \, dt = \sum_{n=0}^{k} \frac{a_n}{n+1} x^{n+1}$$

converges uniformly to $F(x) = \int_0^x f(t) \, dt$. Thus,

$$\sum_{n=0}^{\infty} \frac{a_n}{n+1} x^{n+1} = \int_0^x f(t) \, dt$$

as required. ∎

8.5.4. EXAMPLE. We return to the series $f(x) = \sum_{n\geq 0} \frac{x^n}{n!}$ of Example 8.4.3. It was shown using the M-test that this series has infinite radius of convergence, and converges uniformly on $[-A, A]$ for all finite A. Use term-by-term differentiation to compute

$$f'(x) = \sum_{n=1}^{\infty} \frac{x^{n-1}}{(n-1)!} = \sum_{k=0}^{\infty} \frac{x^k}{k!} = f(x).$$

The differential equation $f'(x) = f(x)$ may be rewritten as

$$1 = \frac{f'(x)}{f(x)} = (\log f)'(x).$$

Integrating from 0 to t, we obtain

$$t = \int_0^t 1\,dx = \int_0^t (\log f)'(x)\,dx = \log f(t) - \log f(0).$$

It is evident that $f(0) = 1$ and therefore $\log f(t) = t$, whence $f(t) = e^t$.

8.5.5. EXAMPLE. Consider the power series $\sum_{n\geq 1} n^2 x^n$. Since $\lim_{n\to\infty} \frac{(n+1)^2}{n^2} = 1$, the Ratio Test tells us that the radius of convergence is 1. When $|x| = 1$, the terms do not tend to 0, and thus the series diverges. So the function $f(x) = \sum_{n\geq 1} n^2 x^n$ is defined for $x \in (-1, 1)$.

Define the function $g(x) = \sum_{n\geq 0} x^n$. This also has radius of convergence 1. Since it is a geometric series, it is easily evaluated as $g(x) = \dfrac{1}{1-x}$ for $|x| < 1$. Apply Theorem 8.5.3 to obtain

$$\frac{1}{(1-x)^2} = g'(x) = \sum_{n\geq 1} n x^{n-1}.$$

This series has the same radius of convergence, 1, as does

$$\frac{x}{(1-x)^2} = x g'(x) = \sum_{n\geq 1} n x^n.$$

A second application of Theorem 8.5.3 yields

$$\sum_{n\geq 1} n^2 x^{n-1} = \big(x g'(x)\big)' = \frac{1}{(1-x)^2} + \frac{2x}{(1-x)^3} = \frac{1+x}{(1-x)^3}.$$

Therefore,

$$f(x) = \sum_{n\geq 1} n^2 x^n = \frac{x(1+x)}{(1-x)^3}.$$

In particular, $\sum_{n\geq 1} \dfrac{n^2}{2^n} = f(\tfrac{1}{2}) = 6$.

8.5.6. EXAMPLE. In this example, we obtain the Binomial Theorem for fractional powers. That is, we derive the power series expansion of $(1+x)^\alpha$ for $\alpha \in \mathbb{R}$. If $g(x) = (1+x)^\alpha$, then $g'(x) = \alpha(1+x)^{\alpha-1}$ and so g satisfies the differential equation (DE)

$$(1+x)g'(x) = \alpha g(x), \quad g(0) = 1.$$

Suppose there is a power series $f(x) = \sum_{n=0}^{\infty} a_n x^n$ that satisfies this DE. Then we have

$$(1+x)\sum_{n=1}^{\infty} n a_n x^{n-1} = \alpha \sum_{n=0}^{\infty} a_n x^n.$$

Collecting terms, we have

$$\sum_{n=0}^{\infty}(na_n + (n+1)a_{n+1})x^n = \sum_{n=0}^{\infty}\alpha a_n x^n$$

and so $na_n + (n+1)a_{n+1} = \alpha a_n$ giving $a_{n+1} = \dfrac{\alpha - n}{n+1}a_n$. As $a_0 = f(0) = 1$, we have $a_1 = \alpha$, $a_2 = \dfrac{\alpha(\alpha-1)}{2}$, $a_3 = \dfrac{\alpha(\alpha-1)(\alpha-2)}{6}$, and so on. In general, we obtain the **fractional binomial coefficients**

$$a_n = \frac{\alpha(\alpha-1)\cdots(\alpha-n+2)(\alpha-n+1)}{n!} = \binom{\alpha}{n}.$$

It remains to show that this series has a positive radius of convergence and that it actually converges to $(1+x)^\alpha$.

If α is a nonnegative integer, then the a_n are eventually zero, and so the series reduces to the usual Binomial Theorem. In this case, the radius of convergence is infinite. Otherwise $a_n \neq 0$ for all n, and we can apply the Ratio Test to obtain

$$\lim_{n\to\infty}\left|\frac{a_{n+1}}{a_n}\right| = \lim_{n\to\infty}\left|\frac{\alpha - n}{n+1}\right| = 1.$$

Hence the series has radius of convergence 1.

To show that $f(x) = (1+x)^\alpha$, consider the ratio $f(x)/(1+x)^\alpha$. Differentiating the ratio with respect to x gives

$$\frac{(1+x)^\alpha f'(x) - \alpha(1+x)^{\alpha-1}f(x)}{(1+x)^{2\alpha}}$$

and since we have shown $(1+x)f'(x) = \alpha f(x)$, it follows that the derivative is zero. However, setting $x = 0$ in $f(x)/(1+x)^\alpha$ gives $1/1 = 1$ and so the ratio is constantly equal to 1, showing that $f(x) = (1+x)^\alpha$.

Thus, for $|x| < 1$ and any real α,

$$(1+x)^\alpha = \sum_{n=0}^{\infty}\binom{\alpha}{n}x^n.$$

Exercises for Section 8.5

A. Determine the interval of convergence of the following power series:

(a) $\displaystyle\sum_{n=0}^{\infty} n^3 x^n$

(b) $\displaystyle\sum_{n=1}^{\infty} \frac{(-1)^n}{n^2} x^n$

(c) $\displaystyle\sum_{n=0}^{\infty} \frac{n^2}{2^n} x^n$

(d) $\displaystyle\sum_{n=0}^{\infty} \sqrt{n}\, x^n$

(e) $\displaystyle\sum_{n=0}^{\infty} (-1)^n \frac{x^{2n}}{(2n)!}$

(f) $\displaystyle\sum_{n=0}^{\infty} x^{n!}$

(g) $\displaystyle\sum_{n=1}^{\infty} \frac{n!}{n^n} x^n$

(h) $\displaystyle\sum_{n=0}^{\infty} \frac{(n!)^2}{(2n)!} x^n$

(i) $\displaystyle\sum_{n=0}^{\infty} \frac{1}{n} x^n$

B. Find a power series $\sum_{n=0}^{\infty} a_n x^n$ that does not have the same *interval* of convergence as

$\sum_{n=0}^{\infty} n a_n x^{n-1}$.

C. Suppose that $\lim_{n \to \infty} \dfrac{a_n}{a_{n+1}} = L$ exists. Find the radius of convergence of the power

series $\sum_{n=1}^{\infty} a_n x^n$. (See Exercise 3.3.F.)

D. Using the method of Example 8.5.6, show that if $f(x) = \sum_{n=0}^{\infty} a_n x^n$ satisfies the DE

$f'(x) = f(x)$ and $f(0) = 1$, then $f(x) = \sum_{n=0}^{\infty} x^n / n!$.

E. Repeat the previous exercise with the conditions $f''(x) = -f(x)$, f is an odd function, and $f(0) = 0$.

F. Prove that if infinitely many of the a_n are nonzero integers, then the radius of convergence of $\sum_{n=1}^{\infty} a_n x^n$ is at most 1.

G. (a) Compute $f(x) = \sum_{n=0}^{\infty} \dfrac{1}{n} x^n$.

  (b) Compute $\sum_{n=0}^{\infty} \dfrac{2^n}{n 5^n}$. Justify your method.

H. (a) Compute $f(x) = \sum_{n=0}^{\infty} (n+1) x^n$.

  (b) Compute $\sum_{n=0}^{\infty} \dfrac{n}{3^n}$. Justify your method.

  (c) Is the substitution of $x = -1$ justified?

I. (a) Compute $g(x) = \sum_{n=0}^{\infty} (n^2 + n) x^n$.

  (b) Compute $\sum_{n=0}^{\infty} \dfrac{n^2 + n}{2^n}$. Justify your method.

J. Using the binomial series for $(1 - x)^{-1/2}$ and the formula

$$\int_0^{\pi/2} \sin^{2n} t \, dt = \frac{1 \cdot 3 \cdot 5 \cdots (2n-1)}{2 \cdot 4 \cdots (2n)} \frac{\pi}{2},$$

show that, for $\kappa \in (0, 1)$, the integral $\displaystyle\int_0^{\pi/2} (1 - \kappa^2 \sin^2 t)^{-1/2} dt$ equals

$$\frac{\pi}{2} \left(1 + \left(\frac{1}{2}\right)^2 \kappa^2 + \left(\frac{1 \cdot 3}{2 \cdot 4}\right)^2 \kappa^4 + \left(\frac{1 \cdot 3 \cdot 5}{2 \cdot 4 \cdot 6}\right)^2 \kappa^6 + \cdots \right).$$

K. Recall the Fibonacci sequence defined recursively by $F(0) = F(1) = 1$ and

  $F(n+2) = F(n) + F(n+1)$ for $n \geq 0$. Set $f(x) = \sum_{n=0}^{\infty} F(n) x^n$.

  (a) Show that $F(n) \leq 2^n$ and hence deduce a positive lower bound for the radius of convergence of this power series.

(b) Compute $\displaystyle\lim_{n\to\infty} \frac{F(n+1)}{F(n)}$ and hence find the radius of convergence R.

HINT: Let $r_n = \dfrac{F(n+1)}{F(n)}$. Show by induction that $r_{n+1} - r_n = -(r_n - r_{n-1})/r_n r_{n-1}$ and that $r_n \geq 3/2$. Hence deduce that the limit R exists. Show that R satisfies a quadratic equation.

(c) Compute $(1 - x - x^2)f(x)$ for $|x| < R$, and justify your steps. Hence compute $f(x)$.

(d) Show that $f(x) = \sum_{n=0}^{\infty} (x + x^2)^n$, and that this converges for $|x| < R$.

8.6. Compactness and Subsets of *C(K)*

We saw in Chapter 4 that compactness is a very powerful property. This showed up particularly in Chapter 5 in the proofs of the Extreme Value Theorem and of uniform continuity for a continuous function on a compact set. In this section, we characterize the subsets of $C(K)$ that are themselves compact. We will need this characterization to prove Peano's Theorem (Theorem 12.8.1), which shows that a wide range of differential equations have solutions.

We restrict our attention to sets of functions in $C(K)$ when K is a compact subset of $\mathbb{R}^n$. As usual, a subset $\mathcal{F}$ of $C(K)$ is called compact if every sequence (f_n) of functions in $\mathcal{F}$ has a subsequence (f_{n_i}) that converges uniformly to a function f in $\mathcal{F}$. The Heine–Borel Theorem showed that a subset of $\mathbb{R}^n$ is compact if and only if it is closed and bounded. However, $\mathbb{R}^n$ is a finite-dimensional space; and this is a critical fact. $C(K)$ is infinite dimensional, so some of those arguments are invalid—as is the conclusion.

The arguments of Lemma 4.4.3 are still valid. If $\mathcal{F}$ is not closed, there is a sequence (f_n) in $\mathcal{F}$ that has a uniform limit $f = \displaystyle\lim_{n\to\infty} f_n$ that is not in $\mathcal{F}$. Every subsequence (f_{n_i}) also has limit f, which is not in $\mathcal{F}$. So $\mathcal{F}$ is not compact.

Likewise, if $\mathcal{F}$ is unbounded, it contains a sequence (f_n) such that $\|f_n\| > n$ for $n \geq 1$. Any subsequence (f_{n_i}) satisfies $\displaystyle\lim_{n\to\infty} \|f_{n_i}\| = \infty$. Consequently, it cannot converge uniformly to any function.

However, this is not the whole story. There are other ways in which a subset of $C(K)$ can fail to be compact, as we now show.

8.6.1. EXAMPLE. Look again at Example 8.1.5. We will show that the set $\mathcal{F} = \{f_n(x) = x^n : n \geq 1\}$ is closed and bounded but not compact. The functions f_n on $[0, 1]$ are all bounded by 1. Suppose that (f_{n_i}) is any subsequence of (f_n). By Example 8.1.5,

$$\lim_{i\to\infty} f_{n_i}(x) = \lim_{n\to\infty} f_n(x) = \begin{cases} 0 & \text{for } 0 \leq x < 1 \\ 1 & \text{for } x = 1. \end{cases}$$

However, this limit $\chi_{\{1\}}$ is not continuous, and the convergence is not uniform. Thus *no* subsequence converges. It follows that the only limit points of $\mathcal{F}$ are the points in $\mathcal{F}$ themselves. In particular, $\mathcal{F}$ contains all of its limit points, and therefore it is closed. On the other hand, $\mathcal{F}$ is not compact because the sequence (f_n) has no convergent subsequence.

8.6.2. EXAMPLE. Consider a sequence (g_n) of continuous functions on K such that g_n converges uniformly to a function g. Then the set

$$\mathcal{G} = \{g_n : n \geq 1\} \cup \{g\}$$

is compact. Indeed, suppose that (f_n) is a sequence in $\mathcal{G}$. Either some element of $\mathcal{G}$ is repeated infinitely often, or infinitely many g_k's are represented in this sequence. In the first case, there is a constant subsequence that evidently converges in $\mathcal{G}$. Otherwise, there is a subsequence (f_{n_i}) such that $f_{n_i} = g_{k_i}$ and $\lim_{i \to \infty} k_i = \infty$. In this case, the subsequence converges uniformly to g.

Now consider a point a in K and an $\varepsilon > 0$. Since g is continuous, there is an $r_0 > 0$ such that

$$\|g(x) - g(a)\| < \frac{\varepsilon}{3} \quad \text{whenever} \quad \|x - a\| < r_0.$$

Since g_n converges uniformly to g, there is an integer N so that

$$\|g - g_n\|_\infty < \frac{\varepsilon}{3} \quad \text{for all} \quad n \geq N.$$

Combining these estimates, we can show that for $n \geq N$ and $\|x - a\| < r_0$,

$$\|g_n(x) - g_n(a)\| \leq \|g_n(x) - g(x)\| + \|g(x) - g(a)\| + \|g(a) - g_n(a)\|$$
$$\leq \frac{\varepsilon}{3} + \frac{\varepsilon}{3} + \frac{\varepsilon}{3} = \varepsilon.$$

Now we can modify this to obtain a statement for all functions in $\mathcal{G}$. Each g_n for $n < N$ is continuous. So there are positive real numbers $r_n > 0$ such that

$$\|g_n(x) - g_n(a)\| < \varepsilon \quad \text{whenever} \quad \|x - a\| < r_n.$$

Set $r = \min\{r_n : 0 \leq n \leq N\}$. We have shown that

$$\|f(x) - f(a)\| < \varepsilon \quad \text{whenever} \quad f \in \mathcal{G} \text{ and } \|x - a\| < r.$$

This example suggests a new variant of continuity in which a whole family of functions satisfy the same inequalities.

8.6.3. DEFINITION. A family of functions $\mathcal{F}$ mapping a set $S \subset \mathbb{R}^n$ into $\mathbb{R}^m$ is **equicontinuous at a point** $a \in S$ if for every $\varepsilon > 0$, there is an $r > 0$ such that

$$\|f(x) - f(a)\| < \varepsilon \quad \text{whenever} \quad \|x - a\| < r \text{ and } f \in \mathcal{F}.$$

The family $\mathcal{F}$ is **equicontinuous on a set** S if it is equicontinuous at every point in S. The family $\mathcal{F}$ is **uniformly equicontinuous** on S if for each $\varepsilon > 0$, there is an $r > 0$ such that

$$\|f(x) - f(y)\| < \varepsilon \quad \text{whenever} \quad \|x - y\| < r, \ x, y \in S \text{ and } f \in \mathcal{F}.$$

Reconsider the previous two examples. In the second example, we established that $\mathcal{G}$ is equicontinuous. However, in the first example, the set $\mathcal{F} = \{x^n : n \geq 1\}$ is *not* equicontinuous if $x = 1$. Indeed, take $\varepsilon = 1/10$ and let $0 < r < 1$ be an arbitrary positive number. Take $x = 1 - r/2$. Since $\lim\limits_{n\to\infty} x^n = 0$, there is an integer N sufficiently large that

$$|1 - x^n| > .5 \quad \text{for all} \quad n \geq N.$$

Hence this r does not work in the definition of equicontinuity as $|1 - x| < r$ and $|1 - x^n| > .1$. Since r is arbitrary, there is no choice of r that will work, and so $\mathcal{F}$ is not equicontinuous at 1.

8.6.4. LEMMA. *Let K be a compact subset of $\mathbb{R}^n$. A compact subset $\mathcal{F}$ of $C(K, \mathbb{R}^m)$ is equicontinuous.*

PROOF. Suppose to the contrary that for a certain point a in K and $\varepsilon > 0$, the definition of equicontinuity is not satisfied for $\mathcal{F}$. This means that for each choice of $r = 1/n$, there is a function $f_n \in \mathcal{F}$ and a point $x_n \in K$ such that

$$\|x_n - a\| < \frac{1}{n} \quad \text{and} \quad \|f_n(x_n) - f_n(a)\| \geq \varepsilon.$$

It is evident that no subsequence of (f_n) can be equicontinuous either.

Now if $\mathcal{F}$ were compact, there would be a subsequence (f_{n_i}) that converges uniformly to some function f. By Example 8.6.2, this subsequence would be equicontinuous. This contradiction shows that $\mathcal{F}$ must be equicontinuous. ∎

8.6.5. PROPOSITION. *If $\mathcal{F}$ is an equicontinuous family of functions on a compact set, then it is uniformly equicontinuous.*

PROOF. This is a modification of Theorem 5.5.9. If the result is false, there is an $\varepsilon > 0$ for which the definition of equicontinuity fails. This means that for each $r = 1/n$, there are points x_n and y_n in K and a function f_n in $\mathcal{F}$ such that

$$\|x_n - y_n\| < \frac{1}{n} \quad \text{and} \quad \|f_n(x_n) - f_n(y_n)\| \geq \varepsilon.$$

Since K is compact, the sequence (x_n) has a convergent subsequence with $\lim\limits_{i\to\infty} x_{n_i} = a$. Hence

$$\lim_{i\to\infty} y_{n_i} = \lim_{i\to\infty} x_{n_i} + \lim_{i\to\infty} y_{n_i} - x_{n_i} = a + 0 = a.$$

By the equicontinuity of $\mathcal{F}$ at a, there is an $r > 0$ so that

$$\|f(x) - f(a)\| < \frac{\varepsilon}{2} \quad \text{for all} \quad f \in \mathcal{F}, \ \|x - a\| < r.$$

There is an integer I so large that

$$\|x_{n_i} - a\| < r \quad \text{and} \quad \|y_{n_i} - a\| < r \quad \text{for all} \quad i \geq I.$$

Therefore

$$\|f_n(x_n) - f_n(y_n)\| \leq \|f_n(x_n) - f_n(a)\| + \|f_n(a) - f_n(y_n)\|$$
$$< \frac{\varepsilon}{2} + \frac{\varepsilon}{2} = \varepsilon.$$

This is a contradiction to the hypothesis that uniform equicontinuity fails; and thus it must hold. ∎

We need a new property, total boundedness, which follows from compactness. Conversely, a *closed* totally bounded set is compact, and we will prove this, as part of the Borel–Lebesgue Theorem (Theorem 9.2.3), in the next chapter. For convenience, we give a simple proof of the direction we need valid in $\mathbb{R}^n$.

8.6.6. DEFINITION. A subset S of K is called an ε-net of K if

$$K \subset \bigcup_{a \in S} B_\varepsilon(a).$$

A set K is **totally bounded** if it has a *finite* ε-net for every $\varepsilon > 0$.

8.6.7. LEMMA. *Let K be a compact subset of $\mathbb{R}^m$. Then K is totally bounded.*

PROOF. Fix $\varepsilon > 0$. Choose N so that $\varepsilon > 1/N$. Since K is closed and bounded, it is contained in a large M-cube

$$C = \{x \in \mathbb{R}^m : |x_i| \leq L,\ 1 \leq i \leq M\}.$$

The finite set of points

$$A = \{a \in \mathbb{R}^m : a_i = \frac{k_i \varepsilon}{2M},\ k_i \in \mathbb{Z},\ |k_i| \leq 2MN\}$$

forms an ε-net for C. For each a in A, either $B_{\varepsilon/2}(a) \cap K$ is empty or it contains a point x_a. Let $B = \{x_a : a \in A\}$.

We will show that B is an ε-net for K. Indeed, if x is any point in K, there is a point $a \in A$ such that $\|x - a\| < \varepsilon/2$. Since $B_{\varepsilon/2}(a) \cap K$ is nonempty, x_a is defined and belongs to this intersection. Therefore,

$$\|x - x_a\| \leq \|x - a\| + \|a - x_a\| < \frac{\varepsilon}{2} + \frac{\varepsilon}{2} = \varepsilon. \qquad\blacksquare$$

8.6.8. COROLLARY. *Let K be a compact subset of $\mathbb{R}^m$. Then K contains a sequence $\{x_i : i \geq 1\}$ that is dense in K. Moreover, for any $\varepsilon > 0$, there is an integer N so that $\{x_i : 1 \leq i \leq N\}$ forms an ε-net for K.*

PROOF. Let B_k be a finite $\frac{1}{k}$-net for K for each integer $k \geq 1$. Form a sequence $\{x_i : i \geq 1\}$ by listing all of the elements of each B_k in turn as the points $x_{N_{k-1}+1}, \ldots, x_{N_k}$. If $x \in K$, then for each k there is an element x_{n_k} corresponding to a point in B_k such that $\|x - x_{n_k}\| < 1/k$. Thus $x = \lim_{k \to \infty} x_{n_k}$. As x is an arbitrary point in K, the sequence is dense. By construction, if $\varepsilon > 1/k$, the set of points $\{x_i : 1 \leq i \leq N_k\}$ forms an ε-net for K. ∎

8.6.9. ARZELA–ASCOLI THEOREM.
Let K be a compact subset of $\mathbb{R}^n$. A subset $\mathcal{F}$ of $C(K, \mathbb{R}^m)$ is compact if and only if it is closed, bounded, and equicontinuous.

PROOF. The easy direction has been established: If $\mathcal{F}$ is compact, then it is closed, bounded and equicontinuous.

So assume $\mathcal{F}$ has these three properties and let (f_n) be a sequence in $\mathcal{F}$. We will construct a convergent subsequence. By Corollary 8.6.8, there is a sequence (x_i) so that for each $r > 0$, there is an integer N so that $\{x_1, \ldots, x_N\}$ forms an r-net for K.

We claim that there is a subsequence of (f_n), call it (f_{n_k}), so that

$$\lim_{k \to \infty} f_{n_k}(x_i) = L_i \quad \text{exists for all } i \geq 1.$$

To prove this, let Λ_0 denote the set of positive integers. Since $(f_n(x_1))$ is a bounded sequence, the Bolzano–Weierstrass Theorem (Theorem 2.6.4) provides a convergent subsequence. That is, there is an infinite subset $\Lambda_1 \subset \Lambda_0$ so that

$$\lim_{n \in \Lambda_1} f_n(x_1) = L_1 \quad \text{exists.}$$

Next, $(f_n(x_2))_{n \in \Lambda_1}$ is bounded sequence, so there is an infinite subset $\Lambda_2 \subset \Lambda_1$ so that

$$\lim_{n \in \Lambda_2} f_n(x_2) = L_2 \quad \text{exists.}$$

Continuing in this way, we obtain a decreasing sequence $\Lambda_0 \supset \Lambda_1 \supset \Lambda_2 \supset \cdots$ of infinite sets so that

$$\lim_{n \in \Lambda_i} f_n(x_i) = L_i \quad \text{converges for each } i \geq 1.$$

We now use a diagonalization method, similar to the proof that $\mathbb{R}$ is uncountable, Theorem 2.8.7. Let n_k be the kth entry of Λ_k; and let $\Lambda = \{n_k : k \in \mathbb{N}\}$. Then for each i, there are at most $i - 1$ entries of Λ that are not in Λ_i. Thus,

$$\lim_{k \to \infty} f_{n_k}(x_i) = \lim_{n \in \Lambda_i} f_n(x_i) = L_i \quad \text{for all} \quad i \geq 1,$$

proving the claim.

For simplicity of notation, we use g_k for f_{n_k} in the remainder of the proof. Now fix $\varepsilon > 0$. By uniform equicontinuity, there is $r > 0$ so that

$$\|f(x) - f(y)\| < \frac{\varepsilon}{3} \quad \text{for all} \quad f \in \mathcal{F} \text{ and } \|x - y\| < r.$$

Choose N so that $\{x_1, \ldots, x_N\}$ is an r-net for K. Since the g_k converge at each of these N points, there is some integer M so that

$$|g_k(x_i) - g_l(x_i)| \leq \frac{\varepsilon}{3} \quad \text{for all} \quad k, l \geq M \text{ and } 1 \leq i \leq N.$$

Let $k, l \geq M$ and pick $x \in K$. Since $\{x_1, \ldots, x_N\}$ is an r-net for K, there is some $i \leq N$ such that $\|x - x_i\| < r$. We need an $\varepsilon/3$-argument to finish the proof.

$$\|g_k(x) - g_l(x)\| \leq \|g_k(x) - g_k(x_i)\| + \|g_k(x_i) - g_l(x_i)\| + \|g_l(x_i) - g_l(x)\|$$
$$\leq \frac{\varepsilon}{3} + \frac{\varepsilon}{3} + \frac{\varepsilon}{3} = \varepsilon$$

Thus g_k is uniformly Cauchy, and so converges uniformly by Theorem 8.2.2. The limit g belongs to $\mathcal{F}$ because $\mathcal{F}$ is closed. Finally, as every sequence in $\mathcal{F}$ has a convergent subsequence, it follows that $\mathcal{F}$ is compact. ∎

In particular, this theorem shows that if a sequence of functions (f_n) in $C[a, b]$ forms a bounded equicontinuous subset, then (f_n) has a subsequence that converges uniformly to some function in $C[a, b]$.

8.6.10. EXAMPLE. Consider the subset K of $C[0, 1]$ consisting of all functions $f \in C[0, 1]$ such that $|f(0)| \leq 5$ and f has Lipschitz constant at most 47.

Notice that K is closed. For if $f_n \in K$ converge uniformly to a function f, then

$$|f(0)| = \lim_{n \to \infty} |f_n(0)| \leq 5,$$

and

$$|f(x) - f(y)| = \lim_{n \to \infty} |f_n(x) - f_n(y)| \leq \lim_{n \to \infty} 47|x - y| = 47|x - y|.$$

In particular,

$$|f(x)| \leq |f(x) - f(0)| + |f(0)| \leq 47 + 5 = 52.$$

So K is bounded.

Finally, observe that K is equicontinuous. For if $\varepsilon > 0$, take $r = \varepsilon/47$. Then if $|x - y| < r$,

$$|f(x) - f(y)| \leq 47|x - y| < 47r = \varepsilon.$$

Therefore, this is a compact subset of $C[0, 1]$.

Exercises for Section 8.6

A. Use $f_n(x) = x^n$ on $[0, 1]$ to show that $B = \{f \in C[0, 1] : \|f\| \leq 1\}$ is not compact.

B. Show that $U = \{f \in C[0, 1] : f(x) > 0 \text{ for all } x \in [0, 1]\}$ is open.
  HINT: You need the Extreme Value Theorem.

C. What is the interior of $V = \{f \in C_b(\mathbb{R}) : f(x) > 0 \text{ for all } x \in \mathbb{R}\}$?
  HINT: Compare with the previous exercise.

D. Prove that the family $\{\sin(nx) : n \geq 1\}$ is not an equicontinuous subset of $C[0, \pi]$.

E. (a) Show that

$$\mathcal{F} = \left\{ F(x) = \int_0^x f(t)\, dt : f \in C[0, 1], \ \|f\|_\infty \leq 1 \right\}$$

  is a bounded and equicontinuous subset of $C[0, 1]$.
  (b) Why is $\mathcal{F}$ not closed?

(c) Show that the closure of $\mathcal{F}$ consists of all functions with Lipschitz constant 1 such that $f(0) = 0$.
HINT: Construct F_n in $\mathcal{F}$ so that $F_n(2^{-n}k) = \frac{n}{n+1}f(2^{-n}k)$ for $0 \le k \le 2^n$.

F. (a) Let $\mathcal{F}$ be a subset of $C[0, 1]$ that is closed, bounded, and equicontinuous. Prove that there is a function $g \in \mathcal{F}$ such that

$$\int_0^1 g(x)\, dx \ge \int_0^1 f(x)\, dx \quad \text{for all} \quad f \in \mathcal{F}.$$

(b) Construct a closed bounded subset $\mathcal{F}$ of $C[0, 1]$ for which the conclusion of the previous problem is false.

G. Let K be a compact subset of $\mathbb{R}^n$. Show that a subset S of $C(K)$ is compact if and only if it is closed and totally bounded.
HINT: Show that totally bounded sets are bounded and equicontinuous.

H. Let $\mathcal{F}$ be an equicontinuous family of functions in $C(K)$, where K is a compact subset of $\mathbb{R}^n$. Suppose that for each $x \in K$,

$$\sup\{f(x) : f \in \mathcal{F}\} = M_x$$

is finite. Prove that $\mathcal{F}$ is bounded.
HINT: Use equicontinuity to bound $\mathcal{F}$ by $M_x + 1$ on a ball about x. Suppose that $|f_n(x_n)|$ tends to $+\infty$. Extract a convergent subsequence of $\{x_n\}$.

I. Let $\mathcal{F}$ be a family of continuous functions defined on $\mathbb{R}$ that is (i) equicontinuous and satisfies (ii) $\sup\{f(x) : f \in \mathcal{F}\} = M_x < \infty$ for every x. Show that every sequence $(f_n)_{n=1}^{\infty}$ has a subsequence that converges uniformly on $[-k, k]$ for every $k > 0$.
HINT: Find a subsequence $(f_{1,n})_{n=1}^{\infty}$ that converges uniformly on $[-1, 1]$. Then extract a subsubsequence $(f_{2,n})_{n=1}^{\infty}$ that converges uniformly on $[-2, 2]$, and so on. Now use a diagonal argument.

CHAPTER 9

Metric Spaces

This text focuses on subsets of a normed space, as this is the natural setting for most of our applications. In this chapter, we introduce an apparently more general framework, metric spaces, and some new ideas that are somewhat more advanced. They play an occasional role in the advanced sections of the applications.

9.1. Definitions and Examples

In a normed vector space, the distance between elements is found using the norm of the difference. However, a distance function can be defined abstractly on any set using the idea of a metric. Most of the arguments we used in the normed context will also work for metric spaces, with only minimal changes. The crucial difference is that in a metric space, we do not work in a vector space, so we cannot use the addition or scalar multiplication.

9.1.1. DEFINITION. Let X be a set. A **metric** on a set X is a function ρ defined on $X \times X$ taking values in $[0, \infty)$ with the following properties:

 (1) (positive definiteness) $\rho(x, y) = 0$ if and only if $x = y$,

 (2) (symmetry) $\rho(x, y) = \rho(y, x)$ for all $x, y \in X$,

 (3) (triangle inequality) $\rho(x, z) \le \rho(x, y) + \rho(y, z)$ for all $x, y, z \in X$.

A **metric space** is a set X with a metric ρ, denoted (X, ρ). If the metric is understood, we use X alone.

9.1.2. EXAMPLES.

(1) If X is a subset of a normed space V, define $\rho(x, y) = \|x - y\|$. This is our standard example.

(2) Put a metric on the surface of the sphere by setting $\rho(x, y)$ to be the length of the shortest path from x to y (known as a **geodesic**). This is the length of the shorter arc of the great circle passing through x and y. More generally, we can define such a metric on any smooth surface.

(3) The **discrete metric** on a set X is given by

$$d(x, y) = \begin{cases} 0 & \text{if } x = y \\ 1 & \text{if } x \neq y. \end{cases}$$

(4) Define a metric on $\mathbb{Z}$ by $\rho_2(n, n) = 0$ and $\rho_2(m, n) = 2^{-d}$, where d is the largest power of 2 dividing $m - n$. It is trivial to verify properties (1) and (2). If $\rho_2(l, m) = 2^{-d}$ and $\rho_2(m, n) = 2^{-e}$, then $2^{\min\{d,e\}}$ divides $l - n$ and so

$$\rho(l, n) \leq 2^{-\min\{d,e\}} = \max\{\rho_2(l, m), \rho_2(m, n)\}.$$

This metric is known as the **2-adic metric**. Replacing 2 with another prime p, we can define similarly the **p-adic metric**.

(5) If X is a closed subset of $\mathbb{R}^n$, let $K(X)$ denote the collection of all nonempty compact subsets of X. If A is a compact subset of X and $x \in X$, we define

$$\text{dist}(x, A) = \inf_{a \in A} \|x - a\|.$$

Then we define the **Hausdorff metric** on $K(X)$ by

$$\begin{aligned} d_H(A, B) &= \max\{\sup_{a \in A} \text{dist}(a, B), \sup_{b \in B} \text{dist}(b, A)\} \\ &= \max\{\sup_{a \in A} \inf_{b \in B} \|a - b\|, \sup_{b \in B} \inf_{a \in A} \|a - b\|\}. \end{aligned}$$

Since A is closed, $\text{dist}(x, A) = 0$ if and only if $x \in A$. In particular, we see that $d_H(A, B) = 0$ if and only if $A = B$. So d_H is positive definite and is evidently symmetric. For the triangle inequality, let A, B, C be three compact subsets of X. For each $a \in A$, the Extreme Value Theorem yields the existence of a closest point $b \in B$, so $\|a - b\| = \text{dist}(a, B)$. Then there is a closest point $c \in C$ to b with $\|b - c\| = \text{dist}(b, C)$. Therefore,

$$\begin{aligned} \text{dist}(a, C) &\leq \|a - c\| \leq \|a - b\| + \|b - c\| \\ &= \text{dist}(a, B) + \text{dist}(b, C) \\ &\leq d_H(A, B) + d_H(B, C). \end{aligned}$$

Therefore, $\sup_{a \in A} \text{dist}(a, C) \leq d_H(A, B) + d_H(B, C)$. Reversing the roles of A and C and combining the two inequalities, we obtain $d_H(A, C) \leq d_H(A, B) + d_H(B, C)$.

Let $A_\varepsilon := \{x \in \mathbb{R}^n : \text{dist}(x, A) \leq \varepsilon\}$. Note that $d_H(A, B) \leq \varepsilon$ if and only if $A \subset B_\varepsilon$ and $B \subset A_\varepsilon$.

This example will be examined in more detail in Section 11.7, when we study fractal sets.

The notions of convergence and open set can be carried over to metric spaces by replacing $\|x - y\|$ with $\rho(x, y)$. Once this is done, most of the other definitions do not need to be changed at all.

9.1.3. DEFINITION. The **ball** $B_r(x)$ of radius $r > 0$ about a point x is defined as $\{y \in X : \rho(x, y) < r\}$. We write $B_r^\rho(x)$ if the metric is ambiguous. A subset U is **open** if for every $x \in U$, there is an $r > 0$ so that $B_r(x)$ is contained in U.

A sequence (x_n) is said to **converge** to x if $\lim_{n \to \infty} \rho(x, x_n) = 0$. A set C is **closed** if it contains all limit points of sequences of points in C.

If X is a subset of a normed space, then these definitions agree with our old definitions; so the language is consistent. It is easy to see that a set is open precisely when the complement is closed (adapt the proof of Theorem 4.3.8 for normed vector spaces).

9.1.4. DEFINITION. A sequence $(x_n)_{n=1}^\infty$ in a metric space (X, ρ) is a **Cauchy sequence** if for every $\varepsilon > 0$, there is an integer N so that $\rho(x_i, x_j) < \varepsilon$ for all $i, j \geq N$.

A metric space X is **complete** if every Cauchy sequence converges (in X).

9.1.5. EXAMPLES.

(1) As in Proposition 2.7.1, every convergent sequence is Cauchy.

(2) It is easy to show that a subset of a complete metric space is complete if and only if it is closed. So one useful way to construct many complete metric spaces is to take a closed subset of a complete normed space. The purpose of Exercise 9.1.L is to show that every metric space arises in this manner, although it is not always a natural context.

(3) If X has the discrete metric, the only way a sequence may converge to x is if the sequence is eventually constant (i.e., $x_n = x$ for all $n \geq N$). This is because the ball $B_{1/2}(x) = \{x\}$. So every subset of X is both open and closed. Also X is complete.

(4) Consider the 2-adic metric of Example 9.1.2(4) again. The balls have the form $B_{2^{-d}}(n) = \{m \in \mathbb{Z} : m \equiv n \pmod{2^d}\}$. The sequence $(2^n)_{n=1}^\infty$ converges to 0 because $\rho_2(2^n, 0) = 2^{-n} \to 0$. Observe that $1 - (-2)^n$ is an odd multiple of 3 for all $n \geq 1$. The sequence $a_n = (1 - (-2)^n)/3$ is Cauchy because if $n > m \geq N$, then $a_n - a_m = (-2)^m a_{n-m}$ and therefore $\rho_2(a_m, a_n) = 2^{-m} \leq 2^{-N}$. This sequence does not converge.

(5) When X is a closed subset of $\mathbb{R}^n$, the metric space $(K(X), d_H)$ is complete. We will prove this in Theorem 11.7.2.

Continuous functions are defined by analogy with the norm case.

9.1.6. DEFINITION. A function f from a metric space (X, ρ) into a metric space (Y, d) is **continuous** if for every $x_0 \in X$ and $\varepsilon > 0$, there is a $\delta > 0$ so that $d(f(x), f(x_0)) < \varepsilon$ whenever $\rho(x, x_0) < \delta$.

The proof of Theorem 5.3.1 goes through without change.

9.1.7. THEOREM. *Let f map a metric space (X, ρ) into (Y, σ). The following are equivalent:*

(1) *f is continuous on X;*

(2) *for every convergent sequence $(x_n)_{n=1}^{\infty}$ with $\lim_{n \to \infty} x_n = a$ in X, we have $\lim_{n \to \infty} f(x_n) = f(a)$; and*

(3) *$f^{-1}(U) = \{x \in X : f(x) \in U\}$ is open in X for every open set U in Y.*

As in the norm case, if the domain of a continuous function is not compact, the function need not be bounded. However, we have the same solution: Consider the normed vector space $C_b(X)$ of all *bounded* continuous functions $f : X \to \mathbb{R}$ with the sup norm $\|f\|_{\infty} = \sup_{x \in X} |f(x)|$.

9.1.8. THEOREM. *The space $C_b(X)$ of all bounded continuous functions on X with the sup norm $\|f\| = \sup\{|f(x)| : x \in X\}$ is complete.*

PROOF. The proof of Theorem 8.2.1 works with S replaced by any metric space. Likewise, Theorem 8.2.2 works with $C_b(X)$ because compactness is used only to ensure that the sup norm is finite. Therefore, $C_b(X)$ is a complete normed vector space. The details are left as an exercise. ∎

Exercises for Section 9.1

A. Show that $\rho(x, y) = |e^x - e^y|$ is a metric on $\mathbb{R}$.

B. Show that every subset of a discrete metric space is both open and closed.

C. Prove that U is open in (X, ρ) if and only if $X \setminus U$ is closed.

D. Prove Theorem 9.1.7.

E. Given a metric space (X, ρ), define a new metric on X by $\sigma(x, y) = \min\{\rho(x, y), 1\}$.
(a) Show that σ is a metric on X. Observe that X has finite diameter in the σ metric.
(b) Show that $\lim_{n \to \infty} x_n = x$ in (X, ρ) if and only if $\lim_{n \to \infty} x_n = x$ in (X, σ).
(c) Show that (x_n) is Cauchy in (X, ρ) if and only if it is Cauchy in (X, σ). Hence completeness is the same for these two metrics.

F. Suppose that V is a normed vector space. If (X, ρ) is a metric space, observe that the space $C_b(X, V)$ of all bounded continuous functions from X to V is a vector space. Show that $\|f\|_{\infty} = \sup_{x \in X} \|f(x)\|$ is a norm on $C_b(X, V)$.

G. Two metrics ρ and σ on a set X are **equivalent** if for each $x \in X$ and $r > 0$, there is an $s > 0$ so that $B_s^{\rho}(x) \subset B_r^{\sigma}(x)$ and $B_s^{\sigma}(x) \subset B_r^{\rho}(x)$.
(a) Prove that equivalent metrics have the same open and closed sets.
(b) Prove that equivalent metrics have the same convergent sequences and the same Cauchy sequences.

H. Define a function on $\mathcal{M}_n \times \mathcal{M}_n$ by $\rho(A, B) = \text{rank}(A - B)$. Prove that ρ is a metric that is equivalent to the discrete metric.

I. Put a metric ρ on all the words in a dictionary by defining the distance between two distinct words to be 2^{-n} if the words agree for the first n letters and are different at the $(n+1)$st letter. Here we agree that a space is distinct from a letter. For example, $\rho(\text{car}, \text{cart}) = 2^{-3}$ and $\rho(\text{car}, \text{call}) = 2^{-2}$.
   (a) Verify that this is a metric.
   (b) Suppose that words w_1, w_2 and w_3 are listed in alphabetical order. Show that $\rho(w_1, w_2) \le \rho(w_1, w_3)$.
   (c) Suppose that words w_1, w_2 and w_3 are listed in alphabetical order. Find a formula for $\rho(w_1, w_3)$ in terms of $\rho(w_1, w_2)$ and $\rho(w_2, w_3)$.

J. Recall the 2-adic metric of Examples 9.1.2(4) and 9.1.5(4). Extend it to $\mathbb{Q}$ by setting $\rho_2(a/b, c/d) = 2^{-e}$, where e is the unique integer such that $a/b - c/d = 2^e(f/g)$ and both f and g are odd integers.
   (a) Prove that ρ_2 is a metric on $\mathbb{Q}$.
   (b) Show that the sequence of integers $a_n = (1 - (-2)^n)/3$ converges in $(\mathbb{Q}, \rho_2)$.
   (c) Find the limit of $\dfrac{n!}{n! + 1}$ in this metric.

K. Complete the details of Theorem 9.1.8 as follows:
   (a) Prove that Theorem 8.2.1 is valid when S is replaced by a metric space X.
   (b) Prove that $C_b(X)$ is a complete normed vector space.   HINT: Theorem 8.2.2

L. Suppose that (X, ρ) is a nonempty metric space. Let $C_b(X)$ be the normed space of all bounded continuous functions on X with the sup norm $\|f\|_\infty = \sup\{|f(x)| : x \in X\}$.
   (a) Fix x_0 in X. For each $x \in X$, define $f_x(y) = \rho(x, y) - \rho(x_0, y)$ for $y \in X$. Show that f_x is a bounded continuous function on X.
   (b) Show $\|f_x - f_y\|_\infty = \rho(x, y)$.
   (c) Hence deduce that the map that takes $x \in X$ to the function f_x identifies X with a subset F of $C_b(X)$ that induces the same metric.

M. (a) Give an example of a decreasing sequence of closed balls in a complete metric space with empty intersection. Compare with Exercise 7.2.J.
   HINT: Use a metric on $\mathbb{N}$ equivalent to the discrete metric so that $\{n \ge k\}$ are closed balls.
   (b) Show that a metric space (M, d) is complete if and only if every sequence of closed balls with radii going to zero has a nonempty intersection.

9.2. Compact Metric Spaces

As we have mentioned in the previous chapter, in general, a closed and bounded set need not be compact. There are several useful properties that are equivalent to compactness. In this section, we define these properties and prove they are equivalent.

We have to change our language slightly. Our old notion of compactness will be renamed **sequential compactness**. Although we introduce a new notion and call it compactness, we will prove that the two notions are equivalent in normed spaces and in metric spaces. There is a more general setting, topological spaces, where compactness and sequential compactness differ, but we never deal with it in this book. As you will immediately recognize, the new notion also makes sense in normed spaces and could have been introduced there.

9.2.1. DEFINITION. A collection of open sets $\{U_\alpha : \alpha \in A\}$ in X is called an **open cover** of $Y \subset X$ if $Y \subset \bigcup_{\alpha \in A} U_\alpha$. A **subcover** of $\{U_\alpha : \alpha \in A\}$ is just a subcollection $\{U_\alpha : \alpha \in B\}$ for some $B \subset A$ that is still a cover. In particular, it is a **finite subcover** of Y if it is a finite collection of open subsets that covers Y.

A collection of closed sets $\{C_\alpha : \alpha \in A\}$ has the **finite intersection property** if every *finite* subcollection has nonempty intersection.

For example, consider the (finite) cover indicated in Figure 9.1.

FIGURE 9.1. An open cover for $X = [a, b]$.

9.2.2. DEFINITION. A metric space X is **compact** if every open cover of X has a finite subcover.

A metric space X is **sequentially compact** if every sequence of points in X has a convergent subsequence.

A metric space X is **totally bounded** if for every $\varepsilon > 0$, there are finitely many points $x_1, \ldots, x_k \in X$ so that $\{B_\varepsilon(x_i) : 1 \le i \le k\}$ is an open cover.

9.2.3. BOREL–LEBESGUE THEOREM.
For a metric space X, the following are equivalent.

(1) *X is compact.*

(2) *Every collection of closed subsets of X with the finite intersection property has nonempty intersection.*

(3) *X is sequentially compact.*

(4) *X is complete and totally bounded.*

PROOF. First assume (1) that X is compact, and let us prove (2). Suppose that $\{C_\alpha : \alpha \in A\}$ is a collection of closed sets such that $\bigcap_\alpha C_\alpha = \varnothing$. Consider the open sets $U_\alpha = C_\alpha'$, the complements of C_α. Then $\bigcup_\alpha U_\alpha = \left(\bigcap_\alpha C_\alpha\right)' = X$. Thus there is a finite subcover $X = U_{\alpha_1} \cup \cdots \cup U_{\alpha_k}$. Consequently,

$$C_{\alpha_1} \cap \cdots \cap C_{\alpha_k} = \left(U_{\alpha_1} \cup \cdots \cup U_{\alpha_k}\right)' = \varnothing.$$

So no collection of closed sets with empty intersection has the finite intersection property.

Now assume (2) and let (x_i) be a sequence in X. Define $C_n = \overline{\{x_i : i \ge n\}}$. This is a decreasing sequence of closed sets. The collection $\{C_n : n \ge 1\}$ has the finite intersection property since the intersection of $C_{n_1}, \ldots, C_{n_k}$ contains the point x_n where $n = \max\{n_1, \ldots, n_k\}$. By hypothesis, there is a point x in $\bigcap_{n \geq 1} C_n$. Recursively define a sequence as follows. Let $n_0 = 1$. If n_{k-1} is defined, use the fact that $x \in \overline{\{x_i : i > n_{k-1}\}}$ to choose $n_k > n_{k-1}$ with $\rho(x, x_{n_k}) < 1/k$. By construction, $\lim_{k \to \infty} x_{n_k} = x$ converges.

Assume sequential compactness (3), and consider (4). If (x_i) is a Cauchy sequence, select a convergent subsequence, say $\lim_{k \to \infty} x_{n_k} = x$. Given $\varepsilon > 0$, use the Cauchy property to find N so that $i, j \geq N$ implies that $\rho(x_i, x_j) < \varepsilon/2$. Then use the convergence to find $n_k > N$ so that $\rho(x, x_{n_k}) < \varepsilon/2$. Then if $i \geq N$,

$$\rho(x, x_i) \leq \rho(x, x_{n_k}) + \rho(x_{n_k}, x_i) < \frac{\varepsilon}{2} + \frac{\varepsilon}{2} = \varepsilon.$$

So $\lim_{i \to \infty} x_i = x$. Thus X is complete.

If X were not totally bounded, then there would be some $\varepsilon > 0$ so that no finite collection of ε-balls can cover X. Recursively select points $x_k \in X$ so that $x_k \notin B_\varepsilon(x_1) \cup \cdots \cup B_\varepsilon(x_{k-1})$ for all $k \geq 2$. Consider the sequence (x_i). We will obtain a contradiction by showing that there is no convergent subsequence. Indeed, if (x_{n_i}) were a convergent subsequence, then it is Cauchy. So for some N large, all $i > N$ satisfy $\rho(x_i, x_N) < \varepsilon$, contrary to the fact that $x_i \notin B_\varepsilon(x_N)$. Therefore, X must be totally bounded.

Finally, we show that (4) implies (1). For each $k \geq 1$, choose a finite set $x_1^k, \ldots x_{n_k}^k$ so that $B_{1/k}(x_i^k)$ for $1 \leq i \leq n_k$ covers X. Suppose that there is an open cover $\mathcal{U} = \{U_\alpha : \alpha \in A\}$ of X with no finite subcover. We will recursively define a sequence of points $y_k = x_{i_k}^k$ so that $\bigcap_{j=1}^k \overline{B_{1/j}(y_j)}$ does not have a finite subcover from $\mathcal{U}$. At the stage $k = 0$, X does not have a finite subcover. Suppose that we have this property for $k - 1$. If each $\bigcap_{j=1}^{k-1} \overline{B_{1/j}(y_j)} \cap \overline{B_{1/k}(x_i^k)}$ had a finite subcover from $\mathcal{U}$, then combining them for $1 \leq i \leq n_k$ would yield a finite subcover of $\bigcap_{j=1}^{k-1} \overline{B_{1/j}(y_j)}$, contrary to fact. So we may pick $y_k = x_{i_k}^k$ so that $\bigcap_{j=1}^k \overline{B_{1/j}(y_j)}$ has no finite subcover.

Now we show that this leads to a contradiction. The sequence (y_k) is Cauchy. To see this, note that $\bigcap_{j=1}^{k-1} \overline{B_{1/j}(y_j)}$ is nonempty because the empty set is covered. Thus there is a point $x \in B_{1/j}(y_j) \cap B_{1/k}(y_k)$. Hence

$$\rho(y_j, y_k) \leq \rho(y_j, x) + \rho(x, y_k) < \tfrac{1}{j} + \tfrac{1}{k}.$$

Given $\varepsilon > 0$, choose N so large that $N > 2/\varepsilon$. Then for $j, k \geq N$, it follows that $\rho(y_j, y_k) < 2/N < \varepsilon$.

But (4) states that X is complete, and thus $y = \lim_{k \to \infty} y_k$ exists. Hence there is some α so that U_α contains y. Therefore, there is an $\varepsilon > 0$ so that $B_\varepsilon(y)$ is contained in U_α. Choose any $k > 2/\varepsilon$ for which $\rho(y_k, y) < \varepsilon/2$. Then since $1/k < \varepsilon/2$, $\overline{B_{1/k}(y_k)} \subset B_\varepsilon(y) \subset U_\alpha$. This contradicts the fact that $\overline{B_{1/k}(y_k)}$ does not have a finite subcover. Thus it must be the case that X is compact. $\blacksquare$

In general, being complete and bounded is not sufficient to imply that a metric space is compact. For example, by Exercise 9.1.E the real line has a bounded metric equivalent to the usual one. In this new metric it is bounded and complete but is not compact.

Perhaps a more natural example is the closed unit ball B of $C[0,1]$ in the max norm. This is complete because it is a closed subset of a complete space. However, it is not totally bounded. To see this, let $f_n(x)$ be the functions from Example 9.1.2. Since $\|f_n\|_\infty = 1$, they all lie in B. But $\|f_n - f_m\|_\infty = 1$ if $n \neq m$. Thus no $1/2$-ball can contain more than one of them. Hence no finite family of $1/2$-balls covers B. We already have a complete description of the compact subsets of $C[0,1]$, thanks to the Arzela–Ascoli Theorem (Theorem 8.6.9).

All of our basic theorems on continuous functions go through for continuous functions on metric spaces. In particular, Theorem 5.4.3 can be established with the same proof. We will give a proof based on open covers. In Exercise 9.2.H, this will yield the Extreme Value Theorem.

9.2.4. THEOREM. *Let C be a compact subset of a metric space (X, ρ). Suppose that f is a continuous function of X into (Y, σ). Then the image set $f(C)$ is compact.*

PROOF. Let $\mathcal{U} = \{U_\alpha : \alpha \in A\}$ be an open cover of $f(C)$ in Y. Since f is continuous, $V_\alpha := f^{-1}(U_\alpha)$ are open sets in X. The collection $\mathcal{V} = \{V_\alpha : \alpha \in A\}$ is an open cover of C. Indeed, for each $x \in C$, $f(x) \in f(C)$ and thus $f(x) \in U_\alpha$ for some α. Hence x belongs to V_α. By the compactness of C, select a finite subcover $V_{\alpha_1}, \ldots, V_{\alpha_k}$. Then

$$f(C) \subset \bigcup_{i=1}^{k} f(V_{\alpha_i}) \subset \bigcup_{i=1}^{k} U_{\alpha_i}.$$

So $U_{\alpha_1}, \ldots, U_{\alpha_k}$ is a finite subcover of $f(C)$. Therefore, $f(C)$ is compact. ∎

Exercises for Section 9.2

A. Let (X, ρ) be a metric space and let a subset (Y, ρ) be considered as a metric space with the same metric.
   (a) Show that every open set U in Y has the form $V \cap Y$ for some open set V in X.
      HINT: This is easy for balls.
   (b) Show that Y is compact if and only if every collection $\{V_\alpha : \alpha \in A\}$ of open sets in X that covers Y has a finite subcover.

B. Show that if Y is a subset of a complete metric space X, then Y is compact if and only if it is closed and totally bounded.

C. Show that a closed subset of a compact metric space is compact.

D. Show that every compact metric space is **separable** (i.e., it has a countable dense subset).

E. (a) Prove that every open subset U of $\mathbb{R}^n$ is the countable union of compact subsets.
      HINT: Use the distance to U^c and the norm to define the sets.
   (b) Show that every open cover of an open subset of $\mathbb{R}^n$ has a countable subcover.

F. Prove **Cantor's Intersection Theorem**: A decreasing sequence of nonempty compact subsets $A_1 \supset A_2 \supset \cdots$ of a metric space (X, ρ) has nonempty intersection.

G. Show that a continuous function from a compact metric space (X, ρ) into a metric space (Y, d) is uniformly continuous. HINT: Fix $\varepsilon > 0$. For $x \in X$, choose $\delta_x > 0$ so that $\rho(x, t) < 2\delta_x$ implies $d(f(x), f(t)) < \varepsilon/2$. Then $\{B_{\delta_x}(x) : x \in X\}$ covers X.

H. If f is a continuous function from a compact metric space (X, ρ) into $\mathbb{R}$, prove that there is a point $x_0 \in X$ such that $|f(x_0)| = \sup\{|f(x)| : x \in X\}$.
HINT: Compare to the proof of the Extreme Value Theorem.

I. Let S_n for $n \geq 1$ be a finite union of disjoint closed balls in $\mathbb{R}^k$ of radius at most 2^{-n} such that $S_{n+1} \subset S_n$ and S_{n+1} has at least two balls inside each ball of S_n. Prove that $C = \bigcap_{n \geq 1} S_n$ is a perfect, nowhere dense compact subset of $\mathbb{R}^k$.
HINT: Compare with Example 4.4.8.

J. If f is a continuous one-to-one function of a compact metric space X onto Y, show that f^{-1} is continuous. HINT: Theorem 9.2.4

K. Show that the previous exercise is false if X is not compact.
HINT: Map $(0, 1]$ onto a circle with a tail (i.e., the figure 6).

L. We say (X, ρ) is a **second countable metric space** if there is a countable collection $\mathcal{U}$ of open balls in X so that for every $x \in X$ and $r > 0$, there is a ball $U \in \mathcal{U}$ with $x \in U \subset B_r(x)$. Prove that (X, ρ) is second countable if and only if it is separable.
HINT: For $\Rightarrow$, take the centres of balls in $\mathcal{U}$. For $\Leftarrow$, take all balls of radius $1/k$, $k \geq 1$, about each point in a countable dense set.

9.3. Complete Metric Spaces

Now we turn to an important consequence of completeness. Some more terminology is required.

9.3.1. DEFINITION. A subset A of a metric space (X, ρ) is **nowhere dense** if $\operatorname{int}(\overline{A}) = \varnothing$ (i.e., the closure of A has no interior). A subset B is said to be **first category** if it is the *countable* union of nowhere dense sets. A subset Y of a *complete* metric space is a **residual set** if the complement Y' is first category.

Nowhere dense sets are small in a certain sense, and thus sets of first category are also considered to be small. For example, the Cantor set is nowhere dense in $\mathbb{R}$. Our main result, which has surprisingly powerful consequences, is that complete metric spaces are never first category.

9.3.2. BAIRE CATEGORY THEOREM.
Let (X, ρ) be a complete metric space. Then the union of countably many nowhere dense subsets of X has no interior, and in particular is a proper subset of X. Equivalently, the intersection of countably many dense open subsets of X is dense in X.

PROOF. Consider a sequence $(A_n)_{n=1}^{\infty}$ of nowhere dense subsets of X. To show that the complement of $\bigcup_{n\geq1} A_n$ is dense X, take any ball $\overline{B_{r_0}(x_0)}$. We will construct a point in this ball that is not in any $\overline{A_n}$. It will then follow that this union has no interior, whence the complement contains points arbitrarily close to each point $x \in X$, and so is dense.

As $\overline{A_1}$ has no interior, it does not contain $B_{r_0/2}(x_0)$. Pick x_1 in $B_{r_0/2}(x_0) \setminus \overline{A_1}$. Since $\overline{A_1}$ is closed, $\text{dist}(x_1, \overline{A_1}) > 0$. So we may choose an $0 < r_1 < r_0/2$ so that $\overline{B_{r_1}(x_1)}$ is disjoint from $\overline{A_1}$. Note that $\overline{B_{r_1}(x_1)} \subset \overline{B_{r_0}(x_0)}$. Proceed recursively choosing a point $x_{n+1} \in B_{r_n/2}(x_n)$ and an $r_{n+1} \in (0, r_n/2)$ so that $\overline{B_{r_{n+1}}(x_{n+1})}$ is disjoint from $\overline{A_{n+1}}$. Clearly, $\overline{B_{r_{n+1}}(x_{n+1})} \subset \overline{B_{r_n}(x_n)}$.

The sequence $(x_n)_{n=1}^{\infty}$ is Cauchy. Indeed, given $\varepsilon > 0$, choose N so that $2^{-N}r_0 < \varepsilon/2$. Then $r_N < \varepsilon/2$. However, all x_n for $n \geq N$ lie in $\overline{B_{r_N}(x_N)}$. Thus for $n, m \geq N$,

$$\rho(x_n, x_m) \leq \rho(x_n, x_N) + \rho(x_N, x_m) < \frac{\varepsilon}{2} + \frac{\varepsilon}{2} = \varepsilon.$$

Because X is a complete space, there is a limit $x_\infty = \lim_{n\to\infty} x_n$ in X. The point x_∞ belongs to $\bigcap_{n\geq1} \overline{B_{r_n}(x_n)}$. Hence it is disjoint from every $\overline{A_n}$ for $n \geq 1$.

If U_n are dense open subsets of X, their complements A_n are closed and nowhere dense. By the previous paragraphs, $\bigcup_{n\geq1} A_n$ has dense complement. This complement is exactly $\bigcap_{n\geq1} U_n$. ∎

Here is one interesting and unexpected consequence. It says that *most* continuous functions are nowhere differentiable. This is rather nonintuitive, as it was hard work to construct even one explicit example of such a function in Example 8.4.9.

9.3.3. PROPOSITION. *The set of continuous, nowhere differentiable functions on an interval $[a, b]$ is a residual set and in particular is dense in $C[a, b]$.*

PROOF. Say that a function f is Lipschitz at x_0 if there is a constant L so that $|f(x) - f(x_0)| \leq L|x - x_0|$ for all $x \in [a, b]$. Our first observation is that if f is differentiable at x_0, then it is also Lipschitz at x_0. From the definition of derivative, $\lim_{x\to x_0} \frac{f(x) - f(x_0)}{x - x_0} = f'(x_0)$. Choose $\delta > 0$ so that for $|x - x_0| < \delta$,

$$\frac{|f(x) - f(x_0)|}{|x - x_0|} \leq |f'(x_0)| + 1.$$

Then for $|x - x_0| < \delta$, $|f(x) - f(x_0)| \leq (|f'(x_0)| + 1)|x - x_0|$. If $|x - x_0| \geq \delta$,

$$|f(x) - f(x_0)| \leq 2\|f\|_\infty \leq \frac{2\|f\|_\infty}{\delta}|x - x_0|.$$

So $L = \max\left\{|f'(x_0)| + 1, \frac{2\|f\|_\infty}{\delta}\right\}$ is a Lipschitz constant at x_0.

Let A_n consist of all functions $f \in C[a, b]$ such that f has Lipschitz constant $L \leq n$ at some point $x_0 \in [a, b]$. Let us show that A_n is closed. Suppose that

$(f_k)_{k=1}^{\infty}$ is a sequence of functions in A_n converging uniformly to a function f. Each f_k has Lipschitz constant n at some point $x_k \in [a, b]$. Since $[a, b]$ is compact, there is a subsequence x_{k_i} converging to a point $x_0 \in [a, b]$. Then

$$|f(x) - f(x_0)| \leq |f(x) - f_{k_i}(x)| + |f_{k_i}(x) - f_{k_i}(x_{k_i})|$$
$$+ |f_{k_i}(x_{k_i}) - f_{k_i}(x_0)| + |f_{k_i}(x_0) - f(x_0)|$$
$$\leq \|f - f_{k_i}\|_{\infty} + n|x - x_{k_i}| + n|x_{k_i} - x_0| + \|f - f_{k_i}\|_{\infty}$$
$$= 2\|f - f_{k_i}\|_{\infty} + n(|x - x_{k_i}| + |x_{k_i} - x_0|).$$

Take a limit as $i \to \infty$ to obtain $|f(x) - f(x_0)| \leq n|x - x_0|$. Thus A_n is closed.

Next we show that A_n has no interior and hence is nowhere dense. Fix a function $f \in A_n$ and an $\varepsilon > 0$. Let us look for a function g in $B_\varepsilon(f)$ that is not in A_n. By Theorem 5.5.9, f is uniformly continuous. Choose $\delta > 0$ so that $|x-y| < \delta$ implies that $|f(x) - f(y)| < \varepsilon/4$. Construct a piecewise linear continuous function h which agrees with f on a sequence $a = x_0 < x_1 < \cdots < x_N = b$, where $x_k - x_{k-1} < \delta$ for $1 \leq k \leq N$. A simple estimate shows that $\|f - h\|_{\infty} < \varepsilon/2$. The function h is Lipschitz with constant L, say. Choose $M > 4\pi(L + n)/\varepsilon$ and set $g = h + \frac{\varepsilon}{2} \sin Mx$. Then

$$\|f - g\|_{\infty} \leq \|f - h\|_{\infty} + \|h - g\|_{\infty} < \frac{\varepsilon}{2} + \frac{\varepsilon}{2} < \varepsilon.$$

To see that g is not in A_n, take any point x_0. We can always choose a point $x \in [a, b]$ with $|x - x_0| < 2\pi/M$ so that $\sin Mx = \pm 1$ has sign opposite to the sign of $\sin Mx_0$. Therefore,

$$|g(x) - g(x_0)| \geq \frac{\varepsilon}{2} |\sin Mx - \sin Mx_0| - |h(x) - h(x_0)|$$
$$\geq \frac{\varepsilon}{2} - L|x - x_0| \geq \left(\frac{M\varepsilon}{4\pi} - L\right)|x - x_0| > n|x - x_0|.$$

So g does not have Lipschitz constant n at any point $x_0 \in [a, b]$.

Recall from Theorem 8.2.2 that $C[a, b]$ is complete. By the Baire Category Theorem, it is not the union of countably many nowhere dense sets. The first category set $\bigcup_{n \geq 1} A_n$ contains all functions that are differentiable at any single point. Thus the complement, consisting entirely of nowhere differentiable functions, is a residual set. Therefore, nowhere differentiable functions are dense in $C[a, b]$. ∎

Exercises for Section 9.3

A. Show that A is nowhere dense in X if and only if $X \setminus \overline{A}$ is dense in X.

B. Show that $\mathbb{R}^2$ is not the union of countably many lines.

C. Suppose that X is a countable complete metric space. Show that X has isolated points [i.e., points that are open (and closed)].
   HINT: Write X as a union of single points and apply the Baire Category Theorem.

D. Show that a complete metric space containing more than one point that has no isolated points is uncountable. HINT: Use the previous exercise.

E. A G_δ **set** is the intersection of a countable family of open sets.
 (a) Show that if $A \subset \mathbb{R}$ is closed, then A is a G_δ set.
 (b) Show that $\mathbb{Q}$ is not a G_δ subset of $\mathbb{R}$.   HINT: Show that $\mathbb{R} \setminus \mathbb{Q}$ is not first category.

F. (a) If f is a real-valued function on a metric space X, show that the set of points at which f is continuous is a G_δ set.   HINT: Show that the set U_k of points x for which there are $i \in \mathbb{N}$ and $\delta > 0$ so that $|f(y) - \frac{i}{k}| < \frac{1}{k}$ for $y \in B_\delta(x)$ is open.
 (b) Show that no function on $[0, 1]$ can be continuous just on the rational points. Compare this with Example 5.2.10.

G. Suppose that f_n is a sequence of continuous real-valued functions on a complete metric space X which converge pointwise to a function f.
 (a) Prove that there is a constant M and an open set $U \subset X$ such that $\sup_{n \geq 1} |f_n(x)| \leq M$ for all $x \in U$.   HINT: Let $A_k = \{x \in X : \sup_{n \geq 1} |f_n(x)| \leq k\}$.
 (b) If f is continuous and and $\varepsilon > 0$, show that there is an open set U and an integer N so that $|f(x) - f_n(x)| < \varepsilon$ for all $x \in U$ and $n \geq N$.
 HINT: Let $B_k = \{x \in X : \sup_{n \geq k} |f(x) - f_n(x)| \leq \varepsilon/2\}$.

H. (a) Show that the set of compact nowhere dense subsets of $\mathbb{R}^n$ is dense in $K(X)$, where $K(X)$ is equipped with the Hausdorff metric of Example 9.1.2(5).
 HINT: If C is compact and $\varepsilon > 0$, use a finite ε-net. Finite sets are nowhere dense.
 (b) Show that the set A_n of those compact sets in $K(X)$ that contain a ball of radius $1/n$ is a closed set with no interior.

I. **Banach–Steinhaus Theorem.** Suppose that X and Y are complete normed vector spaces, and $\{T_\alpha : \alpha \in A\}$ is a family of continuous linear maps of X into Y such that for each $x \in X$, $K_x = \sup_{\alpha \in A} \|T_\alpha x\| < \infty$.
 (a) Let $A_n = \{x \in X : K_x \leq n\}$. Show that A_n is closed.
 (b) Prove that there is some n_0 so that A_{n_0} has interior, say containing $B_\varepsilon(x_0)$.
 (c) Show that there is a finite constant L so that every T_α has Lipschitz constant L.
 HINT: If $\|x\| < 1$, then $x_0 + \varepsilon x \in A_{n_0}$. Estimate $\|T_\alpha x\|$.

J. Show that $[0, 1]$ is not the disjoint union of a countably infinite family of disjoint nonempty closed sets $\{A_n : n \geq 1\}$.
 HINT: If $U_n = \text{int } A_n$, observe that $X := [0, 1] \setminus \bigcup_{n \geq 1} U_n = \bigcup_{n \geq 1} (A_n \setminus U_n)$ is complete. Find an integer n_0 and an open interval V so that $\varnothing \neq X \cap V \subset A_{n_0}$. Show $U_n \cap V = \varnothing$ for $n \neq n_0$.

K. A function on $[0, 1]$ that is not monotonic on any interval is called a **nowhere monotonic function**. Show that these functions are a residual subset of $C[0, 1]$.
 HINT: Let $A_n = \{\pm f : \text{there is } x \in [0, 1], (f(y) - f(x))(y - x) \geq 0 \text{ for } |y - x| \leq \frac{1}{n}\}$.

9.4. Connectedness

The Intermediate Value Theorem shows that continuous real-valued functions map intervals onto intervals (see Corollary 5.6.2). This should mean that intervals have some special property not shared by other subsets of the line. This fact may appear to be dependent on the order structure. However, this property can be described in topological terms that allows a generalization to higher dimensions.

9.4.1. DEFINITION. A subset A of a metric space X is **not connected** if there are *disjoint* open sets U and V such that $A \subset U \cup V$ and $A \cap U \neq \varnothing \neq A \cap V$. Otherwise, the set A is said to be **connected**.

A set A is **totally disconnected** if for every pair of distinct points $x, y \in A$, there are two *disjoint* open sets U and V so that $x \in U$, $y \in V$ and $A \subset U \cup V$.

9.4.2. EXAMPLES.

(1) $A = [-1, 0] \cup [3, 4]$ is not connected in $\mathbb{R}$ because $U = (-2, 1)$ and $V = (2, 5)$ are disjoint open sets which each intersect A and jointly contain A.

(2) The set $A = \{x \in \mathbb{R} : x \neq 0\}$ is not connected because $U = (-\infty, 0)$ and $V = (0, +\infty)$ provide the necessary separation. The sets U and V need not be a positive distance apart. Even the small gap created by omitting the origin disconnects the set.

(3) $\mathbb{Q}$ is totally disconnected. For if $x < y \in \mathbb{Q}$, choose an irrational number z with $x < z < y$. Then $U = (-\infty, z) \ni x$ and $V = (z, \infty) \ni y$ are disjoint open sets such that $U \cup V$ contains $\mathbb{Q}$.

(4) It is more difficult to construct a closed set that is totally disconnected. However, the Cantor set C is an example. Suppose that $x < y \in C$. Choose N so that $y - x > 3^{-N}$. By Example 4.4.8, C is obtained by repeatedly removing middle thirds from each interval. At the Nth stage, C is contained in the set S_N that consists of 2^N intervals of length 3^{-N}. Thus x and y belong to distinct intervals in S_N. Therefore, there is a point z in the complement of S_N such that $x < z < y$. Then $U = (-\infty, z) \ni x$ and $V = (z, \infty) \ni y$ are disjoint open sets such that $U \cup V$ contains S_N and therefore contains C.

To obtain an example of a connected set, we must appeal to the Intermediate Value Theorem.

9.4.3. PROPOSITION. *The interval $[a, b]$ is connected.*

PROOF. Suppose to the contrary that there are disjoint open sets U and V such that $[a, b] \cap U$ and $[a, b] \cap V$ are both nonempty, and $[a, b] \subset U \cup V$. Define a function f on $[a, b]$ by

$$f(x) = \begin{cases} 1 & \text{if } x \in [a, b] \cap U \\ -1 & \text{if } x \in [a, b] \cap V. \end{cases}$$

We claim that this function is continuous. Indeed, let x be any point in $[a, b]$ and let $\varepsilon > 0$. If $x \in U$, then since U is open, there is a positive number $r > 0$ such that $B_r(x)$ is contained in U. Thus $|x - y| < r$ implies that

$$|f(y) - f(x)| = |1 - 1| = 0 < \varepsilon.$$

So f is continuous at x. Similarly, if $x \in V$, there is some ball $B_r(x)$ contained in V; and thus $|x - y| < r$ implies that

$$|f(y) - f(x)| = |-1 - (-1)| = 0 < \varepsilon.$$

This accounts for every point in $[a, b]$, and therefore f is a continuous function.

Thus we have constructed a continuous function on $[a, b]$ with range equal to $\{-1, 1\}$. This contradicts the Intermediate Value Theorem. So our assumption that $[a, b]$ is not connected must be wrong. Therefore, the interval is connected. ∎

To obtain a useful criterion for connectedness, we need to know that it is a property that is preserved by continuous functions. This is readily established using the topological form of continuity based on open sets.

9.4.4. THEOREM. *The continuous image of a connected set is connected.*

PROOF. This theorem is equivalent to the contrapositive statement: *If the image of a continuous function is not connected, then the domain is not connected.* To prove this, suppose that f is a continuous function from a metric space (X, ρ) into (Y, σ) such that the range $f(X)$ is not connected. Then there are disjoint open sets U and V in Y such that $f(X) \cap U$ and $f(S) \cap V$ are both nonempty, and $f(X) \subset U \cup V$.

Let $S = f^{-1}(U)$ and $T = f^{-1}(V)$. By Theorem 9.1.7, the inverse image of an open set is open; and thus both S and T are open. Since U and V are disjoint and cover $f(X)$, each $x \in X$ has either $f(x) \in U$ or $f(x) \in V$ but not both. Therefore, S and T are disjoint and cover X. Finally, since $U \cap f(X)$ is nonempty, it follows that $S \cap X$ is also nonempty; and likewise, $T \cap X$ is not empty. This shows that X is not connected, as required. ∎

9.4.5. EXAMPLE. In Section 5.6, we defined a path in $\mathbb{R}^n$. Similarly, we define a **path** in any metric space as a continuous image of a closed interval. A path is connected because a closed interval is connected. In particular, if f is a continuous function of $[a, b]$ into $\mathbb{R}^n$, the graph $G(f) = \{(x, f(x)) : a \le x \le b\}$ is connected.

9.4.6. DEFINITION. A subset S of a metric space X is said to be **path connected** if for each pair of points x and y in S, there is a **path** in S connecting x to y, meaning that there is a continuous function f from $[0, 1]$ into S such that $f(0) = x$ and $f(1) = y$.

9.4.7. COROLLARY. *Every path connected set is connected.*

PROOF. Let S be a path connected set, and suppose to the contrary that it is not connected. Then there are disjoint open sets U and V such that

$$S \subset U \cup V \quad \text{and} \quad S \cap U \neq \varnothing \neq S \cap V.$$

Pick points $u \in S \cap U$ and $v \in S \cap V$. Since S is path connected, there is a path P connecting u to v in S. It follows that $P \subset U \cup V$. And $P \cap U$ contains u and

$P \cap V$ contains v. Thus both are nonempty. Consequently, this shows that the path P is not connected. This contradicts the previous example. Therefore, S must be connected. ∎

9.4.8. EXAMPLES.

(1) $\mathbb{R}^n$ is path connected and therefore connected.

(2) Recall that a subset of $\mathbb{R}^n$ is *convex* if the straight line between any two points in the set belongs to the set. Therefore, any convex subset of $\mathbb{R}^n$ is path connected.

(3) The annulus $\{x \in \mathbb{R}^2 : 1 \le \|x\| \le 2\}$ is path connected and hence connected.

(4) There are connected sets that are not path connected. An example is the compact set

$$S = Y \cup G = \{(0,t) : |t| \le 1\} \cup \{(x, \sin \tfrac{1}{x}) : 0 < x \le 1\}.$$

To see that S is connected, suppose that U and V are two disjoint open sets such that S is contained in $U \cup V$. The point $(0,0)$ belongs to one of them, which may be called U. Therefore, U intersects the line Y. Since Y is path connected, it must be completely contained in U (for if V intersects Y, this would show that Y was disconnected).

Since U is open, it contains a ball of positive radius $r > 0$ around the origin. But $(0,0)$ is a limit point of G, and therefore U contains points of G. Since G is a graph, it is path connected. So by the same argument, U contains G. Therefore, $V \cap S$ is empty. This shows that S is connected.

If S were path connected, there would be a continuous function f of $[0,1]$ into S with $f(0) = (0,0)$ and $f(1) = (1, \sin 1)$. Let $a = \sup\{x : f(x) \in Y\}$. Since f is continuous, there is a $\delta > 0$ such that $\|f(x) - f(a)\| \le .5$ if $|x - a| \le \delta$. Set $b = a + \delta$. Let $f(a) = (0,y)$ and $f(b) = (u, \sin \tfrac{1}{u}) =: \mathbf{u}$. A similar argument shows that there is a point $a < c < b$ such that $f(c) = (t, \sin \tfrac{1}{t}) =: \mathbf{t}$, where $t \le \dfrac{u}{1 + 2\pi u}$. Therefore, $f([c,b])$ is a connected subset of S containing both $\mathbf{t}$ and $\mathbf{u}$. The graph $G' = \{(x, \sin \tfrac{1}{x}) : t \le x \le u\}$ is connected, but removing any point will disconnect it. So $f([c,b])$ contains G'. However, $\sin \tfrac{1}{x}$ takes both values ± 1 on $[t,u]$, and so there is a point $\mathbf{v} = (v, \sin \tfrac{1}{v})$ on G' with $\|\mathbf{v} - f(a)\| > |\sin \tfrac{1}{v} - y| \ge 1$. This contradicts the estimate $\|f(x) - f(a)\| \le .5$ for all $x \in [c,b]$. Therefore, S is not path connected.

Exercises for Section 9.4

A. Show that the only connected subsets of $\mathbb{R}$ are intervals (which may be finite or infinite and may or may not include the endpoints).

B. Show that if S is a connected subset of X, then so is $\overline{S}$.

C. Let $A \subset \mathbb{R}^m$ and $B \subset \mathbb{R}^n$ be connected sets. Prove that $A \times B$ is connected in $\mathbb{R}^{m+n}$. HINT: Recall that $A \times B = \{(a,b) : a \in A, \ b \in B\}$.

D. Show that a connected open set $U \subset \mathbb{R}^n$ is path connected.
HINT: Fix $u_0 \in U$. The set V of points in U path connected to u_0 is open, as is the set W of points in U not path connected to u_0.

E. Find an example of a decreasing sequence $A_1 \supset A_2 \supset A_3 \cdots$ of closed connected sets in $\mathbb{R}^2$ such that $\bigcap_{k \geq 1} A_k$ is not connected.

F. Let $A_1 \supset A_2 \supset A_3 \cdots$ be a decreasing sequence of connected compact subsets of $\mathbb{R}^n$. Show that $\bigcap_{k \geq 1} A_k$ is connected.
HINT: Show that if $U \cup V$ is an open set containing the intersection, then it contains some A_n for n sufficiently large.

G. Show that there is no continuous bijection of $[-1, 1]$ onto the circle.
HINT: Such a function would map $[-1, 0) \cup (0, 1]$ onto a connected set.

H. Let (Y, ρ) be a subset of a metric space (X, ρ). Show that Y is connected in itself if and only if it is connected in X.
HINT: If U and V are nonempty disjoint open subsets of Y with $Y = U \cup V$, let $f(y) = \max\{\text{dist}(y, U), \text{dist}(y, V)\}$. Then $U_1 = \bigcup\{B_{f(y)/2}(y) : y \in U\}$ is open in X.

9.5. Metric Completion

Completeness is such a powerful property that we attempt to work in a complete context whenever possible. The purpose of this section is to show that every metric space may be naturally imbedded in a unique complete metric space known as its completion. We do this by an efficient, if sneaky, argument based on the completeness of $C_b(X)$, the space of bounded real functions on X.

In principle, this method could be applied to the rational numbers to obtain the real numbers in a base independent way. You should be concerned that our argument uses the real numbers in its proof. Because of this circularity, we cannot construct the real numbers this way. Nevertheless, we can show that the real line is unique and hence the various different constructions we have mentioned all give the same object. This puts the real numbers on a firmer theoretical footing.

In the next section, the completeness theorem will be applied to $C[a, b]$ with the L^p norms to obtain the L^p spaces. The completeness of these normed spaces allows us to use the full power of analysis in studying them.

9.5.1. DEFINITION. Let (X, ρ) be a metric space. A **completion** of (X, ρ) is a complete metric space (Y, d) together with a map T of X into Y such that $d(Tx_1, Tx_2) = \rho(x_1, x_2)$ for all $x_1, x_2 \in X$ (an **isometry**) and TX is dense in Y.

When the metric space and its completion are concretely represented, often the map T is an inclusion map, as in the next example. By an **inclusion map**, we mean that $T : X \to Y$, where $X \subset Y$ and T is given by $T(x) = x$. Except for changing the codomain, T is the identity map on X.

9.5.2. EXAMPLE. Let A be an arbitrary subset of $\mathbb{R}^n$ with the usual Euclidean distance. Then the closure $\overline{A}$ of A is a complete metric space (see Exercise 4.3.H). Evidently, the distance functions coincide on these two sets, so the inclusion map is trivially an isometry. As A is dense in $\overline{A}$, this is a completion of A.

9.5.3. THEOREM. *Every metric space (X, ρ) has a completion.*

PROOF. Pick a point $x_0 \in X$. For each $x \in X$, define $f_x(y) = \rho(x, y) - \rho(x_0, y)$. If $\rho(y_1, y_2) < \varepsilon/2$, then $|\rho(x, y_1) - \rho(x, y_2)| \le \rho(y_1, y_2) < \varepsilon/2$. So

$$|f_x(y_1) - f_x(y_2)| \le |\rho(x, y_1) - \rho(x, y_2)| + |\rho(x_0, y_1) - \rho(x_0, y_2)| < \varepsilon.$$

Therefore, f_x is (uniformly) continuous.

Observe that $f_{x_1}(y) - f_{x_2}(y) = \rho(x_1, y) - \rho(x_2, y)$. By the triangle inequality, $\rho(x_1, y) \le \rho(x_1, x_2) + \rho(x_2, y)$, and thus $f_{x_1}(y) - f_{x_2}(y) \le \rho(x_1, x_2)$. Interchanging x_1 and x_2 yields $\|f_{x_1} - f_{x_2}\|_\infty \le \rho(x_1, x_2)$. However, substituting $y = x_2$ shows that this is an equality. In particular as $f_{x_0} = 0$, we see that f_x is bounded. Thus $Tx = f_x$ is an isometric imbedding of X into $C_b(X)$.

Let $F = TX = \{f_x : x \in X\}$, and consider $\overline{F}$, the closure of F in $C_b(X)$. This is a closed subset of a complete normed space, and therefore it is complete. Evidently, $\overline{F}$ is a metric completion of X.   ∎

It is important that this completion is unique in a natural sense. This will justify our use of the terminology *the completion* and will allow us to consider X as a subset of C without explicit use of the isometry T.

To prove uniqueness, we first need a significant result about continuous functions on the metric completion. To simplify notation, we drop the map T and consider X as a subset of its completion C and use the same notation for the two metrics. A continuous function g on C is a **continuous extension** of a function f on X if $g|_X = f$.

9.5.4. EXTENSION THEOREM.
Let (X, ρ) be a metric space with metric completion (C, ρ). Let f be a uniformly continuous function from (X, ρ) into a complete metric space (Y, σ). Then f has a unique uniformly continuous extension g mapping C into Y.

PROOF. Let c be a point in C. Since X is dense in C, we may choose a sequence (x_n) in X converging to c. We claim that $(f(x_n))$ is a Cauchy sequence in Y. Indeed, let $\varepsilon > 0$ be given. By the uniform continuity of f, there is a $\delta > 0$ so that $\rho(x, x') < \delta$ implies $\sigma(f(x), f(x')) < \varepsilon$. Since (x_n) is Cauchy, there is an integer N so that $\sigma(x_m, x_n) < \delta$ for all $m, n \ge N$. Hence $\sigma(f(x_m), f(x_n)) < \varepsilon$ for all $m, n \ge N$. Therefore, $(f(x_n))$ is Cauchy and we may define $g(c)$ to be $\lim_{n \to \infty} f(x_n)$. Moreover, this is the *only* possible choice for a continuous extension of f.

We must verify that g is well defined, meaning that any other sequence converging to c will yield the same value for $g(c)$. Consider a second sequence (x'_n)

in X with $\lim_{n \to \infty} x'_n = c$. Then $(x_1, x'_1, x_2, x'_2, \dots)$ is another sequence converging to c. By the previous paragraph, $(f(x_1), f(x'_1), f(x_2), f(x'_2), \dots)$ is a Cauchy sequence in Y. Therefore $\lim_{n \to \infty} f(x'_n) = \lim_{n \to \infty} f(x_n)$; and g is well defined.

To verify that g is continuous, let $\varepsilon > 0$. Let $\delta > 0$ be chosen as before. Suppose that c, d are points in C with $\rho(c, d) < \delta$ and that (x_n) and (x'_n) are sequences in X converging to c and d, respectively. Then $\lim_{n \to \infty} \rho(x_n, x'_n) = \rho(c, d) < \delta$. Thus there is an N so that $\rho(x_n, x'_n) < \delta$ for all $n \geq N$. It follows that

$$\sigma(g(c), g(d)) = \lim_{n \to \infty} \sigma(f(x_n), f(x'_n)) \leq \varepsilon.$$

Consequently, g is uniformly continuous. ∎

9.5.5. COROLLARY. *The metric completion (C, ρ) of (X, ρ) is unique in the sense that if S is an isometry of X into another completion (Y, σ), then there is a unique isometry S' of C onto Y extending S.*

PROOF. As in Theorem 9.5.4, we consider X as a subset of C with the same metric. The map S is an isometry and thus is uniformly continuous. Let S' be the unique continuous extension of S from C into Y. If $c, d \in C$ with $c = \lim_{n \to \infty} x_n$ and $d = \lim_{n \to \infty} y_n$, then

$$\sigma(S'c, S'd) = \lim_{n \to \infty} \sigma(Sx_n, Sy_n)$$
$$= \lim_{n \to \infty} \rho(x_n, y_n) = \rho(c, d).$$

So S' is also an isometry.

Conversely, there is an isometry T' of Y into C extending the map S^{-1} from SX onto X. Notice that $T'S'$ is therefore a continuous extension of the identity map on X and hence equals the identity map on C. Likewise, $S'T'$ is the identity on Y. In particular, for each y in Y, $S'(T'y) = y$ and so S maps C onto Y. ∎

9.5.6. EXAMPLE. Uniqueness of the Real Number System. We apply this theory to the real number system. The real numbers have several important properties, which were discussed at length in Chapter 2. To establish uniqueness in some sense, we need to be explicit about the required structure. First, $\mathbb{R}$ should be a field that contains the rational numbers $\mathbb{Q}$. In addition to the algebraic operations, it is important to recognize that $\mathbb{R}$ has an order extending the order on $\mathbb{Q}$. Second, $\mathbb{R}$ should be complete in some sense. The sense that fits well with analysis is completeness as a metric space and containing $\mathbb{Q}$ as a dense subset. A subtlety that immediately presents itself is, Where does this metric take its values? Fortunately, the metric on $\mathbb{Q}$ takes values in $\mathbb{Q}$, but any completions of $\mathbb{Q}$ will have to take metric values in $\mathbb{R}$! So by our construction, this would still appear to depend on the original construction of $\mathbb{R}$.

Therefore, it seems prudent to introduce a different and more algebraic notion of completeness, using order. A field is an **ordered field** if it contains a subset P of positive elements with the properties that

(i) every element belongs to exactly one of P, $\{0\}$ or $-P$; and

(ii) whenever $x, y \in P$, then $x + y$ and xy belong to P.

Then we define $x < y$ if $y - x \in P$. An ordered field is **order complete** if every nonempty set that is bounded above has a least upper bound. This notion doesn't mention metric and thus is free from the difficulty mentioned previously about where the metric takes its values. The density of the rationals is also addressed by order. The rationals are **order dense** if whenever $x < y$, there is a rational r so that $x < r < y$.

9.5.7. THEOREM. (1) *The rational numbers $\mathbb{Q}$ have a unique metric completion with a unique field structure extending the field operations on $\mathbb{Q}$ making addition and multiplication continuous. The order structure is also unique.*

(2) *The rational numbers are contained as an order dense subfield of a unique order complete field.*

PROOF. (1) The metric space of real numbers $\mathbb{R}$ is a metric completion of $\mathbb{Q}$, by the completeness of $\mathbb{R}$ (Theorem 2.7.4. By Corollary 9.5.5, it follows that if $\mathbb{S}$ is any metric space completion of $\mathbb{Q}$, then there is an isometry T of $\mathbb{R}$ onto $\mathbb{S}$ that extends the identity map on $\mathbb{Q}$. The Extension Theorem guarantees that there is only one way to extend addition and multiplication on $\mathbb{Q}$ to a continuous function on $\mathbb{S}$. One such way is to transfer the operations on $\mathbb{R}$ by $Tx + Ty = T(x + y)$ and $Tx \cdot Ty = T(xy)$. Uniqueness says that this is the only way, and thus the field operations on $\mathbb{S}$ are equivalent to those on $\mathbb{R}$. The positive elements of $\mathbb{R}$ are exactly those nonzero elements that have a square root. Since this is determined by the algebra structure, we see that order is also preserved.

(2) The real numbers are order complete by the Least Upper Bound Principle (2.5.3), and the rationals are order dense by construction (see Exercise 2.2.F). Suppose that $\mathbb{S}$ is another order complete field containing $\mathbb{Q}$ as an order dense subfield. Define a map T from $\mathbb{R}$ to $\mathbb{S}$ by sending each real number x to the least upper bound in $\mathbb{S}$ of $\{r \in \mathbb{Q} : r < x\}$. Notice that $Tq = q$ for $q \in \mathbb{Q}$. Indeed, q is an upper bound for $\{r \in \mathbb{Q} : r < q\}$. If $Tq < q$, the order density of $\mathbb{Q}$ would imply that there is an $r \in \mathbb{Q}$ such that $Tq < r < q$, contradicting the fact that Tq is the least upper bound for this set.

Let us show that T is a bijection. Clearly, if $x \leq y$, then $Tx \leq Ty$. If $x < y$ in $\mathbb{R}$, then there is a rational r with $x < r < y$. Consequently, $Tx < r < Ty$ and so T is one-to-one and preserves the order. Now suppose that $s \in \mathbb{S}$ and let $S = \{r \in \mathbb{Q} : r < s\}$. This set has a least upper bound x in $\mathbb{R}$. Then Tx is the least upper bound of S in $\mathbb{S}$. Again s is an upper bound by definition and if $Tx < s$, the order density would yield a rational r with $Tx < r < s$. Hence $Tx = s$ and the map T is onto.

Finally, we verify that T preserves addition and multiplication. The fact that $P + P \subset P$ means that addition preserves order: if $x_1 < x_2$ amd $y_1 < y_2$, then $x_1 + y_1 < x_2 + y_2$. Thus in both $\mathbb{R}$ and $\mathbb{S}$, we have that

$$x + y = \sup\{r + s : r, s \in \mathbb{Q}, \ r < x, \ s < y\}.$$

Since T preserves order and therefore sups, it follows that $Tx + Ty = T(x + y)$. Similarly, if x, y are *positive*,

$$xy = \sup\{rs : r, s \in \mathbb{Q} \cap P, \ r < x, \ s < y\}.$$

So $(Tx)(Ty) = T(xy)$ for $x, y \in P$. The rest follows since

$$(-x)y = x(-y) = -(xy) \qquad \text{and} \qquad (-x)(-y) = xy$$

allow the extension of multiplication to the whole field. This establishes that the map T preserves all of the field operations and the order. ∎

So part (2) establishes, without any reference to any properties based on our construction of $\mathbb{R}$, that there is only one field that has the properties we want. This we call the real numbers. We could, for example, define the real numbers using expansions in base 2 or base 3. We have implicitly assumed up to now that these constructions yield the same object. We now know this to be the case.

There are other constructions of the real numbers that do not depend on a base. This has a certain esthetic appeal but does not in itself address the uniqueness question. One approach briefly mentioned in Section 2.2 is Dedekind's construction based on an abstract definition of the sets $\{r \in \mathbb{Q} : r < x\}$ used in the proof here. Another method alluded to at the beginning of this section is to consider Cauchy sequences of rational numbers as representing points. We have to decide when two Cauchy sequences should represent the same point. A little thought shows that if you take (r_k) and (s_k) and combine them as $(r_1, s_1, r_2, s_2, \dots)$, then this new sequence is Cauchy if and only if they have the same limit. Thus we are led to dealing with equivalence classes of Cauchy sequences. Both of these constructions yield a complete structure; Dedekind's is order complete and Cauchy's is metrically complete. But there remains in both cases the tedious, but basically easy, task of defining addition and multiplication and verifying all of the ordered field properties. Indeed, we did not verify all of these details for our decimal construction either.

Another practical approach is to take *any* construction of the real numbers and show that every number has a decimal expansion. This is an alternate route to the uniqueness theorem. The reader interested in the details of these foundational issues should consult [3].

Exercises for Section 9.5

A. Show that the metric completion of a normed vector space is a complete normed vector space. HINT: Use the Extension Theorem to extend the vector space operations.

B. Prove that the map taking each bounded uniformly continuous function on a metric space X to its continuous extension on the completion C is an isometry from $BUC(X)$ *onto* $BUC(C)$, where $BUC(X)$ is the normed space of bounded uniformly continuous functions on X.

C. Show by example that there is a bounded continuous function on $(0, 1)$ that does not extend to a continuous function on the completion.

D. Let (X, ρ) and (Y, σ) be metric spaces with completions (C, ρ) and (D, σ).
  (a) Show that $d\big((x_1, y_1), (x_2, y_2)\big) = \rho(x_1, x_2) + \sigma(y_1, y_2)$ is a metric on the product space $X \times Y$.
  (b) Show that $C \times D$ is the completion of $X \times Y$.

E. Let (X, ρ) be the metric on all finite sequences of letters with the dictionary metric of Example 9.1.I. Describe the metric completion.
  HINT: Consider the natural extension of this metric to all infinite words.

F. Show that the completion of a metric space (X, ρ) is compact if and only if X is totally bounded.

G. The p-adic numbers. Fix a prime number p. For each $n \in \mathbb{Z}$, let $\mathrm{ord}_p(n)$ denote the largest integer d such that p^d divides n. Extend this function to all rational numbers by setting $\mathrm{ord}_p(a/b) = \mathrm{ord}_p(a) - \mathrm{ord}_p(b)$.
  (a) Show that $\mathrm{ord}_p(r)$ is independent of the representation of r as a fraction.
  (b) Set $|r|_p = p^{-\mathrm{ord}_p(r)}$ and $\rho_p(r, s) = |r - s|_p$. Prove that ρ_p is a metric on $\mathbb{Q}$.
  HINT: In fact, $\rho_p(r, t) \le \max\{\rho_p(r, s), \rho_p(s, t)\}$.
  (c) Let $\mathbb{Q}_p$ be the metric completion of $(\mathbb{Q}, \rho_p)$. Prove that addition extends to a continuous operation on $\mathbb{Q}_p$.
  (d) Show that $|xy|_p = |x|_p |y|_p$. Hence show that there is a continuous extension of multiplication to $\mathbb{Q}_p$.
  (e) Verify that every nonzero element of $\mathbb{Q}_p$ has an inverse.
  (f) Why doesn't this complete field containing $\mathbb{Q}$ contradict the uniqueness of the real numbers?

H. We develop further properties of the p-adic numbers, although this requires a bit of number theory.
  (a) If $r \in \mathbb{Q}$ and $|r|_p = p^{-k}$, find $a_k \in \{0, 1, \dots, p-1\}$ so that $|r - a_k p^k|_p \le p^{-k-1}$.
  HINT: Write $r = p^k \frac{a}{b}$ where $\gcd(p, b) = 1$. By the Euclidean algorithm, we can write $1 = mb + np$. Let $a_k \equiv am \pmod{p}$.
  (b) Hence show that every element $r \in \mathbb{Q}$ with $|r|_p \le 1$ is the limit in the ρ_p metric of a sequence of integers of the form $a_0 + a_1 p + \dots + a_n p^n$, where $a_k \in \{0, 1, \dots, p-1\}$.
  (c) Extend (b) to every element $x \in \mathbb{Q}_p$ with $|x|_p \le 1$. Hence deduce that $\mathbb{Z}_p$, the closure of $\mathbb{Z}$ in $\mathbb{Q}_p$, consists of all these points.
  (d) Prove that $\mathbb{Z}_p$ is compact.

9.6. The L^p Spaces and Abstract Integration

In Section 7.7, the L^p norms for $p \ge 1$ were introduced and shown to be norms via the Hölder and Minkowski inequalities. However, these norms were put on $C[a, b]$, and it is easy to show that $C[a, b]$ is not complete in any of these norms. The completion process of the last section may be used to remedy this situation. One caveat to the reader is that there is a better, although more involved, method of defining the L^p spaces using measure theory. Measure theory provides a powerful way to develop not only integration on $\mathbb{R}$, but also integration on many other spaces, and provides a setting for probability theory. Nevertheless, we are able to achieve some useful things by our abstract process, including recognizing L^p as a space of functions (or, more precisely, as equivalence classes of functions).

9.6.1. DEFINITION. The space $L^p[a, b]$ is the completion of $C[a, b]$ in the L^p norm $\|f\|_p = \left(\int_a^b |f(x)|^p \, dx \right)^{1/p}$.

At this point, we know that $L^p[a, b]$ is a complete normed space by Exercise 9.5.A. In particular, $\|f\|_p$ is the distance in $L^p(a, b)$ to the zero function. However, the elements of this space are no longer represented by functions. We will rectify that in this section. The first step is to extend our integration theory to L^p spaces.

Keeping in mind that we expect to represent elements of $L^p(a, b)$ as functions, we will write elements as f, g, and so on. In particular, we shall say that $f \geq 0$ if it is the limit in the L^p norm of a sequence of positive continuous functions f_n. So we say $f \leq g$ if $g - f \geq 0$.

9.6.2. THEOREM. *The Riemann integral has a unique continuous extension to $L^1(a, b)$ denoted by $\int f$. It has the properties*

(1) $\int sf + tg = s \int f + t \int g$ *for all $f, g \in L^1(a, b)$ and $s, t \in \mathbb{R}$.*

(2) $|\int f| \leq \int |f| = \|f\|_1$ *for $f \in L^1(a, b)$.*

(3) *If $f \geq 0$, then $\int f \geq 0$.*

PROOF. The Riemann integral on $C[a, b]$ satisfies the properties (1)–(3). In particular, it is uniformly continuous in the L^1 norm by (2) since for $f, g \in C[a, b]$,

$$\left| \int_a^b f(x) \, dx - \int_a^b g(x) \, dx \right| \leq \int_a^b |f(x) - g(x)| \, dx = \|f - g\|_1.$$

Therefore, by the Extension Theorem, there is a unique continuous extension to $L^1(a, b)$, which we call $\int f$. Properties (1)–(3) follow by taking limits. ∎

The integral defined in this manner is called the **Daniell integral**. The alternative is the **Lebesgue integral**, which is constructed using measure theory. Although the constructions are different, the two integrals have the same theory. They are powerful for two reasons. The first reason is that the L^p spaces of integrable functions are complete. For the Lebesgue integral, this is a theorem, proved using properties of the integral; whereas in our case, the complete space comes first and the properties of the integral are a theorem. The second reason is that there are much better limit theorems than for the Riemann integral. Here is first of these. The best limit theorem, the Dominated Convergence Theorem, we leave as an exercise.

This result, like Theorem 2.5.4, is traditionally called the Monotone Convergence Theorem. It is usually easy to distinguish between the two, as one result applies to sequences of real numbers and the other to sequences of functions.

9.6.3. MONOTONE CONVERGENCE THEOREM.

Suppose $(f_n) \in L^1(a, b)$ is an increasing sequence such that $\sup_{n \geq 1} \int f_n < \infty$. Then f_n converges in the $L^1(a, b)$ norm to an element f and $\int f = \lim\limits_{n \to \infty} \int f_n$.

PROOF. By property (2), $\int f_n$ is an increasing sequence of real numbers that is bounded above. Hence by the Monotone Convergence Theorem for real numbers, $L = \lim\limits_{n \to \infty} \int f_n$ exists. So if $\varepsilon > 0$, there is an integer N so that

$$L - \varepsilon < \int f_N \leq \int f_n \leq L \quad \text{for all} \quad n \geq N.$$

Thus when $N \leq m \leq n$,

$$\|f_n - f_m\| = \int |f_n - f_m| = \int f_n - f_m = \int f_n - \int f_m < \varepsilon.$$

Therefore, (f_n) is Cauchy in $L^1(a, b)$. It follows that $f = \lim\limits_{n \to \infty} f_n$ exists. Since the integral is continuous, $\int f = \lim\limits_{n \to \infty} \int f_n$. ■

The preceding analysis would appear to deal only with $L^1(a, b)$. For $L^p(a, b)$, we obtain much the same result by extending the Hölder inequality. Minkowski's inequality, which is just the triangle inequality, is immediately satisfied because the metric $d(f, g) = \|f - g\|_p$ satisfies the triangle inequality.

9.6.4. THEOREM.

Let $1 < p, q < \infty$ satisfy $\frac{1}{p} + \frac{1}{q} = 1$. If $f \in L^p(a, b)$ and $g \in L^q(a, b)$, then fg 'belongs' to $L^1(a, b)$ and $\left| \int fg \right| \leq \|f\|_p \|g\|_q$.

PROOF. We make $L^p(a, b) \times L^q(a, b)$ into a complete metric space with the metric $d((f_1, g_1), (f_2, g_2)) = \|f_1 - f_2\|_p + \|g_1 - g_2\|_q$. By Exercise 9.5.D, this is the completion of $C[a, b] \times C[a, b]$ with the same metric. By the Hölder inequality (7.7.3), the map $\Phi(f, g) = fg$ maps $C[a, b] \times C[a, b]$ into $L^1(a, b)$ and

$$\|fg\|_1 = \int_a^b |f(x)g(x)| \, dx \leq \|f\|_p \|g\|_q.$$

This map is not uniformly continuous on the whole product, but it is uniformly continuous on balls. So fix $r \geq 0$ and consider

$$X_r = \left\{ (f, g) \in C[a, b] \times C[a, b] : \|f\|_p \leq r \text{ and } \|g\|_q \leq r \right\}.$$

Then for (f_1, g_1) and (f_2, g_2) in X_r,

$$\begin{aligned}
\|\Phi(f_1, g_1) - \Phi(f_2, g_2)\|_1 &\leq \|f_1(g_1 - g_2)\|_1 + \|(f_1 - f_2)g_2\|_1 \\
&\leq \|f_1\|_p \|g_1 - g_2\|_q + \|f_1 - f_2\|_p \|g_2\|_q \\
&\leq r \left(\|f_1 - f_2\|_p + \|g_1 - g_2\|_q \right) \\
&= r \, d((f_1, g_1), (f_2, g_2)).
\end{aligned}$$

So Φ is uniformly continuous on X_r. By the Extension Theorem (Theorem 9.5.4), Φ extends to a uniformly continuous function on the completion of X_r, which is its closure in $L^p(a,b) \times L^q(a,b)$, namely

$$\overline{X_r} = \{(f,g) \in L^p(a,b) \times L^q(a,b) : \|f\|_p \leq r \text{ and } \|g\|_q \leq r\}.$$

In particular, if $f \in L^p(a,b)$ and $g \in L^q(a,b)$, then $\Phi(f,g) = fg$ belongs to $L^1(a,b)$. Moreover, $\|\Phi(f,g)\|_1 \leq \|f\|_p \|g\|_q$ by continuity since this holds on $C[a,b] \times C[a,b]$. Moreover, since $\int$ is uniformly continuous on $L^1(a,b)$, we obtain Hölder's inequality:

$$\left| \int fg \right| = \left| \int \Phi(f,g) \right| \leq \|\Phi(f,g)\|_1 \leq \|f\|_p \|g\|_q. \qquad \blacksquare$$

We turn to the problem of representing each element of L^p as a function. We already know we must allow discontinuous functions, as $C[a,b]$ is not complete in the L^p norm. But this allows the possibility of having a function that is zero except at a countable set of points. Such a function will have integral zero, and then the L^p norm will not be positive definite. So we must identify this function with the zero function.

In general, we must identify functions that are equal on "negligble" sets. That is, each element of L^p will be an equivalence class of functions that are equal almost everywhere, in the sense of Section 6.6. We return to this issue after proving the next theorem.

Measure theory has something useful to add to this picture, since there is a notion of measurable function that provides a natural set to choose the L^p functions from. However, this approach cannot get around the essential difficulty that an element of $L^p(a,b)$ is not actually a function.

9.6.5. THEOREM. *If $f \in L^p(a,b)$, we may choose a sequence (f_n) in $C[a,b]$ converging to f in $L^p(a,b)$ so that $\lim_{n\to\infty} f_n(x)$ converges almost everywhere (a.e.) to a function $f(x)$. Moreover, if (g_n) is another sequence converging to f in $L^1(a,b)$ and pointwise to $g(x)$ a.e., then $g(x) = f(x)$ a.e.*

PROOF. Choose any sequence (f_n) in $C[a,b]$ such that $\|f - f_n\|_p < 4^{-n}$ for $n \geq 1$. Let

$$U_n = \{x \in [a,b] : |f_n(x) - f_{n+1}(x)| > 2^{-n}\}.$$

This is an open set since $|f_n - f_{n+1}|$ is continuous. Hence there are disjoint intervals (c_i, d_i) so that $U_n = \bigcup_{i \geq 1}(c_i, d_i)$. Now

$$\|f_n - f_{n+1}\|_p \leq \|f_n - f\|_p + \|f - f_{n+1}\|_p < 2 \cdot 4^{-n}.$$

On the other hand,

$$\|f_n - f_{n+1}\|_p^p \geq \sum_{i \geq 1} \int_{c_i}^{d_i} |f_n(x) - f_{n+1}(x)|^p \, dx \geq 2^{-np} \sum_{i \geq 1} d_i - c_i.$$

Therefore,

$$\sum_{i \geq 1} d_i - c_i < (2 \cdot 4^{-n})^p 2^{np} = 2^{(1-n)p}.$$

Let $E = \bigcap_{k \geq 1} \bigcup_{n \geq k} U_n$. This set may be covered by the open intervals making up $\bigcup_{n \geq k} U_n$, which have total length at most $\sum_{n \geq k} 2^{(1-n)p} = 2^{-kp}\left(\frac{2^p}{1-2^{-p}}\right)$.

This converges to 0 as k increases and thus may be made less than any given $\varepsilon > 0$. Therefore, E has measure zero.

If $x \in [a, b] \setminus E$, then there is an integer $N = N(x)$ so that x is not in $\bigcup_{n \geq N} U_n$. Hence $|f_n(x) - f_{n+1}(x)| \leq 2^{-n}$ for all $n \geq N$. Thus if $N \leq n \leq m$,

$$|f_n(x) - f_m(x)| \leq \sum_{k=n}^{m-1} |f_k(x) - f_{k+1}(x)| \leq \sum_{k=n}^{m-1} 2^{-k} < 2^{1-n}.$$

Therefore $(f_n(x))$ is a Cauchy sequence, and $\lim_{n \to \infty} f_n(x)$ converges, say to $f(x)$. That is, this sequence converges almost everywhere.

Now suppose that (g_n) is another sequence in $C[a, b]$ converging to f in the L^p norm and converging almost everywhere to a function g. Drop to a subsequence (g_{n_k}) so that $\|g_{n_k} - f\|_p < 4^{-k}$ for $k \geq 1$. Let

$$V_k = \{x \in [a, b] : |g_{n_k}(x) - f_k(x)| > 2^{-k}\}.$$

This is an open set because $|g_{n_k} - f_k|$ is continuous. Now

$$\|g_{n_k} - f_k\|_p \leq \|g_{n_k} - f\|_p + \|f - f_n\|_p < 2 \cdot 4^{-k}.$$

Arguing as previously, the intervals making up V_k have length at most $2^{(1-n)p}$. Consequently, as before, the set $F = \bigcap_{k \geq 1} \bigcup_{n \geq k} V_n$ has measure zero.

If $x \in [a, b] \setminus F$, then there is an $N = N(x)$ so that x is not in $\bigcup_{n \geq N} V_n$. Hence $\lim_{k \to \infty} |g_{n_k}(x) - f_k(x)| = 0$. Now $f_k(x)$ converges to $f(x)$ for $x \in [a, b] \setminus E$. Therefore, $g(x) = f(x)$ for every $x \in [a, b] \setminus (E \cup F)$. Thus $g = f$ a.e.   ∎

One of the subtleties of L^p spaces is that the elements of $L^p[a, b]$ are not really functions. The preceding theorem shows that they may be represented by functions but that representation is not unique. Two functions that differ on a set of measure zero both represent the same element of $L^p[a, b]$. The correct way to handle this is to identify elements of $L^p[a, b]$ with an equivalence class of functions that agree almost everywhere. (See Appendix 1.6.)

Another point that bears repeating is that in this approach, the integral of these limit functions is determined abstractly. The Lebesgue integral develops a method of integrating these functions in terms of their values. This is an important viewpoint that we do not address here. We refer the interested reader to [17] or [15].

Exercises for Section 9.6

A. Show that every piecewise continuous function on $[a, b]$ is simultaneously the limit in $L^1(a, b)$ and the pointwise limit everywhere of a sequence of continuous functions.

B. (a) Show that the characteristic function χ_U of any open set $U \subset [a, b]$ is the increasing limit of a sequence of continuous functions.

    (b) Hence show that χ_U belongs to $L^1(a, b)$. If U is the disjoint union of open intervals $U_n = (c_n, d_n)$, show that $\int \chi_U = \|\chi_U\|_1 = \sum_n |U_n|$.

C. Suppose that $f \in L^1[a, b]$ and $f \geq 0$ and $\int f = 0$. Prove that $f = 0$ a.e.
HINT: Apply the Monotone Convergence Theorem (MCT) with $f_n = nf$.

D. (a) Suppose that $A \subset [a, b]$ has measure zero. Show that there is a decreasing sequence of open sets U_n containing A such that $\lim_{n \to \infty} \int \chi_{U_n} = 0$.

(b) If f is a positive element of $L^1(a, b)$ that is represented by a function $f(x)$ such that $f(x) = 0$ a.e., prove that $\|f\|_1 = 0$.
HINT: Let $A_n = \{x : 0 < f(x) < n\}$. Consider $f \chi_{A_n}$.

(c) Suppose $f \in L^1(a, b)$ is represented by a function $f(x)$ such that $f(x) = 0$ a.e. Prove that $\int |f| = 0$ (i.e. $f = 0$ in $L^1(a, b)$).

E. Suppose that $f_n \in C[a, b]$ is a decreasing sequence of continuous functions converging pointwise to 0. Prove that $\lim_{n \to \infty} \int_a^b f_n(x)\, dx = 0$.
HINT: Exercise 8.1.I

F. (a) Show that if $f, g, h \in L^1[a, b]$, then $f \vee g := \frac{1}{2}(f + g + |f - g|)$ satisfies $f \leq f \vee g$ and $g \leq f \vee g$.

(b) If $f \leq h$ and $g \leq h$, show that $f \vee g \leq h$. That is, $f \vee g = \max\{f, g\}$.

(c) Show that $f \wedge g = \min\{f, g\}$ may be defined as $f \wedge g = -(-f \vee -g)$.

G. **Dominated Convergence Theorem.** Suppose that $f_n, g \in L^1[a, b]$ and $|f_n| \leq g$ for all $n \geq 1$. Also suppose that $\lim_{n \to \infty} f_n(x) = f(x)$ a.e.

(a) Define $u_{mn} = f_m \vee f_{m+1} \vee \cdots \vee f_n$ and $l_{mn} = f_m \wedge f_{m+1} \wedge \cdots \wedge f_n$. Show that $u_m = \lim_{n \to \infty} u_{mn}$ belongs to $L^1(a, b)$, as does $l_m = \lim_{n \to \infty} l_{mn}$.   HINT: MCT

(b) Show that u_m is a decreasing sequence of functions converging to f almost everywhere, and l_m is an increasing sequence converging to f almost everywhere.

(c) Show that $\lim_{m \to \infty} \int u_m = \lim_{m \to \infty} \int l_m = \int f$.   HINT: MCT

(d) Prove that $\lim_{n \to \infty} \int f_n = \int f$.   HINT: $l_n \leq f_n \leq u_n$

H. **Lebesgue Measure.** Let Σ consist of all subsets A of $[a, b]$ such that χ_A belongs to $L^1(a, b)$ in the sense that there is an $f \in L^1(a, b)$ with $f(x) = \chi_A$ a.e. Define a function $m : \Sigma \to [0, +\infty)$ by $m(A) = \int \chi_A$.

(a) Show that if $A, B \in \Sigma$, then $A \cap B$, $A \cup B$ and $A \setminus B$ belong to Σ.

(b) Show that $m(A \cap B) + m(A \cup B) = m(A) + m(B)$ for all $A, B \in \Sigma$.

(c) Show that $m(A) = 0$ if and only if A has measure zero.

(d) **Countable Additivity.** If A_n are disjoint sets in Σ, show that $\bigcup_{n \geq 1} A_n$ belongs to Σ, and $m\left(\bigcup_{n \geq 1} A_n\right) = \sum_{n \geq 1} m(A_n)$.

(e) Show that Σ contains all open and all closed subsets of $[a, b]$; and verify that $m((c, d)) = m([c, d]) = d - c$.

Part B

Applications

CHAPTER 10

Approximation by Polynomials

This chapter introduces some of the essentials of approximation theory, in particular approximating functions by "nice" ones such as polynomials. In general, the intention of approximation theory is to replace some complicated function with a new function, one that is easier to work with, at the price of some (hopefully small) difference between the two functions. The new function is called an approximation. There are two crucial issues in using an approximation: first, how much simpler is the approximation? and second, how close is the approximation to the original function? Deciding which approximation to use means looking at the trade-off between these two issues.

Of course, the answers to these two questions depend on the exact meanings of *simpler* and *close*, which vary according to the context. In this chapter, we study approximations by polynomials, so here a simpler function means a polynomial. *Close* usually refers to some norm. We focus on the uniform norm so that the polynomial is close to the function everywhere on a given interval.

Approximations are closely tied to the notions of limit and convergence, as a sequence of functions approximating a function f to greater and greater accuracy might be expected to converge to f in some sense. Different approximation schemes correspond to different notions of convergence.

10.1. Taylor Series

The first approximation taught in Calculus and the most often used approximation is the the tangent line approximation: If $f : \mathbb{R} \to \mathbb{R}$ is differentiable at $a \in \mathbb{R}$, then for x near a, we have

$$f(x) \approx f(a) + f'(a)(x - a).$$

However, you should learn early that an approximation is only as good as the error estimate that can be verified. Unless we can estimate the error, the difference between $f(x)$ and its approximation $f(a) + f'(a)(x - a)$, it is impossible to say whether or not the approximation is worth the trouble. For example, why not just approximate $f(x)$ with the constant $f(a)$, since x is "near" a and f is continuous?

We start with the error estimate for the constant approximation $f(a)$. This error estimate comes from the Mean Value Theorem (Theorem 6.2.4), which gives us the estimate $|f(x) - f(a)| = |f'(c)(x-a)| \leq C|x-a|$, where c is some point between a and x and $C = \sup\{|f'(c)| : c \text{ between } x \text{ and } a\}$. When C is finite, we obtain a useful error estimate for this constant approximation. Notice that this estimate does not require us to find $f(x)$ exactly. If we could easily find $f(x)$, we wouldn't bother with the approximation.

A more sophisticated use of the Mean Value Theorem shows that the tangent line has an error of the form $M(x - a)^2$ for a constant M that depends on f''. For x very close to a, this is a considerable improvement on $C|x - a|$. See Exercise 10.1.B. Very similar calculations were done in Example 6.2.6.

In this section, we generalize these two approximations and their error estimates to take account of higher derivatives—in other words, we generalize the Mean Value Theorem. As this method requires many derivatives, and because it only uses information at one point, it will not be an ideal method for uniform approximation over an interval. Nevertheless, it is works very well in certain instances of great importance, and it is easier to understand than the alternative methods. Later in this chapter we will explore other methods for finding polynomial approximants which are uniformly close over an interval.

The role of the tangent line to f is replaced by a polynomial $P_n(x)$ of degree at most n that has the same derivatives at a as f up to the nth degree. This is all that the parameters of a polynomial of degree at most n permit.

10.1.1. DEFINITION. If f has n derivatives at a point $a \in [A, B]$, the **Taylor polynomial** of order n for f at a is

$$P_n(x) = f(a) + f'(a)(x - a) + \frac{f''(a)}{2}(x - a)^2 + \cdots + \frac{f^{(n)}(a)}{n!}(x - a)^n$$

$$= \sum_{k=0}^{n} \frac{f^{(k)}(a)}{k!}(x - a)^k.$$

10.1.2. LEMMA. *Let $f(x)$ belong to $C^n[A, B]$ (i.e., f has n continuous derivatives), and let $a \in [A, B]$. The Taylor polynomial $P_n(x)$ of order n for f at a is the unique polynomial $p(x)$ of degree at most n such that $p^{(k)}(a) = f^{(k)}(a)$ for $0 \leq k \leq n$.*

PROOF. Every polynomial of degree at most n has the form $p(x) = \sum_{j=0}^{n} a_j(x - a)^j$. We may differentiate this k times to obtain

$$p^{(k)}(x) = \sum_{j=k}^{n} j(j - 1) \ldots (j + 1 - k)(x - a)^{j-k}.$$

Substituting $x = a$ yields $p^{(k)}(a) = k!a_k$. Therefore, we must choose the coefficients $a_k = f^{(k)}(a)/k!$, which yields the Taylor polynomial $P_n(x)$. ∎

The preceding simple lemma established that the Taylor polynomial is the appropriate analogue of the tangent line for higher order polynomials. The hard work, and indeed the total content of this approximation, comes from the error estimate. In this case, the estimate is only good for points sufficiently close to a when there is reasonable control on the size of the $(n+1)$st derivative. The case $n = 0$ is a direct consequence of the Mean Value Theorem.

10.1.3. TAYLOR'S THEOREM.

Let $f(x)$ belong to $C^n[A, B]$, and furthermore assume that $f^{(n+1)}$ is defined and $|f^{(n+1)}(x)| \leq M$ for $x \in [A, B]$. Let $a \in [A, B]$, and let $P_n(x)$ be the Taylor polynomial of order n for f at a. Then for each $x \in [A, B]$, the error of approximation $R_n(x) = f(x) - P_n(x)$ satisfies

$$|R_n(x)| \leq \frac{M|x - a|^{n+1}}{(n + 1)!}.$$

PROOF. Notice that for $0 \leq k \leq n$,

$$R_n^{(k)}(a) = f^{(k)}(a) - P_n^{(k)}(a) = 0,$$

and because P_n is a polynomial of degree at most n,

$$R_n^{(n+1)}(x) = f^{(n+1)}(x) - P_n^{(n+1)}(x) = f^{(n+1)}(x).$$

We recover R_n by repeated integration.

$$|R_n^{(n)}(x)| = \left| R_n^{(n)}(a) + \int_a^x R_n^{(n+1)}(t)\, dt \right| \leq \left| 0 + \int_a^x M\, dt \right| = M|x - a|$$

Suppose that for some k with $0 \leq k < n$ we have shown that

$$|R_n^{(n-k)}(x)| \leq \frac{M|x - a|^{k+1}}{(k + 1)!}.$$

Then we integrate again to obtain

$$|R_n^{(n-k-1)}(x)| = \left| R_n^{(n-k-1)}(a) + \int_a^x R_n^{(n-k)}(t)\, dt \right|$$

$$\leq \left| 0 + \int_a^x \frac{M|x - a|^{k+1}}{(k + 1)!}\, dt \right| = \frac{M|x - a|^{k+2}}{(k + 2)!}.$$

We established the formula for $k = 0$, and have now completed the induction step. Eventually we obtain the desired formula where $k = n$,

$$|R_n(x)| \leq \frac{M|x - a|^{n+1}}{(n + 1)!}. \qquad \blacksquare$$

When f is C^∞, the **Taylor series** of f about a is $\sum_{k=0}^{\infty} \frac{f^{(k)}(a)}{k!}(x - a)^k$. This is a power series, and so we must watch out for problems with convergence.

10.1.4. EXAMPLE. Consider the function $f(x) = e^x$. This function has the very nice property that $f' = f$. Thus $f^{(n)}(x) = e^x$ for all $n \geq 0$. So expanding around $a = 0$, we obtain Taylor polynomials $P_n(x) = \sum_{k=0}^{n} \dfrac{x^k}{k!}$. We readily see that $|f^{(n+1)}(t)| = e^t \leq \max\{1, e^x\}$ if t lies between 0 and x. Thus Taylor's Theorem for the interval $[0, x]$ (or $[x, 0]$) tells that the error has the form

$$\left| e^x - \sum_{k=0}^{n} \frac{x^k}{k!} \right| \leq \max\{1, e^x\} \frac{|x|^{n+1}}{(n+1)!}.$$

Even for low values of n, $P_n(x)$ appears quite close to e^x around the origin; see Figure 10.1.

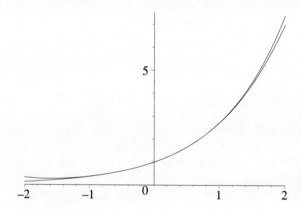

FIGURE 10.1. e^x and $P_4(x)$ on $[-2, 2]$.

The ratio test shows that $\lim_{n \to \infty} \dfrac{|x|^{n+1}}{(n+1)!} = 0$. Thus the Taylor series converges to e^x for every $x \in \mathbb{R}$. Moreover, this series converges uniformly on any interval $[-A, A]$. To see this, notice that the error estimate at any point $x \in [-A, A]$ is greatest for $x = A$. Hence

$$\sup_{|x| \leq A} \left| e^x - \sum_{k=0}^{n} \frac{x^k}{k!} \right| \leq \frac{e^A A^{n+1}}{(n+1)!}.$$

To compute e, we could use this formula to compute

$$\left| e - \sum_{k=0}^{n} \frac{1}{k!} \right| \leq \frac{e}{(n+1)!} \leq \frac{3}{(n+1)!}.$$

So to obtain e to 10 decimal places, we need

$$\frac{3}{(n+1)!} < 5(10)^{-11} \qquad \text{or} \qquad (n+1)! > 6(10)^{10}.$$

A calculation shows that we need $n = 13$.

This is not too bad, yet we can significantly increase the rate of convergence by using a smaller value of x. For example, suppose that we use $x = 1/16$ to compute

$e^{1/16}$. We can then square this number 4 times to obtain e. If we use just the first 10 terms, we have

$$\left| e^{1/16} - \sum_{k=0}^{10} \frac{1}{(16)^k k!} \right| \leq \frac{e^{1/16}}{(16)^{11}(11)!} < 1.6(10)^{-21}.$$

Then we take the number $a = \sum_{k=0}^{10} \frac{1}{(16)^k k!}$ and square it 4 times to obtain a^{16} as an approximation to e. Since we know that $e^{1/16} - \varepsilon < a < e^{1/16}$, where $\varepsilon = 1.6(10)^{-21}$, we have

$$e > a^{16} > (e^{1/16} - \varepsilon)^{16} > e - 16e^{15/16}\varepsilon > e - 7(10)^{-20}$$

So roughly the same number of calculations yields almost double the number of digits of accuracy.

Consider the power series $\sum_{n=0}^{\infty} \frac{x^n}{n!}$. The ratio test shows that

$$\lim_{n \to \infty} \frac{|x|^{n+1}/(n+1)!}{|x|^n/n!} = \lim_{n \to \infty} \frac{|x|}{n+1} = 0$$

for every real x. Thus this power series has an infinite radius of convergence. Moreover, as we showed previously, this series converges to the function e^x; and this convergence is uniform on each bounded interval.

A similar situation occurs for $\sin x$ and $\cos x$.

Many functions in common use are C^∞, meaning that they have continuous derivatives of all orders. For such functions, the Taylor polynomials of all orders are defined. Thus it is natural to consider the convergence of the Taylor series of f around $x = a$. Recall from Section 8.5 that every power series has a radius of convergence. In the previous example, the best possible result occurred—the power series of e^x had an infinite radius of convergence, and the limit of the series was the function itself. Unfortunately, things are not always so good.

10.1.5. EXAMPLE. Consider the function $f(x) = \dfrac{1}{1 + x^2}$. The formulas for the derivatives are a bit complicated, so we use a trick. Consider the geometric series

$$\sum_{k=0}^{n} (-x^2)^k = 1 - x^2 + x^4 + \cdots + (-x^2)^n = \frac{1 - (-x^2)^{n+1}}{1 + x^2}.$$

If we set $P_{2n}(x) = \sum_{k=0}^{n} (-x^2)^k$, we have the estimate

$$\left| \frac{1}{1 + x^2} - P_{2n}(x) \right| = \frac{x^{2n+2}}{1 + x^2}.$$

Thus

$$\lim_{x \to 0} \frac{|f(x) - P_{2n}(x)|}{x^{2n+1}} = \lim_{x \to 0} \frac{|x|}{1 + x^2} = 0.$$

By Exercise 10.1.C, it follows that this is indeed the Taylor polynomial for f not only of order $2n$, but also of order $2n + 1$.

So the Taylor series is $\sum_{k=0}^{\infty} (-x^2)^k$. The radius of convergence is readily seen to be 1, since for $|x| \geq 1$ the terms to do not go to 0; while for $|x| < 1$, the geometric series does converge. Moreover, the limit is our function $f(x)$. In Example 8.4.8, we saw that convergence is uniform on each set $[-r, r]$, for $r < 1$.

We conclude that even if the function is defined and C^∞ on the whole real line, the Taylor series may only converge on a finite interval.

10.1.6. EXAMPLE. In the last example, we showed that for any $r < 1$, the series for $1/(1 + x^2)$ converges uniformly on $[-r, r]$. Thus we may integrate this series term by term by Theorem 8.3.1. For $|x| = r < 1$,

$$\tan^{-1}(x) = \int_0^x \frac{1}{1 + t^2} \, dt = \int_0^x \lim_{n \to \infty} \sum_{k=0}^{n} (-x^2)^k \, dt$$

$$= \lim_{n \to \infty} \int_0^x \sum_{k=0}^{n} (-x^2)^k \, dt$$

$$= \lim_{n \to \infty} \sum_{k=0}^{n} \frac{(-1)^k}{2k + 1} x^{2k+1} = \sum_{k=0}^{\infty} \frac{(-1)^k}{2k + 1} x^{2k+1}.$$

This is the Taylor series for $\tan^{-1}(x)$. It also has radius of convergence 1 and converges uniformly on $[-r, r]$ for any $r < 1$. This series also converges at $x = \pm 1$ by the alternating series test.

The next thing to notice is that this series does converge to $\tan^{-1}(x)$ uniformly on $[-1, 1]$. To see this, we use the alternating series test at each point $x \in [-1, 1]$. We have $P_{2n}(x) = \sum_{k=0}^{n-1} \frac{(-1)^k}{2k + 1} x^{2k+1}$. This is a sequence of polynomials of degree $2n - 1$ that converges to $\tan^{-1}(x)$ at each point in $[-1, 1]$. The corresponding series is alternating in sign with terms of modulus $|x|^{2k+1}/(2k + 1)$ tending monotonely to 0. Thus the error is no greater than the modulus of the next term.

$$\left| \tan^{-1}(x) - P_{2n}(x) \right| < \frac{|x|^{2n+1}}{2n + 1} \leq \frac{1}{2n + 1}.$$

Therefore,

$$\sup_{|x| \leq 1} \left| \tan^{-1}(x) - P_{2n}(x) \right| \leq \frac{1}{2n + 1}.$$

So P_{2n} converges uniformly to $\tan^{-1}(x)$ on $[-1, 1]$.

However, this sequence converges very slowly. Indeed, by the triangle inequality for the max norm in $C[-1, 1]$,

$$\frac{1}{2n+1} = \|P_{2n} - P_{2n+2}\|_\infty \leq \|P_{2n} - f\|_\infty + \|f - P_{2n+2}\|_\infty.$$

So $\max\{\|P_{2n} - f\|_\infty, \|f - P_{2n+2}\|_\infty\} \geq \dfrac{1}{4n+2}$, which is a rather slow rate of convergence.

On the other hand, this estimate shows that the error on $[-r, r]$ is no more than $r^{-2n-1}/(2n+1)$, which goes to zero quite quickly as $n \to \infty$, if r is small. So Taylor series can sometimes be a good approximation in a limited range. See Exercise 10.1.E for a method of rapidly computing π using these polynomials.

Taylor polynomials can be used to evaluate limits. Perhaps you have seen L'Hôpital's Rule in calculus. Taylor polynomials are equally powerful and provide a better method because their application follows naturally from an understanding of approximation rather than blindly following a rule.

For convenience, we introduce the **big O notation** (pronounced big oh). A function $f(x)$ is $O(g)$ near $x = a$ if there is a constant M and a $\delta > 0$ so that

$$|f(x)| \leq M|g(x)| \quad \text{for all} \quad 0 < |x - a| < \delta.$$

For example, Taylor's Theorem concludes that $|f(x) - P_n(x)| = O\big((x - a)^{n+1}\big)$. The advantage of big O notation is that it allows you to forget about the precise constants involved in the inequality. The reader can check that the following simple arithmetic rules are valid.

(1) $O(f) \pm O(g) = O(\max\{f, g\})$
In particular, $O\big((x - a)^m\big) \pm O\big((x - a)^n\big) = O\big((x - a)^{\min\{m,n\}}\big)$

(2) $O(f)O(g) = O(fg)$
In particular, $O\big((x - a)^m\big)O\big((x - a)^n\big) = O\big((x - a)^{m+n}\big)$

(3) $\dfrac{O\big((x - a)^m\big)}{(x - a)^n} = O\big((x - a)^{m-n}\big)$

10.1.7. EXAMPLE. Consider $\lim\limits_{x \to 0} \dfrac{e^x + e^{-x} - 2}{x^2}$. We saw in Example 10.1.4 that $e^x = 1 + x + \frac{1}{2}x^2 + O(x^3)$ near $x = 0$. Hence

$$\frac{e^x + e^{-x} - 2}{x^2} = \frac{1 + x + \frac{1}{2}x^2 + O(x^3) + 1 - x + \frac{1}{2}x^2 + O(x^3) - 2}{x^2}$$

$$= \frac{x^2 + O(x^3)}{x^2} = 1 + O(x).$$

Hence $\lim\limits_{x \to 0} \dfrac{e^x + e^{-x} - 2}{x^2} = \lim\limits_{x \to 0} 1 + O(x) = 1.$

10.1.8. EXAMPLE. Now consider a more complicated limit $\lim\limits_{x\to 0} \cot^2 x - \dfrac{1}{x^2}$.
We will use the Taylor polynomials for sin and cos:

$$\sin x = x - x^3/6 + O(x^5) \quad \text{and} \quad \cos x = 1 - x^2/2 + O(x^4).$$

Then

$$\cot^2 x - \frac{1}{x^2} = \frac{\cos^2 x}{\sin^2 x} - \frac{1}{x^2} = \frac{\left(1 - x^2/2 + O(x^4)\right)^2}{x^2\left(1 - x^2/6 + O(x^4)\right)^2} - \frac{1}{x^2}$$

$$= \frac{\left(1 - x^2 + O(x^4)\right) - \left(1 - x^2/3 + O(x^4)\right)}{x^2\left(1 - x^2/3 + O(x^4)\right)}$$

$$= \frac{-2/3 + O(x^2)}{1 - x^2/3 + O(x^4)}.$$

Hence

$$\lim_{x\to 0} \cot^2 x - \frac{1}{x^2} = \lim_{x\to 0} \frac{-2/3 + O(x^2)}{1 - x^2/3 + O(x^4)} = -\frac{2}{3}.$$

10.1.9. EXAMPLE. Even if f has derivatives of all orders that we can evaluate accurately *and* the Taylor series converges uniformly, the Taylor polynomials may not converge to the right function! The classic example of this is the function

$$f(x) = \begin{cases} e^{-1/x^2} & \text{if} \quad x \neq 0 \\ 0 & \text{if} \quad x = 0. \end{cases}$$

Figure 10.2 shows just how flat this function is near zero.

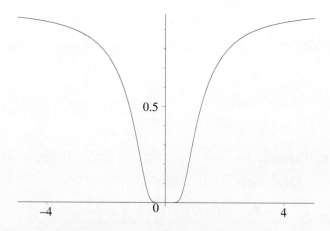

FIGURE 10.2. The graph of e^{-1/x^2}.

We will show that f is C^∞ on all of $\mathbb{R}$ and $f^{(n)}(0) = 0$ for all n. We claim that there is a polynomial $q_n(x)$ of degree at most $2n$ so that

$$f^{(n)}(x) = \frac{q_n(x)}{x^{3n}} e^{-1/x^2} \quad \text{for} \quad x \neq 0.$$

Indeed, this is obvious for $n = 0$, where $q_0 = 1$. We proceed by induction. If it is true for n, then by the product rule

$$f^{(n+1)}(x) = f^{(n)\prime}(x) = e^{-1/x^2}\left(\frac{q_n'(x)}{x^{3n}} - \frac{3nq_n(x)}{x^{3n+1}} + \frac{q_n(x)}{x^{3n}}\left(\frac{-2}{x^3}\right)\right)$$

$$= \frac{x^3 q_n'(x) - 3nx^2 q_n(x) - 2q_n(x)}{x^{3n+3}}e^{-1/x^2}.$$

To check the degree of

$$q_{n+1}(x) = x^3 q_n'(x) - 3nx^2 q_n(x) - 2q_n(x),$$

note that

$$\deg(x^3 q_n') \le 3 + \deg q_n' \le 2 + \deg q_n \le 2n + 2$$

and

$$\deg(3nx^2 q_n) \le 2 + \deg q_n \le 2n + 2.$$

Clearly this nice algebraic formula for the derivatives is continuous on both $(-\infty, 0)$ and $(0, \infty)$. So f is C^∞ everywhere except possibly at $x = 0$. Also,

$$\lim_{x\to 0}\frac{e^{-1/x^2}}{x^k} = \lim_{t\to\pm\infty}\frac{t^k}{e^{t^2}}$$

where we substitute $t = 1/x$. Since

$$e^{t^2} = \sum_{n=0}^{\infty}\frac{1}{n!}t^{2n} \ge \frac{1}{k!}t^{2k},$$

we see that

$$\lim_{t\to\pm\infty}\left|\frac{t^k}{e^{t^2}}\right| \le \lim_{t\to\pm\infty}\frac{k!}{|t|^k} = 0.$$

Therefore, if $q_n = \sum_{j=0}^{2n} a_j x^j$, we obtain

$$\lim_{x\to 0}f^{(n)}(x) = \lim_{x\to 0}\sum_{j=0}^{2n}a_j x^j\frac{e^{-1/x^2}}{x^{3n}} = \sum_{j=0}^{2n}a_j\lim_{x\to 0}\frac{e^{-1/x^2}}{x^{3n-j}} = 0.$$

We use the same fact to show that $f^{(n)}(0) = 0$ for $n \ge 1$. Indeed,

$$f^{(n+1)}(0) = \lim_{h\to 0}\frac{f^{(n)}(h) - f^{(n)}(0)}{h} = \lim_{h\to 0}\frac{q_n(h)e^{-1/h^2}}{h^{3n+1}} = 0.$$

So $f^{(n)}$ is defined on the whole line and is continuous for each n. Therefore, f is C^∞.

Because all of the derivatives of f vanish at $x = 0$, *all* of the Taylor polynomials are $P_n(x) = 0$. While this certainly converges rapidly, it converges to the wrong function! The Taylor polynomials completely fail to approximate f anywhere except at the one point $x = 0$.

Exercises for Section 10.1

A. Find the Taylor polynomials of order 3 for each of the following functions at the given point a, and estimate the error at the point b.
(a) $f(x) = \tan x$ about $a = \frac{\pi}{4}$ and $b = .75$
(b) $g(x) = \sqrt{1 + x^2}$ about $a = 0$ and $b = .1$
(c) $h(x) = x^4$ about $a = 1$ and $b = 0.99$
(d) $k(x) = \sinh x$ about $a = 0$ and $b = .003$

B. Let $a \in [A, B]$, $f \in C^2[A, B]$, and let $P_1(x) = f(a) + f'(a)(x - a)$ be the first-order Taylor polynomial. Fix a point x_0 in $[A, B]$.
(a) Define $h(t) = f(t) + f'(t)(x_0 - t) + A(x_0 - t)^2$. Find the constant A that makes $h(a) = h(x_0)$.
(b) Apply Rolle's Theorem to h to obtain a point c between a and x_0 so that
$$f(x_0) - P_1(x_0) = f''(c)\frac{(x_0 - a)^2}{2}.$$
(c) Find a constant M so that $|f(x) - f(a)| \leq M(x - a)^2$ for all $x \in [A, B]$.

C. Let f satisfy the hypotheses of Taylor's Theorem at $x = a$.
(a) Show that $\lim\limits_{x \to a} \dfrac{f(x) - P_n(x)}{(x - a)^n} = 0$.
(b) If $Q(x) \in \mathbb{P}_n$ and $\lim\limits_{x \to a} \dfrac{f(x) - Q(x)}{(x - a)^n} = 0$, prove that $Q = P_n$.

D. (a) Find the Taylor series for $\sin x$ about $x = 0$, and prove that it converges to $\sin x$ uniformly on any bounded interval $[-N, N]$.
(b) Find the Taylor expansion of $\sin x$ about $x = \pi/6$. Hence show how to approximate $\sin(31°)$ to 10 decimal places. Do careful estimates.

E. (a) Verify that $4\tan^{-1}(\frac{1}{5}) - \tan^{-1}(\frac{1}{239}) = \frac{\pi}{4}$.   HINT: Take the tan of both sides.
(b) Using the estimates for $\tan^{-1}(x)$ derived in Example 10.1.6, compute how many terms are needed to approximate π to 1000 decimals accuracy using this formula.
(c) Calculate π to 6 decimals of accuracy using this method.

F. Let $f(x) = \log x$.
(a) Find the Taylor series of f about $x = 1$.
(b) What is the radius of convergence of this series?
(c) What happens at the two endpoints of the interval of convergence? Hence find a series converging to $\log 2$.
(d) By observing that $\log 2 = \log 4/3 - \log 2/3$, find another series converging to $\log 2$. Why is this series more useful?
(e) Show that $\log 3 = 3\log .96 + 5\log \frac{81}{80} - 11\log .9$. Find a finite expression that estimates $\log 3$ to 50 decimal places.

G. Compute the following limits using Taylor polynomials.
(a) $\lim\limits_{x \to 0} \dfrac{2^x - 3^x}{x}$.   HINT: $a^x = e^{x \log a}$
(b) $\lim\limits_{x \to 0} \dfrac{x \cos x - \sin x}{x \sin^2 x}$
(c) Compute $\lim\limits_{x \to 0} \left(\dfrac{\tan x}{x}\right)^{1/x^2}$.   HINT: Take the log, and use Taylor polynomials for both $\tan(x)$ and $\log(1 + x)$ about 0.

H. Suppose that f and g have $n + 1$ continuous derivatives on $[a - \delta, a + \delta]$ and that $f^{(k)}(a) = g^{(k)}(a) = 0$ for $0 \le k < n$ and $g^{(n)}(a) \ne 0$. Use Taylor polynomials to show that $\displaystyle\lim_{x \to a} \frac{f(x)}{g(x)} = \frac{f^{(n)}(a)}{g^{(n)}(a)}$.

I. Show that $\displaystyle\int_{-\infty}^{+\infty} \log\left|\frac{1+x}{1-x}\right| \frac{dx}{x} = \pi^2$ as follows:

(a) Reduce the integral to $4 \displaystyle\int_0^1 \log\left|\frac{1+x}{1-x}\right| \frac{dx}{x}$.

(b) Use the Taylor series for $\log x$ about $x = 1$ found in Exercise 10.1.F. Use your knowledge of convergence and integration to evaluate $\displaystyle\int_0^r \log\left|\frac{x+1}{x-1}\right| \frac{dx}{x}$ as a series when $r < 1$.

(c) Justify the improper integral obtained by letting r go to 1.

(d) Use the famous identity $\displaystyle\sum_{n=1}^{\infty} \frac{1}{n^2} = \frac{\pi^2}{6}$ to complete the argument.

J. Let $f(x) = (1+x)^{-1/2}$.

(a) Find a formula for $f^{(k)}(x)$. Hence show that

$$f^{(k)}(0) = \binom{-\frac{1}{2}}{k} := \frac{-\frac{1}{2}(-\frac{1}{2} - 1) \cdots (-\frac{1}{2} + 1 - k)}{k!}$$

$$= \frac{(-1)(-3) \cdots (1 - 2k)}{2^k k!} = \frac{(-1)^k (2k)!}{2^{2k} (k!)^2} = \left(\frac{-1}{4}\right)^k \binom{2k}{k}.$$

(b) Show that the Taylor series for f about $x = 0$ is $\displaystyle\sum_{k=0}^{\infty} \binom{2k}{k} \left(\frac{-x}{4}\right)^k$, and compute the radius of convergence.

(c) Show that $\sqrt{2} = 1.4 f(-.02)$. Hence compute $\sqrt{2}$ to 8 decimal places.

(d) Express $\sqrt{2} = 1.415 f(\varepsilon)$ where ε is expressed as a fraction in lowest terms. Use this to obtain an alternating series for $\sqrt{2}$. How many terms are needed to estimate $\sqrt{2}$ to 100 decimal places?

K. Let a be the decimal number with 198 ones (i.e., $a = \underbrace{11 \ldots 11}_{198 \text{ ones}}$). Find $\sqrt{a}$ to 500 decimal places. HINT: $a = \left(\frac{10^{99}}{3}\right)^2 (1 - 10^{-198})$. Your decimal expansion should end in 97916.

10.2. How Not to Approximate a Function

Given a continuous function $f : [a, b] \to \mathbb{R}$, can we approximate f by a polynomial? For example, suppose you need to write a computer program to evaluate f. Since some round-off errors are inevitable, why not replace f with a polynomial that is close to f, as the polynomial will be easy to evaluate?

What precisely do we mean by *close*? This depends on the context, but in the context of the preceding programming example, we mean a polynomial p so that

$$\|f - p\|_\infty = \max_{x \in [a,b]} |f(x) - p(x)|$$

is small. This is known as **uniform approximation**. Such approximations are important both in practical work and in theory. Later in this chapter, there are several methods for computing such approximations, including methods well adapted to programming.

We start by looking at several plausible ways to answer this problem that do not work.

It might seem that the Taylor polynomials of the previous section are a good answer to this problem. However, there are several serious flaws. First, and most important, the function f may not be differentiable at all and f must have at least n derivatives in order to compute P_n. Moreover, the bound on the $(n+1)$st derivative is the crucial factor in the error estimate, so f must have $n + 1$ derivatives. The bounds may be so large that this estimate is useless. Second, even for very nice functions, the Taylor series may converge only on a small interval about the point a. Recall Example 10.1.6, which shows that the Taylor series for $\tan^{-1}(x)$ only converges on the small interval $[-1, 1]$ even though the function is C^∞ on all of $\mathbb{R}$. Moreover, the convergence is very slow unless we further restrict the interval to something like $[-.5, .5]$. Worse yet was $f(x) = e^{-1/x^2}$ of Example 10.1.9, which is C^∞ and for which the Taylor series about $x = 0$ converges everywhere, but to the wrong function. Because the Taylor series only uses information at one point, it cannot be expected always to do a good job over an entire interval.

Differentiation is a very unstable process when the function is known only approximately in the uniform norm—a small error in evaluating the function can result in a huge error in the derivative (see Example 10.2.1). So even when the Taylor series does converge, it can be difficult to compute the coefficients numerically.

10.2.1. EXAMPLE. One reason that Taylor polynomials fail is that they only use information available at one point. Another failing is that they try too hard to approximate derivatives at the same time. Here is an example where convergence for the function is good, but not for the derivative. Let

$$f_n(x) = x + \frac{1}{\sqrt{n}} \sin nx \quad \text{for} \quad -\pi \le x \le \pi.$$

It is easy to verify that f_n converges uniformly on $[-\pi, \pi]$ to the function $f(x) = x$. Indeed,

$$\|f - f_n\|_\infty = \max_{-\pi \le x \le \pi} \frac{1}{\sqrt{n}} |\sin nx| = \frac{1}{\sqrt{n}} \quad \text{for} \quad n \ge 1.$$

However $f'(x) = 1$ everywhere, while $f'_n(x) = 1 + \sqrt{n} \cos nx$. Therefore,

$$\|f' - f'_n\|_\infty = \max_{-\pi \le x \le \pi} \sqrt{n} |\cos nx| = \sqrt{n} \quad \text{for} \quad n \ge 1.$$

As the approximants f_n get closer to f, they oscillate more and more dramatically. So the derivatives are very far from the derivative of f.

Another possible method for finding good approximants is to use polynomial interpolation. Pick $n + 1$ points distributed over $[a, b]$, for example,

$$x_i = a + \frac{i(b - a)}{n} \quad \text{for} \quad i = 0, 1, \ldots, n.$$

There is a unique polynomial p_n of degree at most n that goes through the $n + 1$ points $(x_i, f(x_i))$, $i = 0, 1, \ldots, n$. One way to find this polynomial is to solve the system of linear equations in $n + 1$ variables $a_0, a_1, \ldots, a_n$

$$a_0 + x_0 a_1 + x_0^2 a_2 + \cdots + x_0^n a_n = f(x_0)$$

$$a_0 + x_1 a_1 + x_1^2 a_2 + \cdots + x_1^n a_n = f(x_1)$$

$$\vdots$$

$$a_0 + x_n a_1 + x_n^2 a_2 + \cdots + x_n^n a_n = f(x_n).$$

This determines the **van der Monde matrix** X and vectors $\mathbf{a}$ and $\mathbf{f}$:

$$X = \begin{bmatrix} 1 & x_0 & x_0^2 & \cdots & x_0^n \\ 1 & x_1 & x_1^2 & \cdots & x_1^n \\ \vdots & \ddots & \ddots & & \vdots \\ 1 & x_n & x_n^2 & \cdots & x_n^n \end{bmatrix} \qquad \mathbf{a} = \begin{bmatrix} a_0 \\ a_1 \\ \vdots \\ a_n \end{bmatrix} \quad \text{and} \quad \mathbf{f} = \begin{bmatrix} f(x_0) \\ f(x_1) \\ \vdots \\ f(x_n) \end{bmatrix}.$$

The system becomes $X\mathbf{a} = \mathbf{f}$. This always has a unique solution because the matrix X is invertible (see Exercise 10.2.N). This determines the polynomial p_n, which agrees with f at the $n + 1$ points $x_0, \ldots, x_n$.

It seems reasonable to suspect that as n increases, p_n will converge uniformly to f. However, this is not true. In 1901, Runge showed that the polynomial interpolants on $[-5, 5]$ to the function

$$f(x) = \frac{1}{1 + x^2}$$

do not converge to f. In fact, if p_n is polynomial interpolant of degree at most n, then

$$\lim_{n \to \infty} \|f - p_n\|_\infty = \infty.$$

Proving this is more than a little tricky. If instead of choosing $n + 1$ equally spaced points, we choose the points $x_i = 5\cos(i\pi/n)$, $i = 0, 1, \ldots, n$, then the interpolating polynomials $p_n(x)$ *will* converge uniformly to this particular function.

However, if you specify the points in advance, then no matter which points you choose at each stage, there is some continuous function $f \in C[a, b]$ so that the interpolating polynomials of degree n do not converge uniformly to f. This was proved by Bernstein and Faber independently in 1914.

There are algorithms using interpolation that involve varying the points of interpolation strategically. There are also ways of making interpolation into a practical method by using splines instead of polynomials. We'll return to the latter idea in the last two sections of this chapter.

After all of these negative results, we might wonder if it is possible to approximate an arbitrary continuous function by a polynomial. It is a remarkable and important theorem, proved by Weierstrass in 1885, that this is possible.

10.2.2. WEIERSTRASS APPROXIMATION THEOREM.

Let f be any continuous real-valued function on $[a, b]$. Then there is a sequence of polynomials p_n that converges uniformly to f on $[a, b]$.

In the language of normed vector spaces, this theorem says that the polynomials are dense in $C[a, b]$ in the max norm.

In fact, this theorem is sufficiently important that many different proofs have been found. The proof we give was found in 1912 by Bernstein, a Russian mathematician. It explicitly constructs the approximating polynomial. This algorithm is not the most efficient, but the problem of finding efficient algorithms can wait until we have proved that the theorem is true.

Exercises for Section 10.2

A. Assume the Weierstrass Theorem is true for $C[0, 1]$, and then prove it is true for $C[a, b]$, for an arbitrary interval $[a, b]$.
HINT: For $f \in C[a, b]$, consider $g(t) := f\big(a + (b - a)t\big)$ in $C[0, 1]$.

B. Let $\alpha > 0$. Using the Weierstrass Theorem, prove that every continuous function $f : [0, +\infty] \to \mathbb{R}$ with $\lim_{n \to \infty} f(x) = 0$ can be uniformly approximated as closely as we like by a function of the form $q(x) = \sum_{n=1}^{N} C_n e^{-n\alpha x}$.
HINT: Consider $g(y) = f(-\log(y)/\alpha)$ on $(0, 1]$.

C. (a) Show that every continuous function f on $[a, b]$ is the uniform limit of polynomials of the form $p_n(x^3)$.
  (b) Describe the subspaces of $C[-1, 1]$ which are uniform limits of polynomials of the form $p_n(x^2)$.

D. Suppose that f is a continuous function on $[0, 1]$ such that $\int_0^1 f(x) x^n \, dx = 0$ for all $n \geq 0$. Prove that $f = 0$.
HINT: Use the Weierstrass Theorem to show that $\int_0^1 |f(x)|^2 \, dx = 0$.

E. Let X be a compact subset of $[-N, N]$.
  (a) Show that every continuous function f on X may be extended to a continuous function g defined on $[-N, N]$ with $\|g\|_\infty = \|f\|_\infty$.
  (b) Show that every continuous function on X is the uniform limit of polynomials.

F. Show that if f is continuously differentiable on $[0, 1]$, then there is a sequence of polynomials p_n converging uniformly to f such that p_n' converge uniformly to f' as well. HINT: Approximate f' first.

G. Show that if f is in $C^\infty[0, 1]$, then there is a sequence p_n such that the kth derivatives $p_n^{(k)}$ converge uniformly to $f^{(k)}$ for every $k \geq 0$.
HINT: Adapt Exercise 10.2.F to find p_n with $\|f^{(k)} - p_n^{(k)}\|_\infty < \frac{1}{n}$ for $0 \leq k \leq n$.

H. Prove that e^x is not a polynomial. HINT: Consider behaviour at $\pm\infty$.

I. (a) If $0 \notin [a, b]$, show that every continuous function f on $[a, b]$ is the uniform limit of a sequence of polynomials (q_n), where $q_n(x) = x^n p_n(x)$ for polynomials p_n.

(b) If $0 \in [a, b]$, show that a continuous function f on $[a, b]$ is the uniform limit of a sequence of polynomials (q_n), where $q_n(x) = x^n p_n(x)$ for polynomials p_n, if and only if $f(0) = 0$.

J. (a) If $x_0, \ldots, x_n$ are points in $[a, b]$ and $\mathbf{a} = (a_0, \ldots, a_n) \in \mathbb{R}^{n+1}$, show that there is a unique polynomial $p_{\mathbf{a}}$ in $\mathbb{P}_n$ such that $p(x_i) = a_i$ for $0 \le i \le n$.
HINT: Find polynomials q_j such that $q_j(x_i)$ is 1 if $i = j$ and is 0 if $0 \le i \ne j \le n$.
(b) Show that there is a constant M so that $\|p_{\mathbf{a}}\|_\infty \le M \|\mathbf{a}\|_2$.

K. Suppose that $f \in C[a, b]$, $\varepsilon > 0$ and $x_1, \ldots, x_n$ are points in $[a, b]$. Prove that there is a polynomial p such that $p(x_i) = f(x_i)$ for $1 \le i \le n$ and $\|f - p\|_\infty < \varepsilon$.
HINT: First approximate f closely by some polynomial. Then use the previous exercise to adjust the difference.

L. Show that every continuous function $h(x, y)$ on $[a, b] \times [c, d]$ is the uniform limit of polynomials in two variables as follows:
(a) Show that every function of the form $f(x)g(y)$, where $f \in C[a, b]$ and $g \in C[c, d]$ is the uniform limit of polynomials.
(b) Let $k(x) = \max\{1 - |x|, 0\}$. Set $k_{ij}^{(n)}(x, y) = k(2^n x - i)k(2^n y - j)$ for $i, j \in \mathbb{Z}$ and $n \ge 0$. Show that $\sum_i \sum_j k_{ij}^{(n)}(x, y) = 1$.
(c) Define $h_n(x, y) = \sum_i \sum_j h(2^{-n}i, 2^{-n}j)k_{ij}^{(n)}(x, y)$. Show that h_n is the uniform limit of polynomials.
(d) Use the uniform continuity of h to prove that h_n converges to h uniformly.

M. Let $\alpha > 0$ and $\varepsilon > 0$. Using the following outline, show by induction that for each positive integer n, there is a polynomial p so that $|e^{-n\alpha x} - e^{-\alpha x}p(x)| < \varepsilon$ for all $x \in [0, +\infty)$.
(a) Show that for each n, it suffices to prove the result for $\alpha = 1$.
(b) For $n = 2$, first show that for $f(x) = e^{-2x} - e^{-x}\sum_{k=0}^{N-1}(-x)^k/k!$ satisfies $\|f\|_\infty \le \frac{1}{N!} \sup\{e^{-x}x^N : x \ge 0\}$.   HINT: Example 10.1.4.
(c) For fixed N, let $g(x) = e^{-x}x^N$, show that g has a unique maximum at $x = N$ and hence $|f(x)| \le K/\sqrt{N}$ for some constant K.   HINT: Stirling's formula.
(d) Deduce that the result holds for $n = 2$.
(e) For the inductive step, find a polynomial q with $|e^{-(n+1)x} - e^{-nx/2-x/2}q(x)| \le \varepsilon/2$ for all $x \in [0, +\infty)$.   HINT: Use part (d) for $e^{-2\alpha x}$ with $\alpha = (n+1)/2$.
(f) Next, find a polynomial r so that $|e^{-nx/2-x/2}q(x) - e^{-x}q(x)r(x)| \le \varepsilon/2$ for all $x \in [0, +\infty)$.   HINT: Note that $h(x) = e^{-x/2}|q(x)|$ is bounded on $[0, +\infty)$ and use the inductive result for $e^{-n\alpha x}$ with $\alpha = 1/2$.
(g) Deduce that the result holds for $e^{-(n+1)x}$.
(h) Finally, use this result and Exercise 10.2.B to show that every continuous function $f : [0, +\infty) \to \mathbb{R}$ with $\lim_{n \to \infty} f(x) = 0$ can be uniformly approximated as closely as we like by a function of the form $e^{-\alpha x}p(x)$ for some polynomial p.

N. (a) Show that the determinant of the van der Monde matrix X is a polynomial of degree $n(n + 1)/2$ in the variables $x_0, \ldots, x_n$.
(b) Show that the determinant is 0 if $x_i = x_j$ for some $0 \le i < j \le n$. Hence show that it is a scalar multiple of $\prod_{0 \le i < j \le n} (x_j - x_i)$. By looking at the coefficient of $x_1 x_2^2 \ldots x_n^n$ in both the determinant and the product, show that the scalar is 1.
(c) Show that X is invertible when $x_0, \ldots, x_n$ are distinct.

10.3. Bernstein's Proof of the Weierstrass Theorem

Recall the binomial formula, $(x + y)^n = \sum_{k=0}^{n} \binom{n}{k} x^k y^{n-k}$. If we set $y = 1 - x$, then we obtain

$$1 = \sum_{k=0}^{n} \binom{n}{k} x^k (1 - x)^{n-k}.$$

Bernstein started by considering the functions

$$P_k^n(x) = \binom{n}{k} x^k (1 - x)^{n-k} \quad \text{for} \quad k = 0, 1, \ldots, n,$$

now called **Bernstein polynomials**. They have several virtues. They are all polynomials of degree n. They take only nonnegative values on $[0, 1]$. And they add up to 1. Moreover, P_k^n is a "bump" function with a maximum at k/n, as a routine calculus calculation shows. For example, the four functions P_k^3 for $0 \le k \le 3$ are given in Figure 10.3.

Given a continuous function f on $[0, 1]$, define a polynomial $B_n f$ by

$$(B_n f)(x) = \sum_{k=0}^{n} f\left(\tfrac{k}{n}\right) P_k^n(x) = \sum_{k=0}^{n} f\left(\tfrac{k}{n}\right) \binom{n}{k} x^k (1 - x)^{n-k}.$$

This is a linear combination of the polynomials P_k^n; and so $B_n f$ is a polynomial of degree at most n. We think of B_n as a function from the vector space $C[0, 1]$ into itself. This map has several easy but important properties. If $f, g \in C[0, 1]$, we say that $f \ge g$ if $f(x) \ge g(x)$ for all $0 \le x \le 1$.

10.3.1. PROPOSITION. *The map B_n is linear and monotone. That is, for all $f, g \in C[0, 1]$ and $\alpha \in \mathbb{R}$,*

(1) $B_n(f + g) = B_n f + B_n g$

(2) $B_n(\alpha f) = \alpha B_n f$

(3) $B_n f \ge 0 \quad \text{if} \quad f \ge 0$

(4) $B_n f \ge B_n g \quad \text{if} \quad f \ge g$

(5) $|B_n f| \le B_n g \quad \text{if} \quad |f| \le g.$

The only part that requires any cleverness is the monotonicity. However, since each $P_k^n \ge 0$, it follows that when $f \ge 0$, then $B_n f$ is also positive. In particular, $|f| \le g$ means that $-g \le f \le g$; and hence $-B_n g \le B_n f \le B_n g$. The details are left to the reader.

Next let us compute $B_n f$ for three basic polynomials: 1, x, and x^2.

10.3.2. LEMMA. $B_n 1 = 1, \quad B_n x = x, \quad$ *and*

$$B_n x^2 = \frac{n - 1}{n} x^2 + \frac{1}{n} x = x^2 + \frac{x - x^2}{n}.$$

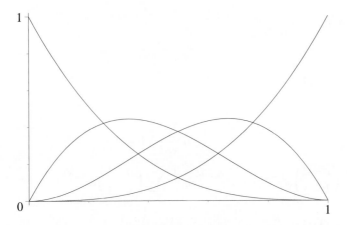

FIGURE 10.3. The Bernstein polynomials of degree 3.

PROOF. For the first equation, we use the Binomial Theorem to get

$$B_n 1 = \sum_{k=0}^{n} 1 \binom{n}{k} x^k (1-x)^{n-k} = 1.$$

Next, notice that

$$\frac{k}{n} \binom{n}{k} = \frac{k}{n} \frac{n!}{k!(n-k)!} = \frac{(n-1)!}{(k-1)!(n-k)!} = \binom{n-1}{k-1}.$$

Using this result and the Binomial Theorem, we have

$$
\begin{aligned}
B_n x &= \sum_{k=0}^{n} \frac{k}{n} \binom{n}{k} x^k (1-x)^{n-k} \\
&= x \sum_{k=1}^{n} \binom{n-1}{k-1} x^{k-1} (1-x)^{n-k} \\
&= x \big(x + (1-x) \big)^{n-1} = x.
\end{aligned}
$$

Finally, notice that

$$
\begin{aligned}
\frac{k^2}{n^2} \binom{n}{k} &= \frac{k^2}{n^2} \frac{n!}{k!(n-k)!} \\
&= \frac{(k-1)+1}{n} \frac{(n-1)!}{(k-1)!(n-k)!} \\
&= \frac{n-1}{n} \frac{k-1}{n-1} \frac{(n-1)!}{(k-1)!(n-k)!} + \frac{1}{n} \frac{(n-1)!}{(k-1)!(n-k)!} \\
&= \frac{n-1}{n} \binom{n-2}{k-2} + \frac{1}{n} \binom{n-1}{k-1}.
\end{aligned}
$$

Note that $\binom{m}{-2} = \binom{m}{-1} = 0$. Hence

$$B_n x^2 = \sum_{k=0}^{n} \frac{k^2}{n^2} \binom{n}{k} x^k (1-x)^{n-k}$$

$$= \frac{n-1}{n} \sum_{k=2}^{n} \binom{n-2}{k-2} x^k (1-x)^{n-k} + \frac{1}{n} \sum_{k=1}^{n} \binom{n-1}{k-1} x^k (1-x)^{n-k}$$

$$= \frac{n-1}{n} x^2 \sum_{k=2}^{n} \binom{n-2}{k-2} x^{k-2}(1-x)^{n-k} + \frac{x}{n} \sum_{k=1}^{n} \binom{n-1}{k-1} x^{k-1}(1-x)^{n-k}$$

$$= \frac{n-1}{n} x^2 \left(x + (1-x)\right)^{n-2} + \frac{1}{n} x \left(x + (1-x)\right)^{n-1}$$

$$= \frac{n-1}{n} x^2 + \frac{1}{n} x = x^2 + \frac{x - x^2}{n}. \qquad \blacksquare$$

PROOF OF WEIERSTRASS'S THEOREM. By Exercise 10.2.A, it suffices to prove the theorem for the interval $[0, 1]$. Fix a continuous function f in $C[0, 1]$. We will prove that for each $\varepsilon > 0$, there is some $N > 0$ so that

$$\|f(x) - B_n f(x)\| < \varepsilon \quad \text{for all} \quad n \geq N.$$

Since $[0, 1]$ is compact, f is uniformly continuous on $[0, 1]$ by Theorem 5.5.9. Thus for our given $\varepsilon > 0$, there is some $\delta > 0$ so that

$$|f(x) - f(y)| \leq \frac{\varepsilon}{2} \quad \text{for all} \quad |x - y| \leq \delta, \ x, y \in [0, 1].$$

Also, f is bounded on $[0, 1]$ by the Extreme Value Theorem (Theorem 5.4.4). So let

$$M = \|f\|_\infty = \sup_{x \in [0,1]} |f(x)|.$$

Fix any point $a \in [0, 1]$. We claim that

$$|f(x) - f(a)| \leq \frac{\varepsilon}{2} + \frac{2M}{\delta^2} (x - a)^2.$$

Indeed, if $|x - a| \leq \delta$, then

$$|f(x) - f(a)| \leq \frac{\varepsilon}{2} \leq \frac{\varepsilon}{2} + \frac{2M}{\delta^2} (x - a)^2$$

by our estimate of uniform continuity. And if $|x - a| \geq \delta$, then

$$|f(x) - f(a)| \leq 2M \leq 2M \left(\frac{x - a}{\delta}\right)^2 \leq \frac{\varepsilon}{2} + \frac{2M}{\delta^2} (x - a)^2.$$

Notice that by linearity and the fact that $B_n 1 = 1$, we obtain

$$B_n(f - f(a))(x) = B_n f(x) - f(a).$$

Now use the positivity of our map B_n to obtain

$$|B_n f(x) - f(a)| \leq B_n \left(\frac{\varepsilon}{2} + \frac{2M}{\delta^2} (x-a)^2 \right)$$

$$= \frac{\varepsilon}{2} + \frac{2M}{\delta^2} \left(x^2 + \frac{x - x^2}{n} - 2ax + a^2 \right)$$

$$= \frac{\varepsilon}{2} + \frac{2M}{\delta^2} (x-a)^2 + \frac{2M}{\delta^2} \frac{x - x^2}{n}.$$

Evaluate this at $x = a$ to obtain

$$|B_n f(a) - f(a)| \leq \frac{\varepsilon}{2} + \frac{2M}{\delta^2} \frac{a - a^2}{n} \leq \frac{\varepsilon}{2} + \frac{M}{2\delta^2 n}.$$

We use the fact that $\max\{a - a^2 : 0 \leq a \leq 1\} = \frac{1}{4}$.
This estimate does not depend on the point a. So we have found

$$\|B_n f - f\|_\infty \leq \frac{\varepsilon}{2} + \frac{M}{2\delta^2 n}.$$

So now choose $N \geq \dfrac{M}{\delta^2 \varepsilon}$ so that $\dfrac{M}{2\delta^2 N} < \dfrac{\varepsilon}{2}$. Then for all $n \geq N$,

$$\|B_n f - f\|_\infty \leq \frac{\varepsilon}{2} + \frac{\varepsilon}{2} = \varepsilon. \qquad \blacksquare$$

As was already mentioned, using Bernstein polynomials is not an efficient way of finding polynomial approximations. However, Bernstein polynomials have other advantages, which are developed in the Exercises.

Exercises for Section 10.3

A. Show that $P_k^n(x) = \binom{n}{k} x^k (1-x)^{n-k}$ attains its maximum at $\frac{k}{n}$.

B. Show that $\|B_n f\|_\infty \leq \|f\|_\infty$.   HINT: Use monotonicity.

C. Prove that $B_n(f)^2 \leq B_n(f^2)$.   HINT: Expand $B_n((f-a)^2)$.

D. (a) Compute $B_n x^3$.   HINT: $\left(\frac{k}{n}\right)^3 = \frac{k(k-1)(k-2) + 3k(k-1) + k}{n^3}$
(b) Compute $\lim\limits_{n \to \infty} n(B_n x^3 - x^3)$.

E. Work through our proof of the Weierstrass theorem with the function $f(x) = |x - \frac{1}{2}|$ on $[0, 1]$ to obtain an estimate for the degree of a polynomial p needed to ensure that $\|f - p\|_\infty < .0005$.

F. (a) Show that $B_n(e^x) = (1 + (e^{1/n} - 1)x)^n$.
(b) Show that this may be rewritten as $(1 + \frac{x}{n} + \frac{c_n}{n^2})^n$, where $0 \leq c_n \leq 1$.
(c) Hence prove directly that $B_n(e^x)$ converges uniformly to e^x on $[0, 1]$.

G. (a) Show that the derivative of $B_{n+1} f$ is

$$(B_{n+1} f)'(x) = \sum_{k=0}^{n} \frac{f\left(\frac{k+1}{n+1}\right) - f\left(\frac{k}{n+1}\right)}{\frac{1}{n+1}} \binom{n}{k} x^k (1-x)^{n-k}.$$

(b) If f has a continuous first derivative, use the Mean Value Theorem and the uniform continuity of f to show that $\lim\limits_{n \to \infty} \|(B_n f)' - f'\|_\infty = 0$.

H. (a) Fix $m \geq 0$ and set $f_{nm}(x) = x\left(x - \frac{1}{n}\right)\left(x - \frac{2}{n}\right)\ldots\left(x - \frac{m-1}{n}\right)$ for $n \geq 1$. Show that $B_n f_{nm} = f_{nm}(0)x^m$.

   (b) Hence show that as n tends to infinity, both sequences (f_{nm}) and $(B_n f_{nm})$ converge uniformly to x^m.

   (c) Show that $B_n x^m$ converges to x^m by estimating
   $$\|B_n x^m - x^m\|_\infty \leq \|B_n(x^m - f_{nm})\|_\infty + \|B_n f_{nm} - x^m\|_\infty.$$

   (d) Use this to give another proof that $B_n p$ converges uniformly to p for every polynomial p.

10.4. Accuracy of Approximation

In this section, we try to measure the speed of convergence of polynomial approximations. Let us define the best possible error. The aim is to get a reasonable idea of what it is for a given function, and how well a given approximation compares with it.

Let $\mathbb{P}_n$ denote the vector space of polynomials of degree at most n. We will write $\mathbb{P}_n[a, b]$ to mean that $\mathbb{P}_n$ is considered as a subspace of $C[a, b]$ with norm given by the maximum modulus over the interval $[a, b]$.

10.4.1. DEFINITION. If $f \in C[a, b]$, then we define **error function** $E_n(f)$ by

$$E_n(f) = \inf\{\|f - q\|_\infty : q \in \mathbb{P}_n\}.$$

Likewise, if $\mathcal{F}$ is a set of functions, we let

$$E_n(\mathcal{F}) = \sup_{f \in \mathcal{F}} E_n(f).$$

We will know how good a polynomial approximation $p \in \mathbb{P}_n$ to f is by how close $\|f - p\|_\infty$ is to $E_n(f)$.

A little thought reveals that wildly oscillating functions will not be well approximated by polynomials of low degree. For example, the function $f(x) = \cos(n\pi x)$ in $C[0, 1]$ alternately takes the extreme values ± 1 at $\frac{k}{n}$ for $0 \leq k \leq n$. Any function close to f (within 1) will have to switch signs between these points. This suggests that in order to get a reasonable estimate, we must measure how quickly f varies.

10.4.2. DEFINITION. The **modulus of continuity** of $f \in C[a, b]$ is defined for each $\delta > 0$ by

$$\omega(f; \delta) = \sup\{|f(x_1) - f(x_2)| : |x_1 - x_2| < \delta, \ x_1, x_2 \in [a, b]\}.$$

In other words, $\omega(f, \delta)$ is the smallest choice of ε for which δ "works" in the definition of uniform continuity. This is closely related to the notion of oscillation used in the proof of Lebesgue's Theorem (Theorem 6.6.6).

By Theorem 5.5.9, every continuous function on the compact set $[0, 1]$ is uniformly continuous. Therefore, for each $\varepsilon > 0$, there is a $\delta > 0$ such that

$$|f(x) - f(y)| < \varepsilon \quad \text{for all} \quad |x - y| < \delta, \ x, y \in [0, 1].$$

Restating this with our new terminology, we see that for every $\varepsilon > 0$, there is a $\delta > 0$ such that $\omega(f; \delta) < \varepsilon$. Thus the uniform continuity of f is equivalent to

$$\lim_{\delta \to 0^+} \omega(f; \delta) = 0.$$

10.4.3. EXAMPLE. Consider $f(x) = \sqrt{x}$ on $[0, 1]$. Fix $\delta \geq 0$ and look at

$$\sup_{0 \leq t \leq \delta} f(x + t) - f(x) = \sqrt{x + \delta} - \sqrt{x}$$

$$= \frac{\delta}{\sqrt{x + \delta} + \sqrt{x}} \leq \sqrt{\delta}.$$

This inequality is sharp at $x = 0$. Thus, $\omega(f; \delta) = \sqrt{\delta}$.

The class of functions f with $\omega(f; \delta) \leq \delta$ for all $\delta > 0$ are precisely the functions satisfying

$$|f(x) - f(y)| \leq |x - y|,$$

namely the functions with Lipschitz constant 1. Denote by $\mathcal{S}$ the class of functions in $C[0, 1]$ with Lipschitz constant 1. We will prove our results first for the class $\mathcal{S}$.

A good lower bound for the error is obtained using an idea due to Chebychev. We will use the idea behind the next proof repeatedly, so examine it carefully.

10.4.4. PROPOSITION. $E_n(\mathcal{S}) \geq \dfrac{1}{2n + 2} \quad \text{for} \quad n \geq 0.$

PROOF. Fix $n \geq 0$. Consider the sawtoothed function f that takes the values

$$f\left(\tfrac{k}{n+1}\right) = \frac{(-1)^k}{2n + 2} \quad \text{for} \quad 0 \leq k \leq n + 1$$

and is linear in between with slope ± 1. Clearly, f belongs to $\mathcal{S}$.

We will show that the closest polynomial to f in $\mathbb{P}_n$ is the zero polynomial, which is clearly distance $1/(2n + 2)$ from f. To this end, suppose that p is a polynomial with $\|p - f\|_\infty < \frac{1}{2n+2}$. Then

$$\left| p\left(\tfrac{k}{n+1}\right) - \frac{(-1)^k}{2n + 2} \right| < \frac{1}{2n + 2}.$$

It follows that $\operatorname{sign} p\left(\tfrac{k}{n+1}\right) = (-1)^k$. Consequently, p changes sign between $\tfrac{k}{n+1}$ and $\tfrac{k+1}{n+1}$ for each $0 \leq k \leq n$. By the Intermediate Value Theorem (Theorem 5.6.1), p has a root in the open interval $\left(\tfrac{k}{n+1}, \tfrac{k+1}{n+1}\right)$. So p is a nonconstant polynomial with at least $n + 1$ roots, and thus is not in $\mathbb{P}_n$. Consequently,

$$E_n(\mathcal{S}) \geq E_n(f) = \|f\|_\infty = \frac{1}{2n + 2}. \qquad \blacksquare$$

To obtain an upper bound, let us look carefully at the estimate that comes out of Bernstein's proof of the Weierstrass Theorem.

10.4.5. PROPOSITION. *If $f \in C[0,1]$ has Lipschitz constant 1, then*

$$\|B_n f - f\| \leq \frac{1}{\sqrt{n}}.$$

PROOF. We recall the details of the proof of the Weierstrass Approximation Theorem in our context. Let ε be any positive number. We claim that the Lipschitz condition gives the strong inequality

$$|f(x) - f(a)| \leq |x - a| \leq \varepsilon + \frac{(x-a)^2}{\varepsilon}.$$

To check this, consider the cases $|x - a| \leq \varepsilon$ and $|x - a| > \varepsilon$ separately.

Now apply the Bernstein map B_n, which by monotonicity yields

$$|B_n f(x) - f(a)| \leq \varepsilon + \frac{B_n\big((x-a)^2\big)}{\varepsilon} = \varepsilon + \frac{(x-a)^2}{\varepsilon} + \frac{x - x^2}{n\varepsilon}.$$

Substituting $x = a$ and maximizing over $[0, 1]$, we obtain

$$\|B_n f - f\|_\infty \leq \varepsilon + \frac{1}{n\varepsilon}\|x - x^2\|_\infty = \varepsilon + \frac{1}{4n\varepsilon}.$$

Minimizing this leads to the choice of $\varepsilon = \frac{1}{2\sqrt{n}}$. Thus

$$\|B_n f - f\|_\infty \leq \frac{1}{\sqrt{n}}. \qquad \blacksquare$$

There is quite a gap between our upper and lower bounds when n is large. In fact, the lower bound has the correct order of growth. In order to obtain superior upper bounds, we need to replace Bernstein approximations $B_n f$ with a better method of polynomial approximation. We do this in Section 14.9 using Fourier series.

Exercises for Section 10.4

A. Show that $\omega(f; \delta_1) \leq \omega(f; \delta_2)$ if $\delta_1 \leq \delta_2$.

B. If f is C^1 on $[a, b]$, show that $\omega(f; \delta) \leq \|f'\|_\infty \delta$.

C. Show that a function f on $\mathbb{R}$ is uniformly continuous if and only if $\lim_{\delta \to 0^+} \omega(f; \delta) = 0$.

D. Show that f is Lipschitz with constant L if and only if f satisfies $\omega(f, \delta) \leq L\delta$.

E. If f is Lipschitz with constant L, prove that $\|B_n f - f\| \leq \dfrac{L}{\sqrt{n}}$.

F. For $f, g \in C[a, b]$ and $\alpha, \beta \in \mathbb{R}$,
(a) Show that $E_n(\alpha f + \beta g) \leq |\alpha| E_n(f) + |\beta| E_n(g)$.
(b) Show that $E_{m+n}(fg) \leq \|f\|_\infty E_n(g) + \|g\|_\infty E_m(f)$.

G. Show that if $\lim_{\delta \to 0^+} \dfrac{\omega(f; \delta)}{\delta} = 0$, then f is constant.

H. (a) In $C[0, 1]$, show that $E_n(\cos m\pi x) = 1$ for $n < m$.

(b) Use the Taylor series about $a = 1/2$ to show that $E_{10n}(\cos n\pi x) < 10^{-3n}$.

I. Let $f(x) = |2x - 1|$ on $[0, 1]$.

(a) Show that $B_n f(\frac{1}{2}) = 2^{-2n} \binom{2n}{n}$.

(b) Compute $\lim\limits_{n \to \infty} \sqrt{n} B_n f(\frac{1}{2})$.   HINT: Use Stirling's formula (6.5.2).

(c) Hence show that Proposition 10.4.5 is the right order of magnitude.

10.5. Existence of Best Approximations

Suppose that f is a continuous function on $[a, b]$. We may search for the optimal polynomial approximation of given degree. The analysis tools that we have developed will allow us to show that such an optimal approximation always exists. Moreover, in the next section, the best approximation will be shown to be unique.

A polynomial $p(x) = a_0 + a_1 x + \cdots + a_n x^n$ of degree at most n is determined by the $n + 1$ coefficients $a_0, \ldots, a_n$. Moreover, a nonzero polynomial of this form has at most n zeros and thus is not equal to the zero function on $[a, b]$. Hence $\mathbb{P}_n[a, b]$ is an $(n + 1)$-dimensional vector subspace of $C[a, b]$. It may be identified with $\mathbb{R}^{n+1}$ by associating p to the vector $(a_0, \ldots, a_n)$. The norm is quite different from the Euclidean norm. However, the results of Section 7.6 are precisely what we need to solve our problem.

First, Lemma 7.6.1 shows that there are constants $0 < c < C$ (depending on a, b and n) such that every polynomial $p(x) = a_0 + a_1 x + \cdots + a_n x^n$ in $\mathbb{P}_n[a, b]$ satisfies

$$c\left(\sum_{k=0}^{n} |a_k|^2\right)^{1/2} \leq \|p\|_\infty = \sup_{a \leq x \leq b} |p(x)| \leq C\left(\sum_{k=0}^{n} |a_k|^2\right)^{1/2}.$$

This lemma allows us to transfer our convergence results for $\mathbb{R}^{n+1}$ over to $\mathbb{P}_n$. It is important to note that these results depend on having a fixed bound on the degree of the polynomials. They are false for polynomials of unbounded degree.

10.5.1. COROLLARY. $\mathbb{P}_n[a, b]$ *has the same convergent sequences as* $\mathbb{R}^{n+1}$ *in the sense that sequence* $p_i = \sum\limits_{k=0}^{n} a_{ik} x^k$ *in* $\mathbb{P}_n[a, b]$ *converges uniformly on* $[a, b]$ *to a polynomial* $p = \sum\limits_{k=0}^{n} a_k x^k$ *if and only if* $\lim\limits_{i \to \infty} a_{ik} = a_k$ *for* $0 \leq k \leq n$.

PROOF. Let p_i correspond to the vector $\mathbf{a}_i = (a_{i0}, \ldots, a_{in})$ for $i \geq 1$, and let p correspond to the vector $\mathbf{a}$. Then if $\mathbf{a}_i$ converges $\mathbf{a}$, then

$$\lim_{i \to \infty} \|p - p_i\|_\infty \leq \lim_{i \to \infty} C\|\mathbf{a} - \mathbf{a}_i\|_2 = 0.$$

Hence p_i converges uniformly to p on $[a, b]$. Conversely, if p_i converges uniformly to p on $[a, b]$, then

$$\lim_{i \to \infty} \|\mathbf{a} - \mathbf{a}_i\|_2 \leq \lim_{i \to \infty} \frac{1}{c} \|p - p_i\|_\infty = 0.$$

So $\mathbf{a}_i$ converges to $\mathbf{a}$ in the Euclidean norm.

The second statement follows from Lemma 4.2.3, which shows that a sequence converges in $\mathbb{R}^{n+1}$ if and only if each coefficient converges. ∎

Second, Corollary 7.6.3 applies directly. Again this is false if the degree of the polynomials is not bounded.

10.5.2. COROLLARY. *A subset of $P_n[a, b]$ is compact if and only if it is closed and bounded.*

Finally, an immediate consequence of Theorem 7.6.5 is the result we are looking for.

10.5.3. THEOREM. *Let f be a continuous function on $[a, b]$. For each integer $n \geq 0$, there exists a closest polynomial of degree at most n to f in the max norm on $C[a, b]$.*

We consider an example that shows that a best approximation in certain more general circumstances may not exist; and when it does exist, it may not be unique.

10.5.4. EXAMPLE. Consider the subspace

$$S = \{h \in C[0, 1] : h(0) = 0\}$$

of $C[0, 1]$. Note that if f is any function in $C[0, 1]$, then $f - f(0)$ belongs to S. This shows that the linear span of S and the constant function 1 is all of $C[0, 1]$. We therefore say that S is a subspace of **codimension one**. In particular, it is infinite-dimensional since $C[0, 1]$ is infinite dimensional. So the type of arguments we used for $\mathbb{P}_n$ do not apply.

Consider the function $f = 1$. What are the best approximations to f in S? Clearly, for any $h \in S$,

$$\|f - h\|_\infty \geq |f(0) - h(0)| = 1.$$

On the other hand, $\|f - h\|_\infty = 1$ is equivalent to the inequalities

$$0 \leq h(x) \leq 2 \quad \text{for all} \quad 0 \leq x \leq 1.$$

There are many functions $h \in S$ within these constraints. For example, $h_0 = 0$, $h_1(x) = x/2$, $h_2(x) = 2x$, and $h_3(x) = \frac{\pi}{2} \sin^2(6\pi x)$. We see that there are (infinitely) many closest points.

Now consider the subspace

$$T = \{h \in S : \int_0^1 h(x) \, dx = 0\}.$$

This is a subspace because if g and h belong to T and α and β are in $\mathbb{R}$, then $(\alpha g + \beta h)(0) = 0$ and

$$\int_0^1 (\alpha g + \beta h)(x)\, dx = \alpha \int_0^1 g(x)\, dx + \beta \int_0^1 h(x)\, dx = 0.$$

The subspace T has codimension 2 because $C[0,1]$ is spanned by T, 1 and x. Moreover, it is closed. For if $h_n \in T$ converge uniformly to a function h, then $h(0)$ equals the limit of $h_n(0)$, which is 0, and by Theorem 8.3.1,

$$\int_0^1 h(x)\, dx = \lim_{n \to \infty} \int_0^1 h_n(x)\, dx = 0.$$

So h belongs to T.

Let $g(x) = x$ and consider the distance of g to T. Note that $g(0) = 0$ but

$$\int_0^1 g(x)\, dx = \frac{1}{2}.$$

Suppose that $h \in T$, and compute that

$$\frac{1}{2} = \int_0^1 g(x) - h(x)\, dx \leq \int_0^1 \|g - h\|_\infty\, dx = \|g - h\|_\infty.$$

If $\|g - h\|_\infty = \frac{1}{2}$, then this inequality must be an equality. This can only occur if

$$g(x) - h(x) = \|g - h\|_\infty = \frac{1}{2} \quad \text{for all} \quad 0 \leq x \leq 1.$$

This implies that $h(x) = x - \frac{1}{2}$. Note that h does not lie in T because $h(0) \neq 0$. So the distance $1/2$ is not attained.

However, we can easily come arbitrarily close to this distance. Indeed, we will show that for any integer n, there will be a continuous function h_n in T so that $\|g - h_n\|_\infty \leq \frac{1}{2} + \frac{1}{n}$. The idea is to make $h_n(x) = x - \frac{1}{2} - \frac{1}{n}$ on $[a_n, 1]$, $h_n(0) = 0$ and linear in between, with a_n chosen so that the integral zero. It is easy to check that the function with these properties does the job. A calculation shows that $a_n = \frac{4}{n+2}$. We find

$$h_n(x) = \begin{cases} -\dfrac{(n-2)^2}{8n}\, x & \text{for} \quad 0 \leq x \leq \dfrac{4}{n+2} \\[2mm] x - \dfrac{1}{2} - \dfrac{1}{n} & \text{for} \quad \dfrac{4}{n+2} \leq x \leq 1. \end{cases}$$

This shows that when infinite-dimensional subspaces are involved, there need not be a closest point.

Exercises for Section 10.5

A. Suppose that $f \in C[0,1]$ satisfies $f(0) = f(1) = 0$.
  (a) Show that f is a limit of polynomials such that $p(0) = p(1) = 0$.
  (b) Show that there is a closest polynomial of degree at most n with this property.

B. Let $f \in C^1[0, 1]$. Show that there is a closest polynomial of degree at most n to f in the $C^1[0, 1]$ norm, analogous to the $C^3[a, b]$ norm defined in Example 7.1.4.

C. Find *all* closest lines $p(x) = ax + b$ to $f(x) = x^2$ in the $C^1[0, 1]$ norm. Note that the best approximation is not unique.

D. Find the closest polynomial to $\sin x$ on $\mathbb{R}$.

E. (a) Show that for every bounded function on $[a, b]$, there is a closest polynomial $p \in \mathbb{P}_n$ in the max norm.
  (b) Show by example that a closest polynomial need not be unique.

F. Recall that a norm is strictly convex if $\|x\| = \|y\| = \|(x + y)/2\|$ implies that $x = y$.
  (a) Suppose that V has a strictly convex norm and M is a finite-dimensional subspace of V. Prove that each $v \in V$ has a unique closest point in M.
  (b) Prove that an inner product norm is strictly convex.
  (c) Show by example that $C[0, 1]$ is not strictly convex.

10.6. Characterizing Best Approximations

Perhaps in view of the previous examples, it is surprising that the best polynomial approximant of degree n to any continuous function f is uniquely determined. However, it is unique. We are able to show this because there is an interesting condition that characterizes this best approximation. This result was established by Borel in 1905, building on work of Chebychev.

10.6.1. EXAMPLE. Consider any continuous function f in $C[0, 1]$. What is the best approximation by a polynomial of degree 0 (i.e., a constant)? We want to make $\|f - c\|_\infty$ is as small as possible. By the Extreme Value Theorem, there are two points x_{min} and x_{max} in $[0, 1]$ so that

$$f(x_{min}) \leq f(x) \leq f(x_{max}) \quad \text{for all} \quad 0 \leq x \leq 1.$$

Clearly, $\|f - c\|_\infty$ is the maximum of $|f(x_{min}) - c|$ and $|f(x_{max}) - c|$. To make both as small as possible, we must take

$$c = \frac{x_{min} + x_{max}}{2}.$$

With this choice, the error $r(x) = f(x) - c$ satisfies

$$r(x_{max}) = \|r\|_\infty = \frac{x_{min} - x_{max}}{2}$$

and

$$r(x_{min}) = -\|r\|_\infty = -\frac{x_{min} - x_{max}}{2}.$$

10.6.2. EXAMPLE. Consider the continuous function $f(x) = x^2$ in $C[0, 1]$. What is the best linear approximation? One approach is to find the maximum modulus of $x^2 - ax - b$ and then minimize over choices of a and b. This is a calculus problem that is not too difficult. However, our approach will be to "guess" the answer and to verify it by geometric means.

First subtract x from f to get $g(x) = x^2 - x$. This function is symmetric about the line $x = \frac{1}{2}$. It takes its maximum value 0 at both 0 and 1, while its minimum is $-\frac{1}{4}$ at $x = \frac{1}{2}$. From the previous example, we know that to minimize $\|g(x) - b\|_\infty$ we should set the constant b equal to $-\frac{1}{8}$ so that the maximum and minima have the same absolute value, $\frac{1}{8}$. This intuitive approach yields a guess that the best linear approximation is $x - \frac{1}{8}$. The error is $r(x) = x^2 - x + \frac{1}{8}$. We know that

$$r(0) = \tfrac{1}{8}, \qquad r(-\tfrac{1}{2}) = -\tfrac{1}{8}, \quad \text{and} \quad r(1) = \tfrac{1}{8}.$$

Now we will show that $y = x - \frac{1}{8}$ is indeed the closest line to x^2 on $[0, 1]$. Equivalently, it suffices to show that $y = 0$ is the closest line to $y = r(x)$ on $[0, 1]$. Suppose that some linear function g satisfies $\|r - g\| < \frac{1}{8}$. Then

$$g(0) \in (0, \tfrac{1}{4}), \quad g(\tfrac{1}{2}) \in (-\tfrac{1}{4}, 0), \quad \text{and} \quad g(1) \in (0, \tfrac{1}{4}).$$

Therefore,

$$g(0) > 0 > g(\tfrac{1}{2}) < 0 < g(1).$$

By the Intermediate Value Theorem, g has a zero between 0 and $\frac{1}{2}$ and another zero between $\frac{1}{2}$ and 1. But g is linear, and thus it has at most one root. This contradiction shows that no better linear approximation exists.

Notice that the strategy we used in this example is essentially the same as that used in proving Proposition 10.4.4.

In the first example, the best approximation yields an error function r that achieves the values $\pm\|r\|_\infty$. In the case of our linear approximation, we found three points at which r alternately achieved the values $\pm\|r\|_\infty$. This notion generalizes to give a condition that is sufficient to be the best approximation.

10.6.3. DEFINITION. We say a function $g \in C[a, b]$ satisfies the **equioscillation condition** of degree n if there are $n + 2$ points $x_1 < x_2 < \cdots < x_{n+2}$ in $[a, b]$ so that either

$$g(x_i) = (-1)^i \|g\|_\infty \quad \text{for} \quad 1 \le i \le n + 2$$

or

$$g(x_i) = (-1)^{i+1} \|g\|_\infty \quad \text{for} \quad 1 \le i \le n + 2.$$

In other words, g attains its maximum absolute value at $n+2$ points and it alternates in sign between these points.

Figure 10.4 shows a function that satisfies the equioscillation condition.

10.6.4. THEOREM. *Suppose $f \in C[a, b]$ and $p \in \mathbb{P}_n$. If $r = f - p$ satisfies the equioscillation condition of degree n, then*

$$\|f - p\|_\infty = \inf\{\|f - q\|_\infty : q \in \mathbb{P}_n\}.$$

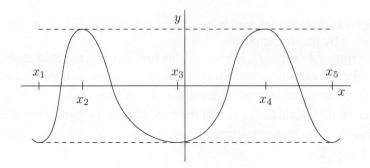

FIGURE 10.4. A function satisfying the equioscillation condition

PROOF. If the equality were not true, then there would be some nonzero $q \in \mathbb{P}_n$ so that $p + q$ is a better approximation to f; that is,

$$\|f - (p + q)\|_\infty < \|f - p\|_\infty$$

or, equivalently,

$$\|r - q\|_\infty < \|r\|_\infty.$$

In particular, if $x_1, \ldots, x_{n+2}$ are the points from the equioscillation condition for r, then

$$|r(x_i) - q(x_i)| < \|r\|_\infty = |r(x_i)| \quad \text{for} \quad 1 \le i \le n + 2.$$

It follows that $q(x_i) \ne 0$ and has the same sign as $r(x_i)$ for $1 \le i \le n + 2$.

Therefore, q changes sign between x_i and x_{i+1} for $1 \le i \le n + 1$. By the Intermediate Value Theorem, q has a root between x_i and x_{i+1} for $1 \le i \le n + 1$. Hence it has at least $n + 1$ zeros. But q is a polynomial of degree at most n, so the only way it can have $n + 1$ zeros is if q is the zero polynomial. This is false, and thus no better approximation exists. ∎

The important insight of Chebychev and Borel is that this condition is not only sufficient, it is also necessary. The argument is more subtle. When the function r fails to satisfy the equioscillation condition, a better approximation of degree n needs to be found.

10.6.5. THEOREM. *If $f \in C[a, b]$ and $p \in \mathbb{P}_n$ satisfy*

$$\|f - p\|_\infty = \inf\{\|f - q\|_\infty : q \in \mathbb{P}_n\},$$

then $f - p$ satisfies the equioscillation condition of degree n.

PROOF. Let $r = f - p \in C[a, b]$ and set $R = \|r\|_\infty$. By Theorem 5.5.9, r is uniformly continuous on $[a, b]$. Thus there is a $\delta > 0$ so that

$$|r(x) - r(y)| < \frac{R}{2} \quad \text{for all} \quad |x - y| < \delta, \ x, y \in [a, b].$$

Partition $[a, b]$ into disjoint intervals of length at most δ. Let $I_1, I_2, \ldots, I_l$ denote those intervals of the partition on which $|r(x)|$ attains the value R (including at the endpoints). Notice that since each I_j has length less than δ, if $r(x) = R$ for some $x \in I_j$, then for all $y \in I_j$ we have

$$r(y) \geq r(x) - |r(x) - r(y)| \geq R/2.$$

Similarly, if $r(x) = -R$ for some $x \in I_j$, then $r(y) \leq -R/2$ for all $y \in I_j$. Let ε_j be $+1$ or -1 according to whether $r(x)$ is positive or negative on I_j.

CLAIM: The sequence $(\varepsilon_1, \ldots, \varepsilon_l)$ has at least $n + 1$ changes of sign.

Accepting this claim for a moment, we produce the points required in the definition of the equioscillation condition. Group together adjacent intervals with the same sign, and label them $J_1, J_2, \ldots, J_k$, where $k \geq n + 2$. For each i between 1 and $n + 2$, pick a point $x_i \in J_i$ so that $|r(x_i)| = R$. By the choice of J_i, the signs alternate and so the equioscillation condition holds.

Thus, it remains only to prove the claim. Suppose the claim is false; that is, there are at most n changes of sign in $(\varepsilon_1, \ldots, \varepsilon_l)$. We will construct a better approximating polynomial, contradicting the choice of p.

Again, group together adjacent intervals with the same sign, and label them $J_1, J_2, \ldots, J_k$, where $k \leq n + 1$. Because f changes sign between J_i and J_{i+1}, it is possible to pick a point $a_i \in \mathbb{R}$ that lies in between. Define a polynomial q of degree $k - 1 \leq n$ by

$$q(x) = \prod_{i=1}^{k-1} x - a_i.$$

As $q \in \mathbb{P}_n$ and q changes sign at each a_i, either q or $-q$ agrees in sign with $r(x)$ on each set J_i, $i = 1, \ldots, k$. If necessary, replace q with $-q$ so that this agreement holds.

Let $L_0 = \overline{\bigcup_{j=1}^m I_j}$ and $L_1 = \overline{[a,b] \backslash L_0}$. Since L_0 is compact and q is never zero on L_0, the minimum,

$$m = \min\{|q(x)| : x \in L_0\}$$

is strictly positive by the Extreme Value Theorem. Let $M = \|q\|_\infty$.

Since L_1 is compact and each I_j has been removed from L_1, $|r(x)|$ does not attain the value R on L_1. Again using the Extreme Value Theorem, there is some $d > 0$ so that

$$\max\{|r(x)| : x \in L_1\} = R - d < R.$$

We will show that the polynomial

$$s(x) = p(x) - \frac{d}{2M} q(x)$$

is a better approximation to f than $p(x)$, contradicting the choice of p. Notice that

$$f(x) - s(x) = r(x) - \frac{d}{2M} q(x).$$

Because r and q have the same sign on each I_j,

$$\max_{x \in L_0} |f(x) - s(x)| \leq R - \frac{dm}{2M}.$$

On the remainder, we have

$$\max_{x \in L_1} |f(x) - s(x)| \le \max_{x \in L_1} |f(x) - p(x)| + \max_{x \in L_1} |p(x) - s(x)|$$

$$\le R - d + \frac{dm}{2M} \|q\| = R - \frac{d}{2}.$$

Now $[a, b] = L_0 \cup L_1$. Thus

$$\|f - s\| = \max_{a \le x \le b} |f(x) - s(x)| \le \max\left\{ R - \frac{dm}{2M}, R - \frac{d}{2} \right\} < R.$$

This contradicts the minimality of p and so proves the claim. ∎

Let us put these results together with one more idea to complete the main result.

10.6.6. CHEBYCHEV APPROXIMATION THEOREM.

For each continuous function f in $C[a, b]$, there is a unique polynomial p of degree at most n so that

$$\|f - p\|_\infty = \inf\{\|f - q\|_\infty : q \in \mathbb{P}_n[a, b]\}.$$

This best approximant is characterized by the fact that $f - p$ either is 0 or satisfies the equioscillation condition of degree n.

PROOF. By Theorem 10.5.3, there is at least one closest polynomial to f in $\mathbb{P}_n$. Suppose $p, q \in \mathbb{P}_n$ are both closest polynomials in $\mathbb{P}_n$, and let

$$R = \|f - p\|_\infty = \|f - q\|_\infty.$$

Then the midpoint $(p+q)/2$ is also a polynomial in $\mathbb{P}_n$ which is closest to f because of the triangle inequality:

$$R \le \left\| f - \frac{p + q}{2} \right\|_\infty \le \tfrac{1}{2}\|f - p\|_\infty + \tfrac{1}{2}\|f - q\|_\infty = R.$$

Thus by Theorem 10.6.5, $r = f - \frac{1}{2}(p + q)$ satisfies the equioscillation condition. Let $x_1 < x_2 < \cdots < x_{n+2}$ be the required points such that

$$|f(x_i) - r(x_i)| = R \quad \text{for} \quad 1 \le i \le n + 2.$$

Another use of the triangle inequality yields

$$R = \left| f(x_i) - \frac{p(x_i) + q(x_i)}{2} \right|$$

$$\le \tfrac{1}{2}|f(x_i) - p(x_i)| + \tfrac{1}{2}|f(x_i) - q(x_i)| \le R.$$

Consequently, $f(x_i) - p(x_i)$ and $f(x_i) - q(x_i)$ both have absolute value R. Since there is no cancellation when they are added, they must have the same sign. Therefore, $f(x_i) - p(x_i) = f(x_i) - q(x_i)$, and hence

$$p(x_i) = q(x_i) \quad \text{for} \quad 1 \le i \le n + 2.$$

Therefore, $p - q$ is a polynomial of degree at most n with $n + 2$ roots; and so is identically equal to zero. In other words, the closest point is unique. ∎

10.6.7. EXAMPLE. Chebychev's characterization sometimes allows exact calculation of the best polynomial. Consider the question: Find the closest cubic to $f(x) = \cos x$ on $[-\frac{\pi}{2}, \frac{\pi}{2}]$.

Since f is an even function, we expect that the best approximation will be even. This is indeed the case. Let $p \in \mathbb{P}_3$ be the closest cubic, and let $\tilde{p}(x) = p(-x)$. Then

$$\|f - \tilde{p}\|_\infty = \max_{-1 \le x \le 1} |f(x) - p(-x)|$$

$$= \max_{-1 \le x \le 1} |f(-x) - p(-x)| = \|f - p\|_\infty.$$

As the closest polynomial is unique, it follows that $p = \tilde{p}$, namely $p(-x) = p(x)$. So we are looking for $p(x) = ax^2 + b$.

From Chebychev's Theorem, we are looking for a polynomial that differs from f by $\pm d$ at 5 points with alternating signs, where $d = \|f - p\|_\infty$. Consider the derivatives of $r(x) = \cos x - ax^2 - b$.

$$r'(x) = -\sin x - 2ax \qquad \text{and} \qquad r''(x) = -\cos x - 2a$$

Since $-\cos x$ is concave on $[-\frac{\pi}{2}, \frac{\pi}{2}]$, the second derivative r'' has at most two zeros. This happens only if $-\frac{1}{2} < a < 0$; and these zeros are at $\pm z = \pm \cos^{-1}(2a)$. So r' is decreasing on $[-1, -z]$, then increasing on $[-z, z]$, and then decreasing again on $[z, 1]$. Thus r' can have at most three zeros. But r achieves its extreme values five times. So r' has exactly three zeros corresponding to extrema of r; and the other two extrema must be at the endpoints. One extremum is at 0, and the other two critical points will be called $\pm x_0$. Moreover, $r'(0) = 0$ and $r''(0) = -1 - 2a < 0$; so this is a maximum.

So far, we have

$$d = r(0) \quad = 1 - b$$

$$d = r(\pm \frac{\pi}{2}) = -a\left(\frac{\pi}{2}\right)^2 - b$$

$$-d = r(\pm x_0) = \cos x_0 - ax_0^2 - b$$

$$0 = r'(x_0) \quad = -\sin x_0 - 2ax_0.$$

Solving the first two equations for a yields

$$a = -\frac{4}{\pi^2}.$$

Plugging this into the fourth yields

$$\sin x_0 = \frac{8}{\pi^2} x_0.$$

Since $\sin x$ is concave on $[0, \frac{\pi}{2}]$, this equation has a unique positive solution. It may be found numerically to be approximately $x_0 := 1.0988243$. From the third equation (and the first), we obtain

$$b = \frac{1}{2}\left(1 + \cos x_0 + \frac{4x_0^2}{\pi^2}\right) := 0.9719952.$$

So the closest cubic to $\cos x$ on $[-\frac{\pi}{2}, \frac{\pi}{2}]$ is $p(x) = -\frac{4}{\pi^2}x^2 + 0.9719952$ and the error is $d = 0.0280048$.

Exercises for Section 10.6

A. Find the closest line to e^x on $[0, 1]$.

B. Find the cubic polynomial which best approximates $|x|$ on the interval $[-1, 1]$. HINT: Use symmetry first.

C. Suppose that $f \in C[a, b]$ is twice continuously differentiable function with $f''(x) > 0$ on $[a, b]$. Show that the best linear approximation to f has slope $\dfrac{f(b) - f(a)}{b - a}$.

D. Apply the previous exercise to find the closest line to $f(x) = \sqrt{1 + 3x^2}$ on $[0, 1]$, and compute the error.

E. If f in $C[-1, 1]$ is an even (odd) function, show that the best approximation of degree n is also even (odd).

F. Let p be the best polynomial approximation of degree n to $\sqrt{x}$ on $[0, 1]$. Show that $q(x) = p(x^2)$ is the best polynomial approximation of degree $2n + 1$ to $|x|$ on $[-1, 1]$. HINT: How does the equioscillation condition on $\sqrt{x} - p(x)$ translate to the approximation of $|x|$?

10.7. Expansions Using Chebychev Polynomials

Ideally, we would like to find the polynomial that is exactly the best approximation to a given continuous function f. There is an algorithm that constructs a sequence of polynomials converging uniformly to the best approximating polynomial of degree n known as **Remes's algorithm**. Roughly, it works as follows: Pick $n + 2$ points $x_1 < x_2 < \ldots x_{n+2}$ in $[a, b]$. These points might be equally spaced, but foreknowledge of the function could lead you to pick points clustered in regions where f behaves more wildly. Then solve the linear equations for $a_0, a_1, \ldots, a_n$ and d:

$$
\begin{array}{llllll}
a_0 + & x_1 a_1 + & x_1^2 a_2 + \ldots + & x_1^n a_n - & d & = f(x_1) \\
a_0 + & x_2 a_2 + & x_2^2 a_2 + \ldots + & x_2^n a_n + & d & = f(x_2) \\
& & \vdots & & & \\
a_0 + & x_{n+2} a_1 + & x_{n+2}^2 a_2 + \ldots + & x_{n+2}^n a_n + & (-1)^{n+2} d & = f(x_{n+2}).
\end{array}
$$

This method attempts to find a polynomial that satisfies Chebychev's Theorem. However, the function may well take its extrema on other points. So the algorithm proceeds to choose new points $x_1', \ldots, x_{n+2}'$ by selecting points where the error is largest (or close to it) near each point. Eventually, this procedure converges to the nearest polynomial of degree n.

However, each step of this process involves many calculations, so convergence to the optimal polynomial is slow. In practice, it is better to find quickly an approximating polynomial that is not quite the best. It is often easier to find a polynomial of say double the degree with half the accuracy, rather than seeking the optimal solution. Solutions that are less than optimal, but still quite good, can be found very efficiently. We develop such an algorithm in this section using Chebychev polynomials. Chebychev polynomials are also useful in numerical analysis, algebra, and other areas.

10.7.1. DEFINITION. For $n \geq 0$, define the **Chebychev polynomial** of degree n in $\mathbb{P}_n[-1, 1]$ by

$$T_n(x) = \cos(n \cos^{-1} x).$$

It is not immediately obvious that T_n is a polynomial, much less a polynomial of degree n. Figure 10.5 gives the graphs of T_1 through T_5 on $[-1, 1]$. The T in T_n comes from Tchebycheff, an earlier transliteration of Chebychev from the original Russian.

10.7.2. LEMMA. $T_0(x) = 1$, $T_1(x) = x$ and

$$T_n(x) = 2xT_{n-1}(x) - T_{n-2}(x) \quad \text{for} \quad n \geq 2.$$

For each $n \geq 1$, $T_n(x)$ is a polynomial of degree n with leading coefficient 2^{n-1}. Also, $\|T_n\|_\infty = 1$, and

$$T_n\left(\cos\left(\tfrac{k}{n}\pi\right)\right) = (-1)^k \quad \text{for} \quad 0 \leq k \leq n.$$

PROOF. Recall the sum and difference of angles formulas for cosine:

$$\cos(A + B) = \cos A \cos B - \sin A \sin B$$

and

$$\cos(A - B) = \cos A \cos B + \sin A \sin B.$$

Let $A = n\theta$ and $B = \theta$, and add these formulas to get

$$\cos(n + 1)\theta + \cos(n - 1)\theta = 2 \cos n\theta \cos \theta.$$

Substituting $\theta = \cos^{-1} x$ gives

$$T_{n+1}(x) + T_{n-1}(x) = 2xT_n(x) \quad \text{for all} \quad n \geq 1.$$

The first few terms are

$$T_2(x) = 2xT_1(x) - T_0(x) = 2x^2 - 1,$$
$$T_3(x) = 2xT_2(x) - T_1(x) = 4x^3 - x,$$
$$T_4(x) = 2xT_3(x) - T_2(x) = 8x^4 - 8x^2 + 1,$$
$$T_5(x) = 2xT_4(x) - T_3(x) = 16x^5 - 20x^3 + 5x.$$

By induction, it follows that that $T_n(x)$ is a polynomial of degree n with leading coefficient 2^{n-1}.

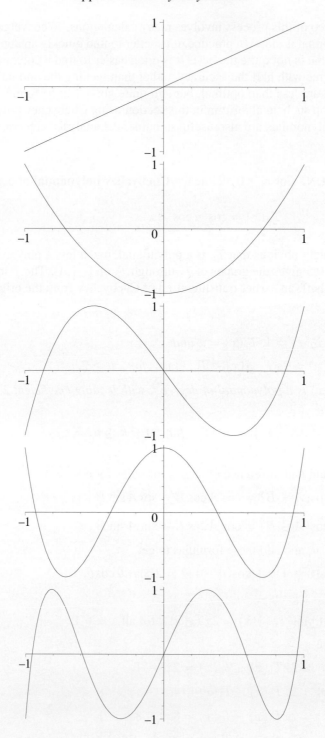

FIGURE 10.5. The Chebychev polynomials T_1 through T_5.

Since $|\cos\theta| \leq 1$ for all values of θ, it follows that $\|T_n\|_\infty \leq 1$. Now $|\cos(\theta)| = 1$ only when θ is an integer multiple of π. It follows that T_n attains its maximum modulus when $n\cos^{-1}x = k\pi$ for some integer k. Solving, we obtain

$$x_k = \cos\left(\tfrac{k}{n}\pi\right) \quad \text{for} \quad 0 \leq k \leq n.$$

Other choices of k just repeat these values. Finally,

$$T_n(x_k) = (-1)^k \quad \text{for} \quad 0 \leq k \leq n. \qquad \blacksquare$$

10.7.3. COROLLARY. *The unique polynomial of degree at most $n-1$ that best approximates x^n on $[-1, 1]$ is $p_n(x) = x^n - 2^{1-n}T_n(x)$, and $E_{n-1}(x^n) = 2^{1-n}$.*

PROOF. Since the leading term of T_n is $2^{n-1}x^n$, $p_n(x)$ is a polynomial of degree at most $n-1$. The difference

$$x^n - p_n(x) = 2^{1-n}T_n(x)$$

has maximum modulus 2^{1-n}, and it attains this maximum modulus at the $n+1$ points $x_k = \cos(k\pi/n)$ for $0 \leq k \leq n$ with alternating sign. Hence it satisfies the equioscillation condition of degree $n-1$. By Chebychev's Theorem (Theorem 10.6.4), this is the unique closest polynomial of degree $n-1$. $\qquad \blacksquare$

Without developing any further results, we can already use Chebychev polynomials to find good approximations.

10.7.4. EXAMPLE. We will approximate $f(x) = \sin(x)$ on the interval $[-1, 1]$ by modifying the Taylor approximations. The Taylor polynomial of degree 10 is

$$p(x) = x - \frac{1}{3!}x^3 + \frac{1}{5!}x^5 - \frac{1}{7!}x^7 + \frac{1}{9!}x^9.$$

For $x \in [-1, 1]$, the error term is given by Taylor's Theorem,

$$|\sin(x) - p(x)| \leq \frac{|x|^{11}}{11!}\|f^{(11)}\|_\infty \leq \frac{1}{11!} < 2.506 \times 10^{-8}.$$

The idea is to replace the term $x^9/9!$ with the best approximation of degree less than 9, which we have seen is $(x^9 - T_9(x)/2^8)/9!$. This increases the error by at most

$$\left\|\frac{x^9}{9!} - \frac{x^9 - T_9(x)/2^8}{9!}\right\|_\infty = \frac{1}{2^89!}\|T_9(x)\|_\infty = \frac{1}{2^89!} \leq 1.077 \times 10^{-8}.$$

Using the three-term recurrence relation (or looking it up in a computer algebra package), we find that

$$T_9(x) = 2^8x^9 - 576x^7 + 432x^5 - 120x^3 + 9x.$$

Thus

$$p(x) = x - \frac{1}{3!}x^3 - \frac{1}{5!}x^5 + \frac{1}{7!}x^7 + \frac{x^9 - T_9(x)/2^8}{9!}$$

has degree 7. This polynomial approximates $\sin(x)$ on $[-1, 1]$ with error at most 3.6×10^{-8}.

For comparison, the Taylor polynomial of degree 7 gives an error of about 2.73×10^{-6}. Thus, $p(x)$ is 75 times as accurate as the Taylor polynomial of the same degree.

In practice, we want to do away with ad hoc methods and find an algorithm that yields reasonable good approximations quickly. We need the following inner product for $f, g \in C[-1, 1]$:

$$\langle f, g \rangle_T = \frac{1}{\pi} \int_{-1}^{1} f(x)g(x) \frac{dx}{\sqrt{1-x^2}}.$$

It is easy to verify that this is an inner product on $C[-1, 1]$ (i.e., it is linear in both variables, positive definite and symmetric). The crucial property we need is that the Chebychev polynomials are orthogonal with respect to this inner product. The constant $1/\pi$ makes the constant function 1 have norm 1, which is computationally convenient.

10.7.5. LEMMA.

$$\langle T_n, T_m \rangle_T = \begin{cases} 0 & \text{if } m \neq n \\ \frac{1}{2} & \text{if } m = n \neq 0 \\ 1 & \text{if } m = n = 0 \end{cases}$$

PROOF. Make the substitution $\cos \theta = x$ in the integral, so that $-\sin \theta \, d\theta = dx$, whence $d\theta = -dx/\sqrt{1-x^2}$. We have

$$\langle T_n, T_m \rangle_T = \frac{1}{\pi} \int_{-1}^{1} T_n(x) T_m(x) \frac{dx}{\sqrt{1-x^2}}$$

$$= \frac{1}{\pi} \int_{0}^{\pi} \cos n\theta \cos m\theta \, d\theta$$

$$= \frac{1}{2\pi} \int_{0}^{\pi} \cos(m+n)\theta + \cos(m-n)\theta \, d\theta,$$

where again we have used the identity $2 \cos A \cos B = \cos(A+B) + \cos(A-B)$.

There are three different integrals, depending on the values of m and n. If $m \neq n$, then both $m + n$ and $m - n$ are not zero and the integral is

$$\frac{1}{2\pi} \left(\frac{\sin(m+n)\theta}{m+n} + \frac{\sin(m-n)\theta}{m-n} \right) \Big|_{0}^{\pi} = 0.$$

If $m = n \neq 0$, then $m - n$ is zero and $m + n$ is not, so the integral is

$$\frac{1}{2\pi} \left(\frac{\sin(m+n)\theta}{m+n} + x \right) \Big|_{0}^{\pi} = \frac{1}{2}.$$

Finally, if $m = n = 0$, then the integral is 1. ∎

Now suppose for a moment that a function $f \in C[-1, 1]$ can be expressed as an infinite sum of Chebychev polynomials. That is, there is a sequence of constants $(a_n)_{n=1}^{\infty}$ so that

$$f(x) = \sum_{n=1}^{\infty} a_n T_n(x)$$

for all $x \in [-1, 1]$. Then, willfully ignoring the issue of convergence, we can write

$$\langle f, T_k \rangle_T = \left\langle \sum_{n=1}^{\infty} a_n T_n, T_k \right\rangle_T = \sum_{n=1}^{\infty} \langle a_n T_n, T_k \rangle_T = a_k \langle T_k, T_k \rangle_T.$$

Solving for a_k in the preceding equation and using the definition of the inner product, we have a possible formula for the coefficients:

$$a_k = 2 \langle f, T_k \rangle_T = \frac{2}{\pi} \int_{-1}^{1} f(x) T_k(x) \frac{dx}{\sqrt{1 - x^2}} \quad \text{for} \quad k \geq 1$$

and

$$a_0 = \langle f, T_0 \rangle_T = \frac{1}{\pi} \int_{-1}^{1} f(x) \frac{dx}{\sqrt{1 - x^2}}.$$

10.7.6. DEFINITION. We define the **Chebychev series** for f in $C[-1, 1]$ to be $\sum_{n=1}^{\infty} a_n T_n(x)$, where the sequence (a_n) is given by the preceding formulas.

There are a host of questions about this series. For which x does this infinite series converge? Is the resulting function continuous? Does it equal f or not?

Under some conditions, the expansion does converge. The underlying reason is a connection between Chebychev series and Fourier series, which we explore at the end of Chapter 14. Rather than be too repetitive, we limit ourselves here to one of the easier results. See Theorem 14.8.2 for a much better result that requires a more subtle analysis.

10.7.7. THEOREM. *If $f \in C[-1, 1]$ has a continuous second derivative, then the Chebychev series of f converges uniformly to f.*

PROOF. We wish to obtain a good estimate on the size of the coefficients a_k for $k \geq 1$. Again make the change of variables $x = \cos \theta$; then

$$a_n = \frac{2}{\pi} \int_{-1}^{1} f(x) T_n(x) \frac{dx}{\sqrt{1 - x^2}}$$

$$= \frac{2}{\pi} \int_{0}^{\pi} g(\theta) \cos n\theta \, d\theta,$$

where $g(\theta) = f(\cos\theta)$. Using integration by parts,

$$= \frac{2}{\pi}\left(g(\theta)\frac{\sin n\theta}{n}\bigg|_0^\pi - \int_0^\pi g'(\theta)\frac{\sin n\theta}{n}\,d\theta\right)$$

$$= \frac{2}{\pi}\left(-g'(\theta)\frac{\cos n\theta}{n^2}\bigg|_0^\pi - \int_0^\pi g''(\theta)\frac{\cos n\theta}{n^2}\,d\theta\right).$$

Now $g'(\theta) = -f'(\cos\theta)\sin\theta$ vanishes at 0 and π, so the integral simplifies to

$$= \frac{-2}{\pi n^2}\int_0^\pi g''(\theta)\cos n\theta\,d\theta.$$

Since f has a continuous (and thus bounded) second derivative, it follows that g does also. If $M = \sup\{|g''(\theta)| : \theta \in [0,\pi]\}$, then $a_n \le 2M/n^2$. So, by the Comparison Test, $\sum\limits_{n=1}^{\infty} a_n$ converges.

Since $\|a_nT_n\|_\infty = a_n$, the Weierstrass M-test (8.4.7) shows that $\sum\limits_{n=1}^{\infty} a_nT_n$ converges uniformly to some continuous function $F \in C[-1,1]$. It remains only to show that $F = f$.

Observe that for all $n \ge 1$,

$$\langle f - F, 2T_n\rangle_T = a_n - \lim_{k\to\infty}\left\langle\sum_{j=0}^k a_jT_j(x), 2T_n(x)\right\rangle_T$$

$$= a_n - a_n = 0.$$

Likewise, $\langle f - F, T_0\rangle_T = 0$. Using the integral definition of the inner product, we have

$$\int_{-1}^1 \frac{f(x) - F(x)}{\sqrt{1-x^2}}T_n(x)\,dx = 0 \quad\text{for all}\quad n \ge 0.$$

Since T_n is a polynomial of degree n, a routine induction argument shows that

$$\int_{-1}^1 \frac{f(x) - F(x)}{\sqrt{1-x^2}}x^n\,dx = 0 \quad\text{for all}\quad n \ge 0,$$

and therefore

$$\int_{-1}^1 (f(x) - F(x))\sqrt{1-x^2}x^n\,dx = \int_{-1}^1 \frac{f(x) - F(x)}{\sqrt{1-x^2}}(x^n - x^{n+2})\,dx = 0.$$

By Exercise 10.2.D, it follows that $(f(x) - F(x))\sqrt{1-x^2} = 0$ for all $x \in [-1,1]$. Since f and F are continuous, we deduce that $f(x) = F(x)$ on $[-1,1]$. $\blacksquare$

Exercises for Section 10.7

A. Verify the following properties of the Chebychev polynomials $T_n(x)$.
    (a) If m is even, then T_m is an even function [i.e., $T_m(-x) = T_m(x)$], and if m is odd, then T_m is odd [i.e., $T_m(-x) = -T_m(x)$].
    (b) Show that every polynomial p of degree n has a unique representation using Chebychev polynomials [i.e., there are constants $a_0, a_1, \ldots, a_n$, so that $p(x) = a_0 T_0(x) + a_1 T_1(x) + \cdots + a_n T_n(x)$].
    (c) $T_m(T_n(x)) = T_{mn}(x)$
    (d) $(1 - x^2)T_n''(x) - xT_n'(x) + n^2 T_n(x) = 0$

B. Show by induction that $T_n(x) = \dfrac{\left(x + \sqrt{x^2 - 1}\right)^n + \left(x - \sqrt{x^2 - 1}\right)^n}{2}$.

C. Find a sequence of polynomials that converges uniformly to $f(x) = |x|^3$ on $[-1, 1]$.
    HINT: f is C^2.

D. Suppose that $f \in C[-1, 1]$ has a Chebychev series $\sum\limits_{n=0}^{\infty} a_n T_n$. If $\sum\limits_{n=0}^{\infty} |a_n| < \infty$, show that the Chebychev series converges uniformly to f.
    HINT: Study the proof of Theorem 10.7.7.

E. Verify the following expansions in Chebychev polynomials:
    (a) $|x| = \dfrac{2}{\pi} - \dfrac{4}{\pi} \sum\limits_{j=1}^{\infty} \dfrac{(-1)^j}{4j^2 - 1} T_{2j}(x)$

    (b) $\sqrt{1 - x^2} = \dfrac{2}{\pi} - \dfrac{4}{\pi} \sum\limits_{j=1}^{\infty} \dfrac{1}{4j^2 - 1} T_{2j}(x)$

    HINT: Substitute $x = \cos\theta$ in computing the coefficients. Apply the previous exercise.

F. Suppose that $f \in C[-1, 1]$ has a Chebychev series $\sum\limits_{n=0}^{\infty} a_n T_n$.

    (a) Show that $E_n(f) \le \sum\limits_{k=n+1}^{\infty} |a_k|$.

    (b) Show that $E_n(T_{n+1}) = 1$.   HINT: Theorem 10.6.4

    (c) Show that $\left| E_n(f) - |a_{n+1}| \right| \le \sum\limits_{k=n+2}^{\infty} |a_k|$.

        HINT: Show that $E_n(f) \ge E_n(|a_{n+1}|T_{n+1}) - \sum\limits_{k=n+2}^{\infty} |a_k|$.

    (d) Show that if $\lim\limits_{n \to \infty} \dfrac{\sum_{k=n+1}^{\infty} |a_k|}{|a_n|} = 0$, then $\lim\limits_{n \to \infty} \dfrac{E_n(f)}{|a_{n+1}|} = 1$.

G. Let a_n be a sequence of real numbers monotone decreasing to 0. Define the sequence of polynomials $p_n(x) = \sum\limits_{k=1}^{n} (a_k - a_{k+1})T_{3^k}(x)$.
    (a) Show that this sequence converges uniformly on $[-1, 1]$ to a continuous function $f(x)$.   HINT: Weierstrass M-test
    (b) Evaluate $(f - p_n)\left(\cos(3^{-n-1}k\pi)\right)$ for $0 \le k \le 3^{n+1}$.
    (c) Show that $E_{3^n}(f) = a_{n+1}$. Conclude that there are continuous functions for which the optimal sequence of polynomials converges exceedingly slowly.

10.8. Splines

Splines are smooth piecewise polynomials. They are well adapted to use on computers, and so they are much used in practical work. Because they are closely related to polynomials, we give a brief treatment of splines here, concentrating on issues related to real analysis. For algorithmic and implementation issues, we refer the reader to [18].

To motivate the idea behind splines, observe that approximation by polynomials can be improved either by increasing the degree of the polynomial or by decreasing the size of the interval on which the approximation is used. Splines take the latter approach, successively chopping the interval into small pieces and approximating the function on each piece by a polynomial of fixed small degree, such as a cubic.

We search for a relatively smooth function that is piecewise a polynomial of low degree but is not globally a polynomial at all. This turns out to be worth the additional theoretical complications. Why? First, evaluating the approximation will be easier on each subinterval because it is a polynomial of small degree. Instead of having to do some multiplications, we have several comparisons to decide which interval we are in. Comparison is much simpler than multiplication, so evaluation can be much faster, even if we have to do many comparisons.

Second, since the degree is small, we can use simple methods like interpolation to find the polynomial on each subinterval. Interpolation is both easy to implement on the computer and (mostly) easy to understand.

Third, local irregularities of the function only affect the approximation locally, in contrast to polynomial approximation. This means that if a function f is not differentiable at one point of the interval, then this affects the polynomial approximation over the whole interval. Working with polynomials on subintervals, the lack of differentiability at one point affects the polynomial approximation on that subinterval only.

The discussion so far has pretended that we are free to choose completely different polynomials on each subinterval. In fact, we would like the polynomials to fit together smoothly. To start, we begin with the revealing special case of approximation by a piecewise linear continuous function.

Choose a partition Δ of the interval $[a, b]$ into k subintervals with endpoints $a = x_0 < x_1 < \cdots < x_k = b$. We define $\mathbb{S}_1(\Delta)$ to be the subspace of $C[a, b]$ given by

$$\mathbb{S}_1(\Delta) = \{g \in C[a, b] : g|_{[x_i, x_{i+1}]} \text{ is linear for } 0 \le i < k\}.$$

Clearly, a function $g \in \mathbb{S}_1(\Delta)$ is uniquely determined by its values $g(x_i)$ at the **nodes** x_i for $0 \le i \le k$. Indeed, we just construct the line segments between the points $(x_i, g(x_i))$. Thus $\mathbb{S}_1(\Delta)$ is a finite-dimensional subspace of dimension $k + 1$. The elements of this space are called **linear splines**. Figure 10.6 shows an element of $\mathbb{S}_1(\Delta)$ approximating a given continuous function.

Now take a continuous function f in $C[a, b]$. By Theorem 7.6.5, there is some g in $\mathbb{S}_1(\Delta)$ so that

$$\|f - g\|_\infty = \inf\{\|f - h\|_\infty : h \in \mathbb{S}_1(\Delta)\}.$$

Instead of trying to find this optimal choice, we choose the function h in $\mathbb{S}_1(\Delta)$ such that $h(x_i) = f(x_i)$ for $0 \le i \le k$. Define $J_1 : C[a,b] \to \mathbb{S}_1(\Delta)$ by letting $J_1 f$ be this function h. Notice that our characterization of functions in $\mathbb{S}_1(\Delta)$ shows that $J_1 g = g$ for $g \in \mathbb{S}_1(\Delta)$. Also, J_1 is linear:

$$J_1(af + bg) = aJ_1 f + bJ_1 g \quad \text{for} \quad f, g \in C[a,b] \text{ and } a, b \in \mathbb{R}.$$

The following lemma shows that choosing $J_1 f$ instead of the best approximant does not increase the error too much.

10.8.1. LEMMA. *If $f \in C[a,b]$, then*

$$\|f - J_1 f\|_\infty \le 2 \inf\{\|f - g\|_\infty : g \in \mathbb{S}_1(\Delta)\}.$$

PROOF. Notice that for any $f \in C[a,b]$,

$$\|J_1 f\|_\infty = \max\{|f(x_i)| : 0 \le i \le k\} \le \|f\|_\infty.$$

If $g \in \mathbb{S}_1(\Delta)$ is the closest point to f, we use linearity and $J_1 g = g$ to obtain

$$\|f - J_1 f\|_\infty = \|f - g + J_1(f - g)\|_\infty$$
$$\le \|f - g\|_\infty + \|J_1(f - g)\|_\infty \le 2\|f - g\|_\infty. \qquad \blacksquare$$

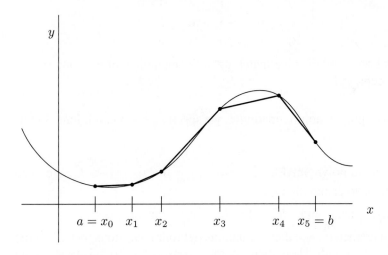

FIGURE 10.6. Approximation in $\mathbb{S}_1(\Delta)$.

10.8.2. DEFINITION. A **cubic spline** for a partition Δ of $[a,b]$ is a C^2 function h such that $h|_{[x_i, x_{i+1}]}$ a polynomial of degree at most 3 for $0 \le i < k$. Let $\mathbb{S}(\Delta)$ denote the vector space of all cubic splines for the partition Δ.

Cubic spline interpolation is popular in practice. It may seem rather surprising that it is possible to fit cubics together and remain twice continuously differentiable and still have the flexibility to approximate functions well.

10.8.3. EXAMPLE. Consider

$$h(x) = \begin{cases} 2x^3 + 12x^2 + 24x + 16 & \text{if } -2 \le x \le -1 \\ -7x^3 - 15x^2 - 3x + 7 & \text{if } -1 \le x \le 0 \\ 9x^3 - 15x^2 - 3x + 7 & \text{if } 0 \le x \le 1 \\ -5x^3 + 27x^2 - 45x + 21 & \text{if } 1 \le x \le 2 \\ x^3 - 9x^2 + 27x - 27 & \text{if } 2 \le x \le 3. \end{cases}$$

We readily compute the following.

$h'(x)$	$h''(x)$	interval
$6x^2 + 24x + 24$	$12x + 12$	$-2 \le x \le -1$
$-21x^2 - 30x - 3$	$-42x - 15$	$-1 \le x \le 0$
$27x^2 - 30x - 3$	$54x - 15$	$0 \le x \le 1$
$-15x^2 + 54x - 45$	$-30x + 27$	$1 \le x \le 2$
$3x^2 - 18x + 27$	$6x - 9$	$2 \le x \le 3$

We can now verify the following table of values.

x_i	-2	-1	0	1	2	3
$h(x_i)$	0	2	7	-2	-1	0
$h'(x_i)$	0	6	-3	-6	3	0
$h''(x_i)$	0	12	-30	24	-6	0

Since the first and second derivative match up at the endpoints of each interval, h is C^2 and so is a cubic spline.

To find a cubic spline h approximating f, we specify certain conditions. Let us demand first that

$$h(x_i) = f(x_i) \quad \text{for} \quad 0 \le i \le k.$$

Let's write each cubic polynomial as $h_i = h|_{[x_i, x_{i+1}]}$ for $0 \le i < k$. We need additional conditions to ensure that h is C^2:

$$h_i'(x_i) = h_{i+1}'(x_i), \quad \text{and} \quad h_i''(x_i) = h_{i+1}''(x_i) \quad \text{for} \quad 1 \le i \le k - 1.$$

A cubic has four parameters, and these equations put four conditions on each cubic except for the two on the ends where there are three constraints. To finish specifying the spline, we add two endpoint conditions

$$h_1'(x_0) = f'(x_0) \quad \text{and} \quad h_k'(x_k) = f'(x_k),$$

assuming that these derivatives exist. (If they do not, we may set them equal to 0.) For convenience, we shall assume that f is C^2, which ensures that these data are defined and allows some interesting theoretical consequences.

We shall see that a cubic spline h in $\mathbb{S}(\Delta)$ is uniquely determined by these equations. There are $k + 3$ data conditions determined by f, namely $f(x_0), \ldots, f(x_k)$ and $f'(x_0)$ and $f'(x_k)$. Hence we expect to find that $\mathbb{S}(\Delta)$ is a finite-dimensional subspace of dimension $k + 3$. This will allow us to define a map J from $C^2[a, b]$ to $\mathbb{S}(\Delta)$ by setting Jf to be the function h specified previously.

10.8.4. LEMMA. *Given $c < d$ and real numbers a_1, a_2, s_1, s_2, there is a unique cubic polynomial p satisfying*

$$p(c) = a_1 \qquad\qquad p(d) = a_2$$
$$p'(c) = s_1 \qquad\qquad p'(d) = s_2.$$

Setting $\Delta = d - c$, we obtain

$$p''(c) = \frac{6(a_2 - a_1)}{\Delta^2} - \frac{4s_1 + 2s_2}{\Delta}$$

and

$$p''(d) = -\frac{6(a_2 - a_1)}{\Delta^2} + \frac{2s_1 + 4s_2}{\Delta}.$$

PROOF. Consider the cubics

$$p_1(x) = \frac{(x - d)^2}{(d - c)^2}\left(1 + 2\frac{x - c}{d - c}\right)$$

$$p_2(x) = \frac{(x - c)^2}{(d - c)^2}\left(1 - 2\frac{x - d}{d - c}\right)$$

$$q_1(x) = \frac{(x - c)(x - d)^2}{(d - c)^2}$$

$$q_2(x) = \frac{(x - c)^2(x - d)}{(d - c)^2}.$$

For example, $p_1(c) = 1$ and $p_1'(c) = p_1'(d) = p_1(d) = 0$. The reader can verify that $p(x) = a_1 p_1(x) + a_2 p_2(x) + s_1 q_1(x) + s_2 q_2(x)$ is the desired cubic.

For uniqueness, we can note that the difference of two such cubics is a cubic q such that $q(c) = q(d) = q'(c) = q'(d) = 0$. The first two conditions show that c and d are roots of q; and the second two conditions then imply that they are double roots. So $(x - c)^2(x - d)^2$ divides q. As q has degree at most 3, this forces $q = 0$.

Finding the value of p'' at c and d is a routine calculation. ∎

10.8.5. THEOREM. *Given a partition $\Delta : a = x_0 < x_1 < \cdots < x_k = b$ of the interval $[a, b]$ and real numbers $a_0, \ldots, a_k$, s_0 and s_k, there is a unique cubic spline $h \in \mathbb{S}(\Delta)$ such that $h(x_i) = a_i$ for $0 \le i \le k$ and $h'(a) = s_0$ and $h'(b) = s_k$.*

PROOF. If such a spline exists, we could define $s_i = h'(x_i)$ for $1 \le i \le k - 1$. We search for such values of s_i that allow a spline. Given the values a_i of h and s_i of h' at the points x_{i-1} and x_i, the previous lemma determines a unique cubic h_i on the interval $[x_{i-1}, x_i]$. So for each choice of $(s_1, \ldots, s_{k-1})$, there is one piecewise cubic function on $[a, b]$ that interpolates the values a_i and derivatives s_i at each point x_i for $0 \le i \le k$. However, in general this will not be C^2. There are $k - 1$ conditions that must be satisfied:

$$h_i''(x_i) = h_{i+1}''(x_i) \quad \text{for} \quad 1 \le i \le k - 1.$$

Our job is to compute a formula for these second derivatives to obtain conditions on the hypothetical data $s_1, \ldots, s_{k-1}$.

Let us write $\Delta_i = x_i - x_{i-1}$ for $1 \leq i \leq k$. By the previous lemma, the second derivative conditions at x_i for $1 \leq i \leq k - 1$ are

$$(10.8.6) \qquad h''(x_i) = -\frac{6(a_i - a_{i-1})}{\Delta_i^2} + \frac{2s_{i-1} + 4s_i}{\Delta_i}$$

$$= \frac{6(a_{i+1} - a_i)}{\Delta_{i+1}^2} - \frac{4s_i + 2s_{i+1}}{\Delta_{i+1}}.$$

Rearranging this yields a linear system of $k - 1$ equations in the $k - 1$ unknowns $s_1, \ldots, s_{k-1}$. For $1 \leq i \leq k - 1$,

$$\Delta_{i+1} s_{i-1} + 2(\Delta_i + \Delta_{i+1})s_i + \Delta_i s_{i+1} = \frac{3\Delta_i(a_{i+1} - a_i)}{\Delta_{i+1}} + \frac{3\Delta_{i+1}(a_i - a_{i-1})}{\Delta_i}.$$

The terms involving s_0 and s_n may be moved to the right-hand side.

It now remains to show that this system has a unique solution. This will follow if we can show that the matrix

$$X = \begin{bmatrix} 2(\Delta_1 + \Delta_2) & \Delta_1 & 0 & 0 & \cdots & & 0 \\ \Delta_3 & 2(\Delta_2 + \Delta_3) & \Delta_2 & 0 & \cdots & & 0 \\ \vdots & \vdots & \vdots & \ddots & & \vdots & \vdots \\ 0 & \cdots & 0 & \Delta_{k-1} & 2(\Delta_{k-2} + \Delta_{k-1}) & \Delta_{k-2} \\ 0 & \cdots & 0 & 0 & \Delta_k & 2(\Delta_{k-1} + \Delta_k) \end{bmatrix}$$

is invertible. The property of this system that makes it possible is that the matrix is **diagonally dominant**, which means that the diagonal entries are greater than the sum of all other entries in each row.

To see this, suppose that $y = (y_1, \ldots, y_{k-1})$ is in the kernel of X. Choose a coefficient i_0 so that $|y_{i_0}| \geq |y_j|$ for $1 \leq j \leq k - 1$. Then looking only at the i_0th coefficient of $0 = Xy$, we obtain

$$0 = \left| \Delta_{i_0+1} y_{i_0-1} + 2(\Delta_{i_0} + \Delta_{i_0+1})y_{i_0} + \Delta_{i_0} y_{i_0+1} \right|$$

$$\geq 2(\Delta_{i_0} + \Delta_{i_0+1})|y_{i_0}| - \Delta_{i_0+1}|y_{i_0-1}| - \Delta_{i_0}|y_{i_0+1}|$$

$$\geq (\Delta_{i_0} + \Delta_{i_0+1})|y_{i_0}|.$$

Hence $y = 0$ and so X has trivial kernel. Hence X is invertible. So for each set of data $a_0, \ldots, a_k$, s_0 and s_k, there is a unique choice of points $s_1, \ldots, s_{k-1}$ solving our system. Thus there is a unique spline $h \in \mathbb{S}(\Delta)$ satisfying these data. ∎

Thus if f is a C^2 function on $[a, b]$, there is a unique cubic spline h such that $h(x_i) = a_i := f(x_i)$ for $0 \leq i \leq k$, $h'(a) = s_0 := f'(a)$ and $h'(b) = s_k := f'(b)$. We denote the function h by Jf. Let us show that J is linear. If f_1 and f_2 are functions in $C^2[a, b]$ with $h_i = Jf_i$, then $h = b_1 h_1 + b_2 h_2$ is a spline such that

$$h(x_i) = (b_1 h_1 + b_2 h_2)(x_i) = (b_1 f_1(x_i) + b_2 f_2)(x_i)$$

and

$$h'(a) = \big(b_1 h_1' + b_2 h_2'\big)(a)$$
$$= b_1 f_1'(a) + b_2 f_2'(a) = \big(b_1 f_1(x_i) + b_2 f_2\big)'(a).$$

Similarly, this holds at b. By the uniqueness of the spline, it follows that

$$J(b_1 f_1 + b_2 f_2) = b_1 h_1 + b_2 h_2 = b_1 J f_1 + b_2 J f_2.$$

In particular, we may find specific splines c_i satisfying

$$c_i'(a) = c_i'(b) = 0 \quad \text{and} \quad c_i(x_j) = \begin{cases} 1 & \text{if } j = i \\ 0 & \text{if } j \neq i \end{cases}$$

for $0 \le i \le k$. The linear space $\mathbb{S}(\Delta)$ is spanned by $\{x, x^2, c_i : 0 \le i \le k\}$. To see this, let h be the spline with data $a_0, \dots, a_k$, s_0 and s_k be given. Let q be the unique quadratic $q(x) = cx + dx^2$ such that $q'(a) = s_1$ and $q'(b) = s_k$. Then $g = h - q$ is a spline with $g'(a) = g'(b) = 0$. Form the spline

$$s(x) = q(x) + \sum_{i=0}^{n} g(x_i) c_i(x).$$

It is easy to check that s is another spline with the same data as h. Since this uniquely determines the spline, $h = s$ has the desired form. This exhibits a specific basis for $\mathbb{S}(\Delta)$. Figure 10.7 shows c_0 and c_2 for a paricular partition Δ.

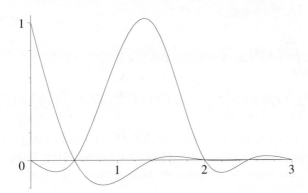

FIGURE 10.7. Graphs of c_0 and c_2 for $\Delta = \{0, .5, 1.4, 2, 2.5, 3\}$.

Now that we have shown that cubic splines are plentiful, we investigate how well Jf approximates the original function f. We need the following:

10.8.7. LEMMA. *If $f \in C^2[a, b]$, then for all $\varphi \in \mathbb{S}_1(\Delta)$,*

$$\int_a^b \varphi(x)(f - Jf)''(x)\, dx = 0.$$

PROOF. We use integration by parts twice. Letting $du = (f - Jf)''(x)\, dx$ and $v = \varphi(x)$, we have

$$\int_a^b \varphi(x)(f - Jf)''(x)\, dx = \varphi(x)(f - Jf)'(x)\Big|_a^b + \int_a^b \varphi'(x)(f - Jf)'(x)\, dx.$$

Observe that $f'(a) = Jf'(a)$ and $f'(b) = Jf'(b)$, so the first term above is zero. Using integration by parts again with $du = (f - Jf)'(x)\, dx$ and $v = \varphi'(x)$, the integral equals

$$\varphi'(x)(f - Jf)(x)\Big|_a^b + \int_a^b \varphi''(x)(f - Jf)(x)\, dx.$$

Now, $f(a) = Jf(a)$ and $f(b) = Jf(b)$, so the first term is zero. For the second term, we observe that since φ is piecewise linear, φ'' is equal to zero except for the points $x_0, x_1, \ldots, x_n$, where it is not defined. Thus the integral is zero. ∎

10.8.8. THEOREM. *If $f \in C^2[a, b]$, then*

$$\int_a^b \left(f''(x)\right)^2 dx = \int_a^b \left((Jf)''(x)\right)^2 dx + \int_a^b \left((f - Jf)''(x)\right)^2 dx.$$

PROOF. Let $g = f - Jf$. We have

$$\int_a^b \left(f''(x)\right)^2 dx = \int_a^b \left(Jf''(x) + g''(x)\right)^2 dx$$

$$= \int_a^b \left(Jf''(x)\right)^2 dx + 2\int_a^b (Jf)''(x)g''(x)\, dx + \int_a^b \left(g''(x)\right)^2 dx.$$

However, since $Jf \in \mathbb{S}(\Delta)$ is piecewise cubic and C^2, it follows that $(Jf)''$ is piecewise linear, whence it belongs to $\mathbb{S}_1(\Delta)$. Hence by Lemma 10.8.7, the second term is zero. This gives the required equality, so we're done. ∎

This allows a characterization of the cubic spline approximating f as optimal in a certain sense.

10.8.9. COROLLARY. *Fix $f \in C^2[a, b]$. Among all functions $g \in C^2[a, b]$ such that $g(x_i) = f(x_i)$ for $0 \leq i \leq k$, $g'(a) = f'(a)$ and $g'(b) = f'(b)$, the cubic spline interpolant Jf minimizes the energy integral*

$$\int_a^b \left(g''(x)\right)^2 dx.$$

PROOF. For any such function g, we have $Jg = Jf$. So by the previous theorem,

$$\int_a^b \left(g''(x)\right)^2 dx = \int_a^b \left((Jf)''(x)\right)^2 dx + \int_a^b \left((g - Jf)''(x)\right)^2 dx$$

$$\geq \int_a^b \left((Jf)''(x)\right)^2 dx.$$

This inequality becomes an equality only if $g'' = (Jf)''$. Since we also have $g(a) = Jf(a)$ and $g'(a) = (Jf)'(a)$ by hypothesis, this implies that $g = Jf$ by integrating twice. ∎

This property is called the **smoothest interpolation property** of cubic spline interpolation. Minimizing $\int_a^b \left(g''\right)^2(x)\, dx$ is roughly equivalent to minimizing the strain energy. Historically, flexible thin strips of wood called splines were used in drafting to approximate curves through a set of points. In 1946, when Schoenberg introduced spline curves, he observed that they represent the curves drawn by means of wooden splines, hence the name. Splines appear to be smooth since they avoid discontinuous first derivatives, which people recognize as "spikes," and avoid discontinuous second derivatives, which are recognized as sudden changes in curvature. Discontinuous third derivatives are not visible in any obvious geometric way.

Exercises for Section 10.8

A. Fill in the details of the proof of Lemma 10.8.4.

B. Find a nice explicit basis for $\mathbb{S}_1(\Delta)$.

C. Show that if $f \in \mathbb{S}_1(\Delta)$ and f' is continuous on $[a, b]$, then f is a straight line.

D. Show that if $f \in \mathbb{S}(\Delta)$ and f''' is continuous on $[a, b]$, then f is a cubic polynomial.

E. Show that for $f \in C^2[a, b]$ that $\|f - Jf\|_\infty \leq 2\inf\{\|f - g\|_\infty : g \in \mathbb{S}(\Delta)\}$.
   HINT: Compare with Lemma 10.8.1.

F. Suppose that Δ has $k > 4$ intervals. Let $1 \leq i \leq k - 4$. If $h \in \mathbb{S}(\Delta)$ is 0 everywhere except on (x_i, x_{i+3}), show that $h = 0$. HINT: What derivative conditions are forced?

G. Let x_+^3 denote the function $\max\{x^3, 0\}$. Show that every cubic spline in $\mathbb{S}(\Delta)$ has the form $p(x) + \sum_{i=1}^{k-1} c_i(x - x_i)_+^3$, where $p(x)$ is a cubic polynomial and $c_i \in \mathbb{R}$.
   HINT: Given h, let $c_i = \delta_i/6$, where δ_i is the change in the third derivative h at x_i.

H. Find a nonzero spline h for the partition $\{-1, 0, 1, 2, 3, 4, 5\}$ such that h is 0 on $[-1, 0] \cup [4, 5]$. HINT: Use the previous exercise.

10.9. Uniform Approximation by Splines

To complete our analysis of cubic splines, we will obtain an estimate for the error of approximation. This is a rather delicate argument that combines a generalized mean value theorem with another system of linear equations. Our goal is to establish the following theorem:

10.9.1. THEOREM. *Let Δ be a partition $a = x_0 < x_1 < \cdots < x_k = b$ of the interval $[a, b]$ and set $\delta = \max\{x_i - x_{i-1} : 1 \leq i \leq k\}$. Let $f \in C^2[a, b]$ and let $h = Jf$ be the cubic spline in $\mathbb{S}(\Delta)$ approximating f. Then*

$$\|f - h\|_\infty \leq \frac{5}{2}\delta^2\omega(f''; \delta)$$

$$\|f' - h'\|_\infty \leq 5\delta\omega(f''; \delta)$$

$$\|f'' - h''\|_\infty \leq 5\omega(f''; \delta).$$

Since the proof is long and computational, we give an overview first. Following the algebra of the last section, we obtain a system of $k + 1$ linear equations satisfied by $h''(x_0), \ldots, h''(x_k)$. The constant terms in this system are estimated using a second-order Mean Value Theorem. Then the equations are used to show that $|h''(x_i) - f''(x_i)| \leq 4\omega(f''; \delta)$. It is then straightforward to bound $\|h'' - f''\|_\infty$, and then integratation gives bounds for $\|f' - h'\|_\infty$ and $\|f - h\|_\infty$.

We begin with a second-order Mean Value Theorem.

10.9.2. LEMMA. *Suppose that $f \in C^2[a, c]$ and $a < b < c$. There is a point ξ in (a, c) such that*

$$f(b) - \left(\frac{c - b}{c - a}f(a) + \frac{b - a}{c - a}f(c)\right) = \frac{-(c - b)(b - a)}{2}f''(\xi).$$

PROOF. Let $L(x)$ be the straight line through $(a, f(a))$ and $(c, f(c))$, namely

$$L(x) = \frac{c - x}{c - a}f(a) + \frac{x - a}{c - a}f(c).$$

Consider the function

$$g(x) = (c - b)(b - a)\big(f(x) - L(x)\big) - (c - x)(x - a)\big(f(b) - L(b)\big).$$

Notice that $g(a) = g(b) = g(c) = 0$. So by Rolle's Theorem, there are points $\xi_1 \in (a, b)$ and $\xi_2 \in (b, c)$ such that $g'(\xi_1) = g'(\xi_2) = 0$. Applying Rolle's Theorem to g' now yields a point ξ in (ξ_1, ξ_2) such that

$$0 = g''(\xi) = (c - b)(b - a)f''(\xi) + 2\big(f(b) - L(b)\big).$$

This is just a rearrangement of the desired formula.   ∎

Notice that there is a limiting situation where b equals a or c. Take $b = a$, for example. Divide both sides by $b - a$ and take the limit, ignoring the important point that ξ depends on b. Then we obtain

$$-\frac{c-a}{2}f''(\xi) = \frac{f(a) - f(c)}{c - a} + \lim_{b \to a} \frac{f(b) - f(a)}{b - a}$$

$$= \frac{f(a) - f(c)}{c - a} + f'(a).$$

Rearranging, this becomes

$$f(c) = f(a) + f'(a)(c - a) + \frac{(c-a)^2}{2}f''(\xi) \quad \text{for some } \xi \in (a, c).$$

We could make the limit argument correct, but we do not need to do so because this is just a consequence of the order one Taylor Theorem (see Exercise 10.1.B).

PROOF OF THEOREM 10.9.1. We need to show that h'' is close to f'' at the points x_i. We rewrite the formula (10.8.6) as

$$h''(x_i) = \frac{6(a_{i+1} - a_i)}{\Delta_{i+1}^2} - \frac{4s_i + 2s_{i+1}}{\Delta_{i+1}} \qquad \text{for} \quad 0 \le i \le k - 1$$

$$h''(x_i) = \frac{-6(a_i - a_{i-1})}{\Delta_i^2} + \frac{2s_{i-1} + 4s_i}{\Delta_i} \qquad \text{for} \quad 1 \le i \le k.$$

The idea is to eliminate the unknown $s_1, \ldots, s_{k-1}$ from these $2k$ equations to yield $k + 1$ equations for the $h''(x_i)$, $0 \le i \le k$.

We save some time by presenting the list of equations and ask the interested reader to verify that they are correct.

$$2\Delta_1 h''(x_0) + \Delta_1 h''(x_1) = 6\left(\frac{a_1 - a_0}{\Delta_1} - s_0\right)$$

$$\Delta_i h''(x_{i-1}) + 2(\Delta_i + \Delta_{i+1})h''(x_i) + \Delta_{i+1}h''(x_{i+1}) = 6\left(\frac{a_{i+1} - a_i}{\Delta_{i+1}} - \frac{a_i - a_{i-1}}{\Delta_i}\right)$$

$$\text{for} \quad 1 \le i \le k - 1$$

$$\Delta_k h''(x_{k-1}) + 2\Delta_k h''(x_k) = 6\left(s_k - \frac{a_k - a_{k-1}}{\Delta_k}\right)$$

Let us define the matrix

$$Y = \begin{bmatrix} 2\Delta_1 & \Delta_1 & 0 & 0 & \cdots & 0 & 0 & 0 \\ \Delta_1 & 2(\Delta_1 + \Delta_2) & \Delta_2 & 0 & \cdots & 0 & 0 & 0 \\ 0 & \Delta_2 & 2(\Delta_2 + \Delta_3) & \Delta_3 & \cdots & 0 & 0 & 0 \\ \vdots & \vdots & \vdots & \vdots & \ddots & \vdots & \vdots & \vdots \\ 0 & 0 & 0 & \cdots & \Delta_{k-2} & 2(\Delta_{k-2} + \Delta_{k-1}) & \Delta_{k-1} & 0 \\ 0 & 0 & 0 & \cdots & 0 & \Delta_{k-1} & 2(\Delta_{k-1} + \Delta_k) & \Delta_k \\ 0 & 0 & 0 & \cdots & 0 & 0 & \Delta_k & 2\Delta_k \end{bmatrix}.$$

So if we set

$$
\mathbf{h}'' = \begin{bmatrix} h''(x_0) \\ h''(x_1) \\ \vdots \\ h''(x_{k-1}) \\ h''(x_k) \end{bmatrix}, \quad \mathbf{f}'' = \begin{bmatrix} f''(x_0) \\ f''(x_1) \\ \vdots \\ f''(x_{k-1}) \\ f''(x_k) \end{bmatrix}, \text{ and } \mathbf{z} = \begin{bmatrix} 6\left(\frac{a_1-a_0}{\Delta_1} - s_0\right) \\ 6\left(\frac{a_2-a_1}{\Delta_2} - \frac{a_1-a_0}{\Delta_1}\right) \\ \vdots \\ 6\left(\frac{a_k-a_{k-1}}{\Delta_k} - \frac{a_{k-1}-a_{k-2}}{\Delta_{k-1}}\right) \\ 6\left(s_k - \frac{a_k-a_{k-1}}{\Delta_k}\right) \end{bmatrix},
$$

then the equation becomes $Y\mathbf{h}'' = \mathbf{z}$.

Now we apply Lemma 10.9.2 to approximate $\mathbf{z}$. Use x_{i-1}, x_i, x_{i+1} for a, b, c. There is a point ξ_i in $[x_{i-1}, x_{i+1}]$ so that

$$
6\left(\frac{a_{i+1} - a_i}{\Delta_{i+1}} - \frac{a_i - a_{i-1}}{\Delta_i}\right)
$$
$$
= 6\frac{(x_i - x_{i-1})f(x_{i+1}) - (x_{i+1} - x_{i-1})f(x_i) + (x_{i+1} - x_i)f(x_{i-1})}{(x_{i+1} - x_i)(x_i - x_{i-1})}
$$
$$
= 3(x_{i+1} - x_{i-1})f''(\xi_i) \quad = \quad 3(\Delta_i + \Delta_{i+1})f''(\xi_i)
$$

for $1 \le i \le k - 1$. The two end terms are approximated using the limit version, namely Taylor's formula of order 2

$$
6\left(\frac{a_1 - a_0}{\Delta_1} - s_0\right) = 6\left(\frac{f(x_1) - f(x_0)}{x_1 - x_0} - f'(x_0)\right)
$$
$$
= 3(x_1 - x_0)f''(\xi_0) \quad = \quad 3\Delta_1 f''(\xi_0)
$$

for some ξ_0 in $[x_0, x_1]$. Similarly, there is a point ξ_k in $[x_{k-1}, x_k]$ such that

$$
6\left(s_k - \frac{a_k - a_{k-1}}{\Delta_k}\right) = 6\left(f'(x_k) - \frac{f(x_k) - f(x_{k-1})}{x_k - x_{k-1}}\right)
$$
$$
= 3(x_k - x_{k-1})f''(\xi_k) \quad = \quad 3\Delta_k f''(\xi_k).
$$

Now we approximate $\mathbf{h}'' - \mathbf{f}''$ by evaluating $Y(\mathbf{h}'' - \mathbf{f}'')$. The first coefficient is estimated by

$$
\left|3\Delta_1 f''(\xi_0) - 2\Delta_1 f''(x_0) - \Delta_1 f''(x_1)\right|
$$
$$
\le 2\Delta_1 |f''(\xi_0) - f''(x_0)| + \Delta_1 |f''(\xi_0) - f''(x_1)|
$$
$$
\le 3\Delta_1 \omega(f''; \delta).
$$

Recall from Lemma 14.9.10 that $\omega(f''; 2\delta) \le 2\omega(f''; \delta)$. For $1 \le i \le k - 1$, we obtain

$$
\left|3(\Delta_i + \Delta_{i+1})f''(\xi_i) - \Delta_i f''(x_{i-1}) - 2(\Delta_i + \Delta_{i+1})f''(x_i) - \Delta_{i+1} f''(x_{i+1})\right|
$$
$$
\le 2(\Delta_i + \Delta_{i+1})|f''(\xi_i) - f''(x_i)| + \Delta_i |f''(\xi_i) - f''(x_{i-1})|
$$
$$
+ \Delta_{i+1} |f''(\xi_i) - f''(x_{i+1})|
$$
$$
\le 2(\Delta_i + \Delta_{i+1})\omega(f''; \delta) + (\Delta_i + \Delta_{i+1})\omega(f''; 2\delta)
$$
$$
\le 4(\Delta_i + \Delta_{i+1})\omega(f''; \delta).
$$

Finally, the last term is estimated

$$\left|3\Delta_k f''(\xi_k) - \Delta_k f''(\xi_{k-1}) - 2\Delta_k f''(x_k)\right|$$
$$\leq \Delta_k |f''(\xi_k) - f''(\xi_{k-1})| + 2\Delta_k |f''(\xi_k) - f''(x_k)|$$
$$\leq 3\Delta_k \, \omega(f''; \delta).$$

Let $A = \max\{|h''(x_i) - f''(x_i)| : 0 \leq i \leq k\}$ occur at i_0. Then looking at the i_0th coefficient of $Y(\mathbf{h}'' - \mathbf{f}'')$, we obtain

$$4(\Delta_{i_0} + \Delta_{i_0+1})\omega(f''; \delta)$$

$$\geq \left|\Delta_{i_0}\big(h''(x_{i_0-1}) - f''(x_{i_0-1})\big) + 2(\Delta_{i_0} + \Delta_{i_0+1})\big(h''(x_{i_0}) - f''(x_{i_0})\big)\right.$$

$$\left. + \Delta_{i_0+1}\big(h''(x_{i_0+1}) - f''(x_{i_0+1})\big)\right|$$

$$\geq 2\left|(\Delta_{i_0} + \Delta_{i_0+1})\big(h''(x_{i_0}) - f''(x_{i_0})\big)\right| - \left|\Delta_{i_0}\big(h''(x_{i_0-1}) - f''(x_{i_0-1})\big)\right|$$

$$- \left|\Delta_{i_0+1}\big(h''(x_{i_0+1}) - f''(x_{i_0+1})\big)\right|$$

$$\geq (\Delta_{i_0} + \Delta_{i_0+1})A.$$

So $A \leq 4\omega(f''; \delta)$. When the $i_0 = 0$ or k, we obtain $3\omega(f''; \delta)$ instead.

We are almost done with the estimate for the second derivative. Notice that on $[x_{i-1}, x_i]$, $h''(x) = h_i''(x)$ is linear. So $h''(x)$ lies between $h''(x_{i-1})$ and $h''(x_i)$. For convenience, suppose that $h''(x_{i-1}) \leq h''(x_i)$. The other case is similar. Then

$$-5\omega(f''; \delta) \leq \big(h''(x_{i-1}) - f''(x_{i-1})\big) + \big(f''(x_{i-1}) - f''(x)\big)$$
$$\leq h''(x) - f''(x)$$
$$\leq \big(h''(x_i) - f''(x_i)\big) - \big(f''(x_i) - f''(x)\big)$$
$$\leq 5\omega(f''; \delta).$$

Thus $\|h'' - f''\|_\infty \leq 5\omega(f''; \delta)$.

The rest is easy. Since $f(x_{i-1}) - h(x_{i-1}) = 0 = f(x_i) - h(x_i)$, Rolle's Theorem provides a point ζ_i in $[x_{i-1}, x_i]$ such that $f'(\zeta_i) - h'(\zeta_i) = 0$. Hence for any point x in $[x_{i-1}, x_i]$,

$$|f'(x) - h'(x)| = \left|\int_{\zeta_i}^x f''(t) - h''(t)\, dt\right|$$
$$\leq 5\omega(f''; \delta)|x - \zeta_i| \leq 5\delta\omega(f''; \delta).$$

Therefore $\|h' - f'\|_\infty \leq 5\delta\omega(f''; \delta)$. Now pick the nearest partition point x_i to x, so that $|x - x_i| \leq \delta/2$. Since $f(x_i) = h(x_i)$,

$$|f(x) - h(x)| = \left|\int_{x_i}^x f'(t) - h'(t)\, dt\right|$$
$$\leq 5\delta\omega(f''; \delta)|x - x_i| \leq \frac{5}{2}\delta^2\omega(f''; \delta).$$

So $\|h - f\|_\infty \leq \frac{5}{2}\delta^2\omega(f''; \delta).$ ∎

Exercises for Section 10.9

A. Show that $\mathbb{S}(\Delta)$ has dimension $k + 3$.

B. A second-order Mean Value Theorem (Lemma 10.9.2) suggests the possibility of a third-order Mean Value Theorem. Suppose that $f \in C^3[a, d]$ and b, c in (a, d) with $b \neq c$. If P is the unique quadratic polynomial through $(a, f(a))$, $(c, f(c))$ and $(d, f(d))$, show that there is a point $\xi \in [a, d]$ with

$$f(b) - P(b) = \frac{(b - c)(b - a)(b - d)}{6} f'''(\xi).$$

C. Prove that every continuous function on $[0, 1]$ is the uniform limit of the sequence of cubic splines h_k with nodes at $\{j2^{-k} : 0 \leq j \leq 2^k\}$.

10.10. Appendix: The Stone–Weierstrass Theorem

We conclude this chapter with a very general approximation theorem that has many applications to approximation problems. It provides a very simple, easy to check criterion for when all continuous functions on a compact metric space can be approximated by some element of a subalgebra of functions. In particular, we shall see immediate consequences for approximation by polynomials in several variables and by trigonometric polynomials.

10.10.1. DEFINITION. A subset A of $C(X)$, the space of continuous real-valued functions on a compact metric space X, is an **algebra** if it is a subspace of $C(X)$ that is closed under multiplication (i.e., if $f, g \in A$, then $fg \in A$).

For $f, g \in C(X)$, define new elements of $C(X)$, $f \vee g$ and $f \wedge g$, by

$$(f \vee g)(x) = \max\{f(x), g(x)\} \quad \text{and} \quad (f \wedge g)(x) = \min\{f(x), g(x)\}.$$

A subset L of $C(X)$ is a **vector lattice** if it is a subspace that is closed under these two operations, that is, f, g both in L imply $f \vee g$ and $f \wedge g$ are in L.

It is easy to verify the two identities $f \vee g = 1/2(f + g) + 1/2|f - g|$ and that $f \wedge g = 1/2(f + g) - 1/2|f - g|$. It follows that an algebra, A, is a vector lattice if and only if $|f| \in A$ for each $f \in A$.

10.10.2. DEFINITION. A set S of functions on X **separates points** if for each pair of points $x, y \in X$, there is a function $f \in S$ such that $f(x) \neq f(y)$. Say that S **vanishes** at x_0 if $f(x_0) = 0$ for all $f \in S$.

In order to approximate arbitrary continuous functions on X from elements of A, a moment's thought shows that A must separate points. Moreover, A cannot vanish at any point, for then we could not approximate the constant function 1. These rather modest requirements, combined with the algebraic structure of an algebra, yield the following beautiful result.

10.10.3. STONE–WEIERSTRASS THEOREM.
An algebra A of continuous real-valued functions on a compact metric space X that separates points and does not vanish at any point is dense in $C(X)$.

We break the proof into several parts. Because norms over several domains occur in this proof, let us write $\|f\|_X$ for the uniform norm over X, and write $\|p\|_\infty$ for the uniform norm of a function on a real interval $[a, b]$, if the interval is understood.

10.10.4. LEMMA. *If A is an algebra of real-valued continuous functions on X, then its closure $\overline{A}$ is a closed algebra and a vector lattice.*

PROOF. It is easy to check that the closure of a subspace is still a subspace. The verification is left as an exercise. To see that $\overline{A}$ is closed under multiplication, take $f, g \in \overline{A}$. Choose sequences f_n and g_n in A that converge uniformly to f and g, respectively. Since A is an algebra, $f_n g_n$ belongs to A. Then the sequence $(f_n g_n)$ converges uniformly to fg (see Exercise 8.2.D). So $\overline{A}$ is closed under multiplication, and thus is an algebra.

Fix a function $f \in \overline{A}$. By the Weierstrass Approximation Theorem, the function $h(t) = |t|$ for $t \in [-\|f\|_X, \|f\|_X]$ is the uniform limit of a sequence p_n of polynomials. We can arrange that $p_n(0) = 0$. Indeed, $0 = h(0) = \lim_{n \to \infty} p_n(0)$. Thus $q_n(x) = p_n(x) - p_n(0)$ satisfies $q_n(0) = 0$ and

$$\|h - q_n\|_\infty \le \|h - p_n\|_\infty + |p_n(0)|.$$

The right-hand side converges to 0, and thus (q_n) converges uniformly to h.

We will show that $|f|$ belongs to $\overline{A}$ also. Since $\overline{A}$ is an algebra, all linear combinations of $f, f^2, f^3, \ldots$ belong to $\overline{A}$. So if q is a polynomial with $q(0) = 0$, say $q(x) = a_1 x + a_2 x^2 + \cdots + a_k x^k$, then

$$q(f) = a_1 f + a_2 f^2 + \cdots + a_k f^k$$

belongs to $\overline{A}$. Moreover, if p, q are two such polynomials, then

$$\begin{aligned}
\|p(f) - q(f)\|_X &= \sup_{x \in X} |p(f(x)) - q(f(x))| \\
&\le \sup_{t \in [-\|f\|_X, \|f\|_X]} |p(t) - q(t)| = \|p - q\|_\infty.
\end{aligned}$$

Since $\|q_n(f) - q_m(f)\|_X \le \|q_n - q_m\|_\infty$ and (q_n) is a Cauchy sequence, we conclude that $(q_n(f))$ is also a Cauchy sequence. Thus the limit g belongs to $\overline{A}$. Then $g(x) = \lim_{n \to \infty} q_n(f(x)) = h(x) = |f(x)|$. So $|f|$ belongs to $\overline{A}$ for each $f \in \overline{A}$. As we remarked previously, this implies that $\overline{A}$ is a vector lattice. ∎

10.10.5. LEMMA. *If A is an algebra on X that separates points and never vanishes, then for any $x, y \in X$ and $\alpha, \beta \in \mathbb{R}$, there is a function $h \in A$ such that $h(x) = \alpha$ and $h(y) = \beta$.*

PROOF. There is a function $f \in A$ such that $f(x) \neq f(y)$. We may assume that $f(y) \neq 0$. If $f(x) \neq 0$ also, try $h = f^2 - tf$. We require

$$f(x)^2 - tf(x) = \alpha$$
$$f(y)^2 - tf(y) = \beta.$$

Solving yields $t = \dfrac{\alpha - \beta}{f(y) - f(x)} + f(x) + f(y)$. If $f(x) = 0$, there is a function $g \in A$ with $g(x) \neq 0$. In this case,

$$h = \frac{\alpha}{g(x)} g + \frac{\beta g(x) - \alpha g(y)}{g(x) f(y)} f$$

will suffice. ∎

PROOF OF THE STONE–WEIERSTRASS THEOREM. Fix a function $f \in C(X)$ and $\varepsilon > 0$. We will approximate f within ε by functions in $\overline{A}$. For each pair of points $x, y \in X$, use Lemma 10.10.5 to find functions $g_{x,y} \in A$ such that $g_{x,y}(x) = f(x)$ and $g_{x,y}(y) = f(y)$.

Fix y. For each $x \neq y$,

$$U_x = \{z \in X : g_{x,y}(z) > f(x) - \varepsilon\} = (g_{x,y} - f)^{-1}(-\varepsilon, \infty)$$

is an open set containing x and y. Thus $\{U_x : x \in X \setminus \{y\}\}$ is an open cover of X. By the Borel–Lebesgue Theorem (Theorem 9.2.3), this cover has a finite subcover $U_{x_1}, \ldots, U_{x_k}$. Let $g_y = g_{x_1,y} \vee g_{x_2,y} \vee \ldots g_{x_k,y}$. By Lemma 10.10.4, g_y belongs to $\overline{A}$. By construction, $g_y(y) = f(y)$ and $g_y(x) > f(x) - \varepsilon$ for all $x \in X$.

Now define $V_y = \{x \in X : g_y(x) < f(x) + \varepsilon\}$, which is an open set containing y. Then $\{V_y : y \in X\}$ is an open cover of X. By the Borel–Lebesgue Theorem, this cover has a finite subcover $V_{y_1}, \ldots, V_{y_l}$. Let $g = g_{y_1} \wedge g_{y_2} \wedge \ldots g_{y_l}$. By Lemma 10.10.4, g belongs to $\overline{A}$. By construction, $g(x) < f(x) + \varepsilon$ for all $x \in X$. Moreover, since $g_{y_j} > f(x) - \varepsilon$ for $1 \leq j \leq l$,

$$f(x) - \varepsilon < g(x) < f(x) + \varepsilon \quad \text{for all} \quad x \in X.$$

Thus $\|f - g\|_X < \varepsilon$ as desired. ∎

10.10.6. COROLLARY. *Let X be a compact subset of $\mathbb{R}^n$. The algebra of all polynomials $p(x_1, x_2, \ldots, x_n)$ in the n coordinates is dense in $C(X)$.*

PROOF. It is immediately clear that the set $\mathbb{P}$ of all polynomials in n variables is an algebra. The constant function 1 does not vanish at any point. Finally, any two distinct points are separated by at least one of the coordinate functions x_i. Therefore, $\mathbb{P}$ satisfies the hypotheses of the Stone–Weierstrass Theorem and hence is dense in $C(X)$. ∎

Another application of the Stone–Weierstrass Theorem yields an abstract proof of the following corollary, which will be given a different and more direct proof in Fejér's Theorem (Theorem 14.6.4). See also Corollary 13.6.6 for yet another proof.

10.10.7. COROLLARY. *The set of all trigonometric polynomials is dense in* $C_*[-\pi, \pi]$, *the space of* 2π-*periodic functions on* $\mathbb{R}$.

PROOF. The set $\mathbb{TP}$ of all trig polynomials $f(t) = a_0 + \sum_{i=1}^{n} a_k \cos kt + b_k \sin kt$ (see Section 7.4) is not so obviously an algebra. The crucial identities are

$$\sin kt \, \sin lt = \tfrac{1}{2} \cos(k - l)t - \tfrac{1}{2} \cos(k + l)t$$
$$\sin kt \, \cos lt = \tfrac{1}{2} \sin(k + l)t + \tfrac{1}{2} \sin(k - l)t$$
$$\cos kt \, \cos lt = \tfrac{1}{2} \cos(k - l)t + \tfrac{1}{2} \cos(k + l)t.$$

This shows that the product of two trig functions is a trig polynomial. As the set of trig polynomials is a vector space, the general multiplication of trig polynomials will also be a trig polynomial. So $\mathbb{TP}$ is an algebra.

To put this problem into the context of the Stone–Weierstrass Theorem, let T be the unit circle. Consider the map $P(t) = (\cos t, \sin t)$ of $\mathbb{R}$ onto T. Observe that $P(s) = P(t)$ if and only if $s - t$ is an integer multiple of 2π. In particular, $[-\pi, \pi]$ is wrapped around the circle with only the endpoints $\pm\pi$ coinciding. A continuous function $g \in C(T)$ may be identified with the function $f(t) = g(P(t))$. It is easy to see that $f(-\pi) = f(\pi)$. If we define $C_*[-\pi, \pi]$ to be the closed subspace given by

$$\{f \in C[-\pi, \pi] : f(-\pi) = f(\pi)\},$$

then it is also easy to see that every function in $C_*[-\pi, \pi]$ is obtained as $g(P(t))$ for some $g \in C(T)$. Moreover, this correspondence between $C(T)$ and $C_*[-\pi, \pi]$ is isometric, linear, and preserves multiplication.

The two coordinate functions x_1 and x_2 on T send the point $P(t)$ to $\cos t$ and $\sin t$, respectively. So the first paragraph shows that polynomials in x_1 and x_2 are actually trig polynomials of the variable t. So the algebra $\mathbb{TP}$ of all trig polynomials in $C_*[-\pi, \pi]$ is identified with the algebra of all usual polynomials on T. By Corollary 10.10.6, the algebra of polynomials is dense in $C(T)$. Therefore, the algebra of trig polynomials is dense in $C_*[-\pi, \pi]$. ∎

Exercises for Section 10.10

A. Show that the closure of a subspace of $C(X)$ is also a subspace of $C(X)$.

B. Let X and Y be compact metric spaces. Show that the set of all functions of the form $\sum_{i=1}^{k} f_i(x)g_i(y)$ for $k \geq 1$ and $f_i \in C(X)$ and $g_i \in C(Y)$ is dense in $C(X \times Y)$.

C. Let $h \in C[0, 1]$. Show that every $f \in C[0, 1]$ is a limit of polynomials in h if and only if h is strictly monotone.

D. Let X be a compact metric space. Suppose that A is a subalgebra of $C(X)$ that separates the points of X. If $\overline{A} \neq C(X)$, show that there is a point $x_0 \in X$ such that $\overline{A} = \{f \in C(X) : f(x_0) = 0\}$.
HINT: Show that $A + \mathbb{R}1$ is an algebra that does not vanish. Can A vanish at more than one point?

E. Let X be a compact metric space. A subset J of $C(X)$ is an **ideal** if it is a vector space with the property that if $p \in J$ and $f \in C(X)$, then $fp \in J$.

(a) Let $E = \{x \in X : J$ vanishes at $x\}$. Show that E is closed.

(b) Show that J separates points of $X \setminus E$.

(c) Show that J is dense in the set of continuous functions on X that vanish on E.

HINT: Fix f and $\varepsilon > 0$, and set $F = \{x \in X : |f(x)| \geq \varepsilon\}$. For $x \in X$, find $p_x \in J$ such that $p_x(x) = 2$. Hence find a finite set of elements $p_i \in J$ such that
$$p(x) = \sum_{i=1}^{k} p_i(x)^2 \geq 1 \text{ for } x \in F. \text{ Let } q = \frac{fnp}{1 + np} \text{ for } n \text{ sufficiently large.}$$

F. Let A be a complex algebra of complex-valued continuous functions on X. That is, A is a vector space over $\mathbb{C}$ and is closed under multiplication. Suppose that A separates points and does not vanish at any point and further if $f \in A$, then $\bar{f}$ belongs to A. Then show that A is dense in the space $C_{\mathbb{C}}(X)$ of all complex-valued continuous functions on X.

HINT: Consider the set A_r of all real-valued functions in A.

CHAPTER 11

Discrete Dynamical Systems

Suppose we wish to describe some physical system. The dynamical systems approach starts with a space X of all possible states of the system—think of a point x in X as representing a state for some physical situation at some time. We will assume that X is a subset of some normed linear space. Often X will be a subset of $\mathbb{R}$. The evolution of the system over time determines a function T mapping X into itself that takes each state x to a new state, one unit of time later.

Suppose that x_0 is the state of the system at time zero and x_n is the state of the system at time n. If $T^{n+1}(x) := T(T^n(x))$ for $n \geq 1$, then we have

$$x_n = Tx_{n-1} = T^2 x_{n-2} = \cdots = T^n x_0,$$

Notice that in this chapter, $T^2 x$ virtually always means $T(T(x))$ and not $(T(x))^2$.

11.0.8. DEFINITION. Suppose that X is a subset of a normed linear space, and T is a continuous map from X into itself. The pair (X, T) is called a **discrete dynamical system**. For each point $x \in X$, the **forward orbit** of x is the sequence $\mathcal{O}(x) := \{T^n x : n \geq 0\}$.

To determine the behaviour of this system, we study the forward orbit $\mathcal{O}(x)$. What is the limit of this orbit for each x, as n goes to $+\infty$? If x is changed a little (in some sense), then how does this limit change? This turns out to be an interesting problem, and there are systems where the limit can change dramatically for tiny changes in the starting point. This leads to the idea of chaos, which we study in Section 11.5.

We concentrate first on fixed points, that is, points $x_0 \in X$ with $x_0 = Tx_0$. These are interesting for their own sake, since analysis problems can sometimes be formulated so that the solution is the fixed point of some dynamical system. Viewing a problem this way can give important results that are otherwise hard to obtain. In this chapter, we will take this approach to Newton's method for solving equations numerically. In Chapter 12, we will use it to study differential equations.

Later in this chapter, we describe other types of orbits and then look at the notion of chaos for dynamical systems. The final section of the chapter considers iterated function schemes, which provide a means of generating fractals. It requires only Section 11.1 and the first section of Chapter 9.

11.1. Fixed Points and the Contraction Principle

Fixed points are so important in analysis that we will discuss two methods for ensuring their existence. Our first method has the added advantage that it comes equipped with a natural algorithm for computing the limit point. This has both theoretical and computational advantages over purely existential arguments.

As motivation, we start with an example, the function graphed in Figure 11.1, namely

$$Tx = 1.8(x - x^3).$$

Factoring shows that this cubic has roots at -1, 0 and 1. It has a local maximum at the point $(\frac{1}{\sqrt{3}}, \frac{2\sqrt{3}}{5})$. Since T is an odd function, it has a local minimum at $(-\frac{1}{\sqrt{3}}, -\frac{2\sqrt{3}}{5})$. To find the fixed points, we solve the equation

$$x = Tx = 1.8x - 1.8x^3.$$

This has three solutions, $x = -\frac{2}{3}$, 0, and $\frac{2}{3}$.

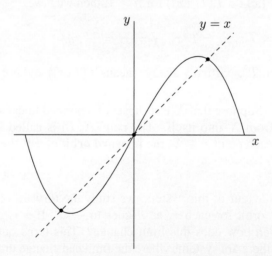

FIGURE 11.1. Graph of $Tx = 1.8(x - x^3)$ showing fixed points.

We classify fixed points based on the orbits of nearby points. This behaviour is crucial to understanding how to approximate a fixed point.

11.1.1. DEFINITION. A fixed point x^* is called an **attractive fixed point** or a **sink** is there is a open neighbourhood $U = (a, b)$ containing x^* so that for every point x in (a, b), the orbit $\mathcal{O}(x)$ converges to x^*. A fixed point x^* is called a **repelling fixed point** or a **source** if there is a neighbourhood $U = (a, b)$ containing

x^* so that for every point x in (a, b) except for x^* itself, the orbit $\mathcal{O}(x)$ always leaves the interval U.

Notice that for an attractive fixed point x^*, the interval U around x^* may be quite small. Also, for a repelling fixed point x^*, the orbit $\mathcal{O}(x)$ may return to the interval U after it leaves. However, it must then leave the interval again eventually.

We shall see that if T is differentiable, the difference between attractive and repelling fixed points comes down to the size of the derivative at x^*. A fixed point x^* is attracting if $|T'(x^*)| < 1$ and repelling if $|T'(x^*)| > 1$. The case $|T'(x^*)| = 1$ is ambiguous and might be one, the other, or neither.

In our example,

$$T'(x) = 1.8 - 5.4x^2.$$

So $T'(0) = 1.8 > 1$. The tangent line at the origin is $L(x) = 1.8x$. Since the tangent line is a good approximation to T near $x = 0$, it follows that T roughly multiplies x by the factor 1.8 when x is small. So repeated application of this to a very small nonzero number will eventually move the point far from 0. We will make this precise in Lemma 11.1.2 by using the Mean Value Theorem to show that 0 is a repelling point in the interval $\left(-\frac{1}{3}, \frac{1}{3}\right)$.

On the other hand, at $x = \pm\frac{2}{3}$, $T'(x) = -.6$. This has absolute value less than 1. So near $x = \frac{2}{3}$, the function is approximated by the tangent line

$$L(x) = \tfrac{2}{3} - .6(x - \tfrac{2}{3}).$$

This decreases the distance to $\frac{2}{3}$ by approximately a factor of .6 each iteration. That is,

$$T^{n+1}x - \tfrac{2}{3} \approx .6(T^n x - \tfrac{2}{3}).$$

So $T^n x$ converges to $\frac{2}{3}$. Again, we will obtain a precise inequality in Lemma 11.1.2. So the points $\pm\frac{2}{3}$ are attractive fixed points.

Consider the graph of the function given in Figure 11.2. Fixed points correspond to the intersection of the graph of T with the line $y = x$. Starting with any point x_0, mark the point (x_0, x_0) on the diagonal. A vertical line from this point meets the graph of T at $(x_0, Tx_0) = (x_0, x_1)$. Then a horizontal line from here meets the diagonal at (x_1, x_1). Repeated application yields a graphical picture of the dynamics. Note that starting near a fixed point, the slope of the graph determines whether the points approach or move away from the fixed point.

In our example, another typical behaviour is exhibited by all $|x|$ sufficiently large. For the sake of simplicity, consider $|x| > 2$. Then

$$|Tx| = 1.8(x^2 - 1)|x| \geq 5.4|x|.$$

It is clear then that $\lim_{n\to\infty} |T^n x| = +\infty$. So all of these orbits go off to infinity. Usually we will try to restrict our domain to a bounded region that is mapped back into itself by the transformation T.

We now connect our classification of fixed points with derivatives. Say that T is a C^1 **dynamical system** on $X \subset \mathbb{R}$ if the function T is C^1, that is, has a continuous derivative.

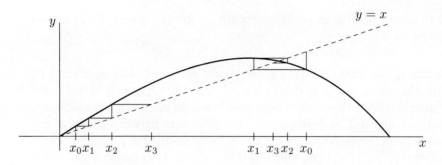

FIGURE 11.2. Fixed points for $Tx = 1.8(x - x^3)$.

11.1.2. LEMMA. *Suppose that T is a C^1 dynamical system with a fixed point x^*. If $|T'(x^*)| < c < 1$, then x^* is an attractive fixed point. Moreover, there is an interval $U = (x^* - r, x^* + r)$ about x^* so that for every $x_0 \in U$, the sequence $x_n = T^n x_0$ satisfies*

$$|x_n - x^*| \le c^n |x_0 - x^*| \le \frac{c^n}{1 - c} |x_1 - x_0|.$$

If $|T'(x^*)| > 1$, then x^* is an repelling fixed point.

PROOF. Suppose that $|T'(x^*)| < c < 1$, and let $\varepsilon = c - |T'(x^*)| > 0$. By the continuity of T', there is an $r > 0$ so that

$$|T'(x) - T'(x^*)| < \varepsilon \quad \text{for all} \quad x^* - r < x < x^* + r.$$

Hence for x in the interval $U = (x^* - r, x^* + r)$,

$$|T'(x)| \le |T'(x^*)| + |T'(x) - T'(x^*)| < (c - \varepsilon) + \varepsilon = c.$$

Let x_0 be an arbitrary point in U, and consider the sequence $x_n = T^n x_0$. Applying the Mean Value Theorem to the points x_n and x^*, there is a point z between them so that

$$\frac{T x_n - T x^*}{x_n - x^*} = T'(z).$$

Rewriting this using $T x_n = x_{n+1}$ and $T x^* = x^*$, we obtain

$$|x_{n+1} - x^*| = |T'(z)| \, |x_n - x^*|.$$

Provided that x_n belongs to the interval U, we obtain that

$$|x_{n+1} - x^*| < c|x_n - x^*|.$$

In particular, x_{n+1} is closer to x^* than x_n is; and therefore x_{n+1} also belongs to U. By induction, we obtain (verify this!) that

$$|x_n - x^*| < c^n |x_0 - x^*| \quad \text{for all} \quad n \ge 1.$$

Hence $\lim_{n \to \infty} |x_n - x^*| = 0$ by the Squeeze Theorem. That is, $\lim_{n \to \infty} x_n = x^*$. So x^* is an attractive fixed point.

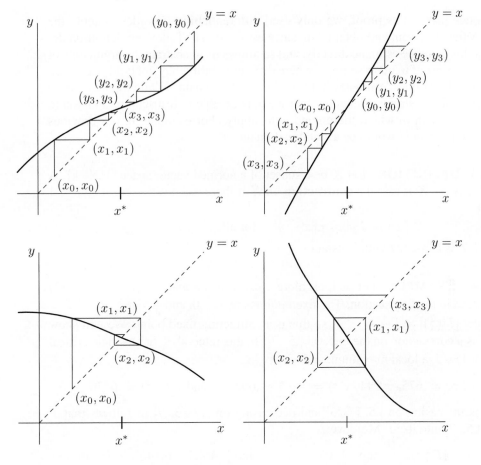

FIGURE 11.3. Iteration of a function g.

Similarly, suppose that $|T'(x^*)| > c > 1$, and let $\varepsilon = |T'(x^*)| - c > 0$. By the continuity of T', there is an $r > 0$ so that

$$|T'(x) - T'(x^*)| < \varepsilon \quad \text{for all} \quad x^* - r < x < x^* + r.$$

Hence for x in the interval $U = (x^* - r, x^* + r)$,

$$|T'(x)| \geq |T'(x^*)| - |T'(x) - T'(x^*)| > (c + \varepsilon) - \varepsilon = c.$$

The Mean Value Theorem argument works the same way as long as x_n belongs to the interval U. However, in this case, repeated iteration eventually moves x_n outside of U, at which point we have almost no information about the dynamics of T. So as long as x_n is in U, we obtain a point z between x_n and x^* so that

$$|x_{n+1} - x^*| = |T'(z)|\,|x_n - x^*| > c|x_n - x^*|.$$

This can be repeated as long as x_n remains inside U to obtain

$$|x_n - x^*| > c^n|x_0 - x^*|.$$

As this distance to x^* is tending to $+\infty$, repeated iteration eventually will move x_{n+1} outside of U. Therefore, x^* is a repelling fixed point. ∎

Notice that in this proof, we only used differentiability in order to apply the Mean Value Theorem and obtain a distance estimate. The following definition describes this distance estimate directly, and so allows us to abstract the arguments of the previous proof to other settings, where differentiability may not hold. However, we demand that these estimates hold on the whole domain. While this may seem to be excessively strong, remember that we may be able to restrict our attention to a smaller domain in which these conditions apply. For example, in our previous proof, the interval U would be a suitable domain.

11.1.3. DEFINITION. Let X be a subset of a normed vector space $(V, \|\cdot\|)$. A map $T : X \to X$ is called a **contraction** on X if there is a positive constant $c < 1$ so that
$$\|Tx - Ty\| \le c\|x - y\| \quad \text{for all} \quad x, y \in X.$$
That is, T is Lipschitz with constant $c < 1$.

11.1.4. EXAMPLE. Let us look more closely at the map $Tx = 1.8(x - x^3)$ introduced in the first section. The fixed points are $-\frac{2}{3}$, 0, and $\frac{2}{3}$.

Now $|T'(\frac{2}{3})| = |-.6| < 1$, so this is an attracting fixed point. We will show that T is a contraction on the interval $[.5, .7]$. In this interval, T has a single critical point at $1/\sqrt{3}$, a local maximum, and
$$T(.5) = .675, \quad T(1/\sqrt{3}) = .4\sqrt{3} \approx .6928, \quad \text{and} \quad T(.7) = .6426.$$
As T is increasing on $[.5, 1/\sqrt{3}]$ and decreasing on $[1/\sqrt{3}, .7]$, it follows that T maps $[.5, .7]$ into itself. Moreover,
$$\sup_{.5 \le x \le .7} |T'(x)| = \sup_{.5 \le x \le .7} 1.8|1 - 3x^2| = \max\{|.45|, |-0.846|\} = 0.846.$$
Therefore, T is a contraction on $[.5, .7]$ with contraction constant $c = .846$.

Now consider the fixed point 0. Since $T'(0) = 1.8 > 1$, this is a repelling fixed point. On the interval $[-\frac{1}{3}, \frac{1}{3}]$, we have
$$\inf_{|x| \le \frac{1}{3}} T'(x) = \inf_{|x| \le \frac{1}{3}} 1.8(1 - 3x^2) = 1.2.$$
So the proof of Lemma 11.1.2 shows that
$$|Tx| \ge 1.2|x| \quad \text{for all} \quad x \in [-\tfrac{1}{3}, \tfrac{1}{3}].$$
So the sequence $T^n x$ moves away from 0 until it leaves this interval.

11.1.5. EXAMPLE. Consider the linear function $T(x) = mx + b$ for $x \in \mathbb{R}$. Then
$$|Tx - Ty| = |m||x - y|.$$
Hence T is a contraction on $\mathbb{R}$ provided that $|m| < 1$. This map has a fixed point if there is a solution to
$$x = Tx = mx + b.$$

It is easy to compute that $x^* = \frac{b}{1-m}$ is the unique solution provided that $m \neq 1$. What happens when $m = 1$?

We may think of T as the dynamical system on $\mathbb{R}$ that maps each point x to Tx. Consider the forward orbit $\mathcal{O}(x) = \{T^n x : n \geq 0\}$ of a point x. We obtain a sequence defined by the recurrence

$$x_{n+1} = Tx_n \quad \text{for} \quad n \geq 0.$$

A simple calculation shows that

$$x_1 = mx_0 + b$$
$$x_2 = m^2 x_0 + (1 + m)b$$
$$x_3 = m^3 x_0 + (1 + m + m^2)b$$
$$x_4 = m^4 x_0 + (1 + m + m^2 + m^3)b.$$

It appears that there is a general formula

$$x_n = m^n x_0 + (1 + m + m^2 + \cdots + m^{n-1})b.$$

We may verify this by induction. It evidently holds true for $n = 1$. Suppose that it is valid for a given n. Then

$$x_{n+1} = mx_n + b = m\big(m^n x_0 + (1 + m + m^2 + \cdots + m^{n-1})b\big) + b$$
$$= m^{n+1} x_0 + (m + m^2 + \cdots + m^n)b + b$$
$$= m^{n+1} x_0 + (1 + m + m^2 + \cdots + m^n)b.$$

Hence the formula follows for $n + 1$, and so for all positive integers by induction.

When $|m| < 1$, the contraction case, this sequence has a limit that we obtain by summing an infinite geometric series.

$$\lim_{n \to \infty} x_n = \lim_{n \to \infty} m^n x_0 + (1 + m + m^2 + \cdots + m^{n-1})b$$

$$= 0 + b \sum_{k=0}^{\infty} m^k = \frac{b}{1 - m}$$

Hence this sequence converges to the fixed point x^*. Therefore, x^* is an attractive fixed point that may be located by starting *anywhere* and iterating T. This means that an approximate solution to $Tx = x$ will be **stable**, meaning that the orbit of any point close to x^* will remain close to x^*. In fact, it was not even necessary to start close to x^* to obtain convergence.

On the other hand, when $|m| > 1$, $\lim_{n \to \infty} |x_n| = \infty$ and so this sequence diverges. There is still a fixed point, but it is repelling. In this case, the answer is in the past. The map T is one to one and thus we may solve for

$$T^{-1}x = (x - b)/m = \tfrac{1}{m}x - \tfrac{b}{m}.$$

This map T^{-1} is a contraction, and we can apply the previous analysis to it. Each point x comes from the point $x_{-1} = T^{-1}x$. Going "back into the past" by setting $x_{-n-1} = T^{-1}x_{-n}$ converges to the fixed point x^*. As points close to x^* move outward and eventually go off to infinity, x^* is a source.

Finally, when $m = -1$, there is a unique fixed point. But this point cannot be located as the limit of an orbit. The reason is that

$$T^2x = -(Tx) + b = -(-x + b) + b = x.$$

That is, T^2 equals the identity map on $\mathbb{R}$. So with the exception of the fixed point $x^* = b/2$, every point has a period 2 orbit.

11.1.6. THE BANACH CONTRACTION PRINCIPLE.
Let X be a closed subset of a complete normed vector space $(V, \| \cdot \|)$. If T is a contraction map of X into X, then T has a unique fixed point x^*. Furthermore, if x is any vector in X, then $x^* = \lim\limits_{n \to \infty} T^n x$ and

$$\|T^n x - x^*\| \le c^n \|x - x^*\| \le \frac{c^n}{1 - c}\|x - Tx\|,$$

where c is the Lipschitz constant for T.

PROOF. The statement of the theorem suggests how the proof should proceed. Pick any point x_0 in X and form the sequence (x_n) given by $x_{n+1} = Tx_n$ for $n \ge 0$.

CLAIM: This sequence is Cauchy. To see this, first observe that

$$\begin{aligned}
\|x_{n+1} - x_n\| &= \|Tx_n - Tx_{n-1}\| \\
&\le c\|x_n - x_{n-1}\| \\
&\le c^2\|x_{n-1} - x_{n-2}\| \\
&\le c^n\|x_1 - x_0\| = c^n D,
\end{aligned}$$

where $D = \|x_1 - x_0\|$ is a finite number. Using this fact and the triangle inequality, we compute

$$(11.1.7) \qquad \|x_{n+m} - x_n\| \le \sum_{i=0}^{m-1} \|x_{n+i+1} - x_{n+i}\| \le \sum_{i=0}^{m-1} c^{n+i} D$$

$$< \sum_{i=0}^{\infty} c^{n+i} D = \frac{c^n}{(1 - c)}D.$$

Given $\varepsilon > 0$, choose N so large that $c^N < \varepsilon(1 - c)/D$, which is possible since $\lim\limits_{n \to \infty} c^n = 0$. Hence for $n \ge N$ and $m \ge 0$, we have $\|x_{n+m} - x_n\| < \varepsilon$. So the sequence (x_n) is Cauchy, as claimed.

Because V is complete, the sequence (x_n) converges to a point in V, say

$$\lim_{n \to \infty} x_n = x^*.$$

And since X is closed, this limit point belongs to X. Observe that T is uniformly continuous because it is Lipschitz (Proposition 5.5.4). Using continuity, we have

$$Tx^* = T\Big(\lim_{n \to \infty} x_n\Big) = \lim_{n \to \infty} Tx_n = \lim_{n \to \infty} x_{n+1} = x^*.$$

Hence x^* is a fixed point.

Suppose that $y \in X$ is also a fixed point: so $Ty = y$. Then

$$\|x^* - y\| = \|Tx^* - Ty\| \le c\|x^* - y\|.$$

Since $c < 1$, this implies that $\|x^* - y\| = 0$, whence $x^* = y$. Therefore, x^* is the unique fixed point.

Now using the estimate (11.1.7), we obtain that

$$\begin{aligned}
\|T^n x_0 - x^*\| &= \|T^n x_0 - T^n x^*\| \\
&\le c^n \|x_0 - x^*\| \\
&= c^n \lim_{m \to \infty} \|x_0 - x_m\| \\
&\le c^n (1 - c)^{-1} \|Tx_0 - x_0\|.
\end{aligned}$$

Finally, for an arbitrary point x in X, the preceding estimates yield the same story. Since the fixed point is unique, the limit x^* is obtained as the limit of every orbit independent of the starting point. ∎

11.1.8. EXAMPLE. Let $V = \mathbb{R}$, $X = [-1, 1]$ and $Tx = \cos x$. By the Mean Value Theorem, for any $x, y \in X$, there is a point z between x and y so that

$$|Tx - Ty| = |x - y| \, |\sin z|.$$

In particular, $|z| < 1$. Since $\sin x$ is increasing on $[-1, 1]$,

$$\max_{|z| \le 1} |\sin z| = |\sin \pm 1| = \sin 1 < 1.$$

Thus T is a contraction. To find the fixed point experimentally, type any value into your calculator and repeatedly hit the $\boxed{\cos}$ button. If your calculator is set for radians, the sequence will converge rapidly to $0.73908513321516064\ldots$.

What happens when you do the same for $Tx = \sin x$? It is not a contraction. Nevertheless, there is a unique fixed point $\sin 0 = 0$, and the iterated sequence converges. But it converges at a painfully slow rate. Try it on your calculator. It is slow because the derivative at the fixed point is $\cos 0 = 1$.

Indeed, the inequality $0 < \sin \theta < \theta$ for θ in $(0, \pi/2)$ shows that the only fixed point in $[0, \infty)$ is $x = 0$. As $\sin x$ is odd, the same is true on $(-\infty, 0]$. The first iteration $x_1 = \sin x_0$ lies in $[-1, 1]$; and thereafter this inequality shows that if $0 < x_1 \le 1$, then $0 \le x_{n+1} < x_n < 1$ for $n \ge 1$. Likewise, if $-1 \le x_1 < 0$, then the sequence x_n is monotone increasing and bounded above by 0. Hence the limit exists and must be the unique fixed point of $\sin x$. We can use the Taylor series of $\sin x$ (see Exercise 10.1.D) to show that

$$|\sin x - x| \le \frac{|x|^3}{6}.$$

Thus if $x_0 = .1$, it follows that

$$|x_n - x_{n+1}| \le \frac{x_0^3}{6} = 10^{-3}/6.$$

Since each step moves us at most $10^{-3}/6$, it will take over $\dfrac{.1 - .01}{10^{-3}/6} = 540$ iterations before $x_n \le .01$. After that, it will take at least 54,000 iterations to get below .001 and 5.4 million more steps to obtain a fourth decimal of accuracy in the approximation of the fixed point.

One moral of this example is that for maps T that are *not* contractions, even if applying T moves a point x very little, x need not be very close to a fixed point x^*. For contractions, it is true and we have a numerical estimate for $\|x - x^*\|$ in terms of $\|Tx - x\|$ and the Lipschitz constant.

11.1.9. EXAMPLE. If, in the definition of contraction, the condition is weakened to $\|Tx - Ty\| \le \|x - y\|$, then we cannot conclude that there is a fixed point. For example, take $X = \mathbb{R}$ and $Tx = x + 1$. Clearly, T has no fixed points. But since $\|Tx - Ty\| = \|x - y\|$, distance is preserved.

Even the strict inequality $\|Tx - Ty\| < \|x - y\|$ is not sufficient. Consider $X = [1, \infty)$ and $Sx = x + x^{-1}$. Then

$$|Sx - Sy| = (x + \tfrac{1}{x}) - (y + \tfrac{1}{y}) = |x - y|(1 - \tfrac{1}{xy}) < |x - y|$$

for all $x, y \ge 1$. However, $Sx > x$ for all x, so S has no fixed point.

11.1.10. REMARK. The first part of Lemma 11.1.2 may be proved as a simple consequence of the Banach Contraction Principle. Indeed, suppose that T is a C^1 dynamical system with a fixed point x^* such that $|T'(x^*)| < c < 1$. As before, use the continuity of T' to find an interval $I = [x^* - r, x^* + r]$ so that $|T'(x)| \le c$ for all $x \in I$. The only difference here is that we are using a closed interval and a $\le$ sign instead of an open interval and strict inequality. Apply the Mean Value Theorem to any two points x, y in I. For such points, there is a point z between them so that $|Tx - Ty| = |T'(z)|\,|x - y| \le c|x - y|$.

In particular,

$$|Tx - x^*| = |Tx - Tx^*| \le c|x - x^*| \le cr.$$

Therefore, T maps the interval I into itself, and it is a contraction with contraction constant c. By the Contraction Principle, we see that x^* is the unique fixed point in I. Moreover, we obtain the desired distance estimates.

11.1.11. EXAMPLE. This example deals with a system of linear equations and looks for a condition that guarantees an attractive fixed point. Let $A = \begin{bmatrix} a_{ij} \end{bmatrix}$ be an $n \times n$ matrix, and let $\mathbf{b} = (b_1, \dots, b_n)$ be a (column) vector in $\mathbb{R}^n$. Consider the system of equations

(11.1.12) $\mathbf{x} = A\mathbf{x} + \mathbf{b}.$

We will try to analyze this problem by studying the dynamical system given by the map $T : \mathbb{R}^n \to \mathbb{R}^n$ according to the rule

$$T\mathbf{x} = A\mathbf{x} + \mathbf{b}.$$

A solution to (11.1.12) corresponds to a fixed point of T.

There are various norms on $\mathbb{R}^n$, and different norms lead to different criteria for T to be a contraction. In this example, we will use the max norm,

$$\|\mathbf{x}\|_\infty = \max_{1 \le i \le n} |x_i|.$$

If we think of points in $\mathbb{R}^n$ as real-valued functions on $\{1, \ldots, n\}$, say $\mathbf{x}(i) = x_i$ for $1 \le i \le n$, then this is the uniform norm on a space of continuous functions. We will show that T is a contraction on $\mathbb{R}^n$ in this norm if and only if

$$c := \max_{1 \le i \le n} \sum_{j=1}^n |a_{ij}| < 1.$$

First suppose that $c \ge 1$. There is some integer i_0 so that

$$c = \sum_{j=1}^n |a_{i_0 j}|.$$

Set $x_j = \text{sign}(a_{i_0 j})$. Then

$$\|\mathbf{x} - 0\|_\infty = \|\mathbf{x}\|_\infty = \max_{1 \le i \le n} |x_i| = 1$$

while

$$\|T\mathbf{x} - T0\|_\infty = \|A\mathbf{x}\|_\infty \ge |(Ax)_{i_0}|$$
$$= \sum_{j=1}^n a_{i_0 j} x_j = \sum_{j=1}^n |a_{i_0 j}|$$
$$= c \ge 1 = \|\mathbf{x} - 0\|_\infty.$$

So T is not contractive.

On the other hand, if $c < 1$, then we compute for $\mathbf{x}, \mathbf{y} \in \mathbb{R}^n$,

$$\|T\mathbf{x} - T\mathbf{y}\|_\infty = \|A(\mathbf{x} - \mathbf{y})\|_\infty$$
$$= \max_{1 \le i \le n} \Big| \sum_{j=1}^n a_{ij}(x_j - y_j) \Big|$$
$$\le \max_{1 \le i \le n} \sum_{j=1}^n |a_{ij}| \, |x_j - y_j|$$
$$\le \Big(\max_{1 \le i \le n} \sum_{j=1}^n |a_{ij}| \Big) \Big(\max_{1 \le i \le n} \sum_{j=1}^n |x_j - y_j| \Big)$$
$$= c \|\mathbf{x} - \mathbf{y}\|_\infty.$$

So T is a contraction.

Thus by the Banach Contraction Principle, we know that there is a unique solution that can be computed iteratively from any starting point. Choose $\mathbf{x}_0 = 0$.

Then

$$\mathbf{x}_1 = T\mathbf{x}_0 = \mathbf{b}$$
$$\mathbf{x}_2 = T\mathbf{x}_1 = \mathbf{b} + A\mathbf{b} = (I + A)\mathbf{b}$$
$$\mathbf{x}_3 = T\mathbf{x}_2 = \mathbf{b} + A(\mathbf{b} + A\mathbf{b}) = (I + A + A^2)\mathbf{b}.$$

We will show that

$$\mathbf{x}_n = (I + A + \cdots + A^{n-1})\mathbf{b}.$$

This is evident for $n = 0, 1, 2, 3$ by the previous calculations. Assume it is true for some integer n. Then

$$\mathbf{x}_{n+1} = T\mathbf{x}_n = \mathbf{b} + A(I + A + \cdots + A^{n-1})\mathbf{b} = (I + A + \cdots + A^n)\mathbf{b}.$$

So the formula is established by induction.

The solution to (11.1.12) is the unique fixed point

$$\mathbf{x}^* = \lim_{n \to \infty} \mathbf{x}_n$$

$$= \lim_{n \to \infty} (I + A + \cdots + A^{n-1})\mathbf{b} = \sum_{k=0}^{\infty} A^k \mathbf{b}.$$

The important factor that makes this infinite sum convergent is that, because T and A are contraction,

$$\|A^k \mathbf{b}\|_\infty = \|A^k \mathbf{b} - A^k 0\| \le c^k \|\mathbf{b} - 0\|_\infty = c^k \|\mathbf{b}\|_\infty.$$

So this series is dominated by a convergent geometric series and thus converges by the comparison test. Indeed, the same argument shows that the series $\sum_{k=0}^{\infty} A^k \mathbf{x}$ converges for *every* vector $\mathbf{x} \in \mathbb{R}^n$. So the sum $C = \sum_{k=0}^{\infty} A^k$ makes sense as a linear transformation. We note that, in particular, if $\mathbf{x}$ is any vector in $\mathbb{R}^n$, then $\lim_{n \to \infty} A^n \mathbf{x} = 0$. The solution to our problem is $\mathbf{x}^* = C\mathbf{b}$.

We know from linear algebra how to solve the equation $\mathbf{x} = A\mathbf{x} + \mathbf{b}$. This leads to $(I - A)\mathbf{x} = \mathbf{b}$. When $I - A$ is invertible, there is a unique solution $\mathbf{x} = (I - A)^{-1}\mathbf{b}$. This suggests that our contractive condition (11.1.11) leads to the conclusion that $I - A$ is invertible with inverse $C = \sum_{k=0}^{\infty} A^k$. To see that this is the case, compute

$$(I - A)C\mathbf{x} = \lim_{n \to \infty} (I - A)(I + A + \cdots + A^{n-1})\mathbf{x}$$

$$= \lim_{n \to \infty} (I + A + \cdots + A^{n-1})\mathbf{x} - (A + A^2 + \cdots + A^n)\mathbf{x}$$

$$= \lim_{n \to \infty} \mathbf{x} - A^n \mathbf{x} = \mathbf{x}.$$

Therefore, $I - A$ has inverse C.

This formula $(I - A)^{-1} = \sum_{k=0}^{\infty} A^k$ should be seen as parallel to the power series identity

$$\frac{1}{1-x} = \sum_{k=0}^{\infty} x^k \quad \text{for} \quad |x| < 1.$$

The condition $|x| < 1$ guarantees that the series converges. The same role is played for matrices by the contractive condition (11.1.11).

Exercises for Section 11.1

A. Let $Tx = 1.8(x - x^3)$. Find the smallest real number R so that $\lim_{n \to \infty} |T^n x| = +\infty$ for all $|x| > R$.

B. Show that $Tx = \sin x$ is not a contraction on $[-1, 1]$.

C. Give an example of a differentiable map T from $\mathbb{R}$ to $\mathbb{R}$ whose fixed points are exactly the set of integers. Find points where $|T'(x)| > 1$.

D. Explain why, in the previous example, points with $|T'(x)| > 1$ necessarily exist.

E. Suppose that S and T are contraction with Lipschitz constants s and t, respectively. Prove that the composition ST is a contraction with Lipschitz constant st.

F. Consider a case of Example 11.1.11 for $A = \begin{bmatrix} .5 & .4 \\ 0 & .8 \end{bmatrix}$ and $b = \begin{bmatrix} .1 \\ .2 \end{bmatrix}$. Explicitly compute the infinite sum $\sum_{k=0}^{\infty} A^k$ in order to solve for the fixed point of $Tx = Ax + b$.

G. Redo Example 11.1.11 using the 1-norm $\|\mathbf{x}\|_1 = \sum_{i=1}^{n} |x_i|$ on $\mathbb{R}^n$ in place of $\|\mathbf{x}\|_\infty$.

HINT: Show that T is a contraction if and only if $\max_{1 \le j \le n} \sum_{i=1}^{n} |a_{ij}| < 1$.

H. Define T on $C[-1, 1]$ by $Tf(x) = f(x) + \dfrac{x^2 - f(x)^2}{2}$. Set $f_0 = 0$ and $f_{n+1} = Tf_n$ for $n \ge 0$. Prove that f_n is a monotone increasing sequence of functions such that

$$0 \le x^2 - f_{n+1}(x)^2 \le (1 - \tfrac{1}{4}x^2)(x^2 - f_n(x)^2).$$

Hence show that f_n is a sequence of polynomials converging uniformly to $|x|$.

I. Define a map $\mathcal{D}$ on $C[0, 1]$ as follows:

$$\mathcal{D}f(x) = \begin{cases} \frac{2}{3} + \frac{1}{3}f(3x) & \text{for} \quad 0 \le x \le \frac{1}{3} \\ (2 + f(1))(\frac{2}{3} - x) & \text{for} \quad \frac{1}{3} \le x \le \frac{2}{3} \\ x - \frac{2}{3} & \text{for} \quad \frac{2}{3} \le x \le 1. \end{cases}$$

(a) Sketch the graph of some function f and $\mathcal{D}f$.
(b) Show that $\mathcal{D}$ is a contraction.
(c) Describe the fixed point. HINT: f_∞ of Exercise 11.4.H

J. Suppose that S and T are contractions on X with Lipschitz constant $c < 1$ and fixed points x_s and x_t respectively. Prove that $\|x_s - x_t\| \leq (1 - c)^{-1}\|S - T\|_\infty$, where $\|S - T\|_\infty = \sup_{x \in X} \|Sx - Tx\|$.
HINT: Estimate $\|x_s - x_t\|$ in terms of $\|x_s - Tx_s\|$.

K. Suppose that for $0 \leq s \leq 1$, T_s is a contraction of a complete normed space X with Lipschitz constant $c < 1$. Moreover, assume that this is a continuous path of contractions. That is, $\lim_{s \to s_0} \|T_s - T_{s_0}\|_\infty = 0$. Prove that the fixed points x_s of T_s form a continuous path.
HINT: Use the previous exercise.

11.2. Newton's Method

Newton's method is an iterative method for rapidly computing the zeros of differentiable functions. The Contraction Principle gives a nice proof that this method works.

Suppose we have a function f on $\mathbb{R}$ and a reasonable guess x_0 for the zero (or root) of f. If f is differentiable at x_0, then we can draw the tangent line at x_0. It seems plausible that *if x_0 is close to the root of the function, the root of the tangent line will be even closer.* This uses one of the basic ideas of Differential Calculus: The tangent line to f at x_0 is a good approximation to f near x_0.

The equation of the tangent line is

$$y = f(x_0) + f'(x_0)(x - x_0).$$

To find the root, we set $y = 0$ and solve for x, to obtain

$$x = x_0 - \frac{f(x_0)}{f'(x_0)}.$$

Call this root x_1. By repeating the same calculation for x_1, we obtain a sequence $(x_n)_{n=1}^\infty$, where

$$x_{n+1} = x_n - \frac{f(x_n)}{f'(x_n)} \quad \text{for} \quad n \geq 1.$$

Our hope is that the sequence converges to a point x^* satisfying $f(x^*) = 0$. See Figure 11.4 for an example of (the first few terms) of such a sequence.

To prove this, we reformulate the problem using fixed points. Suppose $f(x)$ is twice differentiable and $f(x^*) = 0$ for some point x^*. Define a dynamical system T by

$$Tx = x - \frac{f(x)}{f'(x)}.$$

It is easy to see that if $f(x^*) = 0$, then $Tx^* = x^*$ and vice versa. So we are looking for a fixed point for the function T. Notice that for this definition to make sense, we require $f'(x) \neq 0$ on our domain. This will show up in our hypotheses as the condition $f'(x^*) \neq 0$.

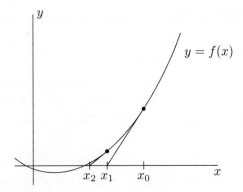

FIGURE 11.4. Example of a function f and sequence (x_n).

We compute the derivative

$$T'(x) = 1 - \frac{f'(x)^2 - f(x)f''(x)}{f'(x)^2} = \frac{f(x)f''(x)}{f'(x)^2}.$$

Notice that $T'(x^*) = 0$. So x^* is an attractive fixed point by Lemma 11.1.2. Moreover, the constant c used may be any small positive number. Indeed, the closer we get to x^*, the smaller the value of c that may be used. This will be important in obtaining very rapid convergence.

In addition to verifying that Newton's method works, we can estimate how quickly the error decreases. This kind of error analysis is fundamental to deciding the effectiveness of an algorithm. It is considered minimally acceptable if the error tends to zero **geometrically** in the sense that

$$|x_{n+1} - x^*| \le c|x_n - x^*| \quad \text{for} \quad n \ge N$$

for some constant $c < 1$. For if $c^s < .1$, we obtain an additional digit of accuracy every s iterations. Compare Example 11.1.8.

However, if great accuracy is desired, this is not all that fast. Some algorithms, such as Newton's method, converge **quadratically**, meaning that there is a constant M so that

$$|x_{n+1} - x^*| \le M|x_n - x^*|^2 \quad \text{for} \quad n \ge N.$$

Once $|x_n - x^*| < .1/M$, the number of digits of accuracy doubles every iteration. Thus, once we get sufficiently close to the solution, the sequence approaches x^* very rapidly.

11.2.1. NEWTON'S METHOD.

Suppose that f is twice continuously differentiable, and there is a point $x^* \in \mathbb{R}$ such that $f(x^*) = 0$ and $f'(x^*) \ne 0$. There is an $r > 0$ so that if x_0 belongs to $[x^* - r, x^* + r]$, then the iterates $x_{n+1} = x_n - f(x_n)/f'(x_n)$ converge to x^*. Moreover, there is a constant M such that

$$|x_{n+1} - x^*| < M|x_n - x^*|^2 \quad \text{for} \quad n \ge 1;$$

so the iterates converge quadratically.

PROOF. Let $Tx = x - f(x)/f'(x)$. Then $Tx^* = x^*$. As computed previously,

$$T'(x) = \frac{f(x)f''(x)}{f'(x)^2} .$$

In particular, $T'(x^*) = 0$. Also, $T'(x)$ is defined for x near x^* since $f'(x) \neq 0$ for x near x^*. Choose $r > 0$ sufficiently small so that $|T'(x)| \leq \frac{1}{2}$ on the interval $[x^* - r, x^* + r]$. By Lemma 11.1.2, for any x_0 in this range, the iterates $x_{n+1} = Tx_n$ converge to x^*, and

$$|x_n - x^*| \leq 2^{-n}|x_0 - x^*| \leq 2^{1-n}|x_0 - x_1|.$$

In particular, x_n converge at least geometrically to the fixed point, and each x_{n+1} is closer to this point than x_n.

Quadratic convergence is a consequence of using the Mean Value Theorem in a more precise way. The estimate from Lemma 11.1.2 only used the fact that $|T'(x)| \leq c$ near the fixed point x^*. We will exploit the fact that $T'(x^*) = 0$. Here are the details. Let

$$A = \sup_{|x-x^*| \leq r} |f''(x)| \qquad \text{and} \qquad B = \inf_{|x-x^*| \leq r} |f'(x)|,$$

and set $M = A/B$. By the Mean Value Theorem, there is a point a_n between x_n and x^* such that

$$f(x_n) = f(x_n) - f(x^*) = f'(a_n)(x_n - x^*).$$

Solve for $x_n - x^*$ and substitute into the following:

$$x_{n+1} - x^* = (x_n - x^*) + (x_{n+1} - x_n)$$

$$= \frac{f(x_n)}{f'(a_n)} - \frac{f(x_n)}{f'(x_n)}$$

$$= \frac{f(x_n)}{f'(a_n)f'(x_n)}\big(f'(x_n) - f'(a_n)\big)$$

$$= \frac{x_n - x^*}{f'(x_n)}\big(f'(x_n) - f'(a_n)\big).$$

Applying the Mean Value Theorem again, this time to f', provides a point b_n between x_n and a_n such that

$$|f'(x_n) - f'(a_n)| = |f''(b_n)(x_n - a_n)| \leq A|x_n - x^*|.$$

Combining this with $|f'(x_n)| \geq B$ and substituting again yields

$$|x_{n+1} - x^*| \leq \frac{|x_n - x^*|}{B}A|x_n - x^*| = M|x_n - x^*|^2.$$

This establishes the quadratic convergence. ∎

11.2.2. EXAMPLE. COMPUTATION OF SQUARE ROOTS. A long time ago in places far far away, everyone had to compute square roots by hand. Today, a few people have to program calculators to compute them. Newton's method is an excellent way for a person or a computer to find square roots rapidly to high accuracy.

To illustrate the value of the method and of quadratic convergence, we'll compute a few square roots. Unlike a calculator, we also give an upper bound for the error between our computed value and the true value.

Finding the square root of $a > 0$ means finding the positive root of the function $f(x) = x^2 - a$. Applying Newton's method, a simple computation gives

$$Tx = x - \frac{x^2 - a}{2x} = \frac{1}{2}\left(x + \frac{a}{x}\right).$$

In fact, we will see that any initial positive choice of x_1 will converge to $\sqrt{a}$. Try this on your calculator as a method of computing $\sqrt{2}$ or $\sqrt{71}$.

Suppose that you are asked to compute $\sqrt{149}$ to 20 digits of accuracy. Since $12^2 = 144 < 149 < 169 = 13^2$, let $x_0 = 12$ be our first approximation. Clearly, the error is at most .5. Moreover,

$$T'(x) = \frac{1}{2}\left(1 - \frac{149}{x^2}\right).$$

This is evidently monotone. Thus on $[12, 13]$, the derivative is bounded by

$$\max\{|T'(12)|, |T'(13)|\} = \max\left\{\frac{5}{288}, \frac{10}{169}\right\} = \frac{10}{169} < .06.$$

Therefore, Newton's method guarantees that the sequence

$$x_0 = 12, \qquad x_{n+1} = \frac{1}{2}\left(x_n + \frac{149}{x_n}\right) \quad \text{for} \quad n \geq 0$$

converges quadratically to $\sqrt{149}$.

Let us compute the other constants involved. As $f'(x) = 2x$ and $f''(x) = 2$, we obtain

$$A = \sup_{12 \leq x \leq 13} |f''(x)| = 2$$
$$B = \inf_{12 \leq x \leq 13} |f'(x)| = 24$$
$$M = \frac{A}{B} < .084.$$

Again using the Mean Value Theorem, note that

$$|f(x_0)| = |f(x_0) - f(x^*)| = |f'(c)| \, |x_0 - x^*| \geq B|x_0 - x^*|.$$

Thus

$$|x_0 - x^*| < |f(x_0)|/24 = 5/24 < .21.$$

From the error estimate for Newton's method, we obtain

$$|x_{n+1} - x^*| \leq .084|x_n - x^*|^2.$$

Starting with $x_0 = 12$, we have the following table of terms and bounds on the error.

n	x_n	Bound on $\lvert x_n - x^* \rvert$
0	12	.21
1	12.2083	3.704×10^{-3}
2	12.206555745165	1.153×10^{-6}
3	12.2065556157337036	3.572×10^{-13}
4	12.20655561573370295189	1.117×10^{-26}
5	12.20655561573370295189	1.048×10^{-53}

We already have 20 digits of accuracy at $n = 4$. In fact, to progress further, we need to worry more about the round-off error of our calculations than with Newton's method.

Look at the global aspects of this example. It is easy to see from the graph that $f(x) = x^2 - 149$ is concave, much like Figure 11.4. If x_0 lies in $(0, \sqrt{149})$, then $x_1 > \sqrt{149}$. However, taking x_0 very small will result in a huge second term. After that, it is also apparent from the graph that x_n decreases monotonely and quickly to $\sqrt{149}$. Similarly, starting with $x_0 < 0$, the same procedure follows from reflection of the whole picture in the y-axis—the sequence converges rapidly to $-\sqrt{149}$. The point $x_0 = 0$ is a nonstarter because $f'(0) = 0$.

Exercises for Section 11.2

A. Show that $f(x) = x^3 + x + 1$ has exactly one real root. Use Newton's method to approximate it to eight decimal places. Show your error estimates.

B. The equation $\sin x = x/2$ has exactly one positive solution. Use Newton's method to approximate it to eight decimal places. Show your error estimates.

C. (a) Set up Newton's method for computing cube roots.
(b) Show by hand that $\sqrt[3]{2} - 1.25 < .01$.
(c) Compute $\sqrt[3]{2}$ to eight decimal places.

D. Let $f(x) = (\sqrt{2})^x$ for $x \in \mathbb{R}$. Sketch $y = f(x)$ and $y = x$ on the same graph. Given $x_0 \in \mathbb{R}$, we define a sequence by $x_{n+1} = f(x_n)$ for $n \geq 0$.
(a) Find all fixed points of f.
(b) Show that the sequence $(x_n)_{n=1}^{\infty}$ is monotone.
(c) If $x^* = \lim_{n \to \infty} x_n$ exists, prove that $f(x^*) = x^*$.
(d) For which $x_0 \in \mathbb{R}$ does the sequence (x_n) converge, and what is the limit?

E. Find the largest critical point of $f(x) = x^2 \sin(1/x)$ to four decimal places.

F. Find the minimum value of $f(x) = (\log x)^2 + x$ on $(0, \infty)$ to four decimal places.

G. Apply Newton's method to find the root of $f(x) = (x - r)^{1/3}$. Start with any point $x_0 \neq r$, and compute $\lvert x_n - r \rvert$. Explain what went wrong here.

H. **Modified Newton's method**. With the same setup as for Newton's method, show that
the sequence $x_{n+1} = x_n - \dfrac{f(x_n)}{f'(x_0)}$ for $n \geq 0$ converges to x^*.

I. (a) Let $a > 0$. Show that Newton's formula for solving $xa = 1$ yields the iteration
$x_{n+1} = 2x_n - ax_n^2$.

    (b) Suppose that $x_0 = \dfrac{1 - \varepsilon}{a}$ for some $|\varepsilon| < 1$. Derive the formula for x_n.

    (c) Do the same analysis for the iteration scheme $x_{n+1} = x_n(3 - ax_n - (ax_n)^2)$.
Explain why this is a superior algorithm.

J. Let $h(x) = x^{1/3} e^{-x^2}$.

    (a) Set up Newton's method for this function.

    (b) If $0 < |x_n| < 1/\sqrt{6}$, then $|x_{n+1}| > 2|x_n|$.

    (c) Show that if $x_n > 1/\sqrt{6}$, then $x_n < x_{n+1} < x_n + \frac{1}{2x_n}$.

    (d) Hence show that Newton's method never works unless $x_0 = 0$. However, given
$\varepsilon > 0$, there will be an N so large that $|x_{n+1} - x_n| < \varepsilon$ for $n \geq N$.

    (e) Sketch h and try to explain this nasty behaviour.

K. Three towns, Alphaville, Betatown, and Gammalot, are situated around the shore of a
circular lake of radius 1 km. The largest town Alphaville claims one half of the area of
the lake as its territory. The town mathematician is charged with computing the radius
r of a circle from the town hall (which is right on the shore) that will cut off half the
area of the lake. Compute r to seven decimal places.

HINTS: Let T denote the town hall, and O the centre of the lake. The circle of radius
r meets the shoreline at X and Y. The area enclosed is the union of two segments of
circles cut off by the chord XY. Express r and the area A as functions of the angle
$\theta = \angle OTX$.

MORE HINTS: (1) Show that (i) $1 < r < 2$; (ii) $\angle TOX = \pi - 2\theta$; (iii) the area
of the segment of a circle of radius ρ cut off by a chord that subtends an angle α is
$\rho^2(\alpha - \sin\alpha)/2$. (2) Show that $r = 2\cos\theta$ and $A(\theta) = 2\pi - 4\theta \sin^2\theta - 2\sin(2\theta)$. (3)
Solve $A(\theta) = \pi$ using Newton's method. Show error estimates.

11.3. Orbits of a Dynamical System

There are several possibilities for the structure of the orbit of a point x_0. Fixed
points, which we've discussed in detail, have the simplest possible orbits, namely
$\mathcal{O}(x_0) = \{x_0\}$. Almost as good as fixed points from the point of view of dynamics,
and certainly more common, are periodic points. We say that x^* is a **periodic point**
if there is a positive integer n such that $T^n x^* = x^*$. The smallest positive n for
which this holds is called the **period**. Notice that x^* is a fixed point of T^n. We can
therefore call x^* an **attractive periodic point** or a **repelling periodic point** for T
if it is an attractive or repelling fixed point of T^n. Because T maps an open set
around x^* to an open set around Tx^*, it is easy to check that points in the same
periodic orbit are either all attractive or all repelling.

Let us discuss the terminology of dynamical systems by examining a particular
map, namely the map $Tx = 1.8(x - x^3)$ that we discussed in Section 11.1. Our

example contains an orbit of period 2, namely $\{\pm\sqrt{14}/3\}$. Indeed, it is an easy calculation to see that

$$T(\sqrt{14}/3) = -\sqrt{14}/3 \quad \text{and} \quad T(-\sqrt{14}/3) = \sqrt{14}/3.$$

See Figure 11.5. This is a repelling orbit since

$$(T^2)'(\sqrt{14}/3) = T'(T\sqrt{14}/3)T'(\sqrt{14}/3)$$

$$= (1.8 - 5.4\frac{14}{3})^2 = 547.56 > 1.$$

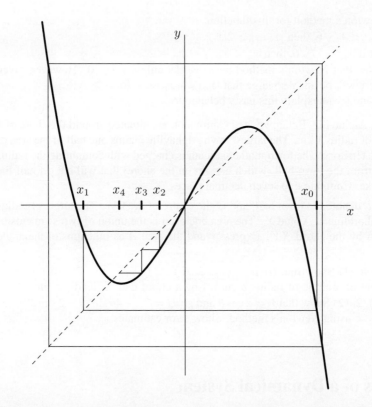

FIGURE 11.5. Graph of $Tx = 1.8(x-x^3)$ showing period 2 orbit and a nearby orbit.

A point x is called an **eventually periodic point** if $T^n x$ eventually belongs to a period. Consider our example again. Notice that $T(1) = 0$. So $T^n(1) = 0$ for all $n \geq 1$. The equation $Tx = 1$ has a solution that is the root of the cubic

$$1 = 1.8x - 1.8x^3.$$

This has a solution r since every real polynomial of odd degree has a root by Exercise 5.6.D and we may calculate $r \approx -1.20822631883$. Then with $x_0 = r$, we obtain $x_1 = Tx_0 = 1$ and $x_2 = T(1) = 0$, and so $x_0 = T^n x_0 = 0$ for all $n \geq 2$. Similarly, we have a sequence (r_n) of points so that $T^n(r_n) = 0$, such as $r_0 = 0$,

$r_1 = 1$, $r_2 \approx -1.20823$, $r_3 \approx 1.24128$. Observe that r_n is close to $(-1)^{n-1}\sqrt{14}/3$ for large n, in the sense that $r_n - (-1)^{n-1}\sqrt{14}/3$ converges to zero.

Likewise there are points that are eventually mapped onto the two attractive fixed points $\pm\frac{2}{3}$.

Figure 11.6, called a **phase portrait**, attempts to show graphically the behaviour of our mapping T. Rather than using a graph, it represents the space as a line and the function as arrows taking a point x to Tx. In particular, you should observe that except for a sequence of points that eventually map to 0, every other point in $(-\sqrt{14}/3, \sqrt{14}/3)$ has an orbit that converges to one of the two attractive fixed points $\pm 2/3$. Notice that, except for the two arrows for the period 2 orbit, arrows above the line show the first few terms of the sequence (r_n) mentioned previously, while those below the line show the "nearby" orbit from Figure 11.5.

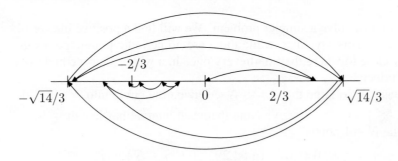

FIGURE 11.6. Phase portrait of $Tx = 1.8(x - x^3)$

Of course, this example does not exhaust the possibilities for the behaviour of an orbit. It can happen that the orbit of a single point is dense in the whole space. A point x is called a **transitive point** if $\overline{\mathcal{O}(x)} = X$. The following examples are more complicated than our first one and exhibit this new possibility as well as having many periodic points.

11.3.1. EXAMPLE. Let the space be the unit circle $\mathbb{T}$. We can describe a typical point on the circle by the angle θ it makes to the positive real axis in radians. If θ is changed to $\theta + 2n\pi$, the point determined remains the same. So this angle is determined "up to a multiple of 2π." We may add two angles, or multiply them by an integer. The resulting point does not depend on how we initially measure the angle. (Notice that when multiplying by a fraction such as $\frac{1}{2}$, the answer does depend on which way the angle is represented. Half of 1 does not determine the same point as half of $1 + 2\pi$.) We call this arithmetic **modulo** 2π. We will write

$$\theta \equiv \varphi \pmod{2\pi}$$

to mean that $\theta - \varphi$ is an integer multiple of 2π, which is to say that θ and φ represent the same point on the circle. In this way, we may think of the unit circle $\mathbb{T}$ as the real line wrapping around the circle infinitely often, meeting up at the same point every 2π. This will be convenient for calculation.

Consider the rotation map R_α through an angle α given by

$$R_\alpha \theta \equiv \theta + \alpha \pmod{2\pi}.$$

It is rather easy to analyze what happens here. The only way to obtain a periodic point is to have

$$\theta \equiv R_\alpha^n \theta = \theta + n\alpha \pmod{2\pi}.$$

This requires $n\alpha$ to be an integer multiple of 2π. When this occurs and n is as small as possible, it follows that *every* point in $\mathbb{T}$ has period n. Indeed, $R_\alpha^n = \mathrm{id}$, the identity map. This is the case for those α that are a rational multiple of 2π. Equivalently, R_α is periodic if and only if $\alpha/2\pi \in \mathbb{Q}$.

On the other hand, for all irrational values of $\alpha/2\pi$, R_α has no periodic points. We will show that the orbit of every point is dense in the circle. Indeed, suppose that $\varepsilon > 0$ and that θ and φ are two points on the circle. We wish to find an integer n so that $|R_\alpha^n \theta - \varphi| < \varepsilon$.

To do this, we first solve a simpler problem. We will find a positive integer m so that $|R_\alpha^m 0| < \varepsilon$. This says that, while there are no periodic points, points do come back very close to where they start every once in a while. Our method uses the pigeonhole principle.

Pick an integer N so large that $2\pi < N\varepsilon$. Divide the circle into N intervals $I_k = \left[\frac{(k-1)2\pi}{N}, \frac{k2\pi}{N} \right)$ for $1 \leq k \leq N$. Note that each interval has length $\frac{2\pi}{N} < \varepsilon$. Now consider the $N + 1$ points

$$\{ x_j \equiv R_\alpha^j 0 = j\alpha \pmod{2\pi} : 0 \leq j \leq N \}.$$

These $N + 1$ points are distributed in some way among the N intervals I_k. As there are more points than intervals, the pigeonhole principle merely asserts that some interval contains at least two points. Let $i < j$ be two integers in our set so that x_i and x_j both lie in some interval I_k. These two points are therefore close to each other. Precisely,

$$|x_j - x_i| < \frac{2\pi}{N} < \varepsilon.$$

Now we use the fact that R_α is isometric, meaning that

$$|R_\alpha \theta - R_\alpha \varphi| = |\theta - \varphi|$$

for any two points $\theta, \varphi \in \mathbb{T}$. This just says that the map R_α is a rigid rotation that does not change the distance between points. Let $m = j - i$. Hence

$$|R_\alpha^m 0| = |x_{j-i} - x_0|$$
$$= |R_\alpha^i x_{j-i} - R_\alpha^i x_0| = |x_j - x_i| < \varepsilon.$$

It is also important that x_m is not 0. This follows because $\alpha/2\pi$ is irrational. So we observe that

$$R_\alpha^{km} 0 \equiv k x_m \pmod{2\pi} \quad \text{for} \quad k \geq 0$$

forms a sequence of points that move around the circle in steps smaller than ε. So it is possible to choose k so that $k x_m$ is close to $\varphi - \theta$, and thus

$$|\varphi - R_\alpha^{km} \theta| = |(\varphi - \theta) - R_\alpha^{km} 0| = |(\varphi - \theta) - k x_m| < \varepsilon.$$

So every orbit is dense in the whole circle.

11.3.2. EXAMPLE. We will now look at an example that has both interesting periodic points and transitive orbits. Some of the proofs must be left until later (see Examples 11.5.5, 11.5.10, and 11.5.14). We shall see that this example is chaotic. While this word is suggestive of wild behaviour, it actually has a precise mathematical meaning which we will explore in Section 11.5.

Consider the map T from $\mathbb{T}$ into itself given by $T\theta = 2\theta$. This is called the **doubling map** on the circle. Essentially this map wraps the circle twice around itself. That is, the top semicircle $[0, \pi)$ is mapped one-to-one and onto the whole circle; and the bottom semicircle $[\pi, 2\pi)$ is also mapped one-to-one and onto the whole circle. Thus this map is two-to-one.

A point θ is periodic of period $n \geq 1$ if

$$\theta \equiv T^n\theta = 2^n\theta \pmod{2\pi}.$$

This happens if and only if $(2^n - 1)\theta$ is an integer multiple of 2π. The period of $\mathcal{O}(\theta)$ will be the smallest positive integer k such that $(2^k - 1)\theta$ is an integer multiple of 2π. Thus the point $\frac{2\pi}{2^n-1}$ is periodic of period n, and

$$\mathcal{O}\left(\tfrac{2\pi}{2^n-1}\right) = \{\tfrac{2^j\pi}{k-1} : 0 \leq j \leq 2^n - 1\}.$$

Indeed, every point $\frac{2\pi s}{2^n-1}$ for every $n \geq 1$ and $1 \leq s \leq 2^n - 1$ is a periodic point, although the period will possibly be a proper divisor of n rather than n itself. These points are dense in the whole circle. Because the derivative of T is 2, as a map from $\mathbb{R}$ to $\mathbb{R}$, it follows that every periodic point is repelling.

Also, there are eventually periodic points, namely $\frac{2\pi s}{2^p(2^n-1)}$ for $p \geq 1$. After p iterations, these points join the periods identified previously. On the other hand, it is not difficult to see that this is a complete list of all the periodic and eventually periodic points. So every other point has infinite orbit. Unlike our first example, these orbits will not converge to some period, as every periodic point is repelling.

This example also has a dense set of transitive points, although we only outline the argument. Write a point θ as $2\pi t$ for $0 \leq t < 1$. Then write t in binary as $t = (0.\varepsilon_1\varepsilon_2\varepsilon_3 \dots)_{\text{base 2}}$. Then $T^k\theta \equiv 2\pi t_k \pmod{2\pi}$, and the binary expansion is $t_k = (0.\varepsilon_{k+1}\varepsilon_{k+2}\varepsilon_{k+3} \dots)_{\text{base 2}}$. The possible limit points of this orbit has little to do with the first few (say) billion coefficients. So we may use these to specify θ close to any point in the circle. Now arrange the tail of the binary expansion to include all possible finite sequences of 0s and 1s. Then by applying T repeatedly, each of these finite sequences eventually appears as the initial part of the binary expansion of t_k. This shows that the orbit is dense in the whole circle.

Exercises for Section 11.3

A. Suppose that x^* is a point of period n. Show that if x^* is attracting (or repelling), then each $T^i x^*$ is also an attracting (or repelling) periodic point.

B. Draw a phase diagram of the dynamics of $Tx = .5(x - x^3)$ for $x \in \mathbb{R}$.

C. Find the periodic points of the **tripling map** on the circle: $T : \mathbb{T} \to \mathbb{T}$ given by $T\theta = 3\theta$.

D. Consider $Tx = a(x - x^3)$ for $x \in \mathbb{R}$ and $a > 0$.
   (a) Find all fixed points. Decide if they are attracting or repelling.
   (b) Find all points of period 2.
   HINT: First look first for solutions of $Tx = -x$. To factor $T^2 x - x$, use the fact that each fixed point is a root to factor out a cubic, and factor out a quadratic corresponding to the period 2 cycle already found.
   (c) Decide if the period 2 points are attracting or repelling.
   (d) Find the three bifurcation points corresponding to the changes in the period 1 and 2 points (i.e., at which values of the parameter a do changes in the dynamic occur?).
   (e) Draw a phase diagram of the dynamics for $a = 2.1$.

E. Consider the **tent map** T of $[0, 1]$ onto itself by $Tx = \begin{cases} 2x & \text{if } 0 \le x \le \frac{1}{2} \\ 2 - 2x & \text{if } \frac{1}{2} \le x \le 1. \end{cases}$
   (a) Graph T^n for $n = 1, 2, 3, 4$.
   (b) Using the graph, show that there are exactly 2^n points which are fixed for T^n. How are they distributed?
   (c) Use (b) to show that the periodic points are dense in X.
   (d) Show that there are two distinct orbits of period 3.
   HINT: Solve $T^3 x = x$ for $x \in [\frac{1}{8}, \frac{1}{4}]$ and for for $x \in [\frac{1}{4}, \frac{3}{8}]$.
   (e) Show that there are points of period n for every positive integer n.
   (f) Find all points that are not fixed but are eventually fixed. Show that they are also dense in X.

F. Let $\omega(x) = \bigcap_{n \ge 0} \overline{\mathcal{O}(T^n x)}$ be the **cluster set** of the forward orbit of T.
   (a) Show that T maps $\omega(x)$ into itself.
   (b) Show that $\overline{\mathcal{O}(x)} = \mathcal{O}(x) \cup \omega(x)$.
   (c) Show by example that $\omega(x)$ can be empty.
   (d) If $\mathcal{O}(x)$ is compact, show that $\omega(x)$ is a nonempty subset of $\mathcal{O}(x)$.

G. Suppose that $\mathcal{O}(x)$ is compact.
   (a) Let n_0 be the smallest integer such that $T^{n_0} x \in \omega(x)$. Prove that
   $$\mathcal{O}(T^{n_0} x) = \omega(x) = \omega(T^{n_0} x).$$
   (b) If $\omega(x)$ is infinite, show that it must be perfect. This contradicts Exercise 4.4.L.
   (c) Show that $\mathcal{O}(x)$ is a compact set if and only if there is an n_0 so that $T^{n_0} x$ is periodic.

H. A dynamical system (X, T) is **minimal** if the only nonempty closed subset F of X such that $TF \subset F$ is X itself.
   (a) Show that (X, T) is minimal if and only if every point in X is transitive.
   (b) Show that the rotation R_α on the circle $\mathbb{T}$ is minimal if and only if $\alpha/2\pi$ is irrational.

I. Let C be the Cantor set, and represent each point $x \in C$ in its ternary expansion using only 0s and 2s. Define $T : C \to C$ by

$$T \, 0.\overbrace{2 \ldots 2}^{k-1} 0 \varepsilon_{k+1} \varepsilon_{k+2} \cdots = 0.\overbrace{0 \ldots 0}^{k-1} 2 \varepsilon_{k+1} \varepsilon_{k+2} \cdots$$

   and $T \, 0.22222 \cdots = 0.00000 \ldots$.
   (a) Prove that T is continuous and bijective.
   (b) Show that $\mathcal{O}(0)$ is dense in C.
   (c) Prove that (C, T) is minimal.

11.4. Periodic Points

In this section, we will show that the Intermediate Value Theorem is a powerful tool for establishing the *existence* of fixed points and periodic points. This theorem does not generally yield a computational scheme. So the conclusions of the Contraction Principle are stronger. However, this new technique applies more widely.

First, we look at a couple of situations that imply a fixed point.

11.4.1. LEMMA. *Suppose that T is a continuous function of a closed bounded interval $I = [a, b]$ into itself. Then T has a fixed point.*

PROOF. Consider the function

$$f(x) = Tx - x \quad \text{for} \quad a \le x \le b.$$

Notice that

$$f(a) = Ta - a \ge 0$$
$$f(b) = Tb - b \le 0.$$

By the Intermediate Value Theorem (Theorem 5.6.1), it follows that there is a point x^* in $[a, b]$ such that

$$0 = f(x^*) = Tx^* - x^*.$$

Thus x^* is the desired fixed point. ∎

The second result is very similar, but instead of mapping an interval *into* itself, we have an interval mapping *onto* itself. The proof is very similar.

11.4.2. LEMMA. *Let T be a continuous function on a closed bounded interval $I = [a, b]$ such that $T(I)$ contains I. Then T has a fixed point.*

PROOF. Again consider the function $f(x) = Tx - x$ for $a \le x \le b$. By hypothesis, there are points c and d such that $Tc = a$ and $Td = b$. Thus

$$f(c) = Tc - c = a - c \ge 0$$
$$f(d) = Td - d = b - d \le 0.$$

Again the Intermediate Value Theorem implies that there is a point x^* in $[c, d]$ such that

$$0 = f(x^*) = Tx^* - x^*.$$

So x^* is the desired fixed point. ∎

11.4.3. EXAMPLE. Consider the family of quadratic maps $Q_a x = a(x - x^2)$ for $a > 1$ known as the **logistic functions**. These maps are inverted parabolas with zeros at 0 and 1 and a maximum at $(\frac{1}{2}, \frac{a}{4})$. Each map Q_a takes positive values on $[0, 1]$ and negative values elsewhere. The derivative is $Q'_a(x) = a(1 - 2x)$. It is evident that $|Q'_a(x)| > 1$ on $\mathbb{R} \setminus [0, 1]$. So it is not difficult to show that if $x < 0$ or $x > 1$, then $Q^n_a x$ converges to $-\infty$. For this reason, we restrict our domain to the interval $I = [0, 1]$.

There are two cases. Suppose that $a \leq 4$. Then Q_a maps I into itself. Thus it has a fixed point by Lemma 11.4.1. On the other hand, if $a \geq 4$, then since $Q_a 0 = 0$ and $Q_a \frac{1}{2} = \frac{a}{4} \geq 1$, it follows from the Intermediate Value Theorem that $Q_a(I)$ contains I. So Q_a has a fixed point by Lemma 11.4.2.

It will be an added convenience, when the image of one interval contains another, to find a smaller interval that exactly maps onto the target interval. While this is intuitively clear, the details need to be checked. The proof is left as an exercise.

11.4.4. LEMMA. *Let T be a continuous function on a closed bounded interval $I = [a, b]$ such that $T(I)$ contains an interval $J = [c, d]$. Then there is a (smaller) interval $I' = [a', b']$ contained in I such that $T(I') = J$ and $T(\{a', b'\}) = \{c, d\}$.*

We will write $I \to J$ to indicate that $T(I)$ contains J. Let us see how this can be used to discover periodic points.

11.4.5. EXAMPLE. Consider the family of logistic functions $Q_a x = a(x - x^2)$ acting on $I = [0, 1]$ for $a \geq 4$. Write $I_0 = [0, \frac{1}{2}]$ and $I_1 = [\frac{1}{2}, 1]$. By the argument in the previous example, $Q_a(I_0)$ and $Q_a(I_1)$ both contain I (i.e., $I_0 \to I_0 \cup I_1$ and $I_1 \to I_0 \cup I_1$).

In particular, $I_0 \to I_0 \to I_1 \to I_0$. Use Lemma 11.4.4 repeated as follows. First find an interval $J_2 \subset I_1$ such that $Q_a(J_2) = I_0$. Then find $J_1 \subset I_0$ such that $Q_a(J_1) = J_2$. Finally, find $J_0 \subset I_0$ such that $Q_a(J_0) = J_1$. Then

$$Q^3_a(J_0) = Q^2_a(J_1) = Q_a(J_2) = I_0.$$

By Lemma 11.4.1, Q^3_a has a fixed point in J_0, say x_0. We will show that this point has period 3. Indeed, let $x_1 = Q_a x_0$ and $x_2 = Q_a x_1$. By construction, $Q_a x_2 = Q^3_a x_0 = x_0$. Thus x_0 is either a period 3 point or a fixed point. Now x_0 and x_1 belong to I_0 and x_2 belongs to I_1. If x_0 were a fixed point, it would belong to $I_0 \cap I_1 = \{\frac{1}{2}\}$. But $Q_a \frac{1}{2} = \frac{a}{4} \geq 1$ is not fixed or even periodic. Thus x_2 is different from x_0 and x_1, and consequently Q_a has an orbit of length 3.

This *proof* requires $a \geq 4$ to work. However, it is actually the case that period 3 orbits begin to appear when $a \approx 3.8284$.

11.4.6. EXAMPLE. Period doubling. Let f be a map from $I = [0, 1]$ into itself. Define a map $\mathcal{D}f$ as follows:

$$\mathcal{D}f(x) = \begin{cases} \frac{2}{3} + \frac{1}{3}f(3x) & \text{for} \quad 0 \leq x \leq \frac{1}{3} \\ (2 + f(1))(\frac{2}{3} - x) & \text{for} \quad \frac{1}{3} \leq x \leq \frac{2}{3} \\ x - \frac{2}{3} & \text{for} \quad \frac{2}{3} \leq x \leq 1. \end{cases}$$

We claim that with the single exception of a repelling fixed point in the interval $I_2 = [\frac{1}{3}, \frac{2}{3}]$, the periodic orbits of $\mathcal{D}f$ correspond to the periodic orbits of f, the periods are exactly double, and the dynamics (attracting or repelling) of these orbits are preserved.

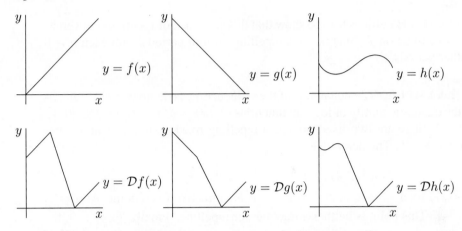

FIGURE 11.7. Applying $\mathcal{D}$ to various functions.

Indeed, it is clear from the graphs in Figure 11.7 that there is (an easily computed) fixed point in the interval I_2. As the function $\mathcal{D}f$ has slope $-2 - f(1) \leq -2$, this evidently is a repelling fixed point. Any other point in I_2 is mapped further and further from this point until it leaves this middle third.

The interval $I_1 = [0, \frac{1}{3}]$ is mapped into $I_3 = [\frac{2}{3}, 1]$ and I_3 is mapped bijectively onto I_1. These orbits map back and forth between I_1 and I_3, never intersecting I_2. Since no orbit, other than the fixed point, stays in I_2, eventually every orbit (other than the fixed point) alternates between I_1 and I_3.

Now notice that if $x \in [0, 1]$, then

$$(\mathcal{D}f)^2(x/3) = \mathcal{D}f(\tfrac{2}{3} + \tfrac{1}{3}f(x)) = \tfrac{1}{3}f(x).$$

This means that the graph of $(\mathcal{D}f)^2$ on I_1 is identical to the graph of f on I except that it is scaled (in both the x-direction and the y-direction) by a factor of one third. To make this precise, let $\sigma(x) = x/3$ be a map of I onto I_1. This map is a continuous bijection with continuous inverse $\sigma^{-1}(x) = 3x$. The relation between f and $(\mathcal{D}f)^2$ can be expressed as

$$\sigma f(x) = (\mathcal{D}f)^2 \sigma \quad \text{or} \quad (\mathcal{D}f)^2 = \sigma f \sigma^{-1}.$$

In Section 11.6, we will see that this relationship is a special case of an important notion, topological conjugacy.

So if x is a periodic point for f of period n, then $x/3$ has period n for $(\mathcal{D}f)^2$ and vice versa. Since $\mathcal{D}f$ flips back and forth between I_1 and I_3, it follows that $x/3$ has period $2n$ for $\mathcal{D}f$. Moreover, it will be attracting or repelling as x is.

Now consider a map f_0 with a unique attracting fixed point like $f_0(x) = \frac{1}{3}$. Define a sequence of functions by

$$f_{n+1} = \mathcal{D}f_n \quad \text{for all} \quad n \geq 1.$$

Then f_1 has an attracting orbit of period 2 and has a repelling fixed point in between. The function f_2 will have an attracting orbit of period 4, and in between, there will be a repelling orbit of period 2 and a repelling fixed point. Recursively we find that f_n has an attracting orbit of period 2^n and in between it has repelling orbits of lengths $1, 2, \ldots, 2^{n-1}$.

Exercise 11.4.H outlines how to show that this sequence of functions converges uniformly to a function f_∞ that has one repelling orbit of period 2^n for each $n \geq 0$ and no other periods.

11.4.7. EXAMPLE. The behaviour of the sequence f_n is demonstrated dramatically by the quadratic family of logistic functions $Q_a(x) = a(x - x^2)$ on $I = [0, 1]$. When $a > 1$, there are two fixed points: a repelling fixed point at 0 and another fixed point at $1 - \frac{1}{a}$. The derivative

$$Q_a'\left(1 - \tfrac{1}{a}\right) = a\left(1 - 2\left(1 - \tfrac{1}{a}\right)\right) = 2 - a$$

lies in $(-1, 1)$ for $1 < a < 3$, and so $1 - \frac{1}{a}$ is an attractive fixed point. At $a = 3$, $Q_3'(\frac{2}{3}) = -1$. This point is neither attracting nor repelling. Finally, for $a > 3$, the fixed point $1 - \frac{1}{a}$ is repelling since $Q_a'(1 - \frac{1}{a}) < -1$.

Exercise 11.4.F considers points of period 2. Notice that

$$Q_a^2 x = a\big(ax(1 - x)\big)\big(1 - ax + ax^2\big)$$

is a polynomial of degree 4. To find these points of period 2, solve the identity $Q_a^2 x = x$. This is made easier because we already know that 0 and $1 - \frac{1}{a}$ are solutions. So we may factor x and $a\big(x - (1 - \frac{1}{a})\big) = \big(ax - (a-1)\big)$ out of $Q_a^2 x - x$ to obtain

$$Q_a^2 x - x = x\big(ax - (a - 1)\big)\big(a^2 x^2 - (a^2 + a)x + (a + 1)\big).$$

The quadratic factor has discriminant

$$(a^2 + a)^2 - 4a^2(a + 1) = a^4 - 2a^3 - 3a^2 = a^2\big((a - 1)^2 - 4\big).$$

This is positive precisely when $a > 3$. So for $1 < a < 3$, there are no period 2 points. At $a = 3$, $Q_a^2 x - x = x(3x - 2)^3$ also has no period 2 orbits. However, once $a > 3$, a period 2 orbit appears consisting of the points

$$p_\pm = \frac{a^2 + a \pm \sqrt{(a - 1)^2 - 4}}{2a^2}.$$

When a parametric family of maps changes its dynamical behaviour, this is called a **bifurcation**.

To compute the derivative of Q_a^2 at $p_\pm$, we use the chain rule

$$\begin{aligned}
(Q_a^2)'(p_+) &= Q_a'(Q_a p_+)\, Q_a'(p_+) = Q_a'(p_-)\, Q_a'(p_+) \\
&= a(1 - 2p_-)a(1 - 2p_+) \\
&= a^2 - 2a^2(p_- + p_+) + 4a^2(p_- p_+) \\
&= a^2 - 2(a^2 + a) + 4(a + 1) \\
&= 5 - (a - 1)^2.
\end{aligned}$$

It follows that this period is attracting for $3 < a < 1+\sqrt{6}$. Then another bifurcation occurs at $a = 1 + \sqrt{6}$ when this becomes a repelling orbit. It turns out that at this point, an attracting orbit of period 4 appears.

However, the story does not stop here. An infinite sequence of bifurcations occurs at which point an attractive period of length 2^n appears and the period of length 2^{n-1} becomes repelling. This is sometimes called the period doubling route to chaos. The limit of this procedure is a point $a_\infty \approx 3.5699$. For every $a \geq a_\infty$, Q_a has repelling orbits of period 2^n for all $n \geq 0$. However, the story continues and yet more bifurcations happen between a_∞ and 4. For example, period 3 orbits first appear at about 3.8284.

There is an ordering among the possible periods of orbits that was discovered by Sharkovskii. In this ordering, the existence of a period n orbit implies the existence of all periods that occur later in the ordering:

$$3 \triangleright 5 \triangleright 7 \triangleright 9 \triangleright \dots \quad \triangleright 6 \triangleright 10 \triangleright 14 \triangleright \dots \quad \triangleright 12 \triangleright 20 \triangleright 28 \triangleright \dots$$

$$\triangleright 3 \cdot 2^n \triangleright 5 \cdot 2^n \triangleright 7 \cdot 2^n \triangleright \dots \quad \triangleright 3 \cdot 2^{n+1} \triangleright 5 \cdot 2^{n+1} \triangleright 7 \cdot 2^{n+1} \triangleright \dots$$

$$\triangleright \dots \quad \dots \triangleright 2^n \triangleright 2^{n-1} \dots \triangleright 4 \triangleright 2 \triangleright 1.$$

We will prove the special case of this result, which shows that period 3 is preeminent among all periods. The more complicated general argument follows the same basic lines.

11.4.8. LEMMA. *Let T be a continuous map from an interval I into itself. Suppose that there are intervals such that*

$$I_1 \to I_2 \to \cdots \to I_n.$$

Then there are intervals $J_k \subset I_k$ for $1 \leq k \leq n-1$ so that $T(J_k) = J_{k+1}$ for $1 \leq k \leq n-2$ and $TJ_{n-1} = I_n$.

PROOF. This is an easy application of Lemma 11.4.4. Using this lemma, find $J_{n-1} \subset I_{n-1}$ such that $T(J_{n-1}) = I_n$. Then use the lemma again to obtain an interval $J_{n-2} \subset I_{n-2}$ such that $T(J_{n-2}) = J_{n-1}$. Proceed in this way to define the whole sequence. ∎

11.4.9. SHARKOVSKII'S THEOREM.
Suppose that T is a continuous map of an interval $I = [a,b]$ into itself that has an orbit of period 3. Then T has an orbit of period n for every $n \geq 1$.

PROOF. Let $x_1 < x_2 < x_3$ be the period 3 orbit. Either $Tx_1 = x_2$ or $Tx_1 = x_3$. But in the second case, we may consider the interval with order reversed, in which case x_3 is the smallest, and it maps to the second point x_2. So the argument for the first case must equally apply in the second. Thus we may assume that

$$x_1 < x_2 = Tx_1 < x_3 = Tx_2 \quad \text{and} \quad Tx_3 = x_1.$$

Let $I_0 = [x_1, x_2]$ and $I_1 = [x_2, x_3]$. Then it is evident that $T I_0$ contains I_1 and $T I_1$ contains $I_0 \cup I_1$. That is, $I_0 \to I_1$ and $I_1 \to I_0$ and $I_1 \to I_1$. By Lemma 11.4.4, there is an interval J_0 contained in I_0 and there are intervals J_1 and J_2 contained in I_1 such that

$$T(J_0) = I_1 \qquad T(J_1) = I_0 \quad \text{and} \quad T(J_2) = I_1.$$

For $n = 1$, we may invoke Lemma 11.4.1 to obtain a fixed point a_1.

And for $n = 2$, we may use the fact that $J_0 \to J_1 \to J_0$. From this it follows that $T^2(J_0)$ contains J_0. So by Lemma 11.4.2, there is a fixed point a_2 of T^2 in J_0. We note that if $J_0 \cap J_1$ is nonempty, it must consist only of the point x_2. Since $T^2 x_2 = x_1$, it is not fixed for T^2. Thus a_2 is some other point in J_0. This means that $T a_2 \neq a_2$. Hence a_2 has period 2.

Now consider $n \geq 4$. We will proceed as in Example 11.4.5. Notice that

$$J_0 \to \underbrace{J_2 \to \cdots \to J_2}_{n-2 \text{ copies}} \to J_1 \to J_0.$$

Use Lemma 11.4.8 to find intervals $K_0 \subset J_0$, $K_i \subset J_2$ for $1 \leq i \leq n-2$ and $K_{n-1} \subset J_1$ so that

$$T(K_i) = K_{i+1} \quad \text{for} \quad 0 \leq i \leq n-2 \quad \text{and} \quad T(K_{n-1}) = J_0.$$

In particular, $T^n(K_0) = J_0$ contains K_0. Applying Lemma 11.4.2 again yields a fixed point a_n of T^n in K_0.

We must verify that a_n has no smaller period. But this is a consequence of a fact guaranteed by our construction:

$$T^i a_n \in J_2 \quad \text{for} \quad 1 \leq i \leq n-2 \quad \text{and} \quad T^{n-1} a_n \in J_1.$$

CLAIM: None of these points is equal to a_n. Indeed, as in the period 2 case, the only possible intersection of J_0 and $J_1 \cup J_2$ is the point x_2. However, were $a_n = x_2$, it would follow that J_2 would be an interval containing $T^2 a_n = x_1$, which is not possible. Hence the period of a_n is exactly n. ∎

Exercises for Section 11.4

A. Find a continuous function from $(0, 1)$ onto itself with no fixed points. Why does this not contradict Lemma 11.4.1 or 11.4.2?

B. Prove Lemma 11.4.4. HINT: Pick a_0 and b_0 in $[a, b]$ such that $T a_0 = c$ and $T b_0 = d$. If $a_0 < b_0$, let $a' = \sup\{x \in [a_0, b_0] : Tx = c\}$ and $b' = \inf\{x \in [a', b_0] : Tx = d\}$. Consider the case $a_0 > b_0$ separately.

C. Consider the function T mapping $I = [0, 4]$ onto itself by

$$Tx = \begin{cases} 2x + 2 & \text{for} \quad 0 \leq x \leq 1 \\ 5 - x & \text{for} \quad 1 \leq x \leq 2 \\ 7 - 2x & \text{for} \quad 2 \leq x \leq 3 \\ 4 - x & \text{for} \quad 3 \leq x \leq 4. \end{cases}$$

Figure 11.8 gives the graph of T. Note that $\{0, 2, 3, 1, 4\}$ is a period 5 orbit.

(a) Sketch the graphs of T^2 and T^3.

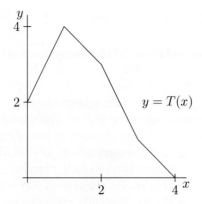

FIGURE 11.8. The graph of T.

   (b) Show that T has one period 2 orbit.
   (c) Show that T has no period 3 orbit.   HINT: Show that $T^3x > x$ on $[0, 2]$, $T^3x < x$
   on $[3, 4]$ and T^3 is monotone decreasing on $[2, 3]$.

D. Suppose that T is a continuous map from an interval I into itself. Moreover, suppose
that there are points $x_1 < x_2 < x_3 < x_4$ such that $Tx_1 = x_2$, $Tx_2 = x_3$, $Tx_3 = x_4$
and $Tx_4 \leq x_1$. Show that T has an orbit of period 3.
HINT: Let $I_k = [x_k, x_{k+1}]$. Show that $I_1 \to I_2 \to I_3 \to I_1$.

E. Give an example of a map with an orbit of period 6, but no odd orbits.

F. Consider the logistic map $Q_a x = a(x - x^2)$ for $a > 1$.
   (a) Show that Q_a has two fixed points in $[0, 1]$. Are they are attracting or repelling?
   (b) What is the orbit behaviour of points in $\mathbb{R} \setminus [0, 1]$?
   (c) Compute $Q_a^2 x$. Look for an orbit of period 2 by solving the quartic polynomial
   equation $Q_a^2 x = x$. Use the fact that fixed points of Q_a are also fixed points of Q_a^2
   to find two of the roots. For which values of a does Q_a have points of period 2?
   (d) Show that if the period 2 orbit is $\{c, d\}$, then $(Q_a^2)'(c) = (Q_a^2)'(d) = Q_a'(c)Q_a'(d)$.
   For which values of a is this period attracting? or repelling?

G. This is a computer experiment for the family of logistic maps Q_a.
   (a) Let $a = 3.46$. Use a computer to calculate $x = Q_a^{100}(.5)$. Then compute $Q_a x$,
   $Q_a^2 x, Q_a^3 x, \ldots, Q_a^{10} x$. What do you observe? Why did this happen?
   (b) Try this for $a = 3.55$. What is different now? What bifurcation occurred?
   (c) Do the same for $a = 3.83$. What do you observe? Try this for $a = 3.8$ and $a = 3.9$.
   The two sequences do not behave in the same way, but the reasons are different.

H. Consider the period doubling method of Example 11.4.6. Start with the constant function $f_0(x) = \frac{1}{3}$. Define a sequence of functions by $f_{n+1} = Df_n$ for $n \geq 0$.
   (a) Show that $f_{n+1}(x) = f_n(x)$ for all $3^{-n} \leq x \leq 1$ and $1 - 3^{-n} \leq f_n(x) \leq 1$ for all
   $0 \leq x \leq 3^{-n}$.
   (b) Use part (a) to show that the sequence f_n converges uniformly to a continuous limit
   function f_∞.
   (c) Calculate the point x_n of intersection between the line $y = 1 - 3^{-n} + x$ and the
   graph of $f_{n+1}(x)$. Show that this is a point of period 2^n for f_∞.
   (d) Show that these are the only periods of the function f_∞.

11.5. Chaotic Systems

In this section, we will define and examine chaotic systems, which are systems of striking complexity with seemingly "wild" behaviour. The surprise is that this complexity arises in seemingly simple situations, as the examples of this section will show. In mathematical physics, it was an important insight that very simple, commonly occurring differential equations exhibit chaotic behaviour. Part of the definition of chaos is that very small perturbations in initial conditions lead to wildly different orbits. For example, this phenomenon makes detailed weather prediction over the long term impossible, even if it is weather produced in a laboratory using an apparently simple model. This is also the reason that water flowing in a river produces complicated eddying that is constantly changing and unpredictable.

The mathematical notion of chaos depends on three things. The first is a dense set of periodic points. The other two items are new, and we study each of them in turn.

11.5.1. DEFINITION. A dynamical system T mapping a set X into itself is **topologically transitive** if, for any two nonempty open sets U and V in X, there is an integer $n \geq 1$ such that $T^n U \cap V$ is nonempty.

11.5.2. PROPOSITION. *For a dynamical system T mapping a set X into itself, topological transitivity is equivalent to the following: For each $x, y \in X$ and $\varepsilon > 0$, there is a point $z \in X$ and an integer $n \geq 1$ such that*

$$\|x - z\| < \varepsilon \quad \text{and} \quad \|y - T^n z\| < \varepsilon.$$

PROOF. To see this, first assume T is topologically transitive. Given $x, y \in X$ and $\varepsilon > 0$, take $U = B_\varepsilon(x)$ and $V = B_\varepsilon(y)$. Transitivity provides n so that $T^n U \cap V$ is nonempty. Pick $z \in U$ so that $T^n z \in V$, and we are done.

Conversely, let nonempty open sets U and V be given. Pick points $x \in U$ and $y \in V$. Since U and V are open, there is an $\varepsilon > 0$ so that

$$B_\varepsilon(x) \subset U \quad \text{and} \quad B_\varepsilon(y) \subset V.$$

Let z and $n \geq 1$ be chosen so that $z \in B_\varepsilon(x)$ and $T^n z \in B_\varepsilon(y)$. Then $T^n z$ belongs to $T^n U \cap V$. ∎

If there is a **transitive point** x_0, meaning that $\mathcal{O}(x_0)$ is dense in X, then T is topologically transitive. For if x and y and $\varepsilon > 0$ are given, pick m so that $\|x - T^m x_0\| < \varepsilon$. Notice that the orbit $\mathcal{O}(T^m x_0)$ is the same as $\mathcal{O}(x_0)$ except for the first m points, and hence it is also dense (explain this). So there is another integer $n \geq 1$ so that $\|y - T^{m+n} x_0\| < \varepsilon$. So $z = T^m x_0$ does the job.

It is perhaps a surprising fact that the converse is true. We require that X is infinite just to avoid the trivial case when X consists of a single finite orbit. The proof depends on the Baire Category Theorem, from Section 9.3.

11.5.3. THE BIRKHOFF TRANSITIVITY THEOREM.
If a mapping T is topologically transitive on an infinite closed subset of $\mathbb{R}^k$, then it has a dense set of transitive points.

PROOF. Let $\{V_n : n \geq 1\}$ be a collection of open sets with the property that every open set V contains one of these V_n. For example, let $\{x_n : n \geq 1\}$ be a dense subset of X in which every point in this set is repeated infinitely often. Take the sets $V_n = B_{1/n}(x_n)$ (verify!).

For each V_n, the set $U_n = \{x \in X : T^k x \in V_n \text{ for some } k \geq 1\}$ is the union of the open sets $T^{-k}(V_n)$ for $k \geq 1$, and thus is open. Since T is topologically transitive, given any open set U, there is some $k \geq 1$ so that $T^k U \cap V_n$ is nonempty. Therefore, $U_n \cap U \neq \varnothing$, and thus U_n is dense.

Consider $R = \bigcap_{n \geq 1} U_n$. Take any point x_0 in R. For each $n \geq 1$, there is an integer k so that $T^k x_0 \in V_n$. Therefore, $\mathcal{O}(x_0)$ intersects every V_n. This shows that $\mathcal{O}(x_0)$ is dense. So R is the set of transitive points of T. Since R is the intersection of countably many dense open sets, the Baire Category Theorem (Theorem 9.3.2) shows that R is dense in X. ∎

11.5.4. EXAMPLE. In Example 11.3.1, if $\alpha/2\pi$ is not rational, then the irrational rotation R_α of the circle $\mathbb{T}$ has transitive points. Hence it is topologically transitive. Indeed, every point is transitive.

11.5.5. EXAMPLE. In Example 11.3.2, the map $T\theta = 2\theta \pmod{2\pi}$ was shown to have a dense set of repelling periodic points, and we outlined how to show that it has a dense set of transitive points. This would imply that it is topologically transitive. We will verify this again directly from the definition.

Let U and V be nonempty open subsets of the circle. Then U contains an interval I of length $\varepsilon > 0$. It follows that $T^n U$ contains $T^n I$, which is an interval of length $2^n \varepsilon$. Eventually $2^n \varepsilon > 2\pi$, at which point $T^n I$ must contain the whole circle. In particular, the intersection of $T^n U$ with V is V itself.

11.5.6. EXAMPLE. Again we consider the quadratic family of logistic maps $Q_a x = a(x - x^2)$ on the unit interval I for large a. Our arguments will work for $a > 2 + \sqrt{5} \approx 4.2361$. However more delicate arguments work for any $a > 4$.

The first thing to notice about the case $a > 4$ is that Q_a does not map I into itself. Notice that once $Q_a^k x$ is mapped outside of $[0, 1]$, it remains outside since Q_a maps $(-\infty, 0) \cup (1, \infty)$ into $(-\infty, 0)$. We recall from Example 11.4.3 that once a point is outside $[0, 1]$, the orbit goes off to $-\infty$.

There is an open interval

$$J_1 = \{x \in [0, 1] : Q_a x > 1\}$$

centred around $x = \frac{1}{2}$. The remainder I_1 consists of two closed intervals, and each is mapped one-to-one and onto $[0, 1]$. In particular, in the middle of each of these closed intervals is an open interval that is mapped onto J_1. Hence

$$J_2 = \{x \in I_1 : Q_a^2 x > 1\}$$

is the union of these two intervals. What remains is the union of four intervals that Q_a^2 maps one-to-one and onto $[0, 1]$.

Proceeding in this way, we may define

$$I_n = \{x \in [0, 1] : Q_a^n x \in [0, 1]\}$$

and

$$J_n = \{x \in I_{n-1} : Q_a^n x > 1\}.$$

See Figure 11.9 for an example. Notice that $I_n = [0, 1] \setminus \bigcup_{k=1}^n J_k$ consists of the union of 2^n disjoint intervals and Q_a^n maps each of these intervals one-to-one and onto $[0, 1]$. We call these 2^n intervals the component intervals of I_n.

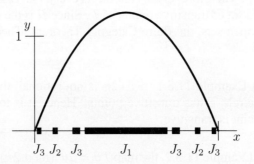

FIGURE 11.9. The graph of Q_5, showing J_1, J_2, and J_3.

We are interested in the set

$$X_a = \{x \in [0, 1] : Q_a^n x \in [0, 1] \text{ for all } n \geq 1\}.$$

If $x \in X_a$, then it is clear that $Q_a x$ remains in X_a. So this set is mapped into itself, making (X_a, Q_a) a dynamical system.

From our construction, we see that $X_a = \bigcap_{n \geq 1} I_n$. In fact, this looks a lot like the construction of the Cantor set C (Example 4.4.8) and X_a has many of the same properties. By Cantor's Intersection Theorem (Theorem 4.4.7), it follows that X_a is nonempty and compact. We will show that it is perfect (no point is isolated) and nowhere dense (it contains no intervals). A set with these properties is often called a **generalized Cantor set**, or sometimes just a Cantor set.

To simplify the argument, we will assume that $a > 2 + \sqrt{5} \approx 4.236$.

11.5.7. LEMMA. *If $a > 2 + \sqrt{5}$, then $c := \min_{x \in I_1} |Q_a'(x)| > 1$. Thus each of the 2^n component intervals of I_n has length at most c^{-n}.*

PROOF. The graph of Q_a is symmetric about the line $x = \frac{1}{2}$. The set I_1 consists of two intervals $[0, s]$ and $[1 - s, 1]$, where s is the smaller root of $a(x - x^2) = 1$, namely

$$s = \frac{a - \sqrt{a^2 - 4a}}{2a} = \frac{1}{2} - \frac{\sqrt{a^2 - 4a}}{2a}.$$

Note that s is a decreasing function of a for $a \geq 4$. Also, $|Q_a'(x)| = a|1 - 2x|$ is decreasing on $[0, \frac{1}{2}]$. So the minimum value is taken at s, which is

$$c = Q_a'(s) = \sqrt{a^2 - 4a} = \sqrt{(a - 2)^2 - 4}.$$

This is an increasing function of a and takes the value 1 when $a^2 - 4a = 1$. Rearranging, we have $(a - 2)^2 = 5$, so that $a = 2 + \sqrt{5}$. Any larger value of a yields a value of c greater than 1.

We will verify that the intervals in I_n have length at most c^{-n} by induction. For $n = 0$, this is clear. Suppose that the conclusion is valid for $n - 1$. Notice that Q_a maps each component interval $[p, q]$ of I_n onto an interval of I_{n-1}. The Mean Value Theorem implies that there is a point r between p and q so that

$$\left| \frac{Q_a(q) - Q_a(p)}{q - p} \right| = |Q_a'(r)| \geq c.$$

Hence

$$|q - p| \leq c^{-1}|Q_a(q) - Q_a(p)| \leq c^{-1}c^{1-n} = c^{-n}. \qquad \blacksquare$$

We can immediately apply this to any interval contained in X_a. As it would also be an interval contained in I_n for all $n \geq 1$, it must have zero length. So X_a has no interior.

Now let x be a point in X_a. It is clear from the construction of X_a that the endpoints of each component interval of I_n belongs to X_a. (In fact, these are eventually fixed points whose orbits end up at 0.) If x is not the left endpoint of one of the intervals in some I_n, let x_n be the left endpoint of the component interval of I_n that contains x. By Lemma 11.5.7, it follows that $|x - x_n| \leq c^{-n}$ and so $x = \lim_{n \to \infty} x_n$. If x happens to be a left endpoint, then use the right endpoints instead. Hence X_a is perfect. This verifies our claim that X_a is a Cantor set.

Now we are ready to establish topological transitivity.

11.5.8. PROPOSITION. *If $a > 2 + \sqrt{5}$, the quadratic map $Q_a = a(x - x^2)$ is topologically transitive on the generalized Cantor set X_a.*

PROOF. Suppose that $x, y \in X_a$ and $\varepsilon > 0$. Choose n so large that $c^{-n} < \varepsilon$, and let J be the component interval of I_n containing x. Then since J has length at most c^{-n}, it is contained in $(x - \varepsilon, x + \varepsilon)$. Now $Q_a^n J$ is the whole interval I. Pick z to be the point in J such that $Q_a^n z = y$. Since y belongs to X_a, it is clear that the orbit of z consists of a few points in $[0, 1]$ together with the orbit of y, which also remains in I. Therefore, z belongs to X_a. We have found a point z in X_a near x that maps precisely onto y via Q_a^n. Therefore, Q_a is topologically transitive on X_a. $\blacksquare$

The third notion we need is the crucial one of sensitive dependence on initial conditions. Roughly, it says that for every point x we can find a point y, as close as we like to x, so that the orbits of x and y are eventually far apart. This means that no measurement of initial conditions, however accurate, can predict the long-term behaviour of the orbit of a point.

11.5.9. DEFINITION. A map T mapping X into itself exhibits **sensitive dependence on initial conditions** if there is a real number $r > 0$ so that for every point $x \in X$ and any $\varepsilon > 0$, there is a point $y \in X$ and $n \geq 1$ so that

$$\|x - y\| < \varepsilon \quad \text{and} \quad \|T^n x - T^n y\| \geq r.$$

11.5.10. EXAMPLE. Consider the circle doubling map $T\theta \equiv 2\theta \pmod{2\pi}$ again. It is easy to see that this map has sensitive dependence on initial conditions. Indeed, let $r = 1$. For any $\varepsilon > 0$ and any $\theta \in \mathbb{T}$, pick any other point $\varphi \neq \theta$ with $|\theta - \varphi| < \varepsilon$. Choose n so that $1 \leq 2^n |\theta - \varphi| \leq 2$. Then it is clear that

$$|T^n \theta - T^n \varphi| = 2^n |\theta - \varphi| \geq 1.$$

11.5.11. EXAMPLE. On the other hand, the rotation map R_α of the circle $\mathbb{T}$ through an angle α is rigid. $|T^n \theta - T^n \varphi| = |\theta - \varphi|$ for all $n \geq 1$. So this map is not sensitive to initial conditions.

11.5.12. PROPOSITION. *When $a > 2 + \sqrt{5}$, then the quadratic logistic map $Q_a x = a(x - x^2)$ exhibits sensitive dependence on initial conditions on the generalized Cantor set X_a.*

PROOF. Set $r = \frac{1}{2}$. Given $x \in X_a$ and $\varepsilon > 0$, we find as before an integer n and a component interval J of I_n which is contained in $(x - \varepsilon, x + \varepsilon)$. Then Q_a^n maps J one-to-one and onto $[0, 1]$. In particular, the two endpoints y and z of J are mapped to 0 and 1. So

$$|Q_a^n z - Q_a^n x| + |Q_a^n x - Q_a^n y| = 1.$$

So $\max \left\{ |Q_a^n z - Q_a^n x|, |Q_a^n x - Q_a^n y| \right\} \geq \frac{1}{2}$ as desired. ∎

Now we can define chaos.

11.5.13. DEFINITION. We call (X, T) a **chaotic dynamical system** if

(1) The set of periodic points is dense in X.

(2) T is topologically transitive on X.

(3) T exhibits sensitive dependence on initial conditions.

This definition demands lots of wild behaviour. In order for the periodic points to be dense, there need to be infinitely many distinct periods. The existence of transitive points already means that orbits are distributed everywhere throughout

X. Sensitive dependence on initial conditions means that orbits that start out nearby can be expected to diverge eventually.

These notions are interrelated. For any metric space, the conditions of dense periodic points and topological transitivity together imply sensitive dependence on initial conditions. The proof is elementary, but delicate; see [42]. However, (2) and (3) do not imply (1), nor do (1) and (3) imply (2). But if the space X is an interval in $\mathbb{R}$, then (2) implies both (1) and (3); a simple proof of this result is given in [47]. Some authors drop condition (1), arguing that it is the other two conditions that are paramount.

11.5.14. EXAMPLE. We have shown that the circle doubling map has a dense set of periodic points in Example 11.3.2. In Example 11.5.5, it was shown to be topologically transitive. And in Example 11.5.10, it was seen to have sensitive dependence on initial conditions. Hence this system is chaotic.

11.5.15. EXAMPLE. The quadratic family $Q_a x = a(x - x^2)$ of logistic maps is chaotic for $a > 2 + \sqrt{5}$. Indeed, Proposition 11.5.8 established topological transitivity and Proposition 11.5.12 established sensitive dependence on initial conditions. In Example 11.4.5, it was established that Q_a has orbits of period 3. Hence by Sharkovskii's Theorem (Theorem 11.4.9), there are orbits of every possible period. But this does not show that they are dense.

It suffices to show that each component interval J of I_n contains periodic points, since as we have argued before, every interval $(x - \varepsilon, x + \varepsilon)$ contains such an interval. Now Q_a^n maps J onto I, which contains J. Therefore, by Lemma 11.4.2, there is a point $y \in J$ that is a fixed point for Q_a^n. So y is a periodic point (whose period is a divisor of n). Moreover, y must belong to X_a since the whole orbit of y remains in $[0, 1]$. It follows that periodic points are dense in X_a and that Q_a is chaotic.

In fact, all of this analysis remains valid for $a > 4$. But because the Mean Value Theorem argument based on Lemma 11.5.7 is no longer valid, the proof is different.

For our last example in this section, we will do a complete proof of chaos for a new system that will be useful in the next section for understanding the relationship between the quadratic maps Q_a for large a.

11.5.16. EXAMPLE. Recall from Example 4.4.8 that the middle thirds Cantor set C can be described as the set of all points x in $[0, 1]$ that have a ternary expansion (base 3) using only 0s and 2s. It is a compact set that is nowhere dense (contains no intervals) and perfect (has no isolated points). It was constructed by removing, in succession, the middle third of each interval remaining at each stage. The endpoints of the removed intervals belong to C and consist of those points that have two different ternary expansions. However, only one of these expansions consists of 0s and 2s alone.

Define the shift map on the Cantor set C by

$$Sy = 3y \pmod{1} = (.y_2 y_3 y_4 \dots)_{\text{base 3}} \quad \text{for} \quad y = (.y_1 y_2 y_3 \dots)_{\text{base 3}} \in C.$$

It is easy to see that

$$Sy = \begin{cases} 3y & \text{for} \quad y \in C \cap [0, 1/3] \\ 3y - 2 & \text{for} \quad y \in C \cap [2/3, 1]. \end{cases}$$

So it follows that S is a continuous map. Moreover, the range is contained in C because every point in the image has a ternary expansion with only 0s and 2s. In fact, it is easy to see that S maps each of the sets $C \cap [0, 1/3]$ and $C \cap [2/3, 1]$ bijectively onto C.

Let us examine the dynamics of the shift map.

First look for periodic points. A moment's reflection shows that $S^n y = y$ if and only if $y_{k+n} = y_k$ for all $k \geq 1$. That is, y has period n exactly when the ternary expansion of y is periodic of period n. There are precisely 2^n points such that $S^n y = y$. Indeed, the first n ternary digits $a_1, \dots, a_n$ are an arbitrary finite sequence of 0s and 2s, and this forces

$$y = (.a_1 \dots a_n a_1 \dots a_n a_1 \dots a_n \dots)_{\text{base3}}$$

$$= \sum_{k=1}^{n} a_k 3^{-k} \left(1 + 3^{-n} + 3^{-2n} + \dots \right)$$

$$= \frac{1}{1 - 3^{-n}} \sum_{k=1}^{n} a_k 3^{-k}.$$

From this, it is evident that the set of periodic points is dense in C. Indeed, given $y = (.y_1 y_2 y_3 \dots)_{\text{base3}}$ in C and $\varepsilon > 0$, choose N so large that $3^{-N} < \varepsilon$. Then let x be the periodic point determined by the sequence $y_1, \dots, y_N$. Then x and y both belong to the interval $[(.y_1 y_2 \dots y_N)_{\text{base3}}, (.y_1 y_2 \dots y_N)_{\text{base3}} + 3^{-N}]$, which has length 3^{-N}. Hence $|x - y| \leq 3^{-N} < \varepsilon$.

It is also easy to see that the set of nonperiodic points that are eventually fixed are also dense. The points in C that have a finite ternary expansion,

$$y = (.y_1 \dots y_n)_{\text{base3}} = (.y_1 \dots y_n 000 \dots)_{\text{base3}},$$

are eventually mapped to 0. These are the left endpoints of all the intervals $T_{\alpha_1 \dots \alpha_n}$.

Next we will show that the set of transitive points are dense. The hard part is to describe one such point. Make a list of all finite sequences of 0s and 2s by first listing all sequences of length 1 in increasing order, then those of length 2, and length 3, and so on:

0, 2,   00, 02, 20, 22,

000, 002, 020, 022, 200, 202, 220, 222,

0000, 0002, 0020, 0022, 0100, 0102, 0120, 0122, ...

String them all together to give the infinite ternary expansion of a point

$$a = (.0200022022000002020022200202220222000000020020002 \dots)_{\text{base3}}.$$

Suppose that y is any point in C and $\varepsilon > 0$ is given. Determine an integer N so that $3^{-N} < \varepsilon$. Somewhere in the expansion of a are the first N digits of y, say starting in the $(p+1)$st place of a. Then $S^p a$ starts with these same N digits. Hence $|y - S^p a| \leq 3^{-N} < \varepsilon$.

To see that the transitive points are dense, first notice that if $S^N x = a$, then x is also transitive. So let x be the point beginning with the first N digits of y followed by the digits of a from the beginning. Then x is transitive. As before, we obtain that $|x - y| < \varepsilon$.

Finally, we need to verify that S has sensitive dependence on initial conditions. This is easy. Let $r = 1/4$. If x and $\varepsilon > 0$ are given, choose $N > 1$ so that $3^{-N} < \varepsilon$. Let y be the point in C obtained by changing the ternary expansion of x only in the Nth digit from a 0 to a 2, or vice versa. Then $|x - y| < \varepsilon$. Also $S^{N-1} x$ and $S^{N-1} y$ differ in the first ternary digit. So they lie in T_0 and T_1, respectively (or vice versa). In particular,

$$\left| S^{N-1} x - S^{N-1} y \right| \geq \frac{1}{3} > r.$$

We conclude that the shift map S is chaotic.

Exercises for Section 11.5

A. Show that if T is topologically transitive on X, then either X is infinite or X consists of a single orbit.

B. Consider the tent map of Exercise 11.3.E, where this map was shown to have a dense set of periodic points.
   (a) What is the slope of the function $T^n(x)$? Use this to establish sensitive dependence on initial conditions.
   (b) Show that T^n maps each interval $[k2^{-n}, (k+1)2^{-n}]$ onto $[0, 1]$. Use this to establish topological transitivity.
   (c) Hence conclude that the tent map is chaotic.

C. Consider the **big tent map** $Sx = \begin{cases} 3x & \text{for} \quad x \leq \frac{1}{2} \\ 3(1-x) & \text{for} \quad x \geq \frac{1}{2} \end{cases}$.
   (a) Sketch the graphs of S, S^2 and S^3.
   (b) What are the dynamics for point outside of $[0, 1]$?
   (c) Describe the set $I_n = \{x \in [0, 1] : S^n x \in [0, 1]\}$.
   (d) Describe the set $X = \bigcap_{n \geq 1} I_n$.
   (e) Show that T^n has exactly 2^n fixed points, and they all belong to X. Hence show that the periodic points are dense in X.
   (f) Show that S is chaotic on X.   HINT: Use the idea of the previous exercise.

D. Let f_∞ be the function constructed in Exercise 11.4.H.
   (a) Show that the middle thirds Cantor set C is mapped into itself by f_∞.
       HINT: Let I_n denote the nth stage consisting of 2^n intervals of length 3^{-n} whose intersection is C. Show that $f_\infty(I_n) = I_n$.
   (b) Show that if x is not periodic for f_∞, then $f_\infty^k(x)$ eventually belongs to each I_n. Hence the distance from $f_\infty^k(x)$ to C tends to zero.
   (c) Show that there are no periodic points in C.

(d) Show that f_∞ maps permutes the 2^n intervals of I_n in a single cycle, so that the orbit of a point $x \in I_n$ intersects all 2^n of these intervals.

(e) Use (d) to show that the orbit of every point in C is dense in C. In particular, f_∞ is topologically transitive on C.

(f) Use (d) to show that f_∞ does *not* have sensitive dependence on initial conditions.

11.6. Topological Conjugacy

In this section, we will discuss how to show that two dynamical systems, possibly on different spaces, are essentially the same. By *essentially the same*, we mean that they have the same dynamical system properties. It is convenient to introduce two new notions that allow us to express the fact that two dynamical systems are the same map up to a reparametrization.

The notion of homeomorphism encodes the fact that two spaces have the same topology, meaning roughly that convergent sequences correspond but distances between points need not correspond.

11.6.1. DEFINITION. Two subsets of normed vector spaces X and Y are said to be **homeomorphic** if there is a continuous, one-to-one, and onto map $\sigma : X \to Y$ such that the inverse map σ^{-1} is also continuous. The map σ is called a **homeomorphism**.

11.6.2. EXAMPLE. Let f be a continuous map from $[0, 1]$ into itself, and consider when this is a homeomorphism. To be onto, there must be points a and b such that $f(a) = 0$ and $f(b) = 1$. By the Intermediate Value Theorem (Theorem 5.6.1), f maps $[a, b]$ onto $[0, 1]$. If $[a, b]$ were a proper subset of $[0, 1]$, then the remaining points would have to be mapped somewhere and f would fail to be one-to-one. Hence we have either $f(0) = 0$ and $f(1) = 1$ or $f(0) = 1$ and $f(1) = 0$. For convenience, let us assume that it is the former for a moment. By the same token, f must be strictly increasing. Indeed, if there were $x < y$ such that $f(y) \le f(x)$, then the Intermediate Value Theorem again yields a point z such that $0 \le z \le x$ such that $f(z) = f(y)$, destroying the one-to-one property.

Conversely, if f is a continuous strictly increasing function such that $f(0) = 0$ and $f(1) = 1$, then the same argument shows that f is one-to-one and onto. So the inverse function f^{-1} is well defined. Moreover, it is evident that f^{-1} is also strictly increasing and maps $[0, 1]$ onto itself. By Corollary 5.7.3, the only discontinuities of monotone functions are jump discontinuities. Hence f^{-1} is also continuous. So f is a homeomorphism of $[0, 1]$. Likewise, if f is a continuous strictly decreasing function such that $f(0) = 1$ and $f(1) = 0$, then it is a homeomorphism.

This example makes it look as though the order on the real line is crucial to establishing the continuity of the inverse. However, this result is actually more basic and depends crucially on compactness. There are two natural proofs of this result based on the equivalent characterizations of continuous functions in Theorem 5.3.1; see Exercise 11.6.A for the other approach.

11.6.3. THEOREM. *Let X and Y be compact subsets of $\mathbb{R}^n$. Suppose that f is a continuous bijection of X onto Y. Then f is a homeomorphism (i.e., f^{-1} is also continuous).*

PROOF. Since f is a bijection, the map f^{-1} is well defined and is a bijection of Y onto X. We need to establish the continuity of f^{-1}. By Theorem 5.3.1, a function g is continuous if and only if $g^{-1}(U)$ is open for every open set U. Let U be an open set in X. Then its complement $C := X \setminus U$ is a closed subset of the compact set X, and therefore C is compact by Lemma 4.4.4.

Since f is one-to-one and onto, we see that

$$\left(f^{-1}\right)^{-1}(U) = f(U) = Y \setminus f(C).$$

By Theorem 5.4.3, we know that $f(C)$ is compact and hence closed. Therefore, its complement $Y \setminus f(C) = f(U)$ must be open. Hence f^{-1} is continuous. ∎

This result is also true if X and Y are compact subsets of a normed vector space, with the same proof; all we need do is show that each of the theorems and lemmas in the proof hold for any normed vector space.

11.6.4. EXAMPLE. Let X be a generalized Cantor set in $\mathbb{R}$ and let C be the standard middle thirds Cantor set, both given as the intersection of sets I_n and S_n, respectively, which are the disjoint union of 2^n intervals with lengths tending to zero: $X = \bigcap_{n \geq 0} I_n$ and $C = \bigcap_{n \geq 0} S_n$, where each component interval of I_n contains two component intervals of $\bar{I}_{n+1}$. We shall show that X is homeomorphic to C. Moreover, this homeomorphism may be constructed to be monotone increasing.

For notational convenience, we let the component intervals of S_n be labeled as follows:

$$T_0 = [0, \tfrac{1}{3}]$$

$$T_{00} = [0, \tfrac{1}{9}]$$

$$T_{01} = [\tfrac{2}{9}, \tfrac{1}{3}]$$

$$T_{000} = [0, \tfrac{1}{27}]$$
$$T_{001} = [\tfrac{2}{27}, \tfrac{1}{9}]$$
$$T_{010} = [\tfrac{2}{9}, \tfrac{7}{27}]$$
$$T_{011} = [\tfrac{8}{27}, \tfrac{1}{3}]$$

$$T_1 = [\tfrac{2}{3}, 1]$$

$$T_{10} = [\tfrac{2}{3}, \tfrac{7}{9}]$$

$$T_{11} = [\tfrac{8}{9}, 1]$$

$$T_{100} = [\tfrac{2}{3}, \tfrac{19}{27}]$$
$$T_{101} = [\tfrac{20}{27}, \tfrac{7}{9}]$$
$$T_{110} = [\tfrac{8}{9}, \tfrac{25}{27}]$$
$$T_{111} = [\tfrac{26}{27}, 1].$$

A component interval of S_n is denoted by a finite sequence of 0s and 1s. When it is split into two intervals of S_{n+1} by removing the middle third, the new intervals are labeled by adding a 0 to the label of the first interval and a 1 to the second. So, for example, when $T_{101} = [20/27, 7/9]$ is split, we label the new intervals as

$T_{1010} = [20/27, 31/81]$ and $T_{1011} = [32/81, 7/9]$. The formula is more transparent in base 3:

$$T_{1010} = [.2020_{\text{base 3}}, .2021_{\text{base 3}}] \quad \text{and} \quad T_{1011} = [.2022_{\text{base 3}}, .2100_{\text{base 3}}].$$

So the label $\alpha_1 \ldots \alpha_n$ specifies the first digits in the ternary expansion of the points in the interval $T_{\alpha_1 \ldots \alpha_n}$ by converting 0s and 1s to 0s and 2s in base 3.

Recall that each point y of C is determined by the sequence of component intervals of S_n that contains it. Indeed, the typical point of C is given in base 3 as

$$y = (.y_1 y_2 y_3 \cdots)_{\text{base 3}} = \sum_{k \geq 1} y_k 3^{-k},$$

where (y_k) is a sequence of 0s and 2s. If we set $\alpha_k = y_k/2$, then y belongs to the component intervals $T_{\alpha_1 \alpha_2 \ldots \alpha_n}$ for each $n \geq 1$. Moreover,

$$\bigcap_{n \geq 1} T_{\alpha_1 \alpha_2 \ldots \alpha_n} = \{y\}.$$

Indeed, since the length of the intervals goes to zero, the intersection can contain at most one point. It is easy to show that the one point must be y.

We now describe X in the same manner. Let the interval components of I_n be denoted as $J_{\alpha_1 \alpha_2 \ldots \alpha_n}$ for each finite sequence $\alpha_1 \alpha_2 \ldots \alpha_n$ of 0s and 1s. When this interval is split into two parts by removing an open interval from the interior, the leftmost remaining interval will be denoted by $J_{\alpha_1 \alpha_2 \ldots \alpha_n 0}$ and the rightmost by $J_{\alpha_1 \alpha_2 \ldots \alpha_n 1}$. By hypothesis, each interval $J_{\alpha_1 \alpha_2 \ldots \alpha_n}$ is nonempty, and the lengths tend to 0 as n goes to $+\infty$.

For each infinite sequence $\mathbf{a} = (\alpha_k)_{n=1}^{\infty}$ of 0s and 1s, define a point $x_{\mathbf{a}}$ in X by

$$\{x_{\mathbf{a}}\} = \bigcap_{n \geq 1} J_{\alpha_1 \alpha_2 \ldots \alpha_n}.$$

Since the lengths of the intervals tends to 0, the intersection may contain at most one point. On the other hand, because of compactness, Cantor's Intersection Theorem (Theorem 4.4.7) guarantees that this intersection in nonempty. So it consists of a single point denoted $x_{\mathbf{a}}$.

Conversely, each point x in X determines a unique sequence $\mathbf{a} = (\alpha_k)_{n=1}^{\infty}$ of 0s and 1s because there is exactly one component interval of I_n containing x, which we denote by $J_{\alpha_1 \alpha_2 \ldots \alpha_n}$. So there is a bijective correspondence between points x in X and the associated symbol sequences $\mathbf{a}$ of 0s and 1s.

Define a function τ from X to C by

$$\tau(x_{\mathbf{a}}) = \sum_{k \geq 1} \frac{2\alpha_k}{3^k}.$$

This is well defined because the sequence $\mathbf{a}$ is uniquely determined by the point x. The range is contained in C because $\tau(x_{\mathbf{a}})$ has a ternary expansion consisting entirely of 0s and 2s, which describes the points of the Cantor set. The map is one-to-one because τ maps $X \cap J_{\alpha_1 \alpha_2 \ldots \alpha_n}$ into $C \cap T_{\alpha_1 \alpha_2 \ldots \alpha_n}$. Different points x_1 and x_2 of X are distinguished at some level n by belonging to different component intervals, and thus have different images in C. This map is also onto because each point $y = (.y_1 y_2 y_3 \ldots)_{\text{base 3}}$ is the image $\tau(x_{\mathbf{a}})$, where $\alpha_k = y_k/2$. Thus τ is a bijection.

Next we establish the continuity of τ. Let $x_{\mathbf{a}} \in X$ and $\varepsilon > 0$ be given. Choose N so large that $3^{-N} < \varepsilon$. Now $x_{\mathbf{a}}$ belongs to $J_{\alpha_1 \alpha_2 \ldots \alpha_N}$. Note that $J_{\alpha_1 \alpha_2 \ldots \alpha_N}$ and $I_N \setminus J_{\alpha_1 \alpha_2 \ldots \alpha_N}$ are disjoint closed sets. Let δ be the positive distance between them. Suppose that $x \in X$ and $|x - x_{\mathbf{a}}| < \delta$. Then x also belongs to $J_{\alpha_1 \alpha_2 \ldots \alpha_N}$. Hence $\tau(x)$ belongs to $T_{\alpha_1 \alpha_2 \ldots \alpha_N}$. This is an interval of length 3^{-N} containing $\tau(x_{\mathbf{a}})$ as well. Hence

$$|\tau(x) - \tau(x_{\mathbf{a}})| \le 3^{-N} < \varepsilon.$$

Finally, we use Theorem 11.6.3 to conclude that τ is a homeomorphism. Alternatively, the continuity of τ^{-1} can be proved in the same way as for τ. Note that the map τ preserves the order on the 2^n intervals in I_n for every n. Hence it follows easily that τ is monotone increasing.

Now we study those homeomorphisms between two spaces that carry a dynamical system on one space to a different system on the other.

11.6.5. DEFINITION. Let S be a dynamical system on a set X and let T be a dynamical system on a set Y. These two systems are said to be **topologically conjugate** if there is a homeomorphism σ from X onto Y such that $\sigma S = T \sigma$ or, equivalently, $T = \sigma S \sigma^{-1}$. The map σ is called a **topological conjugacy** between S and T.

It is clear that if σ is a topological conjugacy between S and T, then

$$\sigma S^n = T^n \sigma \quad \text{for all} \quad n \ge 1.$$

Hence if $x \in X$ is a periodic point for S with period n, then $y = \sigma(x)$ will be periodic for T of the same order. Moreover, the fact that σ is a homeomorphism means that convergent sequences in X correspond exactly to convergent sequences in Y under this map. Hence a periodic point $y = \sigma(x)$ will be attracting or repelling exactly as x is. Indeed, we have $\mathcal{O}_T(\sigma(x)) = \sigma(\mathcal{O}_S(x))$ for every point $x \in X$.

Topological conjugacy is an equivalence relation. First, if σ conjugates S onto T, then σ^{-1} conjugates T back onto S. If R is conjugate to S and S is conjugate to T, then R and T are conjugate. Evidently, the identity map id $: S \to S$ is a topological conjugacy from S to itself.

We will study topological conjugacy by examining a few examples.

11.6.6. PROPOSITION. *The tent map of Exercise 11.3.E and the quadratic map $Q_4 x = 4x - 4x^2$ on $[0, 1]$ are topologically conjugate. Hence Q_4 is chaotic on $[0, 1]$.*

PROOF. We will pull the appropriate homeomorphism out of the air. So the rest of this proof will be easy to understand, but it won't explain how to choose the homeomorphism.

Let $\sigma(x) = \sin^2(\frac{\pi}{2}x)$. It is easily checked that σ is strictly increasing and continuous on $[0, 1]$ and that $\sigma(0) = 0$ and $\sigma(1) = 1$. Hence by Example 11.6.2, it follows that σ is a homeomorphism of $[0, 1]$. Now using $Q_4 x = 4x(1 - x)$, compute

$$Q_4 \tau(x) = 4 \sin^2(\tfrac{\pi}{2}x) \cos^2(\tfrac{\pi}{2}x) = \sin^2(\pi x).$$

Likewise,

$$\tau(Tx) = \begin{cases} \sin^2(\tfrac{\pi}{2}(2x)) = \sin^2(\pi x) & \text{if } 0 \le x \le \tfrac{1}{2} \\ \sin^2(\tfrac{\pi}{2}(2 - 2x)) = \sin^2(\pi - \pi x) & \text{if } \tfrac{1}{2} \le x \le 1. \end{cases}$$

Thus $\tau T = Q_4 \tau$; and so τ is a topological conjugacy intertwining T and Q_4.

By Exercise 11.5.B, the tent map is chaotic. Hence Q_4 is also. $\blacksquare$

The goal of the rest of this section is to continue our analysis of the quadratic maps Q_a for $a > 2 + \sqrt{5}$. We will establish a topological equivalence with the shift on the Cantor set. Hence these quadratic maps are all topologically equivalent to each other, so that dynamically they all behave in exactly the same way.

11.6.7. THEOREM. *For $a > 2 + \sqrt{5}$, the quadratic maps Q_a on the set X_a is topologically conjugate to the shift S on the Cantor set C.*

PROOF. Recall from Example 11.5.6 that X_a is a Cantor set. We will construct a homeomorphism along the lines of Example 11.6.4, except that the ordering will be determined by the dynamics rather than by the usual order on the line. Recall the notation from that example.

The first step in the construction of the Cantor set X_a is the set

$$I_1 = J_0 \cup J_1 = \{x \in [0, 1] : Q_a x \in [0, 1]\}.$$

For each point x in X_a, $Q_a^{n-1} x$ belongs to X_a, and thus to either J_0 or J_1. Define the **itinerary** of x to be the sequence $\Gamma x = \gamma_1 \gamma_2 \ldots$ of 0s and 1s defined by the condition that

$$Q_a^{n-1} x \in J_{\gamma_n} \quad \text{for all} \quad n \ge 1.$$

The interval $J_{\alpha_1 \ldots \alpha_{n-1}}$ is mapped bijectively by Q_a^{n-1} onto the whole unit interval. And $X_a \cap J_{\alpha_1 \ldots \alpha_{n-1}}$ is mapped into X_a, and in particular into $I_1 = J_0 \cup J_1$. This dichotomy determines the sets $J_{\alpha_1 \ldots \alpha_{n-1} 0}$ and $J_{\alpha_1 \ldots \alpha_{n-1} 1}$, as one is mapped onto J_0 by Q_a^{n-1} and the other is mapped onto J_1. The order we need to keep track of is this itinerary ordering, not the usual order on $\mathbb{R}$. Notice that this discussion shows that if x and y both belong to $X_a \cap J_{\alpha_1 \ldots \alpha_n}$, then the itineraries Γx and Γy agree for the first n terms.

Define a map σ from X_a to C by

$$\sigma(x) = y_{\Gamma x} := \sum_{k \ge 1} 2\gamma_k 3^{-k}.$$

Let us verify that σ is a homeomorphism.

For any $x \in X_a$ and $\varepsilon > 0$, choose N so that $3^{-N} < \varepsilon$. Then x belongs to one of the Nth level intervals $J_{\alpha_1 \ldots \alpha_N}$. Let δ be the positive distance between this interval and the remaining $I_N \setminus J_{\alpha_1 \ldots \alpha_N}$. Then any $y \in X_a$ with $|x - y| < \delta$ also belongs to $J_{\alpha_1 \ldots \alpha_N}$. Hence the itinerary of y agrees with x for the first N terms. This means that $\sigma(x)$ and $\sigma(y)$ belong to the same Nth-level interval for C. Hence

$$|\sigma(x) - \sigma(y)| \leq 3^{-N} < \varepsilon.$$

So σ is continuous.

To see that σ is a bijection, consider any point $y \in C$. As usual, we write $y = .y_1 y_2 \cdots {}_{\text{base }3}$ in ternary using a sequence of 0s and 2s. This is the image of all points $x \in X_a$ with itinerary $\Gamma = \gamma_1 \gamma_2 \ldots$ given by $\gamma_k = y_k / 2$. However, Γ determines another unique sequence $\mathbf{a} = \alpha_1 \alpha_2 \ldots$ by the relation

$$Q_a^{n-1} J_{\alpha_1 \ldots \alpha_n} = J_{\gamma_n} \quad \text{for all} \quad n \geq 1.$$

So the points x with itinerary Γ are the points in

$$\bigcap_{n \geq 1} J_{\alpha_1 \ldots \alpha_n} = \{x_{\mathbf{a}}\}.$$

As we have noted before, this intersection consists of exactly one point. So σ is onto because this set is nonempty for each Γ; and it is one-to-one because the set is always a singleton.

Now we may apply Theorem 11.6.3 to see that σ is a homeomorphism.

We must show that σ intertwines Q_a and S. Suppose that $x \in X_a$ has itinerary $\Gamma = \gamma_1 \gamma_2 \gamma_3 \ldots$. Then the itinerary of $Q_a x$ is evidently $\gamma_2 \gamma_3 \gamma_4 \ldots$ because

$$Q_a^{n-1} Q_a x = Q_A^n x \in J_{\gamma_{n+1}} \quad \text{for all} \quad n \geq 1.$$

This is just saying that

$$\sigma(Q_a x) = S \sigma(x).$$

Therefore, Q_a is topologically conjugate to the shift. ∎

11.6.8. COROLLARY. *The quadratic maps Q_a for $a > 2 + \sqrt{5}$ have a dense set of transitive points in X_a.*

PROOF. This follows from our discussion of the shift in Example 11.5.16. The shift has a dense set of transitive points. So any map topologically conjugate to the shift must have such a set as well. ∎

11.6.9. REMARK. We have been studying the quadratic logistic maps in detail throughout this chapter. Our early arguments depended on specific calculations for these functions. However, all of the arguments for chaos depend only on a few fairly general properties.

Suppose that f is a function on $[0, 1]$ with $f(0) = f(1) = 0$ that is **unimodal**, meaning that f increases to a maximum at a point (x_0, y_0) and then decreases back down to $(1, 0)$. In order for our arguments to work, all we need is that $y_0 > 1$ and $|f'(x)| \geq c > 1$ for all x in

$$I_1 = \{x \in [0, 1] : f(x) \in [0, 1]\}.$$

Then the argument of Example 11.5.6 would apply to show that the set

$$X = \{x \in [0,1] : f^n(x) \in [0,1] \text{ for all } n \geq 1\}$$

is a Cantor set.

The preceding proof showing that Q_a is topologically conjugate to the shift only relied on the fact that $Q_a^n J = [0,1]$ for every component interval of the nth-level set I_n for each $n \geq 1$. It is easy to see that this property also holds in the generality of the unimodal function f. Thus, in particular, f is chaotic on X. This enables us to recognize chaos in many situations.

A simple example of this which is very similar to the quadratic family is the function $f(x) = 3x - 3x^3$. However, it is more instructive to look at the quadratic maps again.

Consider the graph of $Q_{3.75}^2$ in Figure 11.10. Notice that $Q_{3.75}$ has a fixed point at $1 - 1/3.75 = \frac{11}{15}$; and $Q_{3.75}\frac{4}{11} = \frac{11}{15}$ as well. On the interval $J = [\frac{4}{15}, \frac{11}{15}]$, $Q_{3.75}^2$ decreases from $(\frac{4}{15}, \frac{11}{15})$ to a local minimum at $(\frac{1}{2}, \frac{225}{1024})$, and increases again to the point $(\frac{11}{15}, \frac{11}{15})$. Since $\frac{225}{1024} \approx .22 < .267 \approx \frac{4}{15}$, this graph "escapes" the square $J \times J$. The qualitative behaviour of this part of the graph is just like that of Q_a for $a > 2 + \sqrt{5}$.

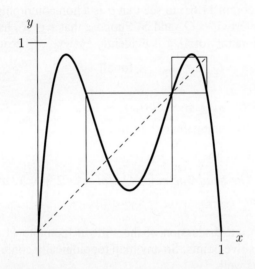

FIGURE 11.10. The graph of $Q_{3.75}^2$, with $J \times J$ and $K \times K$ marked.

To verify that our proof applies, we need to see that the absolute value of the derivative is greater than 1 on

$$J_1 = \{x \in J : Q_{3.75}^2(x) \in J\}.$$

It is notationally easier to use the generic parameter a and substitute 3.75 for a later. What exactly is J_1? To compute it, we first solve the quadratic $Q_a(x) = \frac{1}{a}$. This has solutions

$$\frac{1}{2} \pm \frac{\sqrt{a^2 - 4}}{2a}.$$

These are roughly .077 and .923 for $a = 3.75$. The points that Q_a^2 maps to the endpoints of J_1 are seen from the graph to be solutions of

$$\frac{1}{2} + \frac{\sqrt{a^2 - 4}}{2a} = Q_a x = ax - ax^2.$$

This quadratic has solutions

$$x_\pm := \frac{1}{2} \pm \frac{\sqrt{a^2 - 2a - 2\sqrt{a^2 - 4}}}{2a}.$$

For $a = 3.75$, we obtain $J_1 = [.2667, .4377] \cup [.5623, .7333]$. Now the derivative of Q_a^2 is monotone increasing on J_1, changing sign at $x = .5$, and is symmetric about the midpoint. So the minimal slope is obtained at the two interior endpoints

$$(Q_a^2)'(x_+) = Q_a'(Q_a x_+) \, Q_a'(x_+)$$
$$= a^2(1 - 2Q_a x_+)(1 - 2x_+)$$
$$= a^2 \frac{\sqrt{a^2 - 4}}{a} \frac{\sqrt{a^2 - 2a - 2\sqrt{a^2 - 4}}}{a}$$
$$= \sqrt{(a^2 - 4)(a^2 - 2a - 2\sqrt{a^2 - 4})} \approx 1.482.$$

Thus our earlier arguments apply to show that there is a Cantor set X contained in J on which $Q_{3.75}^2$ acts chaotically, and in fact is topologically conjugate to the shift map.

Now $Q_{3.75}$ maps J_1 onto the interval $K = [.733, .923]$. The restriction of $Q_{3.75}^2$ to $K \times K$ behave in exactly the same way, and there is another Cantor set Y on which $Q_{3.75}^2$ behaves chaotically. Moreover, $Q_{3.75}$ maps the Cantor set X into Y and vice versa. From this, it is not difficult to see that $Q_{3.75}$ acts chaotically on the Cantor set $X \cup Y$. See the Exercises.

Just as Q_a is actually chaotic for $a > 4$ with a more delicate proof, the same is true for this analysis of Q_a^2. It can be shown that the preceding argument works whenever the graph in the interval $J \times J$ escapes in the middle. This occurs at about $a = 3.6786$. So once $a > 3.6786$, the quadratic map Q_a is chaotic on a Cantor set.

Exercises for Section 11.6

A. Use the sequential characterization of continuity Theorem 5.3.1 (2) to provide another proof of Theorem 11.6.3.   HINT: Let $(x_n)_{n=1}^\infty$ be a convergent sequence in Y with $\lim\limits_{n \to \infty} x_n = a$. Use Exercise 4.4.F on $(f(x_n))_{n=1}^\infty$.

B. Consider the map of $[0, 2\pi)$ onto the circle $\mathbb{T}$ by wrapping around exactly once. Show that this map is continuous, one-to-one, and onto. Show that this is *not* a homeomorphism. Explain why this does not contradict Theorem 11.6.3.

C. Show that the circle $\mathbb{T}$ is not homeomorphic to the interval $[0, 1]$.   HINT: If σ maps $\mathbb{T}$ onto $[0, 1]$, show that there are at least two points mapping to each $0 < y < 1$.

D. Consider the map of the circle to itself given by $T\theta \equiv \theta + \frac{2\pi}{n} + \varepsilon \sin(2\pi n x)$, where $0 < \varepsilon < 1/2\pi n$.

(a) Compute $T'(\theta)$ and deduce that T is a homeomorphism.

(b) Show that 0 and $\frac{\pi}{n}$ are periodic points.

(c) If $x \notin \mathcal{O}(0)$, prove that $\text{dist}(T^k x, \mathcal{O}(\frac{\pi}{n}))$ is strictly decreasing.

(d) Show that $\mathcal{O}(0)$ is a repelling orbit and $\mathcal{O}(\frac{\pi}{n})$ is attracting, and that $\omega(x) = \mathcal{O}(\frac{\pi}{n})$ except for $x \in \mathcal{O}(0)$.

E. Show that $f(x) = 1 - 2|x|$ and $g(x) - 1 - 2x^2$ as dynamical systems on $[-1, 1]$ are topologically conjugate as follows:

(a) If φ is a homeomorphism of $[-1, 1]$ such that $\varphi(f(x)) = g(\varphi(x))$, show that φ is an odd function such that $\varphi(-1) = -1$ and $\varphi(0) = 0$.

(b) Use fixed points to show that $\varphi(1/3) = 1/2$. Deduce that $\varphi(2/3) = \sqrt{3}/2$.

(c) Guess a trig function with the properties of φ and verify that it works.

F. Let f be a homeomorphism of $[0, 1]$ with no fixed points in $(0, 1)$.

(a) Show that f is strictly monotone increasing, and either $f(x) < x$ for all x in $(0, 1)$ or $f(x) > x$ for all x in $(0, 1)$.

(b) If $f(x) > x$, prove that the orbit of x under f converges to 1 and orbit under f^{-1} converges to 0.

(c) Show that $(0, 1)$ is the disjoint union of the intervals $[f^k(.5), f^{k+1}(.5))$ for $k \in \mathbb{Z}$.

(d) Let f and g be two homeomorphisms of $[0, 1]$ with no fixed points in $(0, 1)$. Prove that they are topologically conjugate.

HINT: Assume first that $f(x) > x$ and $g(x) > x$ for all x in $(0, 1)$. Define φ from $[.5, f(.5)]$ onto $[.5, g(.5)]$. Extend this to the whole interval to obtain a conjugacy.

G. (a) Show that every quadratic function $p(x) = ax^2 + bx + c$ on $\mathbb{R}$ is topologically conjugate to some $q(x) = x^2 + d$.   HINT: Use a linear map $\tau(x) = mx + e$. Compute $p(\tau(x))$ and $\tau(q(x))$ and equate coefficients to solve for m, e, and d.

(b) For which values of d is $q(x) = x^2 + d$ topologically conjugate to one of the logistic maps Q_a for $a > 0$? What are the dynamics of q when d is outside this range?

H. Suppose that $T : I \to I$ is given, and T^2 maps an infinite compact subset X into itself and is chaotic on X. Show that T is chaotic on $X \cup TX$.

11.7. Iterated Function Systems

An **iterated function system**, or **IFS**, is a multivariable discrete dynamical system. Under reasonable hypotheses, these systems have a unique compact invariant set. This invariant set exhibits certain self-similarity properties that have become known as **fractals**.

We begin with a finite set $\mathcal{T} = \{T_1, \ldots, T_r\}$ of contractions on a closed subset X of $\mathbb{R}^n$. This family of maps determines a multivariable dynamical system. The orbit of a point x will consist of the set of all points obtained by repeated application of the maps T_i in any order with arbitrary repetition. That is, for each finite *word* $i_1 i_2 \ldots i_k$ in the alphabet $\{1, \ldots, r\}$, the point $T_{i_1} T_{i_2} \ldots T_{i_k} x$ is in the orbit $\mathcal{O}(x)$. We wish to find a compact set A with the property that

$$A = T_1 A \cup T_2 A \cup \cdots \cup T_r A.$$

Surprisingly this set turns out to be unique!

A **similitude** is a map T that is a scalar multiple of an isometry. That is, there is a constant $r > 0$ so that $\|Tx - Ty\| = r\|x - y\|$ for all $x, y \in X$. Such maps are obtained as rotations, translations, and scalings (see Exercise 11.7.F). In particular, these maps are a similarity in the geometric sense that they map sets to similar sets and so preserve shape, up to a scaling factor.

In the event that each T_i is a similitude, the fact that $A = T_1 A \cup T_2 A \cup \cdots \cup T_r A$ means that each $T_i A$ is similar to A. This will be especially evident in examples in which these r sets are disjoint. The process repeats and $T_i A$ decomposes as $T_i A = T_i T_1 A \cup T_i T_2 A \cup \cdots \cup T_i T_r A$. After k steps, A is decomposed into r^k similar pieces. This symmetry property is called **self-similarity** and is characteristic of fractals arising from iterated function systems.

11.7.1. EXAMPLES.

(1) Let $X = \mathbb{R}^2$, and consider three affine maps $T_1 \mathbf{x} = \frac{1}{2}\mathbf{x}$, $T_2 \mathbf{x} = \frac{1}{2}\mathbf{x} + (2, 0)$, and $T_3 \mathbf{x} = \frac{1}{2}\mathbf{x} + (1, \sqrt{3})$. Notice that each T_i is a similitude with scaling factor $\frac{1}{2}$. It is easy to verify that the fixed points of these three maps are $\mathbf{v}_1 = (0, 0)$, $\mathbf{v}_2 = (4, 0)$, and $\mathbf{v}_3 = (2, 2\sqrt{3})$, respectively.

Let Δ be the solid equilateral triangle with these three vertices. A computation shows that $T_i \Delta$ for $i = 1, 2, 3$ are the three equilateral triangles with half the dimensions of the original that lie inside Δ and share the vertex $\mathbf{v}_i$ with Δ. So $\Delta_1 = T_1 \Delta \cup T_2 \Delta \cup T_3 \Delta$ equals Δ with the middle triangle removed.

Since $\Delta_1 \subset \Delta$, it follows fairly easily (see Corollary 11.7.5) that when we iterate the procedure by setting

$$\Delta_{k+1} = T_1 \Delta_k \cup T_2 \Delta_k \cup T_3 \Delta_k$$

a decreasing sequence of compact sets is obtained. The intersection Δ_∞ of these sets is the Sierpinski snowflake of Exercise 4.4.J (see Figure 4.4). It has the property that we are looking for,

$$\Delta_\infty = T_1 \Delta_\infty \cup T_2 \Delta_\infty \cup T_3 \Delta_\infty.$$

(2) Not all fractals are solid figures. The **von Koch curve** is obtained from an IFS using the following four similitudes:

$$T_1 \begin{bmatrix} x \\ y \end{bmatrix} = \begin{bmatrix} \frac{1}{3} & 0 \\ 0 & \frac{1}{3} \end{bmatrix} \begin{bmatrix} x \\ y \end{bmatrix}, \qquad T_2 \begin{bmatrix} x \\ y \end{bmatrix} = \begin{bmatrix} \frac{1}{3} \\ 0 \end{bmatrix} + \begin{bmatrix} \frac{1}{6} & \frac{-\sqrt{3}}{6} \\ \frac{\sqrt{3}}{6} & \frac{1}{6} \end{bmatrix} \begin{bmatrix} x \\ y \end{bmatrix},$$

$$T_3 \begin{bmatrix} x \\ y \end{bmatrix} = \begin{bmatrix} \frac{1}{2} \\ \frac{\sqrt{3}}{6} \end{bmatrix} + \begin{bmatrix} \frac{1}{6} & \frac{\sqrt{3}}{6} \\ -\frac{\sqrt{3}}{6} & \frac{1}{6} \end{bmatrix} \begin{bmatrix} x \\ y \end{bmatrix}, \qquad T_4 \begin{bmatrix} x \\ y \end{bmatrix} = \begin{bmatrix} 0 \\ \frac{2}{3} \end{bmatrix} + \begin{bmatrix} \frac{1}{3} & 0 \\ 0 & \frac{1}{3} \end{bmatrix} \begin{bmatrix} x \\ y \end{bmatrix}.$$

Let $B_0 = \{(x, 0) : x \in [0, 1]\}$ and define $B_{k+1} = T_1 B_k \cup T_2 B_k \cup T_3 B_k \cup T_4 B_k$. Graphing these figures shows an increasingly complex curve emerging in which the previous curve is scaled by $1/3$ and used to replace the four line segments of B_1. See Figure 11.11.

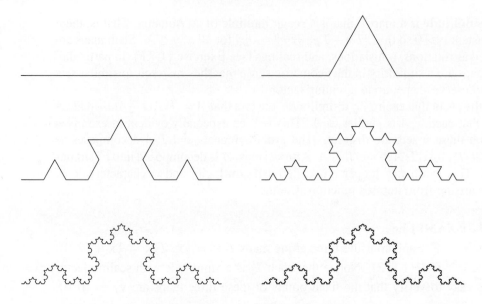

FIGURE 11.11. The sets B_0 through B_5.

As we will see, this construction works in great generality, and many sets can be obtained as the invariant sets for iterated function systems. To establish these facts, we need a framework, in this case, a metric space. Let $K(X)$ denote the collection of all nonempty compact subsets of X. This is a metric space with respect to the Hausdorff metric of Example 9.1.2(5),

$$d_H(A, B) = \max \Big\{ \sup_{a \in A} \mathrm{dist}(a, B), \sup_{b \in B} \mathrm{dist}(b, A) \Big\}.$$

Our first result is the completeness of $K(X)$ in the Hausdorff metric.

11.7.2. THEOREM. *If X is a closed subset of $\mathbb{R}^n$, the metric space $K(X)$ of all compact subsets of X with the Hausdorff metric is complete.*

PROOF. Let A_n be a Cauchy sequence of compact sets in $K(X)$. Define

$$A = \bigcap_{k \geq 1} \overline{\bigcup_{i \geq k} A_i}.$$

Observe that for any $\varepsilon > 0$, there is an integer N so that $d_H(A_i, A_j) < \varepsilon$ for all $i, j \geq N$. In particular, it follows from the definition of the Hausdorff metric that $A_i \subset (A_N)_\varepsilon$ for all $i \geq N$. Consequently, $A \subset (A_N)_\varepsilon$. Now, $(A_N)_\varepsilon$ is a closed and bounded subset of $\mathbb{R}^n$ and so, by the Heine–Borel Theorem, is compact. It follows that A is the decreasing intersection of nonempty compact sets; by Cantor's Intersection Theorem, A is a nonempty compact set.

Having shown that $A \in K(X)$, it remains only to prove that (A_i) converges to A. We also have $A \subset (A_N)_\varepsilon \subset ((A_i)_\varepsilon)_\varepsilon = (A_i)_{2\varepsilon}$ for all $i \geq N$. Conversely, fix $a_i \in A_i$ with $i \geq N$. For each $j > i$, $A_i \subset (A_j)_\varepsilon$ and thus there is a point $a_j \in A_j$

with $\|a_i - a_j\| < \varepsilon$. Each a_j lies in the bounded set $(A_N)_\varepsilon$. By compactness, there is a convergent subsequence $\lim_{l \to \infty} a_{j_l} = a$. Clearly, $\|a_i - a\| \le \varepsilon$.

We must show that a belongs to A. However, all but the first few terms of (a_{j_l}) lie in $\bigcup_{j \ge k} A_j$, and so a belongs to $\overline{\bigcup_{j \ge k} A_j}$. Thus a also lies in the intersection of these sets, A. We deduce that $\text{dist}(a_i, A) \le \varepsilon$ for each point in A_i and therefore $A_i \subset \overline{A_\varepsilon} \subset A_{2\varepsilon}$. Combining the two estimates, $d_H(A_i, A) < 2\varepsilon$ for all $i \ge N$. Therefore, A_i converges to A. ∎

Next we define a map from $K(X)$ into itself by

$$TA = T_1 A \cup T_2 A \cup \cdots \cup T_r A.$$

We need to verify that TA is in $K(X)$ (i.e., that TA is compact). Each T_i is continuous, and the continuous image of a compact set is compact (Theorem 5.4.3). Also, the finite union of compact sets is compact, and therefore TA is a compact set. Our goal is to show that this is a contraction. First we need an easy lemma whose proof is left as Exercise 11.7.A.

11.7.3. LEMMA. *Let $A_1, \ldots, A_r$ and $B_1, \ldots, B_r$ be compact subsets of $\mathbb{R}^n$. Then*

$$d_H(A_1 \cup \cdots \cup A_r, B_1 \cup \cdots \cup B_r) \le \max \{d_H(A_1, B_1), \ldots, d_H(A_r, B_r)\}.$$

11.7.4. THEOREM. *Let X be a closed subset of $\mathbb{R}^n$ and let $T_1, \ldots, T_r$ be contractions of X into itself. Let s_i be the Lipschitz constants for each T_i and set $s = \max\{s_1, \ldots, s_r\}$. Then T is a contraction of $K(X)$ into itself with Lipschitz constant s. Hence there is a unique compact subset A of X such that*

$$A = T_1 A \cup T_2 A \cup \cdots \cup T_r A.$$

Moreover, if B is any compact set, we have the estimates

$$d_H(T^k B, A) \le s^k d_H(B, A) \le \frac{s^k}{1 - s} d_H(B, TB).$$

PROOF. Let A and B be any two compact subsets of X. Observe that

$$d_H(T_i A, T_i B) \le s_i d_H(A, B).$$

Indeed, if $a \in A$, then there is a $b \in B$ with $\|a - b\| \le d_H(A, B)$. Hence $\|T_i a - T_i b\| \le s_i d_H(A, B)$. So $\sup_{a \in A} \text{dist}(T_i a, T_i B) \le s_i d_H(A, B)$. Reversing the roles of A and B, we arrive at the desired estimate. By Lemma 11.7.3, it follows that $d_H(TA, TB) \le s d_H(A, B)$.

Theorem 11.7.2 shows that $(K(X), d_H)$ is a complete metric space. As T is a contraction, we may apply the Banach Contraction Principle (11.1.6). It follows that there is a unique fixed point $A = TA$. By definition of T, this is the unique compact set such that $A = T_1 A \cup T_2 A \cup \cdots \cup T_r A$. It is also an immediate consequence of the Contraction Principle that A is obtained as the limit of iterates of T applied to any initial set B, and the estimates follow directly from the estimates in the Contraction Principle. ∎

11.7.5. COROLLARY. *Suppose that T is a contraction of $K(X)$ into itself with Lipschitz constant s. If B is a compact set such that $TB \subset B$, then the fixed point is given by $A = \bigcap\limits_{k \geq 0} T^k B$.*

PROOF. We show by induction that $T^{k+1}B \subset T^k B$ for $k \geq 0$. If $k = 0$, this is true by hypothesis. Assuming that $T^k B \subset T^{k-1}B$, we have

$$T^{k+1}B = T(T^k B) \subset T(T^{k-1}B) = T^k B.$$

By Theorem 11.7.4, the fixed point A is the limit of the sequence $T^k B$. From the *proof* of Theorem 11.7.2, this limit is given by

$$A = \bigcap_{k \geq 1} \overline{\bigcup_{i \geq k} T^i B} = \bigcap_{k \geq 1} T^k B. \qquad \blacksquare$$

An excellent choice for the compact set B to use in computing the limit set is modeled by the Sierpinski snowflake, Example 11.7.1(1). Another good choice for an initial compact set B is a single point $\{x\}$ that happens to belong to A. Such points are easy to find. Each T_i is a contraction on X and thus has a unique fixed point x_i that may be found by iteration of T_i applied to any initial point. We give a significant strengthening of this fact, which hints at the dynamical properties of the iterated function system.

11.7.6. THEOREM. *For each word $w = i_1 i_2 \ldots i_l$ in the alphabet $\{1, \ldots, r\}$, there is a unique fixed point a_w of $T_w = T_{i_1} T_{i_2} \ldots T_{i_l}$. Each a_w belongs to the fixed set A, and the set of all of these fixed points is dense in A.*

If a is any point in A, the orbit $\mathcal{O}(a) = \{T_w a : w \text{ is a word in } \{1, \ldots, r\}\}$ is dense in A.

PROOF. First observe that the composition of contractions is a contraction (see Exercise 11.1.E). Therefore, each T_w is a contraction and hence has a unique fixed point a_w. Moreover, starting with any point x, the iterates $T_w^k x$ converge to a_w. Take x to be any point in A. Since each T_i maps A into itself, $T_w^k x$ is in A for all $k \geq 0$. As A is closed, the limit a_w belongs to A.

Next we prove that the set of these fixed points $\{a_w : w \text{ is a word in } \{1, \ldots, r\}\}$ is dense in A. Now $A = TA = T_1 A \cup \cdots \cup T_r A$. So

$$A = T^2 A = T_1 T_1 A \cup T_1 T_2 A \cup \ldots \cup T_1 T_r A \cup$$
$$T_2 T_1 A \cup T_2 T_2 A \cup \ldots \cup T_2 T_r A \cup$$
$$\ldots \cup$$
$$T_r T_1 A \cup T_r T_2 A \cup \ldots \cup T_r T_r A.$$

Repeating this N times, we obtain

$$A = T^N A = \bigcup_{\text{words } w \text{ of length } N} T_w A.$$

Fix $\varepsilon > 0$ and choose N so large that $s^N \operatorname{diam}(A) < \varepsilon$. By Exercise 11.1.E, each T_w is a contraction with Lipschitz constant no greater than s^N. Consequently, $\operatorname{diam}(T_w A) \le s^N \operatorname{diam}(A) < \varepsilon$. If $a \in A$, choose a word w of length N so that $a \in T_w A$. Since $a_w = T_w a_w$, it is clear that $a_w \in T_w A$ as well. Therefore, $\|a - a_w\| < \varepsilon$. Thus the set of these fixed points is dense in A.

Now if a is an arbitrary point in A, it follows that $T_w a$ belongs to A for every finite word w. Hence $\mathcal{O}(a) \subset A$. Observe that $T_i T_w a = T_{iw} a$ is another point in $\mathcal{O}(a)$. Therefore, $T\mathcal{O}(a) \subset \mathcal{O}(a)$. Let $B = \overline{\mathcal{O}(a)}$. The continuity of T implies that $TB \subset B$. By Corollary 11.7.5, $A = \bigcap_{k \ge 1} T^k B \subset B$. In particular $A \subset B \subset A$, and so $\overline{\mathcal{O}(a)} = A$. ∎

This result allows us graph the fractal approximately as follows. Pick any point a. This may not lie in A. However, $b = T_1^{100} a$ will be very close to the fixed point of T_1 assuming reasonable constants. Use a computer to calculate b and then recursively plot the sets $T^k\{b\}$ for sufficiently many k. This will frequently give an excellent picture of the fractal. You can make explicit estimates to ensure good convergence.

11.7.7. EXAMPLE. Consider the maps

$$T_1 \begin{bmatrix} x \\ y \end{bmatrix} = \begin{bmatrix} .5 & .5 \\ -.5 & .5 \end{bmatrix} \begin{bmatrix} x \\ y \end{bmatrix} + \begin{bmatrix} 1 \\ 5 \end{bmatrix} \qquad T_2 \begin{bmatrix} x \\ y \end{bmatrix} = \begin{bmatrix} .5 & .5 \\ -.5 & .5 \end{bmatrix} \begin{bmatrix} x \\ y \end{bmatrix} + \begin{bmatrix} -1 \\ 3 \end{bmatrix}.$$

A simple matrix calculation shows that $(-4, 6)$ is the fixed point of T_1 and $(-4, 2)$ is the fixed point of T_2. We use a computer to plot the sets $A_0 = \{(-4, 6), (-4, 2)\}$ and $A_{k+1} = T_1 A_k \cup T_2 A_k$ for $1 \le k \le 11$. Since A equals the closed union of all the A_k's, this yields a reasonable approximation. Look for the self-symmetry in Figure 11.12.

We finish with a simple result that shows that the sets that are fixed for iterated function schemes are extremely plentiful.

11.7.8. PROPOSITION. *Let C be a compact subset of $\mathbb{R}^n$, and let $\varepsilon > 0$. Then there is an IFS $T = \{T_1, \ldots, T_r\}$ with fixed set A such that $d_H(A, C) < \varepsilon$.*

PROOF. Since C is compact, we can find a fintite set of points, $C_0 = \{c_1, \ldots, c_r\}$, so that the union of the balls $B_\varepsilon(c_i)$, call it B, contains C. Observe that B equals $(C_0)_\varepsilon$. Let R be large enough so that the $B_R(0)$ contains B. Define

$$T_i x = \frac{\varepsilon}{2R}(x - c_i) + c_i \quad \text{for} \quad 1 \le i \le r.$$

Since $T_i c_i = c_i$ are fixed points, C_0 is contained in the fixed set A of T. Also, $T_i B \subset T_i B_R(0) \subset B_\varepsilon(c_i) \subset B$ for each i. Therefore by Corollary 11.7.5, $A \subset B = (C_0)_\varepsilon \subset C_\varepsilon$. Plus, $C \subset (C_0)_\varepsilon \subset A_\varepsilon$. So, $d_H(A, C) < \varepsilon$ as required. ∎

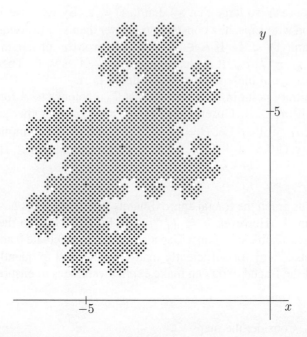

FIGURE 11.12. Pointillist picture of "the twin dragon" set.

Exercises for Section 11.7

A. (a) Let A_1, A_2, B_1 and B_2 be compact subsets of $\mathbb{R}^n$. Show that
$$d_H(A_1 \cup A_2, B_1 \cup B_2) \le \max\{d_H(A_1, B_1), d_H(A_2, B_2)\}.$$
(b) Use induction to prove Lemma 11.7.3.

B. (a) Let $\mathcal{T} = \{T_1, \dots, T_r\}$ be an IFS on $\mathbb{R}^n$. Suppose that B is a compact set. Prove that there is a unique compact set C such that $C = B \cup TC$.
HINT: Add a constant map $T_0(X) = B$ for $X \in K(\mathbb{R}^n)$.
(b) Hence prove that there is always a compact set C containing B so that $TC \subset C$.

C. Consider the four maps on $\mathbb{R}^2$ given by $T_i \mathbf{x} = A\mathbf{x} + \mathbf{b}_i$ for $i = 1, 2, 3, 4$, where $A = \frac{1}{2} \begin{bmatrix} 1 & 1 \\ -1 & 1 \end{bmatrix}$ and the vectors $\mathbf{b}_i$ are $(0, 0), (1, 0), (0, 1)$ and $(1, 1)$.
(a) Show that the fixed points of the T_i's form the vertices of a square S.
(b) Compute TS and $T^2 S$.
(c) Use a computer to generate a picture of the fixed set.

D. (a) Find an IFS on $\mathbb{R}$ that generates the Cantor set.
HINT: Identify two self maps of C with disjoint union equal to C.
(b) Find a different IFS that also generates C.

E. Consider the maps $\mathcal{T} = \{T_1, T_2, T_3, T_4\}$ given by
$$T_1 \begin{bmatrix} x \\ y \end{bmatrix} = \begin{bmatrix} .8 & .0 \\ 0 & .8 \end{bmatrix} \begin{bmatrix} x \\ y \end{bmatrix} + \begin{bmatrix} .1 \\ .04 \end{bmatrix} \qquad T_2 \begin{bmatrix} x \\ y \end{bmatrix} = \begin{bmatrix} .5 & 0 \\ 0 & .5 \end{bmatrix} \begin{bmatrix} x \\ y \end{bmatrix} + \begin{bmatrix} .25 \\ .4 \end{bmatrix}$$
$$T_3 \begin{bmatrix} x \\ y \end{bmatrix} = \begin{bmatrix} .35 & -.35 \\ .35 & .35 \end{bmatrix} \begin{bmatrix} x \\ y \end{bmatrix} + \begin{bmatrix} .27 \\ .08 \end{bmatrix} \qquad T_4 \begin{bmatrix} x \\ y \end{bmatrix} = \begin{bmatrix} .35 & .35 \\ -.35 & .35 \end{bmatrix} \begin{bmatrix} x \\ y \end{bmatrix} + \begin{bmatrix} .38 \\ .43 \end{bmatrix}.$$

(a) Use a computer to plot the maple leaf pattern A fixed by T.

(b) Plot $T_i A$ for $1 \le i \le 4$ to see the self-symmetries.

F. (a) Show that an isometry T with $T\mathbf{0} = \mathbf{0}$ is linear.

(b) Show that every similitude T of $\mathbb{R}^n$ has the form $T\mathbf{x} = rU\mathbf{x} + \mathbf{a}$, where $r > 0$, $\mathbf{a} \in \mathbb{R}^n$, and U is a unitary matrix.

HINT: Set $g(\mathbf{x}) = r^{-1}(T\mathbf{x} - T\mathbf{0})$. Verify that g is isometric and apply (a).

G. The **fractal dimension** of a bounded subset A of $\mathbb{R}^n$ is computed by counting, for $\varepsilon > 0$, the smallest number of cubes of side length ε that cover A, call it $N(A, \varepsilon)$. The fractal dimension is then the limit $\lim\limits_{\varepsilon \to 0^+} \dfrac{\log N(A, \varepsilon)}{\log \varepsilon^{-1}}$, if it exists.

(a) Compute the fractal dimension of the unit n-cube in $\mathbb{R}^n$.

(b) Show that the fractal dimension is not affected by scaling.

(c) If $A \subset \mathbb{R}^n$ has interior, show that the fractal dimension is n.

(d) Compute the fractal dimension of the Cantor set C.

(e) Compute the fractal dimension of the Sierpinski snowflake.

CHAPTER 12

Differential Equations

In this chapter, we apply analysis to the study of ordinary differential equations, generally called DEs or ODEs. *Ordinary* is used to indicate differential equations of a single variable, in contrast with partial differential equations (PDEs), where several variables, and hence partial derivatives, appear. We will see some PDEs in the chapters on Fourier series, Chapters 13 and 14.

Most introductory courses on differential equations present methods for solving DEs of various special types. We will not be concerned with those techniques here except to give a few pertinent examples. Rather we are concerned with why differential equations have solutions, and why these solutions are or are not unique. This topic, crucial to a full understanding of differential equations, is often omitted from introductory courses because it requires the tools of real analysis.

12.1. Integral Equations and Contractions

We consider an example that motivates the approach of the next section. Start with an initial value problem, which consists of two parts:

$$f'(x) = \varphi(x, f(x)) \quad \text{for} \quad a \leq x \leq b$$
$$f(c) = y_0.$$

The first equation is the DE and the second is an initial value condition. The function $\varphi(x, y)$ is a continuous function of two variables defined on $[a, b] \times \mathbb{R}$ and c is a given point in $[a, b]$. By solving the DE or, equivalently, solving the initial value problem, we mean finding a function $f(x)$ that is defined and differentiable on the interval $[a, b]$ and satisfies both the differential equation and the initial value condition.

This DE is of first order, as the equation involves only the first derivative. In general, the **order of a differential equation** is the highest-order derivative of the unknown function that appears in the equation.

Because of the subject's connections to physics, chemistry, and engineering, it is common in differential equations to suppress the dependence of the function f on x [i.e., to write f instead of $f(x)$]. Typically in the sciences, each variable has a physical significance, and it can be a matter of choice about which is the independent variable and which is dependent. We will sometimes do this in examples, where simplifying otherwise complicated expressions seems to be worth the extra demands this notation makes.

Our first step is to turn this problem into a fixed-point problem by integration. Indeed, from the fundamental theorem of calculus, our solution must satisfy

$$f(x) = f(c) + \int_c^x f'(t)\, dt = y_0 + \int_c^x \varphi(t, f(t))\, dt.$$

Conversely, a continuous solution of this **integral equation** is automatically differentiable by the Fundamental Theorem of Calculus, and

$$f'(x) = \frac{d}{dx}\left(y_0 + \int_c^x \varphi(t, f(t))\, dt\right) = \varphi(x, f(x))$$

and

$$f(c) = y_0 + \int_c^c \varphi(t, f(t))\, dt = y_0.$$

Thus f satisfies the DE, including the initial value condition.

This integral equation suggests studying a map from $C[a, b]$ into itself defined by

$$Tf(x) = y_0 + \int_c^x \varphi(t, f(t))\, dt.$$

The solutions to the integral equation, if any, correspond precisely to the fixed points of T. The Contraction Principle (11.1.6) is well suited to this kind of problem. There are also more sophisticated approaches that give weaker conclusions from weaker hypotheses. However, the Contraction Principle gives both existence and uniqueness of a solution, when it can be applied. Consider the following specific example.

12.1.1. EXAMPLE. We will solve the initial value problem

$$f'(x) = 1 + x - f(x) \quad \text{for} \quad -\tfrac{1}{2} \le x \le \tfrac{1}{2}$$
$$f(0) = 1.$$

First convert it to the integral equation

$$f(x) = 1 + \int_0^x 1 + t - f(t)\, dt$$
$$= 1 + x + \tfrac{1}{2}x^2 - \int_0^x f(t)\, dt$$

for f in $C[-\tfrac{1}{2}, \tfrac{1}{2}]$.

Define a map T on $C[-\tfrac{1}{2}, \tfrac{1}{2}]$ by sending f to the function Tf given by

$$Tf(x) = 1 + x + \tfrac{1}{2}x^2 - \int_0^x f(t)\, dt.$$

The solution of the integral equation is a fixed point of T.

To use the Banach Contraction Principle, we must show that T is a contraction. We have

$$|Tf(x) - Tg(x)| = \left| \int_0^x f(t) - g(t)\, dt \right|$$

$$\leq \left| \int_0^x |f(t) - g(t)|\, dt \right|$$

$$\leq \left| \int_0^x \|f - g\|_\infty\, dt \right|$$

$$= \|f - g\|_\infty \int_0^{|x|} dt \leq \frac{1}{2}\|f - g\|_\infty.$$

This estimate is independent of x in $[-\frac{1}{2}, \frac{1}{2}]$, and thus we obtain

$$\|Tf - Tg\|_\infty \leq \frac{1}{2}\|f - g\|_\infty.$$

Hence T is a contraction.

By the Contraction Principle (11.1.6), there is a unique fixed point f_∞ that will solve our DE. Moreover, any sequence of functions (f_n) with $f_{n+1} = Tf_n$ will converge to f_∞ in $C[-\frac{1}{2}, \frac{1}{2}]$. For example, let us take f_0 to be the constant function 1. Then

$$f_1(x) = Tf_0(x) = 1 + x + \frac{1}{2}x^2 - \int_0^x 1\, dt = 1 + \frac{1}{2}x^2.$$

Similarly,

$$f_2(x) = Tf_1(x) = 1 + x + \frac{1}{2}x^2 - \int_0^x 1 + \frac{1}{2}t^2\, dt$$

$$= 1 + \frac{1}{2}x^2 - \frac{1}{6}x^3.$$

And

$$f_3(x) = Tf_2(x) = 1 + x + \frac{1}{2}x^2 - \int_0^x 1 + \frac{1}{2}t^2 - \frac{1}{6}t^3\, dt$$

$$= 1 + \frac{1}{2}x^2 - \frac{1}{6}x^3 + \frac{1}{24}x^4.$$

In general, we can establish by induction (do it yourself!) that

$$f_n(x) = 1 + \frac{1}{2}x^2 - \frac{1}{3!}x^3 + \frac{1}{4!}x^4 - \frac{1}{5!}x^5 + \cdots + \frac{1}{(n+1)!}(-x)^{n+1}.$$

This sequence evidently consists of the partial sums of an infinite series. The new term added at the nth stage is $\frac{1}{(n+1)!}(-x)^{n+1}$, which on $[-\frac{1}{2}, \frac{1}{2}]$ has max norm

$$\max_{|x|\leq 1/2} \left| \frac{(-x)^{n+1}}{(n+1)!} \right| = \left(2^{n+1}(n+1)!\right)^{-1}.$$

Therefore, this power series converges uniformly on $[-\frac{1}{2}, \frac{1}{2}]$ by the Weierstrass M-test. Figure 12.1 gives the graphs of f_0, f_1, f_2, and f_∞ on $[-1, 1]$. The astute reader should notice that these are the Taylor polynomials for e^{-x} except that the term in x is missing.

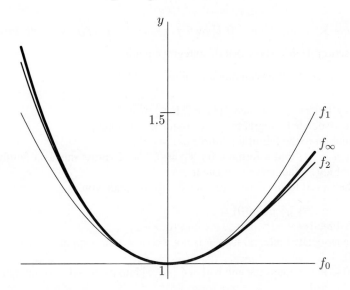

FIGURE 12.1. The first three approximants and f_∞ on $[-1, 1]$.

We obtain

$$f_\infty(x) = x + \sum_{k=0}^{\infty} \frac{1}{k!}(-x)^k = e^{-x} + x.$$

This shows that $e^{-x} + x$ is the unique solution to our integral equation. Indeed,

$$(e^{-x} + x)' = -e^{-x} + 1 = 1 + x - (e^{-x} + x) \quad \text{and} \quad e^{-0} + 0 = 1.$$

Now the reader may notice that we only found a solution valid on $[-\frac{1}{2}, \frac{1}{2}]$, but in fact $f_\infty(x) = e^{-x} + x$ is a valid solution on all of $\mathbb{R}$. One way we can deal with this is to reconsider our problem beginning at the point $x = \frac{1}{2}$, and try to extend to $[\frac{1}{2}, 1]$. We have the DE

$$f'(x) = 1 + x - f(x) \quad \text{for} \quad 0 \le x \le 1$$
$$f(\tfrac{1}{2}) = e^{-1/2} - \tfrac{1}{2}.$$

The exact same argument on this new interval will yield the unique attractive fixed point $f_\infty(x) = e^{-x} + x$ valid on $[0, 1]$. It is then easy to see that we can bootstrap our way to the unique solution on the whole line.

In the next two sections, we show how the solutions to a large family of DEs can formulated as fixed-point problems. Then we may use the Contraction Principle to show that these DEs always have solutions.

Exercises for Section 12.1

A. Use the method of this section to solve $f'(x) = 1 + \frac{1}{2}f(x)$ for $0 \le x \le 1$ and $f(0) = 1$. You should be able to recognize the series and successfully find a closed form for the solution. Show that, in fact, this solution is valid for all real numbers.

B. For $b > 0$ and $a \in \mathbb{R}$, define T on $C[0, b]$ by $Tf(x) = a + \int_0^x f(t)xe^{-xt}\, dt$. Prove that T is a contraction. Hence show that the integral equation

$f(x) = a + \int_0^x f(t)xe^{-xt}\, dt$ has a unique solution $f \in C[0, \infty)$.

C. Consider the DE $f'(x) = \frac{-2}{x}f(x)$ and $f(1) = 1$ for $\frac{2}{3} \le x \le \frac{3}{2}$.
(a) Prove that the associated integral map is a contraction mapping.
(b) Look for a solution to the DE of the form $f(x) = ax^b$.
(c) Starting with $f_0 = 1$, find a formula for $f_n = T^n f_0$. Express this as a familiar power series of $\log x$, and hence evaluate it.
    HINT: Use the closed form of the solution from (b) to guide you.

D. Consider the DE $y' = 1 + y^2$ and $y(0) = 0$.
(a) Solve the DE directly.
(b) Show that the associated integral map T is not a contraction mapping on $C[-r, r]$ for any $r > 0$.
(c) Find an $r > 0$ so that T maps the unit ball of $C[-r, r]$ into itself and is a contraction mapping on this ball.
(d) Hence show that there is a unique solution on $[-r, r]$.

E. Consider a ball falling to the ground. The downward force of gravity is counteracted by air resistance proportional to the velocity. Find a formula for the velocity of the ball if it is at rest at time 0 [i.e., $v(0) = 0$] as follows:
(a) Show that the velocity satisfies $v'(t) = g - cv(t)$, where g is the gravity constant and c is the constant of air resistance. What is the initial condition?
(b) Construct the associated integral map T. Starting with $v_0(t) = 0$, iterate T and obtain a formula for the solution v_∞.
(c) Show that this solution is valid on $[0, \infty)$ (or at least until the ball hits the ground). Compute $\lim_{t \to \infty} v_\infty(t)$. This is known as the **terminal velocity**.

12.2. Calculus of Vector-Valued Functions

In this section, we develop differentiation and integration for vector-valued functions. We will need this material in the next section, to convert an nth-order DE for a real-valued function into a first-order DE for a vector-valued function.

Consider a function $f : [a, b] \to \mathbb{R}^n$, where $f(x) = (f_1(x), f_2(x), \dots, f_n(x))$ and each coordinate function f_i maps $[a, b]$ to $\mathbb{R}$. Although much of this is done by looking at the coordinate functions, there are some crucial differences between $\mathbb{R}^n$ and $\mathbb{R}$. Most notably, we do not have a total order in $\mathbb{R}^n$, so we cannot take the supremum or infimum of the set $f([a, b]) \subset \mathbb{R}^n$.

12.2.1. DEFINITION. We say that a vector-valued function $f : [a, b] \to \mathbb{R}^n$ is **differentiable** at a point $x_0 \in (a, b)$ if

$$\lim_{h \to 0} \frac{f(x_0 + h) - f(x_0)}{h}$$

exists in $\mathbb{R}^n$. As usual, we write $f'(x_0)$ for the limit and call it the **derivative** of f at x_0. Notice that the numerator is in $\mathbb{R}^n$ while the denominator is a scalar, so $f'(x_0)$ is in $\mathbb{R}^n$.

We can define **left differentiable** and **right differentiable** in the natural way, and we say f is **differentiable** on the interval $[a, b]$ if it is differentiable at every point of (a, b) in the sense just stated and left or right differentiable at the endpoints.

We leave it for the reader to verify that if $f : [a, b] \to \mathbb{R}$ is differentiable at $x_0 \in [a, b]$, then f is continuous at x_0.

12.2.2. PROPOSITION. *Suppose that $f : [a, b] \to \mathbb{R}^n$ given by $f(x) = (f_1(x), \ldots, f_n(x))$, where each f_i maps $[a, b]$ into $\mathbb{R}$. Then f is differentiable at $x_0 \in [a, b]$ if and only if each f_i is differentiable at x_0. Moreover, $f'(x_0) = (f_1'(x_0), \ldots, f_n'(x_0))$.*

PROOF. Fix $x_0 \in [a, b]$ and consider the function

$$g(h) = \frac{f(x_0 + h) - f(x_0)}{h},$$

defined for those h so that $x_0 + h \in [a, b]$. If $g(h) = (g_1(h), \ldots, g_n(h))$, then we have that $g_i(h) = (f_i(x_0 + h) - f_i(x_0))/h$ for each i. Thus f is differentiable x_0 if and only if the limit of $g(h)$ exists as $h \to 0$ and each f_i is differentiable at x_0 if and only if the limit of $g_i(h)$ exists as $h \to 0$. So it suffices to show that the limit of $g(h)$ exists and equals $(u_1, \ldots, u_n)$ if and only if the limits of each of its components g_i exists and equals u_i. But this is established in Exercise 5.3.D. ∎

We leave it to the reader to verify that that sums and scalar multiples of differentiable functions are differentiable. Since products of real-valued differentiable functions are differentiable, it follows easily from the previous theorem that if f and g are vector-valued differentiable functions, then $f \cdot g$ is differentiable. Notice that $f \cdot g$ is real valued.

On the other hand, not all results carry over from the real-valued setting. For example, consider the function $f : [0, 2\pi] \to \mathbb{R}^2$ given by $f(x) = (\cos x, \sin x)$. It is easy to see that $f(2\pi) = f(0)$, but there is no $x \in [0, 2\pi]$ so that $f'(x)$ is the zero vector. However, we do have the following result. The analaguous fact for real-valued functions is a corollary of the Mean Value Theorem.

12.2.3. THEOREM. *Suppose that $f : [a, b] \to \mathbb{R}^n$ is continuous on $[a, b]$ and differentiable on (a, b). Then there is $c \in (a, b)$ so that*

$$\|f(b) - f(a)\| \le (b - a)\|f'(c)\|.$$

PROOF. Define $\mathbf{v} = f(b) - f(a)$ and a function $g : [a, b] \to \mathbb{R}$ by $g(x) = \mathbf{v} \cdot f(x)$. Notice that if $\mathbf{v}$ is the zero vector, then we are done. So we may assume $\|\mathbf{v}\| \ne 0$.

A brief calculation shows that g is differentiable and

$$g'(x) = \mathbf{v} \cdot f'(x).$$

Applying the Mean Value Theorem to g shows that there is $c \in (a, b)$ so that

$$g(b) - g(a) = (b - a)(\mathbf{v} \cdot f'(c)).$$

Using the definition of g and rearranging shows that $g(b) - g(a) = \|\mathbf{v}\|^2$. The Schwarz inequality now gives

$$\|\mathbf{v}\|^2 = (b - a)\mathbf{v} \cdot f'(c) \le (b - a)\|\mathbf{v}\|\|f'(c)\|.$$

Dividing by the nonzero quantity $\|\mathbf{v}\|$ gives the inequality. ∎

12.2.4. COROLLARY. *For a function $f : [a, b] \to \mathbb{R}^n$ that is C^1 on $[a, b]$, we have $\|f(b) - f(a)\| \le (b - a)\|f'\|_\infty$.*

Next, we develop integration for vector-valued functions. Following the notation of Section 6.3, given $P = \{x_0 < \ldots < x_n\}$, a partition of $[a, b]$, and a set of points $X = \{x'_j : 1 \le j \le n\}$ with $x'_j \in [x_{j-1}, x_j]$, define the Riemann sum

$$I(f, P, X) = \sum_{j=1}^{n} f(x'_j)\Delta_j,$$

where the vector $f(x'_j)$ is multiplied by the the scalar $\Delta_j = x_j - x_{j-1}$. We say that the set of points X is **subordinate** to the partition P.

12.2.5. LEMMA. *Given a function $f : [a, b] \to \mathbb{R}^n$ with $f = (f_1, \ldots, f_n)$, a partition P of $[a, b]$, and a set of points X subordinate to P, we have*

$$I(f, P, X) = \big(I(f_1, P, X), I(f_2, P, X), \ldots, I(f_n, P, X)\big).$$

We leave the proof as an exercise. The next definition says, in essence, that a function is integrable provided the Riemann sums "eventually" converge. However, making this precise requires several quantifiers. You should compare this definition to Theorem 6.3.8.

12.2.6. DEFINITION. We say a bounded function $f : [a, b] \to \mathbb{R}^n$ is **Riemann integrable** if there is a vector, $\mathbf{v}$, so that for every $\varepsilon > 0$, there is a $\delta > 0$ so that every partition Q of $[a, b]$ with mesh$(Q) < \delta$ and every choice of points X subordinate to Q,

$$\Big\|I(f, Q, X) - \mathbf{v}\Big\| < \varepsilon.$$

As usual, we denote $\mathbf{v}$ by $\int_a^b f(x)\, dx$.

Theorem 6.3.8 shows that this definition agrees with our previous definition of Riemann integrable function if $n = 1$.

12.2.7. PROPOSITION. *Fix* $f : [a, b] \to \mathbb{R}^n$ *with* $f(x) = (f_1(x), \ldots, f_n(x))$ *for all* $x \in [a, b]$. *Then* f *is Riemann integrable if and only if each coordinate function* f_i *is Riemann integrable. In this case,*

$$\int_a^b f(x) \, dx = \left(\int_a^b f_1(x) \, dx, \int_a^b f_2(x) \, dx, \ldots, \int_a^b f_n(x) \, dx \right).$$

PROOF. Suppose that f is Riemann integrable. To show that f_j is Riemann integrable, for some j between 1 and n, we use condition (d) of Theorem 6.3.8. That is, there is a real number, call it M, so that for every $\varepsilon > 0$, there is $\delta > 0$ so that for every partition Q with mesh$(Q) < \delta$ and every choice of points X subordinate to Q, we have $|I(f_j, Q, X) - M| < \varepsilon$.

Since f is Riemann integrable, we have $(u_1, \ldots, u_n) \in \mathbb{R}^n$ with

$$\int_a^b f(x) \, dx = (u_1, \ldots, u_n).$$

Moreover, there is $\delta > 0$ so that for every partition Q with mesh$(Q) < \delta$ and every set of points X subordinate to Q, $\left\| I(f, Q, X) - \int_a^b f(x) \, dx \right\| < \varepsilon$.

For any vector $\mathbf{x} = (x_1, \ldots, x_n)$, recall that $|x_i| \leq \|\mathbf{x}\|$ for $1 \leq i \leq n$. Applying this fact to the vector $I(f, Q, X) - \int_a^b f(x) \, dx$ and using Lemma 12.2.5, we have $|I(f_j, Q, X) - u_j| < \varepsilon$. Thus, if we choose $M = u_j$, it follows that f_j is Riemann integrable and, moreover, $\int_a^b f_j(x) \, dx = u_j$.

Conversely, suppose that each f_j is Riemann integrable. Let $\varepsilon > 0$ and let $\mathbf{u} \in \mathbb{R}^n$ be given by $\mathbf{u} = (\int_a^b f_1(x) \, dx, \ldots, \int_a^b f_n(x) \, dx)$. Find $\delta_j > 0$ satisfying condition (d) of Theorem 6.3.8 so that $|\int_a^b f_j(x) \, dx - I(f_j, Q, X)| < \varepsilon/\sqrt{n}$. Let δ be the minimum of $\delta_1, \ldots, \delta_n$. Then for any partition Q with mesh$(Q) < \delta$ and any set of points X subordinate to Q, we have

$$\|\mathbf{u} - I(f, Q, X)\|^2 \leq \sum_{j=1}^n \left| \int_a^b f_j(x) \, dx - I(f_j, Q, X) \right|^2 \leq \sum_{j=1}^n \varepsilon^2/n = \varepsilon^2.$$

Taking square roots, we are done. ∎

Using this proposition, we can carry over many properties of integration for real-valued functions to vector-valued functions. For example, linearity of integration and the Fundamental Theorem of Calculus carry over in this way.

12.2.8. FUNDAMENTAL THEOREM OF CALCULUS II.
Let $f : [a, b] \to \mathbb{R}^n$ *be a bounded Riemann integrable function and define*

$$F(x) = \int_a^x f(x) \, dx \quad \text{for } a \leq x \leq b.$$

Then $F : [a, b] \to \mathbb{R}^n$ *is a continuous function. If* f *is continuous at a point* x_0, *then* F *is differentiable at* x_0 *and* $F'(x_0) = f(x_0)$.

Finally, we have the following estimate, which says that the norm of an integral is less than the integral of the norm. This is the higher-dimensional analogue of taking the absolute value inside the integral. While the proof is similar in spirit to the scalar case, the argument requires inner products and the Schwarz inequality.

12.2.9. LEMMA. *Let $F : [a, b] \to \mathbb{R}^n$ be continuous. Then*

$$\left\| \int_a^b F(x)\, dx \right\| \leq \int_a^b \|F(x)\|\, dx.$$

PROOF. Let $\mathbf{y} = \int_a^b F(x)\, dx$. If $\|\mathbf{y}\| = 0$, then the inequality is trivial. Otherwise, using the standard inner product on $\mathbb{R}^n$,

$$\|\mathbf{y}\|^2 = \left\langle \int_a^b F(x)\, dx, \mathbf{y} \right\rangle = \int_a^b \langle F(x), \mathbf{y} \rangle\, dx$$

$$\leq \int_a^b |\langle F(x), \mathbf{y} \rangle|\, dx \leq \int_a^b \|F(x)\|\, \|\mathbf{y}\|\, dx$$

$$= \|\mathbf{y}\| \int_a^b \|F(x)\|\, dx.$$

The second line follows by the Schwarz inequality. Dividing through by $\|\mathbf{y}\|$ gives the result. ∎

Exercises for Section 12.2

A. Prove Lemma 12.2.5.

B. Show that if α, β are real numbers and f, g are Riemann integrable functions from $[a, b]$ to $\mathbb{R}^n$, then $\alpha f + \beta g$ is Riemann integrable. Moreover, show that

$$\alpha \int_a^b f(x)\, dx + \beta \int_a^b g(x)\, dx = \int_a^b (\alpha f(x) + \beta g(x))\, dx.$$

C. Prove Theorem 12.2.8.

D. Every continuous function $f : [a, b] \to \mathbb{R}^n$ is Riemann integrable.
  (a) Show this using Proposition 12.2.7.
  (b) Show this directly from the definition, without using Proposition 12.2.7.
      HINT: Adapt Theorem 6.3.10 to use Riemann sums.

E. Define a **regular curve** to be a differentiable function $f : [a, b] \to \mathbb{R}^n$ so that $f'(x)$ is never the zero vector.
  (a) Given two regular curves $f : [a, b] \to \mathbb{R}^n$ and $g : [c, d] \to \mathbb{R}^n$, we say that g is a **reparametrization** of f if there is a differentiable function $h : [c, d] \to [a, b]$ so that $h'(t) \neq 0$ for all t and $g = f \circ h$. Show that reparametrization is an equivalence relation and that $g'(t) = f'(h(t))h'(t)$.

  (b) Define the **length of a regular curve** f to be $L(f) = \int_a^b \|f'(x)\|\, dx$. Show that the length is not changed by reparametrization. HINT: Consider reparametrizations where $h(t)$ is always positive or always negative.

(c) For a regular curve $f : [a, b] \to \mathbb{R}^n$ and a partition $P = \{x_0 < \cdots < x_n\}$ of $[a, b]$, define $\ell(f, P) = \sum_{i=1}^{n} \|f(x_i) - f(x_{i-1})\|$. Show that for each $\varepsilon > 0$, there is some $\delta > 0$ such that for all partitions P with $\text{mesh}(P) < \delta$, $|\ell(f, P) - L(f)| < \varepsilon$.

(d) Given a regular curve f, show there is a reparametrization g with $\|g'(t)\| = 1$ for all t. Such a curve is said to have **unit speed**.

HINT: Show that $x \mapsto \int_a^x \|f(t)\| \, dt$ has an inverse function.

(e) Suppose that $f : [a, b] \to \mathbb{R}^n$ and $g : [c, d] \to \mathbb{R}^n$ are regular curves and both have unit speed. If g is a reparametrization of f, show that either $g(x) = f(x + (a - c))$ for all $x \in [c, d]$ or $g(x) = f(-x + (c - b))$ for all $x \in [c, d]$. In either case, show that $d - c = b - a$.

12.3. Differential Equations and Fixed Points

The goal of this section is to start with a DE of order n, and convert it to the problem of finding a fixed point of an associated integral operator.

The first step is to take a fairly general form of a higher-order differential equation and turn it into a first-order DE at the expense of making the function vector valued. We define an **initial value problem**, for functions on $[a, b]$ and a point $c \in [a, b]$, as follows:

(12.3.1)
$$f^{(n)}(x) = \varphi(x, f(x), f'(x), \ldots, f^{(n-1)}(x))$$
$$f(c) = \gamma_0$$
$$f'(c) = \gamma_1$$
$$\vdots$$
$$f^{(n-1)}(c) = \gamma_{n-1},$$

where φ is a real-valued continuous function on $[a, b] \times \mathbb{R}^n$. This is not quite the most general situation, but it includes many important examples. The first equation is referred to as a **differential equation of nth order**, and the second set is called the **initial conditions**.

A standard trick reduces this to a first order differential equation with values in $\mathbb{R}^n$. As we shall see, this has computational advantages. Replace the function f by the vector valued function $F : [a, b] \to \mathbb{R}^n$ given by

$$F(x) = (f(x), f'(x), \ldots, f^{(n-1)}(x)) \, .$$

Then the differential equation becomes

$$F'(x) = \left(f'(x), \ldots, f^{(n-1)}(x), \varphi(x, f(x), \ldots, f^{(n-1)}(x)) \right),$$

and the initial data become

$$F(c) = (\gamma_0, \gamma_1, \ldots, \gamma_{n-1}).$$

We further simplify the notation by introducing a function Φ from $[a, b] \times \mathbb{R}^n$ to $\mathbb{R}^n$ by

$$\Phi(x, y_0, \ldots, y_{n-1}) = (y_1, y_2, \ldots, y_{n-1}, \varphi(x, y_0, \ldots, y_{n-1})),$$

and the vector

$$\Gamma = (\gamma_0, \gamma_1, \ldots, \gamma_{n-1}).$$

Note that Φ is continuous since φ is continuous. Then (12.3.1) becomes the first-order initial value problem with vector values

(12.3.2)
$$F'(x) = \Phi(x, F(x))$$
$$F(c) = \Gamma.$$

It is easy to see that a solution of (12.3.1) gives a solution of (12.3.2). To go the other way, suppose (12.3.2) has a solution

$$F(x) = (f_0(x), f_1(x), \ldots, f_{n-1}(x)).$$

Then (12.3.2) means

$$\begin{aligned}
F'(x) &= (f_0'(x), f_1'(x), \ldots, f_{n-2}'(x), f_{n-1}'(x)) \\
&= \Phi(x, f_0(x), f_1(x), \ldots, f_{n-1}(x)) \\
&= \left(f_1(x), \ldots, f_{n-1}(x), \varphi(x, f(x), \ldots, f^{(n-1)}(x)) \right).
\end{aligned}$$

By identifying each coordinate, we obtain

$$\begin{aligned}
f_0'(x) &= f_1(x) \\
f_1'(x) &= f_2(x)
\end{aligned}$$

$$\vdots$$

$$\begin{aligned}
f_{n-2}'(x) &= f_{n-1}(x) \\
f_{n-1}'(x) &= \varphi(x, f_0(x), f_1(x), \ldots, f_{n-1}(x)).
\end{aligned}$$

Thus $f_0' = f_1$, $f_0^{(2)} = f_1' = f_2$, $\ldots$, $f_0^{(n-1)} = f_{n-2}' = f_{n-1}$, and

$$f_0^{(n)}(x) = f_{n-1}'(x) = \varphi(x, f_0(x), f_0'(x), \ldots, f_0^{(n-1)}(x)).$$

The initial data become $\Gamma = F(c) = (f_0(c), \ldots, f_{n-1}(c))$ so

$$f_0(c) = \gamma_0, \quad f_0'(c) = \gamma_1, \quad \ldots, \quad f_0^{(n-1)}(c) = \gamma_{n-1}.$$

Thus (12.3.1) is satisfied by $f_0(x)$. Consequently, (12.3.2) is an equivalent formulation of the original problem.

12.3.3. EXAMPLE. We will express the unknown function as y, instead of $f(x)$. Consider the differential equation

$$\left(1 + (y')^2\right)y^{(3)} = y'' - xy'y + \sin x \quad \text{for} \quad -1 \le x \le 1$$
$$y(0) = 1$$
$$y'(0) = 0$$
$$y''(0) = 2.$$

It is first necessary to reformulate this DE to express the highest-order derivative, $y^{(3)}$, as a function of lower-order terms. This yields

$$y^{(3)} = \frac{y'' - xy'y + \sin x}{1 + (y')^2}.$$

Then the vector function Φ defined from $[-1, 1] \times \mathbb{R}^3$ into $\mathbb{R}^3$ given by

$$\Phi(x, y_0, y_1, y_2) = \left(y_1, y_2, \frac{y_2 - xy_1y_0 + \sin x}{1 + y_1^2}\right).$$

The initial vector is $\Gamma = (1, 0, 2)$. The DE is now reformulated as a first order vector valued DE looking for a function $F(x) = (f_0(x), f_1(x), f_2(x))$ defined on $[-1, 1]$ with values in $\mathbb{R}^3$ such that

$$F'(x) = \left(f_1(x), f_2(x), \frac{f_2(x) - xf_1(x)f_0(x) + \sin x}{1 + f_1(x)^2}\right) \quad \text{for} \quad -1 \le x \le 1$$
$$F(0) = (1, 0, 2).$$

Now we can integrate this example as before. Define a mapping T from the space $C([-1, 1], \mathbb{R}^3)$ of functions on $[-1, 1]$ with vector values in $\mathbb{R}^3$ into itself by sending the vector valued function $F(x) = (f_0(x), f_1(x), f_2(x))$ to

$$TF(x) = \Gamma + \int_0^x \Phi(t, F(t)) \, dt$$
$$= \left(1 + \int_0^x f_1(t) \, dt, \; \int_0^x f_2(t) \, dt, \; 2 + \int_0^x \frac{f_2(t) - tf_1(t)f_0(t) + \sin t}{1 + f_1(t)^2} \, dt\right).$$

This converts the differential equation into the integral equation $TF = F$. An argument similar to the preceding one shows that this fixed-point problem is equivalent to the differential equation.

Returning to the general case, we need to find a suitable framework (i.e., a complete normed vector space) in which to solve the problem in general. Since the coordinate functions $f_0, \ldots, f_{n-1}$ of F are all continuous functions on $[a, b]$, we can think of F as an element of $C([a, b], \mathbb{R}^n)$, the vector space of continuous functions from $[a, b]$ into $\mathbb{R}^n$. This space has the norm

$$\|F\|_\infty = \max_{a \le x \le b} \|F(x)\| = \max_{a \le x \le b} \left(\sum_{i=0}^{n-1} |f_i(x)|^2\right)^{1/2}.$$

A sequence $F^k = (f_0^k, \ldots, f_{n-1}^k)$ converges to $F^* = (f_0^*, \ldots, f_{n-1}^*)$ in the max norm if and only if each of the coordinate functions f_i^k converges uniformly to f_i^* for $0 \leq i \leq n-1$. To see this, notice that for each coordinate i,

$$\|f_i^k - f_i^*\| = \max_{a \leq x \leq b} |f_i^k(x) - f_i^*(x)|$$

$$\leq \Big(\sum_{j=0}^{n-1} |f_j^k(x) - f_j^*(x)|^2 \Big)^{1/2} = \|F^k - F^*\|_\infty.$$

Therefore, the convergence of F^k to F^* implies that f_i^k converges to f_i^* for each $0 \leq i \leq n-1$. Conversely,

$$\|F^k - F^*\|_\infty = \max_{a \leq x \leq b} \Big(\sum_{j=0}^{n-1} |f_j^k(x) - f_j^*(x)|^2 \Big)^{1/2}$$

$$\leq \Big(\sum_{j=0}^{n-1} \max_{a \leq x \leq b} |f_j^k(x) - f_j^*(x)|^2 \Big)^{1/2} = \Big(\sum_{j=0}^{n-1} \|f_j^k - f_j^*\|_\infty^2 \Big)^{1/2}.$$

So if $\|f_i^k - f_i^*\|_\infty$ tends to zero for each i, then $\|F^k - F^*\|_\infty$ also converges to zero.

We need to know that this space of functions is complete. This is a straightforward consequence of the scalar version Theorem 8.2.2.

12.3.4. THEOREM. $C([a,b], \mathbb{R}^n)$ *is complete.*

PROOF. It is clear that $C([a,b], \mathbb{R}^n)$ is a normed vector space. To prove completeness, we must show that every Cauchy sequence F^k converges to a function F^* in $C([a,b], \mathbb{R}^n)$. We will write these functions in coordinate form as

$$F^k(x) = (f_0^k(x), f_1^k(x), \ldots, f_{n-1}^k(x)).$$

If F^k is Cauchy, the inequalities just established show that each sequence of coordinate functions $(f_i^k)_{k=1}^\infty$ is Cauchy in $C[a,b]$ for each $0 \leq i \leq n-1$. By Theorem 8.2.2, $C[a,b]$ is complete. Therefore, the sequence (f_i^k) converges uniformly to a function f_i^* in $C[a,b]$. By the preceding remarks, F^k converges uniformly to the function $F^* = (f_0^*, \ldots, f_{n-1}^*)$. ∎

We have found a setting for our problem (12.3.2) in which the Contraction Principle is valid, and so we can formulate the problem in terms of a fixed point. To do this, integrate (12.3.2) from a to x to obtain

$$F(x) = F(c) + \int_c^x F'(t)\,dt = \Gamma + \int_c^x \Phi(t, F)\,dt.$$

This suggests defining a map on $C([a,b], \mathbb{R}^n)$ by

$$TF(x) = \Gamma + \int_c^x \Phi(t, F(t))\,dt.$$

A solution of (12.3.2) is clearly a fixed point of T. Conversely, by the Fundamental Theorem of Calculus, a fixed point of T is a solution of (12.3.2). So the problem (12.3.1) can be solved by finding the fixed point(s) of the mapping T from $C([a, b], \mathbb{R}^n)$ into itself.

Exercises for Section 12.3

For each of the following three differential equations, convert the DE into a first-order vector-valued DE and then into a fixed-point problem.

A. $y^{(3)} + y'' - x(y')^2 = e^x$, $\quad y(0) = 1, y'(0) = -1$ and $y''(0) = 0$

B. $f''(x) = (f(x)^2 + x^2)^{1/2} - f'(x)^2$, $\quad f(1) = 0$ and $f'(1) = 1$

C. $\dfrac{d}{dx}\left(x\dfrac{dv}{dx}\right) + 2x^2v = 0$, $\quad v(1) = 1$ and $v'(1) = 0$

D. Let V be a complete normed vector space. (Think of a few examples!)
(a) Let (f_n) be a Cauchy sequence in $C([a, b], V)$. Show that for each $x \in [a, b]$ that $(f_n(x))$ is Cauchy, and so define the pointwise limit $f(x) = \lim_{n \to \infty} f_n(x)$.
(b) Prove that (f_n) converges uniformly. HINT: Use the Cauchy criterion to obtain an estimate for $\|f_n(x) - f(x)\|$ that is independent of the point x.
(c) Prove that f is continuous, and deduce that $C([a, b], V)$ is complete.

E. (a) Solve the DE $y' = xy$ and $y(0) = 1$. Deduce that the solution is valid on the whole line. HINT: Integrate $y'/y = x$.
(b) Define $Tf(x) = 1 + \displaystyle\int_0^x tf(t)\,dt$. Start with $f_0(x) = 1$ and compute $f_n = Tf_{n-1}$ for $n \geq 1$. Prove that this converges to the same solution. Where is this convergence uniform?

F. Let φ be a continuous positive function on $\mathbb{R}$. Consider the DE $f'(x) = \varphi(f(x))$ and $f(c) = \gamma$. Let $F(y) = c + \displaystyle\int_\gamma^y \dfrac{dt}{\varphi(t)}$. Show that $f(x) = F^{-1}(x)$ is the unique solution. HINT: Integrate $f'(t)/\varphi(f(t))$ from c to x.

G. Consider the DE: $xyy' = y^2 - 1$ and $y(1) = 1/\sqrt{2}$.
(a) Solve the DE. HINT: Integrate $\dfrac{yy'}{y^2-1} = \dfrac{1}{x}$.
(b) The solution only exists for a finite interval containing 1. Explain why the solution cannot extend further.

12.4. Solutions of Differential Equations

In this section, we use a modification of the Contraction Principle to demonstrate the existence and uniqueness of solutions to a large class of differential equations. The basic idea of contraction mappings is that, starting with any function, iteration of the map T leads inevitably to the solution. As we have seen for Newton's method in Section 11.2, it may well be the case that there is a contraction provided that we start near enough to the solution. Even when T is not a contraction, it may still have an attractive fixed point. In this case, if we start at a reasonable initial approximation, the structure of the integral mapping will force convergence, which is *eventually* contractive.

To analyze the map T, we need a computational result.

12.4.1. DEFINITION. A function $\Phi(x, y)$ is **Lipschitz in the y variable** if there is a constant L so that

$$\|\Phi(x, y) - \Phi(x, z)\| \le L\|y - z\|$$

for all (x, y) and (x, z) in the domain of Φ.

Note that both the y-variable and the range may be vectors rather than elements of $\mathbb{R}$. While this estimate does not concern variation in the x-variable, it does require that the constant L be independent of x.

12.4.2. LEMMA. *Let Φ be a continuous function from $[a, b] \times \mathbb{R}^n$ into $\mathbb{R}^n$ which is Lipschitz in y with Lipschitz constant L. Let*

$$TF(x) = \Gamma + \int_a^x \Phi(t, F(t)) \, dt.$$

If $F, G \in C([a, b], \mathbb{R}^n)$ satisfy $\|F(x) - G(x)\| \le \dfrac{M(x - a)^k}{k!}$, then

$$\|TF(x) - TG(x)\| \le \frac{LM(x - a)^{k+1}}{(k + 1)!}.$$

In particular, T is uniformly continuous.

PROOF. Compute

$$\|TF(x) - TG(x)\| = \left\| y_0 + \int_a^x \Phi(t, F(t)) \, dt - y_0 - \int_a^x \Phi(t, G(t)) \, dt \right\|$$

$$= \left\| \int_a^x \Phi(t, F(t)) - \Phi(t, G(t)) \, dt \right\|$$

$$\le \int_a^x \|\Phi(t, F(t)) - \Phi(t, G(t))\| \, dt$$

$$\le \int_a^x L\|F(t) - G(t)\| \, dt$$

$$\le \frac{LM}{k!} \int_a^x (t - a)^k \, dt = \frac{LM}{(k + 1)!} (x - a)^{k+1}.$$

In particular, $\|F - G\|_\infty = \|F - G\|_\infty \frac{(x-a)^0}{0!}$. It follows that

$$\|TF - TG\|_\infty \le \|F - G\|_\infty L\|x - a\|_\infty = \|F - G\|_\infty L(b - a).$$

So T has Lipschitz constant $L(b - a)$ and therefore is uniformly continuous. ∎

12.4.3. GLOBAL PICARD THEOREM.

Suppose that Φ is a continuous function from $[a, b] \times \mathbb{R}^n$ into $\mathbb{R}^n$ which is Lipschitz in y. Then the differential equation

$$F'(x) = \Phi(x, F(x)), \quad F(a) = \Gamma$$

has a unique solution.

PROOF. Let T map $C([a, b], \mathbb{R}^n)$ into itself by

$$TF(x) = \Gamma + \int_a^x \Phi(t, F(t)) \, dt.$$

As discussed at the beginning of this section, a function F in $C([a, b], \mathbb{R}^n)$ is a fixed point for T if and only it is a solution of (12.3.2). Define a sequence of functions by

$$F_0(x) = \Gamma \quad \text{and} \quad F_{k+1} = TF_k \quad \text{for} \quad k \geq 0.$$

Let L be the Lipschitz constant of Φ, and set $M = \max\limits_{a \leq x \leq b} \{\|\Phi(x, \Gamma)\|\}$. We have the inequality

$$\|F_1(x) - F_0(x)\| = \left\| \int_a^x \Phi(t, \Gamma) \, dt \right\| \leq \frac{M(x - a)}{1!}.$$

Therefore, by Lemma 12.4.2,

$$\|F_2(x) - F_1(x)\| \leq \frac{ML^2(x - a)^2}{2!}$$

$$\|F_3(x) - F_2(x)\| \leq \frac{ML^3(x - a)^3}{3!}$$

and, by induction, we get

$$\|F_{k+1}(x) - F_k(x)\| \leq \frac{ML^{k+1}(x - a)^{k+1}}{k + 1!}.$$

As in the proof of the Banach Contraction Principle (11.1.6),

$$\|F_{n+m}(x) - F_n(x)\| \leq \sum_{k=n}^{m+n-1} \|F_{k+1}(x) - F_k(x)\|$$

$$\leq \sum_{k=n}^{m+n-1} \frac{ML^{k+1}(x - a)^{k+1}}{k + 1!}$$

$$\leq M \sum_{k=n+1}^{\infty} \frac{\left(L(b - a)\right)^k}{k!}.$$

But the series $M \sum\limits_{k=0}^{\infty} \left(L(b - a)\right)^k / k!$ converges to $Me^{L(b-a)}$. So given any $\varepsilon > 0$, there is an integer N so that the tail of the series satisfies

$$M \sum_{k=N}^{\infty} \frac{\left(L(b - a)\right)^k}{k!} < \varepsilon.$$

Thus if n, $n + m \geq N$, it follows that

$$\|F_{n+m} - F_n\|_\infty < \varepsilon.$$

This means that the sequence (F_k) is Cauchy in $C([a, b], \mathbb{R}^n)$, and hence converges to a limit function, call it F^*.

Since T is continuous,

$$TF^* = \lim_{k \to \infty} TF_k = \lim_{k \to \infty} F_{k+1} = F^*.$$

If G is another solution, then G satisfies $TG = G$. An argument similar to the previous paragraph shows that

$$\|F^* - G\|_\infty = \|T^k F^* - T^k G\|_\infty \leq \|F^* - G\|_\infty \frac{(L(b-a))^k}{k!}.$$

The left-hand side is constant and nonnegative, while the right-hand side converges to 0 as k tends to infinity. Therefore, $\|F^* - G\|_\infty = 0$. In other words, $G = F^*$ and the solution is unique. ∎

12.4.4. EXAMPLE. Consider the initial value problem

$$y'' + y + \sqrt{y^2 + (y')^2} = 0 \qquad y(0) = \gamma_0 \quad \text{and} \quad y'(0) = \gamma_1.$$

We set this up by letting $Y = (y_0, y_1)$ and

$$\Phi(x, y_0, y_1) = \left(y_1, -y_0 - \sqrt{y_0^2 + y_1^2}\right) = \left(y_1, -y_0 - \|Y\|\right).$$

Then the DE becomes

$$Y'(x) = \Phi(x, Y) \qquad \text{and} \qquad Y(0) = \Gamma := (\gamma_0, \gamma_1).$$

Let us verify that Φ is Lipschitz. Let $Z = (z_0, z_1)$. Recall that the triangle inequality implies that $\big|\|Z\| - \|Y\|\big| \leq \|Z - Y\|$.

$$\|\Phi(x, Y) - \Phi(x, Z)\| = \left\|\left(y_1 - z_1, \; z_0 - y_0 + \|Z\| - \|Y\|\right)\right\|$$
$$\leq \left\|\left(y_1 - z_1, \; z_0 - y_0\right)\right\| + \big|\|Z\| - \|Y\|\big| \leq 2\|Z - Y\|$$

Therefore, Picard's Theorem applies, and this equation has a unique solution that is valid on the whole real line.

It may be surprising that this DE actually can be solved explicitly. We may rearrange this equation to look like

$$\frac{y + y''}{\sqrt{y^2 + (y')^2}} = -1.$$

There is no obvious way to integrate the left-hand side. The key observation is that

$$\left(y^2 + (y')^2\right)' = 2yy' + 2y'y'' = 2y'(y + y'').$$

We may multiply both sides by y', known as an **integrating factor**, which makes both sides easily integrable in closed form:

$$\frac{2yy' + 2y'y''}{2\sqrt{y^2 + (y')^2}} = -y'.$$

The left-hand side is the derivative of $\sqrt{y^2 + (y')^2}$. As y is a function of x and we are about to introduce constants independent of x, it is helpful to switch notation and use $y(x)$ and $y'(x)$ in place of y and y'.

Integrate both sides with respect to x from 0 to x to obtain

$$\sqrt{y(x)^2 + y'(x)^2} - \|\Gamma\| = \gamma_0 - y(x).$$

Take $\|\Gamma\|$ to the other side, square, and simplify to obtain

$$y'(x)^2 = c^2 - 2cy(x) \quad \text{where } c = \|\Gamma\| + \gamma_0.$$

Hence

$$\frac{y'(x)}{\sqrt{c^2 - 2cy(x)}} = \pm 1.$$

Integrating again from 0 to x yields

$$\frac{-1}{c}\sqrt{c^2 - 2cy(x)} + \frac{1}{c}\sqrt{c^2 - 2c\gamma_0} = \pm x.$$

It looks like we may get multiple solutions (which we know isn't the case), but let us persevere. Notice that

$$\sqrt{c^2 - 2c\gamma_0} = \sqrt{\|\Gamma\|^2 - \gamma_0^2} = \pm\gamma_1.$$

Use this identity and simplify to obtain

$$c^2 - 2cy(x) = (\pm cx + \gamma_1)^2 = c^2 x^2 \pm 2c\gamma_1 x + c^2 - 2c\gamma_0.$$

Solving for y produces

$$y(x) = \frac{-c}{2}x^2 \pm \gamma_1 x + \gamma_0.$$

The condition that $y'(0) = \gamma_1$ shows that the sign is $+$ and the unique solution is the quadratic

$$y(x) = \left(\frac{-\sqrt{\gamma_0^2 + \gamma_1^2} - \gamma_0}{2}\right)x^2 + \gamma_1 x + \gamma_0.$$

Notice that this solution depends continuously on the initial data (γ_0, γ_1). This is a general phenomenon that we explore in Section 12.7.

Exercises for Section 12.4

A. Consider the DE $y' = 1 + xy$ and $y(0) = 0$ on $[-1, 1]$.
  (a) Show that the associated integral operator is a contraction mapping.
  (b) Find a convergent power series expansion for the unique solution.
  (c) Use the Global Picard Theorem to show that there is a unique solution on $[-b, b]$
      for any $b < \infty$. Hence deduce that there is a unique solution on $\mathbb{R}$.

B. Consider the DE $y'' = y' + xy + 3x^2$, $y(0) = 2$ and $y'(0) = 1$ for $x \in [0, 2]$.
  (a) Find the function Φ and vector Γ to put this DE in standard form.
  (b) Calculate the constants involved in the proof of the Global Picard Theorem, and
      hence find an integer N such that $\|F^* - F_N\|_\infty < 10^{-3}$.

C. Consider the DE: $xyy' = (2 - x)(y + 2)$ and $y(1) = -1$.

  (a) Separate variables and deduce that the solution y satisfies $\dfrac{e^{y/2}}{y + 2} = xe^{x/2}$.

  (b) Prove that both x and y are bounded.
      HINT: In part (a), minimize the left-hand side and maximize the right-hand side.

D. Consider the DE $f'(x) = xf(x) + 1$ and $f(0) = 0$.
  (a) Use the Global Picard Theorem to show that there is a unique solution on $[-b, b]$
      for any $b < \infty$. Hence deduce that there is a unique solution on $\mathbb{R}$.
  (b) Find an explicit power series that solves the DE.

  HINT: Look for a solution of the form $f(x) = \sum\limits_{n=0}^{\infty} a_n x^n$. Plug this into the DE and

  find a recurrence relation for the a_n. Hence obtain a formula for each a_n.
  (c) Show that this series converges uniformly on the whole real line. Hence validate the
      term-by-term differentiation to verify that this power series is the unique solution.

E. (a) Suppose φ is C^∞ function on $[a, b] \times \mathbb{R}$, and $Tf(x) = c + \displaystyle\int_a^x \varphi(t, f(t))\, dt$. Show

  by induction that if $f_0 \in C[a, b]$, then $T^n f_0$ has n continuous derivatives.
  (b) Hence conclude that a fixed point $f = Tf$ must be C^∞.

F. (a) In the previous question, suppose that φ is C^n. Prove that a solution to $f = Tf$
      has $n + 1$ continuous derivatives.
  (b) Let $\varphi(t) = t$ for $t \leq 1$ and $\varphi(t) = 2 - t$ for $t \geq 1$. Solve the DE $y' = g(y)$ and
      $y(0) = 1.5$ on $\mathbb{R}$. Verify that the solution is C^1 but is not twice differentiable.

G. Let p be a positive real number. Suppose that $u(x)$ is a solution of the integral equation
  $$u(x) = \int_0^x \sin(u(t))u(t)^p\, dt.$$

  (a) Fix $a > 0$ and set $M = \sup\{|u(x)| : |x| \leq a\}$. Prove that $|u(x)| \leq \dfrac{M^{pn}|x|^n}{n!}$ on
      $[-a, a]$ for each $n \geq 0$.   HINT: $|\sin y| \leq |y|$
  (b) Hence prove that $u = 0$.

H. Suppose that Φ and Ψ are Lipschitz functions defined on $[a, b] \times \mathbb{R}$. Let f and g
  be solutions of $f' = \Phi(x, f(x))$ and $g' = \Psi(x, g(x))$, respectively. Suppose that
  $f(a) \leq g(a)$ and $\Phi(x, y) \leq \Psi(x, y)$ for all $(x, y) \in [a, b] \times \mathbb{R}$. Show that $f(x) \leq g(x)$
  for all $x \in [a, b]$.
  HINT: If $f(x) = g(x)$, what about $f'(x)$ and $g'(x)$?

12.5. Local Solutions

The stipulation that Φ has to be Lipschitz over all of $\mathbb{R}^n$ is quite restrictive. However, many functions satisfy a Lipschitz condition in y on a set of the form $[a, b] \times \overline{B_R(\Gamma)}$ for $R < \infty$. For example, if Φ has continuous partial derivatives of first order, this follows from the Mean Value Theorem and the Extreme Value Theorem (see the Exercises). In this case, the proof would go through as long as $F_n(x)$ always stays in this ball. While this isn't usually possible for all x, it is possible to verify this condition on a small interval $[a, a + h]$. In this way, we get a **local solution**. It is then often possible to piece these local solutions together to extend the solution to all of $[a, b]$. We shall see in a few examples that such an extension is not always valid.

12.5.1. LOCAL PICARD THEOREM.
Suppose that Φ is a continuous function from $[a, b] \times \overline{B_R(\Gamma)}$ into $\mathbb{R}^n$ satisfying a Lipschitz condition in y. Then the differential equation

$$F'(x) = \Phi(x, F(x)), \quad F(a) = \Gamma$$

has a unique solution on the interval $[a, a + h]$, where

$$h = \min\{b - a, R/\|\Phi\|_\infty\}.$$

PROOF. The proof is the same as for the Global Picard Theorem except that we must ensure that $F_n(x)$ remains in $\overline{B_R(\Gamma)}$ so that the iterations remain defined. See Figure 12.2.

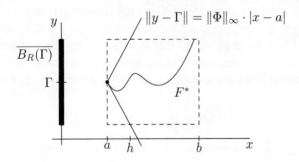

FIGURE 12.2. Schematic setup for Local Picard Theorem.

This follows by induction from the easy estimate for $x \in [a, a + h]$:

$$\|F_{n+1}(x) - \Gamma\| \leq \int_a^x \|\Phi(t, F_n(t))\| \, dt$$

$$\leq \|\Phi\|_\infty |x - a|$$

$$\leq \|\Phi\|_\infty h \leq R.$$

Thus F_n converges uniformly on $[a, a + h]$ to a solution F^* of the differential equation. The uniqueness argument remains the same. $\blacksquare$

12.5.2. EXAMPLE. Consider the differential equation

$$y' = y^2, \qquad y(0) = 1, \qquad 0 \le x \le 2.$$

In this case, the function is $\Phi(x, y) = y^2$. This is not Lipschitz globally because

$$\frac{\Phi(a, n + \frac{1}{n}) - \Phi(a, n)}{(n + \frac{1}{n}) - n} > 2n \quad \text{for all} \quad n \ge 1.$$

However, Φ is continuously differentiable and therefore Lipschitz on any compact set (Exercise 12.5.A); for example, on $[0, 2] \times \overline{B_{100}(1)}$. By the Local Picard Theorem, this has a unique solution beginning at the origin. The maximum of $|\Phi|$ over $[0, 2] \times B_R(1)$ is $(R + 1)^2$, which leads to $h = R/(R + 1)^2$. Our best choice is $R = 1$ and $h = \frac{1}{4}$.

This DE may be solved by **separation of variables**.

$$\frac{y'}{y^2} = 1$$

Integrating from $t = 0$ to $t = x$, we obtain

$$x = \int_0^x 1 \, dt = \int_0^x \frac{y'(t) \, dt}{y(t)^2} = -\frac{1}{y(t)}\Big|_0^x = 1 - \frac{1}{y(x)}.$$

Therefore,

$$y(x) = \frac{1}{1 - x}.$$

Evidently, this is a solution on the interval $[0, 1)$, which has a singularity at $x = 1$. So the solution does not extend in any meaningful way to the rest of the interval. The range $[0, 1)$ is better than our estimate of $[0, \frac{1}{4}]$ but is definitely not a solution on all of $[0, 2]$.

Now let us consider how to improve on this situation. Start over at the point $\frac{1}{4}$ and try to extend the solution some more. More generally, suppose that we have used our technique to establish a unique solution on $[0, a]$, where $a < 1$. Consider the DE

$$y' = y^2, \qquad y(a) = \frac{1}{1 - a}, \qquad 0 \le x \le 2.$$

Again we take a ball $B_R(\frac{1}{1-a})$ and maximize $\Phi(y) = y^2$ over this ball:

$$\|\Phi\|_{B_R(1/1-a)} = \left(R + \frac{1}{1 - a}\right)^2.$$

Thus Theorem 12.5.1 applies and extends the solution to the interval $[a, a + h]$, where $h = R/\|\Phi\|$. A simple calculus argument maximizes h by taking $R = \frac{1}{1-a}$, which yields $h = \frac{1-a}{4}$.

The thrust of this argument is that repeated use of the Local Picard Theorem extends the solution to increasingly larger intervals. Our first step produced a solution on $[0, a_1]$ with $a_1 = \frac{1}{4}$. A second application extends this to $[0, a_2]$, where

$$a_2 = a_1 + \frac{1 - a_1}{4} = \frac{3a_1 + 1}{4} = \frac{7}{16}.$$

Generally, the solution on $[0, a_n]$ is extended to $[0, a_{n+1}]$, where

$$a_{n+1} = a_n + \frac{1 - a_n}{4} = \frac{3a_n + 1}{4}.$$

This is a monotone increasing sequence of real numbers that are readily seen to all be less than 1. Applying the Monotone Convergence Theorem (Theorem 2.5.4) to the sequence of real numbers (a_n), there is a limit $L = \lim_{n \to \infty} a_n$. Therefore,

$$L = \lim_{n \to \infty} a_{n+1} = \lim_{n \to \infty} \frac{3a_n + 1}{4} = \frac{3L + 1}{4}.$$

Solving yields $L = 1$.

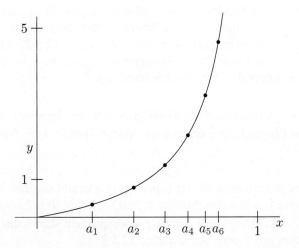

FIGURE 12.3. The solution to $y' = y^2$, $y(0) = 1$ with a_n marked.

The upshot is that repeated use of the Local Picard Theorem did extend the solution until it *blew up* by going off to infinity at $x = 1$. See Figure 12.3. It is not possible to use our method further since we are essentially following along the solution curve, which just carried us off the map.

12.5.3. EXAMPLE. Consider the differential equation

$$y' = y^{2/3}, \qquad y(0) = 0, \qquad 0 \le x \le 2.$$

The function $\Phi(x, y) = y^{2/3}$ is not Lipschitz. Indeed,

$$\lim_{h \to 0} \frac{|\Phi(0, h) - \Phi(0, 0)|}{|h|} = \lim_{h \to 0} \frac{1}{|h|^{1/3}} = +\infty.$$

So the Picard Theorems do not apply.

Nevertheless, we may attempt to solve this equation as before. Separating variables and integrating the equation $y^{-2/3} y' = 1$ from 0 to x yields

$$3y^{1/3} = x + c.$$

The initial data imply that $c = 0$, and thus $y = x^3/27$. This solution is valid on the whole real line.

However, there is another nice solution that stands out, namely $y = 0$. So the solution is not unique. In fact, there are many more solutions as well. For any $a > 0$, let

$$f_a(x) = \begin{cases} 0 & \text{if} \quad x \leq a \\ (x-a)^3/27 & \text{if} \quad x \geq a. \end{cases}$$

Then $y = f_a(x)$ is a C^2 solution for every positive a.

We see that there can be existence of solutions without uniqueness. A result known as Peano's Theorem (see Section 12.8) establishes that the differential equation $F'(x) = \Phi(x, F(x))$ has a solution locally whenever Φ is continuous.

Now we will try to systematize what occurred in Example 12.5.2. The idea is to make repeated use of the Local Picard Theorem to extend the solution until either we reach the whole interval or the solution blows up.

12.5.4. DEFINITION. A function $\Phi(x, y)$ on $[a, b] \times \mathbb{R}^n$ is **locally Lipschitz in the y variable** if it is Lipschitz in y on each compact subset $[a, b] \times \overline{B_R}$ for all positive real numbers R.

Actually a truly local definition of locally Lipschitz in y would say that for each $(x, y) \in [a, b] \times \mathbb{R}^n$, there is a positive number $\varepsilon > 0$ such that Φ is Lipschitz in y on $[x - \varepsilon, x + \varepsilon] \times \overline{B_\varepsilon(y)}$. However, a compactness argument shows that this is equivalent to the definition just given. See Exercise 12.5.J.

12.5.5. CONTINUATION THEOREM.
Suppose that $\Phi : [a, b] \times \mathbb{R}^n \to \mathbb{R}^n$ is a locally Lipschitz function. Consider the differential equation

$$F'(x) = \Phi(x, F(x)), \quad F(a) = \Gamma.$$

Then either

(1) *the DE has a unique solution $F(x)$ on $[a, b]$; or*

(2) *there is a $c \in (a, b)$ so that the DE has a unique solution $F(x)$ on $[a, c)$ and $\lim_{x \to c^-} \|F(x)\| = +\infty$.*

PROOF. The Local Picard Theorem (Theorem 12.5.1) establishes the fact that there is a unique solution on a nontrivial interval $[a, a + h]$. Define c to be the supremum of all values d for which the DE has a unique solution on $[a, d]$. A priori, these might be different solutions for different values of d. However, if F_1 and F_2 are the unique solutions on $[a, d_1]$ and $[a, d_2]$, respectively, for $a < d_1 < d_2$, then the restriction of F_2 to $[a, d_1]$ will also be a solution. Hence by the uniqueness, $F_2(x) = F_1(x)$ when both are defined. Therefore, we may conclude that there is a unique solution $F^*(x)$ of the DE defined on $[a, c)$ by taking the union of the solutions for $d < c$.

If F^* blows up at c, then we satisfy part (2) of the theorem; and if the solution extends to include b, we have part (1). However, it could be the case that the solution F^* remains bounded on some sequence approaching the point c, contradicting (2), yet the solution does not actually extend to or beyond the point c. The proof is complete if we can show that this is not possible. We suppose first that $c < b$.

Let x_n be a sequence increasing to c such that $\|F(x_n)\| \leq K$ for $n \geq 1$. Consider the compact region $D = [a, b] \times \overline{B_{K+1}(0)}$. Since Φ is locally Lipschitz in y, it is Lipschitz on the region D. Set

$$M = \sup_{(x,y) \in D} \|\Phi(x, y)\| \qquad \text{and} \qquad \delta = \min\left\{\frac{b-c}{2}, \frac{1}{2M}\right\}.$$

Choose N large enough that $c - x_N < \delta$. We may apply the Local Picard Theorem 12.5.1 to the DE

$$F'(x) = \Phi(x, F(x)) \qquad F(x_N) = F^*(x_N) \quad \text{for} \quad x \in [x_N, b].$$

The result is a unique solution $F(x)$ on the interval $[x_N, x_N + h]$, where

$$h = \min\{b - x_N, 1/M\} \geq \delta.$$

In particular, $d := x_N + h > c$. The uniqueness guarantees that $F(x)$ agrees with $F^*(x)$ on the interval $[x_N, c)$. Hence extending the definition of F^* by setting $F^*(x) = F(x)$ on $[x_N, x_N + h]$ produces a solution on $[a, d]$. This contradicts the definition of c.

An easy modification of this argument is possible when $c = b$ showing that the solution always extends to the *closed* interval $[a, b]$ if the solution does not blow up. Take $\delta = 1/2M$. We leave the details to the reader. ∎

The solution obtained in the Continuation Theorem is called the **maximal continuation** of the solution to the DE.

12.5.6. EXAMPLE. Not all solutions of differential equations blow up in the manner of the previous theorem. Consider the DE

$$x^4 y'' + 2x^3 y' + y = 0 \qquad \text{and} \qquad y(\tfrac{2}{\pi}) = 1 \quad y'(\tfrac{2}{\pi}) = 0 \quad \text{for} \quad x \in \mathbb{R}.$$

This looks like a reasonably nice linear homogeneous equation (see the next section). However, the coefficient of y'' is not 1, but rather a function of x that vanishes at 0. To put this in our standard form, we divide by x^4 to obtain

$$y'' + \frac{2}{x}y' + \frac{1}{x^4}y = 0.$$

The function Φ is just

$$\Phi(x, y_0, y_1) = \left(y_1, -\frac{2}{x}y_1 - \frac{1}{x^4}y_0\right).$$

This function satisfies

$$\|\Phi(x, y_0, y_1) - \Phi(x, z_0, z_1)\| = \left\|\left(y_1 - z_1, \frac{2}{x}(z_1 - y_1) + \frac{1}{x^4}(z_0 - y_0)\right)\right\|$$

$$\leq |y_1 - z_1|\left(1 + \frac{2}{|x|}\right) + \frac{1}{x^4}|z_0 - y_0|$$

$$\leq \left(1 + \frac{2}{|x|} + \frac{1}{x^4}\right)\|y - z\|.$$

So Φ satisfies a global Lipschitz condition in y provided that x remains bounded away from 0. Thus the Global Picard Theorem (Theorem 12.4.3) applies on the interval $[\varepsilon, R]$ for any $0 < \varepsilon < \frac{1}{\pi} < R < \infty$. It follows that there is a unique solution on $(0, \infty)$.

We will not demonstrate how to solve this equation. However, it is easy to check (do it!) that the solution is $f(x) = \sin(1/x)$. This solution has a very nasty discontinuity at $x = 0$, but it does not blow up—it remains bounded by 1. This does not contradict the Continuation Theorem (Theorem 12.5.5). The reason is that the function Φ is not locally Lipschitz on $\mathbb{R} \times \mathbb{R}^2$. It is not even defined for $x = 0$ and has a bad discontinuity there. This shows why we cannot expect to have a global solution to a DE where the coefficient of the highest-order term $y^{(n)}$ vanishes.

Exercises for Section 12.5

A. Suppose that $\Phi(x, y)$ and $\frac{\partial}{\partial y}\Phi(x, y)$ are both continuous functions on the region $[a, b] \times [c, d]$. Use the Mean Value Theorem to show that Φ is Lipschitz in y.
HINT: Let $L = \left\|\frac{\partial}{\partial y}\Phi(x, y)\right\|_\infty$.

B. Suppose that $\Phi(x, y_0, y_1)$ is C^1 on $[a, b] \times \overline{B_R(0)}$. Show that Φ is Lipschitz in y.
HINT: Let $L = \left\|\frac{\partial}{\partial y_0}\Phi(x, y)\right\|_\infty + \left\|\frac{\partial}{\partial y_1}\Phi(x, y)\right\|_\infty$. Estimate $\Phi(x, y_0, y_1) - \Phi(x, z_0, z_1)$ by subtracting and adding $\Phi(x, z_0, y_1)$.

C. For each of the following DEs, write down the function Φ and decide if it satisfies (a) a global Lipschitz condition in y, (b) a local Lipschitz condition in y, or (c) a Lipschitz condition on a smaller region that allows a local solution.
(a) $y'' = yy'$ and $y(0) = y'(0) = 1$ for $0 \leq x \leq 10$

(b) $y' = \sqrt{1 + y^2}$ and $y(0) = 0$ for $0 \leq x \leq 1$

(c) $f'(x) + f(x)^2 = 4xf(x) - 4x^2 + 2$ and $f(0) = 2$

(d) $(x^2 - x - 2)f'(x) + 3xf(x) - 3x = 0$ and $f(0) = 2$ for $-10 \leq x \leq 10$

D. Solve Exercise C(c) explicitly. Find the maximal continuation of the solution.
HINT: Find the DE satisfied by $g(x) = f(x) - 2x$ and solve it.

E. Solve Exercise C(d) explicitly. Find the maximal continuation of the solution.
HINT: Find the DE satisfied by $g(x) = f(x) - 1$ and solve it.

F. Reformulate Theorem 12.5.1 so that it is valid when the initial conditions apply to a point c in the interior of $[a, b]$.

G. Consider $y' = \sin\left(\dfrac{x^3 + x^2 - 1}{\sqrt{101 - y^2}}\right)$ and $y(2) = 3$. Prove that there is a solution on $[-5, 9]$. HINT: Show that $|y| \leq 10$ first. Then obtain a Lipschitz condition.

H. Provide the details for the proof of Theorem 12.5.5 for the case $c = b$.
HINT: Let $\delta = 1/(2M)$.

I. Consider the DE $y' = 3xy^{1/3}$ for $x \in \mathbb{R}$ and $y(0) = c \geq 0$. Let $A_\varepsilon = \{y : |y| \geq \varepsilon\}$.
  (a) Show that $3xy^{1/3}$ is Lipschitz in y on $[a, b] \times A_\varepsilon$ but not on $[a, b] \times [-1, 1]$.
  (b) Solve the DE when $c > 0$.
  (c) Find at least two solutions when $c = 0$.

J. Show that the two definitions of locally Lipschitz given in Definition 12.5.4 and the subsequent paragraph are equivalent. HINT: Cover $[a, b] \times \overline{B_R(0)}$ by open sets on which Φ is Lipschitz in y. Use the Borel–Lebesgue Theorem (Theorem 9.2.3).

K. Consider the DE $f(x)f'(x) = 1$ and $f(0) = a$ for $x \in \mathbb{R}$.
  (a) Solve this equation explicitly.
  (b) Show that there is a unique solution on an interval about 0 if $a \neq 0$ but that it is only extends to a proper subset of $\mathbb{R}$, even though the solution does not blow up. Why does this not contradict the Continuation Theorem?
  (c) Show that there are two solutions for $a = 0$ valid on $[0, \infty)$. Why does this not contradict the Local Picard Theorem?

L. Show that $f(x) = 1 - \sqrt{1 - x^2}$ for $-1 \leq x \leq 1$ is the unique solution of the DE $y'' = \left(1 + (y')^2\right)^{3/2}$ and $y(0) = y'(0) = 1$. This solution cannot be continued beyond $x = 1$, yet $|f(x)| \leq 1$. Why does this not contradict the Continuation Theorem? HINT: How does $F(x) = \left(f(x), f'(x)\right)$ behave at $x = 1$?

M. Consider the DE: $y' = x^2 + y^2$ and $y(0) = 0$.
  (a) Show that this DE satisfies a local Lipschitz condition but not a global one.
  (b) Integrate the inequality $y' \geq 1 + y^2$ for $x \geq 1$ to prove that the solution must go off to infinity in a finite time. (See Exercise 12.4.H.)

12.6. Linear Differential Equations

In this section, we explore a very important class of differential equations in greater depth. This class occurs frequently in applications and is also especially amenable to analysis.

Consider the differential equation

$$(12.6.1) \quad f^{(n)}(x) = p(x) + q_0(x)f(x) + q_1(x)f'(x) + \cdots + q_{n-1}(x)f^{(n-1)}(x)$$

$$f(c) = \gamma_0, \quad f'(c) = \gamma_1, \quad \cdots f^{(n-1)}(c) = \gamma_{n-1},$$

where $p(x)$ and $q_k(x)$ are continuous functions on $[a, b]$ and c is a point in $[a, b]$. This is called a **linear differential equation** because the function

$$\varphi(x, y) = p(x) + q_0(x)y_0 + q_1(x)y_1 + \cdots + q_{n-1}(x)y_{n-1}$$

defined on $[a, b] \times \mathbb{R}^n$ is linear in the second variable $y = (y_0, y_1, \ldots, y_{n-1})$.

Using the reduction in Section 12.3, we obtain the reformulated first-order differential equation

$$F'(x) = \Phi(x, F(x)) \qquad \text{and} \qquad F(c) = \Gamma = (\gamma_0, \gamma_1, \ldots, \gamma_{n-1}),$$

where

$$\Phi(x, y_0, \ldots, y_{n-1}) = (y_1, \ldots, y_{n-1}, \varphi(x, y)).$$

We will verify that Φ is Lipschitz in y. Set $M = \max\limits_{0 \le k \le n-1} \|q_k\|_\infty$. Then for any $x \in [a, b]$ and $y = (y_0, \ldots, y_{n-1})$ and $z = (z_0, \ldots, z_{n-1})$ in $\mathbb{R}^n$,

$$\|\Phi(x, y) - \Phi(x, z)\| = \left\|\left(y_1 - z_1, \ldots, y_{n-1} - z_{n-1}, \sum_{k=0}^{n-1} q_k(x)(y_k - z_k)\right)\right\|$$

$$= \left(\sum_{i=1}^{n-1} |y_i - z_i|^2 + \left|\sum_{k=0}^{n-1} q_k(x)(y_k - z_k)\right|^2\right)^{1/2}$$

$$\le \left(\|y - z\|^2 + (nM\|y - z\|)^2\right)^{1/2}$$

$$\le (1 + nM)\|y - z\|.$$

So Φ satisfies the Lipschitz condition with $L = 1 + nM$.

Therefore, by the Global Picard Theorem (Theorem 12.4.3), this equation has a unique solution for each choice of initial values Γ.

The most important consequence of linearity is the relationship between solutions of the same DE with different initial values. Suppose that f and g are solutions of (12.6.1) with initial data Γ and $\Delta = (\delta_0, \ldots, \delta_{n-1})$, respectively. Then the function $h(x) = g(x) - f(x)$ satisfies

$$(12.6.2) \qquad h^{(n)}(x) = p(x) + \sum_{k=0}^{n-1} q_k(x)g^{(k)}(x) - p(x) + \sum_{k=0}^{n-1} q_k(x)f^{(k)}(x)$$

$$= q_0(x)h(x) + q_1(x)h'(x) + \cdots + q_{n-1}(x)h^{(n-1)}(x)$$

and satisfies the initial conditions

$$h^{(k)}(c) = \gamma_k - \delta_k \qquad \text{for} \qquad 0 \le k \le n - 1.$$

This is a linear equation with the term $p(x)$ missing. This DE satisfied by h is called a **homogeneous linear DE**, while the equation (12.6.1) with the **forcing term** $p(x) \ne 0$ is called a **nonhomogeneous linear DE**.

Generally the homogeneous DE is much easier to solve. Once the set of solutions for the homogeneous problem is described as an n-dimensional linear space, one searches for a single solution f_p of the nonhomogeneous DE (12.6.1) without regard to the initial conditions. This solution is called a **particular solution**, to distinguish it from the general solution. Then the argument of the previous paragraph shows that the general solution of this DE has the form $f_p + h$ for some solution h of the homogeneous DE (12.6.2).

Let Γ_i denote the initial conditions $\gamma_i = 1$, $\gamma_j = 0$ for $j \ne i$, $0 \le j \le n - 1$. In other words, the vectors Γ_i correspond to the standard basis vectors of $\mathbb{R}^n$. For each $0 \leq i \leq n - 1$, let $h_i(x)$ be the unique solution of the homogeneous DE (12.6.2). Then let $\Gamma = (\gamma_0, \gamma_1, \ldots, \gamma_{n-1})$ be an arbitrary vector. Consider

$$h(x) = \sum_{i=0}^{n-1} \gamma_i h_i(x).$$

It is easy to calculate that $h^{(k)}(c) = \gamma_k$ for $0 \leq k \leq n - 1$ and

$$h^{(n)}(x) = \sum_{k=0}^{n-1} q_k(x) h^{(k)}(x)$$

$$= \sum_{j=0}^{n-1} \gamma_j \sum_{k=0}^{n-1} q_k(x) h_j^{(k)}(x) = 0.$$

In other words, the solutions of (12.6.2) are linear combinations of the special solutions $h_i(x)$, $0 \leq i \leq n - 1$. So these n functions form a basis for the solution space of the homogeneous DE, which is an n-dimensional subspace of $C[a, b]$.

Now suppose that f_0 is the solution of the nonhomogeneous DE (12.6.1) with initial data $\Gamma = 0$. Let g be the solution of (12.6.1) for initial data Δ. Our calculation using $h(x) = g(x) - f_0(x)$ yields a solution of the homogeneous DE with initial data Δ. Thus

$$g(x) = f_0(x) + h(x) = f_0(x) + \sum_{i=0}^{n-1} \delta_i h_i(x).$$

So f_0 is a particular solution of the nonhomogeneous DE. All of the others are obtained by adding this particular solution to a solution of the homogeneous equation. Often one does not look for the particular solution with zero initial conditions, but rather for one with a simple form related to the specific problem.

Summing up, we have the following useful result.

12.6.3. THEOREM. *The homogeneous equation* (12.6.2) *has* n *linearly independent solutions* $h_0, \ldots, h_{n-1}$; *and every solution is a linear combination of these solutions.*

For each set Γ *of initial conditions, the nonhomogeneous linear differential equation* (12.6.1) *has a unique solution. If* f_p *is a particular solution of the nonhomogeneous DE, then every solution (for different initial conditions) is the sum of* f_p *and a solution* h *of the homogeneous DE.*

Some techniques for solving linear DEs will be explored in the Exercises. We now consider a special case in which all the functions q_k are constant.

12.6.4. EXAMPLE. Consider the second-order linear DE with constant coefficients

$$y''(x) - 5y'(x) + 6y(x) = \sin x \quad \text{for} \quad x \in \mathbb{R}$$
$$y(0) = 1 \qquad y'(0) = 0.$$

The first task is to solve the homogeneous equation

$$y'' - 5y' + 6y = 0.$$

It is useful to consider the linear map D, which sends each function to its derivative: $Df = f'$. Our equation may be written as $(D^2 - 5D + 6I)y = 0$, where I is the identity map $If = f$. This quadratic may be factored as

$$D^2 - 5D + 6I = (D - 2I)(D - 3I),$$

where 2 and 3 are the roots of the quadratic equation $x^2 - 5x + 6 = 0$.

Note that the equation $(D - 2I)y = 0$ is just $y' = 2y$. We can recognize by inspection that $f(x) = e^{2x}$ is a solution. Then we may compute

$$(D^2 - 5D + 6I)e^{2x} = (4 - 10 + 6)e^{2x} = 0.$$

Similarly, e^{3x} is a solution of $(D - 3I)y = 0$ and

$$(D^2 - 5D + 6I)e^{3x} = (9 - 15 + 6)e^{3x} = 0.$$

So $h_1(x) = e^{2x}$ and $h_2(x) = e^{3x}$ are both solutions of this equation.

Let $\Gamma_1 = (h_1(0), h_1'(0)) = (1, 2)$ and $\Gamma_2 = (h_2(0), h_2'(0)) = (1, 3)$ be the initial conditions. These two vectors are evidently independent. Thus every possible vector of initial conditions is a linear combination of Γ_1 and Γ_2. From this, we see that every solution of the homogeneous DE is of the form $h(x) = ae^{2x} + be^{3x}$.

Now let us return to the nonhomogeneous problem. A technique called the **method of undetermined coefficients** works well here. This is just a fancy name for good guesswork. It works for forcing functions that are (sums of) exponentials, polynomials, sines, and cosines. We look for a solution of the same type. Here we hypothesize a solution of the form

$$f(x) = c \sin x + d \cos x,$$

where c and d are constants. Plug f into our differential equation:

$$f'' - 5f' + 6f = (-c \sin x - d \cos x) - 5(c \cos x - d \sin x) + 6(c \sin x + d \cos x)$$
$$= (5c + 5d) \sin x + (5d - 5c) \cos x.$$

So we may solve the system of linear equations

$$5c + 5d = 1$$
$$5c - 5d = 0$$

to obtain $c = d = .1$. This is a particular solution.

Now the general solution to the nonhomogeneous equation is of the form

$$f(x) = .1 \sin x + .1 \cos x + ae^{2x} + be^{3x}.$$

We compute the initial conditions

$$1 = f(0) = .1 + a + b$$
$$0 = f'(0) = .1 + 2a + 3b.$$

Solving this linear system yields $a = 2.8$ and $b = -1.9$. Thus the solution is

$$f(x) = .1 \sin x + .1 \cos x + 2.8e^{2x} - 1.9e^{3x}.$$

Exercises for Section 12.6

A. Solve $y'' + 3y' - 10y = 8e^{3x}$, $y(0) = 3$, and $y'(0) = 0$.

B. Consider the homogeneous equation $y'' + by' + cy = 0$ by using the roots r and s of the quadratic $x^2 + bx + c = 0$.
   (a) Solve the DE when r and s are distinct real roots.
   (b) When $r = a + ib$ and $s = a - ib$ are distinct complex roots, show that $e^{ax} \sin bx$ and $e^{ax} \cos bx$ are solutions.
   (c) When r is a double real root, show that e^{rx} and xe^{rx} are solutions.

C. Observe that x is a solution of $y'' - x^{-2}y' + x^{-3}y = 0$. Look for a second solution of the form $f(x) = xg(x)$.   HINT: Find a first-order DE for g'.

D. Let A be an $n \times n$ matrix, and let $y = (y_1, \dots, y_n)$. Consider $y' = Ay$ and $y(a) = \Gamma$.
   (a) Set up the integral equation for this DE.
   (b) Starting with $f_0 = \Gamma$, show that the iterates obtained are $f_k(x) = \sum_{i=0}^{k} \frac{1}{i!}(xA)^i \Gamma$.
   (c) Deduce that this series converges for any matrix A. The limit is $e^{xA}\Gamma$.

E. Solve the DE of the previous exercise explicitly for $A = \begin{bmatrix} 3 & -1 \\ -1 & 3 \end{bmatrix}$ and $\Gamma = (1, 2)$.
   HINT: Find a basis that diagonalizes A.

F. Consider the DE $y'' + xy' + y = 0$.
   (a) Look for a power series solution $y = \sum_{n \geq 0} a_n x^n$. That is, plug this series into the DE and solve for a_n in terms of a_0 and a_1.
   (b) Find the radius of convergence of these solutions.
   (c) Identify one of the resulting series in closed form $h_1(x)$. Look for a second solution of the form $f(x) = h_1(x)g(x)$. Obtain a DE for g and solve it.
   (d) Use the power series expansion of (a) to find a particular *polynomial* solution of $y'' + xy' + y = x^3$.

G. **Bessel's DE.** Consider the DE $x^2 y'' + xy' + (x^2 - N^2)y = 0$, where $N \in \mathbb{N}$.
   (a) Look for a power series solution of the form $y = \sum_{n \geq 0} a_n x^n$.
      HINT: Show $a_n = 0$ for $n < N$ and $n - N$ odd.
   (b) Find the radius of convergence of this power series.
   (c) The Bessel function (of the first kind) of order N, denoted $J_N(x)$, is the multiple of this solution with $a_N = 1/(2^N N!)$. Find a concise expression for the power series expansion J_N in terms of $x/2$ and factorials.

H. **Variation of Parameters.** Let h_1 and h_2 be a basis for the solutions of the homogeneous linear DE $y'' + q_1(x)y' + q_0(x)y = 0$. Define the **Wronskian** determinant to be the function $W(x) = h_1(x)h_2'(x) - h_1'(x)h_2(x)$.
   (a) Show that $W'(x) + q_1(x)W(x) = 0$ and $W(c) \neq 0$.
   (b) Solve for $W(x)$. Hence show that $W(x)$ is never 0.   HINT: Integrate $\dfrac{W'}{W} = -q_1$.
   (c) Let $p(x)$ be a forcing function. Show that
   $$f(x) = -h_1(x) \int_c^x \frac{h_2(t)p(t)}{W(t)}\, dt + h_2(x) \int_c^x \frac{h_1(t)p(t)}{W(t)}\, dt$$
   is a particular solution of $y'' + q_1(x)y' + q_0(x)y = p(x)$.

I. Use variation of parameters to solve $y'' - 5y' + 6y = 4xe^x$, $y(0) = 0$, and $y'(0) = -6$.

12.7. Perturbation and Stability of DEs

Another point of interest that can be readily achieved by our methods is the continuous dependence of the solution F on the initial data $F(a) = \Gamma$. It is important in applications that nearby initial values should lead to nearby solutions, and nearby equations have solutions that are also close. This is a variation on the notion of sensitive dependence on initial conditions, which we studied in Section 11.5.

Since the main theorem of this section is an explicit estimate, it is necessarily a bit complicated. However, you should interpret this theorem as saying something more qualitative: If two DEs are close and at least one is Lipschitz, then their solutions are close. To understand the precise formulation of the estimate, it is important to work through the examples and the Exercises.

12.7.1. PERTURBATION THEOREM.
Let $\Phi(x, y)$ be a continuous function on a region $D = [a, b] \times \overline{B_R(\Gamma)}$ satisfying a Lipschitz condition in y with constant L. Suppose that Ψ is another continuous function on D such that $\|\Psi - \Phi\|_\infty \leq \varepsilon$. (The function Ψ is not assumed to be Lipschitz.) Let F and G be the solutions of the differential equations

$$F'(x) = \Phi(x, F(x)), \qquad F(a) = \Gamma$$

and

$$G(x)' = \Psi(x, G(x)), \qquad G(a) = \Delta$$

such that $(x, F(x))$ and $(x, G(x))$ belong to D for $a \leq x \leq b$. Also suppose that $\|\Delta - \Gamma\| \leq \delta$. Then, for all $x \in (a, b)$,

$$\|G(x) - F(x)\| \leq \delta e^{L|x-a|} + \frac{\varepsilon}{L}(e^{L|x-a|} - 1).$$

Thus

$$\|G - F\|_\infty \leq \delta e^{L|b-a|} + \frac{\varepsilon}{L}(e^{L|b-a|} - 1).$$

PROOF. Define

$$\tau(x) = \|G(x) - F(x)\| = \left(\sum_{i=0}^{n-1}\big(g_i(x) - f_i(x)\big)^2\right)^{1/2}.$$

In particular, $\tau(a) = \|\Delta - \Gamma\| < \delta$. Then by the Cauchy–Schwarz inequality,

$$2\tau(x)\tau'(x) = (\tau(x)^2)' = \sum_{i=0}^{n-1} 2\big(g_i(x) - f_i(x)\big)\big(g_i'(x) - f_i'(x)\big)$$

$$\leq 2\left(\sum_{i=0}^{n-1}\big(g_i(x) - f_i(x)\big)^2\right)^{1/2}\left(\sum_{i=0}^{n-1}\big(g_i'(x) - f_i'(x)\big)^2\right)^{1/2}$$

$$= 2\tau(x)\|G'(x) - F'(x)\|.$$

Now compute

$$\|G'(x) - F'(x)\| = \|\Psi(x, G(x)) - \Phi(x, F(x))\|$$
$$\leq \|\Psi(x, G(x)) - \Phi(x, G(x))\| + \|\Phi(x, G(x)) - \Phi(x, F(x))\|$$
$$\leq \varepsilon + L\|G(x) - F(x)\| = \varepsilon + L\tau(x).$$

Combining these two estimates, we obtain a differential inequality

$$\tau'(x) \leq \|G'(x) - F'(x)\| \leq \varepsilon + L\tau(x).$$

Hence

$$x - a = \int_a^x dt \geq \int_a^x \frac{\tau'(t)}{\varepsilon + L\tau(t)}\, dt$$
$$= \frac{1}{L}\log(L\tau + \varepsilon)\Big|_a^x = \frac{1}{L}\log\Big(\frac{L\tau(x) + \varepsilon}{L\tau(a) + \varepsilon}\Big).$$

Solving for $\tau(x)$ yields

$$L\tau(x) + \varepsilon \leq e^{L(x-a)}(L\tau(a) + \varepsilon) \leq e^{L(x-a)}(L\delta + \varepsilon);$$

whence

$$\|G(x) - F(x)\| = \tau(x) \leq \delta e^{L|x-a|} + \frac{\varepsilon}{L}(e^{L|x-a|} - 1). \qquad \blacksquare$$

An immediate and important consequence of this result is that the solution to this result is continuous dependence of the solution of a DE (with Lipschitz condition) as a function of the parameter Γ. For simplicity only, we assume a global Lipschitz condition.

12.7.2. COROLLARY. *Suppose that Φ satisfies a global Lipschitz condition in y on $[a, b] \times \mathbb{R}^n$. Then the solution F_Γ of*

$$F'(x) = \Phi(x, F(x)), \qquad F(a) = \Gamma$$

is a continuous function of Γ.

PROOF. Let L be the Lipschitz constant. Since the Lipschitz condition is global, there is no need to check whether the values of $F(x)$ remain in the domain. Also, there is no need to keep δ small.

In this application of Theorem 12.7.1, we take $\varepsilon = 0$ since the function Φ is used for both functions. Hence we obtain

$$\|F_\Gamma - F_\Delta\|_\infty \leq \|\Gamma - \Delta\|e^{L|b-a|}.$$

In particular, as Δ converges to Γ, it follows that F_Δ converges uniformly to F_Γ. This is referred to as **continuous dependence on parameters**. $\qquad \blacksquare$

12.7.3. EXAMPLE. Consider a linear DE (12.6.1) of Section 12.6. We showed there that linear DEs satisfy a global Lipschitz condition in y. The solution is a function f_Γ of the initial conditions. However, the estimates are expressed in terms of the vector-valued function $F_\Gamma = \left(f_\Gamma, f_\Gamma', \ldots, f_\Gamma^{(n-1)}\right)$. By Corollary 12.7.2, the solution is a continuous function of the initial data:

$$\|F_\Gamma - F_\Delta\|_\infty \le \|\Gamma - \Delta\| e^{L|b-a|}.$$

Hence we obtain that

$$\|f_\Gamma^{(k)} - f_\Delta^{(k)}\|_\infty \le \|\Gamma - \Delta\| e^{L|b-a|} \quad \text{for all} \quad 0 \le k \le n-1.$$

So the first $n-1$ derivatives also depend continuously on the initial data. Consequently, $f_\Gamma^{(n)} = \varphi(x, F_\Gamma)$ is also a continuous function of Γ. Therefore, f_Γ is a continuous function of Γ in the $C^n[a,b]$ norm.

However, in this case, it is evident from the form of the solution. Recall that we let f_0 be the particular solution with $\Gamma = 0$ and found a basis of solutions h_i for the homogeneous equation (12.6.2) for initial data Γ_i. The general solution is given by

$$f_\gamma(x) = f_0(x) + \sum_{i=0}^{n-1} \gamma_i h_i(x).$$

From this, the continuous dependence of f_γ and its derivatives on Γ is evident.

Theorem 12.7.1 can be interpreted as a stability result. If the differential equation and initial data are measured empirically, then this theorem assures us that the approximate solution based on the measurements remains reasonably accurate. It is rare that differential equations that arise in practice can be explicitly solved in closed form. However, general behaviour can be deduced if the DE is close to a nice one.

12.7.4. EXAMPLE. Suppose that $g(x)$ is a solution of

$$y'' + y = e(x, y, y') \qquad y(0) = 0 \quad \text{and} \quad y'(0) = 1 \quad \text{for} \quad x \in [-2\pi, 2\pi],$$

where $e(x, y, y')$ is a small function bounded by ε. Then g should be close to the solution of

$$y'' + y = 0 \qquad y(0) = 0 \quad \text{and} \quad y'(0) = 1 \quad \text{for} \quad x \in [-2\pi, 2\pi],$$

which is known to be $f(x) = \sin x$. This unperturbed DE corresponds to the function $\Phi(x, y_0, y_1) = (y_1, -y_0)$, which has Lipschitz constant 1. We apply Theorem 12.7.1, with $\delta = 0$, to obtain

$$\left\| (g(x) - \sin x, \; g'(x) - \cos x) \right\| \le \varepsilon(e^{|x|} - 1).$$

It follows that $g(x)$ is bounded between $\sin(x) - \varepsilon(e^{|x|} - 1)$ and $\sin(x) + \varepsilon(e^{|x|} - 1)$; see Figure 12.4.

By using the bound on g', we can describe g more precisely. Let us assume that $\varepsilon < .01$. Since $\cos x > .02$ on $[-1.55, 1.55]$, it follows that $g'(x) > 0$ on this range, and hence g is strictly increasing. So 0 is the only zero of g in this range. Similarly, g is strictly decreasing on $[1.59, 4.69]$. We see that

$$g(2.9) > \sin(2.9) - .01(e^{2.9} - 1) > .06.$$

Similarly, $g(3.5) < 0$. It follows that g has a single zero in the interval $(2.9, 3.5)$. So on $[0, 3.5]$, g oscillates much like the sine function.

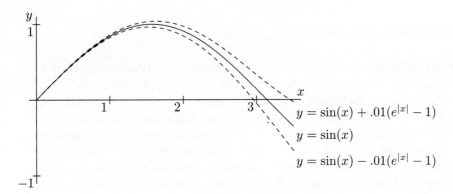

FIGURE 12.4. The bounds for $g(x)$.

Exercises for Section 12.7

A. It is known that $f(x)$ is an exact solution of $y'' + x^2 y' + 2^x y = \sin x$ on $[0, 1]$. However, the initial data must be measured experimentally. How accurate must the measurements of $f(0)$ and $f'(0)$ be in order to be able to predict $f(1)$ and $f'(1)$ to within an accuracy of $.00005$?

B. Let $f(x)$ be the solution of the DE $y' = e^{xy}$ and $y(0) = 1$ for $x \in [-1, 1]$. Suppose that $f_n(x)$ is the solution of $y' = \sum_{k=0}^{n} \frac{(xy)^k}{k!}$ and $y(0) = 1$ for $x \in [-1, 1]$.
(a) Show that $f_n(x)$ converges to $f(x)$ uniformly on $[-1, 1]$.
(b) Find an N so that $\|f - f_N\|_\infty < .0001$.

C. Let $A_\varepsilon = \begin{bmatrix} 1 & \varepsilon \\ \varepsilon & 1 \end{bmatrix}$. Consider the DE: $F'_\varepsilon(x) = A_\varepsilon F(x)$ and $F(0) = (2, 3)$ on $[0, 1]$.
(a) Solve explicitly for F_ε.   HINT: Exercise 12.6.E
(b) Compute $\|F_\varepsilon - F_0\|_\infty$.
(c) Compare (b) with the bound provided by Theorem 12.7.1.

D. For each $n \geq 1$, define a piecewise linear function $g_n(x)$ on $[0, 1]$ by setting $g_n(0) = 1$ and then defining $g_n(x) = g_n(\frac{k}{n})(1 + 3(x - \frac{k}{n}))$ for $\frac{k}{n} < x \leq \frac{k+1}{n}$ and $0 \leq k \leq n-1$.
(a) Sketch $g_{10}(x)$ and e^{3x} on the same graph.
(b) Use the fact that g_n is an approximate solution to the DE $y' = 3y$ and $y(0) = 1$ on $[0, 1]$ to estimate $\|g_n - e^{3x}\|_\infty$.
(c) How does this estimate compare with the exact value?
(d) Show that $g_n(x)$ converges uniformly to e^{3x}.

E. Let F be a solution of $F'(x) = \Phi(x, F(x))$ and $F(a) = \Gamma$, where Φ is a continuous function that is Lipschitz in y with constant L on $[a, b] \times \mathbb{R}^n$. Suppose that G is a differentiable function satisfying $\|G'(x) - \Phi(x, G(x))\| \leq \varepsilon$ and $\|G(a) - \Gamma\| \leq \delta$. Show that $\|G(x) - F(x)\| \leq \delta e^{L|x-a|} + \frac{\varepsilon}{L}(e^{L|x-a|} - 1)$.
HINT: Apply the *proof* of Theorem 12.7.1.

12.8. Existence without Uniqueness

So far, all of our theorems establishing the existence of solutions required a Lipschitz condition. While this is frequently the case in applications, there do exist common situations for which there is no Lipschitz condition near a critical point of some sort. As we saw in Example 12.5.3, this might result in existence of multiple solutions. It turns out that by merely assuming continuity of the function Φ (which is surely not too much to ask), the existence of a solution is guaranteed in a small interval. We could again use continuation methods to obtain solutions on larger intervals. However, we will not do that here. The additional tool we need to proceed is a compactness theorem about functions, the Arzela–Ascoli Theorem (Theorem 8.6.9).

12.8.1. PEANO'S THEOREM.
Suppose that $\Gamma \in \mathbb{R}^n$ and Φ is a continuous function from $D = [a, b] \times \overline{B_R(\Gamma)}$ into $\mathbb{R}^n$. Then the differential equation

$$F'(x) = \Phi(x, F(x)) \qquad f(a) = \Gamma \quad for \quad a \leq x \leq b$$

has a solution on $[a, a + h]$, where $h = \min\{b - a, R/M\}$ and $M = \|\Phi\|_\infty$ is the max norm of Φ over the set D.

PROOF. As in Picard's proof, we convert the problem to finding a fixed point for the integral mapping

$$TF(x) = \Gamma + \int_a^x \Phi(t, F(t)) \, dt.$$

For each $n \geq 1$, we define a function $F_n(x)$ on $[a, a + h]$ as follows:

$$F_n(x) = \Gamma \qquad\qquad \text{for} \quad a \leq x \leq a + \tfrac{1}{n}$$

$$F_n(x) = \Gamma + \int_a^{x-1/n} \Phi(t, F_n(t)) \, dt \qquad \text{for} \quad a + \tfrac{1}{n} \leq x \leq a + h.$$

Notice that the integral defines $F_n(x)$ in terms of the values of $F_n(x)$ in the interval $[a, x - \frac{1}{n}]$. Since F_n is defined to be the constant Γ on $[a, a + \frac{1}{n}]$, the definition of F_n as an integral makes sense on the interval $[a + \frac{1}{n}, a + \frac{2}{n}]$. Once this is accomplished, it then follows that the integral definition makes sense on the interval $[a + \frac{2}{n}, a + \frac{3}{n}]$. Proceeding in this way, we see that the definition makes sense on all of $[a, a + h]$ provided that $F_n(x)$ remains in $\overline{B_R(\Gamma)}$. This is an easy estimate (but it explains the definition of h):

$$\|F_n(x) - \Gamma\| \leq \int_a^{x-1/n} \|\Phi(t, F_n(t))\| \, dt$$
$$\leq M|x - a| \leq Mh \leq R.$$

It is also easy to show that F_n is an approximate solution to the fixed-point problem. For $a \leq x \leq a + \frac{1}{n}$,

$$\|TF_n(x) - F_n(x)\| = \left\| \Gamma + \int_a^x \Phi(t, F_n(t)) \, dt - \Gamma \right\|$$
$$\leq \int_a^x \|\Phi(t, F_n(t))\| \, dt \leq M(x - a) \leq \frac{M}{n}.$$

For $a + \frac{1}{n} \leq x \leq a + h$,

$$\|TF_n(x) - F_n(x)\| = \left\| \int_{x-1/n}^x \Phi(t, F_n(t)) \, dt \right\|$$
$$\leq \int_{x-1/n}^x \|\Phi(t, F_n(t))\| \, dt \leq \frac{M}{n}.$$

So $\|TF_n - F_n\|_\infty \leq M/n$.

We will show that the family $\{F_n : n \geq 1\}$ is equicontinuous. Indeed, given $\varepsilon > 0$, let $\delta = \varepsilon/M$. If $a \leq x_1 < x_2 \leq a + h$ and $|x_2 - x_1| < \delta$, then

$$\|F_n(x_2) - F_n(x_1)\| \leq \int_{x_1-1/n}^{x_2-1/n} \|\Phi(t, F_n(t))\| \, dt$$
$$\leq M\|(x_1 - 1/n) - (x_2 - 1/n)\| < M\delta = \varepsilon.$$

Therefore we may apply the Arzela–Ascoli Theorem. The family of functions $\{F_n : n \geq 1\}$ is bounded by $\|\Gamma\| + R$ and is equicontinuous. So its closure is compact. Thus we can extract an increasing sequence n_k so that F_{n_k} converge uniformly on $[a, a + h]$ to a function $F^*(x)$. We will show that F^* is a fixed point of T, and hence the desired solution. Compute

$$\|F^*(x) - TF^*(x)\|$$
$$\leq \|F^*(x) - F_{n_k}(x)\| + \|F_{n_k}(x) - TF_{n_k}(x)\| + \|TF_{n_k}(x) - TF^*(x)\|$$
$$\leq \|F^* - F_{n_k}\|_\infty + \frac{M}{n_k} + \int_a^{a+h} \|\Phi(t, F^*(t)) - \Phi(t, F_{n_k}(t))\| \, dt.$$

Now Φ is uniformly continuous on the compact set D by Theorem 5.5.9. Since F_{n_k} converges uniformly to F^* on $[a, a + h]$, it follows that $\Phi(x, F_{n_k}(x))$ converges uniformly to $\Phi(x, F^*(x))$. Hence by Theorem 8.3.1,

$$\lim_{k \to \infty} \int_a^{a+h} \|\Phi(t, F^*(t)) - \Phi(t, F_{n_k}(t))\| \, dt = 0.$$

Putting all of our estimates together and letting k tend to ∞, we obtain $TF^* = F^*$, completing the argument. ∎

Exercises for Section 12.8

A. Show that the DE $y^{(4)} = 120y^{1/5}$ and $y(0) = y'(0) = y^{(2)}(0) = y^{(3)}(0) = 0$ has infinitely many solutions on the whole real line.
HINT: Compare with Example 12.5.3.

B. Let $\gamma \in \mathbb{R}$ and let Φ be a continuous real-valued function on $[a, b] \times [\gamma - R, \gamma + R]$. Consider the DE $y'(x) = \Phi(x, y)$ and $y(a) = \gamma$, and suppose that Peano's Theorem guarantees a solution on $[a, a+h]$. If f and g are both solutions on $[a, a+h]$, show that their maximum $f \vee g(x) = \max\{f(x), g(x)\}$ and minimum $f \wedge g(x) = \min\{f(x), g(x)\}$ are also solutions.
HINT: Verify the DE at points x in $U = \{x : f(x) > g(x)\}$, $V = \{x : f(x) < g(x)\}$ and $X = \{x : f(x) = g(x)\}$ separately.

C. Let Φ be a continuous function on $[a, b] \times \mathbb{R}$ such that $\Phi(x, y)$ is a decreasing function of y for each fixed x.
(a) Suppose that f and g are solutions of $y'(x) = \Phi(x, y(x))$. Show that $|f(x) - g(x)|$ is a decreasing function of x.
HINT: suppose that $f(x) > g(x)$ on an interval I and $x_1 < x_2 \in I$. Express $\big(f(x_2) - g(x_2)\big) - \big(f(x_1) - g(x_1)\big)$ as an integral.
(b) Show that this DE has a unique solution for the initial condition $y(a) = \gamma$.

D. Show that the set of all solutions on $[a, a+h]$ to the DE of Peano's Theorem is closed, bounded, and equicontinuous.

E. Consider the setup of Exercise 12.8.B. Prove that the set of all solutions on $[a, a+h]$ has a largest and smallest solution.
HINT: Use Exercise 12.8.D to obtain a countable dense subset $\{f_n\}$ of the set of solutions. Let $g_k = \max\{f_1, \dots, f_k\}$ for $k \geq 1$. Show that g_k converges to the maximal solution $f_{\max}$.

F. Again consider the setup of Exercise 12.8.B, and let $x_0 \in [a, a+h]$. Show that the set $\{f(x_0) : f \text{ is a solution of the DE}\}$ is a closed interval.
HINT: If $c \in [f_{\min}(x_0), f_{\max}(x_0)]$, show that $f'(x) = \Phi(x, f(x))$ and $f(x_0) = c$ has a solution f on $[a, c]$. Consider $g(x) = (f_{\max} \wedge f) \vee f_{\min}$.

CHAPTER 13

Fourier Series and Physics

Fourier series were first developed to solve partial differential equations that arise in physical problems, such as heat flow and vibration. We will look at the physics problem of heat flow to see how Fourier series arise and why they are useful. Then we will proceed with the solution, which leads to a lot of very interesting mathematics. Then we will see that the problem of a vibrating string leads to a different PDE that requires similar techniques to solve.

While this problem sounds very applied, the infinite series that arise as solutions forced mathematicians to delve deeply into the foundations of analysis. When d'Alembert proposed his solution for the motion of a vibrating string in 1754, there were no clear, precise definitions of limit, function, or even of the real numbers— all things taken for granted in most calculus courses today. D'Alembert's solution (which we shall see at the end of this chapter) has a closed form, and thus did not really challenge deep principles. However, the solution to the heat problem that Fourier proposed in 1807 required notions of convergence that mathematicians of that time did not have. Fourier won a major prize in 1812 for this work, but the judges, Laplace, Lagrange, and Legendre, criticized Fourier for lack of rigour. Work in the nineteenth century by many now famous mathematicians eventually resolved these questions by developing the modern definitions of limit, continuity, and uniform convergence. These tools were developed not because of some fetish for finding complicated things, but because they were essential to understanding Fourier series.

13.1. The Steady-State Heat Equation

The purpose of this section is to derive from physical principles the partial differential equation satisfied by heat flow on a surface.

Consider the problem of determining the temperature on a thin metal disk given that the temperature on the boundary circle is fixed. We assume that there is no heat loss in the third dimension. Perhaps this disk is placed between two insulating pads. We also assume that the system is at equilibrium. As a consequence, the temperature at each point remains constant over time. This is known as the **steady-state heat problem**.

It is convenient to work in polar coordinates in order to exploit the symmetry. The disk will be given as the set

$$\overline{\mathbb{D}} = \{(x,y) : x^2 + y^2 \leq 1\} = \{(r,\theta) : 0 \leq r \leq 1, \ -\pi \leq \theta \leq \pi\},$$

where $(0,\theta)$ represents the origin for all values of θ; and $(r,-\pi) = (r,\pi)$ for all $r \geq 0$. More generally, we allow (r,θ) for any real value of θ and make the identification $(r,\theta + 2\pi) = (r,\theta)$.

Let us denote the temperature distribution over the disk by a function $u(r,\theta)$, and let the given function on the boundary circle be $f(\theta)$.

As usual in physical problems such as this, we need to know a mathematical form of the appropriate physical law in order to determine a differential equation that governs the behaviour of the system. In this case, the law is that the heat flow across a boundary is proportional to the temperature difference between the two sides of the curve. Of course, our temperature distribution function will be continuous. So we must deal with the infinitesimal version of temperature change, which is the derivative of the temperature in the direction perpendicular to the boundary, known as the **normal derivative**.

With this assumption, we can write down a mathematical version of the fact that given any region R in the disk with piecewise smooth boundary $\mathcal{C}$, the total amount of heat crossing the boundary $\mathcal{C}$ must be 0. This yields the heat conservation equation

$$0 = \int_{\mathcal{C}} \frac{\partial u}{\partial n} \, ds.$$

Here $\partial u / \partial n$ denotes the normal derivative in the outward direction perpendicular to the tangent, and ds indicates integration over arc length along the curve.

Those students comfortable with multivariable calculus will recognize a version of the Divergence Theorem:

$$\int_{\mathcal{C}} \frac{\partial u}{\partial n} \, ds = \int_{R} \Delta u \, dA,$$

where $\Delta u = u_{xx} + u_{yy}$ is the **Laplacian** and dA represents integration with respect to area. Whenever a continuous function integrates to 0 over *every* nice region (say squares or disks), then the function must be 0 everywhere, which leads to the equation $\Delta u = 0$. This is the desired differential equation, except that it is necessary to express the Laplacian in polar coordinates. A (routine but nontrivial) exercise using the multivariate chain rule shows that

$$\Delta u = u_{rr} + \frac{1}{r} u_r + \frac{1}{r^2} u_{\theta\theta}.$$

For the convenience of students unfamiliar with ideas in the previous paragraph, we show how to derive this equation directly. This has the advantage that we can work in polar coordinates and avoid the need for a messy change of variables. The idea is to take R to be the region graphed in Figure 13.1, namely

$$R = \{(r,\theta) : r_0 \leq r \leq r_1, \ \theta_0 \leq \theta \leq \theta_1\}.$$

Then we will divide the integral by $(\theta_1 - \theta_0)(r_1 - r_0)$ and take the limit as θ_1 decreases to θ_0 and r_1 decreases to r_0, which we evaluate using the Fundamental Theorem of Calculus. This will yield the same result.

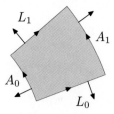

FIGURE 13.1. The region R with boundary and outward normal vectors.

The boundary $\mathcal{C}$ of R consists of two radial line segments

$$L_0 = \{(r, \theta_0) : r_0 \leq r \leq r_1\} \quad \text{and} \quad L_1 = \{(r, \theta_1) : r_0 \leq r \leq r_1\}$$

and two arcs

$$A_0 = \{(r_0, \theta) : \theta_0 \leq \theta \leq \theta_1\} \quad \text{and} \quad A_1 = \{(r_1, \theta) : \theta_0 \leq \theta \leq \theta_1\}.$$

Taking orientation into account, $\mathcal{C} = L_0 + A_1 - L_1 - A_0$. Along L_1, the outward normal at (r, θ_1) is in the θ direction, and arc length along the circle with angle h is rh, thus

$$\frac{\partial u}{\partial n}(r, \theta_1) = \lim_{h \to 0} \frac{u(r, \theta_1 + h) - u(r, \theta_1)}{rh} = \frac{1}{r} u_\theta(r, \theta_1).$$

Along the arc A_1, the outward normal is the radial direction, and thus the normal derivative is $u_r(r_1, \theta)$. The arc length ds along the radii L_0 and L_1 is just dr, while arc length along an arc of radius r is $r\, d\theta$. Thus by adding over the four pieces of the boundary, we obtain the conservation law

$$0 = \int_{r_0}^{r_1} \frac{1}{r}\left(u_\theta(r, \theta_1) - u_\theta(r, \theta_0)\right) dr + \int_{\theta_0}^{\theta_1} u_r(r_1, \theta) r_1\, d\theta - \int_{\theta_0}^{\theta_1} u_r(r_0, \theta) r_0\, d\theta$$

$$= \int_{r_0}^{r_1} \frac{1}{r}\left(u_\theta(r, \theta_1) - u_\theta(r, \theta_0)\right) dr + \int_{\theta_0}^{\theta_1} \left(r_1 u_r(r_1, \theta) - r_0 u_r(r_0, \theta)\right) d\theta.$$

Divide by $\theta_1 - \theta_0$ and take the limit as θ_1 decreases to θ_0. The first term is integrated with respect to r, which is independent of θ, and thus the limit is evaluated using the Leibniz Rule (8.3.4). For the second term, the limit follows from the Fundamental Theorem of Calculus.

$$0 = \int_{r_0}^{r_1} \frac{1}{r} \lim_{\theta_1 \to \theta_0} \frac{u_\theta(r, \theta_1) - u_\theta(r, \theta_0)}{\theta_1 - \theta_0} dr$$

$$+ \lim_{\theta_1 \to \theta_0} \frac{1}{\theta_1 - \theta_0} \int_{\theta_0}^{\theta_1} \left(r_1 u_r(r_1, \theta) - r_0 u_r(r_0, \theta)\right) d\theta$$

$$= \int_{r_0}^{r_1} \frac{1}{r} u_{\theta\theta}(r, \theta_0)\, dr + r_1 u_r(r_1, \theta_0) - r_0 u_r(r_0, \theta_0)$$

Now divide by $r_1 - r_0$ and take the limit as r_1 decreases to r_0. We obtain

$$0 = \lim_{r_1 \to r_0} \frac{1}{r_1 - r_0} \int_{r_0}^{r_1} \frac{1}{r} u_{\theta\theta}(r, \theta_0)\, dr \; + \; \frac{r_1 u_r(r_1, \theta_0) - r_0 u_r(r_0, \theta_0)}{r_1 - r_0}$$

$$= \frac{1}{r_0} u_{\theta\theta}(r_0, \theta_0) + \frac{\partial}{\partial r}(r u_r)(r_0, \theta_0)$$

$$= \frac{1}{r_0} u_{\theta\theta}(r_0, \theta_0) + u_r(r_0, \theta_0) + r_0 u_{rr}(r_0, \theta_0)$$

$$= r_0 \Delta u(r_0, \theta_0).$$

Thus our differential equation with boundary conditions becomes

$$\Delta u := u_{rr} + \frac{1}{r} u_r + \frac{1}{r^2} u_{\theta\theta} = 0 \quad \text{for} \quad 0 \le r < 1,\ -\pi \le \theta \le \pi$$

$$u(1, \theta) = f(\theta) \qquad \text{for} \qquad -\pi \le \theta \le \pi.$$

Exercises for Section 13.1

A. Do the change of variables calculation converting $u_{xx} + u_{yy}$ to polar coordinates.

B. Let $u(r, \theta) = \log r$. Compute Δu and $u(1, \theta)$. Explain why u is not a solution of the heat equation for the boundary function $f(\theta) = 0$.

C. Suppose that u is a solution of the steady-state heat equation for the annulus $A = \{(r, \theta) : r_0 \le r \le r_1\}$.
(a) If u depends only on r, and not on θ, what ODE does u satisfy? Solve the heat equation for the boundary conditions $u(r_0, \theta) = a_0$ and $u(r_1, \theta) = a_1$.
(b) Show that if u depends only on θ, then it is constant.

D. Show that $u(r, \theta) = (3 - 4r^2 + r^4) + (8r^2 - 8r^4)\sin^2 \theta + 8r^4 \sin^4 \theta$ satisfies $\Delta u(r, \theta) = 0$ and $u(1, \theta) = \sin^4 \theta$.

E. Let $S = \{(x, y) : 0 \le x \le 1, y \in \mathbb{R}\}$ and consider the steady-state heat problem on this strip.
(a) Show that $e^{n\pi y} \sin n\pi x$ is a solution for the problem of zero boundary values.
(b) Do you believe that this is a reasonable solution to the physical problem? Discuss.

F. Suppose that an infinite rod has a temperature distribution $u(x, t)$ at the point $x \in \mathbb{R}$ at time $t > 0$. The heat equation is $u_t = u_{xx}$.
(a) Prove that $u(x, t) = \frac{1}{\sqrt{4\pi t}} e^{-x^2/4t}$ is a solution.
(b) Evaluate the total heat at time t: $\int_{-\infty}^{\infty} u(x, t)\, dx$.

HINT: Let $I = \int_{-\infty}^{\infty} e^{-x^2/2}\, dx$. Express I^2 as a double integral over the plane, and convert to polar coordinates. Or use Example 8.3.5.
(c) Evaluate $\lim_{t \to 0} u(x, t)$. Can you give a physical explanation of what this limit represents?

13.2. Formal Solution

The steady-state heat equation is a difficult problem to solve. We approach it first by making the completely unjustified assumption that there will be solutions of a special form. Then we combine the solutions we find to obtain a quite general solution in which all considerations of convergence are ignored. After that, we will work backward and show rigourously that these solutions in fact make good sense and are completely general. So we justify our first steps as *experiments* that lead us to a likely candidate for the solution but do not in themselves constitute a proper derivation of the solution. In subsequent sections, we will use our analysis techniques to justify why it works.

Our method, called **separation of variables**, is to look for solutions of the form $u(r, \theta) = R(r)\Theta(\theta)$, where R is a function only of r and Θ is a function only of θ. The reason for doing this is that it enables us to split the partial differential equation into two ordinary differential equations of a single variable each. Indeed, the DE $\Delta u = 0$ becomes

$$R''(r)\Theta(\theta) + \frac{1}{r}R'(r)\Theta(\theta) + \frac{1}{r^2}R(r)\Theta''(\theta) = 0.$$

Manipulate this by taking all dependence on r to one side of the equation and all dependence on θ to the other to obtain

$$\frac{r^2 R''(r) + rR'(r)}{R(r)} = \frac{-\Theta''(\theta)}{\Theta(\theta)}.$$

The left-hand side does not depend on θ and the right-hand side does not depend on r. As they are equal, they are both independent of all variables and hence are equal to a constant c.

$$\frac{r^2 R''(r) + rR'(r)}{R(r)} = c = \frac{-\Theta''(\theta)}{\Theta(\theta)}.$$

These equations can now be rewritten as

$$\Theta''(\theta) + c\Theta(\theta) = 0$$

and

$$r^2 R''(r) + rR'(r) - cR(r) = 0.$$

The first equation is a well-known linear DE with constant coefficients. We know from Section 12.6 that this DE has a two-parameter space of solutions corresponding to the possible initial values. We can solve the equation $(D^2 + cI)y = 0$ making use of the quadratic equation $x^2 + c = 0$, which has roots $\pm\sqrt{-c}$ if $c < 0$, a double root at 0 for $c = 0$ and two imaginary roots $\pm\sqrt{c}i$ when $c > 0$. Hence by Exercise 12.6.B, we obtain the solutions

$$\Theta(\theta) = A\cos(\sqrt{c}\,\theta) + B\sin(\sqrt{c}\,\theta) \qquad \text{for} \quad c > 0$$
$$\Theta(\theta) = A + B\theta \qquad \text{for} \quad c = 0$$
$$\Theta(\theta) = Ae^{\sqrt{-c}\,\theta} + Be^{-\sqrt{-c}\,\theta} \qquad \text{for} \quad c < 0.$$

However, not all of these solutions fit our problem. Our solutions must be 2π-periodic because $(r, -\pi)$ and (r, π) represent the same point; and more generally (r, θ) and $(r, \theta + 2\pi)$ represent the same point. Hence

$$\Theta(-\pi) = \Theta(\pi) \quad \text{and} \quad \Theta'(-\pi) = \Theta'(\pi).$$

This eliminates the case $c < 0$ and limits the $c = 0$ case to the constant functions. For $c > 0$, this forces $\sqrt{c}$ to be an integer. Hence we obtain

$$\Theta(\theta) = A \cos n\theta + B \sin n\theta \qquad \text{for} \quad c = n^2 \geq 1$$
$$\Theta(\theta) = A \qquad\qquad\qquad\qquad \text{for} \quad c = 0.$$

Now for each $c = n^2$, $n \geq 0$, we must solve the equation

$$r^2 R''(r) + r R'(r) - n^2 R(r) = 0.$$

This is not as easy to solve, but a trick, the substitution $r = e^t$, leads to the answer. Differentiation yields

$$\frac{dR}{dt} = \frac{dR}{dr}\frac{dr}{dt} = R'r$$

and

$$\frac{d^2 R}{dt^2} = \frac{d}{dr}(R'r)\frac{dr}{dt} = (R''r + R')r = r^2 R'' + r R'.$$

Hence our DE becomes $\dfrac{d^2 R}{dt^2} = n^2 R$. This is a linear DE with constant coefficients, which has the solutions

$$R = ae^{nt} + be^{-nt} = ar^n + br^{-n} \qquad \text{for} \quad n \geq 1$$
$$R = a + bt = a + b \log r \qquad\qquad \text{for} \quad n = 0.$$

Again physical considerations demand that R be continuous at $r = 0$. This eliminates the solutions r^{-n} and $\log r$. That leaves the solutions $R(r) = ar^n$ for each $n \geq 0$.

Combining these two solutions for each $c = n^2$ provides the solutions

$$u(r, \theta) = A_n r^n \cos n\theta + B_n r^n \sin n\theta \quad \text{for} \quad n \geq 0.$$

Of course, the case $n = 0$ is special and yields $u(r, \theta) = A_0$. Since the sum of solutions for a homogeneous DE such as ours will also be a solution, we obtain a formal solution (ignoring convergence issues)

$$u(r, \theta) = A_0 + \sum_{n=1}^{\infty} A_n r^n \cos n\theta + B_n r^n \sin n\theta.$$

Continuing to ignore the question of the convergence, we let $r = 1$ and use our boundary condition to obtain

$$f(\theta) = A_0 + \sum_{n=1}^{\infty} A_n \cos n\theta + B_n \sin n\theta.$$

Such a series is called a **Fourier series**, which we have discussed in Section 7.4.

Exercises for Section 13.2

A. (a) Verify that if $\Delta u = 0 = \Delta v$ implies that $\Delta(au + bv) = 0$ for all scalars $a, b \in \mathbb{R}$.
  (b) Solve the DE $y' = -y^2$. Show that the sum of any two solutions can never be a solution.
  (c) Explain the difference in these two situations.

B. Adapt the method of this section (i.e., separation of variables) to find the possible solutions of $\Delta u(r, \theta) = 0$ on the region $U = \{(r, \theta) : r \geq 1\}$ that are continuous on $\overline{U}$ and are **continuous at infinity** in the sense that $\lim_{r \to \infty} u(r, \theta) = L$ exists, independent of θ.

C. Let $HS = \{(x, y) : 0 \leq x \leq 1,\ y \geq 0\}$, and consider the steady-state heat problem on HS with boundary conditions $u(0, y) = u(1, y) = 0$ and $u(x, 0) = x - x^2$.
  (a) Use separation of variables to obtain a family of basic solutions.
  (b) Show that the conditions on the two infinite bounding lines restricts the possible solutions. If in addition, you stipulate that the solution must be bounded, express the resulting solution as a formal series.
  (c) What does the boundary condition on $[0, 1]$ become for this formal series?

D. Consider a circular drum membrane of radius 1. At time t, the point (r, θ) on the surface has a vertical deviation of $u(r, \theta, t)$. The wave equation for the motion is $u_{tt} = c^2 \Delta u$, where c is a constant. In this exercise, we will only consider solutions that have radial symmetry (no dependence on θ).
  (a) What boundary condition should apply to $u(1, \theta, t)$?
  (b) Look for solutions to the PDE of the form $u(r, \theta, t) = R(r)T(t)$. Use separation of variables to obtain ODEs for R and T. An unknown constant must be introduced.
  (c) What conditions on the ODE for T are needed to guarantee that T remains bounded (a reasonable physical hypothesis)?
  (d) The DE for R is called Bessel's DE. What degeneracy of the DE requires us to add another condition that R remain bounded at $r = 0$?

13.3. Orthogonality Relations

The next step in our heuristic development is to determine the coefficients A_n and B_n in the Fourier series given in the last section. To do this, we use the natural inner product on $C[-\pi, \pi]$ given by

$$\langle f, g \rangle = \frac{1}{2\pi} \int_{-\pi}^{\pi} f(\theta)g(\theta)\, d\theta.$$

In Section 7.4, we showed that the functions $\{1, \sqrt{2}\cos n\theta, \sqrt{2}\sin n\theta : n \geq 1\}$ form an orthonormal set in $C[-\pi, \pi]$ with this inner product.

Moreover, we used the orthogonality relations to show that for a trigonometric polynomial $p(\theta) = A_0 + \sum_{k=1}^{n} A_k \cos k\theta + B_k \sin k\theta$, we can recover all of the coefficients from the inner products $A_0 = \langle p, 1 \rangle$, $A_k = \langle p, 2\cos k\theta \rangle$ and $B_k = \langle p, 2\sin k\theta \rangle$ for $k \geq 1$. This was the motivation for defining Fourier series of

$f \in C[-\pi, \pi]$ to be

$$f \sim A_0 + \sum_{k=1}^{\infty} A_k \cos k\theta + B_k \sin k\theta,$$

where $A_0 = \langle f, 1 \rangle = \dfrac{1}{2\pi} \displaystyle\int_{-\pi}^{\pi} f(t)\, dt$, $\quad A_k = \langle f, 2\cos k\theta \rangle = \dfrac{1}{\pi} \displaystyle\int_{-\pi}^{\pi} f(t) \cos kt\, dt$

and $B_k = \langle f, 2\sin k\theta \rangle = \dfrac{1}{\pi} \displaystyle\int_{-\pi}^{\pi} f(t) \sin kt\, dt$ for $k \geq 1$.

We have said nothing yet about the convergence of this series. There are serious difficulties that need to be addressed. We also note that this definition makes sense when f is not continuous provided that

$$\|f\|_1 := \frac{1}{2\pi} \int_{-\pi}^{\pi} |f(\theta)|\, d\theta < \infty.$$

Such functions are called **absolutely integrable functions**. In particular, Fourier series of piecewise continuous functions are defined. We will need the following easy estimate.

13.3.1. LEMMA. *If f is absolutely integrable on $[-\pi, \pi]$, then*

$$|A_0| \leq \|f\|_1, \qquad |A_n| \leq 2\|f\|_1 \quad and \quad |B_n| \leq 2\|f\|_1 \quad for \quad n \geq 1.$$

Since $\|f\|_1 \leq \|f\|_\infty$, it follows that if $f \in C[-\pi, \pi]$, then its Fourier coefficients are bounded.

PROOF. This is routine. For example,

$$|B_n| \leq \frac{1}{\pi} \int_{-\pi}^{\pi} |f(\theta) \sin n\theta|\, d\theta \leq \frac{1}{\pi} \int_{-\pi}^{\pi} |f(\theta)|\, d\theta = 2\|f\|_1.$$

Moreover, it is evident that if f is bounded,

$$\|f\|_1 = \frac{1}{2\pi} \int_{-\pi}^{\pi} |f(\theta)|\, d\theta \leq \frac{1}{2\pi} \int_{-\pi}^{\pi} \|f\|_\infty\, d\theta = \|f\|_\infty.$$

So continuous functions are absolutely integrable, and thus the Fourier coefficients are bounded by $2\|f\|_\infty$. ∎

13.3.2. EXAMPLE. Consider the Fourier series of the function $f(\theta) = |\theta|$ for $-\pi \leq \theta \leq \pi$. First note that f is even, so that $f(\theta) \sin n\theta$ is an odd function for all $n \geq 1$. Hence $B_n = 0$ for all $n \geq 1$ (Exercise 13.3.G). For A_n we compute

$$A_0 = \frac{1}{2\pi} \int_{-\pi}^{\pi} |t|\, dt = \frac{1}{\pi} \int_0^{\pi} t\, dt = \frac{\pi}{2},$$

and, using integration by parts,

$$A_n = \frac{1}{\pi} \int_{-\pi}^{\pi} |t| \cos nt \, dt = \frac{2}{\pi} \int_{0}^{\pi} t \cos nt \, dt$$

$$= \frac{2t}{\pi} \left. \frac{\sin nt}{n} \right|_0^\pi - \frac{2}{\pi} \int_0^\pi 1 \frac{\sin nt}{n} \, dt = 0 + \frac{2 \cos nt}{\pi n^2} \bigg|_0^\pi$$

$$= \frac{2}{\pi n^2} \left((-1)^n - 1 \right) = \begin{cases} 0 & \text{if } n \text{ is even} \\ -\dfrac{4}{\pi n^2} & \text{if } n \text{ is odd.} \end{cases}$$

Thus the Fourier series is

$$|\theta| \sim \frac{\pi}{2} - \frac{4}{\pi} \sum_{k=0}^{\infty} \frac{\cos (2k+1)\theta}{(2k+1)^2}.$$

Exercises for Section 13.3

A. Find the Fourier series of the following functions:
   (a) $f(\theta) = \cos^3(\theta)$
   (b) $f(\theta) = |\sin \theta|$
   (c) $f(\theta) = \theta$ for $-\pi \le \theta \le \pi$

B. Find the Fourier expansion for the function $u(r, \theta)$ of Exercise 13.1.D with boundary function $f(\theta) = \sin^4 \theta$.

C. Verify the remaining parts of Lemma 7.4.5.

D. (a) Suppose that $f(\theta)$ is a 2π-periodic function with known Fourier series. Let α be a real number, and let $g(\theta) = f(\theta - \alpha)$ for $\theta \in \mathbb{R}$. Find the Fourier series of g.
   (b) Combine part (a) with A(b) to find the Fourier series of $|\cos \theta|$.

E. (a) Let $f(\theta)$ be a 2π-periodic function with a given Fourier series. Let $g(\theta) = f(-\theta)$ for $\theta \in \mathbb{R}$. Find the Fourier series of g.
   (b) Suppose that h is a 2π-periodic function such that $h(\pi - \theta) = h(\theta)$. What does this imply about its Fourier series?

F. (a) Compute the Fourier series of $f(\theta) = \begin{cases} 1 - |\theta| & \text{for } -1 \le \theta \le 1 \\ 0 & \text{otherwise} \end{cases}$.

   (b) Compute the Fourier series of $g(\theta) = \begin{cases} 1 & \text{for } -1 \le \theta \le 0 \\ -1 & \text{for } \ \ 0 < \theta \le 1. \\ 0 & \text{otherwise} \end{cases}$

   (c) What relationship do you see between these two functions and series?

G. Show that if $f \in C[-\pi, \pi]$ is an odd function, then the Fourier series of f involves only functions of the form $\sin k\theta$. Similarly, if $f \in C[-\pi, \pi]$ is an even function, then the Fourier series of f involves only the constant term and the cosine functions.

H. For $f \in C[-\pi, \pi]$, define $f_e(x) = \frac{1}{2}(f(\theta) + f(-\theta))$ and $f_o(x) = \frac{1}{2}(f(\theta) - f(-\theta))$. Compute the Fourier series of f_e and f_o in terms of the series for f.

I.   Show that $f_0(\theta) = 1$ and $f_n(\theta) = \sqrt{2}\cos n\theta$ on $0 \le \theta \le \pi$ for $n \ge 1$ is an orthonormal set in $C[0, \pi]$ for the inner product $\langle f, g \rangle = \dfrac{1}{\pi}\displaystyle\int_0^\pi f(\theta)g(\theta)\,d\theta$.

J.   (a) Find an inner product on $C[0, 1]$ so that $\{\sqrt{2}\sin n\pi x : n \ge 1\}$ is an orthonormal set, and verify that this set is orthonormal for your choice of inner product.
   (b) In Exercise 13.2.C, find the coefficients necessary to satisfy the boundary condition on the unit interval.

13.4.   Convergence in the Open Disk

It is now time to return to the original problem and systematically analyze our proposed solutions. The behaviour of the function u on the open disk

$$\mathbb{D} = \{(r, \theta) : 0 \le r < 1,\ -\pi \le \theta \le \pi\}$$

is much easier than the analysis of the boundary behaviour. We deal with that first, and we will find that it leads to a method for understanding the Fourier series of f. Often results about the behaviour of u on $\mathbb{D}$ will follow from a stronger result on each smaller closed disk

$$\overline{\mathbb{D}}_R = \{(r, \theta) : 0 \le r \le R,\ -\pi \le \theta \le \pi\} \quad \text{for} \quad R < 1.$$

13.4.1. PROPOSITION.   *Let f be an absolutely integrable function on $[-\pi, \pi]$ with Fourier series $f \sim A_0 + \sum_{n=1}^{\infty} A_n \cos n\theta + B_n \sin n\theta$. Then the series*

$$A_0 + \sum_{n=1}^{\infty} A_n r^n \cos n\theta + B_n r^n \sin n\theta$$

converges uniformly on $\overline{\mathbb{D}}_R$ for any $R < 1$ and thus converges everywhere on the open disk $\mathbb{D}$ to a continuous function $u(r, \theta)$.

PROOF. This follows from the Weierstrass M-test (8.4.7). Indeed,

$$\left\| A_n r^n \cos n\theta + B_n r^n \sin n\theta \right\|_{\overline{\mathbb{D}}_R} = \max_{(r,\theta) \in \overline{\mathbb{D}}_R} \left| A_n r^n \cos n\theta + B_n r^n \sin n\theta \right|$$

$$\le (|A_n| + |B_n|) R^n \le 4\|f\|_1 R^n.$$

Since

$$\sum_{n=0}^{\infty} 4\|f\|_1 R^n = \frac{4\|f\|_1}{1 - R} < \infty,$$

the M-test guarantees that the series converges uniformly on $\overline{\mathbb{D}}_R$. The uniform limit of continuous functions is continuous by Theorem 8.2.1. Therefore, $u(r, \theta)$ is continuous on $\overline{\mathbb{D}}_R$. Since this is true for each $R < 1$, u is continuous on the whole *open* disk $\mathbb{D}$. ∎

Now we extend this argument to apply to the various partial derivatives of u. This procedure justifies **term-by-term differentiation** under appropriate conditions on the convergence.

13.4.2. LEMMA. *Suppose that $u_n(x,y)$ are C^1 functions on an open set R for $n \geq 0$ such that $\sum_{n=0}^{\infty} u_n(x,y)$ converges uniformly to $u(x,y)$ and $\sum_{n=0}^{\infty} \frac{\partial}{\partial x} u_n(x,y)$ converges uniformly to $v(x,y)$. Then $\frac{\partial}{\partial x} u(x,y) = v(x,y)$.*

PROOF. It is enough to verify the theorem on a small square about an arbitrary point (x_0, y_0) in R. By limiting ourselves to a square, the whole line segment from (x_0, y) to (x, y) will lie in R. We define functions on this small square by

$$w_n(x,y) = \sum_{k=0}^{n} u_k(x,y) = \sum_{k=0}^{n} u_k(x_0, y) + \int_{x_0}^{x} \frac{\partial}{\partial x} \sum_{k=0}^{n} u_k(t, y) \, dt$$

and

$$w(x,y) = u(x_0, y) + \int_{x_0}^{x} v(t, y) \, dt.$$

Then since the integrands $\frac{\partial}{\partial x} \sum_{k=0}^{n} u_n(t, y)$ converge uniformly to $v(t, y)$, Corollary 8.3.2 shows that $w_n(x,y)$ converges uniformly to $w(x,y)$. But this limit is $u(x,y)$. Therefore, by the Fundamental Theorem of Calculus,

$$\frac{\partial}{\partial x} u(x,y) = \frac{\partial}{\partial x} w(x,y)$$
$$= \frac{\partial}{\partial x} \left(u(x_0, y) + \int_{x_0}^{x} v(t, y) \, dt \right)(x, y) = v(x, y). \qquad \blacksquare$$

We may apply this to our function $u(r, \theta)$.

13.4.3. THEOREM. *Let f be an absolutely integrable function with associated function $u(r, \theta) = A_0 + \sum_{n=1}^{\infty} A_n r^n \cos n\theta + B_n r^n \sin n\theta$. Then u satisfies the heat equation $\Delta u(r, \theta) = 0$ in the open disk $\mathbb{D}$.*

PROOF. Let $u_n(r, \theta) = A_n r^n \cos n\theta + B_n r^n \sin n\theta$. Then

$$\frac{\partial}{\partial r} u_n(r, \theta) = n A_n r^{n-1} \cos n\theta + n B_n r^{n-1} \sin n\theta$$
$$\frac{\partial^2}{\partial r^2} u_n(r, \theta) = n(n-1) A_n r^{n-2} \cos n\theta + n(n-1) B_n r^{n-2} \sin n\theta$$
$$\frac{\partial}{\partial \theta} u_n(r, \theta) = -n A_n r^n \sin n\theta + n B_n r^n \cos n\theta$$
$$\frac{\partial^2}{\partial \theta^2} u_n(r, \theta) = -n^2 A_n r^n \cos n\theta - n^2 B_n r^n \sin n\theta.$$

We will apply the M-test to the series $\sum_{n=0}^{\infty} u_n(r, \theta)$ and to each series of partial derivatives. Indeed, on the disk $\overline{\mathbb{D}_R}$ for $0 \leq R < 1$, each of the preceding terms is uniformly bounded by $4\|f\|_1 n^2 R^{n-2}$. Apply the Ratio Test (Exercise 3.2.I) to the series $\sum_{n=0}^{\infty} 4\|f\|_1 n^2 R^{n-2}$:

$$\lim_{n\to\infty} \frac{4\|f\|_1(n+1)^2 R^{n-1}}{4\|f\|_1 n^2 R^{n-2}} = R < 1.$$

Thus this series converges; and therefore it follows from the Weierstrass M-test that each series of partial derivatives converges uniformly on $\overline{\mathbb{D}}_R$.

Therefore, by Lemma 13.4.2, the partial derivative of the sum equal the sum of the partial derivatives. This means that

$$\frac{\partial}{\partial r} u(r,\theta) = \sum_{n=0}^{\infty} \frac{\partial}{\partial r} u_n(r,\theta)$$

$$\frac{\partial^2}{\partial r^2} u(r,\theta) = \sum_{n=0}^{\infty} \frac{\partial^2}{\partial r^2} u_n(r,\theta)$$

$$\frac{\partial}{\partial \theta} u(r,\theta) = \sum_{n=0}^{\infty} \frac{\partial}{\partial \theta} u_n(r,\theta)$$

$$\frac{\partial^2}{\partial \theta^2} u(r,\theta) = \sum_{n=0}^{\infty} \frac{\partial^2}{\partial \theta^2} u_n(r,\theta).$$

This convergence is uniform on every disk $\overline{\mathbb{D}}_R$ for $R < 1$.

Using $\Delta u = u_{rr} + \frac{1}{r} u_r + \frac{1}{r^2} u_{\theta\theta}$, we deduce from Lemma 13.4.2 that

$$\Delta u(r,\theta) = \sum_{n=0}^{\infty} \Delta u_n(r,\theta)$$

and that this convergence is uniform on any disk $\overline{\mathbb{D}}_R$. However, we constructed the functions $r^n \cos n\theta$ and $r^n \sin n\theta$ as solutions of $\Delta u = 0$. Thus the right-hand side of this equation is zero. Hence $\Delta u(r,\theta) = 0$ everywhere in the open disk $\mathbb{D}$. ∎

13.4.4. DEFINITION. A function u such that $\Delta u = 0$ is called a **harmonic function**. The function $u(r,\theta)$ determined by the Fourier series of a 2π-periodic f is called the **harmonic extension** of f.

Exercises for Section 13.4

A. Find the Fourier series of $f(\theta) = \theta^2$ for $-\pi \le \theta \le \pi$.
  (a) Show that the series for $u(r,\theta)$ converges uniformly on the *closed* disk $\overline{\mathbb{D}}$.
  (b) Show that the series for $u_{\theta\theta}$ does not converge *uniformly* on $\mathbb{D}$.

B. Let A_n and B_n be the Fourier coefficients of a continuous function $f(\theta)$.
  (a) Show that $v_t(r,\theta) = u(rt,\theta) = A_0 + \sum_{n=1}^{\infty} A_n(rt)^n \cos n\theta + B_n r^n \sin n\theta$ converges uniformly on the closed disk $\overline{\mathbb{D}}$ for $0 < t < 1$.
  (b) Show that v_t converges uniformly to u on $\mathbb{D}$ as $t \to 1^-$ if and only if u extends to a continuous function on $\overline{\mathbb{D}}$.

C. (a) Suppose that a 2π-periodic function f is C^1 (i.e., f' is continuous), then the Fourier coefficients of f satisfy $|A_n| + |B_n| \le Cn^{-1}$ for some constant C.
HINT: Integrate by parts and compare with Lemma 13.3.1.
(b) Show by induction that if f is C^k, then $|A_n| + |B_n| \le Cn^{-k}$ for some constant C.

D. Suppose $f \sim A_0 + \sum_{n \ge 1} A_n \cos n\theta + B_n \sin n\theta$ is a 2π-periodic function and that $\sum_{n \ge 1} |A_n| + |B_n| < \infty$. Prove that the Fourier series converges uniformly to f.

E. Suppose $f \sim A_0 + \sum_{n \ge 1} A_n \cos n\theta + B_n \sin n\theta$ is a 2π-periodic function and that $\sum_{n \ge 1} n|A_n| + n|B_n| < \infty$. By modifying the proof of Lemma 13.4.2, establish that $\sum_{n \ge 1} nA_n \sin n\theta + nB_n \cos n\theta$ converges uniformly to f'.

F. The Fourier coefficients of a 2π-periodic function f satisfy $|A_n| + |B_n| \le Cn^{-k}$ for an integer $k \ge 2$ and some constant C. Show that f is in the class C^{k-2}.
HINT: Use term-by-term differentiation and the M-test.

G. Show that $u(r, \theta)$ has partial derivatives $\frac{\partial^{j+k}}{\partial r^j \partial \theta^k} u$ of all orders. Hence show that u is C^∞.
HINT: First verify this for the functions $u_n(r, \theta)$. Then show that the sequence of partial derivatives converges uniformly on each $\mathbb{D}_R$.

H. Give necessary and sufficient conditions for $A_0 + \sum_{n=1}^{\infty} A_n \cos n\theta + B_n \sin n\theta$ to be the Fourier series of a C^∞ function.
HINT: Combine Exercises B and E.

13.5. The Poisson Formula

The next step is to find a formula for $u(r, \theta)$ in terms of the boundary function $f(\theta)$. The basic idea is to substitute the formula for each Fourier coefficient and interchange the order of the summation and integration. This again requires uniform convergence.

Compute $u(r, \theta)$ for a *bounded* integrable function f:

$$u(r, \theta) = A_0 + \sum_{n=1}^{\infty} A_n r^n \cos n\theta + B_n r^n \sin n\theta$$

$$= \frac{1}{2\pi} \int_{-\pi}^{\pi} f(t)\, dt + \frac{1}{\pi} \sum_{n=1}^{\infty} \int_{-\pi}^{\pi} f(t) \cos nt\, dt\, r^n \cos n\theta + \int_{-\pi}^{\pi} f(t) \sin nt\, dt\, r^n \sin n\theta$$

$$= \frac{1}{2\pi} \int_{-\pi}^{\pi} f(t)\, dt + \sum_{n=1}^{\infty} \frac{1}{\pi} \int_{-\pi}^{\pi} f(t) r^n \left(\cos nt \cos n\theta + \sin nt \sin n\theta\right) dt$$

$$= \frac{1}{2\pi} \int_{-\pi}^{\pi} f(t)\, dt + \sum_{n=1}^{\infty} \frac{1}{\pi} \int_{-\pi}^{\pi} f(t) r^n \cos n(\theta - t)\, dt.$$

Yet again, we apply the M-test. Since

$$\|f(t)r^n \cos n(\theta - t)\| \le \|f\|_\infty R^n \quad \text{for} \quad 0 \le r \le R,$$

and $\sum_{n=0}^{\infty} \|f\|_\infty R^n = \|f\|_\infty/(1 - R) < \infty$, the series converges uniformly on $\overline{\mathbb{D}}_R$. Therefore, by Theorem 8.3.1, we can interchange the order of the summation and the integral

$$u(r, \theta) = \int_{-\pi}^{\pi} f(t) \frac{1}{2\pi} \left(1 + 2 \sum_{n=1}^{\infty} r^n \cos n(\theta - t) \right) dt$$

$$= \int_{-\pi}^{\pi} f(t) P(r, \theta - t) \, dt = \int_{-\pi}^{\pi} f(\theta - u) P(r, u) \, du.$$

We have introduced the function

$$P(r, \theta) = \frac{1}{2\pi} \left(1 + 2 \sum_{n=1}^{\infty} r^n \cos n\theta \right) \quad \text{for} \quad 0 \le r < 1, \, \theta \in \mathbb{R}.$$

Notice that the last step is a change of variables in which we make use of the fact that both f and P are 2π-periodic in θ.

The function $P(r, \theta)$ is known as the **Poisson kernel**. The purpose of this section is to develop its basic properties. The next theorem is known as the **Poisson formula**.

We will use the fact that exponentiation and trigonometric functions are related using complex variables. (See Appendix 13.10.) In particular, $e^{it} = \cos t + i \sin t$ and consequently

$$\cos nt = \frac{e^{int} + e^{-int}}{2} \quad \text{and} \quad \sin nt = \frac{e^{int} - e^{-int}}{2i}.$$

13.5.1. THEOREM. *Let f be a bounded integrable function on $[-\pi, \pi]$. Then for $0 \le r < 1$ and $-\pi \le \theta \le \pi$, the harmonic extension of f is given by*

$$u(r, \theta) = \int_{-\pi}^{\pi} f(\theta - t) P(r, t) \, dt,$$

where $P(r, t) = \dfrac{1}{2\pi} \dfrac{1 - r^2}{1 - 2r \cos t + r^2}.$

PROOF. From the discussion preceding the theorem, it only remains to evaluate the series for $P(r, t)$. Changing the cosines to complex exponentials allows us to sum a geometric series.

$$2\pi P(r, t) = 1 + 2 \sum_{n=1}^{\infty} r^n \cos nt = 1 + \sum_{n=1}^{\infty} r^n (e^{int} + e^{-int})$$

$$= 1 + \sum_{n=1}^{\infty} (re^{it})^n + \sum_{n=1}^{\infty} (re^{-it})^n$$

$$= 1 + \frac{re^{it}}{1 - re^{it}} + \frac{re^{-it}}{1 - re^{-it}}$$

$$= \frac{(1 - re^{it} - re^{-it} + r^2) + (re^{it} - r^2) + (re^{-it} - r^2)}{(1 - re^{it})(1 - re^{-it})}$$

$$= \frac{1 - r^2}{1 - r(e^{it} + e^{-it}) + r^2}$$

$$= \frac{1 - r^2}{1 - 2r \cos t + r^2} \qquad \blacksquare$$

The Poisson kernel has a number of very nice properties. The most important is that $P(r, t)$ is positive and integrates to 1. Hence $u(r, \theta)$ is a *weighted average* of the values $f(\theta - t)$. The function $P(r, t)$ peaks dramatically at $t = 0$ when r is close to 1, and thus eventually $u(r, \theta)$ depends mostly on the values of $f(u)$ for u close to θ. See Figure 13.2.

13.5.2. PROPERTIES OF THE POISSON KERNEL.
For $0 \le r < 1$ and $-\pi \le t \le \pi$,

(1) $P(r, t) > 0$.

(2) $P(r, -t) = P(r, t)$.

(3) $\displaystyle\int_{-\pi}^{\pi} P(r, t) \, dt = 1$.

(4) $P(r, t)$ *is decreasing in t on $[0, \pi]$ for fixed r.*

(5) *For any $\delta > 0$,* $\displaystyle\lim_{r \to 1^-} \max_{\delta \le |t| \le \pi} P(r, t) = 0$.

PROOF. Statements (1) and (2) are routine. For (3), take $f = 1$. The Fourier series of f is $A_0 = 1$ and $A_n = B_n = 0$ for $n \ge 1$. Hence $u(r, \theta) = 1$. Plugging this into the Poisson formula, we obtain

$$1 = u(r, \theta) = \int_{-\pi}^{\pi} P(r, t) \, dt.$$

For fixed r, the numerator $1 - r^2$ of $P(r, t)$ is constant while the denominator $1 - 2r \cos t + r^2$ is monotone increasing on $[0, \pi]$; whence $P(r, t)$ is monotone decreasing in t. This verifies (4). Finally, from (2) and (4), we see that

$$\lim_{r \to 1^-} \max_{\delta \le |t| \le \pi} P(r, t) = \lim_{r \to 1^-} P(r, \delta) = \lim_{r \to 1^-} \frac{1 - r^2}{1 - 2r \cos \delta + r^2}$$

$$= \frac{0}{2(1 - \cos \delta)} = 0. \qquad \blacksquare$$

In the next chapter, we will develop two other integral kernels, the Dirichlet kernel and the Fejér kernel. The reader may want to compare their properties and compare Figures 13.2, 14.2, and 14.4.

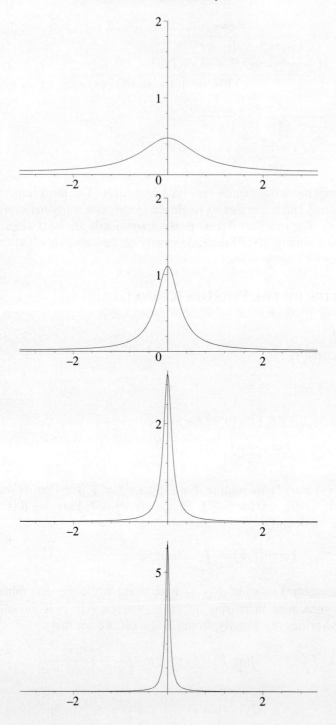

FIGURE 13.2. Graphs of $P(r, t)$ for $r = 1/2, 3/4, 9/10$, and $19/20$.

Exercises for Section 13.5

A. Compute the Fourier series of $f(\theta) = P(s, \theta)$ for $0 \leq s < 1$.

B. Find an explicit value of $r < 1$ for which $\int_{-.01}^{.01} P(r, t) \, dt > 0.999$.

C. Prove that $\dfrac{1 - r}{1 + r} \leq P(r, \theta) \leq \dfrac{1 + r}{1 - r}$.

D. Let f be a *positive* continuous 2π-periodic function with harmonic extension $u(r, \theta)$.
   (a) Prove that $u(r, \theta) \geq 0$.
   (b) Prove **Harnack's inequality**:
$$\frac{1 - r}{1 + r} u(0, 0) \leq u(r, \theta) \leq \frac{1 + r}{1 - r} u(0, 0).$$

   HINT: Use the previous exercise.

E. Suppose that f is a 2π-periodic continuous function such that $L \leq f(\theta) \leq M$ for $-\pi \leq \theta \leq \pi$. Let $u(r, \theta)$ be the Poisson integral of f. Show that $L \leq u(r, \theta) \leq M$ for $0 \leq r \leq 1$ and $-\pi \leq \theta \leq \pi$.

F. Show that if $f(\theta)$ is absolutely integrable on $[-\pi, \pi]$ and $g_n(\theta)$ are continuous functions that converge uniformly to $g(\theta)$ on $[-\pi, \pi]$, then
$$\lim_{n \to \infty} \int_{-\pi}^{\pi} f(\theta) g_n(\theta) \, d\theta = \int_{-\pi}^{\pi} f(\theta) g(\theta) \, d\theta.$$

   HINT: Look at the proof of Theorem 8.3.1.

G. Use the previous exercise to show that the Poisson formula is valid for absolutely integrable functions on $[-\pi, \pi]$.

H. Prove that $\displaystyle\int_{-\pi}^{\pi} P(r, \theta - t) P(s, t) \, dt = P(rs, \theta)$.
   HINT: Use the series expansion of the Poisson kernel.

I. Let $f(\theta)$ be a continuous 2π-periodic function. Let $u(r, \theta)$ be the harmonic extension of f. Let $0 < s < 1$, and define $g(\theta) = u(s, \theta)$. Prove that the harmonic extension of g is $u(rs, \theta)$.
   HINT: Use the Poisson formula twice to obtain the harmonic extension of g as a double integral, and interchange the order of integration.

13.6. Poisson's Theorem

Using the properties of the Poisson kernel, it is now possible to show that $u(r, \theta)$ approaches $f(\theta)$ uniformly as r tends to 1. This means that our proposed solution to the heat problem is continuous on the closed disk and has the desired boundary values. This puts us very close to solving the steady-state heat problem. It also provides a stronger reason for calling $u(r, \theta)$ an extension of f.

13.6.1. POISSON'S THEOREM.

Let f be a continuous 2π-periodic function on $[-\pi, \pi]$ with harmonic extension $u(r, \theta)$. Then $f_r(\theta) := u(r, \theta)$ converges uniformly to f on $[-\pi, \pi]$ as $r \to 1^-$. Consequently, the function $u(r, \theta)$ may be defined on the boundary of the closed disk by $u(1, \theta) = f(\theta)$ to obtain a continuous function on $\overline{\mathbb{D}}$ that is harmonic on $\mathbb{D}$ and agrees with f on the boundary.

PROOF. Let $\varepsilon > 0$ be given. We will find an $r_0 < 1$ so that

$$|f(\theta) - f_r(\theta)| \leq \varepsilon \quad \text{for all} \quad r_0 \leq r < 1, \ -\pi \leq \theta \leq \pi,$$

which will establish uniform convergence.

Let $M = \|f\|_\infty$ and set $\varepsilon_0 = (4\pi M + 1)^{-1}\varepsilon$. By Theorem 5.5.9, f is uniformly continuous. Therefore, there is a $\delta > 0$ such that

$$|f(\theta) - f(t)| \leq \varepsilon_0 \quad \text{for all} \quad |\theta - t| \leq \delta.$$

By property (5), there is an $R < 1$ such that

$$\max_{\delta \leq |t| \leq \pi} P(r, t) \leq \varepsilon_0 \quad \text{for all} \quad R \leq r < 1.$$

Now, using the Poisson formula, compute

$$f(\theta) - f_r(\theta) = f(\theta) \int_{-\pi}^{\pi} P(r, t)\, dt - \int_{-\pi}^{\pi} f(\theta - t) P(r, t)\, dt$$

$$= \int_{-\pi}^{\pi} \big(f(\theta) - f(\theta - t)\big) P(r, t)\, dt.$$

For $R \leq r < 1$, split this integral into two pieces to estimate $|f(\theta) - f_r(\theta)|$

$$\leq \int_{-\delta}^{\delta} |f(\theta) - f(\theta - t)| P(r, t)\, dt + \int_{-\pi}^{-\delta} + \int_{\delta}^{\pi} |f(\theta) - f(\theta - t)| P(r, t)\, dt$$

$$\leq \int_{-\delta}^{\delta} \varepsilon_0 P(r, t)\, dt + \int_{-\pi}^{-\delta} + \int_{\delta}^{\pi} 2M\varepsilon_0\, dt$$

$$\leq \varepsilon_0 + 2\pi(2M\varepsilon_0) = (4\pi M + 1)\varepsilon_0 = \varepsilon.$$

This establishes that $\|f - f_r\|_\infty < \varepsilon$ for $R \leq r < 1$. Hence $f_r(\theta)$ converges uniformly to $f(\theta)$ on $[-\pi, \pi]$.

Extend the definition of $u(r, \theta)$ to the boundary of the closed disk by setting $u(1, \theta) = f(\theta)$. The previous paragraphs show that this function is continuous on the closed disk $\overline{\mathbb{D}}$. By Theorem 13.4.3, u is harmonic on the interior of the disk. ∎

This results yields several important consequences quite easily. The first is the existence of a solution to the heat problem. The question of uniqueness of the solution will be handled in the next section.

13.6.2. COROLLARY. *Let f be a continuous 2π-periodic function. Then the steady-state heat equation $\Delta u = 0$ and $u(1, \theta) = f(\theta)$ has a continuous solution.*

PROOF. By Theorem 13.4.3, the function $u(r, \theta)$ is differentiable on the open disk $\mathbb{D}$ and satisfies $\Delta u = 0$. By Poisson's Theorem, u extends to be continuous on the closed unit disk $\overline{\mathbb{D}}$ and attains the boundary values $u(1, \theta) = f(\theta)$. ∎

This next corollary shows that the Fourier series of a continuous function determines the function uniquely. Continuity is essential for this result. For example, the function that is zero except for a discontinuous point $f(0) = 1$ will have the zero Fourier series but is not the zero function.

13.6.3. COROLLARY. *If two continuous 2π-periodic functions on $\mathbb{R}$ have equal Fourier series, then they are equal functions.*

PROOF. Suppose that f and g have the same Fourier series. Since the harmonic extension $u(r, \theta)$ is defined only in terms of the Fourier coefficients, this is the same function for both f and g. Now, by Poisson's Theorem,

$$f(\theta) = \lim_{r \to 1^-} u(r, \theta) = g(\theta) \quad \text{for} \quad -\pi \le \theta \le \pi.$$

Thus $f = g$. ∎

The final application concerns the case in which the Fourier series itself converges uniformly. This is not always the case, and the delicate question of the convergence of the Fourier series will be dealt with later.

13.6.4. COROLLARY. *Suppose that the Fourier series*

$$f \sim A_0 + \sum_{n=1}^{\infty} A_n \cos n\theta + B_n \sin n\theta$$

of a continuous function f converges uniformly. Then the series in fact converges uniformly to f.

PROOF. Let the uniform limit be g. By Theorem 8.2.1, g is continuous; and it is 2π-periodic as it is the limit of 2π-periodic functions. Compute the Fourier series of g using Theorem 8.3.1:

$$\frac{1}{\pi} \int_{-\pi}^{\pi} g(t) \sin nt \, dt = \lim_{N \to \infty} \frac{1}{\pi} \int_{-\pi}^{\pi} \left(A_0 + \sum_{k=1}^{N} A_k \cos k\theta + B_k \sin k\theta \right) \sin nt \, dt$$

$$= \lim_{N \to \infty} B_n = B_n.$$

The other coefficients are computed in the same way. It follows that f and g have the same Fourier series. Thus by the previous corollary, they are equal. That is, the Fourier series converges uniformly to f. ∎

13.6.5. EXAMPLE. Recall Example 13.3.2. It was shown that

$$|\theta| \sim \frac{\pi}{2} - \frac{4}{\pi} \sum_{k=0}^{\infty} \frac{\cos((2k+1)\theta)}{(2k+1)^2}.$$

Since $\frac{\pi}{2} + \frac{4}{\pi} \sum_{k=0}^{\infty} \frac{1}{(2k+1)^2}$ converges (by the integral test or by comparison with $\frac{1}{n^2}$),
it follows from the Weierstrass M-test (8.4.7) that this series converges uniformly.
By Corollary 13.6.4, it follows that the Fourier series converges uniformly to $|\theta|$.

Let us evaluate this at $\theta = 0$.

$$0 = \frac{\pi}{2} - \frac{4}{\pi} \sum_{k=0}^{\infty} \frac{1}{(2k+1)^2}$$

Therefore,

$$\sum_{k=0}^{\infty} \frac{1}{(2k+1)^2} = \frac{\pi^2}{8}.$$

A little manipulation yields that

$$\sum_{k=0}^{\infty} \frac{1}{k^2} = \sum_{k=0}^{\infty} \frac{1}{(2k+1)^2} + \sum_{k=0}^{\infty} \frac{1}{(2k)^2} = \frac{\pi^2}{8} + \frac{1}{4} \sum_{k=0}^{\infty} \frac{1}{k^2}.$$

Solving, we obtain a famous identity due to Euler (by different methods):

$$\sum_{k=0}^{\infty} \frac{1}{k^2} = \frac{\pi^2}{6}.$$

We conclude this section with an important application of Poisson's Theorem to approximation by trigonometric polynomials. Another proof of this fact will be given in Theorem 14.6.4.

13.6.6. COROLLARY. *Every continuous 2π-periodic functions is the uniform limit of a sequence of trigonometric polynomials.*

PROOF. Let f be a continuous 2π-periodic function. Given $n \in \mathbb{N}$, Poisson's Theorem (Theorem 13.6.1) shows that there is an $r < 1$ such that $\|f - f_r\|_\infty < \frac{1}{2n}$.

Let $M = \|f\|_1$ and choose K so large that $\sum_{k=K+1}^{\infty} 4Mr^k < \frac{1}{2n}$. Then

$$t_n(\theta) = A_0 + \sum_{k=1}^{K} A_k r^k \cos k\theta + B_k r^k \sin k\theta$$

is a trigonometric polynomial such that

$$\|f - t_n\|_\infty \le \|f - f_r\|_\infty + \|f_r - t_n\|_\infty$$

$$\le \frac{1}{2n} + \sum_{k=K+1}^{\infty} \left\| A_k r^k \cos k\theta + B_k r^k \sin k\theta \right\|$$

$$\le \frac{1}{2n} + \sum_{k=N+1}^{\infty} 4M r^k < \frac{1}{n}.$$

Therefore, $(t_n)_{n=1}^{\infty}$ converges uniformly to f. ∎

Exercises for Section 13.6

A. (a) Compute the Fourier series of $f(\theta) = \theta^2$ for $-\pi \le \theta \le \pi$.

   (b) Hence evaluate $\displaystyle\sum_{n\ge 1} \frac{(-1)^n}{n^2}$.

B. (a) Compute the Fourier series of the function $f(\theta) = |\sin\theta|$.

   (b) Hence evaluate $\displaystyle\sum_{n\ge 1} \frac{(-1)^n}{4n^2 - 1}$.

C. Suppose that $Q(r,\theta)$ is any function on $\mathbb{D}$ satisfying properties (1), (3), and (5) of Proposition 13.5.2. If f is a continuous function on the unit circle, define $v(r,\theta) = \int f(\theta - t) Q(r,t)\, dt$. Prove that $f_r(\theta) = v(r,\theta)$ converges to f uniformly on $[-\pi, \pi]$.

D. Use rectangular coordinates for the disk $\mathbb{D} = \{(x,y) : x^2 + y^2 < 1\}$, and define a function $u(x,y) = \tan^{-1}\left(\dfrac{y}{1+x}\right)$.

   (a) Show that $\Delta u = u_{xx} + u_{yy} = 0$.

   (b) Show that u is continuous on $\overline{\mathbb{D}}$ except at $(-1,0)$.

   (c) Show that u is constant on straight line segments through $(-1,0)$. Hence evaluate $f(\theta) := \lim_{r\to 1^-} u(r\cos\theta, r\sin\theta)$.

   (d) Find the Fourier series for f, and hence find an expression for u in polar coordinates.

13.7. The Maximum Principle

The remaining point to be dealt with in the heat problem is the question of uniqueness of solutions. Physically, it is intuitively clear that a fixed temperature distribution on the boundary circle will result in a uniquely determined distribution over the whole disk. We will show this to be the case by establishing a maximal principle showing that a harmonic function must take its maximum value on the boundary circle.

13.7.1. MAXIMUM PRINCIPLE.
Suppose that u is continuous on the closed disk $\overline{\mathbb{D}}$ and $\Delta u = 0$ on the open disk $\mathbb{D}$. Then

$$\max_{(r,\theta)\in\overline{\mathbb{D}}} u(r,\theta) = \max_{-\pi\le\theta\le\pi} u(1,\theta).$$

PROOF. First suppose that $v(r,\theta)$ satisfies $\Delta v \ge \varepsilon > 0$. If v attained its maximum at an interior point (r_0,θ_0), then the first-order partial derivatives are zero,

$$v_r(r_0,\theta_0) = 0 = v_\theta(r_0,\theta_0),$$

and the second-order derivatives are negative,

$$v_{rr}(r_0,\theta_0) \le 0 \quad \text{and} \quad v_{\theta\theta}(r_0,\theta_0) \le 0.$$

Therefore,

$$\varepsilon \le \Delta v(r_0,\theta_0) = v_{rr}(r_0,\theta_0) + \frac{1}{r_0}v_r(r_0,\theta_0) + \frac{1}{r_0^2}v_{\theta\theta}(r_0,\theta_0) \le 0.$$

This contradiction shows that v attains its maximum only on the boundary.

Now consider u. Let $v_n(r,\theta) = u(r,\theta) + \frac{1}{n}r^2$. A simple computation shows that

$$\Delta v_n = \Delta u + \frac{1}{n}\Delta r^2 = \frac{1}{n}\big((r^2)'' + \frac{1}{r}(r^2)' + \frac{1}{r^2}(r^2)_{\theta\theta}\big) = \frac{4}{n} > 0.$$

So v_n attains it maximum only on the boundary circle. Since v_n converges uniformly to u, we obtain

$$\max_{(r,\theta)\in\overline{\mathbb{D}}} u(r,\theta) = \lim_{n\to\infty} \max_{(r,\theta)\in\overline{\mathbb{D}}} v_n(r,\theta)$$

$$= \lim_{n\to\infty} \max_{-\pi\le\theta\le\pi} v_n(1,\theta) = \max_{-\pi\le\theta\le\pi} u(1,\theta). \qquad\blacksquare$$

13.7.2. COROLLARY.
Suppose that u is continuous on the closed disk $\overline{\mathbb{D}}$ and $\Delta u = 0$ on the open disk $\mathbb{D}$ and $u(1,\theta) = 0$ for $-\pi \le \theta \le \pi$. Then $u = 0$.

PROOF. By the Maximum Principle, $u(r,\theta) \le 0$ on $\overline{\mathbb{D}}$. However, $-u$ is also a continuous harmonic function, and thus the Maximum Principle implies that $u(r,\theta) \ge 0$ on $\overline{\mathbb{D}}$. Hence $u = 0$. $\qquad\blacksquare$

All the ingredients are now in place for a complete solution to the heat problem on the disk.

13.7.3. THEOREM.
Let $f(\theta)$ be a continuous 2π-periodic function. There exists a unique solution to the steady-state heat problem

$$\Delta u = 0 \qquad u(1,\theta) = f(\theta)$$

given by the Poisson integral of f.

PROOF. By Corollary 13.6.2, the Poisson integral of f provides a solution $u(r, \theta)$ to the heat problem. It remains to discuss uniqueness. Suppose that $v(r, \theta)$ is another solution. Then consider $w(r, \theta) = u(r, \theta) - v(r, \theta)$. It follows that

$$\Delta w = \Delta u - \Delta v = 0 \quad \text{and} \quad w(1, \theta) = u(1, \theta) - v(1, \theta) = 0.$$

Thus by Corollary 13.7.2, $w = 0$ and so $v = u$ is a only solution. ∎

Exercises for Section 13.7

A. Suppose that $u(x, y)$ is a solution of the heat problem on the disk $\overline{\mathbb{D}}$ written in rectangular coordinates for convenience. Let $D_R(x_0, y_0)$ be a small disk contained inside $\mathbb{D}$. Establish the **mean value property**:

$$u(x_0, y_0) = \frac{1}{2\pi} \int_0^{2\pi} u(x_0 + R\cos\theta, y_0 + R\sin\theta)\, d\theta.$$

HINT: The restriction of u to $\overline{D_R(x_0, y_0)}$ is the solution to the heat problem on this disk. Use the Poisson formula for the value at the centre of the disk.

B. (a) Suppose that $u(x, y)$ is a continuous function on $\overline{\mathbb{D}}$ that satisfies the mean value property of the previous exercise. Prove that u attains its maximum on the boundary.
   (b) Moreover, prove that if u attains its maximum value at a single interior point, then it must be constant.

C. Prove that a continuous function on $\mathbb{D}$ that satisfies the mean value property is harmonic.
   HINT: Fix a point (x_0, y_0) in $\mathbb{D}$ and let $D_R(x_0, y_0)$ be a small disk contained inside $\mathbb{D}$. Let $v(x, y)$ be the solution of the steady-state heat problem on $D_R(x_0, y_0)$ that agrees with u on the boundary circle. Show that $u = v$, and hence deduce that $\Delta u(x_0, y_0) = 0$.

D. Let $u(x, y)$ be a positive harmonic function on an open subset Ω of the plane. Suppose that $\overline{D_R(x_0, y_0)}$ is contained in Ω. Prove that
$$\frac{R - r}{R + r} u(x_0, y_0) \le u(x_0 + r\cos\theta, y_0 + r\sin\theta) \le \frac{R + r}{R - r} u(x_0, y_0) \text{ for } 0 \le r < R.$$
HINT: Compare Exercise 13.5.D.

E. Let Ω be a bounded open subset of the plane with smooth boundary Γ. A function v on Ω is harmonic if $\Delta v = 0$ on Ω.
   (a) Show that if in addition v is continuous on $\overline{\Omega}$, then it must attain its maximum value on Γ.
   (b) Hence show that if f is a continuous function on Γ, there is at most one continuous function on $\overline{\Omega}$ that is harmonic on Ω and with boundary values equal to f.

F. Let $u(r, \theta) = \dfrac{r(1 - r^2)\sin\theta}{(1 - 2r\cos\theta + r^2)^2}$ on $\mathbb{D}$.
   (a) Prove that u is harmonic on $\mathbb{D}$.
   (b) Show that $\lim_{r \to 1^-} u(r, \theta) = 0$ for all values of θ.
   (c) Why does this not contradict the Maximum Principle?

13.8. The Vibrating String (Formal Solution)

The mathematics of a vibrating string was one of the first problems studied using Fourier series. It arose in a discussion on the oscillations of a violin string by d'Alembert in 1747, twenty-two years before Fourier was born.

Most readers will be familiar with swinging a skipping rope. The simplest mode is a single lobe oscillating between the two fixed ends. However, it is possible to set up a wave with two lobes or even three. These vibrations with more lobes are called **harmonics**. They exist in the vibration of any stringed instrument and tend to be characteristic of the instrument, giving it a distinctive sound. For example, violins have significant order five harmonics.

Our problem is to describe the motion of a vibrating string. We imagine a uniform string stretched between two fixed endpoints under tension. We further assume that the oscillations are small compared with the length of the string. This is a reasonable assumption for a stiff string like one found on a violin or piano. This leads to the simplifying assumption that each point on the string moves only in a vertical direction. We ignore all forces other than the effect of string tension, such as the weight of the string and air resistance.

Orient the string along the x-axis of the plane, and choose units so that the endpoints are $(0,0)$ and $(\pi, 0)$. The vertical displacement of the string will be given by a function $y(x,t)$, the function giving the horizontal position at the point $x \in [0, \pi]$ at the time t. For convenience, we assume that time begins at time $t = 0$. Let τ denote the tension force, and let ρ be the density of the string.

Fix x and t and consider the forces acting on the string segment between the nearby points $A = (x, y) = (x, y(x,t))$ and $B = (x + \Delta x, y + \Delta y)$, where $y + \Delta y = y(x + \Delta x, t)$. The tension τ on the string results in forces acting on both ends of the segment in the direction of the tangent, as shown in Figure 13.3.

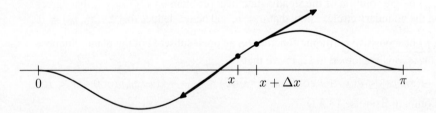

FIGURE 13.3. Forces acting on a segment of the string.

The force at A is

$$-\tau \frac{\left(1, \frac{\partial y}{\partial x}(x,t)\right)}{\sqrt{1 + \frac{\partial y}{\partial x}(x,t)^2}} \approx \left(-\tau, -\tau \frac{\partial y}{\partial x}(x,t)\right).$$

The approximation is reasonable since y and $\frac{\partial y}{\partial x}$ are assumed to be small relative to 1. Likewise, the tensile force at B is approximately

$$\left(\tau, \tau \frac{\partial y}{\partial x}(x + \Delta x, t)\right).$$

The horizontal forces cancel, while the combined vertical force is

$$\Delta V(x,t) = \tau \frac{\partial y}{\partial x}(x + \Delta x, t) - \tau \frac{\partial y}{\partial x}(x, t) \approx \tau \Delta x \frac{\partial^2 y}{\partial x^2}(x, t).$$

By Newton's law, we have $F = ma$, where we have a segment of mass $\rho \Delta x$ and acceleration equal to the second derivative of $y(x,t)$ with respect to t. Substitute this in, divide by $\rho \Delta x$, and take the limit as Δx tends to 0 to obtain the linear partial differential equation

$$\frac{\partial^2 y}{\partial t^2}(x,t) = \frac{\rho}{\tau} \frac{\partial^2 y}{\partial x^2}(x,t).$$

Set $\omega^2 = \rho/\tau$. This is known as the one-dimensional **wave equation**:

(13.8.1)
$$\frac{\partial^2 y}{\partial t^2}(x,t) = \omega^2 \frac{\partial^2 y}{\partial x^2}(x,t).$$

Since the endpoints are fixed, there are boundary conditions

(13.8.2)
$$y(0,t) = y(\pi,t) = 0 \quad \text{for all} \quad t \geq 0.$$

Finally, there are initial conditions: Imagine the string is initially stretched to some (continuous) shape $f(x)$ and moving with initial velocity $g(x)$. This gives the conditions

(13.8.3)
$$y(x,0) = f(x), \quad \frac{\partial y}{\partial t}(x,0) = g(x) \quad \text{for all} \quad x \in [0,\pi].$$

These boundary conditions, together with the wave equation governing subsequent motion of the string, determine a unique solution. We shall see that it can be solved in a manner similar to our analysis of the steady-state heat problem.

As before, we begin by using separation of variables to look for solutions of the special form $y(x,t) = X(x)T(t)$. There is no way to know in advance that there are solutions of this type, but in fact there are many such solutions that can be then be combined to exhaust all possibilities. Substituting $y(x,t) = X(x)T(t)$ into the wave equation gives

$$X(x)T''(t) = \omega^2 X''(x)T(t).$$

Isolating the variables x and t, we obtain

$$\frac{T''(t)}{T(t)} = \omega^2 \frac{X''(x)}{X(x)}.$$

The left-hand side of this equation is independent of x and the right-hand is be independent of t. Thus both sides are independent of both variables and therefore are equal to some constant c.

This results in two ordinary differential equations:

$$X''(x) - \frac{c}{\omega^2}X(x) = 0 \quad \text{and} \quad T''(t) - cT(t) = 0.$$

The boundary condition (13.8.2) simplifies to yield $X(0) = X(\pi) = 0$. At this stage, we must ignore the initial shape conditions (13.8.3).

The equation for X is essentially the same as the equation for Θ in Section 13.2. Depending on the sign of c, the solutions are sinusoidal, linear, or exponential:

$$X(x) = A\cos\left(\frac{\sqrt{|c|}}{\omega}x\right) + B\sin\left(\frac{\sqrt{|c|}}{\omega}x\right) \qquad \text{for} \quad c < 0$$

$$X(x) = A + Bx \qquad\qquad\qquad\qquad\qquad\quad \text{for} \quad c = 0$$

$$X(x) = Ae^{\sqrt{c}x/\omega} + Be^{-\sqrt{c}x/\omega} \qquad\qquad \text{for} \quad c > 0.$$

However, the boundary conditions eliminate both the linear and the exponential solutions. Thus the constant $-\frac{c}{\omega^2}$ is strictly positive, say γ^2. The solutions are

$$X(x) = a\cos\gamma x + b\sin\gamma x.$$

Since $X(0) = 0$, this forces $a = 0$. And $X(\pi) = 0$ yields $b\sin\gamma\pi = 0$. Thus a nonzero solution is possible only if γ is an integer n. Therefore, the possible solutions are

$$X_n(x) = b\sin nx \quad \text{for} \quad n \geq 1.$$

Now return to the equation for T. Since $c = -\gamma^2\omega^2 = -n^2\omega^2$, the DE for T becomes

$$T''(t) + (n\omega)^2 T(t) = 0.$$

Again this has solutions

$$T(t) = A\sin n\omega t + B\cos n\omega t.$$

Putting these together, we obtain solutions for $y(x, t)$ of the form

$$y_n(x, t) = A_n\sin nx\,\cos n\omega t + B_n\sin nx\,\sin n\omega t \quad \text{for} \quad n \geq 1.$$

The functions $y_n(x, t)$ correspond to the modes of vibration of the string. For $n = 1$, we have a string shape of a single sinusoidal loop oscillating up and down. This is the **fundamental vibration** mode of the string with the lowest frequency ω. However, for $n = 2$, we obtain a function that has two "arches" that swing back and forth. The frequency is twice the fundamental frequency. For general n, we have higher frequency oscillations with frequency $n\omega$ that oscillate n times between the two fixed endpoints at n times the rate. As we mentioned before, these higher frequencies are called harmonics.

The differential equation (13.8.1) is linear, so linear combinations of solutions are solutions. Thinking of solutions as waves, this combination of solutions is called **superposition**; see Figure 13.4. Ignoring convergence questions, we have a large family of possible solutions, all of the form

$$y(x, t) = \sum_{n=1}^{\infty} A_n\sin nx\,\cos n\omega t + B_n\sin nx\,\sin n\omega t.$$

Now consider the initial condition (13.8.3). Substituting $t = 0$ into this, we arrive at the boundary condition

$$f(x) = y(x, 0) = \sum_{n=1}^{\infty} A_n\sin nx.$$

Likewise, allowing term-by-term differentiation with respect to t, we obtain

$$\frac{\partial y}{\partial t}(x,t) = \sum_{n=1}^{\infty} -n\omega A_n \sin nx \; \sin n\omega t + n\omega B_n \sin nx \; \cos n\omega t.$$

Substituting in the boundary condition at $t = 0$ yields $g(x) = \sum_{n=1}^{\infty} n\omega B_n \sin nx$.

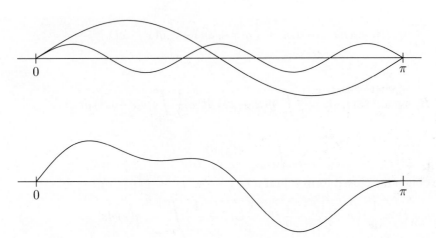

FIGURE 13.4. Superposition of two sine waves.

These are Fourier series. In fact, they are sine series because f and g are only defined on $[0, \pi]$. We have seen that summing them requires a certain amount of delicacy. Notice that since the string is fixed at 0 and π, the boundary functions satisfy

$$f(0) = f(\pi) = g(0) = g(\pi) = 0.$$

Let us extend f and g to $[-\pi, \pi]$ as *odd* functions by setting $f(-x) = -f(x)$ and $g(-x) = -g(x)$. Then extend them to be 2π-periodic functions on the whole real line. The values of the coefficients A_n and B_n for $n \geq 1$ are read off from the Fourier (sine) coefficients of f and g.

$$A_n = \frac{1}{\pi} \int_{-\pi}^{\pi} f(t) \sin nt \, dt = \frac{2}{\pi} \int_0^{\pi} f(t) \sin nt \, dt$$

and

$$B_n = \frac{1}{n\omega\pi} \int_{-\pi}^{\pi} g(t) \sin nt \, dt = \frac{2}{n\omega\pi} \int_0^{\pi} g(t) \sin nt \, dt$$

This form of the solution was proposed by Euler in 1748, and the same idea was advanced by D. Bernoulli in 1753 and Lagrange in 1759. However, we want to find a closed form solution discovered by d'Alembert in 1747.

Returning to our proposed series solution, use the trig identity

$$2 \sin nx \; \cos n\omega t = \sin n(x + \omega t) + \sin n(x - \omega t)$$

and substitute into

$$\sum_{n=1}^{\infty} A_n \sin nx \; \cos n\omega t = \frac{1}{2} \sum_{n=1}^{\infty} A_n \sin n(x + \omega t) + \frac{1}{2} \sum_{n=1}^{\infty} A_n \sin n(x - \omega t)$$

$$= \frac{1}{2} f(x + \omega t) + \frac{1}{2} f(x - \omega t).$$

Similarly,

$$\sum_{n=1}^{\infty} n\omega B_n \sin nx \; \cos n\omega t = \frac{1}{2} g(x + \omega t) + \frac{1}{2} g(x - \omega t).$$

Thus

$$\sum_{n=1}^{\infty} B_n \sin nx \; \sin n\omega t = \frac{1}{2} \int_0^t g(x + \omega t) \, dt + \frac{1}{2} \int_0^t g(x - \omega t) \, dt$$

$$= \frac{1}{2\omega} \int_{x-\omega t}^{x+\omega t} g(s) \, ds.$$

Adding, we obtain the closed form solution

$$y(x, t) = \frac{1}{2} f(x + \omega t) + \frac{1}{2} f(x - \omega t) + \frac{1}{2\omega} \int_{x-\omega t}^{x+\omega t} g(s) \, ds.$$

At this point, we have arrived at d'Alembert's form of the solution. It is worth noting that this form no longer involves a series and takes no notice of the method we used to arrive here.

The role of ω becomes apparent in this formulation of the solution. Notice that $y(x, t)$ depends on the values of f and g only in the range $[x - \omega t, x + \omega t]$. We should think of ω as the speed of propagation of the signal. In particular, if $g = 0$ and f is a small "bump" supported on $[\frac{\pi}{2}, \frac{\pi}{2} + \varepsilon]$, then for $\frac{\varepsilon}{\omega} \leq t \leq 1 - \frac{\varepsilon}{\omega}$, the function $y(x, t)$ consists of two identical bumps moving out toward the ends of the string at speed ω. When they reach the end, they bounce back. When the bumps overlap, they are superimposed (added). But the shape of the bumps is not lost; the message of the bump is transmitted accurately forever. In particular, the wave equation for light has the constant c, the speed of light, in place of ω.

13.9. The Vibrating String (Rigourous Solution)

We may check directly that d'Alembert's formula yields the solution to the vibrating string problem without returning to the Fourier series expansion. Our assumption that y satisfies a second-order PDE forces f to be C^2 and g to be C^1. We will consider more general initial data afterward.

13.9.1. THEOREM. *Consider the vibrating string equation*

$$\frac{\partial^2 y}{\partial t^2}(x, t) = \omega^2 \frac{\partial^2 y}{\partial x^2}(x, t)$$

with initial conditions

$$y(x,0) = f(x) \qquad \text{and} \qquad \frac{\partial y}{\partial t}(x,0) = g(x)$$

such that f is C^2 and g is C^1 and $f(0) = g(0) = f(\pi) = g(\pi) = 0$. This has a unique solution for all $t > 0$ given by

$$y(x,t) = \frac{1}{2}f(x + \omega t) + \frac{1}{2}f(x - \omega t) + \frac{1}{2\omega}\int_{x-\omega t}^{x+\omega t} g(s)\, ds,$$

where f and g have been extended (uniquely) to the whole real line as 2π-periodic odd functions.

PROOF. First, we verify that this is indeed a valid solution. Compute

$$\frac{\partial y}{\partial x}(x,t) = \frac{1}{2}f'(x + \omega t) + \frac{1}{2}f'(x - \omega t) + \frac{1}{2\omega}g(x + \omega t) - \frac{1}{2\omega}g(x - \omega t)$$

$$\frac{\partial^2 y}{\partial x^2}(x,t) = \frac{1}{2}f''(x + \omega t) + \frac{1}{2}f''(x - \omega t) + \frac{1}{2\omega}g'(x + \omega t) - \frac{1}{2\omega}g'(x - \omega t)$$

$$\frac{\partial y}{\partial t}(x,t) = \frac{\omega}{2}f'(x + \omega t) - \frac{\omega}{2}f'(x - \omega t) + \frac{1}{2}g(x + \omega t) + \frac{1}{2}g(x - \omega t)$$

$$\frac{\partial^2 y}{\partial t^2}(x,t) = \frac{\omega^2}{2}f''(x + \omega t) + \frac{\omega^2}{2}f''(x - \omega t) + \frac{\omega}{2}g'(x + \omega t) - \frac{\omega}{2}g'(x - \omega t).$$

Thus $\frac{\partial^2 y}{\partial t^2}(x,t) = \omega^2 \frac{\partial^2 y}{\partial x^2}(x,t)$ and $y(x,0) = f(x)$ and $\frac{\partial y}{\partial t}(x,0) = g(x)$.

The question of uniqueness of solutions remains. As with the heat equation, physical properties yield a clue. The argument in this case is based on conservation of energy. Without actually justifying the physical interpretation, we introduce a quantity called total energy obtained by adding the potential and kinetic energies of the string:

$$E(t) = E(y,t) = \int_0^\pi \left(\frac{\partial y}{\partial x}(x,t)\right)^2 + \frac{1}{\omega^2}\left(\frac{\partial y}{\partial t}(x,t)\right)^2 dx.$$

This is defined for any solution y of our vibration problem. For convenience, we switch to the notation y_x, y_t, and so on for partial derivatives. Use the Leibniz Rule (8.3.4) to compute

$$E'(t) = \frac{\partial}{\partial t}\int_0^\pi \left(y_x(x,t)\right)^2 + \frac{1}{\omega^2}\left(y_t(x,t)\right)^2 dx$$

$$= \int_0^\pi 2y_x y_{xt}(x,t) + \frac{1}{\omega^2}2y_t y_{tt}(x,t)\, dx.$$

Substitute $y_{tt} = \omega^2 y_{xx}$:

$$= \int_0^\pi 2y_x y_{xt}(x,t) + 2y_t y_{xx}(x,t)\, dx$$

$$= \int_0^\pi \frac{\partial}{\partial x}\left(y_x y_t(x,t)\right) dx = y_x y_t(x,t)\Big|_{x=0}^{x=\pi} = 0.$$

The last equality follows since the string is fixed at both endpoints, forcing the relation $y_t(0, t) = y_t(\pi, t) = 0$. This shows that the energy is preserved (constant).

Now suppose that $y_2(x, t)$ is another solution to the wave problem. Then the difference $z(x, t) = y(x, t) - y_2(x, t)$ is also a solution of the wave equation and initial boundary conditions:

$$z_{tt} = \omega^2 z_{xx} \qquad z(x, 0) = z_t(x, 0) = 0.$$

It follows that $z_x(x, 0) = 0$ too, and hence the energy of the system is

$$E = \int_0^\pi z_x(s, 0)^2 + z_t(s, 0)^2 \, ds = 0.$$

By conservation of energy, we deduce that

$$\int_0^\pi z_x(s, t)^2 + z_t(s, t)^2 \, ds = 0 \quad \text{for all} \quad t \ge 0.$$

In other words, $z_x(s, t) = z_t(s, t) = 0$ for all s and t. Therefore, $z = 0$, establishing uniqueness. ∎

We know from real-world experience that a string may be bent into a non-C^2 shape. What happens then? In this case, we may approximate f uniformly by a sequence of C^2 functions f_n. Likewise, g may be approximated uniformly by a sequence g_n of C^1 functions. The wave equation with initial data $y(x, 0) = f_n(x)$ and $y_x(x, 0) = g_n(x)$ yields the solution

$$y_n(x, t) = \tfrac{1}{2} f_n(x + \omega t) + \tfrac{1}{2} f_n(x - \omega t) + \frac{1}{2\omega} \int_{x-\omega t}^{x+\omega t} g_n(s) \, ds.$$

This sequence of solutions converges uniformly to

$$y(x, t) = \tfrac{1}{2} f(x + \omega t) + \tfrac{1}{2} f(x - \omega t) + \frac{1}{2\omega} \int_{x-\omega t}^{x+\omega t} g(s) \, ds.$$

Thus, even though this function y may not be differentiable (so that it does not make sense to plug y into our PDE), we obtain a reasonable solution to the wave problem.

One very interesting aspect of this solution is the fact that $y(x, t)$ is not any smoother for large t than it is at time $t = 0$. Indeed, for every integer $n \ge 0$,

$$y(x, \tfrac{2n\pi}{\omega}) = \tfrac{1}{2} f(x + 2n\pi) + \tfrac{1}{2} f(x - 2n\pi) + \frac{1}{2\omega} \int_{x-2n\pi}^{x+2n\pi} g(s) \, ds$$

$$= f(x).$$

This uses the fact that g has been extended to an odd 2π-periodic function and thus g integrates to 0 over any interval of length (a multiple of) 2π. Similarly,

$$y_t(x, \tfrac{2n\pi}{\omega}) = \frac{\omega}{2} f'(x+2n\pi) - \frac{\omega}{2} f'(x-2n\pi) + \frac{1}{2} g(x+2n\pi) + \frac{1}{2} g(x-2n\pi)$$

$$= g(x).$$

Thus if the initial data f fails to be smooth at x_0, this property persists along the lines $x(t) = x_0 \pm \omega t$. This is known as **propagation of singularities**. And because our solutions are odd 2π-periodic functions, these singularities recur within our range. Following the solution only within the interval $[0, \pi]$, these singularities appear to reflect off the boundary and reenter the interval. This property is distinctly different from the solution of the heat equation, which becomes C^∞ for $t > 0$ (see Exercise 13.4.G) because the initial heat distribution gets averaged out over time.

The lack of averaging or damping in the wave equation is very important in real life. It makes it possible to see, and to transmit radio and television signals over long distances without significant distortion.

The Fourier series approach still has more to tell us. The Fourier coefficients in the expansion of the solution $y(x, t)$ decompose the wave into a sum of harmonics of order n for $n \geq 1$. In fact, the term

$$y_n(x, t) = A_n \sin nx \ \cos n\omega t + B_n \sin nx \ \sin n\omega t$$

may be rewritten as

$$y_n(x, t) = C_n \sin nx \ \sin(n\omega t + \tau_n),$$

where $C_n = \sqrt{A_n^2 + B_n^2}$ and the phase shift τ_n is chosen so that $\sin \tau_n = A_n/C_n$ and $\cos \tau_n = B_n/C_n$. Thus as t increases, y_n modulates through multiples of $\sin nx$ from C_n down to $-C_n$ and back.

The combination of different harmonics gives a wave its shape. In electrical engineering, one often attempts to break down a wave into its component parts or build a new wave by putting harmonics together. This amounts to finding a Fourier series whose sum is a specified function. That is the problem we will investigate further in the next chapter.

Exercises for Section 13.9

A. Using the series for $y(x, t)$ and the orthogonality relations, show that

$$E = \int_0^\pi \frac{1}{2} f'(x)^2 + \frac{1}{2\omega^2} g(x)^2 \ dx = \frac{\pi}{2} \sum_{n=1}^\infty n^2 (|A_n|^2 + |B_n|^2).$$

B. Consider a guitar string that is plucked in the centre to a height h starting at rest. Assume that the initial position is piecewise linear with a sharp cusp in the centre.
(a) What is the odd 2π-periodic extension of the initial position function f? Sketch it.
(b) Plot the graph of the solution $y(x, t)$ for $t = 0, \frac{\pi}{3\omega}, \frac{\pi}{2\omega}, \frac{\pi}{\omega}, \frac{2\pi}{\omega}$.
(c) How does the cusp move? Find a formula.
(d) Compute the energy for this string.

C. Verify that $A_n \sin nx \cos n\omega t + B_n \sin nx \sin n\omega t = C_n \sin nx \sin(n\omega t + \tau_n)$, where $C_n = \sqrt{A_n^2 + B_n^2}$ and τ_n is chosen so that $\sin \tau_n = A_n/C_n$ and $\cos \tau_n = B_n/C_n$.

D. Let F and G be C^2 functions on the line.
(a) Show that $y(x, t) = F(x + \omega t) + G(x - \omega t)$ for $x \in \mathbb{R}$ and $t \geq 0$ is a solution of the wave equation the whole line.
(b) What is the initial position f and velocity g in terms of F and G? Express F and G in terms of f and g.
(c) Show that the value of $y(x, t)$ depends only on the values of f and g in the interval $[x - \omega t, x + \omega t]$.

(d) Explain the physical significance of part (c) in terms of the speed of propagation of the signal.

E. Consider the wave equation on $\mathbb{R}$. Let $u = x + \omega t$ and $v = x - \omega t$.
   (a) Show that

$$\frac{\partial}{\partial u} = \frac{1}{2}\frac{\partial}{\partial x} + \frac{1}{2\omega}\frac{\partial}{\partial t} \qquad \text{and} \qquad \frac{\partial}{\partial v} = \frac{1}{2}\frac{\partial}{\partial x} - \frac{1}{2\omega}\frac{\partial}{\partial t}.$$

Hence deduce that after a change of variables, the wave equation becomes

$$\frac{\partial}{\partial u}\frac{\partial}{\partial v}y = 0.$$

   (b) Hence show that every solution has the form $y = F(u) + G(v)$.
   (c) Combine this with the previous exercise to show that the wave equation has a unique solution on the line.

F. Let $w(x)$ be a strictly positive function on $[0, 1]$, and consider the PDE defined for $0 \le x \le 1$ and $t \ge 0$:

$$\frac{\partial^2 y}{\partial t^2}(x, t) = w(x)^2 \frac{\partial^2 y}{\partial x^2}(x, t) + H(x, t)$$

$$y(x, 0) = f(x) \qquad y_t(x, 0) = g(x).$$

Suppose that $y(x, t)$ and $z(x, t)$ are two solutions, and let $u = y - z$. Consider the quantity

$$E(t) = \int_0^1 u_x^2(x, t) + \frac{u_t^2(x, t)}{w(x)^2}\, dx.$$

Show that E is the zero function, and hence deduce that the solution of the PDE is unique.

13.10. Appendix: The Complex Exponential

Our goal in this section is to extend the definition of the exponential function to all complex numbers. Although we quickly review the basic ideas of complex numbers $\mathbb{C}$, prior experience with complex numbers will be quite helpful.

Complex numbers may be written uniquely as $a + ib$ for $a, b \in \mathbb{R}$. Addition is just given by the rule

$$(a + ib) + (c + id) = (a + c) + i(b + d).$$

Multiplication uses distributivity and the rule $i^2 = -1$. So

$$(a + ib)(c + id) = (ac - bd) + i(ad + bc).$$

The **conjugate** of a complex number $z = a + ib$ is the number $\bar{z} := a - ib$. The absolute value or **modulus** is given by

$$|z| = (z\bar{z})^{1/2} = \sqrt{a^2 + b^2}.$$

The set of all complex numbers is closed under addition and subtraction, multiplication and division by nonzero elements. In particular, if $z = a + ib \ne 0$, then

$$\frac{1}{z} = \frac{\bar{z}}{z\bar{z}} = \frac{a - ib}{a^2 + b^2} = \frac{a}{a^2 + b^2} + i\frac{-b}{a^2 + b^2}.$$

This makes $\mathbb{C}$ into a field containing the reals as a proper subset. We also define the **real part** and **imaginary part** of a complex number by

$$\mathrm{Re}(a + ib) = a \qquad \text{and} \qquad \mathrm{Im}(a + ib) = b.$$

Note that the imaginary part is a real number.

One crucial property of $\mathbb{C}$ is that it is an **algebraically closed field**, meaning that every polynomial with complex coefficients factors as a product of degree one polynomials. This result is called the **Fundamental Theorem of Algebra**, although modern algebra does not accord $\mathbb{C}$ the preeminent role that it had in the nineteenth century, when the theorem was named.

The analysis of complex functions has a lot of interesting aspects. Functions of a complex variable that are differentiable are called **analytic**, and they have many amazing properties. These properties are developed in complex analysis, an important subject that is often quite different from real analysis.

To extend the definition of the exponential function to all complex numbers, we must construct a differentiable function $E : \mathbb{C} \to \mathbb{C}$ such that $E(w + z) = E(w)E(z)$ for all w and z in $\mathbb{C}$ and so that $E(x) = e^x$ for all $x \in \mathbb{R}$. Once we have established the existence of this function, we will write e^z for $E(z)$.

Let us calculate some simple properties that such a function must have. First,

$$E(x + iy) = e^x E(iy).$$

Using the differentiability, we get

$$\begin{aligned} E'(z) &= \lim_{h \to 0} \frac{E(z + h) - E(z)}{h} \\ &= E(z) \lim_{h \to 0} \frac{E(h) - E(0)}{h} \\ &= E(z) \lim_{x \to 0} \frac{e^x - 1}{x} = E(z). \end{aligned}$$

Now concentrate on the function $f(y) = E(iy)$. Split it into its real and imaginary parts as $f(y) = E(iy) = A(y) + iB(y)$. Differentiating with respect to y yields

$$\begin{aligned} f'(y) &= A'(y) + iB'(y) \\ &= E'(iy)\frac{d(iy)}{dy} \\ &= iE(iy) = -B(y) + iA(y). \end{aligned}$$

So we arrive at the system of differential equations

$$\begin{aligned} A'(y) &= -B(y) \\ B'(y) &= A(y). \end{aligned}$$

This leads to the second-order differential equation $A''(y) = -A(y)$. From the identity $1 = E(0) = A(0) + iB(0)$, we also get the initial conditions $A(0) = 1$ and $A'(0) = -B(0) = 0$. We have seen that this system has a unique solution $A(y) = \cos y$ and $B(y) = \sin y$.

Thus we arrive at a unique solution $E(iy) = \cos y + i \sin y$. So

$$E(x + iy) = e^x\big(\cos y + i \sin y\big).$$

13.10.1. THEOREM. *The function $E(x + iy) = e^x\big(\cos y + i \sin y\big)$ satisfies the identities $E(z + w) = E(z)E(w)$ for all complex numbers z, w. Moreover, E is differentiable and $E'(z) = E(z)$ for all $z \in \mathbb{C}$.*

PROOF. Notice that

$$\begin{aligned}
E(iy)E(iv) &= \big(\cos y + i \sin y\big)\big(\cos v + i \sin v\big) \\
&= \big(\cos y \cos v - \sin y \sin v\big) + i\big(\cos y \sin v + \sin y \cos v\big) \\
&= \cos(y + v) + i \sin(y + v) = E(iy + iv).
\end{aligned}$$

Hence

$$\begin{aligned}
E(x + iy)E(u + iv) &= e^x\big(\cos(y) + i\sin(y)\big)e^u\big(\cos(v) + i\sin(v)\big) \\
&= e^{x+u}E(iy + iv) = E((x + iy) + (u + iv)).
\end{aligned}$$

So E satisfies the multiplicative property.

The derivative property is a bit more delicate. The hard part is to show that $E'(0) = 1$. For then, as on the previous page, we obtain

$$E'(z) = E(z)E'(0) = E(z).$$

To verify that $E'(0) = 1$, we must show that

$$0 = \lim_{h \to 0}\left| \frac{E(h) - 1}{h} - 1 \right| = \lim_{h \to 0} \frac{|E(h) - 1 - h|}{|h|}.$$

The complication comes from the fact that h can be a small complex number, not just a real number, as it approaches 0. However, we need only facts from the calculus of real functions to verify this limit. The major tool for making estimates is the Mean Value Theorem. Let us write $h = x + iy$. We may assume that $|h|^2 = x^2 + y^2 < 1$. In particular, $|x| < 1$. Calculate

$$\begin{aligned}
E(h) - 1 - h &= e^x \cos y + ie^x \sin y - 1 - x - iy \\
&= e^x(\cos y - 1) + (e^x - 1 - x) + ie^x(\sin y - y) + iy(e^x - 1).
\end{aligned}$$

Each of these terms can be estimated by the Mean Value Theorem. First, since $f(y) = \cos y$ has derivative $f'(y) = -\sin y$, it follows that there is a value c between 0 and y such that

$$|\cos y - 1| = |-\sin c|\,|y| \le |c|\,|y| \le |y|^2.$$

So $e^x|\cos y - 1| \le e|y|^2 \le e|h|^2$ provided that $|x| \le 1$.

A similar treatment of the function e^x shows that

$$|e^x - 1| \le e|x| \quad \text{for all} \quad |x| \le 1.$$

Now repeat the argument for the function $g(x) = e^x - 1 - x$, which has derivative $g'(x) = e^x - 1$. Again by the Mean Value Theorem, there is a point c between 0 and x so that

$$|e^x - 1 - x| = |e^c - 1|\,|x| \le e|c|\,|x| \le e|h|^2.$$

A third application of the Mean Value Theorem with $k(y) = \sin y - y$ and derivative $k'(y) = \cos y - 1$ yields a point c between 0 and y so that

$$|\sin y - y| = |y|\,|\cos c - 1| \le |y|\,|c|^2 \le |y|^3.$$

Together with the inequality $|e^x| \le e$ for $|x| \le 1$, this yields

$$|ie^x(\sin y - y)| \le e|y|^3 \le e|h|^3.$$

Finally, the fourth term is handled by $2|xy| \le x^2 + y^2 = |h|^2$, so

$$|y(e^x - 1)| \le e|y||x| \le 2|h|^2.$$

Putting it all together yields, for $|h| \le 1$,

$$|E(h) - 1 - h| = e^x\big|\cos y - 1\big| + \big|e^x - 1 - x\big| + \big|ie^x(\sin y - y)\big| + \big|iy(e^x - 1)\big|$$
$$\le e|h|^2 + e|h|^2 + e|h|^3 + 2|h|^2$$
$$= \big(2e + 2 + e|h|\big)|h|^2.$$

Thus

$$\lim_{h \to 0} \frac{|E(h) - 1 - h|}{|h|} = 0. \qquad \blacksquare$$

From now on, we will write e^z instead of $E(z)$, as this theorem justifies calling it the exponential function. If $z = a + ib$ is any complex number, let $r = |z| = \sqrt{a^2 + b^2}$. Then $z/r = a/r + ib/r$ has modulus 1 and hence lies on the unit circle. Therefore there is an angle θ, which is unique up to a multiple of 2π, such that $z/r = \cos\theta + i\sin\theta = e^{i\theta}$. So $z = re^{i\theta}$. This is called the polar form since z is represented as (r, θ) in the polar coordinates of the plane.

The relationship between $e^{\pm ix}$ and $\sin x$ and $\cos x$ is important in Fourier series. In particular, we have

$$\cos nx = \mathrm{Re}\left(e^{inx}\right) = \frac{e^{inx} + e^{-inx}}{2}$$

and

$$\sin nx = \mathrm{Im}\left(e^{inx}\right) = \frac{e^{inx} - e^{-inx}}{2i}.$$

We record some of the facts and ask you to prove them in the Exercises.

Since the pairs $\{\cos nx, \sin nx\}$ and $\{e^{inx}, e^{-inx}\}$ both span the same two-dimensional space (using complex coefficients), it follows that a Fourier series may be written as

(13.10.2)
$$f(\theta) \sim \sum_{n=-\infty}^{\infty} a_n e^{in\theta},$$

where the coefficients are given by

(13.10.3)
$$a_n = \frac{1}{2\pi} \int_{-\pi}^{\pi} f(\theta) e^{-in\theta}\, d\theta.$$

We have already seen that we could sum the Poisson kernel by using complex exponentials and the formula for summing geometric series. This idea has many applications in Fourier series. We give another example.

We begin with a lemma that can be proved using trig identities but is transparent using complex exponentials.

13.10.4. LEMMA.

$$S_n(\theta) := \sum_{k=1}^{n} \sin k\theta = \frac{\cos \theta/2 - \cos(n + \frac{1}{2})\theta}{2\sin \theta/2} \quad \text{for} \quad \theta \neq 2m\pi$$

In particular, $|S_n(\theta)| \leq \pi \varepsilon^{-1}$ *uniformly on* $[\varepsilon, 2\pi - \varepsilon]$ *for all* $n \geq 1$.

PROOF. We use the exponential formula for $\sin kx$ and the formula for summing a geometric series. Compute

$$\sum_{k=1}^{n} \sin k\theta = \frac{1}{2i} \sum_{k=1}^{n} e^{ik\theta} - \frac{1}{2i} \sum_{k=1}^{n} e^{-ik\theta}$$

$$= \frac{e^{i(n+1)\theta} - e^{i\theta}}{2i(e^{i\theta} - 1)} - \frac{e^{-i(n+1)\theta} - e^{-i\theta}}{2i(e^{-i\theta} - 1)}$$

$$= \frac{e^{i(n+1/2)\theta} + e^{-i(n+1/2)\theta} - e^{i\theta/2} - e^{-i\theta/2}}{2i(e^{i\theta/2} - e^{-i\theta/2})}$$

$$= \frac{\left(e^{i\theta/2} + e^{-i\theta/2}\right) - \left(e^{i(n+1/2)\theta} + e^{-i(n+1/2)\theta}\right)}{4} \, \frac{2i}{e^{i\theta/2} - e^{-i\theta/2}}$$

$$= \frac{\cos \theta/2 - \cos(n + \frac{1}{2})\theta}{2\sin \theta/2}.$$

If $\varepsilon \leq \theta \leq 2\pi - \varepsilon$, then

$$|2\sin \theta/2| \geq |2\sin \varepsilon/2| \geq 2\frac{2}{\pi}\frac{\varepsilon}{2} = \frac{2\varepsilon}{\pi}.$$

It follows that in this interval,

$$|S_n(\theta)| \leq \frac{2}{2\varepsilon/\pi} = \frac{\pi}{\varepsilon}. \qquad \blacksquare$$

13.10.5. EXAMPLE.

A Fourier series that has positive coefficients decreasing monotonically to 0 always converges to a function that is continuous on $(0, 2\pi)$ but may have a discontinuity at multiples of 2π. The key tool is Dirichlet's Test (3.4.10) for the convergence of a sequence. We will examine this in a specific instance. Consider the Fourier sine series

$$f(\theta) = \sum_{k=1}^{\infty} \frac{1}{k} \sin k\theta.$$

Notice that

$$\frac{1}{k} = \left(\frac{1}{k} - \frac{1}{k+1}\right) + \left(\frac{1}{k+1} - \frac{1}{k+2}\right) + \cdots + \left(\frac{1}{n-1} - \frac{1}{n}\right) + \frac{1}{n}$$

$$= \sum_{j=k}^{n-1} \frac{1}{j(j+1)} + \frac{1}{n}.$$

Following the proof of the Dirichlet Test (3.4.10), we can rearrange the partial sums of the sine series as

$$\sum_{k=1}^{n} \frac{1}{k} \sin k\theta = \sum_{k=1}^{n} \left(\sum_{j=k}^{n-1} \frac{1}{j(j+1)} + \frac{1}{n} \right) \sin k\theta$$

$$= \sum_{j=1}^{n-1} \frac{1}{j(j+1)} S_j(\theta) + \frac{1}{n} S_n(\theta).$$

On the interval $[\varepsilon, 2\pi - \varepsilon]$, we have the bounds

$$\sup_{\varepsilon \le \theta \le 2\pi - \varepsilon} \left| \frac{1}{j(j+1)} S_j(\theta) \right| \le \frac{\pi}{j(j+1)\varepsilon}.$$

So we can apply the Weierstrass M-test. We have a telescoping sum:

$$\sum_{j=1}^{\infty} \frac{\pi}{j(j+1)\varepsilon} = \frac{\pi}{\varepsilon} \sum_{j=1}^{\infty} \left(\frac{1}{j} - \frac{1}{j+1} \right) = \frac{\pi}{\varepsilon} < \infty.$$

So the series $\sum_{j=1}^{\infty} \frac{1}{j(j+1)} S_j(\theta)$ converges uniformly on $[\varepsilon, 2\pi - \varepsilon]$. Since the uniform limit of continuous functions is continuous by Theorem 8.2.1, the sum is a continuous function on $[\varepsilon, 2\pi - \varepsilon]$ for any $\varepsilon > 0$. Hence the limit exists and is a continuous function $f(\theta)$ on $(0, 2\pi)$.

Therefore, the partial sums also converge:

$$\lim_{n \to \infty} \sum_{k=1}^{n} \frac{1}{k} \sin k\theta = \lim_{n \to \infty} \sum_{j=1}^{n-1} \frac{1}{j(j+1)} S_j(\theta) + \lim_{n \to \infty} \frac{1}{n} S_n(\theta)$$

$$= f(\theta).$$

The second limit is 0 since

$$\sup_{\varepsilon \le \theta \le 2\pi - \varepsilon} \left| \frac{1}{n} S_n(\theta) \right| \le \frac{\pi}{n\varepsilon}.$$

This tends to 0 uniformly on $[\varepsilon, 2\pi - \varepsilon]$ as n tends to infinity.

Now let us attempt to compute $f(\theta)$. It is convenient to work with the harmonic extension

$$u(r, \theta) = \sum_{k=1}^{\infty} \frac{r^n}{n} \sin n\theta$$

as in Section 13.4 because better convergence is obtained inside the disk. Take the partial derivative with respect to r, and sum the resulting geometric series.

$$u_r(r, \theta) = \sum_{k=1}^{\infty} r^{n-1} \sin n\theta$$

$$= \frac{1}{2i} \sum_{k=1}^{\infty} r^{n-1} e^{in\theta} + \frac{1}{2i} \sum_{k=1}^{\infty} r^{n-1} e^{-in\theta}$$

$$= \frac{e^{i\theta}}{2i(1 - re^{i\theta})} + \frac{e^{-i\theta}}{2i(1 - re^{-i\theta})}$$

It is not difficult to integrate this function with respect to r from $(0, \theta)$ to (R, θ) using the initial condition $u(0, \theta) = A_0 = 0$ to obtain

$$u(R, \theta) = \frac{i}{2} \log(1 - Re^{i\theta}) - \frac{i}{2} \log(1 - Re^{-i\theta})$$

$$= \frac{i}{2} \log \frac{1 - Re^{i\theta}}{1 - Re^{-i\theta}}.$$

Notice that $1 - re^{-i\theta}$ is the complex conjugate of $1 - re^{i\theta}$. Thus if we write $1 - re^{i\theta} = se^{i\alpha}$, then

$$u(r, \theta) = \frac{i}{2} \log \frac{se^{i\alpha}}{se^{-i\alpha}} = \frac{i}{2} \log e^{2i\alpha} = -\alpha.$$

Look at Figure 13.5. For θ in $(0, \pi)$, α belongs to $(-\frac{\pi}{2}, 0)$, and it reverses sign on $(\pi, 2\pi)$. So $u(r, \theta) > 0$ for θ in $(0, \pi)$, and it is negative for θ in $(\pi, 2\pi)$.

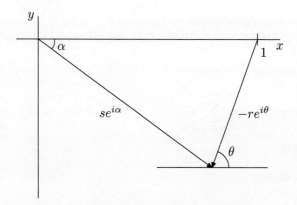

FIGURE 13.5. The vector $se^{i\alpha} = 1 - re^{i\theta}$.

We have a triangle in Figure 13.5 with angle $|\alpha|$ opposite the side of length r and angle $\pi - |\alpha| - |\theta|$ opposite the side of length 1. By the law of sines, we obtain

$$\frac{\sin |\alpha|}{r} = \frac{\sin(\pi - |\alpha| - |\theta|)}{1} = \sin |\alpha| \cos |\theta| + \cos |\alpha| \sin |\theta|.$$

Solving, we obtain

$$\tan |\alpha| = \frac{r \sin |\theta|}{1 - r \cos \theta}.$$

Using the information about the sign of u, we obtain

$$u(r, \theta) = \text{Tan}^{-1} \frac{r \sin \theta}{1 - r \cos \theta}.$$

This definition of u makes sense provided that the denominator $1 - r \cos \theta$ is not 0, which happens on the closed unit disk except for one boundary point $(1, 0)$. Hence $u(r, \theta)$ is defined and continuous on $\overline{\mathbb{D}} \setminus \{(1, 0)\}$. In particular, using trig formulas, we obtain

$$
\begin{aligned}
f(\theta) = u(1, \theta) &= \text{Tan}^{-1} \left(\frac{\sin \theta}{1 - \cos \theta} \right) \\
&= \text{Tan}^{-1} \left(\frac{2 \sin(\theta/2) \cos(\theta/2)}{2 \sin^2(\theta/2)} \right) \\
&= \text{Tan}^{-1} \cot \frac{\theta}{2} = \frac{\pi - \theta}{2}.
\end{aligned}
$$

Note that this is valid for θ in $(0, 2\pi)$, not our usual $(-\pi, \pi)$ interval. For the graph of f, see Figure 13.6.

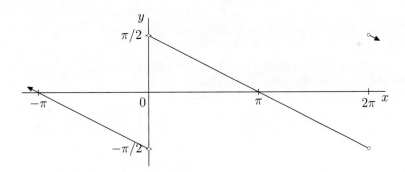

FIGURE 13.6. The graph of f as a 2π-periodic function on $\mathbb{R}$.

As a final check, we compute the Fourier series of f. This is an odd function, so it has a sine series. Use integration by parts:

$$
\begin{aligned}
B_n &= \frac{1}{\pi} \int_0^{2\pi} \frac{\pi - \theta}{2} \sin n\theta \, d\theta \\
&= -\frac{\pi - \theta}{2n\pi} \cos n\theta \Big|_0^{2\pi} - \int_0^{2\pi} \frac{\cos n\theta}{2n\pi} \, d\theta = \frac{1}{n}.
\end{aligned}
$$

Exercises for Section 13.10

A. (a) Graph the image of a line parallel to the y-axis under the exponential map.

(b) Graph the image of a line parallel to the x-axis under the exponential map.

(c) Show that the strip $\{z = x + iy \mid 0 \le y < 2\pi\}$ is mapped by the exponential function one-to-one and onto the whole complex plane except for the point 0.

B. (a) Express $f(\theta) = A_n \cos n\theta + B_n \sin n\theta$ as a linear combination of $e^{in\theta}$ and $e^{-in\theta}$.

(b) Show that the coefficients are given by the formula (13.10.3).

(c) Hence show that the Fourier series of a continuous function $f(\theta)$ is converted to the form (13.10.2).

C. Find a formula for the sum $\sum_{k=1}^{n} \cos k\theta$.

D. Define the **complex Fourier series** of an absolutely integrable 2π-periodic function $f(\theta)$ by $f \sim \sum_{n=-\infty}^{\infty} c_n e^{in\theta}$, where $c_n = \frac{1}{2\pi} \int_{-\pi}^{\pi} f(\theta) e^{-in\theta} \, d\theta$.

(a) Find the relationship between $\{A_k, B_k\}$ and $\{c_k, c_{-k}\}$.

(b) Hence show that $\sum_{k=-n}^{n} c_k e^{ik\theta} = A_0 + \sum_{k=1}^{n} A_k \cos k\theta + B_k \sin k\theta$.

(c) Show that $\{e^{in\theta} : n \in \mathbb{Z}\}$ is an orthonormal set in $L^2(-\pi, \pi)$.

E. Sum the Fourier series $\sum_{n=0}^{\infty} 2^{-n} \cos n\theta$ and $\sum_{n=1}^{\infty} 2^{-n} \sin n\theta$.

HINT: Take the real and imaginary parts of $\sum_{n=0}^{\infty} 2^{-n} e^{in\theta}$.

F. Show that $\sum_{k=1}^{n} \frac{1}{k} \cos k\theta$ converges uniformly on $[\varepsilon, 2\pi - \varepsilon]$ for $\varepsilon > 0$.

G. Define $\cos z = (e^{iz} + e^{-iz})/2$ and $\sin z = (e^{iz} - e^{-iz})/2i$ for all $z \in \mathbb{C}$.

(a) Prove that $\sin(w + z) = \sin w \cos z + \cos w \sin z$.

(b) Find all solutions of $\sin z = 2$.

CHAPTER 14

Fourier Series and Approximation

It is a natural problem to take a wave output and try to decompose it into its harmonic parts. Engineers are able to do this with an oscilloscope. A real difficultly occurs when we try to put the parts back together. Mathematically, this amounts to summing up the series obtained from decomposing the original wave. In this chapter, we examine this delicate question: Under what conditions does a Fourier series converge? We begin with L^2 approximation and convergence, which has a very clean answer. Nice applications of this include the isoperimetric inequality and sums of various interesting series. Then we turn to the more subtle questions of pointwise and uniform convergence. The idea of kernel functions, analogous to the Poisson kernel from the previous chapter, provides an elegant method for understanding these notions of convergence.

14.1. Least Squares Approximations

Approximation in the L^2 norm is important because it is readily computable. Also, the partial sums of the Fourier series are well behaved in this norm, unlike pointwise and uniform convergence, which we study later.

For the purposes of applications, we need to consider piecewise continuous functions on $[a, b]$, as given in Definition 5.2.4. Recall that there is a partition $a = x_0 < x_1 < \cdots < x_N = b$ of $[a, b]$ so that f is continuous on each interval (x_i, x_{i+1}) and one-sided limits exist at each node. Note that a piecewise continuous function is bounded by applying the Extreme Value Theorem on each interval $[x_i, x_{i+1}]$. Since it is continuous on each of these intervals, it is also absolutely integrable. Hence piecewise continuous 2π-periodic functions have Fourier series. The product of two piecewise continuous functions is also piecewise continuous and thus is integrable. So we can compute the L^2 inner product of two piecewise continuous functions by extending the definition for continuous functions given in Example 7.4.4.

Given a piecewise continuous function 2π-periodic f with Fourier series

$$f \sim A_0 + \sum_{n=1}^{\infty} A_n \cos n\theta + B_n \sin n\theta,$$

let us denote the partial sums by $S_N f(\theta) = A_0 + \sum_{n=1}^{N} A_n \cos n\theta + B_n \sin n\theta$.

14.1.1. LEMMA. *Every piecewise continuous function 2π-periodic f is the limit in the $L^2(-\pi, \pi)$ norm of a sequence of trigonometric polynomials.*

PROOF. Let $x_0 = -\pi < x_1 < \cdots < x_N = \pi$ be a partition of f into continuous segments. Fix $n \geq 1$. Let $M = \|f\|_\infty$, and let

$$\delta = \min \left\{ \frac{1}{32N(Mn)^2}, \frac{x_{i+1} - x_i}{2} : 0 \leq i < N \right\}.$$

Define a continuous function g_n on $[-\pi, \pi]$ as follows. Let $g_n(x) = f(x)$ for $x \in [x_i + \delta, x_{i+1} - \delta], 0 \leq i < N$. Also, set $g_n(-\pi) = g(\pi) = 0$. Finally, make g_n linear and continuous on each segment $J_0 = [-\pi, -\pi + \delta], J_i = [x_i - \delta, x_i + \delta], 1 \leq i < N$, and $J_N = [\pi - \delta, \pi]$. See Figure 14.1 for an example.

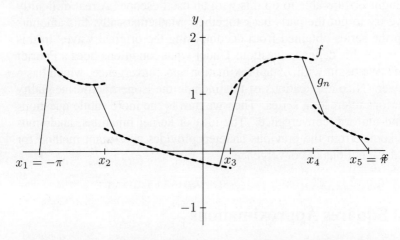

FIGURE 14.1. Piecewise continuous f with continuous approximation g_n.

Observe that $\|g_n\|_\infty \leq \|f\|_\infty = M$ and therefore $|f(x) - g_n(x)| \leq 2M$. Moreover, the two functions agree except on the intervals J_i for $0 \leq i \leq N$. The total length of these intervals is $2N\delta$. Therefore, we can estimate

$$\|f - g_n\|_2^2 \leq \sum_{i=0}^{N} \int_{J_i} (2M)^2 \, dx \leq 8NM^2\delta \leq \frac{1}{4n^2}.$$

By Corollary 13.6.6, the 2π-periodic continuous function g_n is the uniform limit of trig polynomials. So there is a trig polynomial t_n so that $\|g_n - t_n\|_\infty < \frac{1}{2n}$. Then $\|g_n - t_n\|_2 \leq \|g_n - t_n\|_\infty < \frac{1}{2n}$ as well.

$$\|f - t_n\|_2 \leq \|f - g_n\|_2 + \|g_n - t_n\|_2 < \frac{1}{2n} + \frac{1}{2n} = \frac{1}{n}$$

Hence f is an L^2 limit of trig polynomials.   ∎

The main import of the following theorem is part (2), which states that the partial sums $S_N f$ converge to f in the L^2 norm. Part (1) says that $S_N f$ is the best L^2 approximant among all trig polynomials of degree N. Since $S_N f$ is a trigonometric polynomial, it is continuous (and in fact C^∞). But our result applies to piecewise continuous functions. Since the *uniform* limit of continuous functions remains continuous, this L^2 convergence is a weaker notion.

14.1.2. LEAST SQUARES THEOREM.
Suppose that f is a piecewise continuous, 2π-periodic function. Then

(1) *If $t(\theta)$ is a trigonometric polynomial of degree N,*

$$\|f - t\|_2^2 = \|f - S_N f\|_2^2 + \|S_N f - t\|_2^2.$$

(2) $\displaystyle\lim_{N \to \infty} \|f - S_N f\|_2 = 0.$

(3) $\displaystyle\frac{1}{2\pi} \int_{-\pi}^{\pi} |f(\theta)|^2 \, d\theta = \|f\|_2^2 = A_0^2 + \frac{1}{2} \sum_{n=1}^{\infty} A_n^2 + B_n^2.$

PROOF. The L^2 norm comes from an inner product, namely

$$\langle f, g \rangle = \frac{1}{2\pi} \int_{-\pi}^{\pi} f(\theta) g(\theta) \, d\theta.$$

In Lemma 7.4.5, we have already proved that $\{1, \sqrt{2} \cos n\theta, \sqrt{2} \sin n\theta : n \geq 1\}$ form an orthonormal set in $C[-\pi, \pi]$, with this same inner product. Since that result depends only on the inner product, it also shows that the same set of functions is orthonormal in the larger inner product space of piecewise continuous, 2π-periodic functions. Notice that

$$S_N(f) = \langle f, 1 \rangle 1 + \sum_{n=1}^{N} \langle f, \sqrt{2} \cos n\theta \rangle \sqrt{2} \cos n\theta + \langle f, \sqrt{2} \sin n\theta \rangle \sqrt{2} \sin n\theta.$$

Thus, by the Projection Theorem from Section 7.5, $S_N(f)$ is the best approximant to f in the subspace spanned by $\{1, \sqrt{2} \cos n\theta, \sqrt{2} \sin n\theta : 1 \leq n \leq N\}$. Further, the inequality (7.5.4) immediately establishes (1).

By Lemma 14.1.1, f is the limit of trigonometric polynomials in the L^2 norm. Thus given $\varepsilon > 0$, choose a trig polynomial t with $\|f - t\|_2 < \varepsilon$. So it follows from (1) that for $n \geq N = \deg t$,

$$\|f - S_n f\|_2 \leq \|f - S_N f\|_2 \leq \|f - t\|_2 < \varepsilon.$$

Thus (2) holds. Finally,

$$\|f\|_2^2 = \lim_{N \to \infty} \|S_N f\|_2^2 = A_0^2 + \frac{1}{2} \sum_{n=1}^{\infty} A_n^2 + B_n^2. \qquad \blacksquare$$

14.1.3. EXAMPLE. Recall Example 13.3.2. It was shown that

$$|\theta| \sim \frac{\pi}{2} - \frac{4}{\pi} \sum_{k=0}^{\infty} \frac{\cos(2k+1)\theta}{(2k+1)^2}.$$

Compute the L^2 norm using our formula:

$$\frac{1}{2\pi} \int_{-\pi}^{\pi} |\theta|^2 \, d\theta = \left(\frac{\pi}{2}\right)^2 - \frac{1}{2}\left(\frac{4}{\pi}\right)^2 \sum_{k=0}^{\infty} \left(\frac{1}{(2k+1)^2}\right)^2$$

$$= \frac{\pi^2}{4} + \frac{8}{\pi^2} \sum_{k=0}^{\infty} \frac{1}{(2k+1)^4}.$$

The integral is easily found to be $\pi^2/3$, from which we deduce that

$$\sum_{k=0}^{\infty} \frac{1}{(2k+1)^4} = \frac{\pi^2}{8}\left(\frac{\pi^2}{3} - \frac{\pi^2}{4}\right) = \frac{\pi^4}{96}.$$

Hence

$$\sum_{k=1}^{\infty} \frac{1}{k^4} = \sum_{k=0}^{\infty} \frac{1}{(2k+1)^4} + \sum_{k=1}^{\infty} \frac{1}{(2k)^4} = \frac{\pi^4}{96} + \frac{1}{16}\sum_{k=1}^{\infty} \frac{1}{k^4}.$$

Therefore, $\displaystyle\sum_{k=1}^{\infty} \frac{1}{k^4} = \frac{\pi^4}{90}.$

An immediate consequence of this theorem is that the sines and cosines span all of $C[-\pi, \pi]$. Since they are orthogonal by Theorem 7.4.5, it follows that they are an orthonormal basis. The last result of this section requires additional background on Hilbert spaces, from Section 7.5. In Section 9.6, we showed that there is a Hilbert space of functions, $L^2(-\pi, \pi)$, which is the completion of $C[-\pi, \pi]$ in the L^2 norm. Thanks to abstract notion of integration developed there, we can define, for each L^2 function, a Fourier series using the integration formulas given in Definition 7.4.6.

If you have not studied Section 9.6, just consider elements of $L^2(-\pi, \pi)$ as limits, under the L^2 norm, of sequences of continuous functions on $[-\pi, \pi]$. We will not need the exact sense in which these limits are bona fide functions. To define a Fourier series for the limit, use the limit of the Fourier series of the continuous functions. The following theorem shows, among other things, that the limit of the Fourier series exists.

14.1.4. COROLLARY. *The functions* $\{1, \sqrt{2}\cos n\theta, \sqrt{2}\sin n\theta : n \geq 1\}$ *form an orthonormal basis for* $L^2(-\pi, \pi)$. *The map sending a sequence* $\mathbf{a} = (a_n)$ *in* $\ell^2(\mathbb{Z})$ *to* $F\mathbf{a} := f(\theta) = a_0 + \sum_{n=1}^{\infty} \sqrt{2}a_n \cos n\theta + \sqrt{2}a_{-n} \sin n\theta$ *is a unitary map. That is, F maps* $\ell^2(\mathbb{Z})$ *one-to-one and onto* $L^2(-\pi, \pi)$, *and* $\|F\mathbf{a}\|_2 = \|\mathbf{a}\|_2$ *for all* $\mathbf{a} \in \ell^2(\mathbb{Z})$.

PROOF. We first define F just on the space ℓ_0 of all sequences **a** with only finitely many nonzero terms. Then $F\mathbf{a}$ is a trigonometric polynomial, and F maps ℓ_0 onto the set of all trig polynomials. Part (3) of Theorem 14.1.2 shows that $\|F\mathbf{a}\|_2 = \|\mathbf{a}\|_2$ for each $\mathbf{a} \in \ell^2(\mathbb{Z})$.

Theorem 7.5.8 shows that $\ell^2(\mathbb{Z})$ is complete and thus is a Hilbert space. Every vector **a** is a limit of the sequence $P_n\mathbf{a} = \sum_{k=-n}^{n} a_k\mathbf{e}_k$ of vectors in ℓ_0. In particular, this sequence is Cauchy. Therefore, for each $\varepsilon > 0$, there is an integer N so that $\|P_n\mathbf{a} - P_m\mathbf{a}\|_2 < \varepsilon$ for all $n, m \geq N$. Consequently, the sequence of functions $FP_n\mathbf{a} = a_0 + \sum_{k=1}^{n} \sqrt{2}a_k \cos k\theta + \sqrt{2}a_{-k} \sin k\theta$ is also Cauchy because

$$\|FP_n\mathbf{a} - FP_m\mathbf{a}\|_2 = \|P_n\mathbf{a} - P_m\mathbf{a}\|_2 < \varepsilon \quad \text{for all} \quad n, m \geq N.$$

So this sequence converges in the L^2 norm to an element in $L^2(-\pi, \pi)$ (because our definition of L^2 is the set of all such limits).

This function f has a Fourier series, and, for example,

$$A_k = 2\langle f, \cos k\theta \rangle = \lim_{n\to\infty} 2\langle FP_n\mathbf{a}, \tfrac{1}{\sqrt{2}}F\mathbf{e}_k \rangle = \sqrt{2} \lim_{n\to\infty} \langle P_n\mathbf{a}, \mathbf{e}_k \rangle = \sqrt{2}a_k.$$

Hence $f \sim a_0 + \sum_{k=1}^{\infty} \sqrt{2}a_k \cos k\theta + \sqrt{2}a_{-k} \sin k\theta$. Moreover,

$$\|f\|_2^2 = \lim_{n\to\infty} \|FP_n\mathbf{a}\|_2 = \lim_{n\to\infty} \|P_n\mathbf{a}\|_2$$
$$= \sum_{-\infty}^{\infty} |a_n|^2 = |A_0|^2 + 2\sum_{-\infty}^{\infty} |A_n|^2 + |B_n|^2.$$

This establishes a map F which maps $\ell^2(\mathbb{Z})$ into $L^2(-\pi, \pi)$ and preserves the norm. In particular, it takes Cauchy sequences to Cauchy sequences, and vice versa. So the image space is complete. The range has been defined as the completion of the trigonometric polynomials in the L^2 norm, and thus is a subspace of $L^2(-\pi, \pi)$. Part (2) of Theorem 14.1.2 shows that the range of this map contains every continuous function. Since the range is complete, it also contains every L^2 limit of continuous functions. So the range is exactly all of $L^2(-\pi, \pi)$. This completes the proof. ∎

14.1.5. REMARK. It is easy to deduce from Appendix 13.10 that the set of complex exponentials $\{e^{in\theta} : n \in \mathbb{Z}\}$ also forms an orthonormal basis for the Hilbert space $L^2_c(-\pi, \pi)$ of complex-valued L^2 functions. Indeed, because there are no $\sqrt{2}$'s around, the formulas are cleaner. We define a map from the Hilbert space $\ell^2_c(\mathbb{Z})$ of all square summable complex sequences **c** to the L^2 function with complex Fourier series $\sum_{k=-\infty}^{\infty} c_k e^{ik\theta}$. Again, this is a unitary map. In particular, if

$f \sim \sum_{k=-\infty}^{\infty} c_k e^{ik\theta}$, then

$$\|f\|_2^2 + \frac{1}{2\pi} \int_{-\pi}^{\pi} |f(\theta)|^2 \, d\theta = \sum_{k=-\infty}^{\infty} |c_k|^2.$$

The proofs are the same except that we need to use complex inner products.

Exercises for Section 14.1

A. Compute the Fourier series of $f(\theta) = \theta^3 - \pi^2\theta$ for $-\pi \le \theta \le \pi$. Hence evaluate the sums $\sum_{n=0}^{\infty} \frac{(-1)^n}{(2n+1)^3}$ and $\sum_{n=1}^{\infty} \frac{1}{n^6}$.

B. Evaluate $\sum_{n=1}^{\infty} \frac{1}{n^8}$.

C. If $f \sim A_0 + \sum_{n=1}^{\infty} A_n \cos n\theta + B_n \sin n\theta$ is a continuous 2π-periodic function, prove that $\lim_{n\to\infty} A_n = \lim_{n\to\infty} B_n = 0$.

D. Show that $\{e^{in\theta} : n \in \mathbb{Z}\}$ forms an orthonormal basis for $C[-\pi, \pi]$.

E. Use Exercise 13.3.I to find an orthonormal basis for $C[0, \pi]$ in the given inner product.

F. (a) Compute the Fourier series of $f(\theta) = e^{a\theta}$ for $-\pi \le \theta \le \pi$ and $a > 0$.
(b) Evaluate $\|f\|_2$ in two ways, and use this to show that
$$\frac{1}{a^2} + 2\sum_{n=1}^{\infty} \frac{1}{a^2 + n^2} = \frac{\pi}{a}\left(\frac{e^{a\pi} + e^{-a\pi}}{e^{a\pi} - e^{-a\pi}}\right) = \frac{\pi}{a}\coth(a\pi).$$

G. (a) Express Parseval's Theorem in terms of the complex Fourier coefficients given in Exercise 13.10.D.
(b) Let $a \in \mathbb{R} \setminus \mathbb{Z}$, and set $f(\theta) = e^{ia\theta}$ for $\theta \in [-\pi, \pi]$. Evaluate $\|f\|_2^2$ in two ways to deduce that $\sum_{n=-\infty}^{\infty} \frac{1}{(a-n)^2} = \frac{\pi^2}{\sin^2 a\pi}$.

H. Recall the Chebychev polynomials $T_n(x) = \cos(n \cos n\theta)$. Make a change of variables in Exercise E to show that the set $\{T_0, \sqrt{2}T_n : n \ge 1\}$ is an orthonormal basis for $C[-1, 1]$ for the inner product $\langle f, g \rangle_T = \frac{1}{\pi} \int_{-1}^{1} f(x)g(x) \frac{dx}{\sqrt{1-x^2}}$.

14.2. The Isoperimetric Problem

In this section, we provide an interesting and nontrivial application of least squares approximation. The isoperimetric problem asks, What is the largest area that can be surrounded by a continuous closed curve of a given length? The answer is the circle, but a method for demonstrating this rigourously is not at all obvious.

Indeed, the Greeks were aware of the isoperimetric inequality. However, little was done in the way of a rigourous proof until the work of Steiner in 1838. Steiner gave at least five different arguments, but each one had a flaw. He could not establish the *existence* of a curve with the greatest area among all continuous curves of fixed perimeter. This difficulty was not resolved for another 50 years. In 1901, Hurwitz published the first strictly analytic proof. It is this proof that is essentially given here.

For convenience, we shall fix the length of the curve C to be 2π. This is the circumference of the circle of radius 1 and area π. We shall show that the circle is the optimal choice subject to the mild hypothesis that C is piecewise C^1. The argument to remove this differentiability requirement is left to the Exercises.

Points on the curve C may be parametrized by the arc length s as $(x(s), y(s))$ for $0 \leq s \leq 2\pi$. This is a closed curve, and thus $x(2\pi) = x(0)$ and $y(2\pi) = y(0)$. Since the differential of arc length is

$$ds = \left(x'(s)^2 + y'(s)^2\right)^{1/2} ds,$$

we have the condition

$$x'(s)^2 + y'(s)^2 = 1.$$

The area $A(C)$ is given by Green's Theorem (see Exercise 6.4.I) as

$$A(C) = \int_0^{2\pi} x(s)y'(s)\, ds = 2\pi \langle x, y' \rangle.$$

At this stage, we need a simple lemma for computing the Fourier series of a derivative.

14.2.1. LEMMA. *Suppose that* $f \sim A_0 + \sum\limits_{n=1}^{\infty} A_n \cos n\theta + B_n \sin n\theta$ *is a* C^1, 2π-*periodic function with Fourier series. Then* f' *has the Fourier series*

$$f' \sim \sum_{n=1}^{\infty} nB_n \cos n\theta - nA_n \sin n\theta.$$

PROOF. The Fourier coefficients of f' are obtained by integration by parts:

$$\frac{1}{\pi}\int_{-\pi}^{\pi} f'(t)\cos nt\, dt = \frac{1}{\pi}f(t)\cos nt\Big|_{-\pi}^{\pi} + \frac{1}{\pi}\int_{-\pi}^{\pi} f(t)n\sin nt\, dt = nB_n.$$

Similarly,

$$\frac{1}{\pi}\int_{-\pi}^{\pi} f'(t)\sin nt\, dt = \frac{1}{\pi}f(t)\sin nt\Big|_{-\pi}^{\pi} - \frac{1}{\pi}\int_{-\pi}^{\pi} f(t)n\cos nt\, dt = -nA_n.$$

And

$$\frac{1}{2\pi}\int_{-\pi}^{\pi} f'(t)\, dt = \frac{1}{2\pi}f(t)\Big|_{-\pi}^{\pi} = \frac{1}{2\pi}\big(f(\pi) - f(-\pi)\big) = 0. \qquad \blacksquare$$

Since x and y are piecewise C^1, they and their derivatives have Fourier series

$$x(s) = A_0 + \sum_{n=1}^{\infty} A_n \cos ns + B_n \sin ns$$

$$y(s) = C_0 + \sum_{n=1}^{\infty} C_n \cos ns + D_n \sin ns$$

$$x'(s) = \sum_{n=1}^{\infty} -nA_n \sin ns + nB_n \cos ns$$

$$y'(s) = \sum_{n=1}^{\infty} -nC_n \sin ns + nD_n \cos ns.$$

Let us integrate the condition $x'(s)^2 + y'(s)^2 = 1$ to get

$$1 = \frac{1}{2\pi} \int_0^{2\pi} x'(s)^2 + y'(s)^2 \, ds$$

$$= \|x'\|_2^2 + \|y'\|_2^2 = \frac{1}{2} \sum_{n=1}^{\infty} n^2 (A_n^2 + B_n^2 + C_n^2 + D_n^2).$$

The area formula yields

$$A(\mathcal{C}) = 2\pi \langle x, y' \rangle = \pi \sum_{n=1}^{\infty} n(A_n D_n - B_n C_n).$$

Therefore,

$$\pi - A(\mathcal{C}) = \frac{\pi}{2} \sum_{n=1}^{\infty} n^2 (A_n^2 + B_n^2 + C_n^2 + D_n^2) - \pi \sum_{n=1}^{\infty} n(A_n D_n - B_n C_n)$$

$$= \frac{\pi}{2} \sum_{n=1}^{\infty} (n^2 - n)(A_n^2 + B_n^2 + C_n^2 + D_n^2) +$$

$$+ \frac{\pi}{2} \sum_{n=1}^{\infty} n(A_n^2 - 2A_n D_n + D_n^2 + B_n^2 + 2B_n C_n + C_n^2)$$

$$= \frac{\pi}{2} \sum_{n=1}^{\infty} (n^2 - n)(A_n^2 + B_n^2 + C_n^2 + D_n^2) +$$

$$+ \frac{\pi}{2} \sum_{n=1}^{\infty} n(A_n - D_n)^2 + n(B_n + C_n)^2.$$

The right-hand side of this expression is clearly a sum of squares and thus is positive. The minimum value 0 is attained only if

$$D_1 = A_1, \quad C_1 = -B_1, \quad \text{and} \quad A_n = B_n = C_n = D_n = 0 \quad \text{for} \quad n \geq 2.$$

Moreover, the arc length condition gives

$$1 = \frac{1}{2}(A_1^2 + B_1^2 + C_1^2 + D_1^2) = A_1^2 + B_1^2.$$

Therefore there is a real number θ such that $A_1 = \cos\theta$ and $B_1 = \sin\theta$. Thus the optimal solutions are

$$x(s) = A_0 + \cos\theta\,\cos s + \sin\theta\,\sin s = A_0 + \cos(s - \theta)$$
$$y(s) = C_0 - \sin\theta\,\cos s + \cos\theta\,\sin s = C_0 + \sin(s - \theta).$$

Clearly, this is the parametrization of a unit circle centred at (A_0, C_0).

Finally, we should relate this proof to the historical issues discussed at the beginning of this section. Hurwitz's proof, as we just saw, results in an inequality for *all* piecewise smooth curves in which the circle evidently attains the minimum. It does not assume the existence of an extremal curve, avoiding this problematic assumption of earlier proofs.

Exercises for Section 14.2

A. (a) Show that if f is an *odd* 2π-periodic C^1 function, then $\|f\|_2 \le \|f'\|_2$.

(b) Deduce that if f is a C^1 function on $[a, b]$ such that $f(a) = f(b) = 0$, then

$$\int_a^b |f(x)|^2\,dx \le \left(\frac{b-a}{\pi}\right)^2 \int_a^b |f'(x)|^2\,dx.$$

HINT: Build an odd function g on $[-\pi, \pi]$ by identifying $[0, \pi]$ with $[a, b]$.

B. Let f be a C^2 function that is 2π-periodic. Prove that $\|f'\|^2 \le \|f\|\,\|f''\|$.
HINT: Use the Fourier series and Cauchy–Schwarz inequality.

C. (a) Consider the problem of surrounding the maximum area with a curve $\mathcal{C}$ of length 1 mile that begins and ends at points on a straight fence a mile long.
HINT: Reflect the curve in the fence.

(b) Consider the corresponding problem with a fence that makes a right angle and so covers two sides of a large field.

D. Suppose that two rays make an angle $\alpha \in (0, \pi)$. Suppose that a curve of length 1 connects one ray to the other. Find the maximum area enclosed.
HINT: Consider the effect of the transformation in polar coordinates sending (r, θ) to $(r, 2\pi\theta/\alpha)$ on both area and arc length.

E. Use Corollary 13.6.6 to extend the solution of the isoperimetric problem to arbitrary continuous curves.

14.3. The Riemann–Lebesgue Lemma

There are continuous functions with Fourier series that do not converge at every point. Such an example was first found by du Bois Reymond in 1876. Further examples have been found by Fejér and by Lebesgue. Much more recently in 1966, Carleson solved a long-standing problem conjectured 50 years earlier by Lusin. He showed that the Fourier series of a continuous function (and indeed any

L^2 function) converges for all θ except for a set of measure zero. These examples and results are beyond the scope of this course. However, under mild regularity conditions, we can establish convergence of the Fourier series.

We first show that if f is C^2, then the convergence result is easy. With more work, we will be able to handle functions that are only piecewise Lipschitz.

14.3.1. THEOREM. *If f is a C^2, 2π-periodic function, then its Fourier series converges absolutely and uniformly to f.*

PROOF. Let $f \sim A_0 + \sum\limits_{n=1}^{\infty} A_n \cos n\theta + B_n \sin n\theta$. Since both f and f' are C^1, by Lemma 14.2.1 we have

$$f' \sim \sum_{n=1}^{\infty} nB_n \cos n\theta - nA_n \sin n\theta$$

and

$$f'' \sim \sum_{n=1}^{\infty} -n^2 A_n \cos n\theta - n^2 B_n \sin n\theta.$$

By Lemma 13.3.1, the Fourier coefficients are bounded by $2\|f''\|_1$. Thus

$$|A_n| \leq \frac{2\|f''\|_1}{n^2} \quad \text{and} \quad |B_n| \leq \frac{2\|f''\|_1}{n^2} \quad \text{for} \quad n \geq 1.$$

Hence as in Exercise 13.4.D, compute

$$\|A_n \cos n\theta + B_n \sin n\theta\|_\infty \leq \frac{4\|f''\|_1}{n^2}.$$

Since $\sum\limits_{n=1}^{\infty} \dfrac{4\|f''\|_1}{n^2} < \infty$, the Weierstrass M-test shows that this Fourier series converges absolutely and uniformly. Thus by Corollary 13.6.4, it follows that this uniform limit equals f. ∎

14.3.2. EXAMPLE. Consider the function $f(\theta) = \theta^3 - \pi^2\theta$ for $-\pi \leq \theta \leq \pi$. Notice that $f(-\pi) = f(\pi) = 0$, whence f is a continuous 2π-periodic function. Moreover, $f'(\theta) = 3\theta^2 - \pi^2$ and again we have $f'(-\pi) = f'(\pi) = 2\pi^2$. So f is C^1. Finally, $f''(\theta) = 6\theta$. Since $f''(-\pi) \neq f''(\pi)$, this function is not C^2.

Let us compute the Fourier coefficients of f. Since f is odd, we need only compute the sine terms. We integrate by parts three times.

$$B_n = \frac{1}{\pi} \int_{-\pi}^{\pi} (\theta^3 - \pi^2\theta) \sin n\theta \, d\theta$$

$$= \frac{(\theta^3 - \pi^2\theta)}{\pi} \left. \frac{-\cos n\theta}{n} \right|_{-\pi}^{\pi} + \int_{-\pi}^{\pi} \frac{3\theta^2 - \pi^2}{n\pi} \cos n\theta \, d\theta$$

$$= 0 + \frac{3\theta^2 - \pi^2}{n^2\pi} \sin n\theta \left. \right|_{-\pi}^{\pi} - \int_{-\pi}^{\pi} \frac{6\theta}{n^2\pi} \sin n\theta \, d\theta$$

$$= 0 + \frac{6\theta}{n^3\pi} \cos n\theta \left. \right|_{-\pi}^{\pi} - \int_{-\pi}^{\pi} \frac{6}{n^3\pi} \cos n\theta \, d\theta$$

$$= \frac{6\pi}{n^3\pi}(-1)^n - \frac{-6\pi}{n^3\pi}(-1)^n - 0$$

$$= \frac{(-1)^n 12}{n^3}.$$

In particular, we see that these coefficients satisfy

$$\sum_{n=1}^{\infty} \left| \frac{(-1)^n 12}{n^3} \right| = 12 \sum_{n=1}^{\infty} \frac{1}{n^3} < \infty.$$

Therefore, by the Weierstrass M-test, this series converges uniformly to a continuous function. By Corollary 13.6.3, this series must converge to f uniformly on the whole circle. Therefore,

$$\theta^3 - \pi^2\theta = \sum_{n=1}^{\infty} \frac{(-1)^n 12}{n^3} \sin n\theta$$

for all $\theta \in [-\pi, \pi]$.

For example, plug in $\theta = \pi/2$. Note that $\sin(2k + 1)\pi/2 = (-1)^k$ and $\sin(2k)\pi/2 = 0$. So

$$\left(\frac{\pi}{2}\right)^3 - \pi^2 \frac{\pi}{2} = \sum_{k=0}^{\infty} \frac{(-1)^{2k+1} 12}{(2k + 1)^3}(-1)^k.$$

Solving, we find that

$$\sum_{k=0}^{\infty} \frac{(-1)^k}{(2k + 1)^3} = \frac{\pi^3}{32}.$$

Now consider the derivative $f'(\theta) = 3\theta^2 - \pi^2$. The Fourier series may be differentiated term by term. We may apply Exercise 13.4.E since

$$\sum_{n=1}^{\infty} n|B_n| = \sum_{n=1}^{\infty} \frac{12}{n^2} < \infty$$

to obtain another uniformly convergent series

$$3\theta^2 - \pi^2 = \sum_{n=1}^{\infty} \frac{(-1)^n 12}{n^2} \cos n\theta \quad \text{for} \quad -\pi \le \theta \le \pi.$$

Let us substitute $\theta = \frac{\pi}{2}$ here as well. Here $\cos \frac{(2k+1)\pi}{2} = 0$ and $\cos \frac{2k\pi}{2} = (-1)^k$.
So

$$-\frac{\pi^2}{4} = \sum_{k=1}^{\infty} \frac{12}{4k^2}(-1)^k = -3 \sum_{k=1}^{\infty} \frac{(-1)^{k-1}}{k^2}.$$

Therefore,

$$\frac{\pi^2}{12} = \sum_{k=1}^{\infty} \frac{(-1)^{k-1}}{k^2}$$

$$= \sum_{k=1}^{\infty} \frac{1}{k^2} - 2 \sum_{k=1}^{\infty} \frac{1}{(2k)^2} = \frac{1}{2} \sum_{k=1}^{\infty} \frac{1}{k^2}.$$

So we again obtain Euler's sum $\displaystyle\sum_{k=1}^{\infty} \frac{1}{k^2} = \frac{\pi^2}{6}$.

Now we have to work harder to enlarge the class of functions. An important step in establishing convergence is the apparently modest goal of showing that at least the Fourier coefficients converge to 0. This is clearly a necessary condition for any kind of convergence. This was by no means clear in the 1850s when Riemann did his fundamental work. In fact, he introduced the modern notion of integral, which we studied in Chapter 6, in order to address the question of convergence of Fourier series.

14.3.3. THE RIEMANN–LEBESGUE LEMMA.
If f is piecewise continuous on $[a, b]$ and $\tau \in \mathbb{R}$, then

$$\lim_{n \to \infty} \int_a^b f(x) \sin(nx + \tau) \, dx = 0.$$

PROOF. The assumption that f is piecewise continuous is much stronger than is really necessary. The result is true for all absolutely integrable functions. However, the proof is more difficult.

We may assume that $b - a \le \pi$. For otherwise, we just chop the interval $[a, b]$ into pieces of length at most π and prove the lemma on each piece separately. Translation by a multiple of π does not affect anything except possibly the sign of the integral. So we may assume that $[a, b]$ is contained in $[-\pi, \pi]$. Now extend the definition of f to all of $[-\pi, \pi]$ by setting $f(x) = 0$ outside of $[a, b]$. This is still piecewise continuous, and the integrals are unchanged.

We use the fact established in the last section that

$$A_0^2 + \frac{1}{2} \sum_{n=1}^{\infty} A_n^2 + B_n^2 = \|f\|_2^2 < \infty.$$

Since the terms of a convergent series must tend to 0, we have

$$\lim_{n \to \infty} A_n^2 + B_n^2 = 0.$$

So we compute

$$\left| \int_{-\pi}^{\pi} f(x) \sin(nx + \tau)\, dx \right| = \left| \int_a^b f(x) \cos nx\, \sin \tau + \sin nx\, \cos \tau\, dx \right|$$

$$= \left| A_n \sin \tau + B_n \cos \tau \right| \leq \sqrt{A_n^2 + B_n^2}.$$

The last estimate (which is included only for elegance) used the Cauchy–Schwarz inequality. This establishes the desired limit. ∎

14.3.4. COROLLARY. *If f is a piecewise continuous 2π-periodic function with Fourier series*

$$f \sim A_0 + \sum_{n=1}^{\infty} A_n \cos n\theta + B_n \sin n\theta,$$

then $\lim\limits_{n \to \infty} A_n = \lim\limits_{n \to \infty} B_n = 0.$

PROOF. Take the definitions of A_n and B_n and apply the Riemann–Lebesgue Lemma for the interval $[-\pi, \pi]$ and displacements $\tau = \pi/2$ and $\tau = 0$, respectively. ∎

Exercises for Section 14.3

A. (a) Let $0 < a \leq \pi$. Compute the Fourier series for $f(\theta) = \max\{a - |\theta|, 0\}$.

(b) Show that this series converges uniformly to f. Evaluate $f(0)$ in two ways.

(c) Rearrange the series obtained in (b) to evaluate $\sum_{n \geq 1} \dfrac{\cos na}{n^2}$.

B. (a) Find the Fourier series for $f(\theta) = \begin{cases} \sin \theta & \text{if } 0 \leq \theta \leq \pi \\ 0 & \text{if } -\pi \leq \theta \leq 0 \end{cases}.$

(b) Evaluate $\sum_{n \geq 1} \dfrac{(-1)^n}{4n^2 - 1}$.

C. Let $f(x) = \sin \frac{1}{x}$ for $0 < |x| \leq \pi$ and $f(0) = 0$. This is not piecewise continuous. Show that f is the limit of a sequence of continuous functions in the L^2 norm. Hence show that its Fourier coefficients of f tend to 0.

D. Suppose that f is an absolutely integrable function for which there is a sequence of continuous functions g_n so that $\lim\limits_{n \to \infty} \|f - g_n\|_1 = 0$. Prove the Riemann–Lebesgue Lemma for f.

E. Let f be a monotone function on $[-\pi, \pi]$. Prove that the Fourier coefficients satisfy
$$\max\{|A_n|, |B_n|\} \leq \frac{2M}{n\pi}, \text{ where } M = f(\pi) - f(-\pi).$$
HINT: Express B_n as the sum of integrals over intervals on which $\sin n\theta$ has constant sign, and combine into a single integral.

F. Consider the DE $y'' + 4y = g$, where g is an odd C^2 function with Fourier series $\sum_{n \geq 1} B_n \sin nx$.
(a) If $B_2 = 0$, find the Fourier series of the solution.
(b) Verify that this series and its second derivative converge uniformly and thus provide a solution.
(c) Show that $y = -\frac{1}{4}x \cos 2x$ is the solution for $g(x) = \sin 2x$.

G. Let g be an odd function such that $g(\theta) \geq 0$ on $[0, \pi]$.
(a) Show by induction that $\sin n\theta < n \sin \theta$ on $(0, \pi)$.
(b) Prove that the Fourier coefficients satisfy $|B_n| < nB_1$.
(c) Show by a series of examples that the previous inequality cannot be improved in general.

H. (a) Show that for any $a < b$ and any real values of α_n, $n \geq 1$,

$$\liminf_{n \to \infty} \int_a^b |\cos(nx + \alpha_n)| \, dx \geq \frac{b - a}{2}.$$

HINT: $|\cos(nx + \alpha_n)| \geq \cos^2(nx + \alpha_n)$. Express this square in terms of $\cos 2nx$ and $\sin 2nx$.
(b) Show that if r_n are positive numbers such that $\sum_{n=1}^{\infty} r_n |\cos(nx + \alpha_n)| \leq C$ for all $a \leq x \leq b$, then $\sum_{n=1}^{\infty} r_n < \infty$.

14.4. Pointwise Convergence of Fourier Series

In order to establish a more delicate convergence theorem, a better method is needed for computing the partial sums of the Fourier series. It turns out that there is an integral formula using the Dirichlet kernel, which we define in a moment. This new kernel is not nearly as nicely behaved as the Poisson kernel. Nevertheless, it provides a better method for achieving good estimates than crudely estimating individual terms.

14.4.1. DEFINITION. The sequence of functions D_n given by

$$D_n(t) = \begin{cases} \dfrac{2n + 1}{2\pi} & \text{if } t = 2\pi n, \, n \in \mathbb{Z} \\[2mm] \dfrac{\sin(n + \frac{1}{2})t}{2\pi \sin t/2} & \text{if } t \neq 2\pi n, \end{cases}$$

for $n = 1, 2, \ldots$ is called the **Dirichlet kernel**.

To see that D_n is continuous, compute

$$\lim_{t \to 0} D_n(t) = \lim_{t \to 0} \frac{\sin(n + \frac{1}{2})t}{t} \frac{t}{2\pi \sin t/2} = \frac{n + \frac{1}{2}}{2\pi/2} = \frac{2n + 1}{2\pi}.$$

The following theorem connects D_n to Fourier series.

14.4.2. THEOREM. *Let f be a piecewise continuous and 2π-periodic function. Then*

$$S_n f(x) = \int_{-\pi}^{\pi} f(x + t) D_n(t) \, dt.$$

PROOF. If we substitute the formulas for A_n and B_n into the definition of $S_n f$, then we obtain

$$S_n f(x) = A_0 + \sum_{k=1}^{n} A_n \cos nx + B_n \sin nx$$

$$= \frac{1}{2\pi} \int_{-\pi}^{\pi} f(t) \, dt + \sum_{k=1}^{n} \frac{1}{\pi} \int_{-\pi}^{\pi} f(t) \cos kt \, dt \, \cos kx +$$

$$+ \sum_{k=1}^{n} \frac{1}{\pi} \int_{-\pi}^{\pi} f(t) \sin kt \, dt \, \sin kx$$

$$= \frac{1}{2\pi} \int_{-\pi}^{\pi} f(t) \Big(1 + 2 \sum_{k=1}^{n} \cos kt \cos kx + \sin kt \sin kx \Big) \, dt$$

$$= \frac{1}{2\pi} \int_{-\pi}^{\pi} f(t) \Big(1 + 2 \sum_{k=1}^{n} \cos k(t - x) \Big) \, dt$$

using the identity $\cos(A - B) = \cos A \cos B + \sin A \sin B$. Then, substituting $u = t - x$ and using the 2π-periodicity of f, we have

$$= \int_{-\pi}^{\pi} f(x + u) \Big(\frac{1}{2\pi} + \frac{1}{\pi} \sum_{k=1}^{n} \cos ku \Big) \, du.$$

So to complete the proof, it is enough to prove the trig identity

$$\frac{1}{2\pi} \Big(1 + 2 \sum_{k=1}^{n} \cos ku \Big) = \frac{\sin(n + \frac{1}{2})u}{2\pi \sin(u/2)} = D_n(u).$$

This identity may be computed by using complex exponentials as we did in Lemma 13.10.4. Here we give a strictly real variable argument using the trig identity $2 \sin A \cos B = \sin(A + B) - \sin(A - B)$. We set up a telescoping sum, as follows:

$$\sin \tfrac{u}{2}\left(1 + 2 \sum_{k=1}^{n} \cos ku\right) = \sin \tfrac{u}{2} + \sum_{k=1}^{n} 2 \sin \tfrac{u}{2} \cos ku$$

$$= \sin \tfrac{u}{2} + \sum_{k=1}^{n} \left(\sin(k + \tfrac{1}{2})u - \sin(k - \tfrac{1}{2})u\right)$$

$$= \sin(n + \tfrac{1}{2})u$$

$$= 2\pi \sin \tfrac{u}{2} \left(\frac{\sin(n + \tfrac{1}{2})t}{2\pi \sin t/2}\right)$$

provided $u \neq 0$. If $u = 0$, then we have

$$\frac{1}{2\pi}\left(1 + \sum_{k=1}^{n} \cos 0\right) == \frac{2n + 1}{2\pi} = D_n(0),$$

and so the trig identity holds for all u.   ■

We collect here various properties of the functions D_n that we will need later. What makes the Dirichlet kernel significantly inferior to the Poisson kernel is that the Dirichlet kernel is not positive. Indeed, the following properties (2) and (3) together show that significant cancellation must occur in (2). Compare Figure 14.2 with Figure 13.2.

14.4.3. PROPERTIES OF THE DIRICHLET KERNEL.
The Dirichlet kernel D_n satisfies the following properties:

(1) *For each n, D_n is a continuous, 2π-periodic, even function.*

(2) $\displaystyle \int_{-\pi}^{\pi} D_n(t)\, dt = 1.$

(3) *For each n, $\displaystyle .28 + .4 \log n \leq \int_{-\pi}^{\pi} |D_n| \leq 2 + \log n$*

PROOF. Clearly D_n is continuous away from 0. Continuity at 0 follows from the limit computed just after the definition. The identity $D_n(t) = \tfrac{1}{2\pi}\left(1 + 2 \sum_{k=1}^{n} \cos kt\right)$ shows that D_n is even and 2π-periodic.

If $f(\theta) = 1$ for all θ, then the Fourier series of f is just $f \sim 1$. Therefore, $S_n f = 1$ for all $n \geq 1$. Using the integral formula of Theorem 14.4.2, we have

$$1 = \int_{-\pi}^{\pi} 1\, D_n(t)\, dt,$$

which proves (2).

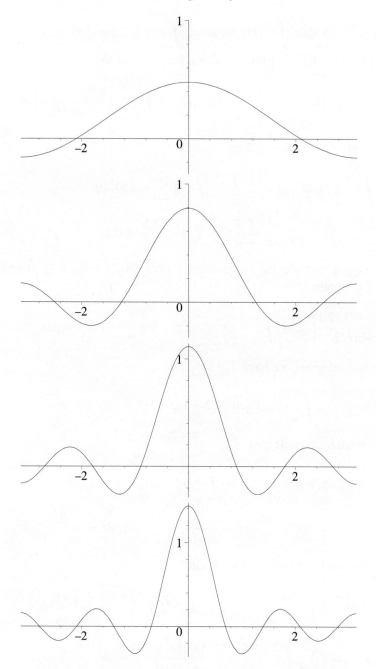

FIGURE 14.2. The graphs of D_1, D_2, D_3, and D_4.

The reader can check that $\int_{-\pi}^{\pi} |D_1(t)|\, dt = \int_{-\pi}^{\pi} |1 + 2\cos t|\, dt < 1.5$. To provide the upper bound on $\int |D_n|$ for $n \geq 2$, we first observe that

$$\int_{-\pi}^{\pi} |D_n(t)|\, dt = 2 \int_0^{\pi} |D_n(t)|\, dt = \int_0^{1/n} 2|D_n(t)|\, dt + \int_{1/n}^{\pi} \left| \frac{\sin(n + \frac{1}{2})t}{\pi \sin t/2} \right| dt.$$

The first integral is estimated for $n \geq 2$ as

$$\int_0^{1/n} 2|D_n(t)|\, dt = \frac{1}{\pi} \int_0^{1/n} \left| 1 + 2\sum_{k=1}^{n} \cos kt \right| dt$$

$$\leq \frac{1}{\pi} \frac{1}{n}(1 + 2n) \leq \frac{2.5}{\pi} < 0.8.$$

For the second integral, we use the inequalities $\pi \sin(t/2) \geq t$ for $t \geq 0$ and $|\sin(n + \frac{1}{2})t| \leq 1$ to obtain

$$\int_{1/n}^{\pi} \left| \frac{\sin(n + 1/2)t}{\pi \sin t/2} \right| dt \leq \int_{1/n}^{\pi} \frac{1}{t}\, dt = \log \pi - \log \frac{1}{n} < 1.2 + \log n.$$

Combining these two integrals, we have

$$\int_{-\pi}^{\pi} |D_n(t)|\, dt \leq 2 + \log n.$$

For the lower bound, first note that

$$\int_{k\pi/(2n+1)}^{(k+1)\pi/(2n+1)} |\sin(n + 1/2)t|\, dt = \int_0^{\pi/(2n+1)} \sin(n + 1/2)t\, dt$$

$$= \frac{2}{2n + 1} \int_0^{\pi/2} \sin t\, dt = \frac{2}{2n + 1}.$$

Now use the fact that $\sin t \leq t$ for $t \geq 0$ to obtain

$$\int_{-\pi}^{\pi} |D_n(t)|\, dt = 2 \int_0^{\pi} \left| \frac{\sin(n + 1/2)t}{2\pi \sin t/2} \right| dt \geq 2 \int_0^{\pi} \left| \frac{\sin(n + \frac{1}{2})t}{\pi t} \right| dt$$

$$= 2 \sum_{k=0}^{2n} \int_{k\pi/(2n+1)}^{(k+1)\pi/(2n+1)} \frac{|\sin(n + 1/2)t|}{\pi t}\, dt$$

$$\geq 2 \sum_{k=0}^{2n} \frac{2n + 1}{(k + 1)\pi^2} \int_{k\pi/(2n+1)}^{(k+1)\pi/(2n+1)} |\sin(n + 1/2)t|\, dt$$

$$\geq 2 \sum_{k=0}^{2n} \frac{2n + 1}{(k + 1)\pi^2} \frac{2}{2n + 1} = \frac{4}{\pi^2} \sum_{k=0}^{2n} \frac{1}{k + 1}.$$

The integral test shows that $\sum\limits_{k=0}^{2n} (k+1)^{-1} \geq \log(2n+2)$, so

$$\int_{-\pi}^{\pi} |D_n(t)|\, dt \geq \frac{4}{\pi^2}(\log 2 + \log(n+1))$$

$$> .28 + .4 \log n.$$

■

This integral formula will now be used to establish convergence for reasonably nice functions. The argument has some similarity to the Poisson Theorem, but property (3) of the previous proposition forces us to be somewhat circumspect. At a certain point, we will need to combine a Lipschitz condition with the Riemann–Lebesgue Lemma to obtain the desired estimate.

14.4.4. DEFINITION. A function f is **piecewise Lipschitz** if f is piecewise continuous, and there is a constant L so that f is has Lipschitz constant L on each interval of continuity.

A function f is **piecewise** C^1 on $[a, b]$ if f is differentiable except at finitely many points, and f' is piecewise continuous.

For example, the Heaviside function H from Example 5.2.2 is piecewise C^1. Another example is $f(x) = x - \lfloor x \rfloor$, where $\lfloor x \rfloor$ indicates the largest integer $n \leq x$. The cubic function f in Example 14.3.2 is piecewise C^2 and f' is piecewise C^1.

Being piecewise C^1 implies piecewise Lipschitz with constant $L = \|f'\|_\infty$. Indeed, observe that f' is bounded on each (closed) interval on which it is continuous because of the Extreme Value Theorem (Theorem 5.4.4); and hence it is bounded on $[-\pi, \pi]$. On any interval of continuity for f', the Mean Value Theorem shows that

$$|f(x) - f(y)| \leq \|f'\|_\infty |x - y|.$$

These estimates can then be spliced together when the function is continuous on an interval even if the derivative is not continuous.

For an example of a Lipschitz function that is not C^1, consider $f(x) = x^2 \sin \frac{1}{x}$ for $x \neq 0$ and $f(0) = 0$. The derivative is defined everywhere:

$$f'(x) = \begin{cases} 2x \sin \frac{1}{x} - \cos \frac{1}{x} & \text{for} \quad x \neq 0 \\ 0 & \text{for} \quad x = 0. \end{cases}$$

This is bounded by 3 on all of $\mathbb{R}$, so the Mean Value Theorem argument is valid. However, f' has a nasty discontinuity at the origin. So f' is not piecewise continuous.

Dirichlet proved a crucial special case of this result in 1829. In his treatise, he introduced the notion of a function which is in use today. Prior to that period, a function was typically assumed to be given by a single analytic formula.

14.4.5. THE DIRICHLET–JORDAN THEOREM.

Suppose that a function f on $\mathbb{R}$ is 2π-periodic and piecewise Lipschitz. If f is continuous at θ, then

$$\lim_{n \to \infty} S_n f(\theta) = f(\theta).$$

If f has a jump discontinuity at θ, then

$$\lim_{n \to \infty} S_n f(\theta) = \frac{f(\theta^+) + f(\theta^-)}{2}.$$

PROOF. By Theorem 14.4.3 (2),

$$f(\theta^+) = f(\theta^+) \int_{-\pi}^{\pi} D_n(t)\,dt = 2 \int_0^{\pi} f(\theta^+) D_n(t)\,dt$$

and a similar equality holds for $f(\theta^-)$. Using Theorem 14.4.2,

$$S_n f(\theta) = \int_{-\pi}^{\pi} f(\theta - t) D_n(t)\,dt$$

$$= \int_0^{\pi} \big(f(\theta + t) + f(\theta - t) \big) D_n(t)\,dt.$$

Using this formula for $S_n f(\theta)$ with the previous two equalities gives

$$S_n f(\theta) - \frac{f(\theta^+) + f(\theta^-)}{2} =$$

$$= \int_0^{\pi} \big(f(\theta + t) + f(\theta - t) \big) D_n(t)\,dt - \int_0^{\pi} \big(f(\theta^+) + f(\theta^-) \big) D_n(t)\,dt$$

$$= \int_0^{\pi} \big(f(\theta + t) - f(\theta^+) \big) D_n(t)\,dt + \int_0^{\pi} \big(f(\theta - t) - f(\theta^-) \big) D_n(t)\,dt.$$

We now consider these two integrals separately. First we prove that

$$\lim_{n \to \infty} \int_0^{\pi} \big(f(\theta + t) - f(\theta^+) \big) D_n(t)\,dt = 0.$$

An entirely similar argument will show the second integral also goes to zero as n goes to infinity. Combining these two results proves the theorem.

Let L be the Lipschitz constant for f, and let $\varepsilon > 0$ be given. Choose a positive $\delta < \varepsilon/L$ that is so small that f is continuous on $[x, x + \delta]$. Hence

$$|f(x + t) - f(x^+)| < Lt \quad \text{for all} \quad x + t \in [x, x + \delta].$$

Since $|\sin(t/2)| \geq |t|/\pi$ for all t in $[-\pi, \pi]$, it follows that

$$|D_n(t)| = \left| \frac{\sin(n + \tfrac{1}{2})t}{2\pi \sin t/2} \right| \leq \frac{1}{2|t|} \qquad -\pi \leq t \leq \pi.$$

Therefore,

$$\left| \int_0^{\delta} \big(f(x + t) - f(x^+) \big) D_n(t)\,dt \right| \leq \int_0^{\delta} Lt \frac{1}{2t}\,dt \leq \frac{L\delta}{2} < \frac{\varepsilon}{2}.$$

For the integral from δ to π, we have

$$\int_\delta^\pi \big(f(x+t) - f(x^+)\big)D_n(t)\,dt = \int_\delta^\pi \frac{f(x+t) - f(x^+)}{2\pi \sin t/2} D_n(t)\,dt$$

$$= \int_\delta^\pi g(t)\sin(n+\tfrac{1}{2})t\,dt,$$

where

$$g(t) = \frac{f(x+t) - f(x^+)}{2\pi \sin t/2}.$$

As g is piecewise continuous on $[\delta, \pi]$, we can apply the Riemann-Lebesgue Lemma to this last integral. Thus, for all n sufficiently large,

$$\left|\int_\delta^\pi \big(f(x+t) - f(x^+)\big)D_n(t)\,dt\right| = \left|\int_\delta^\pi g(t)\sin(n+\tfrac{1}{2})t\,dt\right| < \frac{\varepsilon}{2}.$$

Combining these two estimates, we have

$$\left|\int_0^\pi \big(f(x+t) - f(x^+)\big)D_n(t)\,dt\right| < \varepsilon$$

for all n sufficiently large, and thus the limit is zero. ∎

14.4.6. EXAMPLE. Let h be the following variant on the Heaviside step function:

$$h(x) = \begin{cases} -1 & \text{if} \quad -\pi < x < 0 \\ 1 & \text{if} \quad 0 \le x \le \pi. \end{cases}$$

Evidently, this function is piecewise C^1. Let us compute its Fourier series. Since h is odd, it has only sine terms.

$$B_n = \frac{1}{\pi}\int_{-\pi}^\pi h(x)\sin nx\,dx$$

$$= \frac{2}{\pi}\int_0^\pi \sin nx\,dx$$

$$= \frac{-2}{n\pi}\cos nx\,\Big|_0^\pi = \begin{cases} 0 & \text{if } n \text{ is even} \\ \frac{4}{n\pi} & \text{if } n \text{ is odd} \end{cases}$$

Thus $h \sim \sum\limits_{k=0}^\infty \dfrac{4}{(2k+1)\pi}\sin(2k+1)x$.

The Dirichlet–Jordan Theorem tells us that this series converges to 1 for all $0 < x < \pi$ and to -1 for $-\pi < x < 0$. At the points of discontinuity, it converges to the average 0. This latter fact is clear since $\sin k\pi = 0$ for all integers k. Let us plug in a few points. For example, take $x = \pi/2$. Since $\sin(2k+1)\pi/2 = (-1)^k$,

$$1 = h\Big(\frac{\pi}{2}\Big) = \sum_{k=0}^\infty \frac{4}{(2k+1)\pi}(-1)^k.$$

Therefore,

$$1 - \frac{1}{3} + \frac{1}{5} - \frac{1}{7} + \cdots = \frac{\pi}{4}.$$

Similarly, plugging in $x = 1$, we obtain

$$\sin 1 + \frac{1}{3} \sin 3 + \frac{1}{5} \sin 5 + \cdots = \frac{\pi}{4}.$$

Both of these series converge exceedingly slowly. So they have no real computational value.

We will consider this function again in Example 14.6.5; see Figure 14.5 for a graph of h and $S_{30}h$.

Exercises for Section 14.4

A. (a) Compute the Fourier series for $f(x) = x$ for $-\pi \le x \le \pi$.
   (b) Find the sum of the series.

B. (a) Find the Fourier series for $f(\theta) = \sinh(\theta)$ for $|\theta| \le \pi$.
   (b) Find a constant c so that the Fourier series for $f(\theta) - c\theta$ converges uniformly on $[-\pi, \pi]$.

C. Show that $\sum_{n=2}^{\infty} \frac{(-1)^n 2n^3}{n^4 - 1} \sin nx$ is the Fourier series of a piecewise C^1 function.
   HINT: Use the previous exercise to subtract a multiple of x to leave the uniformly convergent Fourier series of a C^1 function.

D. Sum the Dirichlet kernel using complex exponentials.

E. Sum the series $\sum_{n=1}^{\infty} \frac{\sin n\theta}{n}$.
   HINT: Exercise A and Exercise 13.3.D

F. (a) Show that the function $h(x) = \cos(x/2)$ for $-\pi \le x \le \pi$ is continuous and piecewise C^1.
   (b) Find the Fourier series for h.   HINT: $A_n = \dfrac{4(-1)^{n-1}}{\pi(4n^2 - 1)}$
   (c) Sum the series at the point $x = 0$ in two ways, and show that they yield the same result.
   HINT: $\dfrac{2}{4n^2 - 1} = \dfrac{1}{2n - 1} - \dfrac{1}{2n + 1}$

G. Prove **Dini's Test**: If f is a 2π-periodic function such that

$$\int_0^\pi \left| \frac{f(\theta_0 + t) + f(\theta_0 - t)}{2} - s \right| dt < \infty,$$

then $\lim_{n \to \infty} S_n f(\theta_0) = s$.
   HINT: Study the proof of the Dirichlet–Jordan Theorem. Look for a place where this integral condition may be used in lieu of the Lipshitz condition.

14.5. Gibbs's Phenomenon

In this section, we show that pointwise convergence of Fourier series, which we established in the previous section, is not good enough for many applications. In particular, a sequence of functions can converge pointwise without "looking like" their limit.

We have seen in Section 8.1 that pointwise convergence of functions allows surprisingly bad behaviour. Such a phenomenon arises for the functions $S_n f$ near any jump discontinuity of f. This was first discovered by an English mathematician, Wilbraham, in 1848. Around the turn of the century, it was rediscovered by Michelson and then explained by Gibbs, a (now) famous American physicist, in a letter to the journal *Nature*. For a discussion of this history, see [43]. Put simply, whenever f has a jump discontinuity, the graphs of $S_n f$ overshoot f near the discontinuity and increasing n does *not* reduce the error, it only pushes the overshoot nearer to the discontinuity. See Figure 14.3, for example.

As an example, we demonstrate the phenomenon for the 2π-periodic function given by

$$f(x) = \begin{cases} x & \text{if} \quad x \in (-\pi, \pi), \\ 0 & \text{if} \quad x = \pm\pi. \end{cases}$$

By the Dirichlet–Jordan Theorem (Theorem 14.4.5), $\lim_{n \to \infty} S_n f(x) = f(x)$, for all $x \in \mathbb{R}$. Nonetheless, $S_n f(x)$ always overshoots $f(x)$ at some point near the discontinuity by about 9% of the gap (which is 2π in this case).

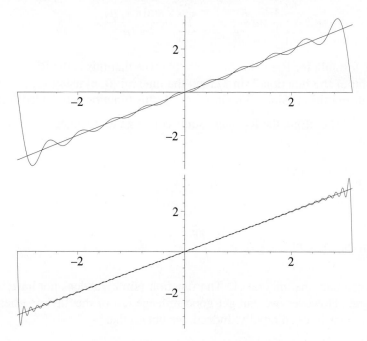

FIGURE 14.3. The graphs of $S_{10} f$ and $S_{50} f$, each plotted with f.

14.5.1. THEOREM. *Let* $A = \dfrac{2}{\pi} \displaystyle\int_0^\pi \dfrac{\sin(x)}{x}\, dx \approx 1.178979744$. *For the function f defined previously, we have*

$$\lim_{n\to\infty} S_n f\big(\pi(1 - \tfrac{1}{n})\big) = A\pi \quad and \quad \lim_{n\to\infty} S_n f\big(-\pi(1 - \tfrac{1}{n})\big) = -A\pi.$$

PROOF. Note that f is an odd function and hence has a sine series. An integration by parts argument (see Exercise 14.4.A) shows that

$$f(x) \sim 2 \sum_{k=1}^\infty \frac{(-1)^{k+1}}{k} \sin kx.$$

Thus,

$$S_n f\big(\pi(1 - \tfrac{1}{n})\big) = 2 \sum_{k=1}^n \frac{(-1)^{k+1}}{k} \sin\left(k\pi - \frac{k\pi}{n}\right)$$

$$= 2 \sum_{k=1}^n \frac{(-1)^{k+1}}{k} \left(\sin k\pi \cos \frac{k\pi}{n} - \cos k\pi \sin \frac{k\pi}{n}\right)$$

and since $\sin k\pi = 0$ and $\cos k\pi = (-1)^k$,

$$= 2 \sum_{k=1}^n \frac{1}{k} \sin \frac{k\pi}{n} = \frac{\pi}{n} \sum_{k=1}^n \frac{2 \sin(k\pi/n)}{k\pi/n}.$$

Remembering the formula for Riemann sums, we observe that this is the Riemann sum for the integral of the function $2\sin x / x$ on the interval $[0, \pi]$ using n evenly spaced points for the partition. Since $\lim_{x\to 0} \sin x / x = 1$, this function is bounded and continuous on $[0, \pi]$. Therefore, the Riemann sums converge to the integral. Thus, we have

$$\lim_{n\to\infty} S_n f\big(\pi(1 - \tfrac{1}{n})\big) = \int_0^\pi \frac{2\sin x}{x}\, dx = \pi A$$

and, similarly,

$$\lim_{n\to\infty} S_n f\big(-\pi(1 - \tfrac{1}{n})\big) = -\int_0^\pi \frac{2\sin x}{x}\, dx = -\pi A.$$

It remains to estimate the integral A. The function $(\sin x)/x$ does not have a closed form integral. However, we can get good mileage out of the Taylor series for $\sin x$ because it converges so rapidly. Indeed, we obtain that

$$\frac{\sin x}{x} = \sum_{k=0}^\infty \frac{(-1)^k}{(2k + 1)!} x^{2k}$$

for all real x. Since this converges uniformly on $[0, \pi]$, we may integrate term by term by Theorem 8.3.1. Therefore,

$$A = \frac{2}{\pi} \int_0^\pi \frac{\sin(x)}{x}\, dx = \frac{2}{\pi} \int_0^\pi \sum_{k=0}^\infty \frac{(-1)^k}{(2k+1)!} x^{2k}\, dx$$

$$= \sum_{k=0}^\infty \frac{2}{\pi} \int_0^\pi \frac{(-1)^k}{(2k+1)!} x^{2k}\, dx = \sum_{k=0}^\infty \frac{2}{\pi} \frac{(-1)^k}{(2k+1)!} \frac{\pi^{2k+1}}{(2k+1)}$$

$$= 2 \sum_{k=0}^\infty \frac{(-1)^k \pi^{2k}}{(2k+1)!(2k+1)} = 2 - \frac{\pi^2}{9} + \frac{\pi^4}{300} - \frac{\pi^6}{17640} + \cdots.$$

Note that this is an alternating series in which the terms decrease monotonely to 0. Therefore,

$$1.17357 \approx 2 - \frac{\pi^2}{9} + \frac{\pi^4}{300} - \frac{\pi^6}{17640} < A$$

$$< 2 - \frac{\pi^2}{9} + \frac{\pi^4}{300} - \frac{\pi^6}{17640} + \frac{\pi^8}{1632960} \approx 1.17938.$$

This is enough for our purposes. However, using *Maple*, we found the integral to be approximately 1.178979744. ∎

Gibbs's phenomenon depends on the finite Fourier series being a best L^2 approximation; see [44] for a related example using using piecewise linear approximations, much like Section 10.8 except that the L^2 norm replaces the uniform norm.

Exercises for Section 14.5

A. (a) Suppose that h is a C^2 function on $[-\pi, \pi]$ with $h(\pi) = h(-\pi)$ but possibly $h'(\pi) \ne h'(-\pi)$. Show that $S_n h$ converges uniformly to h.

(b) Suppose that g is a C^2 function on $[-\pi, \pi]$ but $g(\pi) \ne g(-\pi)$. Subtract a multiple of the function f used in this section from g to obtain a function h as in part (a).

(c) Hence show that g also exhibits Gibbs's phenomenon.

B. (a) Follow through the proof of Gibbs's phenomenon and show that

$$\lim_{n\to\infty} S_n f\left(\pi - \tfrac{a}{n}\right) = \int_0^a \frac{\sin x}{x}\, dx.$$

(b) Let $t_n = \displaystyle\int_{(n-1)\pi}^{n\pi} \frac{\sin x}{x}\, dx$. Show that t_n alternates in sign, $|t_{n+1}| < |t_n|$ and $\displaystyle\lim_{n\to\infty} t_n = 0$.

(c) Hence show that $\displaystyle\sup_{a>0} \left| \int_0^a \frac{\sin x}{x}\, dx \right| = \int_0^\pi \frac{\sin x}{x}\, dx$.

(d) Establish the existence of the improper Riemann integral

$$\int_0^\infty \frac{\sin x}{x}\, dx = \lim_{a\to\infty} \int_0^a \frac{\sin x}{x}\, dx.$$

C. (a) Use the formula for D_n as a sum of cosines to show that

$$\int_0^{\pi - a} 2\pi D_n(x)\, dx = \pi - a + S_n f(a),$$

where $0 < a < \pi$ and $f(x) = x$ for $-\pi \leq x \leq \pi$.

(b) Hence show that

$$|S_n f(a) - a| = \left| \int_0^{(n + \frac{1}{2})(\pi - a)} \frac{2 \sin x}{(2n + 1) \sin \frac{x}{2n+1}}\, dx - \pi \right|.$$

(c) Use the Riemann–Lebesgue Lemma to show that

$$\int_0^{(n + \frac{1}{2})\frac{\pi}{2}} \frac{2 \sin x}{(2n + 1) \sin \frac{x}{2n+1}} - \frac{2 \sin x}{x}\, dx = \int_0^{\frac{\pi}{2}} g(x) \sin(n + \tfrac{1}{2})x\, dx$$

(for a certain continuous function g) tends to 0 as n goes to $+\infty$.

(d) Use the Dirichlet–Jordan Theorem to deduce that $\displaystyle \int_0^\infty \frac{\sin x}{x}\, dx = \frac{\pi}{2}$.

14.6. Cesàro Summation of Fourier Series

Although it is natural to try to recombine the harmonics of a function f simply by adding the first number of terms, we have seen that such functions are not good approximations to f in a number of ways. In this section we consider a new sequence of approximations, built from the Fourier coefficients of f, that converge uniformly to the function f for all continuous functions f. The results of this section were found about 1900 by Fejér, a Hungarian mathematician, at the age of 19.

In order to obtain this better behaviour, we replace the sequence of functions $S_n f$ with their averages, known as **Cesàro means**:

$$\sigma_n f(x) = \frac{1}{n + 1} \sum_{k=0}^{n} S_k f(x).$$

This is defined whenever the Fourier coefficients of f are defined, which includes all absolutely integrable functions. Our primary interest will be for continuous functions. This new sequence of functions also has an associated kernel that is much better behaved than the Dirichlet kernel. It shares many of the good properties of the Poisson kernel, and computing $\sigma_n f$ does not require an infinite sum. Indeed,

$$\sigma_n f(x) = \frac{1}{n + 1} \sum_{k=0}^{n} \left(A_0 + \sum_{j=0}^{k} A_j \cos jx + B_j \sin jx \right)$$

$$= A_0 + \sum_{j=0}^{n} \left(1 - \frac{j}{n + 1} \right) \left(A_j \cos jx + B_j \sin jx \right).$$

Our first result is to turn this summation into an integral formula, using the following sequence of functions.

14.6.1. DEFINITION. The sequence of functions K_n given by

$$
K_n(t) = \begin{cases}
\dfrac{n+1}{2\pi} & \text{if } t = 2n\pi, \ n \in \mathbb{Z}, \\[2ex]
\dfrac{1}{2\pi(n+1)} \left(\dfrac{\sin\frac{n+1}{2}t}{\sin t/2} \right)^2 & \text{if } t \neq 2n\pi,
\end{cases}
$$

for $n = 1, 2, \ldots$, is called the **Fejér kernel**.

14.6.2. THEOREM. *If $f : \mathbb{R} \to \mathbb{R}$ is piecewise continuous and 2π-periodic, then*

$$
\sigma_n f(x) = \int_{-\pi}^{\pi} f(x+t) K_n(t) \, dt.
$$

PROOF. Using Theorem 14.4.2, we have

$$
\sigma_n f(x) = \frac{1}{n+1} \sum_{k=0}^{n} \int_{-\pi}^{\pi} f(t+x) \frac{\sin(k+\frac{1}{2})t}{2\pi \sin t/2} \, dt
$$

$$
= \int_{-\pi}^{\pi} f(t+x) \frac{1}{2\pi(n+1)\sin t/2} \sum_{k=0}^{n} \sin(k+\frac{1}{2})t \, dt.
$$

So all we need to establish is the trigonometric identity

$$
\sum_{k=0}^{n} \sin(k+\tfrac{1}{2})t \, \sin t/2 = \left(\sin \frac{n+1}{2} t \right)^2
$$

as we can then divide through by $\sin^2 t/2$ and substitute into the preceding integral to obtain the equality in the statement of the theorem.

One method for computing this sum is to use complex exponentials to obtain a geometric series as in Lemma 13.10.4. Here we will give a real variable argument using trig identities. Applying the identity $2 \sin A \sin B = \cos(A-B) - \cos(A+B)$ with $A = k + \frac{1}{2}$ and $B = k/2$ to the left-hand side of the identity gives

$$
\sum_{k=0}^{n} \sin(k+\tfrac{1}{2})t \, \sin t/2 = \frac{1}{2} \sum_{k=0}^{n} \left(\cos kt - \cos(k+1)t \right)
$$

$$
= \frac{1}{2}(1 - \cos(n+1)t) = \left(\sin \frac{n+1}{2} t \right)^2,
$$

where we have used the half-angle formula in the last step. ∎

You should compare Figure 14.4 with the graphs of the Poisson and Dirichlet kernels, Figures 13.2 and 14.2. The key difference between the kernels K_n and D_n is that the K_n are positive. Moreover, for $t \neq 0$, $K_n(t) \to 0$ as n goes to infinity, unlike $D_n(t)$. However, $K_n(0) \to \infty$ as n goes to infinity, exactly like D_n. It is helpful to think of the K_n as functions that become more and more like spikes (i.e., large at zero and small elsewhere), as n goes to infinity. All of these properties are shared with the Poisson kernel. This parallel allows us to obtain similar convergence results. As before, we collect the important properties of the K_n in a proposition.

14.6.3. PROPERTIES OF THE FEJÉR KERNEL. (1) *For each n, K_n is a positive, continuous, 2π-periodic, even function.*

(2) $\displaystyle\int_{-\pi}^{\pi} K_n(t)\, dt = \int_{-\pi}^{\pi} |K_n(t)|\, dt = 1.$

(3) *For $\delta \in (0, \pi)$,* $\displaystyle\lim_{n\to\infty} \left(\int_{-\pi}^{-\delta} K_n + \int_{\delta}^{\pi} K_n \right) = 0.$

(4) *Moreover, if $\delta \in (0, \pi)$, then K_n converges uniformly to zero on $[-\pi, -\delta] \cup [\delta, \pi]$.*

PROOF. It is evident from the formula that K_n is positive, even, 2π-periodic, and continuous except possibly at multiples of 2π. Because of the periodicity, it suffices to check continuity at 0. This is a simple consequence of the well-known limit $\displaystyle\lim_{t\to 0} \frac{\sin at}{t} = a.$ Hence

$$\lim_{t\to 0} K_n(t) = \lim_{t\to 0} \frac{1}{2\pi(n+1)} \left(\frac{\sin \frac{n+1}{2} t}{\sin t/2} \right)^2$$

$$= \lim_{t\to 0} \frac{1}{2\pi(n+1)} \left(\frac{\sin \frac{n+1}{2} t}{t} \right)^2 \left(\frac{t}{\sin t/2} \right)^2$$

$$= \frac{1}{2\pi(n+1)} \left(\frac{n+1}{2} \right)^2 2^2 = \frac{n+1}{2\pi}.$$

For (2), taking $f = 1$ in Theorem 14.6.2, we have

$$\int_{-\pi}^{\pi} K_n(t)\, dt = \sigma_n f(0) = 1.$$

For (3), we could use the analogous fact for the D_n and the averaging property. However, (3) is an immediate consequence of (4). For (4), we let $\varepsilon > 0$ and observe that $|\sin t/2| \geq \sin \delta/2$ for t such that $\delta \leq |t| \leq \pi$. Thus,

$$|K_n(t)| \leq \frac{1}{2(n+1)} \frac{1}{|\sin \delta/2|} \quad \text{for all} \quad t \in [-\pi, -\delta] \cup [\delta, \pi].$$

As δ is fixed, if we choose any $N \geq \varepsilon/(2 \sin \delta/2)$, then for all $n \geq N$,

$$|K_n(t)| \leq \varepsilon \quad \text{for all} \quad t \in [-\pi, -\delta] \cup [\delta, \pi].$$

That is, K_n converges uniformly to zero on $[-\pi, -\delta] \cup [\delta, \pi]$. ∎

We can now prove the main result of this section.

14.6.4. FEJÉR'S THEOREM.
If f is continuous and 2π-periodic, then $\sigma_n f$ converges uniformly to f.

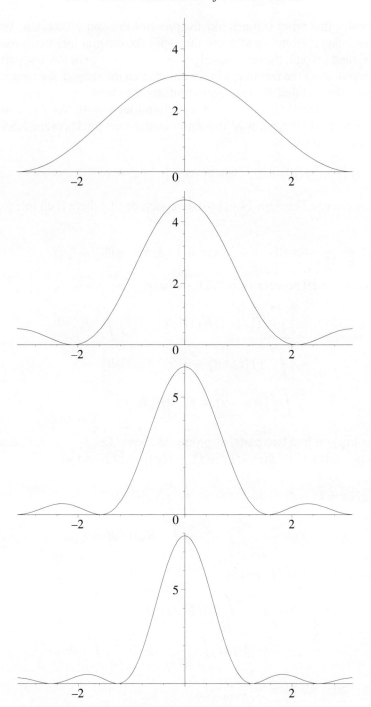

FIGURE 14.4. The graphs of K_1, K_2, K_3, and K_4.

PROOF. Conceptually, this proof is much like the proof of Poisson's Theorem. We write $\sigma_n f(x)$ as an integral from $-\pi$ to π and then split the integral into two parts: the interval $[-\delta, \delta]$, and second, the rest, namely $[-\pi, -\delta] \cup [\delta, \pi]$. On the first part, we control the integral using the fact that f is continuous; on the second, we control the integral by using the fact that K_n converges uniformly to zero.

Let $M = \|f\|_\infty = \max\{|f(x)| : x \in [-\pi, \pi]\}$ and let $\varepsilon > 0$. As f is continuous on the compact set $[-\pi, \pi]$, it is uniformly continuous by Theorem 5.5.9. Hence there is some $\delta > 0$ so that

$$|f(x) - f(y)| < \frac{\varepsilon}{2} \quad \text{whenever} \quad |x - y| < \delta.$$

With this δ fixed, we apply Theorem 14.6.3 (4), to conclude that there is an integer N so that

$$K_n(x) < \frac{\varepsilon}{8\pi M} \quad \text{for all} \quad x \in [-\pi, -\delta] \cup [\delta, \pi] \quad \text{and} \quad n \geq N.$$

Using Theorem 14.6.2 and Theorem 14.6.3 (2), we have

$$|\sigma_n f(x) - f(x)| = \left| \int_{-\pi}^{\pi} f(x+t) K_n(t)\, dt - f(x) \int_{-\pi}^{\pi} K_n\, dt \right|$$

$$= \left| \int_{-\pi}^{\pi} \big(f(x+t) - f(x)\big) K_n(t)\, dt \right|$$

$$\leq \int_{-\pi}^{\pi} |f(x+t) - f(x)| K_n(t)\, dt.$$

Now, we split this integral into two parts, as promised above. Let $I_1 = [-\delta, \delta]$ and $I_2 = [-\pi, -\delta] \cup [\delta, \pi]$. If $t \in I_1$, then $|f(x+t) - f(x)| < \varepsilon/2$, and so

$$\int_{I_1} |f(x+t) - f(x)| K_n(t)\, dt \leq \int_{I_1} \frac{\varepsilon}{2} K_n(t)\, dt$$

$$\leq \frac{\varepsilon}{2} \int_{-\pi}^{\pi} K_n(t)\, dt = \frac{\varepsilon}{2}.$$

If $t \in I_2$, then $|K_n(t)| \leq \varepsilon/(8\pi M)$ and so

$$\int_{I_2} |f(x+t) - f(x)| K_n(t)\, dt \leq \int_{I_2} 2M K_n(t)\, dt$$

$$\leq 2M \int_{I_2} \frac{\varepsilon}{8\pi M}\, dt < \frac{\varepsilon}{4\pi} \int_{-\pi}^{\pi} dt = \frac{\varepsilon}{2}.$$

Adding these two results, we have

$$|\sigma_n f(x) - f(x)| \leq \int_{-\pi}^{\pi} |f(x+t) - f(x)| K_n(t)\, dt \leq \frac{\varepsilon}{2} + \frac{\varepsilon}{2} = \varepsilon$$

for all $x \in [-\pi, \pi]$ and all $n \geq N$. Thus, by the definition of uniform convergence, $\sigma_n f$ converges uniformly to f. ∎

This proof has a strong resemblance to our proof of the Weierstrass Approximation Theorem. The common underlying technique here is controlling the integral of a product by splitting the integral into two parts, where one factor of the product is well-behaved on each part. For another example, look at the proof of one of the inequalities in Theorem 14.4.3 (3).

In fact, it is possible to prove the Weierstrass Approximation Theorem using Fejér's Theorem. One proof is outlined in the Exercises. Another will be given in Section 14.9.

14.6.5. EXAMPLE. Consider the function h introduced in Example 14.4.6,

$$h(x) = \begin{cases} 1 & \text{for} \quad 0 \le x \le \pi \\ -1 & \text{for} \quad -\pi < x < 0. \end{cases}$$

The sequence $S_n h$ will exhibit Gibbs's phenomenon (see Exercise 14.5.A) as shown in Figure 14.5. We will compute the Cesàro means for this function using the Féjer kernel. Since h is an odd function, the approximants $S_n h$ and $\sigma_n h$ are also all odd. Also, $h(\pi - x) = h(x)$, and so $S_n h$ and $\sigma_n h$ also have this symmetry. Consider a point x in $[0, \pi/2]$.

$$\sigma_n h(x) = \int_{-\pi}^{\pi} h(x+t) K_n(t)\, dt$$

$$= -\int_{-\pi}^{-x} K_n(t)\, dt + \int_{-x}^{\pi-x} K_n(t)\, dt - \int_{\pi-x}^{\pi} K_n(t)\, dt$$

$$= \int_{-x}^{x} K_n(t)\, dt - \int_{\pi-x}^{\pi+x} K_n(t)\, dt$$

Here we have exploited the fact that K_n is even to cancel the integral from x to $\pi - x$ with the integral from $x - \pi$ to $-x$.

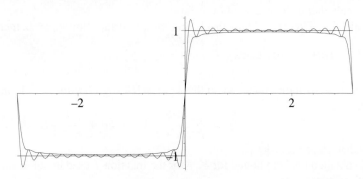

FIGURE 14.5. The graphs of h, $S_{30}h$, and $\sigma_{30}h$ on $[-\pi, \pi]$.

By Proposition 14.6.3, the first integral converges to 1 for x in $(0, \pi/2]$; while the second term tends to 0. On the other hand, $\sigma_n h(0) = 0$ for any n. Rewrite $\sigma_n h(x)$ as $2 \int_0^x K_n(t) - K_n(\pi-t)\, dt$. Since K_n is positive and decreasing on $[0, \pi]$, it follows that $\sigma_n h(x)$ is monotone increasing on $[0, \pi/2]$, then decreases symmetrically back down to 0 at π. Also, $0 < \sigma_n h(x) < 1$ here because $\int_{-x}^{x} K_n(t)\, dt < 1$. Likewise, by symmetry, $\sigma_n h(x)$ converges to -1 on $(-\pi, 0)$ and $\sigma_n h(-\pi) = 0$ for all n.

From the monotonicity, we can deduce that this convergence is uniform on intervals $[\varepsilon, \pi - \varepsilon] \cup [\varepsilon - \pi, -\varepsilon]$ for $\varepsilon > 0$. Since the function h has jump discontinuities at the points 0 or $\pm\pi$, continuous functions cannot converge uniformly to it near these points.

Exercises for Section 14.6

A. Show that if f is an absolutely integrable function with $\lim_{n\to\infty} S_n f(\theta) = a$, then
$$\lim_{n\to\infty} \sigma_n f(\theta) = a.$$

B. Show that if f is a piecewise continuous function with a jump discontinuity at θ, then
$$\lim_{n\to\infty} \sigma_n f(\theta) = \frac{f(\theta^+) + f(\theta^-)}{2}.$$

HINT: Write f as the sum of a continuous function g and a piecewise C^1 function h. Use Exercise A on h.

C. Show that $\|\sigma_n f\|_\infty \le \|f\|_\infty$.

D. Let $\sum_{j \ge 0} a_j$ be an infinite series. Define $s_n = \sum_{j=0}^{n} a_j$ and $\sigma_n = \frac{1}{n} \sum_{j=0}^{n-1} s_j$.
  (a) If $\lim_{n\to\infty} s_n = L$, show that $\lim_{n\to\infty} \sigma_n = L$.
  (b) Show by example that the converse of (a) is false.
  (c) **Hardy's Tauberian Theorem**: Show that if $\lim_{n\to\infty} n a_n = 0$ and $\lim_{n\to\infty} \sigma_n = L$, then
$$\lim_{n\to\infty} s_n = L. \quad \text{HINT: Verify that } s_N - \sigma_{N+1} = \frac{1}{n+1} \sum_{j=1}^{n} j a_j.$$

E. Suppose that f is a 2π-periodic function and that $|A_n| + |B_n| \le C/n$ for $n \ge 1$ and some constant C.
  (a) Find a bound for $S_n f(\theta) - \sigma_n f(\theta) = \sum_{k=1}^{n} \frac{k}{n+1} A_k \cos k\theta + \frac{k}{n+1} B_k \sin k\theta$.
  Hence show that $\|S_n f\|_\infty \le \|f\|_\infty + C$.
  (b) Apply this to obtain a uniform bound for $S_n f$ for the function f used in our example of Gibbs's phenomenon.

F. Prove Weierstrass's Approximation Theorem for a continuous function f on $[0, \pi]$ as follows:
  (a) Set $g(\theta) = f(|\theta|)$ for $\theta \in [-\pi, \pi]$. This is an even, continuous, 2π-periodic function. Use Fejér's Theorem to approximate g within $\varepsilon/2$ by a trig polynomial.
  (b) Use the fact that the Taylor polynomials for $\cos n\theta$ converge uniformly on $[0, \pi]$ to approximate the trigonometric polynomial by actual polynomials within $\varepsilon/2$.

14.7. Best Approximation by Trig Polynomials

We return to the theme of Chapter 10, uniform approximation. Here we are interested in approximating 2π-periodic functions by (finite) linear combinations of trigonometric functions. In this section, we obtain some reasonable estimates. Later we will establish the Jackson and Bernstein Theorems, which yield optimal estimates. It will turn out that there are close connections with approximation by polynomials which we also explore.

The **degree of a trigonometric polynomial** $q(x) = a_0 + \sum_{k=1}^{n} a_k \cos kx + b_k \sin kx$ is n if $|a_n| + |b_n| > 0$ and $a_k = b_k = 0$ for all $k > n$. We let $\mathbb{TP}_n$ denote the subspace of $C[-\pi, \pi]$ consisting of all trigonometric polynomials of degree at most n.

14.7.1. DEFINITION. The **error of approximation** to a 2π-periodic function f by trigonometric polynomials of degree n is

$$\widetilde{E}_n(f) = \inf\{\|f - q\|_\infty : q \in \mathbb{TP}_n\}$$

For example, for any 2π-periodic function $f \in C[-\pi, \pi]$, both $S_n f$ and $\sigma_n f$ are in $\mathbb{TP}_n$. The subspace $\mathbb{TP}_n$ has dimension $2n + 1$ because it is spanned by the linearly independent functions $\{1, \cos kx, \sin kx : 1 \le k \le n\}$. It follows from the compactness argument of Theorem 7.6.5 that there is a best approximation in $\mathbb{TP}_n$ to any function f. That is, given f in $C[-\pi, \pi]$, there is a trig polynomial p in $\mathbb{TP}_n$ so that

$$\|f - p\|_\infty = \inf\{\|f - q\|_\infty : q \in \mathbb{TP}_n\} = \widetilde{E}_n(f).$$

In a certain sense, the functions $S_n f$ and $\sigma_n f$ are natural approximants to f in $\mathbb{TP}_n$. Theorem 14.1.2 shows that $S_n f$ is the best L^2 norm approximant to f in $\mathbb{TP}_n$. However, it has some undesirable wildness when it comes to the uniform norm. Féjer's Theorem (Theorem 14.6.4) suggests that $\sigma_n f$ is a reasonably good approximant in the uniform norm. The following theorem gives bounds on how close $S_n f$ is to the best approximation in $\mathbb{TP}_n$. It says that $S_n f$ can be a relatively bad approximation for large n. Nevertheless, the degree of approximation by $S_n f$ is sufficiently good to yield reasonable approximations if the Fourier series decays at a sufficient rate.

On the other hand, while $\sigma_n f$ is in general a superior approximant, consider $f(x) = \sin nx$. Then $S_n f = f$ is the best approximation with $\widetilde{E}_n(f) = 0$, while $\sigma_n f(x) = \frac{1}{n+1} \sin nx$ is a rather poor estimate. In spite of this, good general results can be obtained from the obvious estimate

$$\widetilde{E}_n(f) \le \|f - \sigma_n f\|_\infty.$$

The reason that S_n works in the following theorem is the fact that $S_n p = p$ for all p in $\mathbb{TP}_n$.

14.7.2. THEOREM. *If $f : \mathbb{R} \to \mathbb{R}$ is a continuous 2π-periodic function, then*

$$\|f - S_n f\|_\infty \le (3 + \log n)\widetilde{E}_n(f).$$

PROOF. By Theorem 14.4.2,

$$S_n f(x) = \int_{-\pi}^{\pi} f(x+t) D_n(t)\, dt.$$

Hence by Theorem 14.4.3 (3), we have

$$\|S_n f\|_\infty \le \|f\|_\infty \int_{-\pi}^{\pi} |D_n(t)|\, dt \le (2 + \log n)\|f\|_\infty.$$

Observe that if $p \in \mathbb{TP}_n$, then $S_n p = p$. In particular, if we let $p \in \mathbb{TP}_n$ be the best approximation to f, then

$$
\begin{aligned}
\|f - S_n f\|_\infty &\le \|(f - p) - S_n(f - p)\|_\infty \\
&\le \|f - p\|_\infty + \|S_n(f - p)\|_\infty \\
&\le \|f - p\|_\infty + (2 + \log n)\|f - p\|_\infty \\
&\le (3 + \log n)\|f - p\|_\infty.
\end{aligned}
$$

As $\|f - p\|_\infty = \widetilde{E}_n(f)$ for our choice of p, we obtain the desired estimate. ∎

We can now apply this estimate to show that the Dirichlet–Jordan Theorem actually yields uniform convergence when the piecewise Lipschitz function is continuous.

14.7.3. THEOREM. *If f is a 2π-periodic Lipschitz function with Lipschitz constant L, then for $n \ge 2$*

$$\|f - \sigma_n f\|_\infty \le \frac{(1 + 2\log n)L}{2n}.$$

Consequently,

$$\|f - S_n f\|_\infty \le \frac{2\pi(1 + \log n)^2 L}{n}.$$

In particular, $S_n f$ converges to f uniformly on $[-\pi, \pi]$.

PROOF. We need a decent estimate for the Féjer kernel. For our purposes, the following is enough:

$$K_n(t) = \frac{1}{2\pi(n+1)}\left(\frac{\sin\frac{n+1}{2}t}{\sin\frac{1}{2}t}\right)^2 \le \min\left\{\frac{n+1}{2\pi},\ \frac{\pi}{2(n+1)t^2}\right\}.$$

Indeed, $K_n(t) \le K_n(0) = \frac{n+1}{2\pi}$ yields the first upper bound. And the inequality $|\sin t/2| \ge (2/\pi)|t/2| = |t|/\pi$ on $[-\pi, \pi]$ is enough to show that

$$\frac{1}{2\pi(n+1)}\left(\frac{\sin\frac{n+1}{2}t}{\sin\frac{1}{2}t}\right)^2 \le \frac{1}{2\pi(n+1)}\left(\frac{1}{|t|/\pi}\right)^2 = \frac{\pi}{2(n+1)t^2}.$$

The first bound is better for small $|t|$, and the second becomes an improvement at the point $\delta = \pi/n + 1$.

This proof follows the proof of Féjer's Theorem using the additional information contained in the Lipschitz condition to sharpen the error estimate. Looking back at that proof, we obtain an estimate by splitting $[-\pi, \pi]$ into two pieces. We use the Lipschitz estimate $|f(x + t) - f(x)| \leq L|t|$, and we take δ as found previously:

$$|\sigma_n f(x) - f(x)| \leq \int_{-\pi}^{\pi} |f(x + t) - f(x)| K_n(t)\, dt$$

$$\leq \int_{-\delta}^{\delta} L|t| \frac{n+1}{2\pi}\, dt + \int_{-\pi}^{-\delta} + \int_{\delta}^{\pi} L|t| \frac{\pi}{2(n+1)t^2}\, dt$$

$$= \frac{(n+1)L}{\pi} \int_{0}^{\pi/n+1} t\, dt + \frac{\pi L}{(n+1)} \int_{\pi/n+1}^{\pi} t^{-1}\, dt$$

$$= \frac{(n+1)L}{\pi} \frac{\pi^2}{2(n+1)^2} + \frac{\pi L}{(n+1)} \log(n+1)$$

$$= \frac{L\pi}{2(n+1)} \big(1 + 2\log(n+1)\big).$$

Finally, a little calculus shows that $f(x) = (1 + 2\log x)/x$ is decreasing for $x > \sqrt{e}$. So for $n \geq 2$ we obtain

$$\|\sigma_n f - f\|_\infty \leq \frac{L\pi}{2n}(1 + 2\log n).$$

Now apply Theorem 14.7.2 to obtain

$$\|f - S_n f\|_\infty \leq (3 + \log n)\, \widetilde{E}_n(f)$$

$$\leq (3 + \log n)\, \|f - \sigma_n f\|_\infty$$

$$\leq (3 + \log n) \frac{L\pi}{2n}(1 + 2\log n) \leq \frac{L\pi}{2n} 4(1 + \log n)^2.$$

Now $\lim\limits_{n\to\infty} \dfrac{2\pi L(1 + \log n)^2}{n} = 0$. Therefore, $S_n f$ converges uniformly to f. ∎

Exercises for Section 14.7

A. Show that if f is a Lipschitz function on $[-\pi, \pi]$ with Lipschitz constant L, then the Fourier coefficients satisfy $|A_n| \leq \frac{2L}{n}$ and $|B_n| \leq \frac{2L}{n}$ for $n \geq 1$.

HINT: Split the integral into n pieces and replace each integral by $\int_{c_k - \pi/n}^{c_k + \pi/n} \big(f(x) - f(c_k)\big) \sin nx\, dx$. Then estimate each piece.

B. For a 2π-periodic continuous function f, define approximants $P_{2n}(f)(\theta) =$

$$A_0 + \sum_{k=1}^{n} A_k \cos k\theta + B_k \sin k\theta + \sum_{k=1}^{n-1} (1 - \tfrac{k}{n})\big(A_{n+k} \cos(n+k)\theta + B_{n+k} \sin(n+k)\theta\big).$$

(a) Show that $P_{2n}(f) = 2\sigma_{2n}(f) - \sigma_n(f)$.
(b) Hence deduce that $\|P_{2n}(f)\|_\infty \leq 3\|f\|_\infty$.
(c) Show that if $p \in \mathbb{TP}_n$, then $P_{2n}(p) = p$.

(d) Hence show that $\|f - P_{2n}(f)\|_\infty \leq 4\widetilde{E}_n(f)$.
   HINT: Follow the method of Theorem 14.7.2.

C. Use the previous exercise to obtain a *lower bound* $\widetilde{E}_n(|\sin\theta|) > C/n$.
   HINT: See Exercise 13.6.B. Show: $\left|P_{2n}(|\sin\theta|)(0)\right| \geq \dfrac{4}{\pi} \sum_{k=n}^{\infty} \dfrac{1}{4k^2 - 1} > \dfrac{1}{\pi n}$.

D. Let $0 < \alpha < 1$. Recall that a 2π-periodic function f is of class Lip α if there is a constant L so that $|f(x) - f(y)| \leq L|x - y|^\alpha$ for all $-\pi \leq y \leq x \leq \pi$.
   (a) If $f \in$ Lip α, show that there is a constant C such that $\|f - \sigma_n f\|_\infty \leq Cn^{-\alpha}$.
   (b) Hence show that $S_n f$ converges uniformly to f.
   HINT: Follow the proof of Theorem 14.7.3 using the new estimate.

E. Let f be a 2π-periodic function with Fourier coefficients A_n and B_n.
   (a) Prove that $A_n^2 + B_n^2 \leq \frac{1}{2}\omega(f; \frac{\pi}{n})$.

   HINT: Show that $A_n = \dfrac{1}{2\pi} \displaystyle\int_{-\pi}^{\pi} f(t) - f(t + \tfrac{\pi}{n})\, dt$.

   (b) If $f \in$ Lip α, prove that there is a constant C so that $A_n^2 + B_n^2 \leq Cn^{-\alpha}$.
   (c) If f is C^p and $f^{(p)} \in$ Lip α, show that $A_n^2 + B_n^2 \leq Cn^{-p-\alpha}$.

F. Let f and g be 2π-periodic with Fourier series $f \sim A_0 + \displaystyle\sum_{n\geq 1} A_n \cos n\theta + B_n \sin n\theta$

   and $g \sim C_0 + \displaystyle\sum_{n\geq 1} C_n \cos n\theta + D_n \sin n\theta$. Suppose that f is absolutely integrable and

   g is Lipschitz. Prove that $\dfrac{1}{2\pi}\displaystyle\int_{-\pi}^{\pi} f(\theta)g(\theta)\, d\theta = A_0 C_0 + \dfrac{1}{2}\sum_{n\geq 1} A_n C_n + B_n D_n$.

   HINT: Use Theorem 14.7.3.

G. (a) Find the Fourier series of $g(x) = \begin{cases} -x - \pi & \text{for } -\pi \leq x < 0 \\ -x + \pi & \text{for } \ \ 0 \leq x \leq \pi \end{cases}$

   (b) For $n \geq 1$, define $g_n(x) = \sin(2nx)\, S_n g(x)$. Find the Fourier series for g_n.
   HINT: $2\sin A \sin B = \cos(A - B) - \cos(A + B)$
   (c) Show that $\|g_n\|_\infty \leq \pi + 2$.   HINT: Use Exercise 14.6.E.
   (d) Show that $S_{2n} g_n(0) > \log n$.

   HINT: $\displaystyle\sum_{k=1}^{n} \frac{1}{n}$ is an upper Riemann sum for $\displaystyle\int_{1}^{n+1} \frac{1}{x}\, dx$.

   (e) Hence prove that $\|g_n - S_{2n}g_n\|_\infty \geq \left(\dfrac{\log n}{\pi + 2} - 1\right)\widetilde{E}_{2n}(g_n)$. This shows that the $\log n$ term is needed in Theorem 14.7.2.
   (f) Show that $h(x) = \displaystyle\sum_{n=1}^{\infty} \frac{1}{n^2} g_{3^{n^3}}(x)$ is a continuous function such that $S_{3^{n^3}} h(0)$ diverges. NOTE: This is difficult.

14.8. Connections with Polynomial Approximation

We will now connect trigonometric polynomial approximation with polynomial approximation on an interval by exploiting an important connection between trig polynomials and the Chebychev polynomials of Section 10.7.

The idea is to relate each function f in $C[-1, 1]$ to an even function Φf in $C[-\pi, \pi]$ in such a way that the set $\mathbb{P}_n$ of polynomials of degree n is carried into $\mathbb{TP}_n$. This map is defined by

$$\Phi f(\theta) = f(\cos\theta) \quad \text{for} \quad -\pi \le \theta \le \pi.$$

Notice immediately that since $\cos\theta$ takes values in $[-1, 1]$, the right-hand side is always defined. Also,

$$\Phi f(-\theta) = f(\cos(-\theta)) = f(\cos\theta) = \Phi f(\theta).$$

Thus Φf is even.

A crucial property of the map Φ is that it is linear. Observe that for f, g in $C[-1, 1]$ and α, β in $\mathbb{R}$,

$$\Phi(\alpha f + \beta g)(\theta) = (\alpha f + \beta g)(\cos\theta) = \alpha f(\cos\theta) + \beta g(\cos\theta)$$
$$= \alpha\Phi f(\theta) + \beta\Phi g(\theta) = (\alpha\Phi f + \beta\Phi g)(\theta).$$

So $\Phi(\alpha f + \beta g) = \alpha\Phi f + \beta\Phi g$, which is linearity.

Recall from Definition 10.7.1 that the Chebychev polynomials are defined on the interval $[-1, 1]$ by $T_n(x) = \cos(n\cos^{-1} x)$. Since T_n is a polynomial of degree n, every polynomial can be expressed as a linear combination of the T_ns, and, in particular, $\mathbb{P}_n$ is spanned by $\{T_0, \ldots, T_n\}$. Also recall that there is an inner product on $C[-1, 1]$ given by

$$\langle f, g \rangle_T = \frac{1}{\pi} \int_{-1}^{1} f(x)g(x)\frac{dx}{\sqrt{1 - x^2}} \quad \text{for} \quad f, g \in C[-1, 1].$$

The Chebychev polynomials form an orthonormal set by Lemma 10.7.5.

Notice that

$$\Phi T_n(\theta) = \cos(n\cos^{-1}(\cos\theta)) = \cos(n\theta).$$

It follows that Φ maps $\mathbb{P}_n$ onto the span of $\{1, \cos\theta, \ldots, \cos n\theta\}$, which consists of the even trig polynomials in $\mathbb{TP}_n$.

This establishes the first parts of the following theorem. Let $E[-\pi, \pi]$ denote the closed subspace of $C[-\pi, \pi]$ consisting of all even continuous functions on $[-\pi, \pi]$.

14.8.1. THEOREM. *The map Φ of $C[-1, 1]$ into $E[-\pi, \pi]$ satisfies the following:*

(1) *Φ is linear, one-to-one, and onto.*

(2) *$\Phi T_n(\theta) = \cos(n\theta)$ for all $n \ge 0$.*

(3) *$\Phi(\mathbb{P}_n) = E[-\pi, \pi] \cap \mathbb{TP}_n$.*

(4) *$\|\Phi f - \Phi g\|_\infty = \|f - g\|_\infty$ for all $f, g \in C[-1, 1]$.*

(5) *$E_n(f) = \widetilde{E}_n(\Phi f)$ for all $f \in C[-1, 1]$.*

(6) *$\langle f, g \rangle_T = \langle \Phi f, \Phi g \rangle$ for all $f, g \in C[-1, 1]$.*

PROOF. For (1), linearity has already been established. To see that Φ is one-to-one, suppose that $\Phi f = \Phi g$ for functions f, g in $C[-1, 1]$. Then $f(\cos \theta) = g(\cos \theta)$ for all θ in $[-\pi, \pi]$. Since $\cos$ maps $[-\pi, \pi]$ onto $[-1, 1]$, it follows that $f(x) = g(x)$ for all x in $[-1, 1]$. Thus $f = g$ as required.

To show that Φ is surjective, we construct the inverse map from $E[-\pi, \pi]$ to $C[-\pi, \pi]$. For each even function g in $E[-\pi, \pi]$, define a function Ψg in $C[-1, 1]$ by

$$\Psi g(x) = g(\cos^{-1} x).$$

(Notice that $\cos^{-1}$ takes all values in $[0, \pi]$. So Ψg depends only on $g(\theta)$ for θ in $[0, \pi]$. This is fine because g is even.) Compute

$$\Phi \Psi g(\theta) = \Psi g(\cos \theta) = g(\cos^{-1}(\cos \theta)) = g(|\theta|) = g(\theta).$$

So Φ maps Ψg back onto g, showing that Φ maps onto $E[-\pi, \pi]$.

We proved (2) and (3) in the discussion before the theorem. Now consider (4). Note that

$$\|\Phi f\|_\infty = \sup_{\theta \in [-\pi, \pi]} |f(\cos \theta)| = \sup_{x \in [-1, 1]} |f(x)| = \|f\|_\infty.$$

Hence by linearity,

$$\|\Phi f - \Phi g\|_\infty = \|\Phi(f - g)\|_\infty = \|f - g\|_\infty.$$

Applying this, we obtain

$$\begin{aligned}
E_n(f) &= \inf\{\|f - p\|_\infty : p \in \mathbb{P}_n\} \\
&= \inf\{\|\Phi f - \Phi p\|_\infty : p \in \mathbb{P}_n\} \\
&= \inf\{\|\Phi f - q\|_\infty : q \in \mathbb{TP}_n \cap E[-\pi, \pi]\}.
\end{aligned}$$

However, since Φf is even, the trig polynomial in $\mathbb{TP}_n$ closest to Φf is also even. To see this, suppose that $r \in \mathbb{TP}_n$ satisfies $\|\Phi f - r\| = \widetilde{E}(\Phi f)$, and let

$$q(\theta) = \frac{r(\theta) + r(-\theta)}{2}.$$

You should verify that if $r(\theta) = a_0 + \sum_{k=1}^n a_n \cos k\theta + b_n \sin \theta$, then $q(\theta) = a_0 + \sum_{k=1}^n a_n \cos k\theta$. Then q belongs to $\mathbb{TP}_n \cap E[-\pi, \pi]$ and

$$\begin{aligned}
|\Phi f(\theta) - q(\theta)| &= \left| \frac{\Phi f(\theta) + \Phi f(-\theta)}{2} - \frac{r(\theta) + r(-\theta)}{2} \right| \\
&\leq \frac{1}{2} |\Phi f(\theta) - r(\theta)| + \frac{1}{2} |\Phi f(-\theta) - r(-\theta)| \\
&\leq \frac{1}{2} \widetilde{E}(\Phi f) + \frac{1}{2} \widetilde{E}(\Phi f) = \widetilde{E}(\Phi f).
\end{aligned}$$

Hence $\|\Phi f - q\|_\infty = \widetilde{E}_n(\Phi f)$. Putting this information into the preceding inequality, we obtain $E_n(f) = \widetilde{E}_n(\Phi f)$.

Finally, to prove (6), we make the substitution $x = \cos\theta$ in the integral.

$$\langle f, g \rangle_T = \frac{1}{\pi} \int_{-1}^{1} f(x)g(x) \frac{dx}{\sqrt{1 - x^2}}$$

$$= \frac{1}{\pi} \int_{\pi}^{0} f(\cos\theta)g(\cos\theta) \frac{-\sin\theta \, d\theta}{\sin\theta}$$

$$= \frac{1}{2\pi} \int_{-\pi}^{\pi} f(\cos\theta)g(\cos\theta) \, d\theta = \langle \Phi f, \Phi g \rangle \qquad \blacksquare$$

In summary, this theorem shows that $C[-1, 1]$ with the inner product $\langle \cdot, \cdot \rangle_T$ and $E[-\pi, \pi]$ with the inner product $\langle \cdot, \cdot \rangle$ are, as inner product spaces, the same.

Approximation questions for Fourier series are well studied, and this allows a transference to polynomial approximation. The reason we obtain estimates more readily in the Fourier series case is that the periodicity allows us to obtain nice integral formulas for our approximations.

Part (6) of this theorem shows that the Chebychev series for $f \in C[-1, 1]$ correspond to the Fourier series of Φf as $f \sim \sum_{k=0}^{\infty} a_k T_k$, where

$$a_0 = \langle f, 1 \rangle_T \quad \text{and} \quad a_n = 2 \langle f, T_n \rangle_T \quad \text{for} \quad n \geq 1.$$

Then define two series corresponding to the Dirichlet and Cesàro series.

$$C_n f = \sum_{k=0}^{n} a_k T_k \quad \text{and} \quad \Sigma_n f = \sum_{k=0}^{n} \left(1 - \frac{k}{n+1}\right) a_k T_k$$

Now it is just a matter of reinterpreting the Fourier series results for polynomials using Theorem 14.8.1.

14.8.2. THEOREM. *Let f be a continuous function on $[-1, 1]$ with Chebychev series $\sum_{k=0}^{\infty} a_k T_k$. Then $(\Sigma_n f)_{n=1}^{\infty}$ converges uniformly to f on $[-1, 1]$. If f is Lipschitz, then $(C_n f)_{n=1}^{\infty}$ also converges uniformly to f. In any event,*

$$\|f - C_n f\|_{\infty} \leq (3 + \log n) E_n(f).$$

PROOF. The map Φ converts the problem of approximating f by polynomials of degree n to the problem of approximating Φf by trig polynomials of degree n. Property (6) of Theorem 14.8.1 shows that $\Phi C_n f = S_n \Phi f$ and $\Phi \Sigma_n f = \sigma_n \Phi f$. So the fact that $\Sigma_n f$ converge uniformly to f on $[-1, 1]$ is a restatement of Féjer's Theorem (Theorem 14.6.4).

If f has a Lipschitz constant L, then

$$|\Phi f(\alpha) - \Phi f(\beta)| = |f(\cos\alpha) - f(\cos\beta)|$$
$$\leq L|\cos\alpha - \cos\beta| \leq L|\alpha - \beta|.$$

The last step follows from the Mean Value Theorem since the derivative of $\cos\theta$ is $\sin\theta$, which is bounded by 1. Thus Φf is Lipschitz with the same constant. (Warning: This step is not reversible.) By Theorem 14.7.3, the sequence $S_n \Phi f$ converges uniformly to Φf, whence $C_n f$ converges uniformly to f by Theorem 14.8.1(4).

Theorem 14.7.2 and Theorem 14.8.1(4) provide the estimate

$$\|f - C_n f\|_\infty = \|\Phi f - S_n f\|_\infty$$
$$\leq (3 + \log n)\tilde{E}_n(\Phi f) = (3 + \log n)E_n(f).$$ ∎

14.8.3. EXAMPLE. Let us try to approximate $f(x) = |x|$ on $[-1, 1]$. We convert this to the function

$$g(\theta) = \Phi f(\theta) = |\cos\theta| \quad \text{for} \quad -\pi \leq \theta \leq \pi.$$

This is an even function and thus has a cosine series. Also, $g(\pi - \theta) = g(\theta)$. The functions $\cos 2n\theta$ have this symmetry, but

$$\cos\big((2n + 1)(\pi - \theta)\big) = -\cos(2n + 1)\theta.$$

So $A_{2n+1} = 0$ for $n \geq 0$. Compute

$$A_0 = \frac{1}{2\pi}\int_{-\pi}^{\pi} |\cos\theta|\, d\theta = \frac{2}{\pi}$$

and for $n \geq 1$,

$$\begin{aligned}
A_{2n} &= \frac{1}{\pi}\int_{-\pi}^{\pi} |\cos\theta|\cos 2n\theta\, d\theta \\
&= \frac{4}{\pi}\int_0^{\pi/2} \cos\theta\cos 2n\theta\, d\theta \\
&= \frac{2}{\pi}\int_0^{\pi/2} \cos(2n - 1)\theta + \cos(2n + 1)\theta\, d\theta \\
&= \frac{2}{\pi}\frac{\sin(2n - 1)\theta}{2n - 1} + \frac{2}{\pi}\frac{\sin(2n + 1)\theta}{2n + 1}\bigg|_0^{\pi/2} \\
&= \frac{2}{\pi}\left(\frac{(-1)^{n-1}}{2n - 1} + \frac{(-1)^n}{2n + 1}\right) = \frac{(-1)^{n-1}4}{\pi(4n^2 - 1)}.
\end{aligned}$$

These coefficients are absolutely summable since they behave like the series $1/n^2$. Thus $S_n g$ converges uniformly to g by the Weierstrass M-test (see Exercise 13.4.D). Hence by Theorem 14.8.1, the sequence of polynomials

$$C_{2n} f(x) = \frac{2}{\pi} - \frac{4}{\pi}\sum_{k=1}^{n} \frac{(-1)^k}{4k^2 - 1}T_{2k}(x)$$

converges to $|x|$ uniformly on $[-1, 1]$.

We can make a crude estimate of the error by summing the remaining terms.

$$\big\| \, |x| - C_{2n}f(x) \big\| \le \frac{4}{\pi} \sum_{k=n+1}^{\infty} \frac{1}{4k^2 - 1} \|T_n\|_\infty$$

$$= \frac{2}{\pi} \sum_{k=n+1}^{\infty} \frac{1}{2k-1} - \frac{1}{2k+1}$$

$$= \frac{2}{\pi(2n+1)} < \frac{2/\pi}{2n}$$

This may not look very good, since $(\pi n)^{-1}$ tends to 0 so slowly. However, in 1913, Bernstein showed that $E_n(|x|) > 0.1n^{-1}$ for all n. So the Chebychev series actually gives an approximation of the correct order of magnitude.

Exercises for Section 14.8

A. Let $f \in C[-1, 1]$ and $g(\theta) = f(\cos\theta)$. Show that $\omega(g; \delta) = \omega(g|_{[0,\pi]}; \delta) \le \omega(f; \delta)$.

B. Find a sequence of polynomials that converges uniformly to

$$f(x) = \begin{cases} -x^2 & \text{for} & -1 \le x \le 0 \\ x^2 & \text{for} & 0 \le x \le 1. \end{cases}$$

C. Find a continuous function $f(x)$ on $[-1, 1]$ that is not Lipschitz but Φf is Lipschitz in $C[-\pi, \pi]$.

D. Show that there is a constant C so that $E_n(|x|) > C/n$.
    HINT: See Exercise 14.7.C. Show that $\tilde{E}_n(|\cos\theta|) > (4\pi n)^{-1}$ by a change of variables. Now use Theorem 14.8.1.

14.9. Jackson's Theorem and Bernstein's Theorem

The goal of this section is to obtain Jackson's Theorem, which provides a good estimate of the error of approximation in terms of the smoothness of the function as measured by the modulus of continuity. First, we will establish a dramatic converse (for trig polynomials) due to Bernstein in 1912 that the growth of the error function provides a good measure of the smoothness of the function.

Recall that, for $\alpha \in (0, 1]$, we defined the class $\text{Lip}\,\alpha$ as the functions f in $C[a, b]$ for which there is a constant C with

$$|f(x) - f(y)| \le C|x - y|^\alpha \quad \text{for all} \quad x, y \in [a, b].$$

In particular, Lip 1 is the class of Lipschitz functions. Observe that $f \in \text{Lip}\,\alpha$ if and only if $\omega(f; \delta) \le C\delta^\alpha$, where $\omega(f; \delta)$ is the modulus of continuity.

We will use this class to illustrate just how tight these two theorems are. The following corollary will be deduced from our two main theorems.

14.9.1. COROLLARY. *Let f be a 2π-periodic function and let $0 < \alpha < 1$. Then f is in* $\mathrm{Lip}\,\alpha$ *if and only if $\widetilde{E}_n(f) \leq Cn^{-\alpha}$ for $n \geq 1$ and some constant C.*

Bernstein's Theorem will be proved first because it is more straightforward. We begin with an easy lemma. This natural proof uses complex numbers. We indicate a strictly real argument in the Exercises.

14.9.2. LEMMA. *Suppose that $f, g \in \mathbb{TP}_n$ and $f(\theta) = g(\theta)$ for $2n+1$ distinct points in $(-\pi, \pi]$. Then $f = g$.*

PROOF. Let $\theta_1, \ldots, \theta_{2n+1}$ be the common points. Using complex exponentials, $(f - g)(\theta)$ may be expressed as

$$(f - g)(\theta) = \sum_{k=-n}^{n} a_k e^{ik\theta} = e^{-in\theta} \sum_{j=0}^{2n} a_{j-n} e^{ij\theta}.$$

Now $p(z) = \sum_{j=0}^{2n} a_{j-n} z^j$ is a polynomial of degree $2n$. Observe that $(f - g)(\theta) = e^{-in\theta} p(e^{i\theta})$. Thus the polynomial p has the roots $z_i = e^{i\theta_k}$ for $1 \leq k \leq 2n+1$. These points are distinct because $|\theta_j - \theta_k| < 2\pi$ if $j \neq k$. Therefore, $p = 0$ and so $f = g$. ∎

The key to Bernstein's Theorem is an elegant inequality. The trig polynomial $p(\theta) = \sin n\theta$ shows that the inequality is sharp (meaning in this case that the constant cannot be improved).

14.9.3. BERNSTEIN'S INEQUALITY.
Let p be a trigonometric polynomial of degree n. Then $\|p'\|_\infty \leq n\|p\|_\infty$.

PROOF. Suppose to the contrary that $p \in \mathbb{TP}_n$ but $\|p'\|_\infty > n\|p\|_\infty$. By replacing p by λp for a carefully chosen scalar λ, we may arrange that $\|p\|_\infty < 1$ yet $n < \|p'\|_\infty = p'(\theta_0)$ for some point θ_0. Choose the angle $\gamma \in [-\frac{\pi}{n}, \frac{\pi}{n}]$ such that $\sin n(\theta_0 - \gamma) = p(\theta_0)$ and the derivative $n \cos n(\theta_0 - \gamma) > 0$.

Define a trigonometric polynomial in $\mathbb{TP}_n$ by $r(\theta) = \sin n(\theta - \gamma) - p(\theta)$. Set $\alpha_k = \gamma + \frac{\pi}{n}(k + \frac{1}{2})$ for $-n \leq k \leq n$. Observe that $r(\alpha_k) = (-1)^k - p(\alpha_k)$. Since $|p(\alpha_k)| < 1$, the sign of $r(\alpha_k)$ is $(-1)^k$. By the Intermediate Value Theorem, there are $2n$ points β_k with $\alpha_k < \beta_k < \alpha_{k+1}$ such that $r(\beta_k) = 0$ for $-n \leq k < n$. The interval (α_s, α_{s+1}) containing θ_0 is special. By choice of γ, $\sin n(\theta - \gamma)$ is increasing from -1 to 1 on this interval. So $r(\alpha_s) < 0 < r(\alpha_{s+1})$. In addition, $r(\theta_0) = 0$ and

$$r'(\theta_0) = n \cos n(\theta_0 - \gamma) - p'(\theta_0) < 0.$$

Therefore there are small positive numbers ε_1 and ε_2 such that $r(\theta_0 - \varepsilon_1) > 0$ and $r(\theta_0 + \varepsilon_2) > 0$. Look at Figure 14.6. Therefore, we may apply the Intermediate Value Theorem three times in this interval. Consequently, we can find two

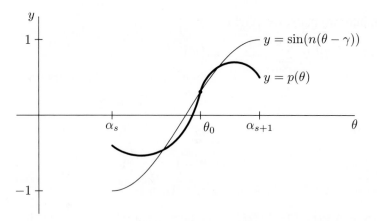

FIGURE 14.6. The graphs of $\sin n(\theta - \gamma)$ and $p(\theta)$ on $[\alpha_s, \alpha_{s+1}]$.

additional zeros, so that r has at least $2n + 2$ zeros in (α_{-n}, α_n), which is an interval of length 2π. Therefore, by Lemma 14.9.2, we reach the absurd conclusion that r is identically 0. We conclude that our assumption was incorrect, and in fact $\|p'\|_\infty \le n\|p\|_\infty$. ∎

We are now in a position to state and prove the desired result. The situation for $\alpha = 1$ is more complicated, and we refer the reader to [19]. The meaning of an error of the form $An^{-\alpha}$ for $\alpha > 1$ will be developed in the Exercises.

14.9.4. BERNSTEIN'S THEOREM.
Let f be a 2π-periodic function such that $\widetilde{E}_n(f) \le An^{-\alpha}$ for $n \ge 1$, where C is a constant and $0 < \alpha < 1$. Then f is in $\mathrm{Lip}\,\alpha$.

PROOF. Choose $p_n \in \mathbb{TP}_n$ so that $\|f - p_n\| \le An^{-\alpha}$ for $n \ge 1$. Define $q_0 = p_1$ and $q_n = p_{2^n} - p_{2^{n-1}}$ for $n \ge 1$. Note that

$$\sum_{n \ge 0} q_n(x) = \lim_{n \to \infty} p_{2^n}(x) = f(x)$$

uniformly on $\mathbb{R}$. Compute for $n \ge 1$,

$$\|q_n\| \le \|p_{2^n} - f\| + \|f - p_{2^{n-1}}\|$$
$$\le A2^{-n\alpha} + A2^{-(n-1)\alpha} \le 3A2^{-n\alpha}.$$

By the Mean Value Theorem and Bernstein's inequality, we can estimate

$$|q_n(x) - q_n(y)| \le \|q_n'\|\,|x - y|$$
$$\le 2^n \|q_n\|\,|x - y| \le 3A2^{n(1-\alpha)}|x - y|.$$

On the other hand, a simple estimate is just

$$|q_n(x) - q_n(y)| \le |q_n(x)| + |q_n(y)| \le 2\|q_n\| \le 6A2^{-n\alpha}.$$

Splitting the sum into two parts, we obtain

$$|f(x) - f(y)| \leq \sum_{n \geq 0} |q_n(x) - q_n(y)|$$

$$\leq \sum_{n=0}^{m-1} 3A2^{n(1-\alpha)}|x - y| + \sum_{n \geq m} 6A2^{-n\alpha}$$

$$\leq 3A|x - y|\frac{2^{m(1-\alpha)} - 1}{2^{1-\alpha} - 1} + 6A\frac{2^{-m\alpha}}{1 - 2^{-\alpha}}.$$

Finally, choose m so that $2^{-m} \leq |x - y| < 2^{1-m}$. Then

$$|f(x) - f(y)| \leq 3A2^{1-m}2^{m(1-\alpha)}(2^{1-\alpha} - 1)^{-1} + 6A2^{-m\alpha}(1 - 2^{-\alpha})^{-1}$$

$$\leq 6A2^{-m\alpha}\left((2^{1-\alpha} - 1)^{-1} + (1 - 2^{-\alpha})^{-1}\right) \leq B|x - y|^{\alpha},$$

where $B = 6A\left((2^{1-\alpha} - 1)^{-1} + (1 - 2^{-\alpha})^{-1}\right)$. ∎

On the other hand, Jackson's Theorem shows that smooth functions have better approximations.

14.9.5. JACKSON'S THEOREM.
Let f belong to $C[-1, 1]$. Then

$$E_n(f) \leq 6\omega(f; \tfrac{1}{n}).$$

Similarly, if g is a continuous 2π-periodic function,

$$\widetilde{E}_n(g) \leq 6\omega(g; \tfrac{1}{n}).$$

Notice that we obtain several interesting consequences immediately. This theorem shows that Proposition 10.4.4 was the correct order of magnitude, and that this result is best possible except possibly for improving the constants.

ANOTHER PROOF OF THE WEIERSTRASS APPROXIMATION THEOREM.
Jackson's Theorem not only proves that every continuous function is the limit of polynomials, it tells you how fast this happens. It suffices to prove the theorem for the interval $[-1, 1]$ (see Exercise 10.2.A). In Section 10.4, we used the *uniform* continuity of f to show that $\lim_{n \to \infty} \omega(f; \tfrac{1}{n}) = 0$. Hence

$$0 \leq \lim_{n \to \infty} E_n(f) \leq \lim_{n \to \infty} 6\omega(f; \tfrac{1}{n}) = 0.$$

This completes the proof. ∎

Let us apply Jackson's Theorem to the classes of functions that we have been discussing.

14.9.6. COROLLARY. *Let S be the class of functions in $C[0, 1]$ with Lipschitz constant 1. Then*

$$E_n(S) \leq \frac{3}{n}.$$

More generally, for Lipschitz constant 1 functions $S[a, b]$ on $[a, b]$,

$$E_n(S[a, b]) \leq \frac{3(b - a)}{n}.$$

PROOF. First consider the interval $[-1, 1]$. For any f in $S[-1, 1]$, we have the inequality $\omega(f; \frac{1}{n}) \leq \frac{1}{n}$. Thus by Jackson's Theorem, $E_n(f) \leq \frac{6}{n}$.

Now the map

$$\gamma(x) = \frac{a + b + (b - a)x}{2} \quad \text{for} \quad -1 \leq x \leq 1$$

maps $[-1, 1]$ onto $[a, b]$. So for any f in $C[a, b]$, the function $\Gamma f = f(\gamma(x))$ belongs to $C[-1, 1]$. Moreover, if $f \in S[a, b]$, then Γf has Lipschitz constant $(b - a)/2$ (see the Exercises). By the first paragraph, choose a polynomial p of degree n so that

$$\|\Gamma f - p\|_\infty \leq 6\frac{b - a}{2} = 3(b - a).$$

Then

$$q(x) = \Gamma^{-1}(q) = q(\gamma^{-1}(x)) = p\left(\frac{2x - a - b}{b - a}\right)$$

is the polynomial of degree n such that $p = \Gamma q$. Thus

$$\begin{aligned}
\|f - q\|_{[a,b]} &= \sup_{x \in [a,b]} |f(x) - q(x)| \\
&= \sup_{t \in [-1,1]} |f(\gamma(t)) - q(\gamma(t))| = \|\Gamma f - p\|_\infty \leq 3(b - a). \quad \blacksquare
\end{aligned}$$

We complete the proof of Corollary 14.9.1 with the following:

14.9.7. COROLLARY. *Suppose that f is a 2π-periodic function of class $\mathrm{Lip}\,\alpha$ for any $0 < \alpha < 1$. Then $\widetilde{E}_n(f) \leq Cn^{-\alpha}$ for $n \geq 1$ and some constant C.*

PROOF. It is immediate from the condition $|f(x) - f(y)| \leq C|x - y|^\alpha$ that $\omega(f; \frac{1}{n}) \leq Cn^{-\alpha}$. Thus Jackson's Theorem yields $\widetilde{E}_n(f) \leq 6Cn^{-\alpha}$. $\blacksquare$

The proof of Jackson's Theorem is difficult, and it requires a better method of approximation that is suited to the specific function. The key idea is convolution: to integrate the function f against an appropriate sequence of polynomials to obtain the desired approximations. These ideas work somewhat better for periodic functions, so we will use the results of the last section and consider approximation by trigonometric polynomials instead. Let $\psi(\theta) = 1 + c_1 \cos\theta + \cdots + c_n \cos n\theta$. It will be important to choose the constants c_i so that ψ is positive. When this is the case, we call ψ a **positive kernel function**. We will then try to make a good choice for these constants. For each 2π-periodic function f, define a function

$$(14.9.8) \qquad\qquad \Psi f(\theta) = \frac{1}{2\pi} \int_{-\pi}^{\pi} f(\theta - t)\psi(t)\, dt.$$

We capture the main properties in the following lemma.

14.9.9. LEMMA. *Suppose that ψ is a trig polynomial of degree n that is a positive kernel function on $[-\pi, \pi]$. Define Ψ as in Equation (14.9.8). Then*

(1) $\Psi 1 = 1$.

(2) Ψ *is linear:* $\Psi(\alpha f + \beta g) = \alpha \Psi f + \beta \Psi g$ *for all* $f, g \in C[-1, 1]$ *and* $\alpha, \beta \in \mathbb{R}$.

(3) Ψ *is monotone:* $f \geq g$ *implies that* $\Psi f \geq \Psi g$.

(4) $\Psi f \in \mathbb{TP}_n$ *for all* $f \in C[-1, 1]$.

PROOF. Part (1) uses the fact that $\cos n\theta$ has mean 0 on $[-\pi, \pi]$.

$$\Psi 1(\theta) = \frac{1}{2\pi} \int_{-\pi}^{\pi} \psi(t)\, dt = \frac{1}{2\pi} \int_{-\pi}^{\pi} 1\, dt + \sum_{k=1}^{n} a_k \int_{-\pi}^{\pi} \cos nt\, dt = 1$$

(2) Linearity follows easily from the linearity of the integral.

(3) If $h \geq 0$, then since $\psi(t) \geq 0$, it follows that $\Psi h(\theta) \geq 0$ as well. Thus if $f \geq g$, then

$$\Psi f - \Psi g = \Psi(f - g) \geq 0.$$

(4) We make a change of variables by substituting $u = \theta - t$ and using the 2π-periodicity to obtain

$$\begin{aligned}
\Psi f(\theta) &= \frac{1}{2\pi} \int_{-\pi}^{\pi} f(\theta - t)\psi(t)\, dt \\
&= \frac{1}{2\pi} \int_{\theta-\pi}^{\theta+\pi} f(u)\psi(\theta - u)\, du = \frac{1}{2\pi} \int_{-\pi}^{\pi} f(u)\psi(\theta - u)\, du \\
&= \frac{1}{2\pi} \int_{-\pi}^{\pi} f(u)\, du + \sum_{k=1}^{n} c_k \int_{-\pi}^{\pi} f(u) \cos k(\theta - u)\, du \\
&= \frac{1}{2\pi} \int_{-\pi}^{\pi} f(u)\, du + \sum_{k=1}^{n} c_k \int_{-\pi}^{\pi} f(u)\big(\cos k\theta \cos ku + \sin k\theta \sin ku\big)\, du \\
&= A_0 + \sum_{k=1}^{n} 2c_k A_k \cos k\theta + 2c_k B_k \sin k\theta,
\end{aligned}$$

where A_k and B_k are the Fourier coefficients of f. This shows that Ψf is a trig polynomial of degree at most N. ∎

Notice that Φf is a trigonometric polynomial that is a weighted linear combination of the Fourier coefficients of f much like the Cesàro means. The positivity of the kernel means that the error estimates can be made in much the same way. However, a more clever choice of the weights c_i will result in a better estimate.

14.9.10. LEMMA. *If f is a continuous function on $[a, b]$, then for any $t > 0$,*

$$\omega(f; t) \le \left(1 + \frac{t}{\delta}\right) \omega(f; \delta).$$

PROOF. Suppose that $(n - 1)\delta < t \le n\delta$. Then if $|x - y| \le t$, we may find points $y = x_0 < x_1 < \cdots < x_n = x$ such that $|x_k - x_{k-1}| \le \delta$ for $1 \le k \le n$. Hence

$$|f(x) - f(y)| \le \sum_{k=1}^{n} |f(x_k) - f(x_{k-1})|$$

$$\le n\omega(f; \delta) \le \left(1 + \frac{t}{\delta}\right)\omega(f; \delta).$$

Taking the supremum over all pairs x, y with $|x - y| \le t$ yields

$$\omega(f; t) \le \left(1 + \frac{t}{\delta}\right) \omega(f; \delta).$$

$\blacksquare$

14.9.11. LEMMA. *Suppose that $f \in C[-\pi, \pi]$ is a 2π-periodic function. Let $\psi = \sum_{k=0}^{n} c_k \cos k\theta$ be a positive kernel function of degree n. Then*

$$\|\Psi f - f\|_\infty \le \omega(f; \tfrac{1}{n})\left(1 + \frac{\pi n}{2} \sqrt{2 - c_1}\right).$$

PROOF. Apply the previous lemma with $\delta = 1/n$ to obtain

$$|f(\theta - t) - f(\theta)| \le \omega(f; |t|) \le (1 + n|t|)\, \omega(f; \tfrac{1}{n}).$$

Now using the integral formula for Ψf,

$$|\Psi f(\theta) - f(\theta)| = \left| \frac{1}{2\pi} \int_{-\pi}^{\pi} f(\theta - t)\psi(t)\, dt - f(\theta)\frac{1}{2\pi} \int_{-\pi}^{\pi} \psi(t)\, dt \right|$$

$$\le \frac{1}{2\pi} \int_{-\pi}^{\pi} |f(\theta - t) - f(\theta)|\psi(t)\, dt$$

$$\le \frac{1}{2\pi} \int_{-\pi}^{\pi} (1 + n|t|)\, \omega(f; \tfrac{1}{n})\psi(t)\, dt$$

$$= \omega(f; \tfrac{1}{n})\left(1 + \frac{n}{2\pi} \int_{-\pi}^{\pi} |t|\psi(t)\, dt\right).$$

To estimate this last term, we need the Cauchy–Schwarz inequality for integrals (Corollary 7.3.5).

$$\frac{1}{2\pi}\int_{-\pi}^{\pi}|t|\psi(t)\,dt = \frac{1}{2\pi}\int_{-\pi}^{\pi}\left(|t|\psi(t)^{1/2}\right)\psi(t)^{1/2}\,dt$$

$$\le \left(\frac{1}{2\pi}\int_{-\pi}^{\pi}t^2\psi(t)\,dt\right)^{1/2}\left(\frac{1}{2\pi}\int_{-\pi}^{\pi}\psi(t)\,dt\right)^{1/2}$$

The second integral is just 1.

Recall the easy estimate $\sin\theta \ge 2\theta/\pi$ for $0 \le \theta \le \pi/2$. This yields

$$1 - \cos t = 2\sin^2\frac{t}{2} \ge 2\frac{4}{\pi^2}\left(\frac{t}{2}\right)^2 = \frac{2}{\pi^2}t^2 \quad \text{for} \quad -\pi \le t \le \pi.$$

Substitute this back into our integral:

$$\frac{1}{2\pi}\int_{-\pi}^{\pi}t^2\psi(t)\,dt \le \frac{1}{2\pi}\int_{-\pi}^{\pi}\frac{\pi^2}{2}(1 - \cos t)\psi(t)\,dt$$

$$= \frac{\pi^2}{2}\left(\frac{1}{2\pi}\int_{-\pi}^{\pi}\psi(t)\,dt - \frac{1}{2\pi}\int_{-\pi}^{\pi}\psi(t)\cos t\,dt\right)$$

$$= \frac{\pi^2}{2}\left(1 - \frac{c_1}{2}\right)$$

Therefore, we obtain

$$\left|\Psi f(\theta) - f(\theta)\right| \le \left(1 + \frac{n\pi}{2}\sqrt{2 - c_1}\right)\omega(f; \tfrac{1}{n}). \qquad\blacksquare$$

PROOF OF JACKSON'S THEOREM. This lemma suggests trying to minimize $2 - c_1$ over all positive kernel functions of degree n. The most straightforward method is to write down a kernel that gets excellent estimates. Consider the complex trig polynomial

$$p(\theta) = \sum_{k=0}^{n}a_k e^{ik\theta} = \sum_{k=0}^{n}\sin\frac{(k+1)\pi}{n+2}e^{ik\theta}.$$

Our kernel will be $\psi(\theta) = c|p(\theta)|^2$. The constant c is chosen to make the constant coefficient equal to 1. Positivity of ψ is automatic since it is the square of the modulus of p. It remains merely to do a calculation to determine the coefficients.

First, we compute

$$|p(\theta)|^2 = \sum_{j=0}^{n}a_j e^{ij\theta}\sum_{k=0}^{n}a_k e^{-ik\theta} = \sum_{j=0}^{n}\sum_{k=0}^{n}a_j a_k e^{i(j-k)\theta}$$

$$= \sum_{j=0}^{n}a_j^2 + \sum_{s=1}^{n}\sum_{k=0}^{n-s}a_k a_{k+s}\left(e^{is\theta} + e^{-is\theta}\right)$$

$$= b_0 + \sum_{s=1}^{n}2b_s\cos s\theta,$$

where $b_s = \sum_{k=0}^{n-s} a_k a_{k+s}$ for $0 \leq s \leq n$. We choose $c = 1/b_0 = \left(\sum_{j=0}^{n} a_j^2 \right)^{-1/2}$. It is clear that ψ is a cosine polynomial of degree n with constant coefficient equal to 1. So ψ is a positive kernel function.

So now we compute the coefficient b_1. We use some clever manipulations and the identity $\sin A + \sin B = 2 \sin \frac{A+B}{2} \cos \frac{A-B}{2}$. Notice that the periodicity of the coefficients is used in the second line.

$$2b_1 = 2 \sum_{k=0}^{n-1} a_k a_{k+1} = \sum_{k=0}^{n-1} 2 \sin \frac{(k+1)\pi}{n+2} \sin \frac{(k+2)\pi}{n+2}$$

$$= \sum_{k=1}^{n} 2 \sin \frac{k\pi}{n+2} \sin \frac{(k+1)\pi}{n+2} = \sum_{k=0}^{n-1} 2 \sin \frac{k\pi}{n+2} \sin \frac{(k+1)\pi}{n+2}$$

$$= \sum_{k=0}^{n-1} \sin \frac{(k+1)\pi}{n+2} \left(\sin \frac{(k+2)\pi}{n+2} + \sin \frac{k\pi}{n+2} \right)$$

$$= 2 \sum_{k=0}^{n-1} \sin^2 \frac{(k+1)\pi}{n+2} \cos \frac{\pi}{n+2} = 2b_0 \cos \frac{\pi}{n+2}$$

Hence

$$c_1 = \frac{2b_1}{b_0} = 2 \cos \frac{\pi}{n+2}.$$

So

$$1 + \frac{n\pi}{2} \sqrt{2 - c_1} = 1 + \frac{n\pi}{2} \sqrt{2 - 2 \cos \frac{\pi}{n+2}}$$

$$= 1 + \frac{n\pi}{2} 2 \sin \frac{\pi}{2(n+2)}$$

$$\leq 1 + n\pi \frac{\pi}{2(n+2)} < 1 + \frac{\pi^2}{2} < 6.$$

The proof is now completed by appealing to Lemma 14.9.11. ∎

Exercises for Section 14.9

A. Suppose that f in $C[a,b]$ has Lipschitz constant L. Let $\gamma(x) = Ax + B$. Show that $f(\gamma(x))$ has Lipschitz constant AL.

B. Suppose that f is a 2π-periodic function such that $\widetilde{E}_n(f) \leq Cn^{-p-\alpha}$, where $p \in \mathbb{N}$ and $0 < \alpha < 1$. Prove that f is C^p and that $f^{(p)}$ is in the class Lip α as follows.
   (a) Write f as a sum of the polynomials q_n as in the proof of Bernstein's Theorem. Apply Bernstein's inequality p times to the q_n's. Show that the resulting series of derivatives still converges, and deduce that f is C^p.
   (b) Use this series to show that $\widetilde{E}_n(f^{(p)}) \leq C'n^{-\alpha}$. Then finish the argument.

C. (a) Suppose that f is C^1 on $[-1, 1]$ and $\|f'\|^\infty = M$. Show that $E_n(f) \leq \frac{6M}{n}$.

   (b) Now choose a polynomial p of degree $n - 1$ so that $\|f' - p\|_\infty = E_{n-1}(f')$. Let
   $$q(x) = \int_0^x p(t)\, dt. \text{ Show that } E_n(f) = E_n(f - q) \leq \frac{6}{n} E_{n-1}(f').$$

D. (a) Use induction on the previous exercise to show that if f has k continuous deriva-tives on $[-1, 1]$, then for $n > k$,
   $$E_n(f) \leq \frac{6^k}{n(n-1)\cdots(n+1-k)} E_{n-k}(f^{(k)}).$$

   (b) Hence show that there is a constant C_k so that for $n > k$.
   $$E_n(f) \leq \frac{C_k}{n^k}\, \omega\left(f^{(k)}; \frac{1}{n-k}\right)$$
   Find this constant explicitly for $k = 2$.

E. Do a change of variables in the previous exercise to show that if f has k continuous derivatives on $[a, b]$, then for $n > k$.
   $$E_n(f) \leq \frac{C_k(b-a)^k}{n^k}\, \omega\left(f^{(k)}; \frac{b-a}{2(n-k)}\right)$$

F. Show that if f is a 2π-periodic function and $\widetilde{E}_n(f) \leq C/n$, then $\omega(f; \delta) \leq B\delta|\log \delta|$.
   HINT: Study the proof of Bernstein's Theorem using $\alpha = 1$.

G. Prove the **Dini–Lipschitz Theorem**: If f is a continuous function on $[-1, 1]$ such that $\lim_{n \to \infty} \omega(f; \frac{1}{n}) \log n = 0$, then the Chebychev series $C_n f$ converges uniformly to f.
   HINT: Combine Jackson's Theorem with Theorem 14.8.2.

H. Let $0 < \alpha < 1$.
   (a) Show that $f(x) = |x|^\alpha$ belongs to Lip α.
   (b) Modify the proof of Proposition 10.4.4 to obtain a lower bound for $E_n(\text{Lip}\,\alpha)$.
   HINT: Piece together translates of $|x|^\alpha$.

CHAPTER 15

Wavelets

15.1. Introduction

In this chapter we develop an important variation on Fourier series, replacing the sine and cosine functions with new families of functions, called wavelets. The strategy is to construct wavelets so that they have some of the good properties of trig functions but avoid the failings of Fourier series that we have seen in previous chapters. With such functions, we can develop new versions of Fourier series methods that will work well for problems where traditional Fourier series work poorly.

What are the good properties of trig functions? First and foremost, we have an orthogonal basis in L^2, namely the set of functions $\sin(nx)$ and $\cos(nx)$ as n runs over $\mathbb{N}_0$. This leads to the idea of breaking up a wave into its harmonic constituents, as the sine and cosine functions appear in the solution of the wave equation. We want to retain some version of this orthogonality.

Fix a positive integer n and consider the span of $\{\sin(nx), \cos(nx)\}$, call it A_n. If $f(t)$ is in the subspace A_n then so is the translated function $f(t-x)$ and, for a positive integer k, the dilated function $f(kt)$ is in A_{kn} (see Exercise 15.1.A). That is, translation leaves each subspace A_n invariant and dilation by k carries A_n to A_{kn} for each n. Moreover, these orthogonal subspaces together span all of $L^2[-\pi, \pi]$. There is a similar decomposition for wavelets, called a multiresolution, and it is central to the study of wavelets.

What are the problems with Fourier series that we would like to fix? Fourier coefficients, and hence the Fourier series approximation, depend on all values of the function. For example, if you change a function f a small amount on the interval $[0, 0.01]$, it is possible that every Fourier coefficient changes. This will then have an effect on the partial sums $S_n f(\theta)$ for all values of θ. Although these changes may be small, there are many subtleties in analyzing Fourier series approximations, as we have seen.

Further, for a badly behaved function, such as a nondifferentiable or discontinuous one, the coefficients decrease slowly. Exercises 13.4.F and 13.4.C show that the Fourier coefficients of a function go rapidly to 0 only when the functions has several continuous derivatives. Thus, we may need many terms to get a close approximation, even at a point relatively far away from the discontinuity, as in Example 14.4.6.

The partial sums $S_n f(\theta)$ do not always converge to $f(\theta)$ when f is merely continuous. Thanks to Gibbs's phenomenon, $S_n f(\theta)$ will always exhibit bad behaviour near discontinuities, no matter how large n is. While we can get better approximations by using $\sigma_n f(\theta)$ instead of $S_n f(\theta)$, this will not resolve such problems as slowly decreasing Fourier coefficients.

This suggests looking for a series expansion with better local properties, meaning that coefficients reflect the local behaviour of the function and a small change on one interval affects only a few of the series coefficients and leaves unchanged the partial sums elsewhere in the domain. It may seem unlikely that there are useful wavelet bases with this local approximation property that still have nice behaviour under translation and dilation. However, they do exist, and they were developed in the 1980s. The discovery has provoked a vast literature of both theoretical and practical importance. No one family of wavelets is ideal for all problems, but we can develop different wavelets to solve specific problems. Developing such wavelets is an important practical problem.

In this chapter, we will illustrate some of the general features of wavelets. The basic example is the Haar wavelet, a rather simple case that is not the best for applications but illuminates the general theory. We construct one of the most used wavelets, the Daubechies wavelet, although we don't prove that it is continuous. This requires tools we don't have; most notably, the Fourier transform. We establish the existence of another continuous wavelet, the Franklin wavelet, but this requires considerable work. Our focus is the use of real analysis in the foundational theory. We leave the development of efficient computational strategies to more specialized treatments, such as those in the bibliography.

Most of the literature deals with bases for functions on the whole real line rather than for periodic functions, so we will work in this context. This means that we will be looking for special orthonormal bases for $L^2(\mathbb{R})$, the Hilbert space of all square integrable functions on $\mathbb{R}$ with the norm

$$\|f\|_2^2 = \int_{-\infty}^{+\infty} |f(x)|^2 \, dx.$$

As in Section 9.6, we define $L^2(\mathbb{R})$ as the completion of $C_c(\mathbb{R})$, the continuous functions of compact support on $\mathbb{R}$, in the L^2 norm.

15.1.1. DEFINITION. A **wavelet** is a function $\psi \in L^2(\mathbb{R})$ such that the set

$$\{\psi_{kj}(x) = 2^{k/2} \psi(2^k x - j)(x) : j, k \in \mathbb{Z}\}$$

forms an orthonormal basis for $L^2(\mathbb{R})$. Sometimes ψ is called the **mother wavelet**.

This is more precisely called a dyadic wavelet to stress that dilations are taken to be powers of 2. This is a common choice but is not the most general one. Notice that the wavelet basis has two parameters, whereas the Fourier basis for $L^2(\mathbb{T})$ has only one, given by dilation alone. From the complex point of view, sines and cosines are written in terms of the exponential function $\psi(\theta) = e^{i\theta}$, and the functions $\psi(k\theta) = e^{ik\theta}$ for $k \in \mathbb{Z}$ form an orthonormal basis for $L^2(-\pi, \pi)$. A singly generated family of this form cannot have the local behaviour we are seeking.

Exercises for Section 15.1

A. (a) Given a Fourier series $f(\theta) \sim A_0 + \sum_{n=1}^{\infty} A_n \cos n\theta + B_n \sin n\theta$, if $f(k\theta) \sim A_0 + \sum_{i=1}^{\infty} C_i \cos i\theta + D_i \sin i\theta$, find the formula for C_i and D_i in terms of the A_n, the B_n, and k.

    (b) Similarly, if $f(\theta - x) \sim A_0 + \sum_{i=1}^{\infty} C_i \cos i\theta + D_i \sin i\theta$, find the formula for C_i and D_i in terms of the A_n, the B_n, and x.

B. Show that if ψ is a function in $L^2(\mathbb{R})$ such that $\{\psi_{0j} : j \in \mathbb{Z}\}$ is an orthonormal set, then $\{\psi_{kj} : j \in \mathbb{Z}\}$ is an orthonormal set for each $k \in \mathbb{Z}$.

C. A map U from a Hilbert space $\mathcal{H}$ to itself is **unitary** if $\|Ux\| = \|x\|$ for all vectors $x \in \mathcal{H}$ and $U\mathcal{H} = \mathcal{H}$. Define linear maps on $L^2(\mathbb{R})$ by $Tf(x) = f(x - 1)$ and $D_f f(x) = \sqrt{2} f(2x)$. Show that these maps are unitary.

D. Let ψ be a wavelet, and let T and D be the unitary maps defined previously. What is the relationship between the subspaces spanned by $\{T^n D\psi : n \in \mathbb{Z}\}$ and $\{DT^n\psi : n \in \mathbb{Z}\}$?

E. Let ψ be a function in $L^2(\mathbb{R})$ such that $\{\psi_{0j} : j \in \mathbb{Z}\}$ is an orthonormal set. Let χ be the characteristic function of the set $\{x \in \mathbb{R} : x - [x] < \frac{1}{2}\}$, where $[x]$ is the greatest integer $n \leq x$. Define $\varphi(x) = \chi(x)\psi(-x - \frac{1}{2}) - (1 - \chi(x))\psi(\frac{1}{2} - x)$.

    (a) Show that φ is orthogonal to ψ_{0j} for all $j \in \mathbb{Z}$.

    (b) Hence deduce that there is no function ψ in $L^2(\mathbb{R})$ such that the set of translates $\{\psi_{0j} : j \in \mathbb{Z}\}$ is an orthonormal basis for $L^2(\mathbb{R})$.

    (c) Show that $\{\psi_{0j}, \varphi_{0j} : j \in \mathbb{Z}\}$ is an orthonormal set.

F. For $t \in \mathbb{R}$, define $T_t f(x) = f(x - t)$ for $f \in L^2(\mathbb{R})$. Show that if $\lim_{n \to \infty} t_n = t$, then $\lim_{n \to \infty} T_{t_n} f = T_t f$ for every $f \in L^2(\mathbb{R})$.

    HINT: If f is continuous with compact support, use the fact that it is uniformly continuous. Next, approximate an arbitrary f.

15.2. The Haar Wavelet

To get started, we describe the **Haar system** for $L^2(0, 1)$. This will then lead to a wavelet basis for $L^2(\mathbb{R})$. For $a < b$, let $\chi_{[a,b)}$ denote the characteristic function of $[a, b)$. Set $\varphi = \chi_{[0,1)}$ and $\psi = \chi_{[0,.5)} - \chi_{[.5,1)}$. Then define

$$\psi_{kj}(x) = 2^{k/2}\psi(2^k x - j)(x) \quad \text{for all} \quad k, j \in \mathbb{Z}.$$

We only use those functions that are supported on $[0, 1)$, namely $0 \leq j < 2^k$ for each $k \geq 0$. The others will be used later. The Haar system is the family

$$\{\varphi, \psi_{kj} : k \geq 0 \text{ and } 0 \leq j < 2^k\}.$$

See Figure 15.1 for examples of elements of the Haar system.

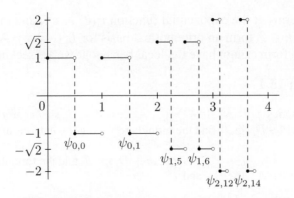

FIGURE 15.1. Some elements of the Haar system.

15.2.1. LEMMA. *The Haar system is orthonormal.*

PROOF. It is straightforward to check that each of these functions has norm 1. Now ψ_{kj} and $\psi_{kj'}$ for $j \neq j'$ have disjoint supports and thus are orthogonal. More generally, if $k < k'$, then φ and ψ_{kj} are constant on the support of $\psi_{k'j'}$. Since $\int_0^1 \psi_{kj}(x)\,dx = 0$ for all j and k, it now follows that these functions are pairwise orthogonal. ∎

We may consider the inner product expansion with respect to this orthonormal set. It is natural to sum all terms of the same order at the same time to obtain a series approximants. Therefore, we define

$$H_n f(x) = \langle f, \varphi \rangle \varphi(x) + \sum_{k=0}^{n-1} \sum_{j=0}^{2^k-1} \langle f, \psi_{kj} \rangle \psi_{kj}(x).$$

The **Haar coefficients** are the inner products $\langle f, \psi_{kj} \rangle$ used in this expansion.

While we have some work yet to see that this orthogonal system spans the whole space, we can see that it has some nice properties. The local character is seen by the fact that these functions have smaller and smaller supports. If f and g agree except on the interval $[3/8, 1/2)$, then the Haar coefficients are the same for about "7/8ths of the terms" in the sense that $\langle f, \psi_{kj} \rangle = \langle g, \psi_{kj} \rangle$ if $k \geq 3$ and $j/2^k \notin [3/8, 1/2)$.

This is the kind of local property we are seeking. The functions ψ_{kj} also have the translation and dilation properties we are want. However, we will have to eliminate φ somehow. We shall see that φ is not needed for a basis of $L^2(\mathbb{R})$ when we add in dilations of ψ by negative powers of 2. On the other hand, φ reappears in a central role in the next section as the scaling function.

We need a more explicit description of $H_n f$. By a **dyadic interval** of length 2^{-n}, we mean one of the form $[j2^{-n}, (j+1)2^{-n})$ for some integer j.

15.2.2. LEMMA. *Let* $f \in L^2(0,1)$. *Then* $H_n f$ *is the unique function that is constant on each dyadic interval of length* 2^{-n} *in* $[0,1]$ *and satisfies*

$$H_n f(x) = 2^n \int_{j2^{-n}}^{(j+1)2^{-n}} H_n f(t)\, dt = 2^n \int_{j2^{-n}}^{(j+1)2^{-n}} f(t)\, dt x$$

for $x \in [j2^{-n}, (j+1)2^{-n})$, $0 \le j < 2^n$. *Moreover,* $\|H_n f\|_2 \le \|f\|_2$.

PROOF. It is easy to see that $\{\varphi, \psi_{00}\}$ span the functions that are constant on $[0, 1/2)$ and on $[1/2, 1)$. By induction, it follows easily that

$$M_n := \operatorname{span}\{\varphi, \psi_{kj} : 0 \le k \le n-1 \text{ and } 0 \le j < 2^k\}$$

is the subspace of all functions that are constant on each of the dyadic intervals $[j2^{-n}, (j+1)2^{-n})$ for $0 \le j < 2^n$. Notice that M_n is also spanned by the characteristic functions $\chi_{n,j} = \chi_{[j2^{-n},(j+1)2^{-n})}$ for $0 \le j < 2^n$.

Now $H_n f$ is contained in the preceding span and therefore is constant on these dyadic intervals. Thus $H_n f$ is the unique function of this form that satisfies $\langle H_n f, \varphi \rangle = \langle f, \varphi \rangle$ and $\langle H_n f, \psi_{kj} \rangle = \langle f, \psi_{kj} \rangle$ for $0 \le k \le n-1$ and $0 \le j < 2^k$. But this basis for M_n may be replaced by the basis of characteristic functions. Since $\|\chi_{n,j}\|_2^2 = 2^{-n}$, $H_n f$ is the unique function in M_n such that

$$H_n f(x) = 2^n \langle H_n f, \chi_{n,j} \rangle = 2^n \langle f, \chi_{n,j} \rangle$$

for all $x \in [j2^{-n}, (j+1)2^{-n})$ and $0 \le j < 2^n$, which is what we wanted.

The map H_n is the orthogonal projection of $L^2(0,1)$ onto M_n. The inequality $\|H_n f\|_2 \le \|f\|_2$ follows from the Projection Theorem (Theorem 7.5.2). An elementary direct argument is outlined in Exercise 15.2.C. ∎

We can now prove that the Haar system is actually a basis. Moreover, we show that it does an excellent job of uniform approximation for continuous functions as well, even though the basis functions are not themselves continuous. In this respect, we obtain superior convergence to the convergence of Fourier series.

15.2.3. THEOREM. *Let* $f \in L^2(0,1)$. *Then* $H_n f$ *converges to* f *in the* L^2 *norm. Consequently, the Haar system is an orthonormal basis for* $L^2(0,1)$. *Moreover, if* f *is continuous on* $[0,1]$, *then* $H_n f$ *converges uniformly to* f.

PROOF. We prove the last statement first. By Theorem 5.5.9, f is uniformly continuous on $[0,1]$. Recall from Definition 10.4.2 that the modulus of continuity is $\omega(f; \delta) = \sup\{|f(x) - f(y)| : |x - y| \le \delta\}$. The remarks there also show that the uniform continuity of f implies that $\lim_{n \to \infty} \omega(f; 2^{-n}) = 0$.

For $x \in [j2^{-n}, (j+1)2^{-n})$, compute

$$|H_n f(x) - f(x)| = \left| 2^n \int_{j2^{-n}}^{(j+1)2^{-n}} f(t)\, dt - 2^n \int_{j2^{-n}}^{(j+1)2^{-n}} f(x)\, dt \right|$$

$$\leq 2^n \int_{j2^{-n}}^{(j+1)2^{-n}} |f(t) - f(x)|\, dt$$

$$\leq 2^n \int_{j2^{-n}}^{(j+1)2^{-n}} \omega(f; 2^{-n})\, dt = \omega(f; 2^{-n}).$$

Hence $\|H_n f - f\|_\infty \leq \omega(f; 2^{-n})$ tends to 0. Therefore, $H_n f$ converges to f uniformly on $[0, 1]$.

Now

$$\|H_n f - f\|_2 \leq \left(\int_0^1 \|H_n f - f\|_\infty\, dt \right)^{1/2} = \|H_n f - f\|_\infty.$$

So we obtain convergence in the $L^2(0, 1)$ norm as well.

Next suppose that f is an arbitrary L^2 function, and let $\varepsilon > 0$ be given. As f is the L^2 limit of a sequence of continuous functions, we may find a continuous function g with $\|f - g\|_2 < \varepsilon$. Now choose n so large that $\|H_n g - g\|_2 < \varepsilon$. Then

$$\|H_n f - f\|_2 \leq \|H_n f - H_n g\| + \|H_n g - g\|_2 + \|g - f\|_2$$

$$\leq \|H_n(f - g)\|_2 + \varepsilon + \varepsilon$$

$$\leq \|f - g\|_2 + 2\varepsilon < 3\varepsilon.$$

So $H_n f$ converges to f in L^2.

Finally, since the orthogonal expansion of f in the Haar system actually sums to f in the L^2 norm, we deduce that this orthonormal set spans all of $L^2(0, 1)$ and thus is a basis. ∎

15.2.4. DEFINITION. The **Haar wavelet** is the function $\psi = \chi_{[0,.5)} - \chi_{[.5,1)}$. The **Haar wavelet basis** is the family $\{\psi_{kj} : k, j \in \mathbb{Z}\}$.

Lemma 15.2.1 can be easily modified to show that the Haar wavelet basis is orthonormal. It remains to verify that it spans $L^2(\mathbb{R})$.

15.2.5. THEOREM. *The Haar wavelet basis spans all of $L^2(\mathbb{R})$.*

PROOF. It is enough to show that any continuous function of bounded support is spanned by the Haar wavelet basis. Each such function is the finite sum of (piecewise) continuous functions supported on an interval $[m, m + 1)$. But our basis is invariant under integer translations. So it is enough to show that a function on $[0, 1)$ is spanned by the Haar wavelet basis. But Theorem 15.2.3 shows that the functions ψ_{kj} supported on $[0, 1)$ together with φ span $L^2(0, 1)$. Consequently, it is enough to approximate φ alone.

Consider the functions $\psi_{-k,0} = 2^{-k/2}\chi_{[0,2^{k-1})} - \chi_{[2^{k-1},2^k)}$ for $k \geq 1$. An easy computation shows that

$$h_N := \sum_{k=1}^{N} 2^{-k/2}\psi_{-k,0} = (1 - 2^{-N})\chi_{[0,1)} - 2^{-N}\chi_{[1,2^N)}.$$

Thus $\|\varphi - h_N\|_2 = \|2^{-N}\chi_{[0,2^N)}\|_2 = 2^{-N/2}$. Hence φ is in the span of the wavelet basis. Therefore, the Haar wavelet basis spans all of $L^2(\mathbb{R})$.   ∎

Exercises for Section 15.2

A. Let $f(x) = \sum\limits_{k=0}^{2^n-1} s_{n,j}\chi_{[j2^{-n},(j+1)2^{-n})}$ and let $\mathbf{s}_n = (s_{n,0}, \ldots, s_{n,2^n-1})$.

    (a) Define $\mathbf{a}_k = (a_{k,0}, \ldots, a_{k,2^k-1})$ and $\mathbf{s}_k = (s_{k,0}, \ldots, s_{k,2^k-1})$ by

$$a_{k,j} = \frac{s_{k+1,2j} - s_{k+1,2j+1}}{\sqrt{2}} \quad \text{and} \quad s_{k,j} = \frac{s_{k+1,2j} + s_{k+1,2j+1}}{\sqrt{2}} \quad \text{for}$$

    $0 \leq k < n$ and $0 \leq j < 2^k$. Show that $f = s_{0,0}\varphi + \sum\limits_{k=1}^{n-1}\sum\limits_{j=0}^{2^k-1} a_{k,j}\psi_{k,j}$.

    (b) Explain how to reverse this process and obtain $\mathbf{s}_n$ from the wavelet expansion of f.

B. Let f be a continuous function with compact support $[0,1]$. Fix $n \geq 1$, and define $s_j = f(j/2^n)$ for $0 \leq j < 2^n$. Show that

$$\left\| H_n f - \sum_{k=0}^{2^n-1} s_j \chi_{[j2^{-n},(j+1)2^{-n})} \right\|_\infty \leq \omega(f; 2^{-n}).$$

C. Prove that $\|H_n f\|_2 \leq \|f\|_2$.

    HINT: Show that it is enough to prove that $\left| \int_a^{a+2^{-k}} f(x)\,dx \right|^2 \leq 2^{-k}\int_a^{a+2^{-k}} |f(x)|^2\,dx$.
Verify this using the Cauchy–Schwarz inequality.

D. Show that $\{\psi_{kj} : k > -N,\ -2^{k+N} \leq j < 2^{k+N}\}$ together with the functions $\varphi_{-N,0} = 2^{-N/2}\varphi(2^{-N}x)$ and $\varphi_{-N,-1} = 2^{-N/2}\varphi(2^{-N}x + 1)$ forms an orthonormal basis for $L^2(-2^N, 2^N)$.

E. Suppose that f is a continuous function with compact support contained in $[-2^N, 2^N]$ for some $N \in \mathbb{N}$. Define $P_n f(x) = \sum\limits_{k=-n}^{n}\sum\limits_{j=-\infty}^{\infty} \langle f, \psi_{kj}\rangle \psi_{kj}(x)$.

    (a) Show that $P_n f$ is the sum of only finitely many nonzero terms.

    (b) If $\int_0^{2^N} f(x)\,dx = 0 = \int_{-2^N}^0 f(x)\,dx$, then the only nonzero terms are for $k \geq -N$.
Verify this. Show that $P_n f$ converges to f uniformly.
HINT: Modify Theorem 15.2.3.

    (c) Show that $P_n f$ converges uniformly to f without the integral conditions.
HINT: Prove uniform convergence for $\varphi_{-N,0}$ and $\varphi_{-N,-1}$.

F. Show that for $f \in L^2(\mathbb{R})$ that $\sum\limits_{k=-\infty}^{\infty}\sum\limits_{j=0}^{\infty} \langle f, \psi_{kj}\rangle \psi_{kj}(x) = \chi_{[0,\infty)}f$.

15.3. Multiresolution Analysis

Motivated by the Haar wavelet, we develop a general framework that applies to a wide range of wavelet systems. We will use this framework to construct other wavelet systems.

As before, $\varphi = \chi_{[0,1)}$ and we define the translations and dilations

$$\varphi_{kj}(x) = 2^{k/2}\varphi(2^k x - j)(x) \quad \text{for all} \quad k, j \in \mathbb{Z}.$$

This is not an orthonormal system. But for each k, the family $\{\varphi_{kj} : j \in \mathbb{Z}\}$ consists of multiples of the characteristic functions of the dyadic intervals of length 2^{-k}. In particular, these families are orthonormal.

Define $V_k = \text{span}\{\varphi_{kj} : j \in \mathbb{Z}\}$. This is the space of L^2 functions that are constant on each dyadic interval of length 2^{-k}. Consequently $V_k \subset V_{k+1}$ for $k \in \mathbb{Z}$. That is, the V_k form a nested sequence of subspaces:

$$\cdots \subset V_{-2} \subset V_{-1} \subset V_0 \subset V_1 \subset V_2 \subset V_3 \subset \cdots.$$

It is also immediate that $f(x)$ belongs to V_k if and only if $f(2x)$ belongs to V_{k+1}. We state and prove the important (but basically easy) properties of this decomposition.

15.3.1. LEMMA. *Let $\varphi = \chi_{[0,1)}$ and define V_k as previously. Then we have*

(1) *orthogonality:* $\{\varphi(x - j) : j \in \mathbb{Z}\}$ *is an orthonormal basis for V_0.*

(2) *nesting:* $V_k \subset V_{k+1}$ *for all $k \in \mathbb{Z}$.*

(3) *scaling:* $f(x) \in V_k$ *if and only if $f(2x) \in V_{k+1}$.*

(4) *density:* $\overline{\bigcup_{k \in \mathbb{Z}} V_k} = L^2(\mathbb{R})$.

(5) *separation:* $\bigcap_{k \in \mathbb{Z}} V_k = \{0\}$.

PROOF. We have already established (1), (2), and (3).

As in the proof of Theorem 15.2.3, we know that every continuous function with compact support $[-N, N]$ is the uniform limit of functions that also have support $[-N, N]$ and are constant on dyadic intervals of length 2^{-k}, (i.e., functions in V_k). These functions therefore converge in L^2 as well. Consequently, the closed union of the V_k's contains all continuous functions of compact support, and thus all of $L^2(\mathbb{R})$.

Notice that ψ_{kj} is orthogonal to $\varphi_{k'j'}$ provided that $k \geq k'$ because $\varphi_{k'j'}$ will be constant on the support of ψ_{kj} and ψ_{kj} integrates to 0. So any function f belonging to the intersection $\bigcap_{k \in \mathbb{Z}} V_k$ must be orthogonal to every ψ_{kj}. By Theorem 15.2.5, it follows that f is orthogonal to every function in $L^2(\mathbb{R})$, including itself. Therefore, $\|f\|_2 = \langle f, f \rangle = 0$, whence $f = 0$. ∎

This leads us to formalize these properties in greater generality.

15.3.2. DEFINITION. A **multiresolution** of $L^2(\mathbb{R})$ with **scaling function** φ is the sequence of subspaces

$$V_j = \text{span}\left\{\varphi_{kj}(x) = 2^{k/2}\varphi(2^k x - j) : j \in \mathbb{Z}\right\}$$

provided that the sequence satisfies the five properties—orthogonality, nesting, scaling, density and separation—described in the preceding lemma.

The function φ is sometimes called a **father wavelet**.

Notice that by a change of variables $t = 2^k x$, we obtain

$$\langle \varphi_{ki}, \varphi_{kj} \rangle = \int_{-\infty}^{\infty} 2^k \varphi(2^k x - i)\varphi(2^k x - j)\,dx$$

$$= \int_{-\infty}^{\infty} \varphi(t - i)\varphi(t - j)\,dx = \delta_{ij}.$$

So $\{\varphi_{kj} : j \in \mathbb{Z}\}$ forms an orthonormal basis of V_k for each $k \in \mathbb{Z}$.

Once we have a nested sequence V_k with these properties, we can decompose $L^2(\mathbb{R})$ into a direct sum of subspaces. Set $W_k = \{f \in V_{k+1} : f \perp V_k\}$. This is the **orthogonal complement** of V_k in V_{k+1}. We write $V_{k+1} = V_k \oplus W_k$, where the $\oplus$ indicates that this is a **direct sum**, that is, a sum of orthogonal subspaces. So each vector $f \in V_{k+1}$ can be written uniquely as $f = g + h$ with $g \in V_k$ and $h \in W_k$. As $\langle f, g \rangle = 0$, we have the Pythagorean identity

$$\|f\|_2^2 = \langle g+h, g+h \rangle = \langle g,g \rangle + \langle g,h \rangle + \langle h,g \rangle + \langle h,h \rangle = \|g\|_2^2 + \|h\|_2^2.$$

Since V_k has an orthonormal basis $\{\varphi_{kj} : j \in \mathbb{Z}\}$, Corollary 7.5.10 of Parseval's Theorem provides an orthogonal projection P_k of $L^2(\mathbb{R})$ onto V_k given by

$$P_k f = \sum_{j=-\infty}^{\infty} \langle f, \varphi_{kj} \rangle \varphi_{kj}$$

and we have the important identity

$$\|f\|_2^2 = \|P_k f\|_2^2 + \|f - P_k f\|_2^2.$$

15.3.3. LEMMA. *$Q_k = P_{k+1} - P_k$ is the orthogonal projection onto W_k.*

PROOF. To verify this, we will show that Q_k is an idempotent with range W_k and kernel $W_k^{\perp}$. Note that $P_k P_{k+1} = P_{k+1} P_k = P_k$ because V_k is contained in V_{k+1}. Hence

$$Q_k^2 = P_{k+1}^2 - P_k P_{k+1} - P_{k+1} P_k + P_k^2 = P_{k+1} - P_k = Q_k.$$

So Q_k is a projection.

We claim that $W_k^{\perp} = V_{k+1}^{\perp} + V_k$. Indeed, $P_{k+1}^{\perp} f$ is orthogonal to V_{k+1} and so is also orthogonal to W_k. So f is orthogonal to W_k if and only if $P_{k+1} f$ is orthogonal to W_k, which is the same as saying that $P_{k+1} f \in V_k$. This latter statement is equivalent to

$$P_{k+1} f = P_k P_{k+1} f = P_k f \qquad \text{or} \qquad (P_{k+1} - P_k) f = 0.$$

So $W_k^{\perp} = \ker Q_k$.

If $f \in W_k$, then $P_{k+1}f = f$ since $f \in V_{k+1}$. Also, $P_k f = 0$ since $f \perp V_k$. Hence $Q_k f = f$. Conversely, if $f = Q_k g$, then

$$P_{k+1}f = P_{k+1}^2 g - P_{k+1}P_k g = (P_{k+1} - P_k)g = f.$$

So f belongs to V_{k+1}. And

$$P_k f = P_k P_{k+1}g - P_k^2 g = (P_k - P_k)g = 0.$$

Thus f is orthogonal to V_k, and so f is in W_k. Therefore, the range of Q_k is exactly W_k. Consequently, Q_k is the orthogonal projection onto W_k. ∎

We may repeat the decomposition $V_{k+1} = V_k \oplus W_k$ finitely often to obtain

$$V_n = V_0 \oplus W_0 \oplus \cdots \oplus W_{n-1} \quad \text{and} \quad V_0 = V_{-n} \oplus W_{-n} \oplus \cdots \oplus W_{-1}.$$

Repetition of this procedure suggests that there is a decomposition of $L^2(\mathbb{R})$ as an infinite direct sum

$$\bigoplus_{k \in \mathbb{Z}} W_k = \cdots \oplus W_{-2} \oplus W_{-1} \oplus W_0 \oplus W_1 \oplus W_2 \oplus \cdots.$$

What we mean by this is that every function f in $L^2(\mathbb{R})$ should decompose uniquely as an infinite sum

$$f = \sum_{k=-\infty}^{\infty} f_k \quad \text{where} \quad f_k \in W_k \quad \text{and} \quad \|f\|_2^2 = \sum_{k \in \mathbb{Z}} \|f_k\|_2^2.$$

We shall prove that this is indeed the case.

15.3.4. LEMMA. *Suppose that $V_k \subset V_{k+1}$ for $k \in \mathbb{Z}$ is the nested sequence of subspaces from a multiresolution of $L^2(\mathbb{R})$. Then*

$$\lim_{k \to \infty} \|f - P_k f\|_2 = 0, \quad \text{and} \quad \lim_{k \to -\infty} \|P_k f\|_2 = 0.$$

PROOF. The limit $\lim_{k \to \infty} \|f - P_k f\|_2$ is a consequence of density. For any $\varepsilon > 0$, there is an integer n and a function $g \in V_n$ such that $\|f - g\|_2 < \varepsilon$. Then for $k \geq n$, Parseval's Theorem (Theorem 7.5.9) shows that

$$\|f - P_k f\|_2 = \|(f - g) - P_k(f - g)\|_2 \leq \|f - g\|_2 < \varepsilon.$$

The second limit $\lim_{k \to -\infty} \|P_k f\|_2 = 0$ is a consequence of separation. We will show that it actually follows from the first part. Let $V_k^\perp$ denote the orthogonal complement of V_k, and note that $I - P_k$ is the orthogonal projection onto it. Notice that these subspaces are also nested in the reverse order $V_{k+1}^\perp \subset V_k^\perp$. We claim that $N = \overline{\bigcup_{k \in \mathbb{Z}} V_k^\perp}$ is all of $L^2(\mathbb{R})$. Indeed, if N were a proper subspace of $L^2(\mathbb{R})$, then there would be a nonzero function $g \perp N$. Thus, in particular, $g \perp V_k^\perp$, so that g belongs to $V_k^{\perp\perp} = V_k$ for every $k \in \mathbb{Z}$. Consequently, g belongs to $\bigcap_{k \in \mathbb{Z}} V_k = \{0\}$. So $N = L^2(\mathbb{R})$. Since $\|P_k f\|_2 = \|f - (I - P_k)f\|_2$, the desired limit follows from the first part. ∎

We now are ready to derive the infinite decomposition.

15.3.5. THEOREM. *Suppose that $V_k \subset V_{k+1}$ for $k \in \mathbb{Z}$ is the nested sequence of subspaces from a multiresolution of $L^2(\mathbb{R})$. Then $L^2(\mathbb{R})$ decomposes as the infinite direct sum $\bigoplus_{k\in\mathbb{Z}} W_k$.*

PROOF. The finite decompositions are valid. So, in particular,

$$V_n = V_{-n} \oplus W_{-n} \oplus \cdots \oplus W_{n-1}.$$

Thus if f belongs to V_n and is orthogonal to V_{-n}, then f decomposes uniquely as

$$f = \sum_{k=-n}^{n-1} f_k \text{ for } f_k \in W_k, \text{ namely } f_k = Q_k f. \text{ Moreover, Parseval's Theorem}$$

shows that $\|f\|_2^2 = \sum_{k=-n}^{n-1} \|f_k\|_2^2$.

If f is an arbitrary function in $L^2(\mathbb{R})$ and $\varepsilon > 0$, then by the lemma we may choose a positive integer n so that $\|(I - P_n)f\|_2^2 + \|P_{-n}f\|_2^2 < \varepsilon^2$. Therefore, $g_n := P_n f - P_{-n} f$ belongs to V_n and is orthogonal to V_{-n}. Consequently, we may write $g_n = \sum_{k=-n}^{n-1} f_k$ for $f_k \in W_k$, where $f_k = Q_k g_n = Q_k f$. By Parseval's Theorem,

$$\|f - g_n\|_2^2 = \|(I - P_n)f + P_{-n}f\|_2^2 = \|(I - P_n)f\|_2^2 + \|P_{-n}f\|_2^2 < \varepsilon^2.$$

Since ε is arbitrary, it follows that g_n converges to f. That is,

$$f = \lim_{n\to\infty} \sum_{k=-n}^{n-1} f_k = \sum_{k=-\infty}^{\infty} f_k$$

and

$$\|f\|_2^2 = \lim_{n\to\infty} \|g_n\|_2^2 = \lim_{n\to\infty} \sum_{k=-n}^{n-1} \|f_k\|_2^2 = \sum_{k=-\infty}^{\infty} \|f_k\|_2^2.$$

To establish uniqueness, suppose that $f = \sum_k f_k = \sum_k h_k$ are two decompositions with f_k and h_k in W_k. Then $0 = \sum_k f_k - h_k$. The norm formula from the previous paragraph shows that $0 = \sum_k \|f_k - h_k\|_2^2$. Therefore, $h_k = f_k$ for all $k \in \mathbb{Z}$. ∎

Exercises for Section 15.3

A. Let V_0 be the span of integer translates of the Haar scaling function φ. Suppose that $f \in V_0$ has bounded support and the set $\{f(x - j) : j \in \mathbb{Z}\}$ is orthonormal. Prove that $f(x) = \pm\varphi(x - n)$ for some integer n.
HINT: Compute $\langle f(x), f(x - j) \rangle$ when the supports overlap on a single interval.

B. Suppose that $\varphi \in L^2(\mathbb{R})$ such that the subspaces V_k satisfy orthogonality, nesting, and scaling. Let $M = \overline{\bigcup_{k\in\mathbb{Z}} V_k}$. Show that if $f \in M$, then $f(x - t) \in M$ for every $t \in \mathbb{R}$.
HINT: First prove this for $t = j2^{-k}$. Then apply Exercise 15.1.F.

C. Suppose that φ is continuous with compact support $[a, a + M]$, and that $\{\varphi_{0j} : j \in \mathbb{Z}\}$ are orthonormal.

(a) Suppose that $f \in V_k$, and express $f(x) = \sum_j c_j 2^{k/2} \varphi(2^k x - j)$. Use the Cauchy–Schwarz inequality to show that $|f(x)| \le 2^{k/2} M \|\varphi\|_\infty \|f\|_2$.

(b) Show that $\bigcap_{k \in \mathbb{Z}} V_k = \{0\}$.

HINT: Take $f \in \bigcap_{k \in \mathbb{Z}} V_k$. Use part (a) to estimate $\int_{-N}^N |f(x)|^2 \, dx$, and let $k \to -\infty$.

15.4. Recovering the Wavelet

Let us look at the decomposition obtained in the previous section in the case of the Haar system. Notice that $\varphi = \chi_{[0,1)}$ satisfies the identity

$$\varphi = \chi_{[0,.5)} + \chi_{[.5,1)} = \tfrac{1}{\sqrt{2}} \varphi_{10} + \tfrac{1}{\sqrt{2}} \varphi_{11}.$$

On the other hand, we can write φ_{10} and φ_{11} in terms of φ and ψ. Recalling that $\psi = \chi_{[0,.5)} - \chi_{[.5,1)}$, we have

$$\varphi_{10} = \tfrac{1}{\sqrt{2}} \varphi + \tfrac{1}{\sqrt{2}} \psi \qquad \text{and} \qquad \varphi_{11} = \tfrac{1}{\sqrt{2}} \varphi - \tfrac{1}{\sqrt{2}} \psi.$$

More generally,

$$\varphi_{kj} = \tfrac{1}{\sqrt{2}} \varphi_{k+1,2j} + \tfrac{1}{\sqrt{2}} \varphi_{k+1,2j+1}$$

and

$$\varphi_{k+1,2j} = \tfrac{1}{\sqrt{2}} \varphi_{kj} + \tfrac{1}{\sqrt{2}} \psi_{kj} \qquad \text{and} \qquad \varphi_{k+1,2j+1} = \tfrac{1}{\sqrt{2}} \varphi_{kj} - \tfrac{1}{\sqrt{2}} \psi_{kj}.$$

The subspace V_k consists of those $L^2(\mathbb{R})$ functions that are constant on the dyadic intervals of length 2^{-k}. Now ψ_{kj} belongs to V_{k+1}, it is supported on one interval of length 2^{-k}, and integrates to 0. Thus $\langle \psi_{kj}, \varphi_{kj'} \rangle = 0$ for all $j, j' \in \mathbb{Z}$. In particular, ψ_{kj} lies in W_k. So $W'_k = \text{span}\{\psi_{kj} : j \in \mathbb{Z}\}$ is a subspace of W_k.

On the other hand, the identities show that every basis vector $\varphi_{k+1,j}$ belongs to $V_k + W'_k$, and thus $V_{k+1} = V_k \oplus W'_k = V_k \oplus W_k$. This forces the identity $W'_k = W_k$. So we have shown that for the Haar system, we have

$$W_k = \text{span}\{\psi_{kj} : j \in \mathbb{Z}\}.$$

There is a systematic way to construct a wavelet from a multiresolution. That is the goal of this section. Let $\{V_k\}$ be a multiresolution with scaling function φ. The construction begins with the fact that $\varphi \in V_0 \subset V_1$. Since φ_{1j} form an orthonormal basis for V_1, we may expand φ as

(15.4.1) $$\varphi(x) = \sum_{j=-\infty}^{\infty} a_j \varphi(2x - j) = \sum_{j=-\infty}^{\infty} \frac{a_j}{\sqrt{2}} \varphi_{1j}(x),$$

where $a_j = 2\langle \varphi(x), \varphi(2x - j) \rangle$. By Parseval's Theorem, $\|\varphi\|_2^2 = \tfrac{1}{2} \sum_{j=-\infty}^{\infty} |a_j|^2$.

Thus (a_j) is a sequence in ℓ^2. Equation (15.4.1) is known as the **scaling relation** for φ.

15.4.2. THEOREM. *Let φ be the scaling function generating a multiresolution $\{V_k\}$ of $L^2(\mathbb{R})$ with scaling relation $\varphi(x) = \sum\limits_{j=-\infty}^{\infty} a_j\varphi(2x - j)$. Define*

$$\psi(x) = \sum_{j=-\infty}^{\infty} (-1)^j a_{1-j}\, \varphi(2x - j).$$

Then ψ is a wavelet generating the wavelet basis $\{\psi_{kj} : k, j \in \mathbb{Z}\}$ such that $W_k = \text{span}\{\psi_{kj} : j \in \mathbb{Z}\}$ for each $k \in \mathbb{Z}$.

PROOF. Since this proof basically consists of several long computations, we provide a brief overview of the plan. The orthonormality of $\{\varphi(x - j) : j \in \mathbb{Z}\}$ will yield conditions on the coefficients a_j. Then we show that $\{\psi(x - k) : k \in \mathbb{Z}\}$ is an orthonormal set that is orthogonal to the $\varphi(x - j)$'s. Finally, we show that V_1 is spanned by V_0 and the $\psi(x - k)$'s.

In this proof, all summations are from $-\infty$ to $+\infty$, but for notational simplicity, only the index will be indicated. We define δ_{0n} to be 1 if $n = 0$ and 0 otherwise. To begin, we have

$$\begin{aligned}
\delta_{0n} &= \langle \varphi(x), \varphi(x - n) \rangle \\
&= \Big\langle \sum_i a_i\varphi(2x - i), \sum_j a_j\varphi(2x - 2n - j) \Big\rangle \\
&= \sum_i \sum_j a_i a_j \langle \varphi(2x - i), \varphi(2x - 2n - j) \rangle \\
&= \frac{1}{2} \sum_j a_{j+2n} a_j.
\end{aligned}$$

The orthonormality of $\{\psi(x - j) : j \in \mathbb{Z}\}$ follows because the coefficients of ψ are obtained from φ by reversing, shifting by one place, and alternating sign. A bit of thought will show that each of these steps preserves the property of orthogonality of translations. Here we provide the direct computation:

$$\begin{aligned}
\langle \psi(x), \psi(x - n) \rangle &= \Big\langle \sum_i (-1)^i a_{1-i}\varphi(2x - i), \sum_j (-1)^j a_{1-j}\varphi(2x - 2n - j) \Big\rangle \\
&= \sum_i \sum_j (-1)^{i+j} a_{1-i} a_{1-j} \langle \varphi(2x - i), \varphi(2x - 2n - j) \rangle \\
&= \frac{1}{2} \sum_j (-1)^{2j+2n} a_{1-j-2n} a_{1-j} \\
&= \frac{1}{2} \sum_i a_i a_{i+2n} = \delta_{0n}.
\end{aligned}$$

So $\{\psi(x - j) : j \in \mathbb{Z}\}$ is orthonormal.

The fact that the ψ's and φ's are orthogonal is more subtle. Calculate

$$\langle \psi(x-m), \varphi(x-n) \rangle = \left\langle \sum_i (-1)^i a_{1-i} \varphi(2x-2m-i), \sum_j a_j \varphi(2x-2n-j) \right\rangle$$

$$= \sum_i \sum_j (-1)^i a_{1-i} a_j \langle \varphi(2x-2m-i), \varphi(2x-2n-j) \rangle$$

but the inner product is 0 unless $2m + i = 2n + j$,

$$= \frac{1}{2} \sum_j (-1)^{j+2n-2m} a_{1-j-2n+2m} a_j$$

$$= \frac{1}{2} \sum_j (-1)^j a_{p-j} a_j,$$

where $p = 2m + 1 - 2n$ is a fixed *odd* integer. Thus by substituting $i = p - j$, we may rearrange this sum:

$$\frac{1}{2} \sum_j (-1)^j a_{p-j} a_j = \frac{1}{2} \sum_i (-1)^{p-i} a_i a_{p-i} = -\left(\frac{1}{2} \sum_i (-1)^i a_i a_{p-i} \right).$$

Thus the sum must be 0. Hence the family $\{\psi(x - k) : k \in \mathbb{Z}\}$ is orthogonal to the family $\{\varphi(x - j) : j \in \mathbb{Z}\}$. Notice that the shift by 1 of the coefficients in the definition of ψ was to make p odd in this calculation.

Now we wish to express $\varphi_{1p}(x) = \sqrt{2}\varphi(2x - p)$ as a linear combination of these two families. To see what the coefficients should be, we compute

$$\langle \varphi_{1p}(x), \varphi(x - n) \rangle = \left\langle \sqrt{2}\varphi(2x - p), \sum_j a_j \varphi(2x-2n-j) \right\rangle = \frac{1}{\sqrt{2}} a_{p-2n}$$

since the inner product is 0 except when $2n + j = p$; and similarly

$$\langle \varphi_{1p}(x), \psi(x - n) \rangle = \left\langle \sqrt{2}\varphi(2x - p), \sum_j (-1)^j a_{1-j} \varphi(2x-2n-j) \right\rangle$$

$$= \frac{(-1)^p}{\sqrt{2}} a_{1-p+2n}.$$

So now it is a matter of adding up the series to see if $\varphi_{1p}(x)$ can be recovered. Compute

$$\sum_n a_{p-2n} \varphi(x - n) = \sum_n \sum_i a_{p-2n} a_i \varphi(2x-2n-i)$$

$$= \sum_k \left(\sum_n a_{p-2n} a_{k-2n} \right) \varphi(2x - k)$$

$$= \sum_k \left(\sum_n a_{p+2n} a_{k+2n} \right) \varphi(2x - k)$$

and

$$\sum_n (-1)^p a_{1-p+2n}\psi(x-n) = \sum_n \sum_i (-1)^p a_{1-p+2n}(-1)^i a_{1-i}\varphi(2x-2n-i)$$

$$= \sum_k \left((-1)^{p+k}\sum_n a_{1+2n-p}a_{1+2n-k}\right)\varphi(2x-k).$$

When $p+k$ is odd,

$$(-1)^{p+k}\sum_n a_{1+2n-p}a_{1+2n-k} = -\sum_m a_{2m+k}a_{2m+p}$$

while if $p+k$ is even,

$$(-1)^{p+k}\sum_n a_{1+2n-p}a_{1+2n-k} = \sum_m a_{1+2m+k}a_{1+2m+p}.$$

When these sums over translates of $\varphi(2x)$ and $\psi(2x)$ are added together, the coefficients of $\varphi(2x-k)$ are canceled when $p+k$ is odd, while for $p+k$ even the two sums conveniently merge to yield the sums from the orthogonality relation for the $\varphi(x-k)$. Hence the sum obtained is

$$\sum_n \tfrac{1}{\sqrt{2}}a_{p-2n}\varphi(x-n) + \sum_n \frac{(-1)^p}{\sqrt{2}}a_{1-p+2n}\psi(x-n)$$

$$= \sum_{k\equiv p \bmod 2}\left(\frac{1}{2}\sum_n a_{p+n}a_{k+n}\right)\sqrt{2}\varphi(2x-k)$$

$$= \sqrt{2}\varphi(2x-p) = \varphi_{1p}(x).$$

Set $W = \text{span}\{\psi(x-j) : j \in \mathbb{Z}\}$. Let us recap what we have established. We have shown that $\{\psi(x-j) : j \in \mathbb{Z}\}$ is an orthonormal basis for W, that W is orthogonal to V_0, and that $\varphi(2x-i)$ belongs to $V_0 + W$ for all $i \in \mathbb{Z}$. Since each $\psi(x-j)$ is expressed in terms of the $\varphi(2x-i)$, it is clear that W is a subspace of V_1. On the other hand, since each $\varphi(2x-i)$ belongs to $V_0 + W$, it follows that $V_1 = V_0 \oplus W$. Hence we deduce that W is the orthogonal complement of V_0 in V_1; that is, $W = W_0$.

It now follows from dilation that

$$\text{span}\{\psi_{kj} : j \in \mathbb{Z}\} = \{2^{k/2}f(2^k x) : f \in W_0\} = W_k.$$

Hence $\{\psi_{kj} : j \in \mathbb{Z}\}$ is an orthonormal basis for W_k for each $k \in \mathbb{Z}$. Since $L^2(\mathbb{R}) = \bigoplus_{k=-\infty}^{+\infty} W_k$, it follows that together the collection $\{\psi_{kj} : k, j \in \mathbb{Z}\}$ is an orthonormal basis for $L^2(\mathbb{R})$. Therefore, ψ is a wavelet. ∎

Exercises for Section 15.4

A. Show that if φ is a scaling function with compact support, then the scaling relation is a finite sum.

B. Let $\{e_k : k \in \mathbb{Z}\}$ be an orthonormal set in a Hilbert space $\mathcal{H}$. Show that the vectors $x = \sum_n a_n e_n$ and $y = \sum_n (-1)^n a_{p-n} e_n$ are orthogonal if p is odd.

C. Given the scaling relation $\varphi(x) = \sum_j a_j \varphi(2x - j)$, we define the **filter** to be the complex function $m_\varphi(\theta) = \sum_j a_j e^{ij\theta}$. Prove that $|m_\varphi(\theta)|^2 + |m_\varphi(\theta + \pi)|^2 = 1$. HINT: Compute the Fourier series of this sum, and compare the sums of coefficients obtained with those that occur in the proof of Theorem 15.4.2.

D. Suppose that φ is a scaling function that is bounded, has compact support, and satisfies $\int_{-\infty}^{\infty} \varphi(x)\, dx \neq 0$. Let $\varphi(x) = \sum_j a_j \varphi(2x - j)$ be the scaling relation.
(a) Show that $\sum_j a_j = 2$.   HINT: Integrate over $\mathbb{R}$.
(b) Show that $\sum_j (-1)^j a_j = 0$.   HINT: Use the previous exercise for $\theta = 0$.

15.5. Daubechies Wavelets

The multiresolution analysis developed in the last two sections can be used to design a continuous wavelet. We start by explaining the properties we want. The only example we have so far of a wavelet system and multiresolution analysis is the Haar wavelet system. The Haar wavelet ψ satisfies

$$\int \psi(x)\, dx = 0$$

and the multiresolution analysis uses subspaces of functions that are constant on dyadic intervals of length 2^k, $k \in \mathbb{Z}$. As a result, Haar wavelets do a good job of approximating functions that are locally constant.

It is possible to do a better job of approximating continuous functions if we use a wavelet that also satisfies

$$\int x\psi(x)\, dx = 0.$$

If you computed moments of inertia in calculus, you won't be surprised to learn that this is called the **first moment** of ψ.

Our goal in this section is to construct a continuous wavelet with this property. To be honest, our construction is not quite complete. At one crucial point, we will *assume* the uniform convergence of a sequence of functions to a continuous function. The full construction of this wavelet requires considerable work, although in the next section we provide a proof that the sequence converges in L^2. Later in this chapter, we give a full proof of the existence of another continuous wavelet, known as the Franklin wavelet.

This is part of a general family of wavelets constructed by Ingrid Daubechies in 1988. Hence these wavelets are called **Daubechies wavelets**.

15.5.1. THEOREM. *There is a continuous function φ of compact support in $L^2(\mathbb{R})$ that generates a multiresolution of $L^2(\mathbb{R})$ so that the associated wavelet ψ is continuous, has compact support, and satisfies $\int \psi(x)\, dx = \int x\psi(x)\, dx = 0$.*

PROOF. As in the last section, all of our summations are from $-\infty$ to $+\infty$; so only the index is given. We will look for a function φ with norm 1 and integral 1, that is,

$$\|\varphi\|_2^2 = \int_{-\infty}^{+\infty} |\varphi(x)|^2 \, dx = 1 \quad \text{and} \quad \int_{-\infty}^{+\infty} \varphi(x) \, dx = 1.$$

Beyond these normalizing assumptions, we use the crucial idea of the previous section by assuming that φ satisfies a scaling relation $\varphi(x) = \sum_j a_j \, \varphi(2x - j)$. Since we wish φ to have compact support, this must be a finite sum.

Compute what follows from our assumptions:

$$1 = \|\varphi\|_2^2 = \int_{-\infty}^{+\infty} |\varphi(x)|^2 \, dx = \int_{-\infty}^{+\infty} \left| \sum_j a_j \, \varphi(2x - j) \right|^2 \, dx$$

$$= \left\langle \sum_j a_j \, \varphi(2x - j), \sum_k a_k \varphi(2x - k) \right\rangle = \frac{1}{2} \sum_j |a_j|^2,$$

where we use the fact that $\{\varphi(2x - j) : j \in \mathbb{Z}\}$ is an orthogonal set of vectors in $L^2(\mathbb{R})$ with norm $1/\sqrt{2}$. In the same way, $\int \varphi(x) \, dx = 1$ implies that $\sum_j a_j = 2$.

From Theorem 15.4.2, if we can find a suitable sequence (a_k), then there is a wavelet ψ, also of norm 1, which is given by

(15.5.2) $$\psi(x) = \sum_j (-1)^j a_{1-j} \, \varphi(2x - j).$$

Consider the consequences of the integral conditions on ψ, namely $\int \psi(x) \, dx = 0$ and $\int x\psi(x) \, dx = 0$, to obtain two more relations that the sequence (a_n) must satisfy.

$$0 = \int \psi(x) \, dx = \int \sum_j (-1)^j a_{1-j} \, \varphi(2x - j) \, dx$$

$$= \sum_j (-1)^j a_{1-j} \int \varphi(2x - j) \, dx = \frac{1}{2} \sum_j (-1)^j a_{1-j}$$

Replace $1 - j$ with j and use $(-1)^{1-j} = -(-1)^j$ to obtain $\sum_j (-1)^j a_j = 0$.

Similarly,

$$0 = \int x\psi(x) \, dx = \sum_j (-1)^j a_{1-j} \int x\varphi(2x - j) \, dx$$

$$= \sum_j (-1)^j a_{1-j} \frac{1}{4} \int (t + j)\varphi(t) \, dt$$

$$= -\left(\frac{1}{4} \sum_j (-1)^j a_{1-j} \right) \int t\varphi(t) \, dt - \left(\frac{1}{4} \sum_j (-1)^j j a_{1-j} \right) \int \varphi(t) \, dt$$

$$= -\frac{1}{4} \sum_j (-1)^j j a_{1-j} = \frac{1}{4} \sum_k (-1)^k (1 - k) a_k$$

$$= \frac{1}{4}\sum_k (-1)^k a_k - \frac{1}{4}\sum_k (-1)^k k a_k = -\frac{1}{4}\sum_k (-1)^k k a_k.$$

Summarizing, we have the following equations:

$$\sum_j |a_j|^2 = \sum_j a_j = 2 \quad \text{and} \quad \sum_j (-1)^j a_j = \sum_j (-1)^j j a_j = 0.$$

As you can verify directly, one solution to these equations is given by

$$a_0 = \frac{1+\sqrt{3}}{4}, \quad a_1 = \frac{3+\sqrt{3}}{4}, \quad a_2 = \frac{3-\sqrt{3}}{4}, \quad a_3 = \frac{1-\sqrt{3}}{4}$$

with $a_j = 0$ for all other $j \in \mathbb{Z}$.

Substituting these values back into the scaling relation, we want the scaling function to satisfy

$$\varphi(x) = \frac{1+\sqrt{3}}{4}\varphi(2x) + \frac{3+\sqrt{3}}{4}\varphi(2x-1) + \frac{3-\sqrt{3}}{4}\varphi(2x-2) + \frac{1-\sqrt{3}}{4}\varphi(2x-3)$$
$$= a_0\varphi(2x) + a_1\varphi(2x-1) + a_2\varphi(2x-2) + a_3\varphi(2x-3).$$

It is not immediately clear why there should be a continuous function satisfying this equation.

We can construct such a function as the limit of a sequence of functions (φ_n), defined by $\varphi_0 = \chi_{[0,1)}$ and for $n \geq 0$,

$$\varphi_{n+1}(x) = a_0\varphi_n(2x) + a_1\varphi_n(2x-1) + a_2\varphi_n(2x-2) + a_3\varphi_n(2x-3)$$

From the first few φ_n, graphed in Figure 15.2, it is plausible that the sequence (φ_n) converges to a continuous function. However, proving this requires careful arguments using the Fourier transform, and so is beyond the scope of this book. We content ourselves with stating the following theorem.

15.5.3. THEOREM. *The sequence of functions (φ_n) converges uniformly to a continuous function φ.*

We will prove convergence in $L^2(\mathbb{R})$ in the next section. The other properties of the Daubechies wavelet can now be deduced. Note that except for the continuity of φ and ψ, all of the other properties follow from convergence in $L^2(\mathbb{R})$.

15.5.4. COROLLARY. *The Daubechies wavelet φ satisfies the properties:*

(1) $\varphi(x) = a_0\varphi(2x) + a_1\varphi(2x-1) + a_2\varphi(2x-2) + a_3\varphi(2x-3)$.

(2) φ *is supported on* $[0,3]$.

(3) $\|\varphi\|_2 = 1$ *and* $\int \varphi(x)\,dx = 1$.

(4) $\psi(x) = -a_0\varphi(2x-1) + a_1\varphi(2x) - a_2\varphi(2x+1) + a_3\varphi(2x+2)$
    is continuous with support in $[-1,2]$.

(5) $\int \psi(x)\,dx = \int x\psi(x)\,dx = 0$.

(6) $\{\varphi(x-j), \psi(x-j) : j \in \mathbb{Z}\}$ *is orthonormal.*

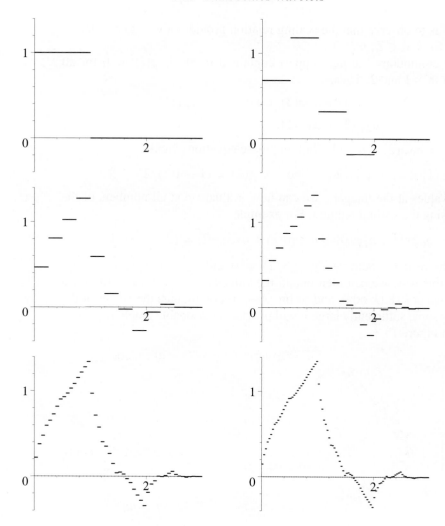

FIGURE 15.2. The graphs of φ_0 through φ_5.

PROOF. The proof will be left as an exercise using the following outline.

From the definition of φ_{n+1} in terms of φ_n and the convergence to φ, it follows immediately that φ satisfies the scaling relation. The Haar function $\varphi_0 = \chi_{[0,1)}$ satisfies (2) and (3) and (6a): $\{\varphi_0(x - j) : j \in \mathbb{Z}\}$ is orthonormal. We also introduce the functions

$$\psi_{n+1}(x) = -a_0\varphi_n(2x - 1) + a_1\varphi_n(2x) - a_2\varphi_n(2x + 1) + a_3\varphi_n(2x + 2)$$

for $n \geq 0$. We show by induction that φ_n and ψ_n satisfy (2)–(6) for all $n \geq 1$ with the exception of continuity for ψ_n, and thus they hold in the limit. The continuity of ψ follows from the continuity of φ. ∎

At this point, a reasonable objection is that we do not have anything resembling a formula for the scaling function φ, much less the wavelet ψ that goes along with it. The remedy is to observe that the scaling relation provides a way to evaluate φ at points $k/2^n$ for $k, n \in \mathbb{Z}$, $n \geq 0$.

Since φ is continuous and has support contained in $[0, 3]$, $\varphi(i) = 0$ for all integers i other than 1 and 2. Thus,

$$\varphi(2) = a_0\varphi(4) + a_1\varphi(3) + a_2\varphi(2) + a_3\varphi(1)$$
$$= a_2\varphi(2) + a_3\varphi(1).$$

Similarly, $\varphi(1) = a_0\varphi(2) + a_1\varphi(1)$. Solving these equations yields

$$\varphi(1) = (1 + \sqrt{3})/2 \quad \text{and} \quad \varphi(2) = (1 - \sqrt{3})/2.$$

From the values at the integers, we can now evaluate φ at all numbers of the form $k/2$ by using the scaling relation. For example,

$$\varphi(\tfrac{1}{2}) = a_0\varphi(1) + a_1\varphi(0) + a_2\varphi(-1) + a_3\varphi(-2) = (2 + \sqrt{3})/4.$$

Likewise, as the reader should verify, $\varphi(3/2) = 0$ and $\varphi(5/2) = (2 - \sqrt{3})/4$. Continuing in this way, we can then obtain the values of φ at points of the form $k/4$ (k odd), then at $k/8$ (k odd), and so on. As φ is continuous, the values at the points $k/2^n$ for some sufficiently large n will provide a reasonable graph of φ, such as that given in Figure 15.3.

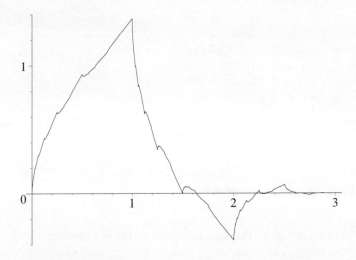

FIGURE 15.3. The Daubechies scaling function.

Similarly, using the relation (15.5.2), we can find the values of ψ at points $k/2^n$ and graph ψ; see Figure 15.4.

Exercises for Section 15.5

A. Prove Corollary 15.5.4 following the outline given there.

B. Evaluate the Daubechies wavelet ψ at the points $k/4$ for $k \in \mathbb{Z}$.

C. Which Daubechies wavelet coefficients are nonzero for the function given by $f(x) = x$ on $[0, 2]$ and 0 elsewhere?

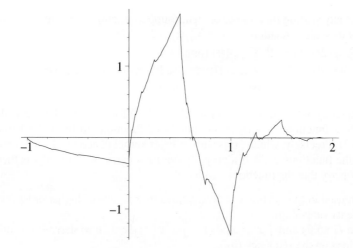

FIGURE 15.4. The Daubechies wavelet.

D. Write out the details of the argument that $\int \varphi(x)\, dx = 1$ implies that $\sum_j a_j = 2$.

E. Define $\mathbb{Q}[\sqrt{3}]$ as the set of numbers $x + \sqrt{3}y$, where $x, y \in \mathbb{Q}$.
  (a) Show that if $a + \sqrt{3}b = x + \sqrt{3}y$, then $a = x$ and $b = y$. Thus we can represent elements of $\mathbb{Q}[\sqrt{3}]$ uniquely as ordered pairs.
  (b) Show that addition, multiplication, and division operations in $\mathbb{Q}[\sqrt{3}]$ are closed and can be expressed using rational operations.
  (c) Explain the significance of this last fact for the efficiency of the algorithm for evaluating φ and ψ given at the end of this section.

F. Let φ be the Daubechies scaling function. Let $\mathbb{D}_n$ be the set $\{a/2^k : k \in \mathbb{N}, a \in \mathbb{Z}\}$ and define the set of dyadic rationals to be $\mathbb{D} = \cup_{n \geq 1} \mathbb{D}_n$. Further, define $\mathbb{D}_n[\sqrt{3}]$ to be the set $\{a + \sqrt{3}b : a, b \in \mathbb{D}_n\}$ and likewise define $\mathbb{D}[\sqrt{3}]$.
  (a) Show that for $d \in \mathbb{D}$, $\varphi(d) \in \mathbb{D}[\sqrt{3}]$.
  (b) If $\overline{a + \sqrt{3}b}$ is defined as $a - \sqrt{3}b$, show that for $d \in \mathbb{D}$, $\varphi(3 - d) = \overline{\varphi(d)}$.
    HINT: Prove it for $\mathbb{D}_n$ using induction on n.
  (c) Show that for $d \in \mathbb{D}$, $1 = \sum_{k \in \mathbb{Z}} \varphi(d - k)$.   HINT: See the previous hint.
  (d) Show that $\sum_{k=0}^{3} k\varphi(k) = \sum_{j=0}^{2} 2j a_{2j} = \dfrac{3 - \sqrt{3}}{2}$.
  (e) Show that for $d \in \mathbb{D}$, $d = \sum_{k \in \mathbb{Z}} \left(\dfrac{3 - \sqrt{3}}{2} + k \right) \varphi(d - k)$.

    HINT: Using part (c), show that it is enough to prove $d - \dfrac{3 - \sqrt{3}}{2} = \sum_{k \in \mathbb{Z}} k\varphi(d - k)$

    Then use induction, part (d), and considerable calculation.
  (f) Using the continuity of φ, deduce that for all $x \in \mathbb{R}$,

$$1 = \sum_{k \in \mathbb{Z}} \varphi(x - k), \qquad x = \sum_{k \in \mathbb{Z}} \left(\frac{3 - \sqrt{3}}{2} + x \right) \varphi(x - k).$$

G. Suppose that φ is any scaling function of compact support arising from a multiresolution analysis and its scaling relation $\varphi(x) = \sum_j a_j \varphi(2x - j)$.

(a) Show that $\sum_n \varphi(2^{-k}n) = 2^k \sum_n \varphi(n)$ for all $k \geq 1$.

(b) Show that $\int \varphi(x)\,dx = \sum_n \varphi(n)$.   HINT: Use Riemann sums and part (a).

(c) Show that $\sum_n \varphi(x - n)$ is constant.

H. Consider the sequence of functions $\varphi_{k0}(x) = 2^{k/2}\varphi(2^k x)$ for $k \geq 0$, where φ is the Daubechies scaling function. Compare this family of functions to the Fejér kernel (Definition 14.6.1). Precisely, which properties of Fejér kernel (Theorem 14.6.3) carry over directly to the functions φ_{k0}? If a property does not carry over directly, is there an analogous property that the functions φ_{k0} have?

I. The construction used in this section can be extended to wavelets of higher order (i.e., with more moments vanishing).

(a) Use Equation (15.5.2) and $\int x^2 \psi(x)\,dx = \int x^3 \psi(x)\,dx = 0$ to derive two additional conditions on the sequence (a_k).

(b) Show that the following values, with all other a_k zero, give an approximate solution to these conditions.

$$a_0 = .470467 \qquad a_1 = 1.141117 \qquad a_2 = .650365$$
$$a_3 = -.190934 \qquad a_4 = -.120832 \qquad a_5 = .049817$$

(c) Using the appropriate scaling relation, plot these higher-order wavelets and scaling functions. Although the graph suggests otherwise, this scaling function is actually differentiable.

15.6. Existence of the Daubechies Wavelet

The purpose of this section is to establish the following:

15.6.1. THEOREM. *The sequence of functions (φ_n) converges in the $L^2(\mathbb{R})$ norm to a function φ.*

We need two computational lemmas that enable us to estimate $\|\varphi_{n+1} - \varphi_n\|_2$. First, we compute $\langle \varphi_{n+1}(x), \varphi_n(x - k)\rangle$.

15.6.2. LEMMA. *Define $c_n(k) = \langle \varphi_{n+1}, \varphi_n(x - k)\rangle$ for $k \in \mathbb{Z}$ and $n \geq 0$. Then $c_n(k) = 0$ if $|k| > 2$, and the sequence of 5-tuples*

$$\mathbf{c}_n = \big(c_n(-2), c_n(-1), c_n(0), c_n(1), c_n(2)\big) \quad \text{for} \quad n \geq 0$$

satisfies $\mathbf{c}_0 = (0, 0, \frac{2+\sqrt{3}}{4}, \frac{2-\sqrt{3}}{4}, 0)$ and $\mathbf{c}_{n+1} = T\mathbf{c}_n$, where

$$T = \begin{bmatrix} 0 & -1/16 & 0 & 0 & 0 \\ 1 & 9/16 & 0 & -1/16 & 0 \\ 0 & 9/16 & 1 & 9/16 & 0 \\ 0 & -1/16 & 0 & 9/16 & 1 \\ 0 & 0 & 0 & -1/16 & 0 \end{bmatrix}.$$

PROOF. Since $\varphi_0 = \chi_{[0,1)}$ and

$$\varphi_1 = \sum_{i=0}^{3} a_i \varphi_0(2x - i) = a_0 \chi_{[0,.5)} + a_1 \chi_{[.5,1)} + a_2 \chi_{[1,1.5)} + a_3 \chi_{[1.5,2)}$$

we easily compute $c_0(0) = \dfrac{a_0 + a_1}{2} = \dfrac{2 + \sqrt{3}}{4}$ and $c_0(1) = \dfrac{a_2 + a_3}{2} = \dfrac{2 - \sqrt{3}}{4}$
and $c_0(k) = 0$ in all other cases. Proceed by induction on n.

$$c_{n+1}(k) = \langle \varphi_{n+1}, \varphi_n(x - k) \rangle$$

$$= \left\langle \sum_{i=0}^{3} a_i \varphi_n(2x - i), \sum_{j=0}^{3} a_j \varphi_{n-1}(2x - 2k - j) \right\rangle$$

$$= \sum_{i=0}^{3} \sum_{j=0}^{3} a_i a_j \langle \varphi_n(2x - i), \varphi_{n-1}(2x - 2k - j) \rangle$$

Notice that making the substitution $y = 2x - i$ in the inner product results in a factor of $\frac{1}{2}$ from the change of variable of integration.

$$= \frac{1}{2} \sum_{i=0}^{3} \sum_{j=0}^{3} a_i a_j \langle \varphi_n(y), \varphi_{n-1}(y + i - 2k - j) \rangle$$

Now set $l = j - i$ to obtain

$$= \frac{1}{2} \sum_{i=0}^{3} \sum_{l=-3}^{3} a_i a_{i+l} \langle \varphi_n(y), \varphi_{n-1}(y - 2k - l) \rangle$$

$$= \sum_{l=-3}^{3} \left(\frac{1}{2} \sum_{i=0}^{3} a_i a_{i+l} \right) c_{n-1}(2k + l).$$

Observe immediately that if $|k| \geq 3$ and $|l| \leq 3$, then $|2k + l| \geq 3$. Therefore, $c_{n+1}(k)$ for $|k| \geq 3$ depend linearly on $c_n(k)$ for $|k| \geq 3$. However, $c_0(k) = 0$ for $|k| \geq 3$, so $c_n(k) = 0$ for all $n \geq 0$ and $|k| \geq 3$. So we need only be concerned with the 5-tuple $\mathbf{c}_n = \big(c_n(-2), c_n(-1), c_n(0), c_n(1), c_n(2) \big)$.

A routine calculation yields

$$\frac{1}{2} \sum_{i=0}^{3} a_i a_{i+l} = \begin{cases} 1 & \text{when} \quad l = 0 \\ 9/16 & \text{when} \quad l = \pm 1 \\ -1/16 & \text{when} \quad l = \pm 3 \\ 0 & \text{otherwise.} \end{cases}$$

Plugging this into our identity and writing the five relations as a matrix, we obtain $\mathbf{c}_{n+1} = T \mathbf{c}_n$. ∎

Next we compute the Jordan form of T. Note immediately that $T\mathbf{e}_0 = \mathbf{e}_0$ is an obvious eigenvector.

15.6.3. LEMMA. *The matrix T factors as $T = VJV^{-1}$, where*

$$
V = \begin{bmatrix} 0 & -1 & 2 & -1 & 4 \\ 0 & 8 & -4 & 4 & 0 \\ 1 & 0 & 0 & -6 & -8 \\ 0 & -8 & 4 & 4 & 0 \\ 0 & 1 & -2 & -1 & 4 \end{bmatrix} \quad and \quad J = \begin{bmatrix} 1 & 0 & 0 & 0 & 0 \\ 0 & 1/2 & 0 & 0 & 0 \\ 0 & 0 & 1/8 & 0 & 0 \\ 0 & 0 & 0 & 1/4 & 1 \\ 0 & 0 & 0 & 0 & 1/4 \end{bmatrix}.
$$

PROOF. We leave it to the reader to show that $VJ = TV$ and to check that V is invertible (see the formula for V^{-1} given in the proof of Theorem 15.6.4). $\blacksquare$

15.6.4. THEOREM. $\sum\limits_{n \geq 0} \|\varphi_{n+1} - \varphi_n\|_2 < \infty$ *and thus* $\lim\limits_{n \to \infty} \varphi_n$ *exists in* $L^2(\mathbb{R})$.

PROOF. The first step is to compute $c_n(0)$. Notice that $\mathbf{c}_n = T^n \mathbf{c}_0 = VJ^nV^{-1}\mathbf{c}_0$. We find that

$$
J^n = \begin{bmatrix} 1 & 0 & 0 & 0 & 0 \\ 0 & 2^{-n} & 0 & 0 & 0 \\ 0 & 0 & 8^{-n} & 0 & 0 \\ 0 & 0 & 0 & 4^{-n} & 4^{1-n}n \\ 0 & 0 & 0 & 0 & 4^{-n} \end{bmatrix} \quad and \quad V^{-1} = \begin{bmatrix} 1 & 1 & 1 & 1 & 1 \\ \frac{1}{6} & \frac{1}{12} & 0 & -\frac{1}{12} & -\frac{1}{6} \\ \frac{1}{3} & \frac{1}{24} & 0 & -\frac{1}{24} & -\frac{1}{3} \\ 0 & \frac{1}{8} & 0 & \frac{1}{8} & 0 \\ \frac{1}{8} & \frac{1}{32} & 0 & \frac{1}{32} & \frac{1}{8} \end{bmatrix}.
$$

Thus a straightforward multiplication yields

$$
\langle \varphi_{n+1}, \varphi_n \rangle = c_n(0) = 1 - 4^{-n-2}(2 - \sqrt{3})(3n + 4).
$$

Set $\varepsilon_n = 4^{-n-2}(2 - \sqrt{3})(3n + 4)$. Notice that

$$
\begin{aligned}
\|\varphi_{n+1} - \varphi_n\|^2 &= \langle \varphi_{n+1} - \varphi_n, \varphi_{n+1} - \varphi_n \rangle \\
&= \|\varphi_{n+1}\|_2^2 - 2\langle \varphi_{n+1}, \varphi_n \rangle + \|\varphi_n\|_2^2 \\
&= 2 - 2(1 - \varepsilon_n) = 2\varepsilon_n.
\end{aligned}
$$

Consequently,

$$
\sum_{n \geq 0} \|\varphi_{n+1} - \varphi_n\|_2 = \sum_{n \geq 0} \frac{\sqrt{2(2 - \sqrt{3})(3n + 4)}}{2^{n+2}} < \sum_{n \geq 0} \frac{n + 2}{2^{n+2}} = \frac{3}{2}.
$$

By Exercise 4.2.B, this implies that (φ_n) is a Cauchy sequence. Thus the limit function φ is defined. $\blacksquare$

Exercises for Section 15.6

A. Verify that $VJ = TV$ and that the formula for V^{-1} is correct.

B. Do the calculation to compute $c_n(0)$ as indicated in Theorem 15.6.4.

C. Verify that $\sum\limits_{n \geq 0} \dfrac{n+2}{2^{n+2}}$ converges. Do not compute its exact value.

D. Observe that the columns of T all sum to 1. Interpret this as saying that a related matrix has a certain eigenvector.

E. Find $\lim\limits_{n \to \infty} T^n$. HINT: Use the Jordan form.

15.7. Approximations Using Wavelets

In this section, we approximate functions using the Daubechies wavelet system. Our goal is to show how properties of the wavelet basis result in better (or worse) approximations. Given that the point of wavelets is to use different kinds of wavelets for different problems, it is worthwhile to see how to use properties of the wavelet and scaling function. We have already devoted Chapter 14 to approximation by Fourier series and Chapter 10 to approximation by polynomials, so we can compare approximation by wavelets to these alternatives.

Recall from the discussion of Haar wavelets that we may construct approximants to functions by first computing the projection $P_k f$ of f onto V_k. Let φ and $\{\psi_{kj}\}$ denote the Daubechies wavelet. Let us define the projections D_n onto $V_n = \text{span}\{\varphi_{n,j} : j \in \mathbb{Z}\}$ by

$$D_n f(x) = \sum_{j \in \mathbb{Z}} \langle f, \varphi_{n,j} \rangle \varphi_{n,j}.$$

For $n \geq 1$, we also realize this as

$$D_n f(x) = D_0 f(x) + \sum_{k=0}^{n-1} \sum_{j \in \mathbb{Z}} \langle f, \psi_{kj} \rangle \psi_{kj}(x).$$

If f has compact support, then only finitely many of these coefficients are nonzero at each level. For example, if the support is $[0, 1]$, then there are at most three nonzero terms in the computation of D_0 and at most $2^{n-1} + 2$ additional terms to compute $D_n f$ knowing $D_{n-1} f$.

For any wavelet arising from a multiresolution analysis, the approximants $P_k f$ converge to f in $L^2(\mathbb{R})$. So, in particular, this is true for the Daubechies wavelets. In this case, we can also establish uniform convergence when f is uniformly continuous. We need a variation on the first part of the proof of Theorem 15.2.3.

15.7.1. LEMMA. *Consider the Daubechies wavelets. Fix $k \in \mathbb{N}$, $j \in \mathbb{Z}$, and $x \in [j/2^k, (j+3)/2^k]$. For any continuous function f on $\mathbb{R}$,*

$$\left| f(x) - 2^{k/2} \langle f, \varphi_{kj} \rangle \right| \leq \sqrt{3}\, \omega(f, 3 \cdot 2^{-k}).$$

PROOF. Using the substitution $z = 2^k t - j$ and $\int \varphi(t)\,dt = 1$, we have

$$
\begin{aligned}
\left| f(x) - 2^{k/2} \langle f, \varphi_{k0} \rangle \right| &= \left| f(x) - 2^k \int f(t)\varphi(2^k t - j)\,dx \right| \\
&= \left| f(x) - \int_0^3 f\!\left(\frac{z+j}{2^k} \right) \varphi(z)\,dz \right| \\
&= \left| \int_0^3 \left[f(x) - f\!\left(\frac{z+j}{2^k} \right) \right] \varphi(z)\,dz \right| \\
&\leq \omega(f; 3 \cdot 2^{-k}) \int_0^3 |\varphi(z)|\,dz
\end{aligned}
$$

since for $z \in [0,3]$, we have $|x - (z+j)/2^k| \leq 3/2^k$. Finally, the Cauchy–Schwarz inequality shows that

$$
\int_0^3 |\varphi(z)|\,dz \leq \left(\int_0^3 |\varphi(z)|^2\,dz \right)^{1/2} \left(\int_0^3 1\,dz \right)^{1/2} = \sqrt{3}. \qquad \blacksquare
$$

15.7.2. THEOREM. *If $f \in L^2(\mathbb{R})$ is uniformly continuous on $\mathbb{R}$, then the approximants $D_k f$ by Daubechies wavelets converge uniformly to f.*

PROOF. From Exercise 15.5.G, we have $\sum_{j \in \mathbb{Z}} \varphi(2^k x - j) = 1$. Multiplying this by $f(x)$, we have

$$
\begin{aligned}
\left| f(x) - D_k f(x) \right| &= \left| \sum_{j \in \mathbb{Z}} f(x)\varphi(2^k x - j) - \sum_{j \in \mathbb{Z}} \langle f, \varphi_{kj} \rangle \varphi_{kj}(x) \right| \\
&\leq \sum_{j \in \mathbb{Z}} \left| f(x) - 2^{k/2} \langle f, \varphi_{kj} \rangle \right| \, \left| \varphi(2^k x - j) \right| \\
&\leq \sqrt{3}\, \omega(f, 3 \cdot 2^{-k}) \sum_{j \in \mathbb{Z}} \left| \varphi(2^k x - j) \right| \\
&\leq 3\sqrt{3} \|\varphi\|_\infty \omega(f, 3 \cdot 2^{-k}),
\end{aligned}
$$

because for any x, there at most three j so that $\varphi(2^k x - j) \neq 0$. By the uniform continuity of f, $\lim_{k \to \infty} \omega(f, 3 \cdot 2^{-k}) = 0$. Thus $\lim_{k \to \infty} \|f - D_k f(x)\|_\infty = 0$. $\qquad \blacksquare$

15.7.3. REMARK. This proof does something even better because of the local nature of the Daubechies wavelets. If f is not continuous everywhere, but is continuous on a neighbourhood $[a - \delta, a + \delta]$, the same argument shows that the series converges uniformly on $[a - \varepsilon, a + \varepsilon]$ for $\varepsilon < \delta$. We will use this in Example 15.7.4.

Theorem 15.2.3 likewise shows that when f is continuous, $H_n f$ converges to f uniformly. In fact, $\|H_n f - f\|_\infty \leq \omega(f; 2^{-n})$. So if f has Lipschitz constant L, the error is at most $2^{-n}L$. Since the number of coefficients in $H_n f$ doubles when n increases by 1, this is not surprising. We now have a comparable rate of convergence for Daubechies wavelets.

In approximating a function, a reasonable measure of the size of the approximant is the number of coefficients used. For $f \in L^2(0,1)$, we have

$$H_n f(x) = \langle f, \varphi \rangle \varphi(x) + \sum_{k=0}^{n-1} \sum_{j=0}^{2^k - 1} \langle f, \psi_{kj} \rangle \psi_{kj}(x)$$

and so $H_n f$ uses 2^n coefficients. If there is a bound on the number of coefficients that we can use, due to storage limitations, for example, one choice is to use the largest value of n so that $H_n f$ does not have too many coefficients. It is frequently better to use a larger value of n and then replace small coefficients with zero. This has the advantage that if f has large irregularities at small resolution, these will appear in the approximation.

15.7.4. EXAMPLE. Consider the function given by

$$f(x) = \begin{cases} x & \text{if} \quad x \in (-\pi, \pi), \\ 0 & \text{if} \quad |x| \geq \pi. \end{cases}$$

In Section 14.5, we analyzed the how the partial sums $S_n f$ of the Fourier series approximated f near the discontinuity at π. The Fourier series for f on $(-\pi, \pi)$ is

$$f(x) \sim 2 \sum_{k=1}^{\infty} \frac{(-1)^{k+1}}{k} \sin kx,$$

which converges very slowly. Even at a point well away from the discontinuity, convergence is slow. For example, if $x = \pi/2$, then we get

$$f(\pi/2) = 2 \sum_{k=0}^{\infty} \frac{(-1)^k}{2k + 1}.$$

To get $2 \sum_{k=0}^{n} (-1)^k / (2k + 1)$ within 10^{-6} of the exact sum of the series, we need $n \geq 500,000$. (To be fair, we can do better with Fejér kernels, but the behaviour is not optimal even then.)

The Haar wavelet approximations $H_n f$ are the step functions taking the average value of f over each interval of length 2^{-n}. It is not difficult to see that except for two intervals about the discontinuities $\pm\pi$, the convergence is uniform. The maximum error is 2^{-n-1}. Since the number of coefficients needed is roughly $2^{n+1}\pi$, we see that this convergence is not much more efficient than the Fourier series *globally*. However, to compute $f(\pi/2)$, note that only n terms in the expansion $H_n f$ are nonzero at $\pi/2$. Thus to compute this value within 10^{-6}, we only need 19 terms since $2^{-20} < 10^{-6}$.

Even better, the vanishing moments of the Daubechies wavelets ensure that so long as the support of ψ_{kj} does not contain $\pm\pi$, the coefficient $\langle f, \psi_{kj} \rangle$ will be zero. Since ψ has support in $[0,3]$, each ψ_{kj} has support in an interval of length $3/2^k$, namely $[j/2^k, (j+3)/2^k]$. So for each k, there are only six nonzero coefficients for ψ_{kj} to contribute to the whole series.

Returning to the point $\pi/2$, notice that for the support of ψ_{kj} to contain both $\pi/2$ and π, we must have $3 \cdot 2^{-k} > \pi/2$ or $k \leq 0$. Thus $D_0 f(\pi/2) = D_n f(\pi/2)$ for all $n \in \mathbb{N}$. Indeed by Remark 15.7.3, this series converges uniformly to f on any interval around $\pi/2$ that is bounded away from $\pm\pi$. In particular, we have $f(\pi/2) = \lim_{n \to \infty} D_n f(\pi/2) = D_0 f(\pi/2)$. So only three nonzero coefficients are involved in recovering this value exactly.

We can use the vanishing of the moments to obtain a better bound when the function f is C^2. Thus as for Fourier series, the Daubechies wavelet coefficients will die off quickly if the function is smooth. Moreover, because of the local nature of these wavelets, if f is smooth on some small interval, the same analysis shows that the wavelet series converges rapidly on that interval.

15.7.5. THEOREM. *If f is twice differentiable on $[(j-2)/2^k, (j+2)/2^k]$, where $j, k \in \mathbb{Z}$, and f'' bounded by B on this interval, then*

$$|\langle f, \psi_{kj} \rangle| \leq \frac{4B}{2^{5k/2}}.$$

PROOF. Substituting $t = 2^k x - j$, we obtain

$$\langle f, \psi_{kj} \rangle = \int_{-\infty}^{+\infty} f(x) 2^{k/2} \psi(2^k x - j)\, dx = 2^{-k/2} \int_{-\infty}^{+\infty} f\left(\frac{t+j}{2^k}\right) \psi(t)\, dt.$$

This integral may be limited to $[-1, 2]$, the support of ψ.

On $[(j-2)/2^k, (j+2)/2^k]$, we have a Taylor series expansion for f centred at the point $b = j/2^k$, namely

$$f(x) = f(b) + f'(b)(x - b) + \frac{f''(c)}{2}(x - b)^2$$

for some point c between x and b. The vanishing moment conditions on ψ imply that $\int (mx + d)\psi(x)\, dx = 0$. As $x - b = (t + j)/2^k - j/2^k = t/2^k$, we end up with

$$\langle f, \psi_{kj} \rangle = 2^{-k/2} \frac{1}{2} f''(c) \int_{-1}^{2} \left(\frac{t}{2^k}\right)^2 \psi(t)\, dt.$$

Since $t^2 \leq 4$, we obtain

$$|\langle f, \psi_{kj} \rangle| \leq \frac{2B}{2^{5k/2}} \int_{-1}^{2} |\psi(t)|\, dt.$$

Finally, the Cauchy–Schwarz inequality shows that

$$\int_{-1}^{2} |\psi(t)|\, dt \leq \left(\int_{-1}^{2} |\psi(t)|^2\, dt \right)^{1/2} \left(\int_{-1}^{2} 1\, dt \right)^{1/2} = \sqrt{3} < 2. \qquad \blacksquare$$

Exercises for Section 15.7

A. Show that $\left\| b_0\varphi + \sum_{k=0}^{n-1}\sum_{j=0}^{2^k-1} a_{kj}\psi_{kj} \right\|_\infty \le |b_0| + \sum_{k=0}^{n-1} 3 \cdot 2^{k/2} \max\left\{ |a_{kj}| : j \in \mathbb{Z} \right\}$.

HINT: For each x and k, how many $\psi_{kj}(x) \neq 0$?

B. Recall the Cantor function f on $[0,1]$ from Example 5.7.8. Find the zero coefficients $\langle f, \psi_{kj}\rangle$ for $k = 1,2,3,4$. What can you conclude about the functions $D_n f$?

C. (a) For the function f in Example 15.7.4, find the least $k \in \mathbb{N}$ so that $f(3) = D_k f(3)$.
(b) In general, find a function $K(\delta)$ so that for $k \ge K(\delta)$, $f(\pi - \delta) = D_k f(\pi - \delta)$.

D. For the higher-order Daubechies wavelets of Exercise 15.5.I, state and prove a version of Theorem 15.7.5.

E. Given a function f, we can represent $P_n f$ in two ways: $b_0\varphi + \sum_{k=0}^{n-1}\sum_{j\in\mathbb{Z}} a_{kj}\psi_{kj}$ or $\sum_{j\in\mathbb{Z}} c_j\varphi_{nj}$.
(a) For the Haar wavelets, what is the significance of the coefficients b_0 and a_{kj}?
(b) For the Haar wavelets, describe how to obtain one set of coefficients from the other.
HINT: Exercise 15.2.A.
(c) For the Daubechies wavelets, describe how to obtain one set of coefficients from the other.

15.8. The Franklin Wavelet

It is not easy to just write down a wavelet or a scaling function. However, it is much easier to find a multiresolution with a scaling function that does not generate an orthonormal basis but does something a bit weaker. The goal is to construct a continuous piecewise linear wavelet by starting with such a system. The technique that we describe here can be adapted and refined to obtain wavelets with greater smoothness and/or with compact support. In this section, we restrict our attention to a single example known as the **Franklin wavelet**.

In these last sections, we need to take a more sophisticated view of linear maps between Hilbert spaces.

Consider the subspaces V_k of $L^2(\mathbb{R})$ consisting of continuous functions in $L^2(\mathbb{R})$ that are linear on each interval $[(j-1)2^{-k}, j2^{-k}]$. These subspaces satisfy most of the requirements of a multiresolution. It is immediately evident that V_k is contained in V_{k+1} for all $k \in \mathbb{Z}$. Also by definition, $f(x)$ belongs to V_k exactly when it is continuous and linear on each dyadic interval of length 2^{-k}, which holds precisely when $f(2x)$ is continuous and linear on each dyadic interval of length 2^{-k-1}, which means that $f(2x)$ belongs to V_{k+1}. So the V_k satisfy scaling.

The union of the V_k's is dense in all of $L^2(\mathbb{R})$. To see this, note that any continuous function g with bounded support, say contained in $[-2^N, 2^N]$, may be uniformly approximated to any desired accuracy by a piecewise linear continuous function f that is linear on dyadic intervals of length 2^{-k} provided that k is sufficiently large. Given $\varepsilon > 0$, choose k and $f \in V_k$ so that f is also supported on $[-2^N, 2^N]$ and $\|g - f\|_\infty < 2^{-(N+1)/2}\varepsilon$. It is easy to see that

$$\|g - f\|_2^2 = \int_{-2^N}^{2^N} |g(x) - f(x)|^2 \, dx \le \int_{-2^N}^{2^N} \left(2^{-(N+1)/2}\varepsilon\right)^2 \, dx = \varepsilon^2.$$

Because $C_c(\mathbb{R})$ is dense in $L^2(\mathbb{R})$, it follows that we have the density property $\overline{\bigcup_{k \in \mathbb{Z}} V_k} = L^2(\mathbb{R})$.

Finally, we will demonstrate the separation property $\bigcap_{k \in \mathbb{Z}} V_k = \{0\}$. Suppose that f belongs to this intersection. Let $f(i) = a_i$ for $i = -1, 0, 1$. Since $f \in V_{-k}$ for each $k > 0$, it is linear on $[0, 2^k]$ and on $[-2^k, 0]$. So $f(x) = a_0 + (a_1 - a_0)x$ on $[0, 2^k]$ and $f(x) = a_0 - (a_{-1} - a_0)x$ on $[-2^k, 0]$. Thus

$$\|f\|_2^2 \ge \int_0^{2^k} |f(x)|^2 \, dx = \tfrac{1}{3}(a_1 - a_0)^2 x^3 + a_0(a_1 - a_0)x^2 + a_0^2 x \Big|_0^{2^n}$$

$$= \tfrac{1}{3}(a_1 - a_0)^2 2^{3n} + a_0(a_1 - a_0)2^{2n} + a_0^2 2^n$$

$$= 2^n \left(\tfrac{1}{3}\left((a_1 - a_0)2^n + \tfrac{3}{2}a_0\right)^2 + \tfrac{1}{4}a_0^2 \right).$$

As n tends to infinity, the right-hand side must remain bounded by $\|f\|_2^2$. This forces $a_0 = a_1 = 0$. Likewise, integration from -2^k to 0 shows that $a_{-1} = 0$. Consequently $f = 0$ on $[-2^k, 2^k]$ for every $k \ge 0$. So $f = 0$ and the intersection is trivial.

We have constructed a multiresolution except for the important scaling function. This example does have a natural scaling function, known as the **hat function** h, which is supported on $[-1, 1]$, has $h(0) = 1$ and $h(-1) = h(1) = 0$, and is linear in between. Figure 15.5 gives its simple graph.

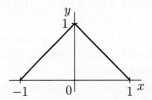

FIGURE 15.5. The graph of the hat function h.

Notice that $f(x) = \sum_{j=-n}^{n} a_j h(x - j)$ is the piecewise linear function in V_0, which is supported on $[-n - 1, n + 1]$ and satisfies $f(j) = a_j$ for $-n \le j \le n$ and $f(j) = 0$ for $|j| > n$. So V_0 is spanned by translates of the hat function. Likewise, V_k is spanned by $\{2^{k/2}h(2^k x - j) : j \in \mathbb{Z}\}$. The problem with h is that these translates are not orthogonal to each other. But it does have a weaker property that serves as a substitute.

15.8.1. DEFINITION. A subset $\{\mathbf{x}_n : n \in \mathbb{Z}\}$ of a Hilbert space $\mathcal{H}$ is a **Riesz basis** if $\overline{\text{span}\{\mathbf{x}_n : n \in \mathbb{Z}\}} = \mathcal{H}$ and there are constants $A > 0$ and $B < \infty$ so that

$$A\left(\sum_n |a_n|^2\right)^{1/2} \leq \left\|\sum_n a_n \mathbf{x}_n\right\| \leq B\left(\sum_n |a_n|^2\right)^{1/2}$$

for all sequences (a_n) with only finitely many nonzero terms.

15.8.2. THEOREM. *The translates $\{h(x-j) : j \in \mathbb{Z}\}$ of the hat function form a Riesz basis for V_0.*

PROOF. Consider the inner product of two compactly supported functions in V_0, say $f(x) = \sum\limits_{j=-n}^{n} a_j h(x-j)$ and $g(x) = \sum\limits_{j=-n}^{n} b_j h(x-j)$. For convenience, set $a_j = b_j = 0$ for $|j| > n$.

$$\langle f, g \rangle = \int_{-\infty}^{+\infty} f(x)g(x)\,dx = \sum_{j=-n-1}^{n} \int_{j}^{j+1} f(x)g(x)\,dx$$

$$= \sum_{j=-n-1}^{n} \int_0^1 \big(a_j + (a_{j+1} - a_j)x\big)\big(b_j + (b_{j+1} - b_j)x\big)\,dx$$

$$= \sum_{j=-n-1}^{n} a_j b_j + \tfrac{1}{2}a_j(b_{j+1} - b_j) + \tfrac{1}{2}(a_{j+1} - a_j)b_j + \tfrac{1}{3}(a_{j+1} - a_j)(b_{j+1} - b_j)$$

$$= \frac{1}{6}\sum_{j=-n-1}^{n} 2a_j b_j + a_j b_{j+1} + a_j b_{j+1} + 2a_{j+1}b_{j+1}$$

It is convenient to rearrange this sum further by moving the term $2a_{j+1}b_{j+1}$ to the next index to obtain

$$\langle f, g \rangle = \frac{1}{6}\sum_{j=-n-1}^{n} 4a_j b_j + a_j b_{j+1} + a_j b_{j+1}.$$

To make sense of this, we introduce two linear transformations.

Recall that $\ell^2(\mathbb{Z})$ is the Hilbert space of all square summable doubly indexed sequences $\mathbf{a} = (a_n)_{-\infty}^{+\infty}$, where we have

$$\langle \mathbf{a}, \mathbf{b} \rangle = \langle (a_n), (b_n) \rangle = \sum_{n=-\infty}^{+\infty} a_n b_n \quad \text{and} \quad \|\mathbf{a}\|_2 = \langle \mathbf{a}, \mathbf{a} \rangle^{1/2} = \left(\sum_{n=-\infty}^{+\infty} a_n^2\right)^{1/2}.$$

Let $\mathbf{e}_k$ for $k \in \mathbb{Z}$ denote the standard basis for $\ell^2(\mathbb{Z})$. Define the **bilateral shift** on $\ell^2(\mathbb{Z})$ by $U\mathbf{e}_k = \mathbf{e}_{k+1}$ or $(U\mathbf{a})_n = a_{n-1}$. It is easy to see that $\|U\mathbf{a}\|_2 = \|\mathbf{a}\|_2$ for all vectors $\mathbf{a} \in \ell^2(\mathbb{Z})$. Also, it is clear that U maps $\ell^2(\mathbb{Z})$ one-to-one and onto itself. Thus, U is a unitary map.

We also recall from linear algebra that the adjoint (or transpose since we are working over the real numbers) of a linear transformation T is the linear map T^*

such that
$$\langle T^*x, y \rangle = \langle x, Ty \rangle \quad \text{for all vectors } x, y \in \mathcal{H}.$$

For the unitary operator U, we have that $U^* = U^{-1}$ is the backward bilateral shift $U^*\mathbf{e}_k = \mathbf{e}_{k-1}$ or $(U^*\mathbf{a})_n = a_{n+1}$.

Second, we define a linear map H from $\ell^2(\mathbb{Z})$ onto V_0 by

$$H\mathbf{a} = \sum_{n=-\infty}^{+\infty} a_n h(x - n).$$

Looking back at our formula for the inner product in V_0, we see that

$$(15.8.3) \quad \langle H\mathbf{a}, H\mathbf{b} \rangle = \frac{1}{6} \sum_{j=-n-1}^{n} 4a_j b_j + a_j b_{j+1} + a_j b_{j+1} = \frac{1}{6}\langle (4I + U + U^*)\mathbf{a}, \mathbf{b} \rangle.$$

By the Cauchy–Schwarz inequality (7.3.4), since $\|U\mathbf{a}\|_2 = \|\mathbf{a}\|_2 = \|U^*\mathbf{a}\|_2$,

$$|\langle U\mathbf{a}, \mathbf{a} \rangle| = |\langle U^*\mathbf{a}, \mathbf{a} \rangle| \leq \|\mathbf{a}\|_2^2.$$

Hence we obtain

$$\|f\|_2^2 = \langle H\mathbf{a}, H\mathbf{a} \rangle = \frac{1}{6}\langle (4I + U + U^*)\mathbf{a}, \mathbf{a} \rangle$$

$$\leq \frac{1}{6}(4\|\mathbf{a}\|_2^2 + |\langle U\mathbf{a}, \mathbf{a} \rangle| + |\langle U^*\mathbf{a}, \mathbf{a} \rangle|) \leq \|\mathbf{a}\|_2^2.$$

Similarly, we obtain a lower bound

$$\|f\|_2^2 = \langle H\mathbf{a}, H\mathbf{a} \rangle = \frac{1}{6}\langle (4I + U + U^*)\mathbf{a}, \mathbf{a} \rangle$$

$$\geq \frac{1}{6}(4\|\mathbf{a}\|_2^2 - |\langle U\mathbf{a}, \mathbf{a} \rangle| - |\langle U^*\mathbf{a}, \mathbf{a} \rangle|) \geq \frac{1}{3}\|\mathbf{a}\|_2^2.$$

So while the translates of h are not an orthonormal set, we find that they do form a Riesz basis for V_0. ∎

Our problem now is to replace the hat function h by a scaling function φ. This function φ must have the property that the translates $\varphi(x - j)$ form an orthonormal set spanning V_0. We make use of another property relating the maps U and H. If $f(x) = H\mathbf{a} = \sum_n a_n h(x - n)$, then

$$HU\mathbf{a} = \sum_n a_{n-1}h(x - n) = \sum_n a_n h(x - n - 1)$$

$$= f(x - 1) = (Tf)(x) = TH\mathbf{a},$$

where $Tg(x) = g(x - 1)$ is the translation operator. So $HU = TH$, namely H carries a translation by U in $\ell^2(\mathbb{Z})$ to translation by 1 in V_0. This suggests that if we can find a vector $\mathbf{c}$ such that the translates $U^n\mathbf{c}$ form an orthonormal basis with respect to the inner product

$$[\mathbf{a}, \mathbf{b}] = \frac{1}{6}\langle (4I + U + U^*)\mathbf{a}, \mathbf{b} \rangle,$$

then $\varphi = H\mathbf{c}$ will be the desired scaling function. Indeed, (15.8.3) becomes $\langle H\mathbf{a}, H\mathbf{b} \rangle = [\mathbf{a}, \mathbf{b}]$.

We make use of the correspondence between $\ell^2(\mathbb{Z})$ and $L^2(-\pi, \pi)$ provided by complex Fourier series (see Remark 14.1.5). We identify the basis $\mathbf{e}_k$ with $e^{ik\theta}$, which is an orthonormal basis for $L^2(-\pi, \pi)$. This identifies the sequence $\mathbf{a}$ in $\ell^2(\mathbb{Z})$ with the function in $L^2(-\pi, \pi)$ given by $f(\theta) = \sum\limits_{n=-\infty}^{+\infty} a_n e^{in\theta}$. Now compute

$$U f(\theta) = U \sum_n a_n e^{in\theta} = \sum_n a_n e^{i(n+1)\theta} = e^{i\theta} f(\theta).$$

Thus U is the operator that multiplies $f(\theta)$ by $e^{i\theta}$. We will write M_g to denote the operator on $L^2(-\pi, \pi)$ that multiplies by g. Such operators are called **multiplication operators**. For example, $U = M_{e^{i\theta}}$. Hence

$$\tfrac{1}{6}(4I + U + U^*) = \tfrac{1}{6}(4I + M_{e^{i\theta}} + M^*_{e^{i\theta}})$$
$$= M_{\frac{1}{6}(4+e^{i\theta}+e^{-i\theta})} = M_{\frac{1}{3}(2+\cos\theta)}.$$

15.8.4. THEOREM. *There is an ℓ^2 sequence (c_j) so that $\varphi(x) = \sum_j c_j h(x-j)$ is a scaling function for V_0.*

PROOF. Define the operator $X = M_g$, where $g(\theta) = \sqrt{3}(2 + \cos\theta)^{-1/2}$. Notice that $X = X^*$, $XU = UX$ and $X^2 = 3M^{-1}_{2+\cos\theta} = 6(4I + U + U^*)^{-1}$. Now define $\mathbf{c} = X\mathbf{e}_0$ and $\varphi = H\mathbf{c}$. Compute

$$\langle \varphi(x-j), \varphi(x-k) \rangle = \langle T^j H\mathbf{c}, T^k H\mathbf{c} \rangle = \langle HU^j\mathbf{c}, HU^k\mathbf{c} \rangle$$
$$= [U^j\mathbf{c}, U^k\mathbf{c}] = \frac{1}{6}\langle (4I + U + U^*)U^j\mathbf{c}, U^k\mathbf{c} \rangle$$
$$= \langle X^{-2}U^j X\mathbf{e}_0, U^k X\mathbf{e}_0 \rangle = \langle X^{-2}XU^j\mathbf{e}_0, XU^k\mathbf{e}_0 \rangle$$
$$= \langle XX^{-2}XU^j\mathbf{e}_0, U^k\mathbf{e}_0 \rangle = \langle U^j\mathbf{e}_0, U^k\mathbf{e}_0 \rangle = \delta_{jk}.$$

This shows that the translates of φ form an orthonormal set in the subspace V_0.

Since $\mathbf{e}_0$ is identified with the constant function 1,

$$\varphi(x) = HX1 = Hg.$$

To compute Hg, we need to find the (complex) Fourier series $g \sim \sum_n c_n e^{in\theta}$. Now g is an even function and thus $c_{-n} = c_n$; and so $g \sim c_0 + \sum\limits_{n=1}^{\infty} 2c_n \cos n\theta$. Moreover,

$$c_n = c_{-n} = \frac{\sqrt{3}}{2\pi} \int_{-\pi}^{\pi} \frac{\cos n\theta}{\sqrt{2 + \cos\theta}} \, d\theta \quad \text{for} \quad n \geq 0.$$

Hence

(15.8.5)
$$\varphi(x) = \sum_{n=-\infty}^{+\infty} c_n h(x-n).$$

This is the continuous piecewise linear function with nodes at the integers taking the values $\varphi(n) = c_n$.

A similar argument shows that the translates of φ span all of V_0. Indeed, note that

$$He^{ij\theta}g = HU^j g = T^j Hg = \varphi(x - j).$$

Now $h = He_0 = HXg^{-1}$. Express $g^{-1}(x) = \sqrt{(2 + \cos\theta)/3}$, which is continuous, as a complex Fourier series $g^{-1} \sim \sum_j b_j e^{ij\theta}$. Then

(15.8.6) $$h = HXg^{-1} = H\sum_j b_j e^{ij\theta} g = \sum_j b_j \varphi(x - j).$$

This expresses h as an ℓ^2 combination of the orthonormal basis of translates of φ, and thus h lies in their span. Evidently, this span also contains all translates of h, and so they span all of V_0. Therefore, φ is the desired scaling function. ∎

Using two formulas from the previous proof, we can write the scaling relation for φ, in terms of the sequences (b_n) and (c_n). Verify that the hat function satisfies the simple scaling relation

$$h(x) = \tfrac{1}{2}h(2x - 1) + h(2x) + \tfrac{1}{2}h(2x + 1).$$

Equation (15.8.6) gives

$$h(2x) = \sum_j b_j \varphi(2x - j)$$

and similar formulas for $h(2x - 1)$ and $h(2x + 1)$. Putting these formulas into the scaling relations gives $h(x)$ as an infinite series involving $h(2x - k)$ as k ranges over the integers. Substituting this series for h in Equation (15.8.5), we obtain

$$\varphi(x) = \sum_{l \in \mathbb{Z}} \left(\sum_{j \in \mathbb{Z}} c_j \left[b_{2j-l} + \tfrac{1}{2}b_{2j-l+1} + \tfrac{1}{2}b_{2j+l+1} \right] \right) \varphi(2x - l).$$

This formula does not appear tractable, but the sequences (c_n) and (b_n) decay quite rapidly, so it is possible to obtain reasonable numerical results by taking sums over relatively small ranges of j, say -10 to 10.

We can then apply Theorem 15.4.2 to obtain a formula for the Franklin wavelet itself. This is plotted in Figure 15.6, along with the scaling relation. Notice that the wavelet is continuous and piecewise linear with nodes at the half integers, as we would expect since Theorem 15.4.2 implies that the wavelet is in V_1. Incidentally, the numerical values of the first few a_n in the scaling relation for φ are

$$a_0 = 1.15633, \quad a_1 = a_{-1} = .56186, \quad a_2 = a_{-2} = -.09772,$$
$$a_3 = a_{-3} = -.07346, \quad a_4 = a_{-4} = -.02400.$$

Exercises for Section 15.8

These exercises are all directed toward the analysis of a different wavelet, known as the **Strömberg wavelet**, that has the same multiresolution subspaces V_k as the Franklin wavelet. See [36] for more details.

Write $PL(X)$ to denote the space of $L^2(\mathbb{R})$ functions that are continuous and piecewise linear with nodes on a discrete subset X of $\mathbb{R}$. Let h be the hat function.

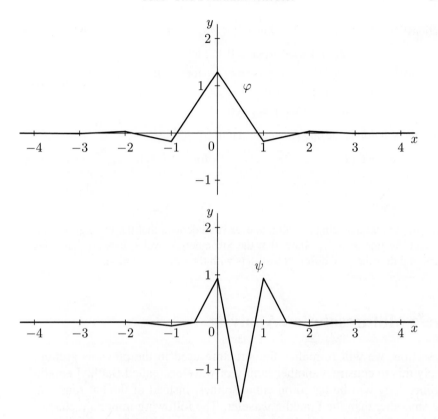

FIGURE 15.6. The Franklin scaling function and wavelet.

A. Show that $PL(-\frac{1}{2}\mathbb{N}_0 \cup \mathbb{N})$ is spanned by $\{h(2x + k + 1),\ h(x - k) : k \geq 0\}$.

B. Show that $PL(-\frac{1}{2}\mathbb{N}_0 \cup \{\frac{1}{2}\} \cup \mathbb{N})$ is spanned by $PL(-\frac{1}{2}\mathbb{N}_0 \cup \mathbb{N})$ and $h(2x)$. Hence show that there is a norm 1 function ψ in $PL(-\frac{1}{2}\mathbb{N}_0 \cup \{\frac{1}{2}\} \cup \mathbb{N})$ that is orthogonal to $PL(-\frac{1}{2}\mathbb{N}_0 \cup \mathbb{N})$.

C. Show that ψ is orthogonal to $2^{-k/2}\psi(2^{-k}x - j)$ for all $k < 0$ and $j \in \mathbb{Z}$ and to $\psi(x + j)$ for $j > 0$. Hence deduce that $\{\psi_{kj} : k, j \in \mathbb{Z}\}$ is orthonormal.
HINT: Some are in $PL(-\frac{1}{2}\mathbb{N}_0 \cup \mathbb{N})$. Do a change of variables for the rest.

D. Let $V_k = PL(2^{-k}\mathbb{Z})$ and let W_k be the orthogonal complement of V_k in V_{k+1}. Show that span$\{\psi_{kj} : j \in \mathbb{Z}\} = W_k$. Hence deduce that ψ is a wavelet.
HINT: Show that the span$\{\psi_{0j} : -n \leq j \leq n\}$ is the orthogonal complement of $PL(\mathbb{Z} \cup \{k/2 : k \leq -2n - 1\})$ in $PL(\mathbb{Z} \cup \{k/2 : k \leq 2n + 1\})$. Let n tend to infinity.

E. The piecewise linear continuous function ψ is determined by its values at the nodes, $\psi(\frac{k}{2}) = a_k$ for $k \leq 0$, $\psi(\frac{1}{2}) = b$ and $\psi(k) = c_k$ for $k \geq 1$. Show that the orthogonality relations coming from the fact that ψ is orthogonal to the basis of $PL(-\frac{1}{2}\mathbb{N}_0 \cup \mathbb{N})$ yield the equations

$$a_{k-1} + a_k + a_{k+1} = 0 \qquad \text{for} \qquad k \leq -1$$
$$c_{k-1} + c_k + c_{k+1} = 0 \qquad \text{for} \qquad k \geq 2$$
$$a_{-2} + 6a_{-1} + 10a_0 + 6b + c_1 = 0$$
$$a_0 + 6b + 13c_1 + 4c_2 = 0.$$

Verify that the solution is the one-parameter family

$$a_k = C(2\sqrt{3} - 2)(\sqrt{3} - 2)^{|k|} \qquad \text{for} \qquad k \leq 0$$
$$b = -C(\sqrt{3} + \tfrac{1}{2})$$
$$c_k = C(\sqrt{3} - 2)^{k-1} \qquad\qquad \text{for} \qquad k \geq 1.$$

F. Show that the Franklin scaling function is even, and deduce that the wavelet is symmetric about the line $x = \frac{1}{2}$. Show that the Strömberg wavelet does not have this symmetry, and thus they are different wavelets with the same resolution.

15.9. Riesz Multiresolution Analysis

In this section, we will formalize the structure used in the previous section. We then apply this to construct another family of wavelets, called Battle–Lemarié wavelets. Since they will be based on cubic splines, instead of the hat function, they will be smoother than the Franklin wavelet. The following important characterization of Riesz bases is our starting point.

15.9.1. THEOREM. *A set of vectors $\{\mathbf{x}_n : n \in \mathbb{Z}\}$ in a Hilbert space $\mathcal{H}$ is a Riesz basis if and only if there is a continuous linear map T from $\ell^2(\mathbb{Z})$ onto $\mathcal{H}$ such that $T\mathbf{e}_n = \mathbf{x}_n$ for $n \in \mathbb{Z}$ and there are constants $0 < A < B < \infty$ such that $A\|\mathbf{a}\|_2 \leq \|T\mathbf{a}\| \leq B\|\mathbf{a}\|_2$ for all $\mathbf{a} \in \ell^2(\mathbb{Z})$.*

PROOF. Let ℓ_0 denote the vector space of all sequences (a_n) with only finitely many nonzero terms. For any set $\{\mathbf{x}_n : n \in \mathbb{Z}\}$, we may define a linear map from ℓ_0 into $\mathcal{H}$ by $T\mathbf{a} = \sum_n a_n \mathbf{x}_n$, which makes sense because the sum is finite.

Suppose that $\{\mathbf{x}_n : n \in \mathbb{Z}\}$ is a Riesz basis. The Riesz condition is readily restated as

$$A\|\mathbf{a}\|_2 \leq \|T\mathbf{a}\| \leq B\|\mathbf{a}\|_2 \quad \text{for all} \quad \mathbf{a} \in \ell_0.$$

In particular, the map T satisfies the Lipschitz condition

$$\|T\mathbf{a} - T\mathbf{b}\| = \|T(\mathbf{a} - \mathbf{b})\| \leq B\|\mathbf{a} - \mathbf{b}\|_2$$

and thus T is uniformly continuous.

Suppose that $\mathbf{a} \in \ell^2(\mathbb{Z})$. Then we may choose a sequence $\mathbf{a}_n$ in ℓ_0 that converges to $\mathbf{a}$ in the ℓ^2 norm. Consequently, $(\mathbf{a}_n)$ is a Cauchy sequence in $\ell^2(\mathbb{Z})$. Therefore, since $\|T\mathbf{a}_n - T\mathbf{a}_m\| \leq B\|\mathbf{a}_n - \mathbf{a}_m\|_2$, it follows that $(T\mathbf{a}_n)$ is a Cauchy sequence in $\mathcal{H}$. Since $\mathcal{H}$ is complete, we may define $T\mathbf{a} = \lim_n T\mathbf{a}_n$. See the Exercises for the argument explaining why this definition does not depend on the choice of the sequence. So the definition of T has been extended to all of $\ell^2(\mathbb{Z})$. Moreover, we obtain that

$$\|T\mathbf{a}\| = \lim_{n \to \infty} \|T\mathbf{a}_n\| \le B \lim_{n \to \infty} \|\mathbf{a}_n\|_2 = B\|\mathbf{a}\|_2.$$

So T is (uniformly) continuous on all of $\ell^2(\mathbb{Z})$.

We similarly obtain

$$\|T\mathbf{a}\| = \lim_{n \to \infty} \|T\mathbf{a}_n\| \ge A \lim_{n \to \infty} \|\mathbf{a}_n\|_2 = A\|\mathbf{a}\|_2.$$

Clearly T maps ℓ_0 onto the set of all finite linear combinations of $\{\mathbf{x}_n : n \in \mathbb{Z}\}$. So the range of T is dense in $\mathcal{H}$ by hypothesis.

Let $\mathbf{y} \in \mathcal{H}$. Choose vectors $\mathbf{y}_n \in \operatorname{span}\{\mathbf{x}_n : n \in \mathbb{Z}\}$ that converge to $\mathbf{y}$. Then $(\mathbf{y}_n)$ is a Cauchy sequence. Since $\mathbf{y}_n$ belongs to the range of T, there are vectors $\mathbf{a}_n \in \ell_0$ with $T\mathbf{a}_n = \mathbf{y}_n$. Therefore,

$$\|\mathbf{a}_n - \mathbf{a}_m\|_2 \le A^{-1}\|T(\mathbf{a}_n - \mathbf{a}_m)\| = A^{-1}\|\mathbf{y}_n - \mathbf{y}_m\|.$$

Consequently, $(\mathbf{a}_n)$ is Cauchy. Since $\ell^2(\mathbb{Z})$ is complete by Theorem 7.5.8, we obtain a vector $\mathbf{a} = \lim_n \mathbf{a}_n$. The continuity of T now ensures that

$$T\mathbf{a} = \lim_{n \to \infty} T\mathbf{a}_n = \lim_{n \to \infty} \mathbf{y}_n = \mathbf{y}.$$

So T maps $\ell^2(\mathbb{Z})$ onto $\mathcal{H}$.

Conversely, if the operator T exists, then the Riesz norm condition holds (by restricting T to ℓ_0). Now because T is continuous and ℓ_0 is dense in $\ell^2(\mathbb{Z})$, it follows that $\overline{\operatorname{span}\{\mathbf{x}_n : n \in \mathbb{Z}\}} = \overline{T\ell_0}$ is dense in $T\ell^2(\mathbb{Z}) = \mathcal{H}$. That is, $\overline{\operatorname{span}\{\mathbf{x}_n : n \in \mathbb{Z}\}} = \mathcal{H}$. ∎

15.9.2. COROLLARY. *If $\{\mathbf{x}_n : n \in \mathbb{Z}\}$ is a Riesz basis for a Hilbert space $\mathcal{H}$, then every vector $\mathbf{y} \in \mathcal{H}$ may be expressed in a unique way as $\mathbf{y} = \sum_n a_n \mathbf{x}_n$ for some $\mathbf{a} = (a_n)$ in $\ell^2(\mathbb{Z})$.*

PROOF. The existence of a vector $\mathbf{a} \in \ell^2(\mathbb{Z})$ such that $T\mathbf{a} = \mathbf{y}$ follows from Theorem 15.9.1. Suppose that $T\mathbf{b} = \mathbf{y}$ as well. Then $T(\mathbf{a} - \mathbf{b}) = 0$. However, $0 = \|T(\mathbf{a}-\mathbf{b})\| \ge A\|\mathbf{a}-\mathbf{b}\|_2$. Hence $\mathbf{b} = \mathbf{a}$. So T is one-to-one and $\mathbf{a}$ is uniquely determined. ∎

15.9.3. REMARK. Note that the norm estimates show that the linear map T has a continuous inverse. Indeed, Corollary 15.9.2 shows that $T^{-1}\mathbf{y} = \mathbf{a}$ is well defined. It is easy to show that the inverse of a linear map is linear. Now Theorem 15.9.1 shows that $A\|\mathbf{a}\|_2 \le \|T\mathbf{a}\|$. Substituting $\mathbf{y} = T\mathbf{a}$, we obtain $\|T^{-1}\mathbf{y}\|_2 \le \frac{1}{A}\|\mathbf{y}\|$ for all $\mathbf{y} \in \mathcal{H}$. This shows that T^{-1} is Lipschitz and hence uniformly continuous.

In fact, a linear map is continuous if and only if it is **bounded**, meaning that $\|T\| = \sup\{\|Tx\| : \|x\| = 1\}$ is finite. (See Exercise 15.9.D.) A basic theorem of functional analysis known as **Banach's Isomorphism Theorem** states that a continuous linear map between complete normed spaces that is one-to-one and onto is invertible. Consequently, the existence of the constants A and B required in Theorem 15.9.1 are automatic if we can verify that T is a bijection.

Now we specialize these ideas to a subspace V_0 of $L^2(\mathbb{R})$ spanned by the translates of a single function h. To obtain a nice condition, we need to know some easy facts about multiplication operators. Let $g \in C[-\pi, \pi]$. Recall that the linear map M_g on $L^2(-\pi, \pi)$ is given by $M_g f(\theta) = g(\theta) f(\theta)$.

15.9.4. PROPOSITION. *Suppose that the complex Fourier series of g is given by $g \sim \sum_k t_k e^{ik\theta}$. Then the matrix $[a_{ij}]$ of M_g with respect to the orthonormal basis $\{e^{ik\theta} : k \in \mathbb{Z}\}$ for $L^2(-\pi, \pi)$ is given by $a_{jk} = t_{j-k}$.*

PROOF. This is an easy computation. Indeed,

$$a_{jk} = \langle M_g e^{ik\theta}, e^{ij\theta} \rangle = \frac{1}{2\pi} \int_{-\pi}^{\pi} g(\theta) e^{ik\theta} \overline{e^{ij\theta}} \, d\theta$$

$$= \frac{1}{2\pi} \int_{-\pi}^{\pi} g(\theta) \overline{e^{i(j-k)\theta}} \, d\theta = t_{j-k}.$$
∎

We also need these norm estimates for M_g.

15.9.5. THEOREM. *If $g \in C[-\pi, \pi]$, then M_g is a continuous linear map on $L^2(-\pi, \pi)$ such that $\|M_g f\|_2 \le \|g\|_\infty \|f\|_2$. Moreover, $\|g\|_\infty$ is the smallest constant B such that $\|M_g f\|_2 \le B\|f\|_2$ for all f.*
Similarly, if $A = \inf\{|g(\theta)| : \theta \in [-\pi, \pi]\}$, then A is the largest constant such that $\|M_g f\|_2 \ge A\|f\|_2$ for all $f \in L^2(-\pi, \pi)$.

PROOF. A straightforward calculation shows that

$$\|M_g f\|_2^2 = \frac{1}{2\pi} \int_{-\pi}^{\pi} |g(\theta)|^2 |f(\theta)|^2 \, d\theta \le \frac{1}{2\pi} \int_{-\pi}^{\pi} \|g\|_\infty^2 |f(\theta)|^2 \, d\theta$$

$$= \|g\|_\infty^2 \frac{1}{2\pi} \int_{-\pi}^{\pi} |f(\theta)|^2 \, d\theta = \|g\|_\infty^2 \|f\|_2^2.$$

In particular, M_g is Lipschitz and hence continuous. Similarly,

$$\|M_g f\|_2^2 = \frac{1}{2\pi} \int_{-\pi}^{\pi} |g(\theta)|^2 |f(\theta)|^2 \, d\theta \ge \frac{1}{2\pi} \int_{-\pi}^{\pi} A^2 |f(\theta)|^2 \, d\theta = A^2 \|f\|_2^2.$$

On the other hand, suppose that $B < C < \|g\|_\infty$. Then there is an nonempty open interval (a, b) on which $|g(\theta)| > C$. Let f be a continuous function on $[-\pi, \pi]$

such that $\|f\|_2 = 1$ and the support of f is contained in $[a, b]$. Then

$$\|M_g f\|_2^2 = \frac{1}{2\pi} \int_{-\pi}^{\pi} |g(\theta)|^2 |f(\theta)|^2 \, d\theta = \frac{1}{2\pi} \int_a^b |g(\theta)|^2 |f(\theta)|^2 \, d\theta$$

$$\geq \frac{1}{2\pi} \int_a^b C^2 |f(\theta)|^2 \, d\theta = C^2 \frac{1}{2\pi} \int_{-\pi}^{\pi} |f(\theta)|^2 \, d\theta > B^2 \|f\|_2^2.$$

Therefore, $B\|f\|_2$ is not an upper bound for $\|M_g f\|$ for all f. So $\|g\|_\infty$ is the optimal choice.

Likewise, if $D > C > A$, there is a continuous function f with $\|f\|_2 = 1$ supported on an interval (c, d) on which $|g(\theta)| < C$. The same calculation shows that $\|M_g f\| \leq C\|f\|_2 < D\|f\|_2$. Therefore, A is the best constant in the lower bound. ∎

We are now ready to obtain a practical characterization of when the translates of h form a Riesz basis.

15.9.6. THEOREM. *Let $h \in L^2(\mathbb{R})$ and $V_0 = \overline{\mathrm{span}\{h(x - j) : j \in \mathbb{Z}\}}$. Set $t_j = \langle h(x), h(x - j)\rangle$ for $j \in \mathbb{Z}$. Assume that there is a continuous function g with Fourier series $t_0 + \sum_{j=1}^{\infty} 2t_j \cos j\theta$. Then $\{h(x - j) : j \in \mathbb{Z}\}$ is a Riesz basis for V_0 if and only if there are constants $0 < A^2 \leq B^2$ such that $A^2 \leq g(\theta) \leq B^2$ for all $-\pi \leq \theta \leq \pi$.*

PROOF. Let $T\mathbf{a} = \sum_n a_n h(x - n)$ for all $\mathbf{a} \in \ell_0$. By Theorem 15.9.1, the set $\{h(x - j) : j \in \mathbb{Z}\}$ is a Riesz basis for V_0 if and only if there are constants $0 < A^2 \leq B^2 < \infty$ such that

$$A^2 \|\mathbf{a}\|_2^2 \leq \|T\mathbf{a}\|^2 = \left\| \sum_n a_n h(x - n) \right\|_2^2 \leq B^2 \|\mathbf{a}\|_2^2.$$

Now

$$\|T\mathbf{a}\|^2 = \langle T\mathbf{a}, T\mathbf{a}\rangle = \langle T^*T\mathbf{a}, \mathbf{a}\rangle$$

The linear map T^*T has a matrix $[t_{ij}]$ with respect to the orthonormal basis $\mathbf{e}_n$ given by

$$t_{ij} = \langle T^*T\mathbf{e}_j, \mathbf{e}_i\rangle = \langle T\mathbf{e}_j, T\mathbf{e}_i\rangle$$
$$= \langle h(x - j), h(x - i)\rangle = \langle h(x), h(x - i + j)\rangle = t_{i-j}.$$

So the matrix of T^*T is constant on diagonals. Note that by symmetry, $t_{-k} = t_k$.

All orthonormal bases are created equal, so we can identify $\mathbf{e}_n$ with $e^{in\theta}$ in $L^2(-\pi, \pi)$, so that ℓ_0 corresponds to all finite *complex* Fourier series. With this identification, we see from Proposition 15.9.4 that $T^*T = M_g$, where

$$g(\theta) = \sum_k t_k e^{ik\theta} = t_0 I + \sum_{k=1}^{\infty} 2t_k \cos k\theta.$$

We used $t_{-k} = t_k$ to obtain a real function g, which returns us to the real domain from this brief foray into complex vector spaces.

Next we observe that $g(\theta) \geq 0$. Indeed, we have

$$\frac{1}{2\pi} \int_{-\pi}^{\pi} g(\theta)|f(\theta)|^2 \, d\theta = \langle T^*Tf, f \rangle = \|Tf\|_2^2 \geq 0.$$

Suppose that g were not positive. Then by the continuity of g, we may choose an interval (a, b) on which $g(\theta) < -\varepsilon < 0$. Then as in the proof of Theorem 15.9.5, we deduce that for any continuous function f supported on $[a, b]$ we have

$$\frac{1}{2\pi} \int_{-\pi}^{\pi} g(\theta)|f(\theta)|^2 \, d\theta \leq -\varepsilon \|f\|_2^2 < 0$$

which contradicts the previous inequality. So g is positive.

This allows us to define the multiplication operator $M_{\sqrt{g}}$. Since $\sqrt{g} \geq 0$, we find that $M_{\sqrt{g}}^* = M_{\sqrt{g}}$.

$$T^*T = M_g = M_{\sqrt{g}}^2 = M_{\sqrt{g}}^* M_{\sqrt{g}}$$

(Be warned that this does *not* show that T is equal to $M_{\sqrt{g}}$. They do not even map into the same Hilbert space.) Consequently,

$$\|Tf\|_2^2 = \langle T^*Tf, f \rangle = \langle M_g f, f \rangle = \langle M_{\sqrt{g}} f, M_{\sqrt{g}} f \rangle = \|M_{\sqrt{g}} f\|_2^2.$$

Finally, an application of Theorem 15.9.5 shows that

$$A^2 \|f\|_2^2 \leq \|M_{\sqrt{g}} f\|_2^2 \leq \|\sqrt{g}\|_\infty^2 \|f\|_2^2 = \|g\|_\infty \|f\|_2^2$$

for all $f \in L^2(-\pi, \pi)$ if and only if

$$A^2 \leq \inf_{\theta \in [-\pi, \pi]} |\sqrt{g}(\theta)|^2 = \inf_{\theta \in [-\pi, \pi]} |g(\theta)|.$$

Thus $A^2 > 0$ is possible only if g is bounded away from 0.   ∎

We are now ready to modify a Riesz basis of translations of h to obtain an orthonormal basis of translates. This is the key to finding wavelets by the machinery we have already developed.

15.9.7. THEOREM. *Let $h \in L^2(\mathbb{R})$ be a function such that the set of translates $\{h(x-j) : j \in \mathbb{Z}\}$ is a Riesz basis for its span V_0. Assume that $t_0 + \sum_{j=1}^{\infty} 2t_j \cos j\theta$ is the Fourier series of a continuous function, where $t_k = \langle h(x), h(x-k) \rangle$. Then there is a function $\varphi \in L^2(\mathbb{R})$ such that $\{\varphi(x-j) : j \in \mathbb{Z}\}$ is an orthonormal basis for V_0.*

PROOF. By Theorem 15.9.1, there is a continuous, invertible linear map T from $\ell^2(\mathbb{Z})$ onto V_0 given by $Ta = \sum_n a_n h(x-n)$. By Theorem 15.9.6, the (continuous) function g with Fourier series $t_0 + \sum_{j=1}^{\infty} 2t_j \cos j\theta$ satisfies $T^*T = M_g$ and there are constants $0 < A^2 \leq B^2 < \infty$ so that

$$A^2 \leq g(\theta) \leq B^2 \quad \text{for all} \quad -\pi \leq \theta \leq \pi.$$

In particular, $g^{-1/2}$ is bounded above by $1/A$.

Now compute the Fourier series of $g^{-1/2}$, say

$$g^{-1/2} \sim \sum_n c_n e^{-in\theta} = c_0 I + \sum_{k=1}^{\infty} 2c_k \cos k\theta,$$

where

$$c_k = c_{-k} = \frac{1}{2\pi} \int_{-\pi}^{\pi} \frac{\cos n\theta}{\sqrt{g(\theta)}} \, d\theta.$$

We claim that the orthogonal generator is obtained by the formula

$$\varphi(x) = Tg^{-1/2} = \sum_{n=-\infty}^{\infty} c_n h(x - n).$$

Indeed,

$$
\begin{aligned}
\langle \varphi(x - j), \varphi(x - k) \rangle &= \langle Te^{-ij\theta} g^{-1/2}(\theta), Te^{-ik\theta} g^{-1/2}(\theta) \rangle \\
&= \langle T^*T e^{-ij\theta} g^{-1/2}(\theta), e^{-ik\theta} g^{-1/2}(\theta) \rangle \\
&= \langle M_g M_g^{-1/2} e^{-ij\theta}, M_g^{-1/2} e^{-ik\theta} \rangle \\
&= \langle M_g^{-1/2} M_g M_g^{-1/2} e^{-ij\theta}, e^{-ik\theta} \rangle \\
&= \langle e^{-ij\theta}, e^{-ik\theta} \rangle = \delta_{ij}.
\end{aligned}
$$

So $\{\varphi(x - k) : k \in \mathbb{Z}\}$ is orthonormal.

It is clear that each $\varphi(x - k)$ belongs to V_0 since they are in the span of the $h(x - j)$'s. Conversely, observe that $\varphi(x - k) = TM_g^{-1/2} e^{ik\theta}$. Thus

$$\text{span}\{\varphi(x - k) : k \in \mathbb{Z}\} = TM_g^{-1/2} L^2(\mathbb{T}) = TL^2(\mathbb{T}) = V_0$$

because M_g is invertible and so $M_g^{-1/2} L^2(\mathbb{T}) = L^2(\mathbb{T})$, and T maps $L^2(\mathbb{T})$ onto V_0 by Corollary 15.9.2. So $\{\varphi(x - k) : k \in \mathbb{Z}\}$ is an orthonormal basis for V_0. ∎

The second notion that arose in our construction of a continuous wavelet was a weaker notion of a multiresolution using Riesz bases.

15.9.8. DEFINITION. A **Riesz multiresolution** of $L^2(\mathbb{R})$ with scaling function h is the sequence of subspaces $V_j = \text{span}\{h(2^k x - j) : j \in \mathbb{Z}\}$ provided that the sequence satisfies the following properties:

(1) Riesz basis: $\{h(x - j) : j \in \mathbb{Z}\}$ is a Riesz basis for V_0.

(2) nesting: $V_k \subset V_{k+1}$ for all $k \in \mathbb{Z}$.

(3) scaling: $f(x) \in V_k$ if and only if $f(2x) \in V_{k+1}$.

(4) density: $\overline{\bigcup_{k \in \mathbb{Z}} V_k} = L^2(\mathbb{R})$.

(5) separation: $\bigcap_{k \in \mathbb{Z}} V_k = \{0\}$.

The main result is now a matter of collecting our results.

15.9.9. THEOREM. *Suppose that h is the scaling function for a Riesz multiresolution $V_j = \text{span}\left\{ h(2^k x - j) : j \in \mathbb{Z} \right\}$. Assume that there is a continuous function with Fourier series $\|h\|_2^2 + \sum_{j=1}^{\infty} 2\langle h(x), h(n-k) \rangle \cos j\theta$. Then there exists a scaling function φ generating the same nested sequence of subspaces. Consequently, there is a wavelet basis for $L^2(\mathbb{R})$ compatible with this decomposition.*

PROOF. Theorem 15.9.7 provides an orthogonal scaling function for this resolution. Then Theorem 15.4.2 provides an algorithm for constructing the corresponding wavelet.   ∎

15.9.10. EXAMPLE. We finish this section by describing some smoother examples of wavelets, known as **Battle–Lemarié wavelets** or **B-spline wavelets**. Let $N_0 = \chi_{[0,1)}$. For each $n \geq 0$, define

$$N_n(x) = N_{n-1} * N_0(x) = \int_{x-1}^{x} N_{n-1}(t)\,dt.$$

For example, $N_1(x) = h(x-1)$ is a translate of the hat function,

$$N_2(x) = \begin{cases} \frac{1}{2}x^2 & \text{for } 0 \leq x \leq 1 \\ \frac{3}{4} - (x - \frac{3}{2})^2 & \text{for } 1 \leq x \leq 2 \\ \frac{1}{2}(3-x)^2 & \text{for } 2 \leq x \leq 3 \end{cases}$$

and

$$N_3(x) = \begin{cases} \frac{1}{6}x^3 & \text{for } 0 \leq x \leq 1 \\ \frac{1}{6}x^3 - \frac{2}{3}(x-1)^3 & \text{for } 1 \leq x \leq 2 \\ \frac{1}{6}(4-x)^3 - \frac{2}{3}(3-x)^3 & \text{for } 2 \leq x \leq 3 \\ \frac{1}{6}(4-x)^3 & \text{for } 3 \leq x \leq 4. \end{cases}$$

Figure 15.7 gives the graphs of N_2 and N_3. Notice that $N_3(x)$ is a cubic spline. That is, N_3 is C^2, has compact support, and on each interval $[k, k+1]$ it is represented by a cubic polynomial. See the Exercises for hints on establishing similar properties for general n.

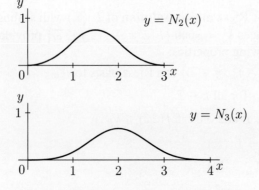

FIGURE 15.7. The graphs of N_2 and N_3.

Let $S_k^{(n)}$ denote the subspace of $L^2(\mathbb{R})$ consisting of **splines of order** n with nodes at the points $2^{-k}\mathbb{Z}$. These are the L^2 functions that have $n-1$ continuous derivatives such that the restriction to each dyadic interval $[2^{-k}j, 2^{-k}(j+1)]$ is a polynomial of degree n. Clearly for each fixed n, the sequence

$$\cdots \subset S_{-2}^{(n)} \subset S_{-1}^{(n)} \subset S_0^{(n)} \subset S_1^{(n)} \subset S_2^{(n)} \subset \cdots$$

is nested and satisfies the scaling property.

In fact, for each n, this forms a Riesz multiresolution of $L^2(\mathbb{R})$ with scaling function $N_n(x)$. We will establish this for $n = 3$.

Theorem 10.9.1 showed that every continuous function f on the closed interval $[-2^N, 2^N]$ is the uniform limit of a sequence of cubic splines h_k with nodes at $2^{-k}\mathbb{Z}$. These cubic splines may be chosen to have support in $[-2^N, 2^N]$ as well. Thus

$$\lim_{k\to\infty} \|f - h_k\|_2^2 = \lim_{k\to\infty} \int_{-2^n}^{2^n} |f(x) - h_k(x)|^2\, dx \le 2^{N+1} \lim_{k\to\infty} \|f - h_k\|_\infty^2 = 0.$$

So f is the limit in L^2 of a sequence of cubic splines. It follows that $\bigcup_k S_k^{(3)}$ is dense in $L^2(\mathbb{R})$.

The separation property is established in much the same way as the piecewise linear case. Suppose that f is a function in $\bigcap_k S_k^{(3)}$. For $k \le 0$, functions in $S_k^{(3)}$ are cubic polynomials on $[0, 2^{|k|}]$ and on $[-2^{|k|}, 0]$. Hence the restrictions of f to $[0, \infty)$ and to $(-\infty, 0]$ agree with cubic polynomials. A nonzero cubic polynomial p has

$$\int_0^\infty |p(x)|^2\, dx = +\infty = \int_{-\infty}^0 |p(x)|^2\, dx.$$

So the only $L^2(\mathbb{R})$ function that is cubic on both half lines is the zero function. Thus $f = 0$ is the only point in the intersection.

Finally, we will show that translates of $N_3(x)$ form a Riesz basis for $L^2(\mathbb{R})$. By Theorem 15.9.6, we must compute $t_j = \langle N_3(x), N_3(x-j)\rangle$. By symmetry, $t_{-k} = t_k$ and the fact that N_3 is supported on $[0, 4]$ means that $t_k = 0$ for $|k| \ge 4$. Therefore, it suffices to compute $t_0, t_1, t_2,$ and t_3. We spare the reader the tedious calculation and use *Maple* to obtain

$$t_0 = \tfrac{151}{315} \quad t_1 = t_{-1} = \tfrac{397}{1680} \quad t_2 = t_{-2} = \tfrac{1}{42} \quad t_3 = t_{-3} = \tfrac{1}{5040}.$$

Thus we are led to consider the function

$$g(\theta) = \tfrac{1208}{2520} + \tfrac{1191}{2520}\cos\theta + \tfrac{60}{2520}\cos 2\theta + \tfrac{1}{2520}\cos 3\theta.$$

An easy calculation shows that this takes its minimum when $\cos\theta = -1$ and the minimum value is $\tfrac{76}{2520} > 0$. Since this function is positive, Theorem 15.9.9 shows that there is a wavelet basis consisting of cubic splines.

Exercises for Section 15.9

A. Let $\varphi = \chi_{[0,2)}$. Show that $\{\varphi(x-j) : j \in \mathbb{Z}\}$ is not a Riesz basis for its span.

B. Show that if $\{h(x-j) : j \in \mathbb{Z}\}$ is a Riesz basis for V_0, then $\{2^{k/2}h(2^kx - j) : j \in \mathbb{Z}\}$ forms a Riesz basis for V_k for each $k \in \mathbb{Z}$.

C. Show that if T is an invertible linear map, then T^{-1} is linear.

D. Let T be a linear map from one Hilbert space $\mathcal{H}$ to itself. Prove that T is continuous if and only if it is bounded.
HINT: If not bounded, find x_n with $\|x_n\| \to 0$ while $\|Tx_n\| \to \infty$.

E. Recall that ℓ_0 is the nonclosed subspace of $\ell^2(\mathbb{Z})$ of elements with only finitely many nonzero entries. Show that if T is a linear map from ℓ_0 into a Hilbert space $\mathcal{H}$ with $\|T\mathbf{a}\| \leq B\|\mathbf{a}\|_2$, then T extends uniquely to a continuous function on $\ell^2(\mathbb{Z})$ into $\mathcal{H}$.
HINT: Fix $\mathbf{a} \in \ell^2(\mathbb{Z})$ and $\varepsilon > 0$. Show that if $\mathbf{b}, \mathbf{c} \in \ell_0$ and both lie in the $\varepsilon/(2B)$ ball about $\mathbf{a}$, then $\|T\mathbf{b} - T\mathbf{c}\| < \varepsilon$. Hence deduce that if $(\mathbf{b}_i)$ and $(\mathbf{c}_j)$ are two sequences in ℓ_0 converging to $\mathbf{a}$, then $\lim_i T\mathbf{b}_i = \lim_j T\mathbf{c}_j$. Consequently, show that setting $T\mathbf{a}$ to be this limit determines a continuous function on $\ell^2(\mathbb{Z})$ extending T.

F. Suppose that $\{\mathbf{x}_n : n \in \mathbb{Z}\}$ is a Riesz basis for $\mathcal{H}$.
(a) Show that there is a unique vector $\mathbf{y}_n$ orthogonal to the subspace
$M_n = \text{span}\{\mathbf{x}_j : j \neq n\}$ such that $\langle \mathbf{x}_n, \mathbf{y}_n \rangle = 1$.
(b) Show that if $\mathbf{z} \in \mathcal{H}$, then $\mathbf{z} = \sum_n \langle \mathbf{z}, \mathbf{y}_n \rangle \mathbf{x}_n$.
(c) Show that $\{\mathbf{y}_n : n \in \mathbb{Z}\}$ is a Riesz basis for $\mathcal{H}$.
HINT: If $(a_n) \in \ell^2$, there is another sequence (b_n) so that $\sum_n a_n\mathbf{y}_n = \sum_n b_n\mathbf{x}_n$. Apply the Cauchy–Schwarz inequality to both $\langle \sum_j a_j\mathbf{x}_j, \sum_k a_k\mathbf{y}_k \rangle$ and $\langle \sum_j b_j\mathbf{x}_j, \sum_k a_k\mathbf{y}_k \rangle$.

G. Show that $\{N_2(x-k) : k \in \mathbb{Z}\}$ is a Riesz basis for its span.

H. Prove by induction on $n \geq 1$ that
(a) N_n is $C^{(n-1)}$.
(b) $N_n|_{[j,j+1]}$ is a polynomial of degree n for each $j \in \mathbb{Z}$.
(c) $\{x \in \mathbb{R} : N_n(x) > 0\} = (0, n+1)$.
(d) $\sum_k N_n(x-k) = 1$ for all $x \in \mathbb{R}$.

CHAPTER 16

Convexity and Optimization

Optimization is a central theme of applied mathematics that involves minimizing or maximizing various quantities. This is an important application of the derivative tests in calculus. In addition to the first and second derivative tests of one-variable calculus, there is the powerful technique of Lagrange multipliers in several variables. This chapter is concerned with analogues of these tests that are applicable to functions that are not differentiable. Of course, some different hypothesis must replace differentiability and this is the notion of convexity. It turns out that many applications in economics, business, and related areas involve convex functions. As in other chapters of this book, we concentrate on the theoretical underpinnings of the subject. The important aspect of constructing algorithms to carry out our program is not addressed. However, the reader will be well placed to read that material. Results from both linear algebra and calculus appear regularly.

The study of convex sets and convex functions is a comparatively recent development. Although convexity appears implicitly much earlier (going back to work of Archimedes, in fact), the first papers on convex sets appeared at the end of the nineteenth century. The main theorems of this chapter, characterizations of solutions of optimization problems, first appeared around the middle of the twentieth century. Starting in the 1970s, there has been considerable work on extending these methods to nonconvex functions.

16.1. Convex Sets

Although convex subsets can be defined for any normed vector space, we concentrate on $\mathbb{R}^n$ with the Euclidean norm. For this space, we have an inner product and the Heine–Borel Theorem (Theorem 4.4.6) characterizing compact sets in $\mathbb{R}^n$ as useful tools.

16.1.1. DEFINITION. A subset A of $\mathbb{R}^n$ is called a **convex set** if

$$\lambda a + (1 - \lambda)b \in A \quad \text{for all} \quad a, b \in A \text{ and } \lambda \in [0, 1].$$

Let $[a, b] = \{\lambda a + (1 - \lambda)b : \lambda \in [0, 1]\}$ denote the line segment joining points a and b in $\mathbb{R}^n$. Define (a, b), $[a, b)$ and $(a, b]$ in the analogous way. You should note that $\lambda \in (0, 1]$ corresponds to $[a, b)$. Notice that A is convex if and only if $[a, b] \subset A$ whenever $a, b \in A$. See Figure 16.1

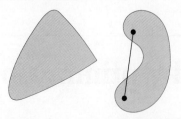

FIGURE 16.1. A convex and a nonconvex set.

16.1.2. DEFINITION. A subset A of $\mathbb{R}^n$ is an **affine set** if

$$\lambda a + (1 - \lambda)b \in A \quad \text{for all} \quad a, b \in A \text{ and } \lambda \in \mathbb{R}.$$

A subset C of $\mathbb{R}^n$ is a **cone** if it is a convex set that contains all of its positive scalar multiples, that is,

$$\lambda a \in C \quad \text{for all} \quad a \in C \text{ and } \lambda > 0.$$

Clearly, affine sets and cones are convex but not conversely.

Now is a good time to mention a related bit of terminology. In this chapter, we reserve **linear function** for functions satisfying $f(\lambda a + \mu b) = \lambda f(a) + \mu f(b)$, for all a, b in the domain and all scalars λ and μ. We use **affine function** for a function g given by $g(x) = f(x) + c$, where f is a linear function and c is a constant. We leave it as an exercise, Exericse 16.1.I, to show that a function $f : \mathbb{R}^n \to \mathbb{R}^m$ is linear or affine if and only if the graph of f is either a subspace or an affine set, respectively.

Convex sets are ubiquitous and we give a few examples, mostly without proof. Proving the following assertions is a useful warm-up exercise.

16.1.3. EXAMPLES.

(1) A subspace of $\mathbb{R}^n$ is both affine and a cone.

(2) Any ball $\overline{B_r(a)} = \{x \in \mathbb{R}^n : \|x - a\| \leq r\}$ in $\mathbb{R}^n$ is convex. Indeed, if $x, y \in \overline{B_r(a)}$ and $\lambda \in [0, 1]$, then

$$\begin{aligned}
\|\lambda x + (1 - \lambda)y - a\| &= \|\lambda(x - a) + (1 - \lambda)(y - a)\| \\
&\leq \lambda\|x - a\| + (1 - \lambda)\|y - a\| \\
&\leq \lambda r + (1 - \lambda)r = r.
\end{aligned}$$

So $\lambda x + (1 - \lambda)y \in \overline{B_r(a)}$. Clearly, the ball is neither affine nor a cone.

(3) The half-space $\{(x, y) : ax + by \geq 0\}$ is a closed convex cone in $\mathbb{R}^2$.

(4) In $\mathbb{R}^n$, the positive orthant $\mathbb{R}^n_+ = \{(x_1, \ldots, x_n) : x_i > 0\}$ is a cone.

(5) If $A \subset \mathbb{R}$, then A is convex if and only if A is an interval, possibly unbounded.

(6) If $A \subset \mathbb{R}^m$ is convex and $T : \mathbb{R}^m \to \mathbb{R}^n$ is linear, then $T(A)$ is convex.

(7) If $A \subset \mathbb{R}^n$ is convex, then any translate of A (i.e., a set of the form $A + x$, $x \in \mathbb{R}^n$), is convex.

We now collect together a number of basic properties of convex and affine sets. The proof of the first lemma is left as an exercise.

16.1.4. LEMMA. *If $\{A_i : i \in I\}$ is a collection of convex subsets of $\mathbb{R}^n$, then $\bigcap_{i \in I} A_i$ is convex. Similarly, the intersection of a collection of affine sets is affine and the intersection of a collection of cones is a cone.*

16.1.5. LEMMA. *If $A \subset \mathbb{R}^n$ is a nonempty affine set, then it is the translate of a unique subspace of $\mathbb{R}^n$.*

PROOF. Fix an element $a_0 \in A$, and let $L = \{a - a_0 : a \in A\}$. If $v = a - a_0 \in L$ and $t \in \mathbb{R}$, then $tv + a_0 = ta + (1 - t)a_0 \in A$. Hence tv lies in $A - a_0 = L$. Suppose that $w = b - a_0$ is another element of L. Then since A is convex, $(a+b)/2$ belongs to A. Now

$$(v + w) + a_0 = a + b - a_0 = 2\left(\tfrac{a+b}{2}\right) + (1 - 2)a_0 \in A.$$

So $v + w$ belongs to L. This shows that L is a subspace.

To see that L is unique, suppose that $A = M + y$, where M is also a subspace of $\mathbb{R}^n$. Then $L = A - a_0 = M + (y - a_0)$. Since 0 is in L, M contains $a_0 - y$ and so $M + (y - a_0) = M$. Therefore, $L = M$. ∎

16.1.6. DEFINITION. Suppose that S is a subset of $\mathbb{R}^n$. The **convex hull** of S, denoted $\mathrm{conv}(S)$, is the intersection of all convex subsets of $\mathbb{R}^n$ containing S.

The **closed convex hull** of S, denoted $\overline{\mathrm{conv}}(S)$, is the intersection of all closed convex subsets of $\mathbb{R}^n$ containing S.

The **affine hull** of S, denoted $\mathrm{aff}(S)$, is the intersection of all affine subsets of $\mathbb{R}^n$ containing S.

Let $L(S)$ denote the unique subspace (as in Lemma 16.1.5) that is a translate of $\mathrm{aff}(S)$. The **dimension** of S, denoted $\dim(S)$, is the dimension of $L(S)$.

Finally, if S is a subset of $\mathbb{R}^n$, we use $\mathrm{cone}(S)$ for the intersection of all cones containing S.

By Lemma 16.1.4, $\mathrm{conv}(S)$ and $\overline{\mathrm{conv}}(S)$ are convex. Hence $\mathrm{conv}(S)$ is the smallest convex set containing S. Therefore, $\mathrm{conv}(\mathrm{conv}(S)) = \mathrm{conv}(S)$. The intersection of closed sets is closed, so $\overline{\mathrm{conv}}(S)$ is the smallest closed convex set containing S. Moreover, $\overline{\mathrm{conv}}(\overline{\mathrm{conv}}(S)) = \overline{\mathrm{conv}}(\mathrm{conv}(S)) = \overline{\mathrm{conv}}(S)$.

Likewise, $\mathrm{aff}(\mathrm{aff}(S)) = \mathrm{aff}(S)$ is the smallest affine set containing S and $\mathrm{cone}(\mathrm{cone}(S)) = \mathrm{cone}(S)$ is the smallest cone containing S. Affine sets are closed because (finite-dimensional) subspaces are closed.

Here is a useful description of the convex hull of an arbitrary set S.

16.1.7. THEOREM. *Suppose that S is a subset of $\mathbb{R}^n$. Then a belongs to* $\mathrm{conv}(S)$ *if and only if there are points $s_1, \ldots, s_r$ in S and scalars $\lambda_1, \ldots, \lambda_r$ in* $[0, 1]$ *with* $\sum_{i=1}^{r} \lambda_i = 1$ *so that* $\sum_{i=1}^{r} \lambda_i s_i = a$.

PROOF. We claim that

$$C = \left\{ \sum_{i=1}^{r} \lambda_i s_i : r \geq 1,\ s_i \in S,\ \lambda_i \in [0,1],\ \sum_{i=1}^{r} \lambda_i = 1 \right\}$$

is a convex set. Consider two points of C, say $a = \sum_{i=1}^{n} \mu_i s_i$ and $b = \sum_{j=1}^{m} \nu_j t_j$, where the s_i and t_j are in S, μ_i and ν_j are in $[0, 1]$, and $\sum_{i=1}^{n} \mu_i = 1 = \sum_{j=1}^{m} \nu_j$. Then $\lambda a + (1 - \lambda)b$ can be written as

$$\sum_{i=1}^{n} \lambda \mu_i s_i + \sum_{j=1}^{m} (1 - \lambda) \nu_j t_j.$$

This is a linear combination of elements of S with coefficients $\lambda \mu_i$ and $\lambda \nu_j$ in $[0, 1]$ such that

$$\sum_{i=1}^{n} \lambda \mu_i + \sum_{j=1}^{m} (1 - \lambda) \nu_j = \lambda + (1 - \lambda) = 1.$$

Thus $\lambda a + (1 - \lambda)b$ also belongs to C.

Since C is convex and contains S, it follows that $\mathrm{conv}(S)$ is contained in C. If we show that C is contained in $\mathrm{conv}(S)$, then it will follow that they are equal.

Suppose that $a = \sum_{i=1}^{r} \lambda_i s_i$, where the λ_i and s_i satisfy the preceding conditions. Set $\Lambda_k = \sum_{i=1}^{k} \lambda_i$ for $1 \leq k \leq r$. Let k_0 be the smallest k for which $\lambda_k > 0$. Inductively define points $a_{k_0}, \ldots, a_r$ in S by $a_{k_0} = s_{k_0}$ and

$$a_k = \frac{\Lambda_{k-1}}{\Lambda_k} a_{k-1} + \frac{\lambda_k}{\Lambda_k} s_k \quad \text{for} \quad k_0 < k \leq r.$$

Since each s_k belongs to S, it follows by induction that these convex combinations all lie in $\mathrm{conv}(S)$. However, we also show by induction that $a_k = \sum_{i=1}^{k} \dfrac{\lambda_i}{\Lambda_k} s_i$ for $k \geq k_0$. This is evident for $k = k_0$. Suppose that it is true for $k - 1$. Then

$$a_k = \frac{\Lambda_{k-1}}{\Lambda_k} \sum_{i=1}^{k-1} \frac{\lambda_i}{\Lambda_{k-1}} s_i + \frac{\lambda_k}{\Lambda_k} s_k = \sum_{i=1}^{k} \frac{\lambda_i}{\Lambda_k} s_i$$

In particular, since $\Lambda_r = 1$, we have $a_r = \sum_{i=1}^{r} \lambda_i s_i = a$ lies in $\mathrm{conv}(S)$. ∎

Since we are working in finite dimensions, this result may be sharpened so that each point in the convex hull is a combination of at most $n + 1$ points.

16.1.8. CARATHÉODORY'S THEOREM.
Suppose that S is a subset of $\mathbb{R}^n$. Then each $a \in \mathrm{conv}(S)$ may be expressed as a convex combination of $n + 1$ elements of S.

PROOF. By the previous proposition, a may be written as a convex combination

$$a = \sum_{i=1}^{r} \lambda_i s_i \quad \text{where} \quad \lambda_i \geq 0, \sum_{i=1}^{r} \lambda_i = 1, \text{ and } s_i \in S, \text{ for } 1 \leq i \leq r.$$

If $r \geq n + 2$, we will construct another representation of a using fewer vectors s_i. Thus we eventually reduce this to a sum over at most $n + 1$ elements. We may suppose that $\lambda_i > 0$ for each i, for otherwise we reduce the list by dropping s_{i_0} if $\lambda_{i_0} = 0$.

Consider $v_i = s_i - s_r$ for $1 \leq i < r$. These are $r - 1 \geq n + 1$ such vectors in an n-dimensional space, and thus they are linearly dependent. Find constants μ_i not all 0 so that

$$0 = \sum_{i=1}^{r-1} \mu_i (s_i - s_r) = \sum_{i=1}^{r} \mu_i s_i,$$

where $\mu_r = -\sum_{i=1}^{r-1} \mu_i$. Let $J = \{i : \mu_i < 0\}$, which is necessarily nonempty. Set $\delta = \min\{\lambda_i / |\mu_i| : i \in J\}$; and pick i_0 so that $\lambda_{i_0} = -\delta \mu_{i_0}$. Then

$$a = \sum_{i=1}^{r} \lambda_i s_i + \sum_{i=1}^{r} \delta \mu_i s_i = \sum_{i=1}^{r} (\lambda_i + \delta \mu_i) s_i.$$

By construction, the constants $\nu_i = \lambda_i + \delta \mu_i \geq 0$ and $\sum_{i=1}^{r} \nu_i = 1$. Moreover, $\nu_{i_0} = 0$. So deleting s_{i_0} from the list represents a as a convex combination of fewer elements of S. ∎

16.1.9. COROLLARY. *If $A \subset \mathbb{R}^n$ is compact, then $\mathrm{conv}(A)$ is compact.*

PROOF. Define a subset $X = A^{n+1} \times \Delta_{n+1}$ of $\mathbb{R}^{(n+1)^2}$ consisting of all points $x = (a_1, a_2, \ldots, a_{n+1}, \lambda_1, \ldots, \lambda_{n+1})$, where $a_i \in A$, $\lambda_i \in [0, 1]$, and $\sum_{i=1}^{n+1} \lambda_i = 1$. It is easy to check that X is closed and bounded and therefore is compact. Consider the function $f(x) = \sum_{i=1}^{n+1} \lambda_i a_i$. This is a continuous function from X into $\mathrm{conv}(A)$. By Carathéodory's Theorem, f maps X onto $\mathrm{conv}(A)$. By Theorem 5.4.3, $f(X)$ is compact. Therefore, $\mathrm{conv}(A)$ is compact. ∎

16.1.10. DEFINITION. A **hyperplane** is an affine set of codimension 1. Thus a hyperplane in $\mathbb{R}^n$ has dimension $n-1$.

Hyperplanes are rather special affine sets, and they serve to split the whole space into two pieces. This is a consequence of the following result.

16.1.11. PROPOSITION. *A subset H of $\mathbb{R}^n$ is a hyperplane if and only if there is a nonzero vector $h \in \mathbb{R}^n$ and a scalar $\alpha \in \mathbb{R}$ so that*

$$H = \{x \in \mathbb{R}^n : \langle x, h \rangle = \alpha\}.$$

PROOF. If H is a hyperplane and $x_0 \in H$, then $L(H) = H - x_0$ is a subspace of dimension $n-1$. Choose a nonzero vector h orthogonal to $L(H)$. This is used to define a linear map from $\mathbb{R}^n$ into $\mathbb{R}$ by $f(x) = \langle x, h \rangle$. Since the set of all vectors orthogonal to h form a subspace of dimension $n-1$ containing $L(H)$, it follows that $L(H) = \{h\}^{\perp} = \ker f$.

Set $\alpha = f(x_0)$. Then $f(x) = \alpha$ if and only if $f(x - x_0) = f(x) - \alpha = 0$, which occurs if and only if $x - x_0 \in L(H)$ or, equivalently, when x belongs to $L(H) + x_0 = H$.

Conversely, the linear map $f(x) = \langle x, h \rangle$ from $\mathbb{R}^n$ into $\mathbb{R}$ maps onto $\mathbb{R}$ since $h \neq 0$. Thus $L(H) := \ker f = \{h\}^{\perp}$ is a subspace of $\mathbb{R}^n$ of dimension $n-1$. Let x_0 be any vector with $f(x_0) = \alpha$. Then following the argument of the previous paragraph, the set $H = \{x \in \mathbb{R}^n : \langle x, h \rangle = \alpha\} = L(H) + x_0$ is a hyperplane. ∎

Notice that the function f (or the vector h) is not unique, but it is determined up to a scalar multiple because the subspace $H^{\perp}$ is one dimensional. When working with a hyperplane, we will usually assume that a choice of this function has been made. This allows us to describe two **half-spaces** associated to H, which we denote by $H^+ = \{x \in \mathbb{R}^n : f(x) \geq \alpha\}$ and $H^- = \{x \in \mathbb{R}^n : f(x) \leq \alpha\}$. These two subsets do not depend on the choice of f except that a sign change can interchange H^+ with H^-.

Exercises for Section 16.1

A. If A is a convex subset of $\mathbb{R}^n$, show that $\overline{A}$ is convex.

B. (a) Prove Lemma 16.1.4: The intersection of convex sets is convex.
   (b) State and prove the analogous result for cones and affine sets.

C. Suppose that A is a closed subset of $\mathbb{R}^n$ and whenever $a, b \in A$, the point $(a+b)/2$ is in A. Show that A is convex.
   HINT: Use induction to show that $\lambda a + (1 - \lambda)b$ is in A if $\lambda = i/2^k$ for $1 \leq i < 2^k$.

D. If A is a convex subset of $\mathbb{R}^n$ such that $\operatorname{aff}(A) \neq \mathbb{R}^n$, show that $\operatorname{int}(A)$ is empty.

E. Let S be a subset of $\mathbb{R}^n$. Show that $\operatorname{aff}(S) = \left\{ \sum_{i=1}^{r} \lambda_i s_i : s_i \in S, \ \lambda_i \in \mathbb{R}, \ \sum_{i=1}^{n} \lambda_i = 1 \right\}$
   and $L(S) = \left\{ \sum_{i=1}^{r} \lambda_i s_i : s_i \in S, \ \lambda_i \in \mathbb{R}, \ \sum_{i=1}^{n} \lambda_i = 0 \right\}.$

F. Suppose that A is a convex subset of $\mathbb{R}^n$ and that T is a linear transformation from $\mathbb{R}^n$ into $\mathbb{R}^m$. Prove that TA is convex and $\text{aff}(TA) = T\,\text{aff}(A)$.

G. If $A \subset \mathbb{R}^m$ and $B \subset \mathbb{R}^n$, then $A \times B := \{(a, b) \in \mathbb{R}^{m+n} : a \in A,\ b \in B\}$. Show that $\text{conv}(A \times B) = \text{conv}(A) \times \text{conv}(B)$.

H. (a) If S is a bounded subset of $\mathbb{R}^n$, prove that $\text{conv}(S)$ is bounded.
(b) Give an example of a closed subset S of $\mathbb{R}^2$ such that $\text{conv}(S)$ is not closed.

I. Recall that the graph of $f : \mathbb{R}^n \to \mathbb{R}^m$ is $G(f) = \{(x, y) \in \mathbb{R}^n \times \mathbb{R}^m : f(x) = y\}$. Show that f is a linear function if and only if $G(f)$ is a linear subspace of $\mathbb{R}^{m+n}$ and f is an affine function if and only if $G(f)$ is an affine set.

J. A function f on $\mathbb{R}$ is **convex** if $f(\lambda x + (1 - \lambda)y) \le \lambda f(x) + (1 - \lambda)f(y)$ for all $x, y \in \mathbb{R}$ and $0 \le \lambda \le 1$. If f is any function on $\mathbb{R}$, define the **epigraph** of f to be $\text{epi}(f) = \{(x, y) \in \mathbb{R}^2 : y \ge f(x)\}$. Show that f is a convex function if and only if $\text{epi}(f)$ is a convex subset of $\mathbb{R}^2$.

K. Let A and B be convex subsets of $\mathbb{R}^n$. Prove that $\text{conv}(A \cup B)$ equals the union of all line segments $[a, b]$ such that $a \in A$ and $b \in B$.

L. **The asymptotic cone.** Let $A \subset \mathbb{R}^n$ be closed and convex, and $x \in A$.
(a) Show that $s(A - x) \subset t(A - x)$ if $0 < s < t$.
(b) Show that $A_\infty(x) = \bigcap_{t>0} t(A - x) := \{d : x + td \in A \text{ for all } t > 0\}$ is a cone.
(c) Show that $A_\infty(x)$ does not depend on the point x.
    HINT: Consider $(1 - \frac{1}{k})y + \frac{1}{k}(x + ktd)$.
(d) Prove that A is compact if and only if $A_\infty = \{0\}$.
    HINT: If $\|a_k\| \to \infty$, find a cluster point d of $a_k/\|a_k\|$. Argue as in (c).

M. A set $S \subset \mathbb{R}^n$ is a **star-shaped set** with respect to $v \in S$ if $[s, v] \subset S$ for all $s \in S$.
(a) Show that S is convex if and only if it is star shaped with respect to every $v \in S$.
(b) Find a set that is star shaped with respect to exactly one of its points.

N. (a) Given a sequence of convex sets $B_1 \subset B_2 \subset B_3 \cdots$ in $\mathbb{R}^n$, show that $\bigcup_{i \ge 1} B_i$ is convex.
(b) For any sequence of convex sets $B_1, B_2, B_3, \ldots$ in $\mathbb{R}^n$, show that $\bigcup_{j \ge 1}(\bigcap_{i \ge j} B_i)$ is convex.

O. A set $S \subset \mathbb{R}^n$ is called **affinely dependent** if there is an $s \in S$ so that $s \in \text{aff}(S \backslash \{s\})$. Show that S is affinely dependent if and only if there are distinct elements $s_1, \ldots, s_r$ of S and scalars $\mu_1, \ldots, \mu_r$, not all zero, so that $\sum_{i=1}^r \mu_i s_i = 0$ and $\sum_{i=1}^r \mu_i = 0$.
    HINT: Solve for some s_i.

P. A subset $S \subset \mathbb{R}^n$ is **affinely independent** if it is not affinely dependent. Show that an affinely independent set in $\mathbb{R}^n$ can have at most $n + 1$ points.
    HINT: If $\{s_0, \ldots, s_n\} \subset S$, show that $\{s_i - s_0 : 1 \le i \le n\}$ is linearly independent.

Q. **Radon's Theorem.** Suppose that $s_1, \ldots, s_r$ are distinct points in $\mathbb{R}^n$, with $r > n + 1$. Show that there are disjoint sets I and J with $I \cup J = \{1, \ldots, r\}$ so that the convex sets $\text{conv}\{s_i : i \in I\}$ and $\text{conv}\{s_j : j \in J\}$ have nonempty intersection.
    HINT: Use the previous two exercises to obtain 0 as a nontrivial linear combination of these elements. Then consider those elements with positive coefficients in this formula.

R. **Helly's Theorem**. Let C_k be convex subsets of $\mathbb{R}^n$ for $1 \le k \le m$. Suppose that any $n + 1$ of these sets have nonempty intersection. Prove that the whole collection has nonempty intersection.

HINT: Use induction on $m \ge n+1$ sets. If true for $m-1$, choose x_j in the intersection of $C_1, \ldots, C_{j-1}, C_{j+1}, \ldots, C_m$. Apply Radon's Theorem.

16.2. Relative Interior

In working with a convex subset A of $\mathbb{R}^n$, the natural space containing it is often $\text{aff}(A)$, not $\mathbb{R}^n$, which may be far too large. The affine hull is a better place to work for many purposes. Indeed, if A is a convex subset of $\mathbb{R}^n$ with $\dim A = k < n$, then thinking of A simply as a subset of $\text{aff}(A)$, which may be identified with $\mathbb{R}^k$, allows us to talk more meaningfully about topological notions such as interior and boundary. One sign that the usual interior is not useful is Exercise 16.1.D, which shows that A has empty interior when $\dim(A) < n$. This section is devoted to developing the properties of the interior relative to $\text{aff}(A)$.

16.2.1. DEFINITION. If A is a convex subset of $\mathbb{R}^n$, then the **relative interior** of A, denoted $\text{ri}(A)$, is the interior of A relative to $\text{aff}(A)$. That is, $a \in \text{ri}(A)$ if and only if there is an $\varepsilon > 0$ so that $B_\varepsilon(a) \cap \text{aff}(A) \subset A$.

Define the **relative boundary** of A, denoted $\text{rbd}(A)$, to be $\overline{A} \setminus \text{ri}(A)$.

16.2.2. EXAMPLE. If $\dim A = k < n$, then $\text{ri}(A)$ is the interior of A when A is considered as a subset of $\text{aff}(A)$, which is identified with $\mathbb{R}^k$. For example, consider the closed convex disk $A = \{(x, y, z) \in \mathbb{R}^3 : x^2 + y^2 \le 1, z = 0\}$. Then $\text{aff}(A) = \{(x, y, z) \in \mathbb{R}^3 : z = 0\}$ is a plane. The relative interior of A is $\text{ri}(A) = \{(x, y, z) \in \mathbb{R}^3 : x^2 + y^2 < 1, z = 0\}$, the interior of the disk in this plane, while the interior as a subset of $\mathbb{R}^3$ is empty. The relative boundary is the circle $\{(x, y, z) \in \mathbb{R}^3 : x^2 + y^2 = 1, z = 0\}$.

Notice that $A \subset B$ does not imply that $\text{ri}(A) \subset \text{ri}(B)$ because an increase in dimension can occur. For example, let $B = \{(x, y, z) : x^2 + y^2 \le 1, z \le 0\}$. Then $\text{ri}(B) = \text{int}(B) = \{(x, y, z) : x^2 + y^2 < 1, z < 0\}$. So $A \subset B$ but $\text{ri}(A)$ and $\text{ri}(B)$ are disjoint.

The following result shows that the phenomenon above occurs precisely because of the dimension shift. Despite its trivial proof, it will be quite useful.

16.2.3. LEMMA. *Suppose that A and B are convex subsets of $\mathbb{R}^n$ with $A \subset B$. If $\text{aff}(A) = \text{aff}(B)$, then $\text{ri}(A) \subset \text{ri}(B)$.*

PROOF. If $a \in \text{ri}(A)$, then there is an $\varepsilon > 0$ with $B_\varepsilon(a) \cap \text{aff}(A) \subset A$. Because $\text{aff}(B) = \text{aff}(A)$, we have $B_\varepsilon(a) \cap \text{aff}(B) \subset A \subset B$, so $a \in \text{ri}(B)$. ∎

A **polytope** is the convex hull of a finite set. Now we obtain an analogue of Theorem 16.1.7 describing the relative interior of a polytope.

16.2.4. LEMMA. *Let $S = \{s_1, \ldots, s_r\}$ be a finite subset of $\mathbb{R}^n$. Then*

$$\text{ri}\left(\text{conv}(S)\right) = \left\{ \sum_{i=1}^{r} \lambda_i s_i : \lambda_i \in (0, 1), \ \sum_{i=1}^{r} \lambda_i = 1 \right\}.$$

PROOF. By Theorem 16.1.7, $\text{conv}(S) = \left\{ \sum_i \lambda_i s_i : \lambda_i \in [0, 1], \sum_i \lambda_i = 1 \right\}$. The subspace $L(S)$ is given by Exercise 16.1.E as $L(S) = \left\{ \sum_i \lambda_i s_i : \sum_i \lambda_i = 0 \right\}$.

Let $e_1, \ldots, e_k$ be an orthonormal basis for $L(S)$. Express each e_j as a combination $e_j = \sum_i \lambda_{ij} s_i$, where $\sum_i \lambda_{ij} = 0$, and define $\Lambda = \max_i \left(\sum_j \lambda_{ij}^2 \right)^{1/2}$. Suppose that x is vector in $L(S)$. Then

$$x = \sum_{j=1}^{k} x_j e_j = \sum_{i=1}^{r} \left(\sum_{j=1}^{k} \lambda_{ij} x_j \right) s_i.$$

By the Schwarz inequality (4.1.1),

$$\left| \sum_{j=1}^{k} \lambda_{ij} x_j \right| \leq \left(\sum_{j=1}^{k} \lambda_{ij}^2 \right)^{1/2} \left(\sum_{j=1}^{k} x_j^2 \right)^{1/2} \leq \Lambda \|x\|.$$

Thus if $\|x\| < \varepsilon/\Lambda$, then x may be expressed as $x = \sum_i \mu_i s_i$, where $\sum_i \mu_i = 0$ and $|\mu_i| < \varepsilon$ for $1 \leq i \leq r$.

Let $a = \sum_i \lambda_i s_i$, where $\lambda_i \in (0, 1)$ and $\sum_i \lambda_i = 1$. Then there is an $\varepsilon > 0$ so that $\varepsilon \leq \lambda_i \leq 1 - \varepsilon$ for $1 \leq i \leq r$. Note that

$$B_{\varepsilon/\Lambda}(a) \cap \text{aff}(S) = a + \left(B_{\varepsilon/\Lambda}(0) \cap L(S) \right).$$

If $x \in L(S)$ and $\|x\| < \varepsilon/\Lambda$, we use the representation above to see that

$$a + x = \sum_{i=1}^{r} (\lambda_i + \mu_i) s_i.$$

Now $\sum_i \lambda_i + \mu_i = 1$ and $\lambda_i + \mu_i \in [\varepsilon, 1 - \varepsilon] + (-\varepsilon, \varepsilon) = (0, 1)$. So $a + x$ belongs to $\text{conv}(S)$. This shows that a belongs to $\text{ri}(\text{conv}(S))$.

Conversely, suppose that $a \in \text{ri}(S)$. Write $a = \sum_i \lambda_i s_i$, where $\lambda_i \in [0, 1]$ and $\sum_i \lambda_i = 1$. We wish to show that it has a possibly different representation with coefficients in $(0, 1)$. Let $\varepsilon > 0$ be given so that $B_{\varepsilon/\Lambda}(a) \cap \text{aff}(S) \subset \text{conv}(S)$. If each $\lambda_i \in (0, 1)$, there is nothing to prove. Suppose that $J = \{j : \lambda_j = 0\}$ is nonempty, and let k be chosen so that $\lambda_k > 0$. Pick a $\delta > 0$ so small that $|J|\delta < \lambda_k$ and $x = \delta \sum_{j \in J} (s_j - s_k)$ satisfies $\|x\| < \varepsilon$. Then $a \pm x$ belong to $\text{conv}(S)$. Write $a - x = \sum_i \mu_i s_i$, where $\mu_i \in [0, 1]$ and $\sum_i \mu_i = 1$. Also,

$$a + x = \sum_{j \in J} \delta s_j + (\lambda_k - |J|\delta) s_k + \sum_{i \in (J \cup \{k\})^c} \lambda_i s_i =: \sum_{i=1}^{r} \nu_i s_i.$$

It is evident from our construction that each $\nu_i > 0$; and since $\sum_i \nu_i = 1$, all are less than 1. Now $a = \sum_i \frac{\mu_i + \nu_i}{2} s_i$ is expressed with all coefficients in $(0, 1)$. ∎

16.2.5. COROLLARY. *If A is a nonempty convex subset of $\mathbb{R}^n$, then $\mathrm{ri}(A)$ is also nonempty. Moreover,* $\mathrm{aff}(\mathrm{ri}(A)) = \mathrm{aff}(A)$.

PROOF. Let $k = \dim \mathrm{conv}(A)$. Then there is a subset S of $k + 1$ points so that $\mathrm{aff}(S) = \mathrm{aff}(A)$. Indeed, fix any element $a_0 \in A$. Since $L(A)$ is spanned by $\{a - a_0 : a \in A\}$ and has dimension k, we may choose k vectors $a_1, \ldots, a_k$ in A such that $a_i - a_0$ are linearly independent. Clearly, they span $L(A)$. So $S = \{a_0, \ldots a_k\}$ will suffice.

Hence, by Lemma 16.2.3, $\mathrm{ri}(\mathrm{conv}\, A)$ contains $\mathrm{ri}(\mathrm{conv}\, S)$. Let $a = \frac{1}{k+1} \sum_{i=0}^{k} a_i$.

By Lemma 16.2.4, $\mathrm{ri}(\mathrm{conv}\, S)$ is nonempty and, in particular, contains the points $b_i = (a + 2a_i)/3$ and $c_i = (2a + a_i)/3$ for $0 \le i \le k$. So $\mathrm{aff}(\mathrm{ri}(\mathrm{conv}\, S))$ contains $2b_i - c_i = a_i$ for $0 \le i \le k$, whence it equals $\mathrm{aff}(A)$ as claimed. ∎

The next theorem will be surprisingly useful. In particular, this theorem applies if b is in the relative boundary, $\mathrm{rbd}(A)$.

16.2.6. ACCESSIBILITY LEMMA.
Suppose that A is a convex subset of $\mathbb{R}^n$. If $a \in \mathrm{ri}(A)$ and $b \in \overline{A}$, then $[a, b)$ is contained in $\mathrm{ri}(A)$.

PROOF. Let $\varepsilon > 0$ be given so that $B_\varepsilon(a) \cap \mathrm{aff}(A) \subset A$. We need to show that $\lambda a + (1 - \lambda)b \in \mathrm{ri}(A)$ for $\lambda \in (0, 1)$. Since $b \in \overline{A}$, pick $c \in A$ so that $x = b - c \in L(A)$ satisfies $\|x\| < \varepsilon\lambda/(2 - 2\lambda)$. Suppose that $z \in L(A)$ with $\|z\| < \varepsilon/2$. Then

$$\lambda a + (1 - \lambda)b + \lambda z = \lambda a + (1 - \lambda)(c + x) + \lambda z$$
$$= \lambda(a + z + (1 - \lambda)x/\lambda) + (1 - \lambda)c.$$

Since $\|z + (1 - \lambda)x/\lambda\| \le \|z\| + (1 - \lambda)\|x\|/\lambda < \varepsilon/2 + \varepsilon/2 = \varepsilon$, this is a vector in $B_\varepsilon(0) \cap L(A)$. Hence $d = a + z + (1 - \lambda)x/\lambda$ belongs to A. Thus $\lambda a + (1 - \lambda)b + \lambda z = \lambda d + (1 - \lambda)c$ also lies in A. So $a + (1 - \lambda)b$ belongs to $\mathrm{ri}(A)$. ∎

Applying the preceding result when $b \in \mathrm{ri}(A)$ shows that $\mathrm{ri}(A)$ is convex. Since $\mathrm{int}(A)$ is either the empty set or equal to $\mathrm{ri}(A)$, we have the following:

16.2.7. COROLLARY. *If A is a convex subset of $\mathbb{R}^n$, so are $\mathrm{ri}(A)$ and $\mathrm{int}(A)$.*

16.2.8. THEOREM. *If A is a convex subset of $\mathbb{R}^n$, then the three sets $\mathrm{ri}(A)$, A, and $\overline{A}$ all have the same affine hulls, closures, and relative interiors.*

PROOF. By Corollary 16.2.5, $\mathrm{aff}(\mathrm{ri}(A)) = \mathrm{aff}(A)$. Since the affine hull is closed, $\overline{A} \subset \mathrm{aff}(A)$ and hence $\mathrm{aff}(A) = \mathrm{aff}(\overline{A})$.

Now $\mathrm{ri}(A) \subset A \subset \overline{A}$, and so $\overline{\mathrm{ri}(A)} \subset \overline{A} = \overline{\overline{A}}$, where the equality follows from Proposition 4.3.5. Suppose that $b \in \overline{A}$. By Corollary 16.2.5, $\mathrm{ri}(A)$ contains a point a. Hence by the Accessibility Lemma (Lemma 16.2.6), $\lambda a + (1 - \lambda)b$ belongs to $\mathrm{ri}(A)$ for $0 < \lambda \le 1$. Letting λ tend to 0 shows that $b \in \overline{\mathrm{ri}(A)}$. Thus $\overline{\mathrm{ri}(A)} = \overline{A}$.

For relative interiors, first observe that since the three sets have the same affine hull, Lemma 16.2.3 shows $\mathrm{ri}(\mathrm{ri}(A)) \subset \mathrm{ri}(A) \subset \mathrm{ri}(\overline{A})$. Now $\mathrm{ri}(\mathrm{ri}(A)) = \mathrm{ri}(A)$. For if $a \in \mathrm{ri}(A)$ and $x \in B_\varepsilon(a) \cap \mathrm{aff}(A) \subset A$, then using $r = \varepsilon - \|x - a\| > 0$, it follows that $B_r(x) \cap \mathrm{aff}(A) \subset B_\varepsilon(a) \cap \mathrm{aff}(A) \subset A$. Hence $B_\varepsilon(a) \cap \mathrm{aff}(A)$ is contained in $\mathrm{ri}(A)$ and so $a \in \mathrm{ri}(\mathrm{ri}(A))$.

Suppose that $a \in \mathrm{ri}(\overline{A})$. Then there is an $\varepsilon > 0$ so that $B_\varepsilon(a) \cap \mathrm{aff}(A) \subset \overline{A}$. Pick any $b \in \mathrm{ri}(A)$, and set $x = (a-b)/\|a-b\|$. Then $c_\pm = a \pm \varepsilon x$ belong to $\overline{A}$. But $c_- = (1 - \varepsilon)a + \varepsilon b \in \mathrm{ri}(A)$ by the Accessibility Lemma. Hence $a = (c_+ + c_-)/2$ belongs to $\mathrm{ri}(A)$, again by the Accessibility Lemma. Therefore, $\mathrm{ri}(\overline{A}) = \mathrm{ri}(A)$. ∎

The next two results will allow us to conveniently compute various combinations of convex sets such as sums and differences. The first is quite straightforward and is left as an exercise.

16.2.9. PROPOSITION. *If $A \subset \mathbb{R}^n$ and $B \subset \mathbb{R}^m$ are convex sets, then $A \times B$ is convex, $\mathrm{ri}(A \times B) = \mathrm{ri}(A) \times \mathrm{ri}(B)$ and $\mathrm{aff}(A \times B) = \mathrm{aff}(A) \times \mathrm{aff}(B)$.*

16.2.10. THEOREM. *If A is a convex subset of $\mathbb{R}^m$ and T is a linear map from $\mathbb{R}^m$ to $\mathbb{R}^n$, then $T\,\mathrm{ri}(A) = \mathrm{ri}(TA)$.*

PROOF. By Exercise 16.1.F, TA is convex. Using Theorem 16.2.8 for the first equality and the continuity of T (see Corollary 5.1.7) for the second containment, we have
$$TA \subset T\overline{A} = T\overline{\mathrm{ri}(A)} \subset \overline{T\,\mathrm{ri}(A)} \subset \overline{TA}.$$
Taking closures, we obtain $\overline{T\,\mathrm{ri}(A)} = \overline{TA}$. Using Theorem 16.2.8 again,
$$\mathrm{ri}(TA) = \mathrm{ri}(\overline{TA}) = \mathrm{ri}(\overline{T\,\mathrm{ri}(A)}) = \mathrm{ri}(T\,\mathrm{ri}(A)) \subset T\,\mathrm{ri}(A).$$

For the reverse containment, let $a \in \mathrm{ri}(A)$. We will show that Ta lies in $\mathrm{ri}(TA)$. By Corollary 16.2.5, $\mathrm{ri}(TA)$ is nonempty. So we may pick $b \in A$ with Tb in $\mathrm{ri}(TA)$. As $a \in \mathrm{ri}(A)$, there is an $\varepsilon > 0$ so that $c = a + \varepsilon(a - b)$ belongs to A. But $a = \frac{1}{1+\varepsilon}c + \frac{\varepsilon}{1+\varepsilon}b$, so $a \in [b, c)$. By linearity, Ta lies in $[Tb, Tc)$. Therefore, by the Accessibility Lemma (Lemma 16.2.6), Ta belongs to $\mathrm{ri}(TA)$. ∎

16.2.11. COROLLARY. *Suppose that A and B are convex subsets of $\mathbb{R}^n$. Then $\mathrm{ri}(A + B) = \mathrm{ri}(A) + \mathrm{ri}(B)$ and $\mathrm{ri}(A - B) = \mathrm{ri}(A) - \mathrm{ri}(B)$.*

PROOF. For the two cases, define maps T from $\mathbb{R}^n \times \mathbb{R}^n$ to $\mathbb{R}^n$ by $T(a, b) = a \pm b$. Then $T(A \times B) = A \pm B$. By the two previous results,

$$\text{ri}(A \pm B) = T\,\text{ri}(A \times B) = T(\text{ri}(A) \times \text{ri}(B)) = \text{ri}(A) \pm \text{ri}(B). \qquad \blacksquare$$

Exercises for Section 16.2

A. Write out a careful proof of Proposition 16.2.9.

B. Explain why the notion of relative closure of a convex set isn't needed.

C. Let C be a convex subset of $\mathbb{R}^n$ and let U be an open set that intersects $\overline{C}$. Show that $U \cap \text{ri}(C)$ is nonempty.

D. Suppose that a convex subset A of $\mathbb{R}^n$ intersects every (affine) line in a closed interval. Show that A is closed.   HINT: Connect $b \in \overline{A}$ to a in $\text{ri}(A)$.

E. Let A be a convex subset of $\mathbb{R}^n$ that is dense in $\text{aff}(A)$. Prove that $A = \text{aff}(A)$.

F. Let A be a convex subset of $\mathbb{R}^n$. Show that if B is a compact subset of $\text{ri}(A)$, then $\text{conv}(B)$ is a compact convex subset of $\text{ri}(A)$.   HINT: Corollary 16.1.9

G. (a) Show that if A and B are closed convex sets and A is compact, then $A + B$ is a closed convex set.
   (b) If B is also compact, show that $A + B$ is compact.
   (c) Give an example of two closed convex sets with nonclosed sum.
   HINT: Look for two closed convex subsets of $\mathbb{R}^2$ that lie strictly above the x-axis but 0 is a limit point of $A + B$.

H. Suppose that C and D are convex subsets of $\mathbb{R}^n$ and that $C \subset \overline{D}$. Show that if $\text{ri}(D) \cap C$ is nonempty, then $\text{ri}(C) \subset \text{ri}(D)$.
   HINT: If $d \in \text{ri}(D) \cap C$ and $c \in \text{ri}(C)$, extend $[d, c]$ beyond c.

I. Suppose that A and B are convex subsets of $\mathbb{R}^n$ with $\text{ri}(A) \cap \text{ri}(B) \neq \varnothing$.
   (a) Show that $\overline{A \cap B} = \overline{A} \cap \overline{B}$.   HINT: Connect $b \in \overline{A} \cap \overline{B}$ to $a \in \text{ri}(A) \cap \text{ri}(B)$.
   (b) Show that $\text{ri}(A \cap B) = \text{ri}(A) \cap \text{ri}(B)$.

16.3. Separation Theorems

The goal of this section is show that we can separate two disjoint convex sets from one another by a hyperplane. Geometrically, it seems obvious that, given a convex set $A \subset \mathbb{R}^2$ and $b \notin A$, there is a line that separates b from A. The appropriate generalization (replacing the line with a hyperplane) is true in $\mathbb{R}^n$. Indeed, there is a general form of this theorem, called the Hahn–Banach theorem, which holds for any normed vector space. We prove this result for $\mathbb{R}^n$, where we can use the Heine–Borel Theorem (Theorem 4.4.6).

We start by showing that a convex set in $\mathbb{R}^n$ always has a *unique* closest point to a given point outside. The *existence* of such a point comes from Exercise 5.4.J for any closed set and does not require convexity. In convexity theory, the map to this closest point is called a **projection**, which differs from the meaning of the term in linear algebra.

16.3.1. CONVEX PROJECTION THEOREM.

Let A be a nonempty closed convex subset of $\mathbb{R}^n$. For each point $x \in \mathbb{R}^n$, there is a unique point $P_A(x)$ in A which is closest to x. The point $P_A(x)$ is characterized by

$$\langle x - P_A(x), a - P_A(x) \rangle \leq 0 \quad \text{for all} \quad a \in A.$$

Moreover, $\|P_A(x) - P_A(y)\| \leq \|x - y\|$ for all $x, y \in \mathbb{R}^n$.

Geometrically, the equation says that the line segment $[x, P_A(x)]$ makes an obtuse angle with $[a, P_A(x)]$ for every point $a \in A$. The last estimate shows that P_A has Lipschitz constant 1 and, in particular, is continuous.

PROOF. Pick any vector a_0 in A, and let $R = \|x - a_0\|$. Then

$$0 \leq \inf\{\|x - a\| : a \in A\} \leq R.$$

The closest point to x in A must belong to $A \cap B_R(x)$. This is the intersection of two closed convex sets, and so is closed and convex. Moreover, $A \cap B_R(x)$ is bounded and thus compact by the Heine–Borel Theorem (Theorem 4.4.6). Consider the continuous function on $A \cap B_R(x)$ given by $f(a) = \|x - a\|$. By the Extreme Value Theorem (Theorem 5.4.4), there is a point $a_1 \in A$ at which f takes its minimum—a closest point.

To see that a_1 is unique, we need to use convexity. Suppose that there is another vector $a_2 \in A$ with $\|x - a_2\| = \|x - a_1\|$. We will show that $b = \frac{1}{2}(a_1 + a_2)$ is closer, contradicting the choice of a_1 as a closest point. See Figure 16.2. One way to accomplish this is to apply the parallelogram law (see Exercise 16.3.A). Instead, we give an elementary argument using Euclidean geometry.

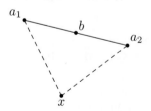

FIGURE 16.2. The points x, a_1, a_2, and b.

Consider the triangle with vertices x, a_1, and a_2. As $\|x - a_1\| = \|x - a_2\|$, this triangle is isosceles. As $\|b - a_1\| = \|b - a_2\|$, the point b is the foot of the perpendicular dropped from x to $[a_1, a_2]$ and therefore $\|x - b\| < \|x - a_i\|$. Indeed, $\angle xba_i$ is a right angle, and the Pythagorean Theorem shows that

$$\|x - a_i\|^2 = \|x - b\|^2 + \frac{1}{4}\|a_1 - a_2\|^2.$$

It now makes sense to define the function P_A by setting it to be this unique closest point to x. Let $a \neq P_A(x)$ be any other point in A. Then for $0 < \lambda \leq 1$,

$$\|x - a_1\|^2 < \|x - (\lambda a + (1 - \lambda)a_1)\|^2$$
$$= \|(x - a_1) - \lambda(a - a_1)\|^2$$
$$= \|x - a_1\|^2 - 2\lambda\langle x - a_1, a - a_1 \rangle + \lambda^2 \|a - a_1\|^2$$

so that $\langle x - a_1, a - a_1 \rangle \leq \frac{\lambda}{2}\|a - a_1\|^2$. Let λ tend to 0 to obtain $\langle x - a_1, a - a_1 \rangle \leq 0$.

This argument is reversible. If a_1 is any point such that $\langle x - a_1, a - a_1 \rangle \leq 0$ for all $a \in A$, then

$$\|x - a\|^2 = \|x - a_1\|^2 - 2\langle x - a_1, a - a_1 \rangle + \|a - a_1\|^2$$
$$\geq \|x - a_1\|^2 + \|a - a_1\|^2.$$

Hence $a_1 = P_A(x)$ is the unique closest point.

Let x and y be points in $\mathbb{R}^n$. Apply the inequality once for each of x and y:

$$\langle x - P_A(x), P_A(y) - P_A(x) \rangle \leq 0$$

and

$$\langle P_A(y) - y, P_A(y) - P_A(x) \rangle = \langle y - P_A(y), P_A(x) - P_A(y) \rangle \leq 0.$$

Adding yields $\langle (x - y) + (P_A(y) - P_A(x)), P_A(y) - P_A(x) \rangle \leq 0$. Hence

$$\|P_A(y) - P_A(x)\|^2 \leq \langle y - x, P_A(y) - P_A(x) \rangle$$
$$\leq \|y - x\| \, \|P_A(y) - P_A(x)\|$$

by the Schwarz inequality (4.1.1). Therefore, $\|P_A(y) - P_A(x)\| \leq \|x - y\|$. ∎

16.3.2. EXAMPLE. Let A be the triangle in $\mathbb{R}^2$ with vertices $(0, 0)$, $(1, 0)$, and $(0, 1)$ as in Figure 16.3. Then

$$P_A((x, y)) = \begin{cases} (x, y) & \text{if } (x, y) \in A \\ (x, 0) & \text{if } 0 \leq x \leq 1 \text{ and } y \leq 0 \\ (0, y) & \text{if } 0 \leq y \leq 1 \text{ and } x \leq 0 \\ (0, 0) & \text{if } x \leq 0 \text{ and } y \leq 0 \\ (1, 0) & \text{if } x \geq 1 \text{ and } y \leq 0 \\ (1, 0) & \text{if } 0 \leq y \leq x - 1 \\ (0, 1) & \text{if } y \geq 1 \text{ and } x \leq 0 \\ (0, 1) & \text{if } 0 \leq x \leq y - 1 \\ (s, 1 - s) & \text{if } x + y \geq 1 \text{ and } |y - x| \leq 1, \text{ where } s = \frac{y + 1 - x}{2}. \end{cases}$$

We can now prove the promised separation theorem, which will be fundamental to all of our later work. We provide only two consequences in this section. There are many more.

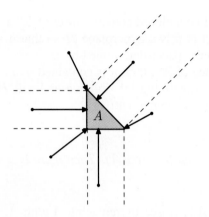

FIGURE 16.3. The triangle A and action of P_A.

16.3.3. SEPARATION THEOREM.
Suppose that A is a closed convex set in $\mathbb{R}^n$ and $b \notin A$. Then there is a vector $h \in \mathbb{R}^n$ and scalar $\alpha \in \mathbb{R}$ so that $\langle a, h \rangle \leq \alpha$ for all $a \in A$ but $\langle b, h \rangle > \alpha$. In particular, h determines a hyperplane H so that A is contained in H^- and b is contained in the interior of H^+.

PROOF. Let $a_1 = P_A(b)$. As $b \notin A$, $h = b - a_1$ is a nonzero vector. Define $\alpha = \langle a_1, h \rangle$, and let $H = \{x : \langle x, h \rangle = \alpha\}$. By Theorem 16.3.1, if $a \in A$, then $\langle b - a_1, a - a_1 \rangle \leq 0$. Rewriting this, we obtain $\langle a, h \rangle \leq \langle a_1, h \rangle = \alpha$. Therefore, A is contained in $H^- = \{x \in \mathbb{R}^n : \langle x, h \rangle \leq \alpha\}$. On the other hand, $\langle b - a_1, b - a_1 \rangle = \|h\|^2 > 0$, which implies that $\langle b, h \rangle > \langle a_1, h \rangle = \alpha$. ∎

16.3.4. COROLLARY. *Let A be a closed convex subset of $\mathbb{R}^n$. Then b belongs to A if and only if $\langle b, x \rangle \leq \sup\{\langle a, x \rangle : a \in A\}$ for all $x \in \mathbb{R}^n$.*

PROOF. If $b \in A$, then the inequality is immediate. For the other direction, suppose that $b \notin A$. Take x to be the vector h given by the Separation Theorem,

$$\langle b, h \rangle > \alpha \geq \langle a, h \rangle \quad \text{for all} \quad a \in A.$$

Thus $\sup\{\langle a, x \rangle : a \in A\} \leq \alpha < \langle b, h \rangle$. Therefore, the inequality of the corollary is valid precisely for $b \in A$. ∎

This corollary has a beautiful geometric meaning. The proof is left as Exercise 16.3.E.

16.3.5. COROLLARY. *A closed convex set in $\mathbb{R}^n$ is the intersection of all the closed half-spaces that contain it.*

16.3.6. DEFINITION. Let A be a closed convex subset of $\mathbb{R}^n$ and $b \in \text{rbd}(A)$. A **supporting hyperplane** to A at b is a hyperplane H so that $b \in H$ and A is contained in one of the closed half-spaces determined by H.

If $\text{aff}(A)$ is a proper subspace of $\mathbb{R}^n$, then it is contained in a hyperplane that supports A. We regard this as a pathological situation and call H a **nontrivial supporting hyperplane** if it does not contain $\text{aff}(A)$.

16.3.7. SUPPORT THEOREM.
Let A be a convex subset of $\mathbb{R}^n$ and $b \in \text{rbd}(A)$. Then there is a nontrivial supporting hyperplane to A at b.

PROOF. We may suppose that A is closed by replacing A with $\overline{A}$. Let a_0 be any point in $\text{ri}(A)$. By the Accessibility Lemma (Lemma 16.2.6), the interval $[a_0, b)$ is contained in $\text{ri}(A)$. By the same token, the point $b_k := b + \frac{1}{k}(b - a_0)$ is not in A for $k \geq 1$, as that would put b into $\text{ri}(A)$. Let $h_k = (b_k - P_A b_k)/\|b_k - P_A b_k\|$. The proof of the Separation Theorem (Theorem 16.3.3) shows that

$$\langle a, h_k \rangle \leq \alpha_k := \langle P_A b_k, h_k \rangle < \langle b_k, h_k \rangle \quad \text{for all} \quad a \in A.$$

By the Heine–Borel Theorem (Theorem 4.4.6), the unit sphere of $\mathbb{R}^n$ is compact. Thus the sequence $\{h_k : k \geq 1\}$ has a convergent subsequence $(h_{k_i})_{i=1}^{\infty}$ with limit $h = \lim_{i \to \infty} h_{k_i}$. Then since P_A is continuous,

$$\lim_{i \to \infty} \alpha_{k_i} = \lim_{i \to \infty} \langle P_A b_{k_i}, h_{k_i} \rangle = \langle P_A b, h \rangle = \langle b, h \rangle =: \alpha.$$

If $a \in A$, then

$$\langle a, h \rangle = \lim_{i \to \infty} \langle a, h_{k_i} \rangle \leq \lim_{i \to \infty} \alpha_{k_i} = \alpha.$$

So A is contained in the half-space $H^- = \{x : \langle x, h \rangle \leq \alpha\}$.

The vectors h_k all lie in the subspace $L(A)$, and thus so does h. Consequently, $b + h$ belongs to $\text{aff}(A)$. Since $\langle b + h, h \rangle = \alpha + 1 > \alpha$, it follows that H does not contain $\text{aff}(A)$. So H is a nontrivial supporting hyperplane. ∎

Exercises for Section 16.3

A. Use the parallelogram law (Exercise 7.4.A) to give a different proof of the uniqueness portion of Theorem 16.3.1.

B. Let $A = \overline{B_r(a)}$. Find an explicit formula for $P_A(x)$.

C. If $A = B_r(a)$ and $b \in \text{rbd}(A)$, show that the *unique* supporting hyperplane to A at b is $H = \{x : \langle x - a, b - a \rangle = r^2\}$.

D. Let A be a subspace. Show that $P_A(x)$ is the (linear) orthogonal projection onto A.

E. Show that a closed convex set in $\mathbb{R}^n$ is the intersection of all the closed half-spaces that contain it.

F. Let $A \subset \mathbb{R}^n$ be a nonempty open convex set. Show that $a \in A$ if and only if for each hyperplane H of $\mathbb{R}^n$ containing a, the two *open* half-spaces $\text{int}\, H^+$ and $\text{int}\, H^-$ both contain elements of A.

G. (a) Suppose $A \subset \mathbb{R}^n$ is convex and $H = \{x \in \mathbb{R}^n : \langle x, h \rangle \leq \alpha\}$ is a nontrivial supporting hyperplane of A. If $a \in A$ and $\langle a, h \rangle = \alpha$, show that $a \in \text{rbd}(A)$.
  (b) Show that an open convex set in $\mathbb{R}^n$ is the intersection of all the open half-spaces that contain it.

H. Suppose that $S \subset \mathbb{R}^n$ and $x \in \text{conv}(S)$ lies in the relative boundary rbd S. Prove that there are n points $s_1, \ldots, s_n$ in S so that $x \in \text{conv}\{s_1, \ldots, s_n\}$.
  HINT: Use a supporting hyperplane and Carathéodory's Theorem.

I. Farkas Lemma. Let $A \subset \mathbb{R}^n$ and let $C = \overline{\text{cone}}(A)$ be the closed convex cone generated by A. Prove that exactly one of the following statements is valid: (1) $x \in C$ or (2) there is an $s \in \mathbb{R}^n$ so that $\langle x, s \rangle > 0 \geq \langle a, s \rangle$ for all $a \in A$.

J. Let $A = \{a_1, \ldots, a_k\}$ be a finite subset of $\mathbb{R}^n$. Prove that the following are equivalent:
  (1) $0 \in \text{conv}(A)$
  (2) There is *no* $y \in \mathbb{R}^n$ so that $\langle a, y \rangle < 0$ for all $a \in A$.
  (3) $f(x) = \log \left(\sum_{i=1}^{n} e^{\langle a_i, x \rangle} \right)$ is bounded below on $\mathbb{R}^n$.
  HINT: (1) $\Leftrightarrow$ (2) separation. Not (2) $\Rightarrow$ not (3), use $f(ty)$. (2) $\Rightarrow$ (3), easy.

K. Suppose that A, B, and C are closed convex sets in $\mathbb{R}^n$ with $A + C = B + C$.
  (a) Is it true that $A = B$?
  (b) What if all three sets are compact?

L. (a) Prove that for any subset S of $\mathbb{R}^n$, a vector b belongs to $\overline{\text{conv}}(S)$ if and only if $\langle b, x \rangle \leq \sup\{\langle s, x \rangle : s \in S\}$ for all $x \in \mathbb{R}^n$.
  (b) Show that the intersection of all closed half-spaces containing S equals $\overline{\text{conv}} \, S$.

M. A hyperplane H **properly separates** A and B if $A \subset H^-$, $B \subset H^+$ and $A \cup B$ is not contained in H. Prove that two convex sets A and B can be properly separated if and only if $\text{ri}(A) \cap \text{ri}(B) = \varnothing$.
  HINT: Let $C = A - B$. Show $0 \notin \text{ri}(C)$, and separate 0 from C.

16.4. Extreme Points

16.4.1. DEFINITION. Let A be a nonempty convex set. A point $a \in A$ is an **extreme point** of A if: Whenever $a_1, a_2 \in A$ and $a = (a_1 + a_2)/2$, then $a_1 = a_2$. The set of all extreme points of A is denoted ext A.

A **face** of A is a (convex) subset F of A such that whenever $a_1, a_2 \in A$ and the open interval (a_1, a_2) intersects F, then $[a_1, a_2]$ is contained in F.

It is evident that a face is convex, and that a one point face is an extreme point. Conversely, an extreme point is a one point face. For if (a_1, a_2) contains a, then a is the average of two points of $[a_1, a_2]$ with one of them being an endpoint. Thus both those points equal a, and hence $[a_1, a_2] = \{a\}$.

Every convex set has two trivial faces, the whole set and the empty set. This could be all of them, but often there is a rich collection.

16.4.2. EXAMPLES.

(1) Consider a cube in $\mathbb{R}^3$. The whole set and $\varnothing$ are faces for trivial reasons. The extreme points are the eight vertices. Note that there are many other boundary

points that are not extreme. The one-dimensional faces are the twelve edges, and the two-dimensional faces are the six sides (which are commonly called faces).

(2) On the other hand, consider the open unit ball U in $\mathbb{R}^3$. Except for the two trivial cases, there are no extreme points or faces at all. For if F is a proper convex subset, let $a \in F$ and $b \in U \setminus F$. Then since a is an interior point, the line segment $[b, a]$ may be extended to some point $c \in U$. Hence $(b, c) \cap F$ contains a but $[b, c]$ is not wholly contained in F.

(3) The closed unit ball B has many extreme points. A modification of the previous argument shows that no interior point is extreme. But every boundary point is extreme. This follows from the proof of Lemma 16.4.3. There are no other faces of the ball except for the two trivial cases.

16.4.3. LEMMA. *Every compact convex set A has an extreme point.*

PROOF. The norm function $f(x) = \|x\|$ is continuous on A, and thus by the Extreme Value Theorem (Theorem 5.4.4) f achieves its maximum value at some point $a_0 \in A$, say $\|a_0\| = R \geq \|a\|$ for all $a \in A$. Suppose that $a_1, a_2 \in A$ and $a_0 = (a_1 + a_2)/2$. Then by the Schwarz inequality (4.1.1),

$$
\begin{aligned}
R^2 = \|a_0\|^2 &= \left\langle \frac{a_1 + a_2}{2}, \frac{a_1 + a_2}{2} \right\rangle \\
&= \tfrac{1}{4}\left(\|a_1\|^2 + 2\langle a_1, a_2\rangle + \|a_2\|^2\right) \\
&\leq \tfrac{1}{4}\left(\|a_1\|^2 + 2\|a_1\|\,\|a_2\| + \|a_2\|^2\right) \\
&\leq \tfrac{1}{4}(R^2 + 2R^2 + R^2) = R^2.
\end{aligned}
$$

Equality at the extremes forces the equalities $\langle a_1, a_2\rangle = \|a_1\|\,\|a_2\|$ in the Schwarz inequality and $\|a_1\| = \|a_2\| = R$. Hence $a_1 = a_2$, and a_0 is extreme. $\blacksquare$

We collect a couple of very easy lemmas that produce faces.

16.4.4. LEMMA. *If F is a face of a convex set A, and G is a face of F, then G is a face of A.*

PROOF. Suppose that $a_1, a_2 \in A$ such that (a_1, a_2) intersects G. Then a fortiori, (a_1, a_2) intersects F. As F is a face of A, it follows that $[a_1, a_2]$ is contained in F. Therefore, $a_1, a_2 \in F$ and since G is a face of F, it follows that $[a_1, a_2]$ is contained in G. So G is a face of A. $\blacksquare$

16.4.5. LEMMA. *If A is a convex set in $\mathbb{R}^n$ and H is a supporting hyperplane, then $H \cap A$ is a face of A.*

PROOF. Let $H = \{x : \langle x, h\rangle = \alpha\}$ be a hyperplane such that A is contained in $H^- = \{x : \langle x, h\rangle \leq \alpha\}$. Let $F = H \cap A$. Suppose that $a_1, a_2 \in A$ such that

$(a_1, a_2) \cap F$ contains a point $a = \lambda a_1 + (1 - \lambda)a_2$ for $\lambda \in (0, 1)$. Then

$$\alpha = \langle a, h \rangle = \lambda \langle a_1, h \rangle + (1 - \lambda)\langle a_2, h \rangle \le \lambda \alpha + (1 - \lambda)\alpha = \alpha.$$

Thus equality holds, so $\langle a_1, h \rangle = \langle a_2, h \rangle = \alpha$. Therefore a_1, a_2 belong to F. ∎

We have set the groundwork for a fundamental result that demonstrates the primacy of extreme points.

16.4.6. MINKOWSKI'S THEOREM.
Let C be a nonempty compact convex subset of $\mathbb{R}^n$. Then $C = \mathrm{conv}(\mathrm{ext}\,C)$.

PROOF. We will prove this by induction on $\dim C$. If $\dim C = 0$, then C is a single point, and it is evidently extreme. Suppose that we have established the result for compact convex sets of dimension at most $k - 1$, and that $\dim C = k$.

Let a be any point in $\mathrm{rbd}(C)$. By the Support Theorem (Theorem 16.3.7), there is a nontrivial supporting hyperplane H to C at a. By Lemma 16.4.5, $F = H \cap C$ is a face of C. Also, F is compact because C is compact and H is closed, and it is nonempty since $a \in F$.

Note that $\mathrm{aff}(F)$ is contained in $\mathrm{aff}(C) \cap H$. This is properly contained in $\mathrm{aff}(C)$ because C is not contained in H. Therefore, $\dim F < \dim C = k$. By the induction hypothesis, F is the convex hull of its extreme points. However, by Lemma 16.4.4, $\mathrm{ext}\,F$ is contained in $\mathrm{ext}\,C$. So $\mathrm{conv}(\mathrm{ext}\,C)$ contains every boundary point of C.

Finally, let $a \in \mathrm{ri}(C)$, and fix another point $b \in C$. Let L be the line passing through a and b. In particular, L is contained in $\mathrm{aff}(C)$. Then $L \cap C$ is a closed bounded convex subset of L and thus is a closed interval that contains a in its relative interior. Let a_1, a_2 be the two endpoints. These points lie in $\mathrm{rbd}(C)$ because any ball about a_i meets L in points outside of C. By the previous paragraph, both a_1, a_2 lie in $\mathrm{conv}(\mathrm{ext}\,C)$. But a belongs to $\mathrm{conv}\{a_1, a_2\}$ and hence is also contained in $\mathrm{conv}(\mathrm{ext}\,C)$. ∎

Exercises for Section 16.4

A. Let $A \subset \mathbb{R}^n$ be convex. Show that no point in $\mathrm{ri}(A)$ is an extreme point.

B. Let $A \subset \mathbb{R}^n$ be convex. Show that $a \in A$ is an extreme point if and only if $A \backslash \{a\}$ is convex.

C. Show that if $B \subset A$ are two convex sets, then any extreme point of A that is contained in B is an extreme point of B.

D. Find a nonempty proper closed convex subset of $\mathbb{R}^2$ with no extreme points.

E. A face of a convex set A of the form $A \cap H$, where H is a hyperplane, is called an **exposed face**. Let $A = \overline{B_1(0)} \cup \{(x, y) : 0 \le x \le 1, |y| \le 1\}$. Show that $(0, 1)$ is an extreme point that is not exposed.

F. Let $A \subset \mathbb{R}^n$ be compact and convex, and let f be an affine map of $\mathbb{R}^n$ into $\mathbb{R}$.
(a) Show that $\{a \in A : f(a) = \sup_{x \in A} f(x)\}$ is an exposed face of A.

(b) Let A be a compact convex set, and let $B \subset \mathbb{R}^n$ such that $A = \overline{\text{conv}}(B)$. Prove that $\overline{B}$ contains $\text{ext}(A)$.

(c) Show that an affine function on A always takes its maximum (and minimum) value at an extreme point.

G. Let $A = [0,1]^n = \{x = (x_1, \ldots, x_n) \in \mathbb{R}^n : 0 \le x_i \le 1, 1 \le i \le n\}$.

(a) Describe $\text{ext}(A)$.

(b) Explicitly show that each $a \in A$ is in the convex hull of $n + 1$ extreme points.
HINT: If $x_1 \le x_2 \le \cdots \le x_n$, consider e^j, where $e_i^j = 1$ if $i > j$ and 0 otherwise, $0 \le j \le n$.

H. A **polyhedral set** is the intersection of a finite number of closed half-spaces. Let A be a closed bounded polyhedral set determined by the intersection of closed half-spaces H_i^- for $1 \le i \le p$.

(a) Show that if $a \in A$ is not in any hyperplane H_i, then a belongs to $\text{ri}(A)$.

(b) Show that every extreme point of A is the intersection of some collection of the hyperplanes H_i.   HINT: Use part (a) and induction on $\dim A$.

(c) Hence deduce that every closed bounded polyhedral set is a polytope, as defined just before Lemma 16.2.4.

I. Let $A = \text{conv}\{a_1, \ldots, a_r\}$ be a polytope.

(a) Show that $\text{ext } A$ is contained in $\{a_1, \ldots, a_r\}$.

(b) Show that every face of A is the convex hull of a subset of $\{a_1, \ldots, a_r\}$.

(c) If F is a face of A, find a hyperplane $H \supset F$ that does not contain all of A.
HINT: Apply the Support Theorem to a point in $\text{ri}(F)$.

(d) Prove that the intersection of half-spaces determined by (c) is a polyhedral set P containing A such that each face of A is contained in $\text{rbd}(P)$.

(e) Show that $P = A$.
HINT: If $p \in P \setminus A$ and $a \in \text{ri}(A)$, consider where $[a, p)$ intersects $\text{rbd}(A)$.

(f) Hence show that every polytope is a closed bounded polyhedral set.

16.5. Convex Functions in One Dimension

Convex functions occur frequently in many applications. Generally, we are interested in minimizing these functions over a convex set determined by a number of constraints, a problem that we will discuss in later sections. The notion of convexity allows us to work with functions that need not be differentiable. The analysis of convex functions can be thought of an an extension of calculus to an important class of nondifferentiable functions.

While a few generalities are introduced here for functions on domains in $\mathbb{R}^n$, most of this section is devoted to convex functions on the line. In the next section, we extend these notions to higher dimensions.

16.5.1. DEFINITION. Suppose that A is a convex subset of $\mathbb{R}^n$. A real-valued function f defined on A is called a **convex function** if

$$f(\lambda x + (1 - \lambda)y) \le \lambda f(x) + (1 - \lambda)f(y) \quad \text{for all} \quad x, y \in A, \ 0 \le \lambda \le 1.$$

A function f is called a **concave function** if $-f$ is convex.

16.5.2. EXAMPLES.

(1) All linear functions are both convex and concave.

(2) If $\| \cdot \|$ is any norm on $\mathbb{R}^n$, the function $f(x) = \|x\|$ is convex. Indeed, the triangle inequality and homogeneity yield

$$\|\lambda x + (1 - \lambda)y\| \le \lambda\|x\| + (1 - \lambda)\|y\| \quad \text{for} \quad x, y \in \mathbb{R}^n \text{ and } 0 \le \lambda \le 1,$$

which is precisely the convexity condition.

(3) $f(x) = e^x$ is convex. This is evident from an inspection of its graph. We will see that any C^2 function g with $g'' \ge 0$ is convex. Since $f''(x) = e^x > 0$, this result applies.

This next result, an easy application of the Mean Value Theorem, provides many examples of convex functions on the line. It also shows that our definition is consistent with the notion introduced in calculus.

16.5.3. LEMMA. *Suppose that f is a differentiable function on (a, b) and f' is monotone increasing. Then f is convex. In particular, this holds if f is C^2 and $f'' \ge 0$.*

PROOF. Suppose that $a < x < y < b$ and $0 < \lambda < 1$. Let $z = \lambda x + (1 - \lambda)y$. Then there are points $c \in (a, z)$ and $d \in (z, y)$ so that

$$\frac{f(z) - f(x)}{z - x} = f'(c) \le f'(d) = \frac{f(y) - f(z)}{y - z}.$$

Substituting $z - x = (1 - \lambda)(y - x)$ and $y - z = \lambda(y - x)$ yields

$$\frac{f(z) - f(x)}{1 - \lambda} \le \frac{f(y) - f(z)}{\lambda}.$$

Just multiply this out to obtain the statement of convexity.

If f is C^2 and $f'' \ge 0$, then f' is an increasing function. ∎

A straightforward induction on r gives the following result from the definition of convexity. The proof is left as an exercise.

16.5.4. JENSEN'S INEQUALITY.

Suppose that $A \subset \mathbb{R}^n$ is convex and f is a convex function on A. If $a_1, \ldots, a_r$ are points in A and $\lambda_1, \ldots, \lambda_r$ are nonnegative scalars that sum to 1, then

$$f(\lambda_1 a_1 + \cdots + \lambda_r a_r) \le \lambda_1 f(a_1) + \cdots + \lambda_r f(a_r).$$

16.5.5. EXAMPLE. In spite of the fact that Jensen's inequality is almost trivial, when it is applied we can obtain results that are not obvious. Consider the exponential function $f(x) = e^x$. Let $t_1, \ldots, t_n$ be positive real numbers and let $a_i = \log t_i$.

Then for positive values λ_i with $\sum_{i=1}^{n} \lambda_i = 1$,

$$e^{\lambda_1 a_1 + \cdots + \lambda_n a_n} \le \lambda_1 e^{a_1} + \cdots + \lambda_n e^{a_n}.$$

In other words,

$$t_1^{\lambda_1} t_2^{\lambda_2} \cdots t_n^{\lambda_n} \le \lambda_1 t_1 + \lambda_2 t_2 + \cdots + \lambda_n t_n.$$

This is the **generalized arithmetic mean–geometric mean inequality**. Setting $\lambda_i = \frac{1}{n}$ for $1 \le i \le n$, we obtain

$$\sqrt[n]{t_1 t_2 \cdots t_n} \le \frac{t_1 + t_2 + \cdots + t_n}{n}.$$

We begin by characterizing a convex function in terms of its graph, or rather its epigraph. This also serves to justify the terminology.

16.5.6. DEFINITION. Let f be a real-valued function on a convex subset A of $\mathbb{R}^n$. The **epigraph** of f is defined to be $\text{epi}(f) = \{(a, y) \in A \times \mathbb{R} : y \ge f(a)\}$.

16.5.7. LEMMA. *Let f be a real-valued function on a convex subset A of $\mathbb{R}^n$. Then f is a convex function if and only if $\text{epi}(f)$ is a convex set.*

PROOF. Suppose $p = (x, t)$ and $q = (y, u)$ belong to $\text{epi}(f)$ and $\lambda \in [0, 1]$. Since A is convex, $z := \lambda x + (1 - \lambda) y \in A$. Consider $\lambda p + (1 - \lambda) q = (z, \lambda t + (1 - \lambda) u)$. If f is convex, then

$$f(z) \le \lambda f(x) + (1 - \lambda) f(y) \le \lambda t + (1 - \lambda) u$$

and thus $\text{epi}(f)$ is convex. Conversely, if $\text{epi}(f)$ is convex, then using $t = f(x)$ and $u = f(y)$ we see that $\text{epi}(f)$ contains $(z, \lambda f(x) + (1 - \lambda) f(y))$ and hence $f(z) \le \lambda f(x) + (1 - \lambda) f(y)$. That is, f is convex. ∎

Now we specialize to functions on the line. We begin with a lemma about secants. In the next section, this will be applied to functions of more variables.

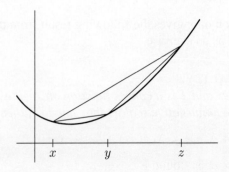

FIGURE 16.4. Secants of a convex function.

16.5.8. SECANT LEMMA.

Let f be a convex function on $[a, b]$, and consider three points $a \leq x < y < z \leq b$. Then

$$\frac{f(y) - f(x)}{y - x} \leq \frac{f(z) - f(x)}{z - x} \leq \frac{f(z) - f(y)}{z - y}.$$

PROOF. See Figure 16.4. Set $\lambda = \dfrac{z - y}{z - x}$, which lies in $(0, 1)$. So $y = \lambda x + (1 - \lambda)z$. By convexity, $f(y) \leq \lambda f(x) + (1 - \lambda)f(z)$. Therefore,

$$f(y) - f(x) \leq (1 - \lambda)\big(f(z) - f(x)\big).$$

Divide by $y - x = (1 - \lambda)(z - x)$ to obtain

$$\frac{f(y) - f(x)}{y - x} \leq \frac{(1 - \lambda)\big(f(z) - f(x)\big)}{(1 - \lambda)(z - x)}$$

as desired. The second inequality is similar. ∎

The main result about convex functions of one real variable is that convex functions are almost differentiable in a certain strong sense. The absolute value function on $\mathbb{R}$ shows that a convex function need not be differentiable, even at interior points of its domain. However, it does have left and right derivatives everywhere.

Recall that a function f has a right derivative at a if $\displaystyle \lim_{h \to 0^+} \frac{f(a + h) - f(a)}{h}$ exists. It is denoted by $D_+ f(a)$. Similarly, we define the left derivative to be the limit $D_- f(a) := \displaystyle \lim_{h \to 0^+} \frac{f(a) - f(a - h)}{h}$ when it exists.

16.5.9. THEOREM.

Let f be a convex function defined on (a, b). Then f has left and right derivatives at every point, and if $a < x < y < b$, then

$$D_- f(x) \leq D_+ f(x) \leq D_- f(y) \leq D_+ f(y).$$

Therefore, f is continuous.

PROOF. Let $0 < h < k$ be small enough that $x \pm k$ belong to the interval (a, b). Apply the Secant Lemma (Lemma 16.5.8) using $x - k < x - h < x < x + h < x + k$,

$$\frac{f(x) - f(x-k)}{k} \leq \frac{f(x) - f(x-h)}{h} \leq \frac{f(x+h) - f(x)}{h} \leq \frac{f(x+k) - f(x)}{k}.$$

Thus the quotient function $d_x(t) = \dfrac{f(x+t) - f(x)}{t}$ is an increasing function of t on an interval $[-r, r]$. In particular, $\{d_x(s) : s < 0\}$ is bounded above by $d_x(t)$ for any $t > 0$. Thus by the Least Upper Bound Principle (2.5.3),

$$D_- f(x) = \lim_{s \to 0^+} \frac{f(x) - f(x-s)}{s} = \sup_{s \to 0^+} \frac{f(x) - f(x-s)}{s}$$

exists. Similarly,

$$D_+f(x) = \lim_{t \to 0^+} \frac{f(x+t) - f(x)}{t} = \inf_{t \to 0^+} \frac{f(x+t) - f(x)}{t}$$

exists. Moreover, $D_-f(x) \le D_+f(x)$ since $d_x(-s) \le d_x(t)$ for all $-s < 0 < t$. Another application of the lemma using $x < x+t < y-s < y$ shows that

$$d_x(t) = \frac{f(x+t) - f(x)}{t} \le \frac{f(y) - f(y-s)}{s} = d_y(-s)$$

if s, t are sufficiently small and positive. Thus $D_+f(x) \le D_-f(y)$.

In particular, since left and right derivatives exist, we have

$$\lim_{t \to 0^+} f(x + t) = \lim_{t \to 0^+} f(x) + t\frac{f(x+t) - f(x)}{t}$$

$$= f(x) + 0D_+f(x) = f(x).$$

Similarly, $\lim_{s \to 0^+} f(x - s) = f(x)$. Therefore f is continuous. ∎

16.5.10. COROLLARY. *Let f be a convex function defined on (a, b). Then f is differentiable except on a countable set of points.*

PROOF. The right derivative $D_+f(x)$ is defined at every point of (a, b) and is a monotone increasing function. By Theorem 5.7.5, D_+f is continuous except for a countable set of jump discontinuities. At every point x where D_+f is continuous, we will show that $D_-f(x) = D_+f(x)$. From the continuity of f, given any $\varepsilon > 0$, there is an $r > 0$ so that $|D_+f(x \pm r) - D_+f(x)| < \varepsilon$. Now if $0 < h < r$,

$$D_+f(x - r) \le D_-f(x - h) \le D_-f(x)$$
$$\le D_+f(x) \le D_-f(x + h) \le D_+f(x + r).$$

Thus $|D_-f(x+h) - D_+f(x)| < \varepsilon$ for all $|h| < r$. As $\varepsilon > 0$ is arbitrary, we deduce that $D_-f(x) = D_+f(x)$ and $\lim_{h \to 0} D_-f(x + h) = D_+f(x)$. So D_-f is continuous at x and agrees with $D_+f(x)$. Consequently, f is differentiable at each point of continuity of D_+f. ∎

16.5.11. DEFINITION. Let f be a convex function on (a, b). For each x in (a, b), the **subdifferential** of f at x is the set $\partial f(x) = [D_-f(x), D_+f(x)]$.

See Figure 16.5. The geometric interpretation is the following pretty result. When the convex function f is differentiable at c, this result says that epi(f) lies above the tangent line to f through $(c, f(c))$, while any other line through this point crosses above the graph.

16.5.12. PROPOSITION. *Let f be a convex function on (a, b) and fix a point $c \in (a, b)$. The line $y = f(c) + m(x - c)$ is a supporting hyperplane of epi(f) at $(c, f(c))$ if and only if $m \in \partial f(c)$.*

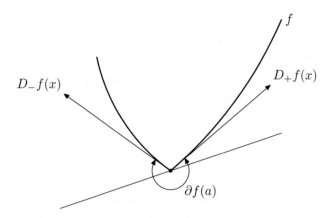

FIGURE 16.5. The subdifferential and a tangent line at a non-smooth point.

PROOF. Suppose that $a < x < c < z < b$ and $m \in \partial f(c)$. From the previous proof,

$$\frac{f(c) - f(x)}{c - x} \leq D_-f(c) \leq m \leq D_+f(c) \leq \frac{f(z) - f(c)}{z - c}.$$

Hence $f(x) \geq f(c) + m(x - c)$ and $f(z) \geq f(c) + m(z - c)$. So epi($f$) lies above the line, and thus $y = f(c) + m(x - c)$ is a support line at $(c, f(c))$.

Conversely, suppose that $m > D_+f(c)$. Since $D_+f(c) = \inf\limits_{z>c} \frac{f(z)-f(c)}{z-c}$, there is some point $z > x$ so that $\frac{f(z)-f(c)}{z-c} < m$. Thus $f(z) < f(c) + m(z - c)$ and the line intersects the interior of epi(f), which is a contradiction. A similar argument deals with $m < D_-f(c)$.   ∎

16.5.13. EXAMPLE. There can be problems at the endpoints when f is defined on a closed interval, even if f is continuous there. For example, consider the function $f(x) = -\sqrt{1 - x^2}$ on $[-1, 1]$. Then on $(-1, 1)$, $f'(x) = \dfrac{x}{\sqrt{1 - x^2}}$ and $f''(x) = (1 - x^2)^{-3/2}$. As f is C^2 and $f'' > 0$ on $(-1, 1)$, this function is convex. It is differentiable at every interior point. However, at the two endpoints, the graph has a vertical tangent. It is for this reason that we did not define the subdifferential at endpoints. If we wished to define them, the only reasonable definition for $\partial f(\pm 1)$ would be the empty set.

Exercises for Section 16.5

A. Show that if f and $-f$ are convex, then f is linear.

B. Show that the function f on $[0, 1]$ given by $f(x) = 0$ for $x < 1$ and $f(1) = 1$ is a discontinuous convex function. Why does this not contradict Theorem 16.5.9?

C. Prove Jensen's inequality (Theorem 16.5.4).

D. Let f be a convex function on $\mathbb{R}$ and let g be any continuous function on $[0, 1]$. Show that $f\left(\int_0^1 g(x)\, dx\right) \leq \int_0^1 f(g(x))\, dx$.
HINT: Approximate the integrals by the Riemann sums.

E. (a) Apply the arithmetic mean–geometric mean inequality to $t_1 = \cdots = t_n = 1 + \frac{1}{n}$ and $t_{n+1} = 1$ and a second application with $t_1 = \cdots = t_{n+1} = \frac{n}{n+1}$ and $t_{n+2} = 1$.
(b) Hence prove that

$$\left(1 + \tfrac{1}{n}\right)^n \leq \left(1 + \tfrac{1}{n+1}\right)^{n+1} \leq \left(1 + \tfrac{1}{n+1}\right)^{n+2} \leq \left(1 + \tfrac{1}{n}\right)^{n+1}.$$

(c) Hence show that $\lim_{n \to \infty} \left(1 + \frac{1}{n}\right)^n$ exists.
HINT: Compare this approach with Proposition 3.3.1.

F. Suppose that $A \subset \mathbb{R}^n$ is a convex set. Show that if $f_1, \ldots, f_k$ are convex functions on A, then the function $f(x) = \max\{f_i(x) : 1 \leq i \leq k\}$ is convex.

G. Suppose that $\{f_i : i \in I\}$ is a collection of convex functions such that for each $x \in \mathbb{R}$, $g(x) = \sup\{f_i(x) : i \in I\}$ is finite. Show that g is convex.

H. Suppose that f is a convex function on $\mathbb{R}$ that is bounded above. Show that f is constant.

I. (a) If f is a C^2 function on (a, b) such that $f''(x) > 0$ for all $x \in (a, b)$, prove that f is strictly convex: $f(\lambda x + (1 - \lambda)y) < \lambda f(x) + (1 - \lambda)f(y)$ for $x \neq y$ and $0 < \lambda < 1$.
(b) Find an example of a strictly convex C^2 function on $\mathbb{R}$ such that $f''(0) = 0$.

J. Let f be a convex function on (a, b). Show that if f attains its minimum at $c \in (a, b)$, then $0 \in \partial f(c)$.

K. **Convex Mean Value Theorem.** Consider a continuous convex function f on $[a, b]$.
Show that there is a point $c \in (a, b)$ so that $\dfrac{f(b) - f(a)}{b - a} \in \partial f(c)$.
HINT: Adapt the proof of the Mean Value Theorem (Theorem 6.2.4). Consider the function $h(x) = f(x) - f(a) - [(f(b) - f(a))/(b - a)](x - a)$ and use the previous exercise.

L. (a) If f is a convex function on $(a, b) \supset [c, d]$, prove that f is Lipschitz on $[c, d]$.
HINT: Theorem 16.5.9.
(b) Show by example that a continuous convex function on $[a, b]$ need not be Lipschitz.

M. Suppose that a function f on (a, b) and satisfies $f\left(\frac{x+y}{2}\right) \leq \frac{1}{2}(f(x) + f(y))$ for all $x, y \in (a, b)$.
(a) Show by induction that $f\left(\frac{x_1 + \cdots + x_{2^k}}{2^k}\right) \leq 2^{-k}\left(f(x_1) + \cdots + f(x_{2^k})\right)$ for $x_i \in (a, b)$.
(b) Prove that if f is continuous, then f is convex.
HINT: Take each x_i to be either x or y.

N. Prove that $f(x) = \log\left(\dfrac{\sinh ax}{\sinh bx}\right)$ is convex if $0 < b \leq a$ as follows:
(a) Show that $f''(x) > 0$ for $x \neq 0$.
HINT: Show that $g(x) = b \sinh ax - a \sinh bx$ is increasing on $[0, \infty)$.
(b) Find the second-order Taylor polynomial of f about 0 from your knowledge of $\sinh x$ and $\log(1 + x)$. Deduce that f is twice differentiable at 0.

O. Suppose that $A \subset \mathbb{R}^n$ is a convex set.

    (a) Show that if f is convex on A and g is convex and increasing on $\mathbb{R}$, then $g \circ f$ is convex on A.

    (b) Give an example of convex functions f, g on $\mathbb{R}$ so that $g \circ f$ is not convex.

16.6. Convex Functions in Higher Dimensions

In this section, we will show that convex functions are continuous on the relative interior of their domain. Then we will investigate certain sets that are defined in terms of convex functions.

This first lemma would be trivial if f were known to be continuous.

16.6.1. LEMMA. *Let f be a convex function defined on a convex set $A \subset \mathbb{R}^n$. If C is a compact convex subset of* $\mathrm{ri}(A)$, *then f is bounded on C.*

PROOF. Since we are only working in A, there is no loss of generality in assuming that $\mathrm{aff}(A) = \mathbb{R}^n$ and so $\mathrm{ri}(A) = \mathrm{int}(A)$.

Fix a point $a \in \mathrm{int}(A)$, and choose $\varepsilon > 0$ so that $\overline{B_\varepsilon(a)} \subset A$. We first show that there is an $r > 0$ so that f is bounded on $B_r(a)$. Let $e_1, \dots, e_n$ be an orthonormal basis for $\mathbb{R}^n$, and define $S = \{a \pm \varepsilon e_i : 1 \le i \le n\}$. Then let $M = \max\{f(s) : s \in S\}$. By Theorem 16.1.7, every point b in $\mathrm{conv}(S)$ has the form $b = \sum_{i=1}^{n} \lambda_i(a - \varepsilon e_i) + \mu_i(a + \varepsilon e_i)$, where $\lambda_i, \mu_i \ge 0$ and $\sum_{i=1}^{n} \lambda_i + \mu_i = 1$. By Jensen's inequality, $f(b) \le M$.

On the other hand, $2a - b = a - (b - a)$ lies in $\mathrm{conv}(S)$. Since $f(2a - b) \le M$,

$$f(a) \le \frac{f(b) + f(2a - b)}{2} \le \frac{f(b) + M}{2}.$$

Thus $f(b) \ge 2f(a) - M$. Therefore, f is bounded on S and

$$\sup_{b \in S} |f(b)| \le \max \left\{ |M|, |2f(a) - M| \right\}.$$

Now $\mathrm{conv}(S)$ contains the ball $B_r(a)$ where $r = \varepsilon/\sqrt{n}$.

To complete the argument, we proceed via proof by contradiction. Suppose that there is a sequence a_k in C such that $|f(a_k)|$ tends to ∞. Since C is compact, this sequence has a convergent subsequence, say $a = \lim_{i \to \infty} a_{k_i}$. Let $r > 0$ and M be chosen as before so that $|f|$ is bounded by M on $B_r(a)$. There is an integer N so that $\|a_{k_i} - a\| < r$ for all $i \ge N$. Hence $|f(a_{k_i})| \le M$ for $i \ge N$, contrary to our hypothesis. Thus f must be bounded on C. $\blacksquare$

16.6.2. THEOREM. *Let f be a convex function defined on a convex set $A \subset \mathbb{R}^n$. If C is a compact convex subset of* $\mathrm{ri}(A)$, *then f is Lipschitz on C. In particular, f is continuous on* $\mathrm{ri}(A)$.

PROOF. The set $X = \text{aff}(A) \setminus \text{ri}(A)$ is closed. Define a function on A by

$$d(a) = \text{dist}(a, X) = \inf\{\|a - x\| : x \in X\}.$$

Evidently, this is a continuous function, and $d(a) = 0$ only if $a \in X$. This is continuous and $d(a) = 0$ only if $a \in X$. Now C is a compact subset of $\text{ri}(A)$, and thus $d(a) > 0$ for all $a \in C$. By the Extreme Value Theorem (Theorem 5.4.4), the minimum value of d on C is attained. So $r = \frac{1}{2}\inf\{d(a) : a \in C\} > 0$.

Let $C_r = \{a \in \text{aff}(A) : \text{dist}(a, C) \leq r\}$. By construction, this is a closed bounded set that is contained in $\text{ri}(A)$. Thus it is compact by the Heine–Borel Theorem (Theorem 4.4.6). The reader should verify that C_r is convex.

By the preceding lemma, $|f|$ is bounded on C_r by some number M. Now fix points $x, y \in C$. Let $\|y - x\| = R$ and $e = (y - x)/R$. Then the interval $[x - re, y + re]$ is contained in C_r. The function $g(t) = f(x + te)$ is a convex function on $[-r, R + r]$. Thus the Secant Lemma (Lemma 16.5.8) applies:

$$\frac{g(0) - g(-r)}{r} \leq \frac{g(R) - g(0)}{R} \leq \frac{g(R + r) - g(R)}{r}.$$

Rewriting and using the bound M yields

$$\frac{-2M}{r} \leq \frac{f(y) - f(x)}{\|y - x\|} \leq \frac{2M}{r}$$

whence $|f(y) - f(x)| \leq L\|y - x\|$ for $L = 2M/r$.

This shows that f is Lipschitz and therefore continuous on each compact subset of $\text{ri}(A)$. In particular, for each $a \in \text{ri}(A)$, there is a ball $\overline{B_r(a)} \cap \text{aff}(A)$ contained in $\text{ri}(A)$, and thus f is continuous on this ball about a. ∎

A basic question in many applications of analysis is how to minimize a function, often subject to various constraints. It is frequently the case in problems from business, economics, and related fields that the functions and the constraints are convex. These functions may be differentiable, in which case the usual multivariable calculus plays a role. However, even in this case, convexity theory provides a useful perspective. When the functions are not all differentiable, this more sophisticated machinery is necessary.

16.6.3. DEFINITION. Consider a convex function f on a convex set $A \subset \mathbb{R}^n$. A point $a_0 \in A$ is a **global minimizer** for f on A if

$$f(a_0) \leq f(a) \quad \text{for all} \quad a \in A.$$

We call $a_0 \in A$ a **local minimizer** for f on A if there is an $r > 0$ so that

$$f(a_0) \leq f(a) \quad \text{for all} \quad a \in B_r(a_0) \cap A.$$

Clearly, a global minimizer is always a local minimizer. The converse is not true for arbitrary functions. For convex functions they are the same, as we prove in the next theorem. Henceforth, we drop the modifier and refer only to **minimizers**.

16.6.4. THEOREM. *Suppose that $A \subset \mathbb{R}^n$ is a convex set, and consider a convex function f on A. If $a_0 \in A$ is a local minimizer for f on A, then it is a global minimizer for f on A. The set of all global minimizers of f on A is a convex set.*

PROOF. Suppose that a_0 minimizes f on $B_r(a_0) \cap A$. Let $a \in A$. As A is convex, the line $[a_0, a]$ is contained in A. Therefore, there is some $\lambda \in (0, 1)$ such that $\lambda a_0 + (1 - \lambda)a \in B_r(a_0) \cap A$. By the convexity of f, we have

$$f(a_0) \leq f(\lambda a_0 + (1 - \lambda)a) \leq \lambda f(a_0) + (1 - \lambda)f(a).$$

Solving this yields $f(a_0) \leq f(a)$. Hence a_0 is a global minimizer.

The last statement follows from Exercise 16.6.A. ∎

A **sublevel set** of a convex function f on A has the form $\{a \in A : f(a) \leq \alpha\}$ for some $\alpha \in \mathbb{R}$. For convenience, we may assume $\alpha = 0$ by the simple device of replacing f by $f - \alpha$. The constraints on the minimization problems which we will consider have this form.

16.6.5. LEMMA. *Let f be a convex function on $\mathbb{R}^n$. Suppose that there is a point $a_0 \in \mathbb{R}^n$ with $f(a_0) < 0$. Then $\operatorname{int}(\{x : f(x) \leq 0\}) = \{x : f(x) < 0\}$ and $\overline{\{x : f(x) < 0\}} = \{x : f(x) \leq 0\}$.*

PROOF. By Theorem 16.6.2, f is continuous. Thus $A := \{x : f(x) \leq 0\}$ is closed and $\{x : f(x) < 0\}$ is an open subset and so is contained in $\operatorname{int} A$.

Let $a \in \operatorname{int}(A)$. Then the line segment $[a_0, a]$ extends beyond a in A to some point, say $b = (1 + \varepsilon)a - \varepsilon a_0$. Thus $a = \lambda a_0 + (1 - \lambda)b$ for $\lambda = \varepsilon/(1 + \varepsilon)$. By the convexity of f,

$$f(a) \leq \lambda f(a_0) + (1 - \lambda)f(b) < 0.$$

Finally, Theorem 16.2.8 shows that $\overline{\operatorname{int} A} = A$. ∎

Exercises for Section 16.6

A. Suppose that $A \subset \mathbb{R}^n$ is convex and f is a convex function on A. Show that the sublevel set $\{a \in A : f(a) \leq \alpha\}$ is convex for each $\alpha \in \mathbb{R}$.

B. Suppose that $A \subset \mathbb{R}^n$ is convex. A real-valued function f on A is called a **strictly convex function** if for all *distinct* points $a, b \in A$ and $0 < \lambda < 1$,

$$f(\lambda a + (1 - \lambda)b) < \lambda f(a) + (1 - \lambda)f(b).$$

If f is strictly convex, show that its set of minimizers is either empty or a singleton.

C. Show that if a convex function f on a convex set $A \subset \mathbb{R}^n$ attains its maximum value at $a \in \operatorname{ri}(A)$, then f is constant.

D. Show that the set $C_r = \{a \in \operatorname{aff}(A) : \operatorname{dist}(a, C) \leq r\}$ used in the proof of Theorem 16.6.2 is convex.

E. A real-valued function f defined on a cone C is called a **positively homogeneous function** if $f(\lambda x) = \lambda f(x)$ for all $\lambda > 0$ and $x \in C$. Show that a positively homogeneous function f is convex if and only if $f(x + y) \leq f(x) + f(y)$ for all $x, y \in C$.

F. A function f on $\mathbb{R}^n$ is a **sublinear function** if $f(\lambda x + \mu y) \leq \lambda f(x) + \mu f(y)$ for all $x, y \in \mathbb{R}^n$ and $\lambda, \mu \in [0, \infty)$.
(a) Prove that sublinear functions are positively homogeneous and convex.
(b) Prove that f is sublinear if and only if epi(f) is a cone.

G. Let f be a sublinear function on $\mathbb{R}^n$.
(a) Prove that $f(x) + f(-x) \geq 0$.
(b) Show that if $f(x) + f(-x) = 0$, then f is linear on the line $\mathbb{R}x$ spanned by x.
(c) If $f(x_j) + f(-x_j) = 0$ for $1 \leq j \leq k$, prove that f is linear on span$\{x_1, \ldots, x_k\}$.
HINT: Consider $f(\pm(x_1 + x_2))$. Use induction.

H. Let f be a bounded convex function on a convex subset $A \times B$ of $\mathbb{R}^m \times \mathbb{R}^n$. Define $g(x) = \inf\{f(x, y) : y \in B\}$. Show that g is convex on A.

I. Suppose that a function $f(x, y)$ defined on $\mathbb{R}^2$ is a convex function of x for each fixed y and is a continuous function of y for each fixed x. Prove that f is continuous.
HINT: Given a point (a, b), find $r, s > 0$ so that $f(x, y)$ is close to $f(a, b)$ on $X = \{(x, b) : x \in [a - r, a + r]\} \cup \{(a \pm r, y) : y \in [b - s, b + s]\}$.

J. Let $f_1, \ldots, f_r$ be convex functions on a convex set $A \subset \mathbb{R}^n$. Suppose that there is no point $a \in A$ satisfying $f_i(a) < 0$ for $1 \leq i \leq k$. Prove that there exist $\lambda_i \geq 0$ so that $\sum_i \lambda_i = 1$ and $\sum_i \lambda_i f_i \geq 0$ on A.
HINT: Separate 0 from $Y = \{y \in \mathbb{R}^r : \text{for some } a \in A, \ y_i > f_i(a) \text{ for } 1 \leq i \leq r\}$.

K. Suppose that f and g are convex functions on $\mathbb{R}^n$. The **infimal convolution** of f and g is the function $h(x) := f \odot g(x) = \inf\{f(y) + g(z) : y + z = x\}$. The value $-\infty$ is allowed.
(a) Suppose that there is an affine function $k(x) = \langle m, x \rangle + b$ for some vector $m \in \mathbb{R}^n$ and $b \in \mathbb{R}$ so that $f(x) \geq k(x)$ and $g(x) \geq k(x)$ for all $x \in \mathbb{R}^n$. Prove that $h(x) > -\infty$ for all x.
(b) Assuming $h(x) > -\infty$ for all x, show that h is convex.

L. Let $h = f \odot g$ denote the infimal convolution of two convex functions f and g on $\mathbb{R}^n$. Prove that ri(epi(h)) = ri(epi(f) + epi(g)) = ri(epi(f)) + ri(epi(g)).

M. Prove that if f is C^2 on a convex set $A \subset \mathbb{R}^n$, then f is convex if and only if the Hessian matrix $\nabla^2 f(x) = \left[\frac{\partial}{\partial x_i} \frac{\partial}{\partial x_j} f(x)\right]$ is positive definite.
HINT: Let $x, y \in A$ and set $g(t) = f(x_t)$, where $x_t = (1 - t)x + ty$. Show that $g'(t) = \langle \nabla f(x_t), y - x \rangle$ and $g''(t) = \langle (\nabla^2 f(x_t))(y - x), y - x \rangle$, where $\nabla f(a)$ is the gradient $\left(\frac{\partial}{\partial x_1} f(a), \ldots, \frac{\partial}{\partial x_n} f(a)\right)$.

N. Prove that $f(x) = -\sqrt[n]{x_1 x_2 \ldots x_n}$ is convex on $\mathbb{R}_+^n$.
HINT: Use the previous exercise.

O. Fix a convex set $A \subset \mathbb{R}^n$. Suppose that f_n are convex functions on A that converge pointwise to a function f.
(a) Prove that f is convex.

(b) If $S \subset \text{ri}(A)$ is compact, show that f_n converges to f uniformly on S.
  HINT: Show that $g = \sup_k f_k$ is bounded above on each ball $B_r(a)$ and use the Lipschitz condition from Theorem 16.6.2.

P.   A function f defined on a convex subset A of $\mathbb{R}^n$ is called a **quasi-convex function** if the sublevel set $\{a \in A : f(a) \leq \alpha\}$ is a convex set for each $\alpha \in \mathbb{R}$.
  (a) Show that f is quasi-convex if and only if
  $$f(\lambda a + (1 - \lambda)b) \leq \max\{f(a), f(b)\} \quad \text{for all } a, b \in A.$$
  (b) By part (a), a convex function is always quasi-convex. Give an example of a function f on $\mathbb{R}$ that is quasi-convex but not convex.
  (c) Suppose that f is differentiable on $\mathbb{R}$. Show that f is quasi-convex if and only if $f(y) \geq f(x) + f'(x)(y - x)$ for all $x, y \in \mathbb{R}$.

16.7. Subdifferentials and Directional Derivatives

We now turn to the notions of derivatives and subdifferentials. For a convex function of one variable, we had two different notions that were useful, the left and right derivatives and the subdifferential set of supporting hyperplanes to $\text{epi}(f)$. Both have natural generalizations to higher dimensions.

16.7.1. DEFINITION. Suppose that A is a convex subset of $\mathbb{R}^n$, $a \in A$, and f is a convex function on A. A **subgradient** of f at a is a vector $s \in \mathbb{R}^n$ such that

$$f(x) \geq f(a) + \langle x - a, s \rangle \quad \text{for all} \quad x \in A.$$

The set of all subgradients of f at a is the **subdifferential** and is denoted by $\partial f(a)$.

These terms are motivated by the corresponding terms for differentiable functions $f : \mathbb{R}^n \to \mathbb{R}$, so we recall them. The **gradient** of f at a, denoted $\nabla f(a)$, is the n-tuple of partial derivatives: $\left(\frac{\partial}{\partial x_1} f(a), \ldots, \frac{\partial}{\partial x_n} f(a)\right)$. The **differential** of f at a is the hyperplane in $\mathbb{R}^{n+1}$ given by those points $(x, t) \in \mathbb{R}^n \times \mathbb{R}$ so that $t = f(a) + \langle x - a, \nabla f(a) \rangle$.

If f is a convex function on A and $a \in A$, then a vector s determines a hyperplane of $\mathbb{R}^n \times \mathbb{R}$ by

$$\begin{aligned} H &= \{(x, t) \in \mathbb{R}^n \times \mathbb{R} : t = f(a) + \langle x - a, s \rangle\} \\ &= \{(x, t) \in \mathbb{R}^n \times \mathbb{R} : \langle (x, t), (-s, 1) \rangle = f(a) - \langle a, s \rangle\} \end{aligned}$$

The condition that s be a subgradient is precisely that the graph of f is contained in the half-space $H^+ = \{(x, t) \in \mathbb{R}^n \times \mathbb{R} : t \geq f(a) + \langle x - a, s \rangle\}$. Clearly, this is equivalent to saying that $\text{epi}(f)$ is contained in H^+. Since H contains the point $(a, f(a))$, we conclude that the subgradients of f at a correspond to the supporting hyperplanes of $\text{epi}(f)$ at $(a, f(a))$. It is important that the vector $(-s, 1)$ determining H has a 1 in the $(n+1)$st coordinate. This ensures that the hyperplane is nonvertical, meaning that it is not of the form $H' \times \mathbb{R}$ for some hyperplane H' of $\mathbb{R}^n$. In the case $n = 1$, this rules out vertical tangents.

An immediate and important fact is that the subdifferential characterizes minimizers for convex functions. This is the analogue of the calculus fact that minima are critical points. But with the hypothesis of convexity, it becomes a sufficient condition as well.

16.7.2. PROPOSITION. *Suppose that f is a convex function on a convex set A. Then $a \in A$ is a minimizer for f if and only if $0 \in \partial f(a)$.*

PROOF. By definition, $0 \in \partial f(a)$ if and only if $f(x) \geq f(a) + \langle x - a, 0 \rangle = f(a)$ for all $x \in A$. ∎

16.7.3. THEOREM. *Suppose that A is a convex subset of $\mathbb{R}^n$, $a \in A$, and f is a convex function on A. Then $\partial f(a)$ is convex and closed. Moreover, if $a \in \mathrm{ri}(A)$, then $\partial f(a)$ is nonempty. If $a \in \mathrm{int}(A)$, then $\partial f(a)$ is compact.*

PROOF. Suppose that $s_1, s_2 \in \partial f(a)$ and $\lambda \in [0, 1]$. Then

$$f(x) \geq f(a) + \langle x - a, s_i \rangle \quad \text{for} \quad i = 1, 2.$$

Hence

$$f(x) \geq \lambda\big(f(a) + \langle x - a, s_1 \rangle\big) + (1 - \lambda)\big(f(a) + \langle x - a, s_2 \rangle\big)$$
$$= f(a) + \langle x - a, \lambda s_1 + (1 - \lambda)s_2 \rangle.$$

Thus $\lambda s_1 + (1 - \lambda)s_2$ is a subdifferential of f at a, and so $\partial f(a)$ is convex.

Since the inner product is continuous, it is easy to check that $\partial f(a)$ is closed. For if (s_i) is a sequence of vectors in $\partial f(a)$ converging to $s \in \mathbb{R}^n$, then

$$f(x) \geq \lim_{i \to \infty} f(a) + \langle x - a, s_i \rangle = f(a) + \langle x - a, s \rangle.$$

Now suppose that $a \in \mathrm{ri}(A)$. The point $(a, f(a))$ lies on the relative boundary of $\mathrm{epi}(f)$ since if $y < f(a)$, the line segment $[(a, f(a)), (a, y)]$ meets $\mathrm{epi}(f)$ in a single point. By the Support Theorem (Theorem 16.3.7), there is a nontrivial supporting hyperplane H for $\mathrm{epi}(f)$ at $(a, f(a))$, say

$$H = \{(x, t) \in \mathbb{R}^n \times \mathbb{R} : \langle (x, t), (h, r) \rangle = \alpha\},$$

where (h, r) is nonzero and $\langle a, h \rangle + r f(a) = \alpha$.

We require a nonvertical hyperplane. Suppose that $r = 0$; so $\langle a, h \rangle = \alpha$. As H is nontrivial, $\mathrm{epi}(f)$ contains a point (b, t) so that

$$\alpha < \langle (b, t), (h, 0) \rangle = \langle b, h \rangle.$$

Because a is in the relative interior, there is an $\varepsilon > 0$ so that $c = a + \varepsilon(a - b)$ belongs to A. But then

$$\alpha \leq \langle (c, f(c)), (h, 0) \rangle = \langle a, h \rangle + \varepsilon \langle a - b, h \rangle = \alpha + \varepsilon(\alpha - \langle b, h \rangle).$$

Thus $\langle b, h \rangle \leq \alpha < \langle b, h \rangle$, which is absurd. Hence $r \neq 0$.

Let $s = -h/r$. Then H is nonvertical and consists of vectors (x, t) such that $rt - \langle x, rs \rangle = \alpha = r f(a) - \langle a, rs \rangle$ or, equivalently, $t = \langle x - a, s \rangle + f(a)$. Since epi$(f)$ is contained in H^+, $f(x) \geq t = f(a) + \langle x - a, s \rangle$ for all $x \in A$. Therefore, $s \in \partial f(a)$, and so the subdifferential is nonempty.

Finally, suppose that $a \in \text{int}(A)$ and choose $r > 0$ so that $\overline{B_r(a)} \subset \text{int}(A)$. By Theorem 16.6.2, f is Lipschitz on the compact set $\overline{B_r(A)}$, say with Lipschitz constant L. If $s \in \partial f(a)$ with $s \neq 0$, then $b = a + rs/\|s\| \in \overline{B_r(A)}$. Moreover,

$$f(b) \geq f(a) + \langle b - a, s \rangle = f(a) + r\|s\|.$$

Hence

$$\|s\| \leq \frac{f(b) - f(a)}{r} \leq \frac{L\|b - a\|}{r} = L.$$

So $\partial f(a)$ is closed and bounded and thus is compact.   ∎

16.7.4. EXAMPLE. Let g be a convex function on $\mathbb{R}$ and set $G(x, y) = g(x) - y$. Consider $\partial G(a, b)$. This consists of all vectors (s, t) such that

$$G(x, y) \geq G(a, b) + \langle (x - a, y - b), (s, t) \rangle \quad \text{for all} \quad (x, y) \in \mathbb{R}^2.$$

Rewrite this as $g(x) \geq g(a) + (x - a)s + (y - b)(t + 1)$ for all $x, y \in \mathbb{R}$. If $t \neq -1$, then fixing x and letting y vary, the right-hand side will take all real values and hence will violate the inequality for certain values of y. Thus $t = -1$. The desired inequality then becomes $g(x) \geq g(a) + (x - a)s$ for all $x \in \mathbb{R}$, which is the condition that $s \in \partial g(a)$. Hence $\partial G(a, b) = \{(s, -1) : s \in \partial g(a)\}$.

Subdifferentials are closely related to directional derivatives. For simplicity, we assume the function is defined on $\mathbb{R}^n$ instead of just a convex subset.

16.7.5. DEFINITION. Suppose that f is a convex function on $\mathbb{R}^n$. For a and $d \in \mathbb{R}^n$, we define the **directional derivative** of f at a in the direction d to be

$$f'(a; d) = \inf_{h \to 0+} \frac{f(a + hd) - f(a)}{h}.$$

It is routine to verify that this function is **positively homogeneous** in the second variable, meaning that $f'(a; td) = tf'(a; d)$ for all $t > 0$.

16.7.6. PROPOSITION. *Suppose that f is a convex function on $\mathbb{R}^n$ and fix a point a in $\mathbb{R}^n$. Then $f'(a; d) = \lim_{h \to 0+} \dfrac{f(a + hd) - f(a)}{h}$ for all $d \in \mathbb{R}^n$, and $f'(a; \cdot)$ is a convex function in the second variable.*

PROOF. Fix d in $\mathbb{R}^n$. Then $g(t) = f(a + td)$ is a convex function on $\mathbb{R}$. Hence by Theorem 16.5.9, $D_+g(0)$ exists. However,

$$D_+g(0) = \lim_{h \to 0+} \frac{g(h) - g(a)}{h} = \inf_{h \to 0+} \frac{g(h) - g(a)}{h}$$

$$= \lim_{h \to 0+} \frac{f(a + hd) - f(a)}{h} = \inf_{h \to 0+} \frac{f(a + hd) - f(a)}{h} = f'(a; d).$$

Fix directions $d, e \in \mathbb{R}^n$ and $\lambda \in [0, 1]$. Let $c = \lambda d + (1 - \lambda)e$. A short calculation using the convexity of f gives

$$\frac{f(a + hc) - f(a)}{h} = \frac{f\big(\lambda(a + hd) + (1 - \lambda)(a + he)\big) - f(a)}{h}$$

$$\leq \lambda \frac{f(a + hd) - f(a)}{h} + (1 - \lambda)\frac{f(a + he) - f(a)}{h}.$$

Taking the limit as h decreases to 0, we obtain

$$f'(a; c) \leq \lambda f'(a; d) + (1 - \lambda)f'(a; e).$$

So $f'(a; \cdot)$ is a convex function. ∎

We are now able to obtain a characterization of the subgradient in terms of the directional derivatives, generalizing Proposition 16.5.12.

16.7.7. THEOREM. *Suppose that f is a convex function on $\mathbb{R}^n$, and $a, s \in \mathbb{R}^n$. Then $s \in \partial f(a)$ if and only if $\langle d, s \rangle \leq f'(a; d)$ for all $d \in \mathbb{R}^n$.*

PROOF. Suppose that $s \in \partial f(a)$. Then for any direction $d \in \mathbb{R}^n$ and $h > 0$, we have $f(a + hd) \geq f(a) + \langle hd, s \rangle$. Rearranging, we have

$$f'(a; d) = \lim_{h \to 0^+} \frac{f(a + hd) - f(a)}{h} \geq \langle d, s \rangle.$$

Conversely, suppose that $s \in \mathbb{R}^n$ satisfies

$$\langle d, s \rangle \leq f'(x; d) = \inf_{h \to 0+} \frac{f(a + hd) - f(a)}{h}$$

for all $d \in \mathbb{R}^n$. Then rearranging yields $f(a) + \langle hd, s \rangle \leq f(a + hd)$. Therefore, s belongs to $\partial f(a)$. ∎

Conversely, we may express the directional derivatives in terms of the subdifferential. This requires a separation theorem. If C is a nonempty compact convex set, the **support function** of C is defined as

$$\sigma_C(x) = \sup\{\langle x, c \rangle : c \in C\}.$$

Note that σ_C is a convex function on $\mathbb{R}^n$ that is positively homogeneous.

It is a consequence of the Separation Theorem (Theorem 16.3.3) that different compact convex sets have different support functions. For if D is another convex set that contains a point $d \notin C$, then there is a vector x so that

$$\sigma_C(x) = \sup\{\langle x, c \rangle : c \in C\} \leq \alpha < \langle x, d \rangle \leq \sigma_D(x).$$

The next lemma shows conversely how to recover the convex set from a support function.

16.7.8. SUPPORT FUNCTION LEMMA.
Suppose that g is a convex, positively homogeneous function on $\mathbb{R}^n$. Let

$$C(g) = \{c \in \mathbb{R}^n : \langle x, c \rangle \le g(x) \quad \text{for all} \quad x \in \mathbb{R}^n\}.$$

Then $C(g)$ is compact and convex, and $\sigma_{C(g)} = g$. Thus if A is a compact convex set, then $C(\sigma_A) = A$.

PROOF. Let $M = \sup\{g(x) : \|x\| = 1\}$, which is finite since g is continuous and the unit sphere is compact. If $c \in C(g)$, then

$$\|c\| = \langle c/\|c\|, c \rangle \le g(c/\|c\|) \le M.$$

So $C(g)$ is bounded. Since the inner product is continuous, $\{c : \langle x, c \rangle \le g(x)\}$ is closed for each $x \in \mathbb{R}^n$. The intersection of closed sets is closed, and hence C is closed. Thus, $C(g)$ is compact by the Heine–Borel Theorem. If $c, d \in C(g)$ and $x \in \mathbb{R}^n$, then

$$\langle x, \lambda c + (1 - \lambda)d \rangle \le \lambda g(x) + (1 - \lambda)g(x) = g(x).$$

So $C(g)$ is convex.

The inequality $\sigma_{C(g)} \le g$ is immediate from the definition.

Fix a vector a in $\mathbb{R}^n$. By Theorem 16.7.3, there is a vector s in $\partial g(a)$. Thus $g(x) \ge g(a) + \langle x - a, s \rangle$ for $x \in \mathbb{R}^n$. Rewriting this yields

$$g(x) - \langle x, s \rangle \ge g(a) - \langle a, s \rangle =: \alpha.$$

Take $x = ha$ for $h > 0$ and use the homogeneity of g to compute

$$\alpha \le g(ha) - \langle ha, s \rangle = h(g(a) - \langle a, s \rangle) = h\alpha.$$

Hence $\alpha = 0$, $g(a) = \langle a, s \rangle$ and s belongs to $C(g)$. So $\sigma_{C(g)}(a) \ge \langle a, s \rangle = g(a)$. Therefore, $\sigma_{C(g)} = g$.

If A is compact and convex, setting $g = \sigma_A$ yields $\sigma_A = \sigma_{C(\sigma_A)}$. By the remarks preceding the proof, this implies that $C(\sigma_A) = A$. ∎

16.7.9. COROLLARY.
Suppose that f is a convex function on $\mathbb{R}^n$. Fix $a \in \mathbb{R}^n$. Then $f'(a; \cdot)$ is the support function of $\partial f(a)$, namely

$$f'(a; d) = \sup\{\langle d, s \rangle : s \in \partial f(a)\} \quad \text{for} \quad d \in \mathbb{R}^n.$$

PROOF. Theorem 16.7.7 shows that $\partial f(a) = C(g)$, where $g(x) = f'(a; x)$. Thus by the Support Function Lemma (Lemma 16.7.8),

$$f'(a; d) = \sigma_{\partial f(a)}(d) = \sup\{\langle d, s \rangle : s \in \partial f(a)\}. \quad ∎$$

16.7.10. EXAMPLE. Consider $f(x) = \|x\|$, the Euclidean norm, on $\mathbb{R}^n$. First look at $a = 0$. A vector s is in $\partial f(0)$ if and only if $\langle x, s \rangle \leq \|x\|$ for all $x \in \mathbb{R}^n$. Thus $\partial f(0) = \overline{B_1(0)}$. Indeed, if $\|s\| \leq 1$, then the Schwarz inequality shows that

$$\langle x, s \rangle \leq \|x\| \, \|s\| \leq \|x\|.$$

If $\|s\| > 1$, take $x = s$ and notice that $\langle x, s \rangle = \|x\|^2 > \|x\|$.

Now we compute the directional derivatives at 0,

$$f'(0; d) = \lim_{h \to 0^+} \frac{\|hd\|}{h} = \|d\| = f(d).$$

In this case, since $f'(0; \cdot) = f$, Theorem 16.7.7 is redundant and provides no new information. Let us verify Corollary 16.7.9 directly:

$$\sup \left\{ \langle d, s \rangle : s \in \overline{B_1(0)} \right\} = \|d\| = f'(0; d)$$

because $\langle d, s \rangle \leq \|d\| \, \|s\| = \|d\|$, while the choice $s = d/\|d\|$ attains this bound.

Now look at $a \neq 0$. We first compute the directional derivatives.

$$f'(a; d) = \lim_{h \to 0^+} \frac{\|a + hd\| - \|a\|}{h} = \lim_{h \to 0^+} \frac{\|a + hd\|^2 - \|a\|^2}{h \big(\|a + hd\| + \|a\| \big)}$$

$$= \lim_{h \to 0^+} \frac{2h \langle d, a \rangle + h^2 \|d\|^2}{h \big(\|a + hd\| + \|a\| \big)} = \frac{2 \langle d, a \rangle}{2 \|a\|} = \left\langle d, \frac{a}{\|a\|} \right\rangle$$

This time Theorem 16.7.7 is very useful. It says that $s \in \partial f(a)$ if and only if $\langle d, s \rangle \leq \langle d, \frac{a}{\|a\|} \rangle$ for all $d \in \mathbb{R}^n$. Equivalently, $\langle d, s - \frac{a}{\|a\|} \rangle \leq 0$ for all d in $\mathbb{R}^n$. But the left-hand side takes all real values except when $s = \frac{a}{\|a\|}$. Therefore, $\partial f(a) = \left\{ \frac{a}{\|a\|} \right\}$.

Observe that $f'(a; d)$ is a linear function of d when $a \neq 0$. Thus f has continuous partial derivatives at a and $\nabla f(a) = \frac{a}{\|a\|}$. So $\partial f(a) = \{\nabla f(a)\}$.

Let's look more generally at the situation that occurs when a convex function f is differentiable at a. Unlike the case of arbitrary functions, the existence of partial derivatives together with convexity implies differentiability.

16.7.11. THEOREM. *Suppose that f is a convex function on an open convex set $A \subset \mathbb{R}^n$. Then the following are equivalent for $a \in A$:*

(1) $\dfrac{\partial f}{\partial x_i}(a)$ *are defined for $1 \leq i \leq n$.*

(2) *f is differentiable at a.*

(3) *$\partial f(a)$ is a singleton.*

In this case, $\partial f(a) = \{\nabla f(a)\}$.

PROOF. Suppose that (1) holds and consider the convex function

$$g(x) = f(a + x) - f(a) - \langle x, \nabla f(a) \rangle.$$

To show that f is differentiable, it suffices to show that $f(a) + \langle x, \nabla f(a) \rangle$ approximates $f(x)$ to first order near a or, equivalently, that $g(x)/\|x\|$ tends to 0 as $\|x\|$ tends to 0. We use $e_1, \ldots, e_n$ for the standard basis of $\mathbb{R}^n$.

The fact that the n partial derivatives exist means that for $1 \le i \le n$,

$$0 = \lim_{h \to 0} \frac{f(a + he_i) - f(a) - h \frac{\partial f}{\partial x_i}(a)}{h}$$

$$= \lim_{h \to 0} \frac{f(a + he_i) - f(a) - \langle he_i, \nabla f(a) \rangle}{h} = \lim_{h \to 0} \frac{g(he_i)}{h}.$$

Fix an $\varepsilon > 0$ and choose r so small that $|g(he_i)| < \varepsilon |h|/n$ for $|h| \le r$ and $1 \le i \le n$.

Take $x = (x_1, \ldots, x_n)$ with $\|x\| \le r/n$. Then

$$g(x) = g\Big(\frac{1}{n} \sum_{i=1}^{n} nx_i e_i\Big) \le \frac{1}{n} \sum_{i=1}^{n} g(nx_i e_i) \le \frac{1}{n} \sum_{i=1}^{n} \varepsilon |x_i| \le \varepsilon \|x\|.$$

Now

$$0 = g(0) = g\Big(\frac{x + (-x)}{2}\Big) \le \tfrac{1}{2} g(x) + \tfrac{1}{2} g(-x).$$

Thus $g(x) \ge -g(-x) \ge -\varepsilon \|x\|$. Therefore, $|g(x)| \le \varepsilon \|x\|$ on $B_r(0)$. As $\varepsilon > 0$ is arbitrary, this proves that f is differentiable at a.

Assuming (2) that f is differentiable at a, the function $f(a) + \langle x, \nabla f(a) \rangle$ approximates $f(x)$ to first order near a with error $g(x)$ satisfying $\lim_{\|x\| \to 0} \frac{g(x)}{\|x\|} = 0$.

Compute

$$f'(a; d) = \lim_{h \to 0^+} \frac{f(a + hd) - f(a)}{h}$$

$$= \lim_{h \to 0^+} \frac{\langle hd, \nabla f(a) \rangle + g(he_i)}{h}$$

$$= \langle d, \nabla f(a) \rangle + \lim_{h \to 0} \frac{g(he_i)}{h} = \langle d, \nabla f(a) \rangle.$$

Theorem 16.7.7 now says that $s \in \partial f(a)$ if and only if $\langle d, s \rangle \le \langle d, \nabla f(a) \rangle$ for all $d \in \mathbb{R}^n$. Equivalently, $\langle d, s - \nabla f(a) \rangle \le 0$ for all $d \in \mathbb{R}^n$. But the left-hand side takes all real values except when $s = \nabla f(a)$. Therefore, $\partial f(a) = \{\nabla f(a)\}$.

Finally, suppose that (3) holds and $\partial f(a) = \{s\}$. Then by Corollary 16.7.9, $f'(a; d) = \langle d, s \rangle$. In particular, $f'(a; -e_i) = -f'(a; e_i)$ and thus

$$\langle e_i, s \rangle = \lim_{h \to 0^+} \frac{f(a + he_i) - f(a)}{h}$$

$$= \lim_{h \to 0^-} \frac{f(a + he_i) - f(a)}{h} = \frac{\partial f}{\partial x_i}(a).$$

Hence the partial derivatives of f are defined at a, which proves (1). ∎

16.7.12. EXAMPLE. Let Q be a positive definite $n \times n$ matrix, and let $q \in \mathbb{R}^n$ and $c \in \mathbb{R}$. Minimize the quadratic function $f(x) = \langle x, Qx \rangle + \langle x, q \rangle + c$.

We compute the differential

$$
\begin{aligned}
f'(x; d) &= \lim_{h \to 0^+} \frac{f(x + hd) - f(x)}{h} \\
&= \lim_{h \to 0^+} \frac{\langle x + hd, Q(x + hd) \rangle + \langle x + hd, q \rangle - \langle x, Qx \rangle - \langle x, q \rangle}{h} \\
&= \lim_{h \to 0^+} \frac{1}{h} \big(\langle hd, Qx \rangle + \langle x, Qhd \rangle + \langle hd, Qhd \rangle + \langle hd, q \rangle \big) \\
&= \lim_{h \to 0^+} \langle d, Qx \rangle + \langle Qx, d \rangle + h \langle d, Qd \rangle + \langle d, q \rangle = \langle d, q + 2Qx \rangle.
\end{aligned}
$$

Thus $\nabla f(x) = q + 2Qx$. We solve $\nabla f(x) = 0$ to obtain the unique minimizer $x = -\frac{1}{2} Q^{-1} q$ with minimum value $f(-\frac{1}{2} Q^{-1} q) = -\frac{1}{4} \langle q, Q^{-1} q \rangle + c$.

This problem can also be solved using linear algebra. The spectral theorem for Hermitian matrices states that there is an orthonormal basis $v_1, \dots, v_n$ of eigenvectors that diagonalizes Q. Thus there are positive eigenvalues $d_1, \dots, d_n$ so that $Qv_i = d_i v_i$. Write q in this basis as $q = q_1 v_1 + \cdots + q_n v_n$. We also write a generic vector x as $x = x_1 v_1 + \cdots + x_n v_n$. Then

$$
f(x) = c + \sum_{i=1}^{n} d_i x_i^2 + q_i x_i = c' + \sum_{i=1}^{n} d_i (x_i - a_i)^2,
$$

where $a_i = -q_i/(2d_i)$ and $c' = c - \sum_{i=1}^{n} d_i a_i^2$.

Now observe by inspection that the minimum is achieved when $x_i = a_i$. To complete the circle, note that

$$
x = \sum_{i=1}^{n} a_i v_i = -\frac{1}{2} \sum_{i=1}^{n} \frac{q_i}{d_i} v_i = -\frac{1}{2} Q^{-1} q
$$

and the minimum value is

$$
f(x) = c' = c - \sum_{i=1}^{n} d_i \left(\frac{q_i}{2d_i} \right)^2 = c - \sum_{i=1}^{n} \frac{q_i^2}{4d_i} = c - \frac{1}{4} \langle q, Q^{-1} q \rangle.
$$

We finish this section with two of the calculus rules for subgradients. The proofs are very similar, so we prove the first and leave the second as an exercise.

16.7.13. THEOREM. *Suppose that $f_1, \dots, f_k$ are convex functions on $\mathbb{R}^n$ and set $f(x) = \max\{f_1(x), \dots, f_k(x)\}$. For $a \in \mathbb{R}^n$, set $J(a) = \{j : f_j(a) = f(a)\}$. Then $\partial f(a) = \operatorname{conv}\{\partial f_j(a) : j \in J(a)\}$.*

PROOF. By Theorem 16.6.2, each f_j is continuous. Thus there is an $\varepsilon > 0$ so that $f_j(x) < f(x)$ for all $\|x - a\| < \varepsilon$ and all $j \notin J(a)$. So for $x \in B_\varepsilon(a)$, $f(x) = \max\{f_j(x) : j \in J(a)\}$. Fix $d \in \mathbb{R}^n$ and note that $f(a + hd)$ depends only on $f_j(a + hd)$ for $j \in J(a)$ when $|h| < \varepsilon/d$. Thus using Corollary 16.7.9,

$$f'(a; d) = \lim_{h \to 0^+} \frac{f(a + hd) - f(a)}{h} = \lim_{h \to 0^+} \max_{j \in J(a)} \frac{f_j(a + hd) - f_j(a)}{h}$$

$$= \max_{j \in J(a)} f'_j(a; d) = \max_{j \in J(a)} \sup\{\langle d, s_j \rangle : s_j \in \partial f_j(a)\}$$

$$= \sup\{\langle d, s \rangle : s \in \cup_{j \in J(a)} \partial f_j(a)\}$$

$$= \sup\{\langle d, s \rangle : s \in \text{conv}\{\partial f_j(a) : j \in J(a)\}\}.$$

This shows that $f'(a; \cdot)$ is the support function of the compact convex set $\text{conv}\{\partial f_j(a) : j \in J(a)\}$. However, by Corollary 16.7.9, $f'(a; \cdot)$ is the support function of $\partial f(a)$. So by the Support Function Lemma (Lemma 16.7.8), these two sets are equal. ∎

16.7.14. THEOREM. *Suppose that f_1 and f_2 are convex functions on $\mathbb{R}^n$ and λ_1 and λ_2 are positive real numbers. Then for $a \in \mathbb{R}^n$,*

$$\partial(\lambda_1 f_1 + \lambda_2 f_2)(a) = \lambda_1 \partial f_1(a) + \lambda_2 \partial f_2(a).$$

Exercises for Section 16.7

A. Show that $f'(a; d)$ is sublinear in d: $f'(a; \lambda d + \mu e) \leq \lambda f'(a; d) + \mu f'(a; e)$ for all $d, e \in \mathbb{R}^n$ and $\lambda, \mu \in [0, \infty)$.

B. Suppose that f is a convex function on a convex subset $A \subset \mathbb{R}^n$. If $a, b \in A$ and $s \in \partial f(a), t \in \partial f(b)$, show that $\langle s - t, a - b \rangle \geq 0$.

C. Give an example of a convex set $A \subset \mathbb{R}^n$, a convex function f on A and a point $a \in A$ such that $\partial f(a)$ is empty.

D. Compute the subdifferential for the norm $\|x\|_\infty = \max\{|x_1|, \ldots, |x_n|\}$.

E. Let g be a convex function on $\mathbb{R}$, and define a function on $\mathbb{R}^2$ by $G(x, y) = g(x) - y$.
   (a) Compute the directional derivatives of the function G at a point (a, b).
   (b) Use Theorem 16.7.7 to evaluate $\partial G(a, b)$. Check you answer against Example 16.7.4.

F. Prove Theorem 16.7.14—compute $\partial f(a)$ when $f = \lambda_1 f_1 + \lambda_2 f_2$ as follows:
   (a) Show that $f'(a; d) = \lambda_1 f'_1(a; d) + \lambda_2 f'_2(a; d)$.
   (b) Apply Corollary 16.7.9 to both sides.
   (c) Show that $\partial(f)(a) = \lambda_1 \partial f_1(a) + \lambda_2 \partial f_2(a)$.

G. Given convex functions f on $\mathbb{R}^n$ and g on $\mathbb{R}^m$, define h on $\mathbb{R}^n \times \mathbb{R}^m$ by $h(x, y) = f(x) + g(y)$. Show that $\partial h(x, y) = \partial f(x) \times \partial g(y)$.
   HINT: $h = F + G$, where $F(x, y) = f(x)$ and $G(x, y) = g(y)$.

H. Suppose S is any nonempty subset of $\mathbb{R}^n$. We may still define the **support function** of S by $\sigma_S = \sup\{\langle s, x \rangle : s \in S\}$, but it may sometimes take the value $+\infty$.
   (a) If $A = \overline{\text{conv}}(S)$, show that $\sigma_S = \sigma_A$.
   (b) Show that σ_S is convex.
   (c) Show that σ_S is finite valued everywhere if and only if S is bounded.

I. Consider a convex function f on a convex set $A \subset \mathbb{R}^n$ and $a \neq b \in \mathrm{ri}(A)$.
  (a) Define g on $[0, 1]$ by $g(\lambda) = f(\lambda a + (1 - \lambda)b)$. If $x_\lambda = \lambda a + (1 - \lambda)b$, then show that $\partial g(\lambda) = \{\langle m, b - a \rangle : m \in \partial f(x_\lambda)\}$.
  (b) Use the Convex Mean Value Theorem (Exercise 16.5.K) and part (a) to show that there are $\lambda \in (0, 1)$ and $s \in \partial f(x_\lambda)$ so that $f(b) - f(a) = \langle s, b - a \rangle$.

J. Define a **local subgradient** of a convex function f on a convex set $A \subset \mathbb{R}^n$ to be a vector s so that $f(x) \geq f(a) + \langle x - a, s \rangle$ for all $x \in A \cap B_r(a)$ for some $r > 0$. Show that if s is a local subgradient, then it is a subgradient in the usual sense.

K. (a) If f_k are convex functions on a convex subset $A \subset \mathbb{R}^n$ converging pointwise to f and $s_k \in \partial f_k(x_0)$ converge to s, prove that $s \in \partial f(x_0)$.
  (b) Show that in general, it is not true that every $s \in \partial f$ is obtained as such a limit by considering $f_k(x) = \sqrt{x^2 + 1/k}$ on $\mathbb{R}$.

L. Let $h = f \odot g$ be the infimal convolution of two convex functions f and g on $\mathbb{R}^n$.
  (a) Suppose that there are points $x_0 = x_1 + x_2$ such that $h(x_0) = f(x_1) + g(x_2)$. Prove that $\partial h(x_0) = \partial f(x_1) \cap \partial g(x_2)$.
    HINT: $s \in \partial h(x_0) \Leftrightarrow f(y) + g(z) \geq f(x_1) + g(x_2) + s(y + z - x_0)$. Take $y = x_1$ or $z = x_2$.
  (b) If (a) holds and g is differentiable, show that h is differentiable at x_0.
    HINT: Theorem 16.7.11

M. Moreau–Yosida. Let f be a convex function on an open convex subset A of $\mathbb{R}^n$. For $k \geq 1$, define $f_k(x) = f \odot (k\|\cdot\|^2)(x) = \inf_y f(y) + k\|x - y\|^2$.
  (a) Show that $f_k \leq f_{k+1} \leq f$.
  (b) Show that x_0 is a minimizer for f if and only if it is a minimizer for every f_k.
    HINT: If $f(x_0) = f(x_1) + \varepsilon$, find $r > 0$ so that $f(x) \geq f(x_1) + \varepsilon/2$ on $B_r(x_0)$.
  (c) Prove that f_k converges to f.
    HINT: If L is a Lipschitz constant on some ball about x_0, estimate f_k inside and outside $B_{L/\sqrt{k}}(x_0)$ separately.
  (d) Prove that f_k is differentiable for all $k \geq 1$.
    HINT: Use the previous exercise.

16.8. Tangent and Normal Cones

In this section, we study two special cones associated to a convex subset of $\mathbb{R}^n$. We develop only a small portion of their theory, since our purpose is to set the stage for our minimization results, and our results are all related to that specific goal.

16.8.1. DEFINITION. Consider a convex set $A \subset \mathbb{R}^n$ and $a \in A$. Define the cone $C_A(a) = \mathbb{R}_+(A - a)$ generated by $A - a$. The **tangent cone** to A at a is the closed cone $T_A(a) = \overline{C_A(a)} = \overline{\mathbb{R}_+(A - a)}$. The **normal cone** to A at a is defined to be $N_A(a) = \{s \in \mathbb{R}^n : \langle s, x - a \rangle \leq 0 \text{ for all } x \in A\}$.

It is routine to verify that $T_A(a)$ and $N_A(a)$ are closed cones. The cone $C_A(a)$ is only used as a tool for working with $T_A(a)$. Notice that $\langle s, x - a \rangle \leq 0$ implies that $\langle s, \lambda(x - a) \rangle \le 0$ for all $\lambda > 0$. Thus $s \in N_A(a)$ satisfies $\langle s, d \rangle \le 0$ for all $d \in C_A(a)$. Since the inner product is continuous, the inequality also holds for $d \in T_A(a)$.

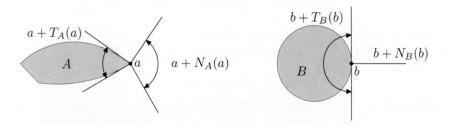

FIGURE 16.6. Two examples of tangent and normal cones.

16.8.2. EXAMPLE. As a motivating example, let

$$A = \{(x, y) : x \ge 0, \ y > 0, \ x^2 + y^2 < 1\} \cup \{(0, 0)\} \subset \mathbb{R}^2.$$

Then $C_A((0,0)) = \{(x, y) : x \ge 0, \ y > 0\} \cup \{(0,0)\}$, so the tangent cone is $T_A((0,0)) = \{(x, y) : x, y \ge 0\}$. At the boundary points $(0, y)$ for $y \in (0, 1)$, $C_A((0, y)) = T_A((0, y)) = \{(x, y) : x \ge 0\}$. Finally, at points $(x, y) \in \text{int } A$, $C_A((x, y)) = T_A((x, y)) = \mathbb{R}^2$.

The normal cone gets smaller as the tangent cone increases in size. Here we have $N_A((0,0)) = \{(a, b) : a, b \le 0\}$, $N_A((0, y)) = \{(a, 0) : a \le 0\}$ for $y \in (0, 1)$ and $N_A((x, y)) = \{0\}$ for $(x, y) \in \text{int } A$.

You may find it useful to draw pictures like Figure 16.6 for various points in A.

Let us formalize the procedure that produced the normal cone.

16.8.3. DEFINITION. Given a nonempty subset A of $\mathbb{R}^n$, the **polar cone** of A, denoted A°, is $A^\circ = \{s \in \mathbb{R}^n : \langle a, s \rangle \le 0 \text{ for all } a \in A\}$.

It is easy to verify that A° is a closed cone. It is evident from the previous definition that $N_A(a) = T_A(a)^\circ$.

We need the following consequence of the Separation Theorem.

16.8.4. BIPOLAR THEOREM.
If C is a closed cone in $\mathbb{R}^n$, then $C^{\circ\circ} = C$.

PROOF. From the definition, $C^{\circ\circ} = \{d : \langle d, s \rangle \le 0 \text{ for all } s \in C^\circ\}$. This clearly includes C. Conversely, suppose that $x \notin C$. Applying the Separation Theorem (Theorem 16.3.3), there is a vector s and scalar α so that $\langle c, s \rangle \le \alpha$ for all $c \in C$ and $\langle x, s \rangle > \alpha$. Since $C = \mathbb{R}_+ C$, the set of values $\{\langle c, s \rangle : c \in C\}$ is a cone in $\mathbb{R}$. Since C is bounded above by α, it follows that $C = \mathbb{R}_-$ or $\{0\}$. Hence

$\langle c, s \rangle \leq 0 \leq \alpha < \langle x, s \rangle$ for $c \in C$. Consequently, s belongs to C°. Therefore, x is not in $C^{\circ\circ}$. So $C^{\circ\circ} = C$. ∎

Since $N_A(a) = T_A(a)^\circ$, we obtain the following:

16.8.5. COROLLARY. *Let A be a convex subset of $\mathbb{R}^n$ and $a \in A$. Then the normal and tangent cones at a are polar to each other, namely $N_A(a) = T_A(a)^\circ$ and $T_A(a) = N_A(a)^\circ$.*

16.8.6. EXAMPLE. Let $s_1, \ldots, s_m$ be vectors in $\mathbb{R}^n$. Consider the convex polyhedron given as $A = \{x \in \mathbb{R}^n : \langle x, s_j \rangle \leq r_j, \ 1 \leq j \leq m\}$. What are the tangent and normal cones at $a \in A$?

Fix $a \in A$. Set $J(a) = \{j : \langle a, s_j \rangle = r_j\}$. For example, this set is empty if $a \in \text{ri}(A)$. Then

$$C_A(a) = \{d = t(x - a) : \langle x, s_j \rangle \leq r_j \text{ and some } t \geq 0\}$$
$$= \{d : \langle \tfrac{d}{t} + a, s_j \rangle \leq r_j \text{ and some } t \geq 0\}$$
$$= \{d : \langle d, s_j \rangle \leq t(r_j - \langle a, s_j \rangle) \text{ and some } t \geq 0\}.$$

If $r_j - \langle a, s \rangle > 0$, this is no constraint; so $C_A(a)$

$$= \{d : \langle d, s_j \rangle \leq 0, \ j \in J(a)\}.$$

This is closed, and thus $T_A(a) = C_A(a)$.

Note that $\{d : \langle d, s \rangle \leq 0\}^\circ = \mathbb{R}_+ s$. Indeed, $(\mathbb{R}_+ s)^\circ = \{d : \langle d, s \rangle \leq 0\}$. So the result follows from the Bipolar Theorem. Now Exercise 16.8.J tells us that

$$N_A(x) = \left(\bigcap_{j \in J(a)} \{d : \langle d, s \rangle \leq 0\} \right)^\circ$$
$$= \overline{\sum_{j \in J(a)} \mathbb{R}_+ s_j} = \text{cone}\{s_j : j \in J(a)\}.$$

Indeed, without the exercise, we can see that

$$\text{cone}\{s_j : j \in J(a)\}^\circ = \{d : \langle d, s_j \rangle \leq 0, \ j \in J(a)\} = T_A(a).$$

Therefore, by Corollary 16.8.5 and the Bipolar Theorem,

$$N_A(x) = T_A(a)^\circ = \text{cone}\{s_j : j \in J(a)\}^{\circ\circ} = \text{cone}\{s_j : j \in J(a)\}.$$

We need to compute the tangent and normal cones for a convex set A given as the sublevel set of a convex function.

16.8.7. LEMMA. *Let A be a compact convex subset of $\mathbb{R}^n$ that does not contain the origin 0. Then the cone $\mathbb{R}_+ A$ is closed.*

PROOF. Suppose that $a_k \in A$ and $\lambda_k \geq 0$ and that $c = \lim\limits_{k\to\infty} \lambda_k a_k$ is a point in $\overline{\mathbb{R}_+ A}$. From the compactness of A, we deduce that there is a subsequence (k_i) so that $a_0 = \lim\limits_{i\to\infty} a_{k_i}$ exists in A. Because $\|a_0\| \neq 0$,

$$\lambda_0 := \frac{\|c\|}{\|a_0\|} = \lim_{i\to\infty} \frac{\|\lambda_{k_i} a_{k_i}\|}{\|a_{k_i}\|} = \lim_{i\to\infty} \lambda_{k_i}.$$

Therefore, $c = \lim\limits_{i\to\infty} \lambda_{k_i} a_{k_i} = \lambda_0 a_0$ belongs to $\mathbb{R}_+ A$. ∎

16.8.8. THEOREM. *Let g be a convex function on $\mathbb{R}^n$, and let A be the convex sublevel set $\{x : g(x) \leq 0\}$. Assume that there is a point x with $g(x) < 0$. If $a \in \mathbb{R}^n$ satisfies $g(a) = 0$, then*

$$T_A(a) = \{d \in \mathbb{R}^n : g'(a; d) \leq 0\} \quad \text{and} \quad N_A(a) = \mathbb{R}_+ \partial g(a).$$

PROOF. Let $C = \{d \in \mathbb{R}^n : g'(a; d) \leq 0\}$, which is a closed cone. Suppose that $d \in A - a$. Then $[a, a + d]$ is contained in A and thus $g(a + hd) - g(a) \leq 0$ for $0 < h \leq 1$. So $g'(a; d) \leq 0$ and hence $d \in C$. As C is a closed cone, it follows that C contains $\overline{\mathbb{R}_+(A - a)} = T_A(a)$.

Choose $x \in A$ with $g(x) < 0$, and set $d = x - a$. Then

$$g'(a; d) = \inf_{h > 0} \frac{g(a + hd) - g(a)}{h}$$
$$\leq \frac{g(a + d) - g(a)}{1} < 0.$$

Hence by Lemma 16.6.5, $\operatorname{int} C = \{d : g'(a; d) < 0\}$ is nonempty and moreover $C = \overline{\operatorname{int} C}$.

Let $d \in \operatorname{int} C$. Since $g'(a; d) < 0$, there is some $h > 0$ so that $g(a + hd) < 0$. Consequently, $a + hd$ belongs to A and $d \in \mathbb{R}_+(A - a) \subset T_A(a)$. So $\operatorname{int} C$ is a subset of $T_A(a)$. Thus $C = \overline{\operatorname{int} C}$ is contained in $T_A(a)$, and the two cones agree.

By Corollary 16.7.9, $g'(a; d) = \sup\{\langle d, s \rangle : s \in \partial g(a)\}$. Thus $d \in T_A(a)$ if and only if $\langle d, s \rangle \leq 0$ for all $s \in \partial g(a)$, which by definition is the polar cone of $\partial g(a)$. Hence by the Bipolar Theorem (Theorem 16.8.4), the polar $N_A(a)$ of $T_A(a)$ is the closed cone generated by $\partial g(a)$. Note that $0 \notin \partial g(a)$ because a is not a minimizer of g (Proposition 16.7.2). Therefore, by Lemma 16.8.7, $N_A(a)$ is just $\mathbb{R}_+ \partial g(a)$. ∎

Exercises for Section 16.8

A. Show that $T_A(a)$ and $N_A(a)$ are closed convex cones.

B. For a point $v \in \mathbb{R}^n$, show that $v \in N_A(a)$ if and only if $P_A(a + v) = a$.

C. If C is a closed cone, show that $T_C(0) = C$ and $N_C(0) = C^\circ$.

D. Suppose that $A \subset \mathbb{R}^n$ is convex and f is a convex function on A. Prove that $x_0 \in A$ is a minimizer for f in A if and only if $f'(x_0, d) \geq 0$ for all $d \in T_A(x_0)$.

E. Let $f(x,y) = (x - y^2)(x - 2y^2)$. Show that $(0,0)$ is not a minimizer of f on $\mathbb{R}^2$ yet $f'((0,0), d) \geq 0$ for all $d \in \mathbb{R}^2$. Why does this not contradict the previous exercise?

F. If $C_1 \subset C_2$, show that $C_2{}^\circ \subset C_1{}^\circ$.

G. If A is a subspace of $\mathbb{R}^n$, show that A° is the orthogonal complement of A.

H. Suppose that $a_1, \ldots, a_r$ are vectors in $\mathbb{R}^n$. Compute $\mathrm{conv}(\{a_1, \ldots, a_r\})^\circ$.

I. If A is a convex subset of $\mathbb{R}^n$, show that $A^\circ = \{0\}$ if and only if $0 \in \mathrm{int}(A)$.
HINT: Use the Separation Theorem and Support Theorem.

J. Suppose that C_1 and C_2 are closed cones in $\mathbb{R}^n$.
(a) Show that $(C_1 + C_2)^\circ = C_1{}^\circ \cap C_2{}^\circ$.
(b) Show that $(C_1 \cap C_2)^\circ = \overline{C_1{}^\circ + C_2{}^\circ}$.
HINT: Use the Bipolar Theorem and part (a).

K. Given a convex function f on $\mathbb{R}^n$, define g on $\mathbb{R}^{n+1}$ by $g(x, r) = f(x) - r$ and $A = \mathrm{epi}(f) = \{(x, r) : f(x) \leq r\}$. Use Theorem 16.8.8 to verify
(a) $T_A\big((x, f(x))\big) = \{(d, p) : f'(x; d) \leq p\}$,
(b) $\mathrm{int}\, T_A\big((x, f(x))\big) = \{(d, p) : f'(x; d) < p\}$, and
(c) $N_A\big((x, f(x))\big) = \mathbb{R}_+[\partial f(x) \times \{-1\}]$.
(d) For $n = 1$, explain the last equation geometrically.

L. For a convex subset $A \subset \mathbb{R}^n$, show that the following are equivalent for $x \in A$:
(1) $x \in \mathrm{ri}(A)$.
(2) $T_A(x)$ is a subspace.
(3) $N_A(x)$ is a subspace.
(4) $y \in N_A(x)$ implies that $-y \in N_A(x)$.

M. (a) Suppose $C \subset \mathbb{R}^n$ is a closed convex cone and $x \notin C$. Show that $y \in C$ is the closest vector to x if and only if $x - y \in C^\circ$ and $\langle y, x - y \rangle = 0$.
HINT: Recall Theorem 16.3.1. Expand $\|x - y\|^2$.
(b) Hence deduce that $x = P_C(x) + P_{C^\circ}x$.

N. Give an example of two closed cones in $\mathbb{R}^3$ whose sum is not closed.
HINT: Let $C_i = \mathrm{cone}\{(x, y, 1) : (x, y) \in A_i\}$, where A_1 and A_2 come from Exercise 16.2.G(c).

O. A **polyhedral cone** in $\mathbb{R}^n$ is a set $A\mathbb{R}_+^m = \{Ax : x \in \mathbb{R}_+^m\}$ for some matrix A mapping $\mathbb{R}^m$ into $\mathbb{R}^n$. Show that $(A\mathbb{R}_+^m)^\circ = \{y \in \mathbb{R}^n : A^t y \leq 0\}$, where A^t is the transpose of A and $z \leq 0$ means $z_i \leq 0$ for $1 \leq i \leq m$.

P. Suppose $A \subset \mathbb{R}^n$ and $B \subset \mathbb{R}^m$ are convex sets. If $(a, b) \in A \times B$, then show that $T_{A \times B}(a, b) = T_A(a) \times T_B(b)$ and $N_{A \times B}(a, b) = N_A(a) \times N_B(b)$.

Q. Suppose that A_1 and A_2 are convex sets and $a \in A_1 \cap A_2$.
(a) Show that $T_{A_1 \cap A_2}(a) \subset T_{A_1}(a) \cap T_{A_2}(a)$.
(b) Give an example where this inclusion is proper.
HINT: Find a convex set A in $\mathbb{R}^2$ such that the positive y-axis Y_+ is contained in $T_A(0)$ but $A \cap Y_+ = \{(0,0)\}$.

16.9. Constrained Minimization

The goal of this section is to characterize the minimizers of a convex function subject to constraints that limit the domain of the function to a convex set. Generally, this convex set is not explicitly described but is given as the intersection of level sets. That is, we are only interested in minimizers in some specified convex set. The first theorem characterizes such minimizers abstractly, using the normal cone of the constraint set and the subdifferentials of the function. If the constraint is given as the intersection of sublevel sets of convex functions, these conditions may be described explicitly in terms of subgradients analogous to the Lagrange multiplier conditions of multivariable calculus. Finally, we present another characterization in terms of saddlepoints.

We will only consider convex functions that are defined on all of $\mathbb{R}^n$, rather than a convex subset. This is not as restrictive as it might seem. Exercise 16.9.H will guide you through a proof that any convex function satisfying a Lipschitz condition on a convex set A extends to a convex function on all of $\mathbb{R}^n$. There are convex functions that cannot be extended. For example, $f(x) = -\sqrt{x - x^2}$ on $[0, 1]$ is convex, but cannot be extended to all of $\mathbb{R}$ because the derivative blows up at 0 and 1.

We begin with the problem of minimizing a convex function f defined on $\mathbb{R}^n$ over a convex subset A. A point x in A is called a **feasible point**.

16.9.1. THEOREM. *Suppose that $A \subset \mathbb{R}^n$ is convex and that f is a convex function on $\mathbb{R}^n$. Then the following are equivalent for $a \in A$:*

(1) *a is a minimizer for $f|_A$.*

(2) *$f'(x; d) \geq 0$ for all $d \in T_A(a)$.*

(3) *$0 \in \partial f(a) + N_A(a)$.*

PROOF. First assume (3) that $0 \in \partial f(a) + N_A(a)$; so there is a vector $s \in \partial f(a)$ such that $-s \in N_A(a)$. Recall that $N_A(a) = \{v : \langle x - a, v \rangle \leq 0 \text{ for all } x \in A\}$. Hence $\langle x - a, s \rangle \geq 0$ for $x \in A$. Now use the fact that $s \in \partial f(a)$ to obtain

$$f(x) \geq f(a) + \langle x - a, s \rangle \geq f(a).$$

Therefore, a is a minimizer for f on A. So (3) implies (1).

Assume (1) that a is a minimizer for f on A. Let $x \in A$ and set $d = x - a$. Then $[a, x] = \{a + hd : 0 \leq h \leq 1\}$ is contained in A. So $f(a + hd) \geq f(a)$ for $h \in [0, 1]$ and thus

$$f'(a; d) = \lim_{h \to 0^+} \frac{f(a + hd) - f(a)}{h} \geq 0.$$

Because $f'(a; \cdot)$ is positively homogeneous and is nonnegative on $A - a$, it follows that $f'(a; d) \geq 0$ for d in the cone $\mathbb{R}_+(A - a)$. But $f'(a; \cdot)$ is defined on all of $\mathbb{R}^n$, and hence is continuous by Theorem 16.6.2. Therefore, $f'(a; \cdot) \geq 0$ on the closure $T_A(a) = \overline{\mathbb{R}_+(A - a)}$. This establishes (2).

By Theorem 16.7.3, since a is in the interior of $\mathbb{R}^n$, the subdifferential $\partial f(a)$ is a nonempty compact convex set. Thus the sum $\partial f(a) + N_A(a)$ is closed and convex by Exercise 16.2.G. Suppose that (3) fails: $0 \notin \partial f(a) + N_A(a)$. Then we may apply the Separation Theorem 16.3.3 to produce a vector d and scalar α so that

$$\sup\{\langle s+n, d\rangle : s \in \partial f(a),\, n \in N_A(a)\} \leq \alpha < \langle 0, d\rangle = 0.$$

It must be the case that $\langle n, d\rangle \leq 0$ for $n \in N_A(a)$, for if $\langle n, d\rangle > 0$, then

$$\langle s + \lambda n, d\rangle = \langle s, d\rangle + \lambda\langle n, d\rangle > 0$$

for very large λ. Therefore, d belongs to $N_A(a)^\circ = T_A(a)$ by Corollary 16.8.5. Now take $n = 0$ and apply Corollary 16.7.9 to compute

$$f'(a; d) = \sup\{\langle s, d\rangle : s \in \partial f(a)\} \leq \alpha < 0.$$

Thus (2) fails. Contrapositively, (2) implies (3). ∎

Theorem 16.9.1 is a fundamental and very useful result. In particular, condition (3) does not depend on where a is in the set A. For example, if a is an interior point of A, then $N_A(a) = \{0\}$ and this theorem reduces to Proposition 16.7.2. Given that all we know about the constraint set is that it is convex, this theorem is the best we can do. However, when the constraints are described in other terms, such as the sublevel sets of convex functions, then we can find more detailed characterizations of the optimal solutions.

16.9.2. DEFINITION. By a **convex program**, we mean the ingredients of a minimization problem involving convex functions. Precisely, we have a convex function f on $\mathbb{R}^n$ to be minimized. The set over which f is to be minimized is not given explicitly but instead is determined by constraint conditions of the form $g_i(x) \leq 0$, where $g_1, \ldots, g_r$ are convex functions. The associated problem is

$$\text{Minimize} \qquad f(x)$$
$$\text{subject to constraints} \quad g_1(x) \leq 0, \ldots, g_r(x) \leq 0.$$

We call $a \in \mathbb{R}^n$ a **feasible vector** for the convex program if a satisfies the constraints, that is, $g_i(a) \leq 0$ for $i = 1, \ldots, r$. A solution $a \in \mathbb{R}^n$ of this problem is called an **optimal solution** for the convex program, and $f(a)$ is the **optimal value**.

The set over which f is minimized is the convex set $A = \bigcap_{1 \leq i \leq r} \{x : g_i(x) \leq 0\}$. The r functional constraints may be combined to obtain $A = \{x : g(x) \leq 0\}$, where $g(x) = \max\{g_i(x) : 1 \leq i \leq r\}$. The function g is also convex. This is useful for technical reasons, but in practice the conditions g_i may be superior (for example, they may be differentiable). It is better to be able to express optimality conditions in terms of the g_i themselves.

In order to solve this problem, we need to impose some sort of regularity condition on the constraints that allows us to use our results about sublevel sets.

16.9.3. DEFINITION. A convex program satisfies **Slater's condition** if there is point $x \in \mathbb{R}^n$ so that $g_i(x) < 0$ for $i = 1, \ldots, r$. Such a point is called a **strictly feasible point** or a **Slater point**.

16.9.4. KARUSH–KUHN–TUCKER THEOREM.
Consider a convex program that satisfies Slater's condition. Then $a \in \mathbb{R}^n$ is an optimal solution if and only if there is a vector $w = (w_1, \ldots, w_r) \in \mathbb{R}^r$ with $w_j \geq 0$ for $1 \leq j \leq r$ so that

$$0 \in \partial f(a) + w_1 \partial g_1(a) + \cdots + w_r \partial g_r(a),$$
(KKT)
$$g_j(a) \leq 0, \quad w_j g_j(a) = 0 \quad for \quad 1 \leq j \leq r.$$

The relations (KKT) are called the **Karush-Kuhn-Tucker conditions**. If a is an optimal solution, the set of vectors $w \in \mathbb{R}_+^r$ that satisfy (KKT) are called the (Lagrange) **multipliers**.

A slight variant on these conditions for differentiable functions was given in a 1951 paper by Kuhn and Tucker and was labeled with their names. Years later, it came to light that they also appeared in Karush's unpublished Master's thesis of 1939, and so Karush's name was added.

This definition of multipliers appears to depend on which optimal point a is used. However, the set of multipliers is in fact independent of a; see Exercise 16.9.E.

PROOF. We introduce the function $g(x) = \max\{g_i(x) : 1 \leq i \leq r\}$ as mentioned previously. Then the feasible set becomes

$$A = \{x \in \mathbb{R}^n : g_i(x) \leq 0, \ 1 \leq i \leq r\} = \{x \in \mathbb{R}^n : g(x) \leq 0\}.$$

Slater's condition guarantees that the set $\{x : g(x) < 0\}$ is nonempty. Hence by Lemma 16.6.5, this is the interior of A.

Assume that $a \in A$ is an optimal solution. In particular, a is feasible, so $g_j(a) \leq 0$ for all j. By Theorem 16.9.1, $0 \in \partial f(a) + N_A(a)$. When a is an interior point, then $N_A(a) = \{0\}$ and so $0 \in \partial f(a)$. Set $w_j = 0$ for $1 \leq j \leq r$ and the conditions (KKT) are satisfied. Otherwise we may suppose that $g(a) = 0$.

The hypotheses of Theorem 16.8.8 are satisfied, and so $N_A(a) = \mathbb{R}_+ \partial g(a)$. When the subdifferential of g is computed using Theorem 16.7.13, it is found to be $\partial g(a) = \text{conv}\{\partial g_j(a) : j \in J(a)\}$, where $J(a) = \{j : g_j(a) = 0\}$. We claim that

$$N_A(a) = \Big\{ \sum_{j \in J(a)} w_j s_j : w_j \geq 0, \ s_j \in \partial g_j(a) \Big\}.$$

Indeed, every element of $\partial g(a)$ has this form, and multiplication by a positive scalar preserves it. Conversely, if $w = \sum_{j \in J(a)} w_j \neq 0$, then $\sum_{j \in J(a)} \frac{w_j}{w} s_j$ belongs to $\text{conv}\{\partial g_j(a) : j \in J(a)\}$, and so $\sum_{j \in J(a)} w_j s_j$ belongs to $\mathbb{R}_+ \partial g(a)$.

Therefore, the condition $0 \in \partial f(a) + N_A(a)$ may be restated as $s \in \partial f(a)$, $s_j \in \partial g_j(a)$, $w_j \geq 0$ for $j \in J(a)$ and $s + \sum_{j \in J(a)} w_j s_j = 0$. By definition, $g_j(a) = g(a) = 0$ for $j \in J(a)$, whence we have $w_j g_j(a) = 0$. For all other j, we set $w_j = 0$ and (KKT) is satisfied.

Conversely suppose that (KKT) holds. Since $g_j(a) \leq 0$, a is a feasible point. The conditions $w_j g_j(a) = 0$ mean that $w_j = 0$ for $j \notin J(a)$, where $J(a)$ is defined as previously. If a is strictly feasible, $J(a)$ is the empty set. In this event, (KKT) reduces to $0 \in \partial f(a) = \partial f(a) + N_A(a)$. On the other hand, when $g(a) = 0$, we saw that in this instance the (KKT) condition is equivalent to $0 \in \partial f(a) + N_A(a)$. In both cases, Theorem 16.9.1 implies that a is an optimal solution. ∎

Notice that if f and each g_i is differentiable, then first part of (KKT) becomes

$$0 = \nabla f(a) + w_1 \nabla g_1(a) + \cdots + w_r \nabla g_r(a),$$

which is more commonly written as a system of linear equations

$$0 = \frac{\partial f}{\partial x_i}(a) + w_1 \frac{\partial g_1}{\partial x_i}(a) + \cdots + w_r \frac{\partial g_r}{\partial x_i}(a) \quad \text{for} \quad 1 \leq i \leq n.$$

This is known as a Lagrange multiplier problem. So we adopt the same terminology here.

These conditions can be used to solve concrete optimization problems in much the same way as in multivariable calculus. Their greatest value for applications is in understanding minimization problems, which can lead to the development of efficient numerical algorithms.

16.9.5. DEFINITION. Given a convex program, define the **Lagrangian** of this system to be the function L on $\mathbb{R}^n \times \mathbb{R}^r$ given by

$$L(x, y) = f(x) + y_1 g_1(x) + \cdots + y_r g_r(x).$$

Next, we recall the definition of a saddlepoint from multivariable calculus. There are several equivalent conditions for saddlepoints given in the Exercises.

16.9.6. DEFINITION. Suppose that X and Y are sets and L is a real-valued function on $X \times Y$. A point $(x_0, y_0) \in X \times Y$ is a **saddlepoint** for L if

$$L(x_0, y) \leq L(x_0, y_0) \leq L(x, y_0) \quad \text{for all} \quad x \in X, \ y \in Y.$$

We shall be interested in saddlepoints of the Lagrangian over the set $\mathbb{R}^n \times \mathbb{R}^r_+$. We restrict the y variables to the positive orthant $\mathbb{R}^r_+$ because the (KKT) conditions require nonnegative multipliers.

16.9.7. THEOREM. *Consider a convex program that admits an optimal solution and satisfies Slater's condition. Then a is an optimal solution and w a multiplier for the program if and only if (a, w) is a saddlepoint for its Lagrangian function $L(x, y) = f(x) + y_1 g_1(x) + \cdots y_r g_r(x)$ on $\mathbb{R}^n \times \mathbb{R}^r_+$. The value $L(a, w)$ at any saddlepoint equals the optimal value of the program.*

PROOF. First suppose that $L(a, y) \leq L(a, w)$ for all $y \in \mathbb{R}^n_+$. Observe that

$$\sum_{j=1}^{r}(y_j - w_j)g_j(a) = L(a, y) - L(a, w) \leq 0.$$

Since each y_j may be taken to be arbitrarily large, this forces $g_j(a) \leq 0$ for each j. So a is a feasible point. Also, taking $y_j = 0$ and $y_i = w_i$ for $i \neq j$ yields $-w_j g_j(a) \leq 0$. Since this quantity is positive, we deduce that $w_j g_j(a) = 0$. So $L(a, w) = f(a) + \sum_j w_j g_j(a) = f(a)$.

Now turn to the condition $L(a, w) \leq L(x, w)$ for all $x \in \mathbb{R}^n$. This means that $h(x) = L(x, w) = f(x) + \sum_j w_j g_j(x)$ has a global minimum at a. By Proposition 16.7.2, $0 \in \partial h(a)$. We may compute $\partial h(a)$ using Theorem 16.7.14. So $0 \in \partial f(a) + \sum_{j=1}^{r} w_j \partial g_j(a)$. This establishes the (KKT) conditions, and thus a is a minimizer for the convex program and w is a multiplier.

Conversely, suppose that a and w satisfy (KKT). Then $g_j(a) \leq 0$ because a is feasible and $w_j g_j(a) = 0$ for $1 \leq j \leq r$. So $w_j = 0$ except for $j \in J(a)$. Thus $L(a, y) - L(a, w) = \sum_{j \notin J(a)} y_j g_j(a) \leq 0$. The other part of (KKT) states that the function $h(x)$ has $0 \in \partial h(a)$. Thus by Proposition 16.7.2, a is a minimizer for h. That is, $L(a, w) \leq L(x, w)$ for all $x \in \mathbb{R}^n$. So L has a saddlepoint at (a, w). ∎

If we had a multiplier w for the convex program, then to solve the convex program it is enough to solve the unconstrained minimization problem:

$$\inf\{L(x, w) : x \in \mathbb{R}^n\}.$$

This shows one important property of multipliers: They turn constrained optimization problems into unconstrained ones. In order to use multipliers in this way, we need a method for finding multipliers without first solving the convex program. This problem is addressed in the next section.

16.9.8. EXAMPLE. Consider the following example. Let g be a convex function on $\mathbb{R}$ and fix two points $p = (x_p, y_p)$ and $q = (x_q, y_q)$ in $\mathbb{R}^2$. Minimize the sum of the distances to p and q over $A = \mathrm{epi}(g) = \{(x, y) : G(x, y) \leq 0\}$, where $G(x, y) = g(x) - y$, as indicated in Figure 16.7.

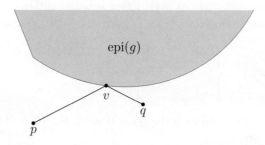

FIGURE 16.7. Minimizing the sum of two distances.

Then the function of $v = (x, y) \in \mathbb{R}^2$ to be minimized is

$$f(v) = \|v-p\| + \|v-q\| = \sqrt{(x-x_p)^2 + (y-y_p)^2} + \sqrt{(x-x_q)^2 + (y-y_q)^2}.$$

Using Example 16.7.10, we may compute that

$$\partial f(v) = \begin{cases} \overline{B_1(0)} + \frac{p-q}{\|p-q\|} & \text{if} \quad v = p \\ \overline{B_1(0)} - \frac{p-q}{\|p-q\|} & \text{if} \quad v = q \\ \frac{v-p}{\|v-p\|} + \frac{v-q}{\|v-q\|} & \text{if} \quad x \neq p, q. \end{cases}$$

Note that $0 \in \partial f(v)$ if $v = p$ or q or the two vectors $v - p$ and $v - q$ point in opposite directions, namely $v \in [p, q]$. This is obvious geometrically.

To make the problem more interesting, let us assume that A does not intersect the line segment $[p, q]$. The (KKT) conditions at the point $v = (x, y)$ become

$$g(x) \leq y \qquad w \geq 0 \qquad w(g(x) - y) = 0$$
$$0 \in \partial f(v) + w \partial G(v).$$

The first line reduces to saying that $v = (x, g(x))$ lies in the boundary of A and $w \geq 0$. Alternatively, we could observe that $N_A(v) = \{0\}$ when $v \in \text{int } A$, and thus $0 \notin \partial f(v) + N_A(v)$, whence the minimum occurs on the boundary.

At a point x, we know that $\partial g(x) = [D_- g(x), D_+ g(x)]$. Thus by Example 16.7.4 for $v = (x, g(x))$,

$$\partial G(v) = \left\{ (s, -1) : s \in [D_- g(x), D_+ g(x)] \right\}.$$

So by Theorem 16.8.8, $N_A(v) = \left\{ (st, -t) : s \in [D_- g(x), D_+ g(x)], \ t \geq 0 \right\}$. Thus the second statement of (KKT) says that the sum of the two unit vectors in the directions $p - v$ and $q - v$ is an element of $N_A(v)$. Now geometrically this means that $p - v$ and $q - v$ make the same angle on opposite sides of some normal vector.

In particular, if g is differentiable at v, then $N_A(v) = \mathbb{R}_+(g'(x), -1)$ is the outward normal to the tangent line at v. So the geometric condition is just that the angles to the tangent (from opposite sides) made by $[p, v]$ and $[q, v]$ are equal. In physics, this is the well-known law: The angle of incidence equals the angle of reflection. This fact for a light beam reflecting off a surface is explained by Fermat's Principle that a beam of light will follow the fastest path (hence the shortest path) between two points.

However, this criterion works just as well when g is not differentiable. For example, take $g(x) = |x|$ and points $p = (-1, -1)$ and $q = (2, 0)$. Then

$$\partial g(x) = \begin{cases} -1 & \text{if} \quad x < 0 \\ [-1, 1] & \text{if} \quad x = 0 \\ +1 & \text{if} \quad x > 0. \end{cases}$$

We will verify the (KKT) condition at the point $0 = (0, 0)$. First observe that $\partial G(0) = [-1, 1] \times \{-1\}$. So

$$N_A(0) = \mathbb{R}_+ (\partial g(0) \times \{-1\}) = \{(s, t) : t \leq 0\}$$

consists of the lower half-plane. Now

$$\partial f(0) = (1,1)/\sqrt{2} + (-2,0)/2 = (\tfrac{1}{\sqrt{2}} - 1, \tfrac{1}{\sqrt{2}})$$

lies in the upper half-plane. In particular,

$$(\tfrac{1}{\sqrt{2}} - 1, \tfrac{1}{\sqrt{2}}) + \tfrac{1}{\sqrt{2}}(1 - \sqrt{2}, -1) = (0,0).$$

So $w = 1/\sqrt{2}$ and $v = (0,0)$ satisfy (KKT), and thus $(0,0)$ is the minimizer.

Exercises for Section 16.9

A. Minimize $x^2 + y^2 - 4x - 6y$ subject to $x \geq 0$, $y \geq 0$, and $x^2 + y^2 \leq 4$.

B. Minimize $ax + by$ subject to $x \geq 1$ and $\sqrt{xy} \geq K$. Here a, b, and K are positive constants.

C. Minimize $\dfrac{1}{x} + \dfrac{4}{y} + \dfrac{9}{z}$ subject to $x + y + z = 1$ and $x, y, z > 0$.
HINT: Find the Lagrange multipliers.

D. Suppose (x_1, y_1) and (x_2, y_2) are saddlepoints of a real valued function p on $X \times Y$.
   (a) Show that (x_1, y_2) and (x_2, y_1) are also saddlepoints.
   (b) Show that p takes the same value at all four points.
   (c) Prove that the set of saddlepoints of p has the form $A \times B$ for $A \subset X$ and $B \subset Y$.

E. (a) Use the previous exercise to show that the set of multipliers for a convex program do not depend on the choice of optimal point.
   (b) Show that the set of multipliers is a closed convex subset of $\mathbb{R}_+^r$.
   (c) Show that the set of saddlepoints for the Lagrangian is a closed convex rectangle $A \times M$, where A is the set of optimal solutions and M is the set of multipliers.

F. Given a real-valued function p on $X \times Y$, define functions α on X and β on Y by $\alpha(x) = \sup\{p(x, y) : y \in Y\}$ and $\beta(y) = \inf\{p(x, y) : x \in X\}$. Show that for $(x_0, y_0) \in X \times Y$ the following are equivalent:
   (1) (x_0, y_0) is a saddlepoint for p.
   (2) $p(x_0, y) \leq p(x, y_0)$ for all $x \in X$ and all $y \in Y$.
   (3) $\alpha(x_0) = p(x_0, y_0) = \beta(y_0)$.
   (4) $\alpha(x_0) \leq \beta(y_0)$.

G. Let g be a convex function on $\mathbb{R}$ and let $p = (x_p, y_p) \in \mathbb{R}^2$. Find a criterion for the closest point to p in $A = \text{epi}(g)$.
   (a) What is the function f to be minimized? Find $\partial f(v)$.
   (b) What is the constraint function G? Compute $\partial G(v)$.
   (c) Write down the (KKT) conditions.
   (d) Simplify these conditions and interpret them geometrically.

H. Suppose that A is an open convex subset of $\mathbb{R}^n$ and f is a convex function on A that is Lipschitz with constant L. Construct a convex function g on $\mathbb{R}^n$ extending f:
   (a) Show if $a \in A$ and $v \in \partial f(a)$, then $\|v\| \leq L$.
      HINT: Check the proof of Theorem 16.7.3.
   (b) For $x \in \mathbb{R}^n$ and $a, b \in A$ and $s \in \partial f(a)$, show that

$$f(b) + L\|x - b\| \geq f(a) + \langle s, x - a \rangle.$$

(c) Define g on $\mathbb{R}^n$ by $g(x) = \inf\{f(a) + L\|x - a\| : a \in A\}$. Show that $g(x) > -\infty$ for $x \in \mathbb{R}^n$ and that $g(a) = f(a)$ for $a \in A$.

(d) Show that g is convex.

I. Let f and $g_1, \ldots, g_m$ be C^1 functions on $\mathbb{R}^n$. The problem is to minimize f over the set $A = \{x : g_j(x) \leq 0, 1 \leq j \leq m\}$. Let $J(x) = \{j : g_j(x) = 0\}$. Prove that a feasible point x_0 is a local minimum if and only if there are constants λ_i not all 0 so that $\lambda_0 \nabla f(x_0) + \sum_{j \in J(x_0)} \lambda_j \nabla g_j(x_0) = 0$.

HINT: Let $g(x) = \max\{f(x) - f(x_0), g_j(x) : j \in J(x_0)\}$. Compute $\partial g(x_0)$ by Theorem 16.7.13. Use Exercise 16.3.J to deduce $g'(x_0; d) \geq 0$ for all d if and only if $0 \in \text{cone}\{\nabla f(x_0), \nabla g_j(x_0) : j \in J(x_0)\}$.

J. **Duffin's duality gap.** Let $b \geq 0$, and consider the convex program:

Minimize $f(x, y) = e^{-y}$ subject to $g(x, y) = \sqrt{x^2 + y^2} - x \leq b$ in $\mathbb{R}^2$.

(a) Find the feasible region. For which b is Slater's condition satisfied?

(b) Solve the problem. When is the minimum attained?

(c) Show that the solution is not continuous in b.

K. An alternative approach to solving minimization problems is to eliminate the constraint set $g_i(x) \leq 0$ and instead modify f by adding a term $h(g_i(x))$, where h is an increasing function with $h(y) = 0$ for $y \leq 0$. The quantity $h(g_i(x))$ is called a **penalty**, and this approach is the **penalty method**. Assume that f and each g_i are continuous functions on $\mathbb{R}^n$ but not necessarily convex. Let $h_k(y) = k(\max\{y, 0\})^2$. For each integer $k \geq 1$, we have the minimization problem: Minimize $F_k(x) = f(x) + \sum_{i=1}^{r} h_k(g_i(x))$ for $x \in \mathbb{R}^n$. Suppose that this minimization problem has a solution a_k and the original has a solution a.

(a) Show that $F_k(a_k) \leq F_{k+1}(a_{k+1}) \leq f(a)$.

(b) Show that $\lim_{k \to \infty} \sum_{i=1}^{r} h_k(g_i(a_k)) = 0$.

(c) If a_0 is the limit of a subsequence of (a_k), show that it is a minimizer.

(d) If $f(x) \to \infty$ as $\|x\| \to \infty$, deduce that the minimization problem has a solution.

16.10. The Minimax Theorem

In addition to the Lagrangian of the previous section, saddlepoints play a central role in various other optimization problems. For example, they arise in game theory and mathematical economics. Our purpose in this section is to examine the mathematics that leads to the existence of a saddlepoint under quite general hypotheses. Examination of a typical saddlepoint in $\mathbb{R}^3$ shows that the cross sections in the xz-plane are convex functions while the cross sections in the yz-plane are concave. See Figure 16.8. It is this trade-off that gives the saddle its characteristic shape. Hence we make the following definition:

16.10.1. DEFINITION. A function $p(x, y)$ defined on $X \times Y$ is called **convex–concave** if $p(\cdot, y)$ is a convex function of x for each fixed $y \in Y$ and $p(x, \cdot)$ is a concave function of y for each fixed $x \in X$.

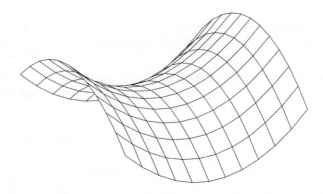

FIGURE 16.8. A typical saddlepoint.

The term **minimax** comes from comparing two interesting quantities:

$$p_* = \sup_{y \in Y} \inf_{x \in X} p(x, y) \quad \text{and} \quad p^* = \inf_{x \in X} \sup_{y \in Y} p(x, y),$$

which are the **maximin** and **minimax**, respectively. These quantities make sense for any function p. Moreover, for $x_1 \in X$ and $y_1 \in Y$,

$$\inf_{x \in X} p(x, y_1) \leq p(x_1, y_1) \leq \sup_{y \in Y} p(x_1, y).$$

Take the supremum of the left-hand side over $y_1 \in Y$ to get $p_* \leq \sup_{y \in Y} p(x_1, y)$, since the right-hand side does not depend on y_1. Then take the infimum over all $x_1 \in X$ to obtain $p_* \leq p^*$.

Suppose that there is a saddlepoint $(\bar{x}, \bar{y})$, that is, $p(\bar{x}, y) \leq p(\bar{x}, \bar{y}) \leq p(x, \bar{y})$ for all $x \in X$ and $y \in Y$. Then

$$p_* \geq \inf_{x \in X} p(x, \bar{y}) \geq p(\bar{x}, \bar{y}) \geq \sup_{y \in Y} p(\bar{x}, y) \geq p^*.$$

Thus the existence of a saddlepoint shows that $p_* = p^*$.

We will use the following variant of Exercise 16.6.J.

16.10.2. LEMMA. *Let $f_1, \ldots, f_r$ be convex functions on a convex subset X of $\mathbb{R}^n$. For $c \in \mathbb{R}$, the following are equivalent:*

(1) *There is no point $x \in X$ satisfying $f_j(x) < c$ for $1 \leq j \leq r$.*

(2) *There exist $\lambda_j \geq 0$ so that $\sum_j \lambda_j = 1$ and $\sum_j \lambda_j f_j \geq c$ on X.*

PROOF. If (1) is false and $f_j(x_0) < c$ for $1 \leq j \leq k$, then $\sum_j \lambda_j f_j(x_0) < c$ for all choices of $\lambda_j \geq 0$ with $\sum_j \lambda_j = 1$. Hence (2) is also false.

Conversely, assume that (1) is true. Define

$$Y = \{y \in \mathbb{R}^r : y_j > f_j(x) \text{ for } 1 \leq j \leq r \text{ and some } x \in X\}.$$

This set is open and convex (left as Exercise 16.10.B). By hypothesis, the point $z = (c, c \ldots, c) \in \mathbb{R}^r$ is not in Y. Depending on whether z belongs to $\overline{Y}$ or not, we apply either the Support Theorem (Theorem 16.3.7) or the Separation Theorem

(Theorem 16.3.3) to obtain a hyperplane that separates z from $\overline{Y}$. That is, there is a nonzero vector $h = (h_1, \ldots, h_r)$ in $\mathbb{R}^r$ and $\alpha \in \mathbb{R}$ so that $\langle y, h \rangle < \alpha \leq \langle z, h \rangle$ for all $y \in \overline{Y}$.

We claim that each coefficient $h_j \leq 0$. Indeed, for any $x \in X$, $\overline{Y}$ contains $(f_1(x), \ldots, f_r(x)) + te_j$ for any $t \geq 0$, where e_j is a standard basis vector for $\mathbb{R}^r$. Thus $\sum_{j=1}^{r} h_j f_j(x) + t h_j \leq \alpha$, which implies that $h_j \leq 0$.

Define $\lambda_j = h_j / H$, where $H = \sum_{j=1}^{r} h_j < 0$. Then $\lambda_j \geq 0$ and $\sum_j \lambda_j = 1$. Restating the separation for $(f_1(x), \ldots, f_r(x)) \in \overline{Y}$, we obtain

$$\sum_{j=1}^{r} \lambda_j f_j(x) \geq \frac{\alpha}{H} \geq \frac{\langle z, h \rangle}{H} = \sum_{j=1}^{r} c\lambda_j = c.$$

So (2) holds. ■

Now we establish our saddlepoint result. First we assume compactness. We will remove it later, at the price of adding a mild additional requirement.

16.10.3. Minimax Theorem (Compact Case).

Let X be a compact convex subset of $\mathbb{R}^n$ and let Y be a compact convex subset of $\mathbb{R}^m$. If p is a convex–concave function on $X \times Y$, then p has a nonempty compact convex set of saddlepoints.

PROOF. For each $y \in Y$, define convex functions on X by $p_y(x) = p(x, y)$. For each $c > p_*$, define $A_{y,c} = \{x \in X : p_y(x) \leq c\}$. Then this is a nonempty compact convex subset of X.

For any finite set of points $y_1, \ldots, y_r$ in Y, we claim that $A_{y_1,c} \cap \cdots \cap A_{y_r,c}$ is nonempty. If not, then there is no point x so that $p_{y_j}(x) < c$ for $1 \leq j \leq r$. So by Lemma 16.10.2, there would be scalars $\lambda_i \geq 0$ with $\sum_i \lambda_i = 1$ so that $\sum_i \lambda_i p_{y_i} \geq c$ on X. Set $\tilde{y} = \sum_{i=1}^{r} \lambda_i y_i$. As p is concave in y,

$$c \leq \sum_{i=1}^{r} \lambda_i p(x, y_i) \leq p(x, \tilde{y}) \quad \text{for all} \quad x \in X.$$

Consequently, $c \leq \min_{x \in X} p(x, \tilde{y}) \leq p_*$, which is a contradiction.

Let $\{y_i : i \geq 1\}$ be a dense subset of Y, and set $c_n = p_* + 1/n$. Then the set $A_n = \bigcap_{i=1}^{n} A_{y_i, c_n}$ is nonempty closed and convex. It is clear that A_n contains A_{n+1}, and thus this is a decreasing sequence of compact sets in $\mathbb{R}^n$. By Cantor's Intersection Theorem (Theorem 4.4.7), the set $A = \bigcap_{n \geq 1} A_n$ is a nonempty compact set. (It is also convex, as the reader can easily verify.)

Let $\bar{x} \in A$. Then $\bar{x} \in A_{y_i, c_n}$ for all $n \geq 1$. Thus $p(\bar{x}, y_i) \leq c_n$ for all $i \geq 1$ and $n \geq 1$. So $p(\bar{x}, y_i) \leq p_*$. But the set $\{y_i : i \geq 1\}$ is dense in Y and p is continuous, so $p(\bar{x}, y) \leq p_*$ for all $y \in Y$. Therefore,

$$p^* = \inf_{x \in X} \max_{y \in Y} p(x, y) \leq \max_{y \in Y} p(\bar{x}, y) \leq p_*.$$

Since $p_* \leq p^*$ is always true, we obtain equality. Choose a point $\bar{y} \in Y$ so that $p(\bar{x}, \bar{y}) = p_*$. Then $p(\bar{x}, y) \leq p(\bar{x}, \bar{y}) \leq p(x, \bar{y})$ for all $x \in X$ and $y \in Y$. That is, $(\bar{x}, \bar{y})$ is a saddlepoint for p.

By Exercise 16.9.D, the set of saddlepoints is a rectangle $A \times B$. Moreover, the same argument required in Exercise 16.9.E shows that this rectangle is closed and convex. ∎

Slippery things can happen at infinity if precautions are not taken. However, the requirements of the next theorem are often satisfied.

16.10.4. MINIMAX THEOREM.

Suppose that X is a closed convex subset of $\mathbb{R}^n$ and Y is a closed convex subset of $\mathbb{R}^m$. Assume that p is convex–concave on $X \times Y$, and in addition assume that

(1) *if X is unbounded, then there is a $y_0 \in Y$ so that $p(x, y_0) \to +\infty$ as $\|x\| \to \infty$ for $x \in X$.*

(2) *if Y is unbounded, then there is an $x_0 \in X$ so that $p(x_0, y) \to -\infty$ as $\|y\| \to \infty$ for $y \in Y$.*

Then p has a nonempty compact convex set of saddlepoints.

PROOF. We deal only with the case in which both X and Y are unbounded. The reader can find a modification that works when only one is unbounded.

By the hypotheses, $\max\limits_{y \in Y} p(x_0, y) = \alpha < \infty$ and $\min\limits_{x \in X} p(x, y_0) = \beta > -\infty$. Clearly, $\beta \leq p(x_0, y_0) \leq \alpha$. Set

$$X_0 = \{x \in X : p(x, y_0) \leq \alpha + 1\} \quad \text{and} \quad Y_0 = \{y \in Y : p(x_0, y) \geq \beta + 1\}.$$

Conditions (1) and (2) guarantee that X_0 and Y_0 are bounded, and thus they are compact and convex. Let $A \times B$ be the set of saddlepoints for the restriction of p to $X_0 \times Y_0$ provided by the compact case.

In particular, let $(\bar{x}, \bar{y})$ be one saddlepoint, and let $c = p(\bar{x}, \bar{y})$ be the critical value. Then

$$\beta \leq p(\bar{x}, y_0) \leq p(\bar{x}, \bar{y}) = c \leq p(x_0, \bar{y}) \leq \alpha.$$

Let $x \in X \setminus X_0$, so that $p(x, y_0) > \alpha + 1$. Now $p(\cdot, y_0)$ is continuous, and hence there is a point x_1 in $[x, \bar{x}]$ with $p(x_1, y_0) = \alpha + 1$. So $x_1 \in X_0$ and $x_1 \neq \bar{x}$. Thus $x_1 = \lambda x + (1 - \lambda)\bar{x}$ for some $0 < \lambda < 1$. As $p(\cdot, \bar{y})$ is convex,

$$c \leq p(x_1, \bar{y}) \leq \lambda p(x, \bar{y}) + (1 - \lambda)p(\bar{x}, \bar{y}) = \lambda p(x, \bar{y}) + (1 - \lambda)c.$$

Hence $p(\bar{x}, \bar{y}) = c \leq p(x, \bar{y})$. Similarly, for every $y \in Y \setminus Y_0$, we may show that $p(\bar{x}, y) \leq c = p(\bar{x}, \bar{y})$. Therefore, $(\bar{x}, \bar{y})$ is a saddlepoint in $X \times Y$. ∎

Now let us see how this applies to the problem of constrained optimization. Consider the convex programming problem: Minimize a convex function $f(x)$ over the closed convex set $X = \{x : g_j(x) \leq 0, 1 \leq j \leq r\}$. Suppose that it satisfies Slater's condition. The Lagrangian $L(x, y) = f(x) + y_1 g_1(x) + \cdots + y_r g_r(x)$ is a convex function of x for each fixed $y \in \mathbb{R}_+^r$, and is a linear function of y for each $x \in \mathbb{R}^n$. So, in particular, L is convex–concave on $\mathbb{R}^n \times \mathbb{R}_+^r$.

Now we also suppose that this problem has an optimal solution. Then we can apply Theorem 16.9.7 and the Karush–Kuhn–Tucker Theorem (Theorem 16.9.4) to guarantee a saddlepoint (a, w) for L; and $L(a, w)$ is the solution of the convex program. By the arguments of this section, it follows that the existence of a saddlepoint means that the optimal value is also obtained as the maximin:

$$f(a) = \min_{x \in X} f(x) = L^* = \max_{y \in \mathbb{R}^r_+} \inf_{x \in \mathbb{R}^n} L(x, y).$$

Define $h(y) = \inf_{x \in \mathbb{R}^n} f(x) + y_1 g_1(x) + \cdots y_r g_r(x)$ for $y \in \mathbb{R}^r_+$. While its definition requires an infimum, h gives a new optimization problem, which can be easier to solve. This new problem is called the **dual program**:

$$\text{Maximize } h(y) \text{ over } y \in \mathbb{R}^r_+.$$

16.10.5. PROPOSITION. *Consider a convex program that admits an optimal solution and satisfies Slater's condition. The solutions of the dual program are exactly the multipliers of the original program, and the optimal value of the dual program is the same.*

PROOF. Suppose a is an optimal solution of the original program and w a multiplier. Then $L(a, y) \le L(a, w) \le L(x, w)$ for all $x \in \mathbb{R}^n$ and $y \in \mathbb{R}^r_+$ since (a, w) is a saddlepoint for the Lagrangian L. In particular,

$$h(w) = \inf\{L(x, w) : x \in \mathbb{R}^n\} = L(a, w).$$

Moreover, for any $y \in \mathbb{R}^r_+$ with $y \ne w$, $h(y) \le L(a, y) \le L(a, w) = h(w)$. So w is a solution of the dual problem, and the value is $L(a, w) = L^*$, which equals the value of the original problem.

Conversely, suppose that w' is a solution of the dual program. Let (a, w) be a saddlepoint. Then

$$L^* = L(a, w) \ge L(a, w') \ge h(w') = L^*.$$

Thus $h(w') = L(a, w') = L^*$. Therefore, $L(a, w') = h(w') \le L(x, w')$ for all $x \in \mathbb{R}^n$. Also, since (a, w) is a saddlepoint, $L(a, y) \le L(a, w) = L(a, w')$ for all $y \in \mathbb{R}^r_+$. Consequently, w' is a multiplier. $\blacksquare$

An important fact for computational purposes is that since these two problems have the same answer, we can obtain estimates for the solution by sampling. Suppose that we have a point $x_0 \in X$ and $y_0 \in \mathbb{R}^m_+$ so that $h(y_0) - f(x_0) < \varepsilon$. Then since we know that the solution lies in $[f(x_0), h(y_0)]$, we have a good estimate for the solution even if we cannot compute it exactly.

16.10.6. EXAMPLE. Consider a **quadratic programming** problem. Let Q be a positive definite $n \times n$ matrix, and let $q \in \mathbb{R}^n$. Also let A be an $m \times n$ matrix and $a \in \mathbb{R}^m$. Minimize the quadratic function $f(x) = \langle x, Qx \rangle + \langle x, q \rangle$ over the region $Ax \le a$.

We can assert before doing any calculation that this minimum will be attained. This follows from the global version, Example 16.7.12, where it was shown that

f may be written as a sum of squares. Thus f tends to infinity as $\|x\|$ goes to infinity. Therefore, the constraint set could be replaced with a compact set. Then the Extreme Value Theorem asserts that the minimum is attained.

The constraint condition is really m linear conditions $\langle x, A^t e_j \rangle - a_j \leq 0$ for $1 \leq j \leq m$, where $a = (a_1, \ldots, a_m)$ with respect to the standard basis $e_1, \ldots, e_m$. If rank $A = m$, then $m \leq n$ and A maps $\mathbb{R}^n$ onto $\mathbb{R}^m$. Thus there are strictly feasible points and so Slater's condition is satisfied. In general this needs to be checked.

The Lagrangian is defined on $\mathbb{R}^n \times \mathbb{R}^m_+$ by

$$L(x, y) = f(x) + \sum_{j=1}^{m} \left(\langle x, A^t e_j \rangle - a_j \right) y_j$$

$$= \langle x, Qx \rangle + \langle x, q + A^t y \rangle - \langle a, y \rangle.$$

To find a solution to the dual problem, we first must compute $h(y) = \inf_{x \in \mathbb{R}^n} L(x, y)$. This was solved in Example 16.7.12, so

$$h(y) = -\tfrac{1}{4} \langle q + A^t y, Q^{-1}(q + A^t y) \rangle - \langle a, y \rangle$$

$$= -\tfrac{1}{4} \langle y, AQ^{-1}A^t y \rangle - \langle y, a + \tfrac{1}{2} AQ^{-1}q \rangle - \tfrac{1}{4} \langle q, Q^{-1}q \rangle.$$

The dual problem is to maximize $h(y)$ over the set $\mathbb{R}^m_+$. This is now a quadratic programming problem with a simpler set of constraints, possibly at the expense of extra variables if $m > n$. The matrix $AQ^{-1}A^t$ is positive semidefinite but may not be invertible. This is not a serious problem.

Now let's look at a specific case.

Minimize $f(x_1, x_2) = 2x_1^2 - 2x_1 x_2 + 2x_2^2 - 6x_1$
subject to $x_1 \geq 0, \; x_2 \geq 0$ and $x_1 + x_2 \leq 2.$

This is a quadratic programming problem with $Q = \begin{bmatrix} 2 & -1 \\ -1 & 2 \end{bmatrix}$, $q = (-6, 0)$,

$A = \begin{bmatrix} -1 & 0 \\ 0 & -1 \\ 1 & 1 \end{bmatrix}$, and $a = (0, 0, 2)$. Note that Slater's condition is satisfied, for example, at the point $(1/2, 1/2)$.

We can compute $Q^{-1} = \tfrac{1}{3} \begin{bmatrix} 2 & 1 \\ 1 & 2 \end{bmatrix}$ and $\tfrac{1}{4} AQ^{-1}A^t = \tfrac{1}{12} \begin{bmatrix} 2 & 1 & -3 \\ 1 & 2 & -3 \\ -3 & -3 & 6 \end{bmatrix}$, and also $a + \tfrac{1}{2} AQ^{-1}q = (2, 1, -1)$ and $\tfrac{1}{4} \langle q, Q^{-1}q \rangle = 6$. Thus $h(y_1, y_2, y_3) =$

$$\frac{1}{12}(2y_1^2 + 2y_2^2 + 6y_3^2 + 2y_1 y_2 - 6y_1 y_3 - 6y_2 y_3) - 2y_1 - y_2 + y_3 - 6.$$

This problem can be solved most easily using the (KKT) conditions. It will be left to Exercise 16.10.G to show that they are satisfied for $x = (3/2, 1/2)$ and $y = (0, 0, 1)$. Notice that $f(3/2, 1/2) = -11/2 = h(0, 0, 1)$. Since the minimum of f and the maximum of h are equal, this is the minimax value. Hence the value of the program must be $-11/2$, the minimizer is $(3/2, 1/2)$ and the multiplier is $(0, 0, 1)$.

Exercises for Section 16.10

A. Compute p_* and p^* for $p(x, y) = \sin(x + y)$ on $\mathbb{R}^2$.

B. Show that the set Y defined in the proof of Lemma 16.10.2 is convex and open.

C. Show that the set A constructed in the proof of the Minimax Theorem (compact case) is compact and convex.

D. Modify the proof of the Minimax Theorem to deal with the case in which X is unbounded and Y is compact.

E. Let $p(x, y) = e^{-x} - e^{-y}$.
  (a) Show that p is convex–concave.
  (b) Show that $p_* = p^*$.
  (c) Show that there are no saddlepoints.
  (d) Why does this not contradict the Minimax Theorem?

F. Suppose that $p(x, y)$ is convex–concave on $X \times X$ for a compact subset X of $\mathbb{R}^n$ and satisfies $p(y, x) = -p(x, y)$. Prove that $p_* = p^* = 0$.

G. (a) Solve the (KKT) equations for the numerical example in Example 16.10.6.
  (b) Write down the Lagrangian and verify the the saddlepoint.

H. Consider the **linear programming** problem: Minimize $\langle x, q \rangle$ subject to $Ax \le a$, where $q \in \mathbb{R}^n$, A is an $m \times n$ matrix, and $a \in \mathbb{R}^m$.
  (a) Express this problem as a convex program and compute the Lagrangian.
  (b) Find the dual program.
  (c) Show that if the original program satisfies Slater's condition and has a solution v, then the dual program has a solution w and $\langle v, q \rangle = \langle w, a \rangle$.

REFERENCES

Foundations and Proofs

[1] P. Halmos, *Naive Set Theory*, Van Nostrand, Princeton, N.J., 1960.

[2] I. Lakatos, *Proofs and Refutations: The Logic of Mathematical Discovery*, Cambridge University Press, Cambridge, U.K., 1977.

[3] E. G. H. Landau, *Foundations of Analysis*, Chelsea Pub. Co., New York, 1951.

[4] G. Polya, *How to Solve It: A New Aspect of Mathematical Method*, Princeton University Press, Princeton, N.J., 1945.

Calculus

[5] G. Klambauer, *Aspects of Calculus*, Springer-Verlag, New York, 1986.

[6] M. Spivak, *Calculus*, 3rd edition, Publish or Perish, Inc., Houston, 1994.

[7] M. Spivak, *Calculus on Manifolds*, W. A. Benjamin, New York, 1965.

Basic Analysis

[8] J. E. Marsden, *Elementary Classical Analysis*, W.H. Freeman and Co., San Francisco, 1974.

[9] A. Mattuck, *Introduction to Analysis*, Prentice Hall, Upper Saddle River, N.J., 1999.

[10] W. Rudin, *Principles of Mathematical Analysis*, 3rd edition, McGraw-Hill, New York, 1976.

[11] W. R. Wade, *An Introduction to Analysis*, Prentice Hall, Englewood Cliffs, N.J., 1995.

Linear Algebra

[12] S. H. Friedberg, A. J. Insel, and L. E. Spence, *Linear Algebra*, 2nd edition, Prentice Hall, Englewood Cliffs, N.J., 1989.

[13] K. Hoffman and R. Kunze, *Linear Algebra*, 2nd edition, Prentice Hall, Englewood Cliffs, N.J., 1971.

[14] G. Strang, *Linear Algebra and Its Applications*, 2nd edition, Academic Press, New York, 1980.

Advanced Analysis (Measure Theory)

[15] A. M. Bruckner, J. B. Bruckner, and B. S. Thomson, *Real Analysis*, Prentice Hall, Upper Saddle River, N.J., 1997.

[16] G. B. Folland, *Real Analysis*, 2nd edition, John Wiley & Sons, New York, 1999.

[17] H. Royden, *Real Analysis*, 3rd edition, Macmillan Pub. Co., New York, 1988.

Approximation Theory

[18] C. deBoor, *A Practical Guide to Splines*, Springer-Verlag, New York, 1978.

[19] W. Cheney, *An Introduction to Approximation Theory*, McGraw-Hill, New York, 1966.

[20] W. Cheney and W. Light, *A Course in Approximation Theory*, Brooks/Cole Pub. Co., Pacific Grove, Calif., 2000.

[21] P. J. Davis, *Interpolation and Approximation*, Blaisdell Pub. Co., New York, 1963.

[22] R. P. Feinerman and D. J. Newman, *Polynomial Approximation*, Williams & Wilkins, Baltimore, 1974.

Dynamical Systems

[23] M. Barnsley, *Fractals Everywhere*, 2nd edition, Academic Press, Boston, 1993.

[24] R. Devaney, *An Introduction to Chaotic Dynamical Systems*, Addison-Wesley, Redwood City, Calif., 1989.

[25] K. Falconer, *Fractal Geometry: Mathematical Foundations and Applications*, John Wiley & Sons, New York, 1990.

[26] R. A. Holmgren, *A First Course in Discrete Dynamical Systems*, Springer-Verlag, New York, 1996.

[27] C. Robinson, *Dynamical Systems*, 2nd edition, CRC Press, Boca Raton, Fla., 1999.

Differential Equations

[28] G. Birkhoff and G. C. Rota, *Ordinary Differential Equations*, 4th edition, John Wiley & Sons, New York, 1989.

[29] G. F. Simmons, *Differential Equations*, 2nd edition, McGraw-Hill, New York, 1991.

[30] W. Walter, *Ordinary Differential Equations*, Grad. Texts in Math., Vol. 182, Springer-Verlag, New York, 1998.

Fourier Series

[31] H. Dym and H. P. McKean, *Fourier Series and Integrals*, Academic Press, New York, 1972.

[32] T. Körner, *Fourier Analysis*, Cambridge University Press, Cambridge, U.K., 1988.

[33] R.T. Seeley, *An Introduction to Fourier Series and Integrals*, W. A. Benjamin, New York, 1966.

Wavelets

[34] G. Bachman, E. Beckenstein, and L. Narici, *Fourier and Wavelet Analysis*, Springer-Verlag, New York, 2000.

[35] E. Hernández and G. Weiss, *A First Course on Wavelets*, CRC Press, Boca Raton, Fla., 1996.

[36] P. Wojtaszczyk, *A Mathematical Introduction to Wavelets*, LMS Student Texts, Vol. 37, Cambridge University Press, Cambridge, U.K., 1997.

Convex Optimization

[37] J. Borwein and A. Lewis, *Convex Analysis and Nonlinear Optimization*, Springer-Verlag, New York, 2000.

[38] J. B. Hiriart-Urruty and C. Lemarichal, *Convex Analysis and Minimization algorithms I*, Springer-Verlag, New York, 1993.

[39] A. L. Peressini, F. E. Sullivan, and J. J. Uhl, *The Mathematics of Nonlinear Programming*, Springer-Verlag, New York, 1988.

[40] J. van Tiel, *Convex Analysis*, John Wiley & Sons, New York, 1984.

[41] R. Webster, *Convexity*, Oxford University Press, New York, 1994.

Articles

[42] J. Banks, J. Brooks, G. Cairns, G. Davis, and P. Stacey, "On Devaney's Definition of Chaos," *Amer. Math. Monthly* **99** (1992), 332–334.

[43] H. Carslaw, "A Historical Note on Gibbs' Phenomenon in Fourier's Series and Integrals," *Bull. Amer. Math. Soc.* (2) **31** (1925), 420–424.

[44] J. Foster and F. B. Richards, "The Gibbs Phenomenon for Piecewise-Linear Approximation," *Amer. Math. Monthly* **98** (1991), 47–49.

[45] J. E. Hutchinson, "Fractals and Self-Similarity," *Indiana Univ. Math. J.* **30** (1981), 713–747.

[46] T. Li, J. Yorke, "Period Three Implies Chaos," *Amer. Math. Monthly* **82** (1975), 985–992.

[47] M. Vellekoop and R. Berglund, "On Intervals, Transitivity = Chaos," *Amer. Math. Monthly* **101** (1994), 353–355.

[48] R. Weinstock, "Elementary Evaluations of $\int_0^\infty e^{-x^2}\,dx$, $\int_0^\infty \cos^2(x^2)\,dx$ and $\int_0^\infty \sin^2(x^2)\,dx$," *Amer. Math. Monthly* **97** (1990), 39–42.

INDEX

A **boldface** page number indicates a definition or primary entry.